Power:

$$\mathbf{1\ hp} = 0.746\ \text{kW} = 550\ \text{ft} \cdot \text{lbf / s} = 33{,}000\ \text{ft} \cdot \text{lbf / min} = 2545\ \text{Btu / h}$$

$$\mathbf{1\ W} = 1.34 \cdot 10^{-3}\ \text{hp} = \text{J / s} = \text{N} \cdot \text{m / s} = \text{V} \cdot \text{A} = 0.239\ \text{cal / s}$$
$$= 9.49 \cdot 10^{-4}\ \text{Btu / s}$$

Pressure:

$$\mathbf{1\ atm} = 101.3\ \text{kPa} = 1.013\ \text{bar} = 14.696\ \text{lbf / in}^2 = 33.89\ \text{ft of water}$$
$$= 29.92\ \text{in of mercury} = 1.033\ \text{kgf / cm}^2 = 10.33\ \text{m of water}$$
$$= 760\ \text{mm of mercury} = 760\ \text{torr}$$

$$\mathbf{1\ psi} = \text{atm} / 14.696 = 6.89\ \text{kPa} = 27.7\ \text{in}\ H_2O = 51.7\ \text{torr}$$

$$\mathbf{1\ Pa} = \text{N / m}^2 = \text{kg / m} \cdot \text{s}^2 = 10^{-5}\ \text{bar} = 1.450 \cdot 10^{-4}\ \text{lbf / in}^2$$
$$= 0.0075\ \text{torr} = 0.0040\ \text{in}\ H_2O = 10\ \text{dyne / cm}$$

Viscosity:

$$\mathbf{1\ cP} = 0.01\ \text{poise} = 0.01\ \text{g / cm} \cdot \text{s} = 0.001\ \text{kg / m} \cdot \text{s} = 0.001\ \text{N} \cdot \text{s / m}^2$$
$$= 0.001\ \text{Pa} \cdot \text{s} = 0.01\ \text{dyne} \cdot \text{s / cm}^2$$
$$= 6.72 \cdot 10^{-4}\ \text{lbm / ft} \cdot \text{s} = 2.42\ \text{lbm / ft} \cdot \text{h} = 2.09 \cdot 10^{-5}\ \text{lbf} \cdot \text{s / ft}^2$$

Kinematic viscosity:

$$\mathbf{1\ cSt} = 0.01\ \text{Stoke} = 0.01\ \text{cm}^2/\text{s} = 10^{-6}\ \text{m}^2/\text{s} = 1\ \text{cP} / (\text{g / cm}^3)$$
$$= 1.08 \cdot 10^{-5}\ \text{ft}^2/\text{s} = \text{cP} / (62.4\ \text{lbm / ft}^3)$$

Temperature:

$$\mathbf{K} = {}^\circ\text{C} + 273.15 = {}^\circ\text{R} / 1.8 \approx {}^\circ\text{C} + 273 \qquad \mathbf{{}^\circ C} = ({}^\circ\text{F} - 32) / 1.8$$

$$\mathbf{{}^\circ R} = {}^\circ\text{F} + 459.67 \approx {}^\circ\text{F} + 460 = 1.8\ \text{K} \qquad \mathbf{{}^\circ F} = 1.8{}^\circ\text{C} + 32$$

Psia, psig:

Psia means pounds per square inch, absolute. Psig means pounds per square inch, gauge, i.e., above or below the local atmospheric pressure.

Force-mass conversion factor, g_c

This factor is equal to dimensionless 1.00. Any dimensioned quantity may be multiplied or divided by g_c without changing the value of that quantity.

$$g_c = \mathbf{1.0} = 32.2 \frac{\text{lbm} \cdot \text{ft}}{\text{lbf} \cdot \text{s}^2} = 1 \frac{\text{slug} \cdot \text{ft}}{\text{lbf} \cdot \text{s}^2} = 1 \frac{\text{lb}}{\text{pou}}$$

$$= 9.81 \frac{\text{kgmass} \cdot \text{m}}{\text{kgforce} \cdot \text{s}^2}$$

McGraw-Hill Chemical Engineering Series

Bailey and Ollis: *Biochemical Engineering Fundamentals*
Bennett and Myers: *Momentum, Heat and Mass Transfer*
Coughanowr and LeBlanc: *Process Systems Analysis and Control*
Davis and Davis: *Fundamentals of Chemical Reaction Engineering*
de Nevers: *Air Pollution Control Engineering*
de Nevers: *Fluid Mechanics for Chemical Engineers*
Douglas: *Conceptual Design of Chemical Processes*
Edgar, Himmelblau, and Lasdon: *Optimization of Chemical Processes*
Gates, Katzer, and Schuit: *Chemistry of Catalytic Processes*
King: *Separation Processes*
Luyben: *Process Modeling, Simulation, and Control for Chemical Engineers*
Marlin: Process Control: *Designing Processes and Control Systems for Dynamic Performance*
McCabe, Smith, and Harriott: *Unit Operations of Chemical Engineering*
Middleman and Hochberg: *Process Engineering Analysis in Semiconductor Device Fabrication*
Perry and Green: *Perry's Chemical Engineers' Handbook*
Peters, Timmerhaus, and West: *Plant Design and Economics for Chemical Engineers*
Reid, Prausnitz, and Poling: *Properties of Gases and Liquids*
Smith, Van Ness, and Abbott: *Introduction to Chemical Engineering Thermodynamics*
Treybal: *Mass Transfer Operations*

The Founding of a Discipline:
The McGraw-Hill Companies, Inc. Series in Chemical Engineering

Over 80 years ago, fifteen prominent chemical engineers met in New York to plan a continuing literature for their rapidly growing profession. From industry came such pioneer practioners as Leo H. Baekeland, Arthur D. Little, Charles L. Reese, John V.N. Dorr, M.C. Whitaker, and R.S. McBride. From the universities came such eminent educators as William H. Walker, Alfred H. White, D.D. Jackson, J.H. James, Warren K. Lewis, and Harry A. Curtis. H.C. Parmlee, then editor of *Chemical and Metallurgical Engineering,* served as chairman and was joined subsequently by S.D. Kirkpatrick as consulting editor.

After several meetings, this committee submitted its report to the McGraw-Hill Book Company in September 1925. In the report were detailed specifications for a correlated series of more than a dozen texts and reference books which became the McGraw-Hill Series in Chemical Engineering—and in turn became the cornerstone of the chemical engineering curricula.

From this beginning, a series of texts has evolved, surpassing the scope and longevity envisioned by the founding Editorial Board. The McGraw-Hill Series in Chemical Engineering stands as a unique historical record of the development of chemical engineering education and practice. In the series one finds milestone of the subject's evolution: industrial chemistry, stoichiometry, unit operations and processes, thermodynamics, kinetics, and transfer operations.

Textbooks such as McCabe et al, *Unit Operations of Chemical Engineering,* Smith et al, *Introduction to Chemical Engineering Thermodynamics,* and Peters et al, *Plant Design and Economics for Chemical Engineers* have taught generations of students principles that are key to success in chemical engineering. Juan de Pablo, Jay Schieber, and Regina Murphy, McGraw-Hill's next band of classic authors, will lead students worldwide toward the latest developments in chemical engineering.

Chemical engineering is a dynamic profession and its literature continues to grow. McGraw-Hill, with its in-house editors and consulting editors Eduardo Glandt (Dean, Univ. of Pennsylvania), Michael Klein (Dean, Rutgers Univ.), and Thomas Edgar (Professor, Univ. of Texas at Austin) remains committed to a publishing policy that will serve the needs of the global chemical engineering profession for the years to come.

FLUID MECHANICS FOR CHEMICAL ENGINEERS

THIRD EDITION

Noel de Nevers

Department of Chemical and Fuels Engineering

University of Utah

FLUID MECHANICS FOR CHEMICAL ENGINEERS, THIRD EDITION
International Edition 2005

Published by McGraw-Hill Education, 2 Penn Plaza, New York, NY 10121.

20 19 18 17 16
20 15 14
CTP SLP

When ordering this title, use ISBN 978-007-123824-3 or MHID 007-123824-7

Printed in Singapore

www.mhhe.com

ABOUT THE AUTHOR

Noel de Nevers received a B.S. from Stanford in 1954, and M.S. and Ph.D. degrees from the University of Michigan in 1956 and 1959, all in chemical engineering.

He worked for the research arms of the Chevron Oil Company from 1958 to 1963 in the areas of chemical process development, chemical and refinery process design, and secondary recovery of petroleum. He has been on the faculty of the University of Utah from 1963 to the present in the Department of Chemical and Fuels Engineering, becoming emeritus in 2002.

He has worked for the National Reactor Testing Site, Idaho Falls, Idaho, on nuclear problems, for the U.S. Army Harry Diamond Laboratory, Washington DC, on weapons, and for the Office of Air Programs of the U.S. EPA in Durham, NC, on air pollution.

He was a Fulbright student of Chemical Engineering at the Technical University of Karlsruhe, Germany, in 1954–1955, a Fulbright lecturer on Air Pollution at the Universidad del Valle, in Cali, Colombia, in the summer of 1974, and at the Universidad de la República, Montevideo Uruguay and the Universidad Naciónal Mar del Plata, Argentina in the Autumn of 1996.

His areas of research and publication are in fluid mechanics, thermodynamics, air pollution, technology and society, energy and energy policy, and explosions and fires. He regularly consults on air pollution problems, explosions, fires and toxic exposures.

In 1993 he received the Corcoran Award from the Chemical Engineering Division of the American Society for Engineering Education for the best paper ("'Product in the Way' Processes") that year in *Chemical Engineering Education*.

In 2000 his textbook, *Air Pollution Control Engineering,* Second Edition, was issued by McGraw-Hill.

In 2002 his textbook, *Physical and Chemical Equilibrium for Chemical Engineers* was issued by John Wiley.

In addition to his serious work he has three "de Nevers's Laws" in the latest "Murphy's Laws" compilation, and won the title "Poet Laureate of Jell-O Salad" at the Last Annual Jell-O Salad Festival in Salt Lake City in 1983. He is the official discoverer of Private Arch in Arches National Park.

CONTENTS

NOTATION

Symbol	Meaning	English units	SI units
a	acceleration	ft / s^2	m / s^2
a	some arbitrary direction or length (Sec. 2.1)	ft	m
a	resistance of filter medium (Sec. 11.4)	1 / ft	1 / m
a_x, a_y, a_z	x, y, and z components of acceleration	ft / s^2	m / s^2
a_c	centrifugal acceleration	ft / s^2	m / s^2
a, b, c, d	exponents in algebraic procedure (Sec. 9.3)	—	—
A	area or cross-sectional area perpendicular to flow	ft^2	m^2
A	independent variable (Sec. 9.3)	various	various
A, B, C, D	arbitrary constants	various	various
b	background concentration (Sec. 3.6)	lbm / ft^3	kg / m^3
B	dependent variable (Sec. 9.3)	various	various
c	volume fraction in hindered settling	—	—
c	speed of light (Chap. 4 only)	ft / s	m / s
c	speed of sound	ft / s	m / s
c	concentration	lbm / ft^3	kg / m^3
C	heat capacity	Btu / lbm · °F or Btu / lbmol · °F	J / kg · K or J / mol · K
C_d	drag coefficient (Sec. 6.13)	—	—
C_i	constants of integration	various	various
C_l	lift coefficient (Sec. 6.13)	—	—
C_f	integrated drag coefficient (Sec. 17.2)	—	—
C'_f	local drag coefficient (Sec. 17.2)	—	—
C_P	heat capacity at constant pressure	Btu / lbm · °F or Btu / lbmol · °F	J / kg · K or J / mol · K
C_v	orifice or venturi coefficient (Sec. 5.8)	—	—
C_V	heat capacity at constant volume	Btu / lbm · °F or Btu / lbmol · °F	J / kg · K or J / mol · K
CC	capital cost factor (Sec. 6.12)	1 / yr	1 / yr
D	diameter	ft	m
D_p	particle diameter	ft	m
D / Dt	Stokes or substantive or convective derivative	1 / s	1 / s
$\mathcal{D}$	diffusivity, molecular or turbulent	ft^2 / s	m^2 / s
erf	gauss error function (see Fig. 19.5)	—	—
E	energy	Btu or equivalent	J
E	voltage	volt	volt
E_s	surface energy	ft · lbf / ft^2	J / m^2
f	Fanning friction factor (Sec. 6.4)	—	—
$f_{\text{P.M.}}$	friction factor for porous medium (Sec. 11.1)	—	—
$f(n)$	spectrum function (Sec. 18.4)	1 / hertz	1 / hertz

°F	temperature or temperature interval, degrees Fahrenheit	°F	
F	force	lbf	N
$\mathscr{F}$	friction heating per lbm	ft · lbf / lbm or equivalent	J / kg
F_x, F_y, F_z	x, y, and z components of force	lbf	N
F_θ	tangential component of force	lbf	N
$F_I, F_V, F_G, F_S, F_E, F_P$	inertia, viscous, gravity, surface, elastic, and pressure forces (Sec. 9.3)	lbf	N
$F(n)$	spectrum function (Sec. 18.4)		—
$\mathscr{F}r$	Froude number	—	—
g	acceleration of gravity	ft / s^2	m / s^2
g_c	conversion factor $= 1 = 32.2$ lbm · ft / lbf · s^2	—	—
h	height or depth	ft	m
h	enthalpy per unit mass or mole ($u + Pv$)	Btu / lbm or Btu / lbmol	J / kg or J / mol
h_c	centroid depth measured from free surface (Prob. 2.26)	ft	m
H	enthalpy ($U + PV$)	Btu	J
H	height	ft	m
H	effective stack height (Chap. 19)	ft	m
H	mixing height (Chap. 3)	ft	m
hp	horsepower	ft · lbf / s	
$\mathscr{H}e$	Hedstrom number (Chap. 13)	—	—
HR	hydraulic radius	ft	m
i, j, k	unit vectors in the x, y, and z directions	—	—
I	electric current (dQ / dt)	amp	amp
I	angular moment of inertia (Chap. 7)	lbm · ft^2	kg · m^2
I_{sp}	specific impulse	lbf · s / lbm	N · s / kg
J_x, J_y, J_z	x, y, and z components of the electric current density (Sec. 16.3)	amp / m^2	amp / m^2
k	number of independent dimensions (Sec. 9.3)	—	—
k	ratio of specific heats, C_P / C_V (Sec. 8.1)	—	—
k	thermal conductivity (Sec. 16.3)	Btu / hr · °F · ft	W / m · K
k	permeability (Sec. 16.3 and Chap. 11)	ft^2	m^2
k	ratio of radii in a Couette viscometer	—	—
k	turbulent ke per unit mass	Btu / lbm	J / kg
ke	kinetic energy per unit mass	Btu / lbm	J / kg
K	arbitrary constant in "power law" (Chap. 13)	lbf · s^n / ft^2	N · s^n / m^2
K	bulk modulus (Sec. 8.1)	lbf / in^2	Pa
K	resistance coefficient (Sec. 6.9)	—	—
K	arbitrary constant in jet equation (Chap. 19)	—	—
KE	kinetic energy	Btu	J
l	length	ft	m
L	length or lever arm	ft	m
L	angular momentum (Chap. 7)	lbm · ft^2 / s	kg · m^2 / s
m	mass	lbm	kg
$\dot{m}$	mass flow rate	lbm / s	kg / s
M	molecular weight	lbm / lbmol	g / mol
$\mathscr{M}$	Mach number	—	—
n	number of independent variables	—	—
n	number of moles	lbmol	mol
n	arbitrary power in "power law" (Chap. 13)	—	—
n	frequency	cyc / s	hertz
n	constant in Chézy Eq. (Chap. 6)	—	—

N	$4f\,\Delta x / D$ (Sec. 8.4)	—	—
N	rotation rate (rpm or rps)	1 / min; 1 / s	1 / min; 1 / s
pe	potential energy per unit mass	Btu / lbm	J / kg
P	pressure	lbf / in^2	Pa
Po	power	ft · lbf / s	W
PE	potential energy	Btu	J
PC	pumping cost (Sec. 6.12)	$ / yr · hp	
PP	purchased price factor for a pipe (Sec. 6.12)	$ / in · ft	
q	emission rate per unit area (Sec. 3.6)	$lbm / s \cdot ft^2$	$kg / s \cdot m^2$
q_x, q_y, q_z	x, y, and z components of heat flux (Sec. 16.3)	$Btu / h \cdot ft^2$	W / m^2
Q	volumetric flow rate	ft^3 / s	m^3 / s
Q	heat	Btu	J
Q	charge	coul	coul
r	radius	ft	m
R	universal gas constant	See inside back cover	See inside back cover
R	correlation coefficient (Sec. 18.5)	—	—
R	radius of curvature (Chap. 14)	ft	m
$\mathcal{R}$	Reynolds number	—	—
$\mathcal{R}_p$	particle Reynolds number	—	—
$\mathcal{R}_{P.M.}$	Reynolds number for porous media	—	—
$\mathcal{R}_x$	Reynolds number based on distance from leading edge	—	—
$\mathcal{R}_{power\ law}$	Reynolds number for power law fluids	—	—
$\mathcal{R}_{Bingham}$	Reynolds number for Bingham plastics	—	—
$\mathcal{R}_{impeller}$	Reynolds number for a mixer impeller	—	—
s	entropy per unit mass or per mole	Btu / lbm · °R or Btu / lbmol · °R	J / kg · K or J / mol · K
s	cake compressibility coefficient (Sec. 11.4)	—	—
SG	specific gravity	—	—
t	time	s	s
t	wall thickness (Sec. 2.4)	ft	m
T	absolute temperature	°R or K	K
T	relative intensity of turbulence (Sec. 18.4)	—	—
u	internal energy per unit mass or per mole	Btu / lbm or Btu / lbmol	J / kg or J / mol
u^*	friction velocity (Sec. 17.4)	ft / s	m / s
u^+	V_x / u^* (Sec. 17.4)	—	—
U	internal energy	Btu	J
v	volume per unit mass	ft^3 / lbm	m^3 / kg
v	fluctuating component of velocity (Chaps. 17 and 18)	ft / s	m / s
V	velocity	ft / s	m / s
V	volume	ft^3	m^3
V_x, V_y, V_z	x, y, and z components of velocity	ft / s	m / s
V_θ	tangential component of velocity	ft / s	m / s
V_r	radial velocity	ft / s	m / s
V_{avg}	average velocity	ft / s	m / s
$V_{centerline}$	centerline velocity in a pipe	ft / s	m / s
V_∞	free-stream velocity	ft / s	m / s
V_s	superficial velocity (Sec. 11.1)	ft / s	m / s
V_I	interstitial velocity (Sec. 11.1)	ft / s	m / s
V_{mf}	minimum fluidizing velocity (Sec. 11.5)	ft / s	m / s
W	work	ft · lbf	J
W	weight	lbf	N
W	width	ft	m

$W_{n.f.}$	non-flow work (excluding injection work)	ft · lbf	J
W	volumetric solids content of slurry (Sec. 11.4)	—	—
x, y, z	directions of coordinate axes, or lengths	ft	m
x	distance	ft	m
y	distance perpendicular to the flow direction	ft	m
y^+	$(r_{wall} - r)u^* / \nu$ (Sec. 17.4)	—	—
z	elevation	ft	m
α	coefficient of thermal expansion	1 / °F	1 / K
α	specific resistance of filter cake (Sec. 11.4)	1 / lbf	1 / N
α	small angle, jet angle (Chap. 19)	rad	rad
α	thermal diffusivity	ft^2 / s	m^2 / s
α	constant in Chézy Eq. (Chap. 6)	—	—
β	isothermal compressibility $= 1 /$ (bulk modulus)	1 / (lbf / in^2)	1 / Pa
γ	specific weight $= \rho g$	lbf / ft^3	N / m^3
Γ	torque	ft · lbf	N · m
δ	boundary-layer thickness (Chap. 17)	ft	m
δ^*	displacement thickness (Sec. 17.2)	ft	m
ε	absolute roughness	ft	m
ε	porosity or void fraction or volume fraction of gas	—	—
ε	eddy (kinematic) viscosity	ft^2 / s	m^2 / s
ε	turbulent dissipation rate	ft^2 / s^3	m^2 / s^3
ζ	vorticity $= 2\omega$	1 / s	1 / s
η	efficiency	—	—
η	$y(V_x / \nu_x)^{1/2}$ (Sec. 17.2)	—	—
η	viscosity (non-Newtonian fluids)	lbm / ft · s or cP	Pa · s
θ	angle	rad	rad
θ	momentum thickness (Sec. 17.2)	ft	m
θ	contact angle (Sec. 17.3)	rad	rad
μ	viscosity	lbm / ft · s or cP	Pa · s
ν	kinematic viscosity (μ / ρ)	ft^2 / s or cSt	m^2 / s
π	number of dimensionless groups (Chap. 9)	—	—
ρ	density	lbm / ft^3	kg / m^3
ρ	resistivity (Sec. 16.3)	—	ohm · m
σ	surface tension	lbf / ft	N / m
σ	stress	lbf / in^2	Pa
σ	shear rate	1 / s	1 / s
$\sigma_x, \sigma_y, \sigma_z$	turbulent dispersion coefficients (Chap. 19)	ft	m
σ_{xx}	normal stress in x direction	lbf / in^2	Pa
τ	shear stress	lbf / in^2	Pa
τ_{xy}	shear stress in the x direction on a face perpendicular to the y axis	lbf / in^2	Pa
τ_{wall}	shear stress at a solid wall	lbf / in^2	Pa
τ_0	shear stress at a solid surface	lbf / in^2	Pa
τ_{yield}	yield stress for a Bingham fluid	lbf / in^2	Pa
ϕ	potential	ft^2 / s for fluid flow	m^2 / s for fluid flow
$\phi(t)$	arbitrary function of time (Sec. 16.2)	—	—
ψ	stream function	ft^2 / s	m^2 / s
ω	angular velocity	rad / s	rad / s

Superscripts

*	sonic condition (Chap. 8)		
$\overline{X}$	time average of X	various	various

Subscripts			
R	reservoir state in Chap. 8		
S	isentropic condition (speed of sound)		
1, 2	arbitrary states		
x, y	conditions before and after a normal shock in Chap. 8		
Vector			
boldface	indicates a vector	various	various
∇	$\nabla = \mathbf{i}\frac{\partial}{\partial x} + \mathbf{j}\frac{\partial}{\partial y} + \mathbf{k}\frac{\partial}{\partial z}$	1 / ft	1 / m

PREFACE

This book presents an introduction to fluid mechanics for undergraduate chemical engineering students.

Throughout the text, emphasis is placed on the connection between physical reality and the mathematical models of reality, which we manipulate. The ultimate test of a mathematical solution is its ability to predict the results of future experiments. Because a mathematically correct consequence of inapplicable assumptions is often simply wrong, the text occasionally offers intentionally wrong solutions to caution the student.

The simplest mathematical approaches are used, consistent with technical vigor.

Considerable attention is paid to the units of quantities in the equations because students usually have trouble with them, and because this reminds them that each symbol in our equations stands for a real physical quantity.

The book is divided into four sections. Section I, preliminaries, provides background for the study of flowing fluids. It includes a separate chapter on the balance equation. One might think that this is such a simple topic that it deserves only a few lines. However, it is a continual source of trouble to students. Furthermore, it is the most all-pervasive concept of chemical engineering, forming the basic mathematical framework for the application of the laws of thermodynamics, Newtonian mechanics, stoichiometry, and for the study of chemically reacting systems. There is also a chapter on the first law of thermodynamics. In the undergraduate program at the University of Utah, the students study basic engineering thermodynamics before they are introduced to fluid mechanics; thus, Chapter 4 is merely a review for them.

Section II discusses flows that are practically one-dimensional or can be treated as such. This organization of the book is radically different from the organization of fluids books written by mechanical and civil engineers, who begin with three-dimensional fluid mechanics and work their way down to one-dimensional fluid mechanics. The reasons for this organization, which fits better with the background of chemical engineers, are spelled out in Section 1.11. Sections I and II are the core of the book, covering all the basic ideas in fluid mechanics, and many of the problems of greatest interest to chemical engineers.

Section III discusses some other topics that can be viewed by the methods of one-dimensional fluid mechanics. These six chapters introduce other areas of fluid mechanics that are of great practical interest to some chemical engineers but that are not covered in an introductory course for want of time. They can be assigned, in any order, as supplementary reading, or covered briefly in class, introducing students to the terminology and basic ideas of these fields and helping them to read related matters in the current literature.

Section IV introduces the student to two- and three-dimensional fluid mechanics. It shows the relations between the methods used for these flows and the simpler approaches used in Sections II and III. It shows what two- and three-dimensional problems can be solved by hand (a small number) and shows the basis on which most such problems are currently solved by Computational Fluid Mechanics programs. A separate chapter introduces the student to mixing, which is basic chemical engineering, but not routinely covered in fluid mechanics texts.

Computers do not make hand calculations unnecessary. No new or unfamiliar computer solution should be believed until manual plausibility checks have shown that the computer is indeed solving the problem we think it is solving and that its solution is physically reasonable. Simply plugging values into available computer packages does not build physical insight, which is one of the most important tools of the successful engineer. Good pedagogy begins with hand solutions of simplified versions of the real problem, which build physical insight and some understanding of physical magnitudes, followed by computer solutions, which can relax the simplifications and cover a wider variety of conditions, followed by manual plausibility checks of the computer solutions.

After an initial rush of enthusiasm for SI, engineering educators seem to be deciding that the English system of units is not likely to vanish overnight. For this reason our students must become like educated Europeans, who speak more than one language fluently and can read and understand one or two additional languages. Our students must be fluent in SI and in the English system of units and must understand traditional metric and cgs, and be able to read and understand texts using the slug and the poundal. This book has a long discussion of these various systems of units. Examples are presented in both SI and English units. This is unlikely to please purists of any persuasion, but it probably serves our students as well as any other approach and better than some.

My goal is to present a text that average chemical engineering undergraduates can read and understand and from which they can attack a variety of meaningful problems. I have tried to help the student develop a physical insight into the processes of fluid mechanics and develop the understanding that the equations on these pages truly describe what nature does. I have tried to choose examples from the student's own experiences, or that relate to things they can observe in their everyday lives. The home is a wonderful place to observe the principles of chemical engineering; good teachers help students interpret what they see in the home in terms of chemical engineering principles.

The true test of the quality of a textbook is whether it becomes the most worn and tattered book on a practicing engineer's bookshelf. Former students tell me that the first two editions of this book pass that test. I hope copies of this edition will become even more worn and tattered.

For instructor resources visit the text's Web site at http://www.mhhe.com/denevers3e

Acknowledgements

I thank the many secretaries who worked on various editions of this book, the faculty who have reviewed earlier editions and the many students who have used it and given me their criticisms and comments. I also thank the following professors who provided helpful comments on drafts of this edition:

Roger T. Bonnecaze, University of Texas at Austin
John R. Grace, University of British Columbia
Neal R. Houze, Purdue University
Christine M. Hrenya, University of Colorado at Boulder
Johannes Khinast, Rutgers University
Hossain Khoroosi, University of Minnesota
David W. Murhammer, University of Iowa
Thuan Nguyen. California State Polytechnic University, Pomona
Russell Ostermann, University of Kansas
Ravi Saraf, Virginia Polytechnic Institute and State University
Karsten E. Thompson, Louisiana State University

Noel de Nevers

CHAPTER 1

INTRODUCTION

1.1 WHAT IS FLUID MECHANICS?

Mechanics is the study of forces and motions. Therefore, fluid mechanics is the study of forces and motions in fluids. But what is a fluid? We all can think of some things that obviously are fluids: air, water, gasoline, lubricating oil, and milk. We also can think of some things that obviously are not fluids: steel, diamonds, rubber bands, and paper. These we call solids. But there are some very interesting intermediate types of matter: gelatin, peanut butter, cold cream, mayonnaise, toothpaste, roofing tar, library paste, bread dough, and auto grease.

To decide what we mean by the word "fluid," we first have to consider the idea of shear stress. It is easiest to discuss shear stress in comparison with tensile stress and compressive stress; see Fig. 1.1.

In Fig. 1.1(*a*) a rope is holding up a weight. The weight exerts a force that tends to pull the rope apart. A stress is the ratio of the applied force to the area over which it is exerted (force/area). Thus, the stress in the rope is the force exerted by the weight divided by the cross-sectional area of the rope. The force that tries to pull things apart is called a *tensile force,* and the stress it causes is called a *tensile stress.*

In Fig. 1.1(*b*) a steel column is holding up a weight. The weight exerts a force that tends to crush the column. This kind of force is called a *compressive force,* and the stress in the column, the force divided by the cross-sectional area of the column, is called a *compressive stress.*

In Fig. 1.1(*c*) some glue is holding up a weight. The weight exerts a force that tends to pull the weight down the walls and thus to *shear* the glue. This force, which tends to make one surface *slide* parallel to an adjacent surface, is called a *shear force,* and the stress in the glue, the force divided by the area of the glue joint, is called a *shear stress.*

A more detailed examination of these examples would show that all three kinds of stress are present in each case, but those we have identified are the main ones. (For more information on this topic, see any text on strength of materials.)

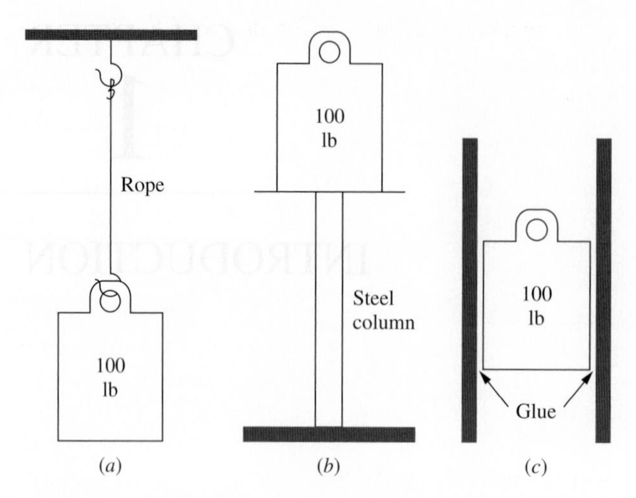

FIGURE 1.1
Comparison of tensile, compressive and shear stresses. (*a*) The rope is in tensile stess; (*b*) the column is in compressive stress; (*c*) the glue is in shear stress.

Solids are substances that can permanently resist very large shear forces. When subject to a shear force, they move a short distance (elastic deformation), thereby setting up internal shear stresses that resist the external force, *and then they stop moving.* Materials that obviously are fluids cannot permanently resist a shear force, no matter how small. When subject to a shear force, they start to move and *keep on moving as long as the force is applied.*

Substances intermediate between solids and fluids can permanently resist a small shear force but cannot permanently resist a large one. For example, if we put a "blob" of any obvious liquid on a vertical wall, gravity will make it run down the wall. If we attach a piece of steel or diamond securely to a wall, it will remain there, no matter how long we wait. If we attach some peanut butter to a wall, it will probably stay, but if we increase the shear stress on the peanut butter by spreading it with a knife, it will flow like a fluid. We cannot spread steel with a knife as we spread peanut butter.

If, as shown above, the relevant difference between peanut butter and steel is the magnitude of the shear stress that the material can resist, then the difference is one of degree, not of kind. At extreme shear stresses steel can be made to "flow like a fluid." In the remainder of this book we will be talking mostly about materials such as air and water, which cannot permanently resist any shear force. However, it is well to keep our minds open to other possibilities of "fluid" behavior [1]. (Numbers in brackets refer to items listed in the References at the end of the chapter.)

1.2 WHAT GOOD IS FLUID MECHANICS?

The problems in fluid mechanics are basically no different from those in "ordinary" mechanics (the mechanics of solids) or in thermodynamics. Therefore, in principle one can solve problems in fluid mechanics with the same methods used to solve problems in mechanics or thermodynamics. However, for many of the problems involving the flow of fluids (or the movement of bodies through fluids), we use a combination of the problem-solving methods of mechanics and thermodynamics. Furthermore, the methods that work for hydraulics problems (dams, canals, locks, river flow, etc.) are applicable, with slight modifications, to aerodynamics problems (airplanes, rockets, wind forces on bridges, etc.) and to problems of special interest to chemical engineers such as the flow in chemical reactors, in distillation columns, or in polymer extrusion dies. Therefore, it makes sense to combine the study of this class of similar problems into one discipline, which we call fluid mechanics.

Consider the important fluids in our lives: the air we breathe, the water we drink, many of the foods we consume, most of the fuels for heating our houses or propelling our vehicles, and the various fluids in our bodies that make up our internal environment. Without some idea of the behavior of fluids, we can have only a very limited understanding of how the world works.

Some of the subdivisions and applications of fluid mechanics are:

1. Hydraulics: the flow of water in rivers, pipes, canals, pumps, turbines.
2. Aerodynamics: the flow of air around airplanes, rockets, projectiles, structures.
3. Meteorology: the flow of the atmosphere.
4. Particle dynamics: the flow of fluids around particles, the interaction of particles and fluids (i.e., dust settling, slurries, pneumatic transport, fluidized beds, air pollutant particles, corpuscles in our blood).
5. Hydrology: the flow of water and water-borne pollutants in the ground.
6. Reservoir mechanics: the flow of oil, gas, and water in petroleum reservoirs.
7. Multiphase flow: coffee percolators, oil wells, carburetors, fuel injectors, combustion chambers, sprays.
8. Combinations of fluid flow: with chemical reactions in combustion, with electromagnetic phenomena in magnetohydrodynamics, with mass transport in distillation or drying.
9. Viscosity-dominated flows: lubrication, injection molding, wire coating, lava, and continental drift.

1.3 BASIC IDEAS IN FLUID MECHANICS

Fluid mechanics is based largely on working out the detailed consequences of four basic ideas:

1. The principle of the conservation of mass.
2. The first law of thermodynamics (the principle of the conservation of energy).
3. The second law of thermodynamics.
4. Newton's second law of motion, which may be summarized in the form $F = ma$.

Each of these four ideas is a generalization of experimental data. None of them can be deduced from the others or from any other prior principle. None of them can be "proven" mathematically. Rather, they stand on their ability to predict correctly the results of any experiment ever run to test them.

Sometimes in fluid mechanics we may start with these four ideas and the measured physical properties of the fluid(s) and proceed directly to solve mathematically for the desired forces, velocities, etc. This is generally possible only in the case of very simple flows. The observed behavior of a great many fluid flows is too complex to be solved directly from these four principles, so we must resort to experimental tests. Through the use of techniques called dimensional analysis (Chap. 9), we often can use the results of one experiment to predict the results of a much different experiment. Thus, careful experimental work is very important in fluid mechanics. With modern computers we can find useful numerical solutions to problems which would previously have required experimental tests. The methods for doing that are outlined in part IV of this book. As computers become faster and cheaper, we will see additional complex fluid mechanics problems solved on computers. Ultimately, though, the computer solutions must be tested experimentally.

These four ideas are applied to fluid mechanical problems as follows. This introductory chapter launches our study and defines some important terms. Then Part I of the book, Chaps. 2–4, deals with preliminaries. We will need these in our study of moving fluids, and they provide direct solutions and/or insight into many practical problems. Parts II and III, Chaps. 5–14, deal with the flow of fluids that are one-dimensional or can be treated as if they were. Part IV, Chaps. 15–20, deals with two- and three-dimensional fluid mechanics. Each of these sections will be described as we begin them.

Students using this book should have previously completed a course in elementary thermodynamics. Chapters 3 and 4 should serve as a review of matter previously covered; they are included because the principles involved are central to fluid mechanics. It is assumed that the student is familiar with the second law of thermodynamics, which is used occasionally. Remember that this entire book is devoted to the application of the four basic ideas and the results of experimental tests to fluid-flow problems. Although the details can become quite involved, the basic ideas are few.

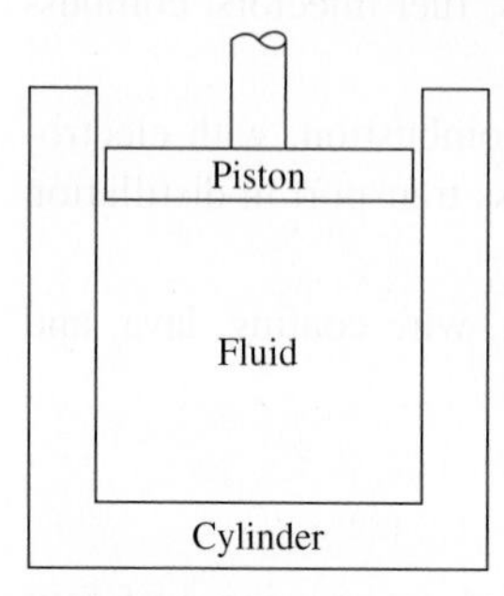

FIGURE 1.2
Piston and cylinder. If the fluid is a gas, we can move the piston up and down as much as we like, and the gas will expand or contract to fill the volume available. If the fluid is a liquid, we can move the piston down very little without producing extreme pressures; if we move it up, the liquid must partly evaporate to produce a gas to fill the space.

1.4 LIQUIDS AND GASES

Fluids are of two types, liquids and gases. On the molecular level these are quite different. In liquids the molecules are close together and are held together by significant forces of attraction; in gases the molecules are relatively far apart and have very weak forces of attraction. As a rule, the specific volumes of gases are ≈1000 times those of liquids, which means that the average intermolecular distance (center to center of the molecules) is roughly 10 times as far in a typical gas as in a typical liquid. As temperature and pressure increase, these differences become less and less, until the liquid and gas become identical at the critical temperature and pressure. The difference between the behavior of liquids and gases is most marked when these fluids are expanded. Suppose that some fluid completely fills the space below the piston in Fig. 1.2. When we raise the piston, the volume occupied by the fluid is increased.

If the fluid is a gas, it will expand readily, filling all the space vacated by the piston; gases can expand without limit to occupy space made available to them. But if the fluid is a liquid, then as the piston is raised, the liquid can expand only a small amount, and then it can expand no more. What fills the space between the piston and the liquid? Part of the liquid must turn into a gas by boiling, and this gas expands to fill the vacant space. This can be explained on the molecular level by saying that there is a maximum distance between molecules over which the attractive forces hold them together to form a liquid and that, when the molecules separate more than this distance, they cease behaving as a liquid and behave as a gas.

Because of their closer molecular spacing, liquids normally have higher densities, viscosities, refractive indices, etc., than gases (see Prob. 1.2). In engineering this frequently leads to quite different behaviors of liquids and gases, as we will see.

1.5 PROPERTIES OF FLUIDS

The physical properties of fluids that will enter our calculations most often are density, viscosity, and surface tension.

1.5.1 Density

The *density* ρ is defined the mass per unit volume:

$$\rho = \frac{m}{V} \tag{1.1}$$

We are all aware of the differences in density between various materials, such as that between lead and wood. How can we measure the density of a material? If we want to know the density of a liquid, we can weigh a bottle of known volume (determine its mass), fill it with the liquid, weigh it again, and compute the density with the aid of Eq. 1.1. (This is one of the standard laboratory methods of determining liquid density; the special weighing bottles designed for this purpose are called *pyncnometers,* Prob. 1.5) If we want to know the density of a cubical solid block, we can measure the length of its sides, compute its volume, weigh it, and apply these results to Eq. 1.1.

Now suppose we are asked to determine the density of a piece of Swiss cheese. If we have a large block of the cheese, we can cut off a cube, measure its sides, compute its volume, weigh it, and then calculate its density. This is an average density, one that includes the density of the air in the holes in the cheese. As long as we are dealing with large pieces of cheese, it is a satisfactory density. Suppose, however, we are asked to find the density at some point inside a large block of the cheese. If we can cut the cheese open, and if we find that the point in question is in the solid cheese and not in one of its holes, we can find the density easily enough or, if the point in question is in a hole, we find the density of the air in the hole. But if the point is on the surface of a hole, the problem is more difficult. Then the density is discontinuous; see Fig. 1.3. There is no meaningful single value of the density at x.

Why this long discussion about the density of Swiss cheese? Because the world is full of holes! Atomic physics tells us that even in a solid bar of steel the space occupied by the electrons, protons, and neutrons is a very small fraction of the total space; the rest presumably is empty. Furthermore, even at the molecular level there

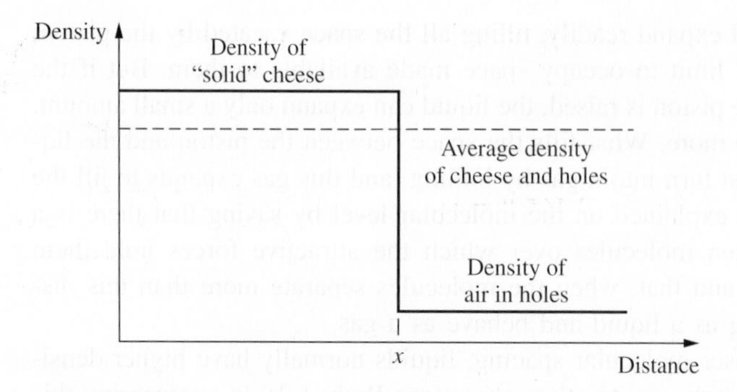

FIGURE 1.3
The density of Swiss cheese is not uniform from point to point, but has local point densities and an average density.

are holes; in a typical gas the space actually occupied by the individual gas molecules at any instant is a small fraction of the total space. Thus, in any attempt to speak of density at a given point we are in the same trouble as with the Swiss cheese. Therefore, we must restrict the definition of density to samples large enough to average out the holes. This causes no problem in fluid mechanics, because of the size of the samples normally used, but it indicates that the concept of density does not readily apply to samples of molecular and subatomic sizes.

In addition, we must be careful in defining the densities of composite materials. For example, a piece of reinforced concrete consists of several parts with different densities. In discussing such materials we must distinguish between the *particle densities* of the individual pebbles or steel-reinforcing bars and the *bulk density* of the mixed mass. When we refer to bulk density, our sample must be large compared with the dimensions of one particle. Some examples of composite solid materials are cast iron, fiberglass-reinforced plastics, and wood. Some examples of composite liquids are slurries, such as muds, milkshakes, and toothpaste, and emulsions, such as homogenized milk, mayonnaise, and cold cream. Smokes and clouds behave as composite gases.

Example 1.1. A typical mud is 70 wt. % sand and 30 wt. % water. What is its density? The sand is practically pure quartz (SiO_2), for which $\rho_{sand} =$ 165 lbm / ft^3 (2.65 g / cm^3). See the inside back cover for the properties of water used in all examples and problems.

Here we assume that there is no volume change on mixing sand and water. There are volume changes on mixing for some substances like ethanol and water, but they are small enough to ignore for most problems, including this one. Then

$$\rho = \frac{m}{V} = \frac{m_{sand} + m_{water}}{V_{sand} + V_{water}} = \frac{m_{sand} + m_{water}}{(m/\rho)_{sand} + (m/\rho)_{water}} \tag{1.A}$$

[Every equation in this book has a number. Those, like this one, that are parts of examples or in other ways specific to some situation are identified with number-letter combinations, such as (1.A). General equations have number-number combinations, such as (1.1).]

We could simplify Eq. 1.A algebraically, but a more intuitive approach is to choose as our *basis* 100 lbm of mud, and substitute into Eq. 1.A, finding

$$\rho = \frac{m_{\text{sand}} + m_{\text{water}}}{\left(\frac{m}{\rho}\right)_{\text{sand}} + \left(\frac{m}{\rho}\right)_{\text{water}}} = \frac{70 \text{ lbm} + 30 \text{ lbm}}{\left(\frac{70 \text{ lbm}}{165 \text{ lbm / ft}^3}\right)_{\text{sand}} + \left(\frac{30 \text{ lbm}}{62.3 \text{ lbm / ft}^3}\right)_{\text{water}}}$$

$$= 110.4 \frac{\text{lbm}}{\text{ft}^3} = 1769 \frac{\text{kg}}{\text{m}^3} \tag{1.B}$$

■

The ■ indicates the end of an example.

1.5.2 Specific Gravity

Specific gravity of liquids and solids (SG) is defined as

$$\text{SG} = \frac{\text{density}}{\text{density of water at some specified temperature and pressure}} \tag{1.2}$$

This definition has the merit of being a ratio and, hence, a pure number, which is independent of the system of units chosen. Occasionally it leads to confusion, because some specific gravities are referred to water at 60°F, some to water at 70°F, and some to water at 39°F = 4°C (all at a pressure of 1 atm). The differences are small but great enough to cause trouble.

If the temperature of the water is specified as 39°F = 4°C, then the density of water is 1.000 g / cm^3. (The gram was defined to make this number come out 1.000). Thus, if this basis of measurement is chosen, then specific gravities become numerically identical with densities expressed in g / cm^3 or kg / L or metric tons / m^3. The mud in Example 1.1 has SG = 1.769.

Many process industries use special scales of fluid density, which are usually referred to as *gravities.* Some of them are the API gravity (American Petroleum Institute) for oil and petroleum products (Prob. 1.6), Brix gravity for the sugar industry, and Baumé gravity for sulfuric acid. Each scale is directly convertible to density; conversion tables and formulae are available in handbooks.

Specific gravities of gases are normally defined as

$$\begin{pmatrix}\text{SG of} \\ \text{a gas}\end{pmatrix} = \left(\frac{\text{density of the gas}}{\text{density of air}}\right)_{\text{Both at the same temperature and pressure}} \tag{1.3}$$

For ideal gases the specific gravity of any gas = $(M_{\text{gas}} / M_{\text{air}})$.

Throughout this text we use liquid and solid specific gravities referred to water at 4°C. Thus a liquid with a specific gravity of 0.8 is a liquid with a density of 0.8 g / cm^3.

1.5.3 Viscosity

Viscosity is a measure of internal, frictional resistance to flow. If we tip over a glass of water on the dinner table, the water will spill out before we can stop it. If we tip over a jar of honey, we probably can set it upright again before much honey flows

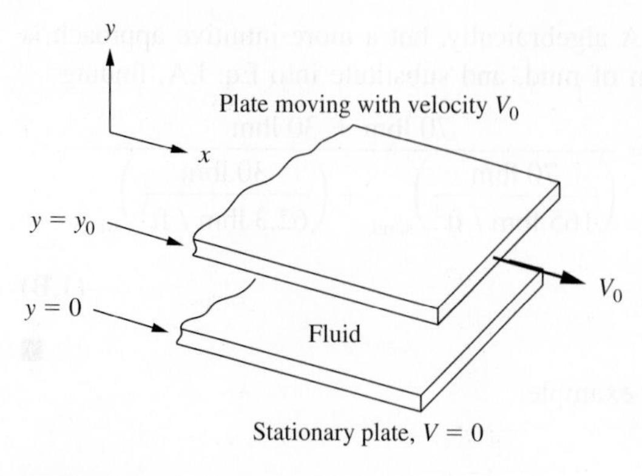

FIGURE 1.4
The sliding-plate experiment.

out; this is possible because the honey has much more resistance to flow, more viscosity, than water. A more precise definition of viscosity is possible in terms of the following conceptual experiment.

Consider two long, solid plates separated by a thin film of fluid (see Fig. 1.4). This apparatus is easy to grasp conceptually and mathematically but difficult to use, because the fluid leaks out at the edges and gravity pulls the two plates together. Other devices that are more complex mathematically but easier to use are actually used to measure viscosities (see Example 1.2 and Chaps. 6 and 13). If we slide the upper plate steadily in the x direction with velocity V_0, a force will be required to overcome the internal friction in the fluid between the plates. This force will be different for different velocities, different plate sizes, different fluids, and different distances between the plates. We can eliminate the effect of different plate sizes, however, by measuring the force per unit area of the plate, which we define as the *shear stress* τ.

It has been demonstrated experimentally that at low values of V_0 the velocity profile in the fluid between the plates is linear, i.e.,

$$V = \frac{V_0 y}{y_0} \tag{1.C}$$

so that

$$\sigma = \begin{pmatrix} \text{shear rate, rectangular} \\ \text{coordinates} \end{pmatrix} = \frac{dV}{dy} = \frac{V_0}{y_0} \tag{1.D}$$

It also has been demonstrated experimentally that for most fluids the results of this experiment can be shown most conveniently on a plot of τ versus dV/dy (see Fig. 1.6). As shown here, dV/dy is simply a velocity divided by a distance. In more complex geometries it is the limiting value of such a ratio at a point. It is commonly called *shear rate, the rate of strain,* and *rate of shear deformation,* all of which mean exactly the same thing.

Example 1.2. Figure 1.5 shows a cutaway photograph of a concentric-cylinder ("cup and bob") viscometer also called a *Couette viscometer.* An inner cylinder (the bob) rotates inside a stationary outer cylinder (the cup). The shaft that drives the bob is instrumented to record both the angular velocity and the applied torque. The solid bob has $D_1 = 25.15$ mm and $L = 92.27$ cm. The surrounding cup has $D_2 = 27.62$ mm and is longer than the bob. When the bob is driven at 10 rpm, the observed torque is $\Gamma = 0.005$ Nm. What are τ and dV/dy?

This viscometer is simply the device in Fig. 1.4, wrapped around a cylinder. In this form, the leakage-at-the-edges problem and the difficulty of

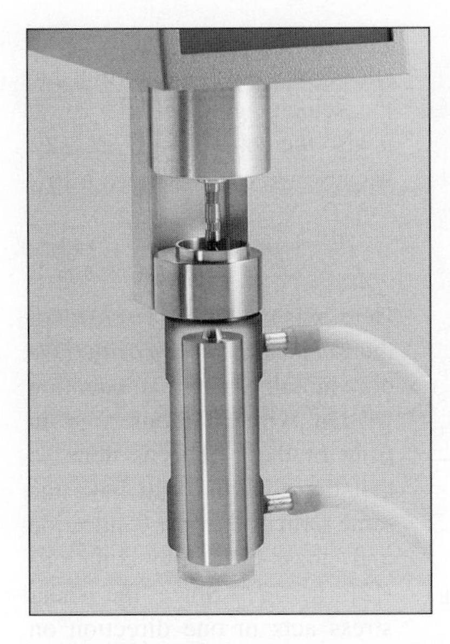

FIGURE 1.5
Cutaway photograph of a concentric-cylinder viscometer. This is simply the sliding-plate arrangement in Fig. 1.4, wrapped around a cylinder, thus eliminating the leaky edges in Fig. 1.4. The drive mechanism at the top holds the outer cylinder fixed and rotates the inner closed cylindrical bob. It provides a measured, controllable rotation rate and simultaneously measures the torque required to produce that rotation. The two flexible hoses circulate constant-temperature water or other fluid, to hold the whole apparatus at a constant temperature. Example 1.2 shows the dimensions of this device. (Courtesy of Brookfield Engineering Company.)

keeping the distance between the two surfaces constant are solved. (Fluid forces hold the rotating inner cylinder properly centered inside the outer cylinder.) Here we must replace the ys in Eq. 1.5 with rs, because the velocity is changing in the radial direction. $\Delta y = y_0$ is replaced by

$$\Delta r = 0.5(D_2 - D_1) = 0.5(27.62 - 25.15) = 1.235 \text{ mm} \quad \text{(1.E)}$$

and

$$V_0 = \pi D_1 \cdot \text{rpm} = \pi \cdot 25.15 \text{ mm} \cdot \frac{10}{\text{min}} = 790.1 \frac{\text{mm}}{\text{min}} = 13.17 \frac{\text{mm}}{\text{s}} \quad \text{(1.F)}$$

Thus,

$$\frac{dV}{dr} = \frac{V_0}{\Delta r} = \frac{13.17 \text{ mm / s}}{1.235 \text{ mm}} = 10.66 \frac{1}{\text{s}} \quad \text{(1.G)}$$

This is a linearized approximation of a cylindrical problem that understates the correct value, which is 12.26 (1 / s), (see Prob. 1.10), a difference of 15%. We will use the correct (cylindrical) value in the rest of this chapter.

The shear stress at the surface of the inner cylinder is

$$\tau = \frac{F}{A} = \frac{\Gamma / r_1}{\pi D_1 L} = \frac{0.005 \text{ Nm}/(0.5 \cdot 25.15 \text{ mm})}{\pi \cdot 25.15 \text{ mm} \cdot 92.37 \text{ mm}} = 5.45 \cdot 10^{-8} \frac{\text{N}}{\text{mm}^2} = 0.0545 \frac{\text{N}}{\text{m}^2} \quad \text{(1.H)}$$

■

This example ignores the stress on the bottom surface of the bob, a small effect, for which a correction is made in real viscosity measurements. The whole device is shown immersed in a constant-temperature bath, because the results are very temperature dependent.

The experiment in Example 1.2 can be repeated at different rotational speeds and the results plotted as shown in Fig. 1.6. Four different kinds of curve are shown as experimental results in the figure. All four of these results are observed in nature. The most common behavior is that represented by the straight line through the origin in the figure. This line is called Newtonian because it is described by Newton's law of viscosity:

$$\tau = \mu \frac{dV}{dy} \qquad \text{[Newtonian fluids]} \qquad (1.4)$$

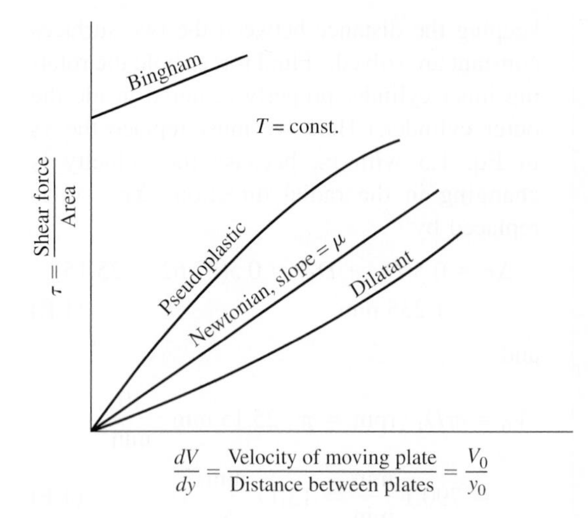

FIGURE 1.6
Possible outcomes of the sliding-plate experiment at constant temperature and pressure.

This equation says that the shear stress τ is linearly proportional to the velocity gradient dV/dy. It is also the definition of viscosity, because we can rearrange it to

$$\mu = \frac{\tau}{dV/dy} \tag{1.5}$$

Here μ is called the *viscosity* or the *coefficient of viscosity.* [We occasionally see this equation written with a minus sign in front of the τ. This is done so that the equation will have the same form as the heat-conduction and mass-diffusion equations ([2], p. 12). Since the shear stress acts in one direction on the rotating cylinder and in the opposite direction on the fluid adjacent to it, we can introduce this minus sign and reverse our idea of the direction of τ so that the result is always the same as in Eq. (1.5).] For Example 1.2, we would calculate

$$\mu = \frac{\tau}{dV/dy} = \frac{0.0.0545 \text{ N/m}^2}{12.26\text{ / s}} = 0.0044 \frac{\text{N} \cdot \text{s}}{\text{m}^2} \tag{1.I}$$

For fluids such as air the value of μ is very low; therefore, their observed behavior is represented in Fig. 1.6 by a straight line through the origin, very close to the dV/dy axis. For fluids such as corn syrup the value of μ is very large, and the straight line through the origin is close to the τ axis.

Fluids that exhibit this behavior in the sliding-plate experiment or its cylindrical equivalent (i.e., fluids that obey Newton's law of viscosity) are called *Newtonian fluids.* All the others are called *non-Newtonian fluids.* Which fluids are Newtonian? All gases are Newtonian. All liquids for which we can write a simple chemical formula are Newtonian, such as water, benzene, ethyl alcohol, carbon tetrachloride, and hexane. Most dilute solutions of simple molecules in water or organic solvents are Newtonian, such as solutions of inorganic salts, or sugar in water, or benzene. Which fluids are non-Newtonian? Generally, non-Newtonian fluids are complex mixtures: slurries, pastes, gels, polymer solutions, etc. (some authors refer to them as *complex fluids*). Most non-Newtonian fluids are mixtures with constituents of very different sizes. For example, toothpaste consists of solid particles suspended in an aqueous solution of various polymers. The solid particles are much, much bigger than water molecules, and the polymer molecules are much bigger than water molecules.

In discussing non-Newtonian fluids we must agree on what we mean by viscosity. If we retain the definition given by Eq. 1.5, then the viscosity can no longer

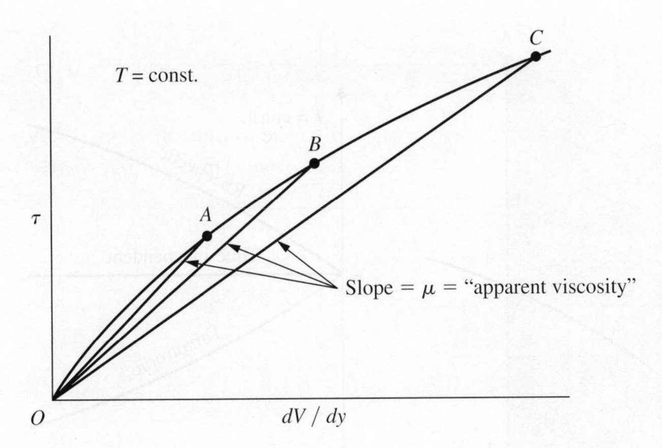

FIGURE 1.7
The "apparent viscosity" of a pseudoplastic fluid decreases as the shear rate increases.

be considered a constant independent of dV/dy for a given temperature, but must be considered a function of dV/dy. This is shown in Fig. 1.7. Here each of the lines *OA*, *OB*, and *OC* have slope μ, so the viscosity is decreasing with increasing dV/dy. (Viscosities defined as the slopes in Fig. 1.7 are often called *apparent viscosities.*) Using this definition, we can observe that there are three common types of non-Newtonian fluid (Fig. 1.6):

1. *Pseudoplastic fluids* show an apparent viscosity that decreases with increasing velocity gradient. Examples are most slurries, muds, polymer solutions, solutions of natural gums, and blood. These fluids are referred to as *shear thinning* fluids. This is the most common type of non-Newtonian behavior.
2. *Bingham fluids,* sometimes called *Bingham plastics,* resist a small shear stress indefinitely but flow easily under larger shear stresses. One may say that at low stresses the viscosity is infinite and at higher stresses the viscosity decreases with increasing velocity gradient. Examples are bread dough, toothpaste, applesauce, some paints, jellies, and some slurries.
3. *Dilatant fluids* show a viscosity that increases with increasing velocity gradient. This behavior is called *shear thickening;* it is uncommon, but starch suspensions and some muds behave this way. For these materials the liquid lubricates the passage of one solid particle over another; at high shear rate the lubrication breaks down, and the particles have more resistance to slipping past each other.

So far, we have assumed that the curve of τ versus dV/dy is not a function of time; i.e., if we move the sliding plate at a constant speed, we will always require the same force. This is true of most fluids, but not of all. A more complete picture

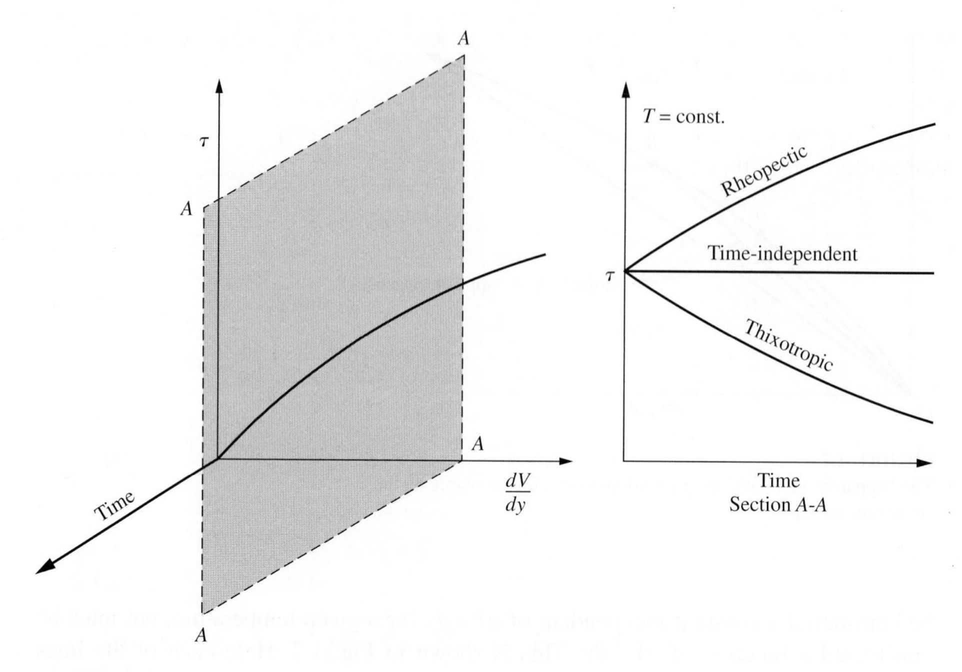

FIGURE 1.8
The viscosity of fluids can be independent of time of shearing or can increase or decrease with time as the fluid is sheared.

is given in Fig. 1.8. In Fig. 1.8 we see a constant dV/dy slice out of the solid constructed of τ versus dV/dy versus time. We see three possibilities:

1. The viscosity can remain constant with time, in which case the fluid is called *time independent.*
2. The viscosity can decrease with time, in which case the fluid is called *thixotropic.*
3. The viscosity can increase with time, in which case the fluid is called *rheopectic.*

All Newtonian fluids are time independent, as are most non-Newtonian fluids. Many thixotropic fluids are known, almost all of which are slurries or solutions of polymers, and a few examples of rheopectic fluids are known.

In addition, some fluids, called *viscoelastic fluids,* can show not only the kinds of behavior represented in Figs. 1.6 and 1.8 but also elastic properties, which allow them to "spring back" when a shear force is released. The most common examples of viscoelastic fluids are egg whites, cookie dough, and the rubber cement sold at stationery stores. Rubber cement's viscoelastic properties can be demonstrated most easily by starting to pour a little out of the bottle and then snapping it back into the bottle with a quick jerk of the hand. The same can be done with egg white. This is quite impossible with any ordinary fluid such as water; try it!

These strange types of fluid behavior are of considerable practical use. A good toothpaste should be a Bingham fluid, so that it can easily be squeezed out of the tube but will not drip off the toothbrush the way water or honey would. A good paint should be a thixotropic Bingham fluid, so that in the can it will be very viscous and the pigment will not settle to the bottom, but when it is stirred, it will become less viscous and can easily be brushed onto a surface. In addition, the brushing should temporarily reduce the viscosity so that the paint will flow sideways (under the influence of surface tension; see below) and fill in the brush marks (called *leveling* in the paint industry); then, as it stands, its viscosity should increase, so that it will not form drops and run down the wall.

Most engineering applications of fluid flow involve water, air, gases, and simple fluids. Therefore, most fluid-flow problems have to do with Newtonian fluids, as do most of the problems in this book. Non-Newtonian fluids are important, however, precisely because of their non-Newtonian behavior; they are discussed in Chap. 13.

The viscosity of simple gases, such as helium, can be calculated for all temperatures and pressures from the kinetic theory of gases using only one experimental measurement for each gas [2]. For the viscosities of most gases and all liquids several experimental data points are required, although ways of predicting viscosity change with changing temperature and pressure are available [3]. As a general rule, the viscosity of gases increases slowly with increasing temperature, and the viscosity of liquids decreases rapidly with increasing temperature. The viscosity of both gases and liquids is practically independent of pressure at low and moderate pressures.

The basic unit of viscosity is the *poise*, where $\text{P} = 1\ \text{g}/(\text{cm}\cdot\text{s}) = 0.1\ \text{Pa}\cdot\text{s} = 6.72 \times 10^{-2}\ \text{lbm}/(\text{ft}\cdot\text{s})$ [See the inside front cover for conversion factors.] The poise is widely used for materials like high-polymer solutions and molten polymers. However, it too large a unit for most common fluids. By sheer coincidence the viscosity of pure water at about 68°F = 20°C is 0.01 poise; for that reason the common unit of viscosity in the United States is the centipoise, $\text{cP} = 0.01\ \text{P} = 0.01\ \text{g}/(\text{cm}\cdot\text{s}) = 0.001\ \text{N}\cdot\text{s}/\text{m}^2 = 0.001\ \text{Pa}\cdot\text{s} = 6.72 \times 10^{-4}\ \text{lbm}/(\text{ft}\cdot\text{s})$. Hence, the viscosity of a fluid expressed in centipoise is the same as the ratio of its viscosity to that of water at room temperature. The viscosities of some common liquids and gases are shown in App. A.1. The computed viscosity of the fluid in Example 1.2 is 4.4 cP.

1.5.4 Kinematic Viscosity

In many engineering problems, viscosity appears only in the relation (viscosity/density). Therefore, to save writing we define

$$\text{Kinematic viscosity} = \nu = \mu/\rho \tag{1.6}$$

The most common unit of kinematic viscosity is the centistoke (cSt):

$$1\ \text{cSt} = \frac{1\ \text{cP}}{1\ \text{g}/\text{cm}^3} = 10^{-6}\frac{\text{m}^2}{\text{s}} = 1.08 \times 10^{-5}\frac{\text{ft}^2}{\text{s}} \tag{1.J}$$

at 68°F = 20°C, water has a kinematic viscosity of $1.004 \approx 1$ cSt. To avoid confusion over which viscosity is being used, some writers refer to the viscosity μ as the *absolute viscosity.* The kinematic viscosity has the same dimension (length^2 / time) as the thermal diffusivity and the molecular diffusivity; in many problems it acts the same way as them. In Chap. 6 we will see some examples of the practical convenience of the kinematic viscosity.

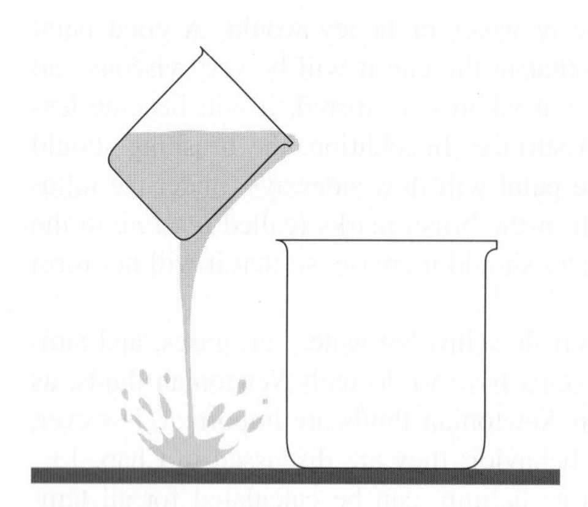

FIGURE 1.9
Disheartening effect of surface tension. The water dribbles down the surface of the container.

1.5.5 Surface Tension

Liquids behave as if they were surrounded by a skin that tends to shrink, or contract, like a sheet of stretched rubber, a phenomenon known as *surface tension.* It is seen in many everyday events, the most disheartening of which is the tendency of water, when poured slowly from a glass, to dribble down the edge of the glass (see Fig. 1.9).

Surface tension is caused by the attractive forces in liquids. All of the molecules attract each other; those in the center are attracted equally in all directions, but those at the surface are drawn toward the center because there are no liquid molecules in the other direction to pull them outward (see Fig. 1.10). The "effort" of each molecule to reach the center causes the fluid to try to take a shape that will have the greatest number of molecules nearest the center, a sphere (Prob. 1.11). Any other shape has more surface per unit volume; therefore, regardless of the shape of a liquid the attractive forces tend to pull the liquid into a sphere. Other forces, such as gravity often oppose surface tension forces, so the spherical shape is only seen for small systems, such as small water drops on a water-repellent surface. The fluid thus tries to decrease its surface area to a minimum. (An analogous situation in two dimensions is observable in the behavior of some army ants. They travel in large groups, and, viewed from above, the swarm often looks like a circle. The reason appears to be that the ants are attracted by the scent of other ants and, hence all try to get to the place where the scent is strongest, the center. The ants all stay in one plane, so the result is the plane figure with the smallest possible ratio of perimeter to area—a circle [4].)

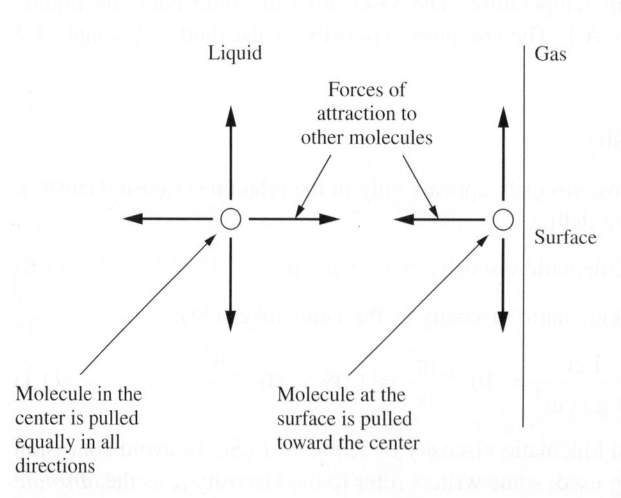

FIGURE 1.10
Surface tension is caused by the attractive forces between molecules.

The tendency of a surface to contract can be measured with the device

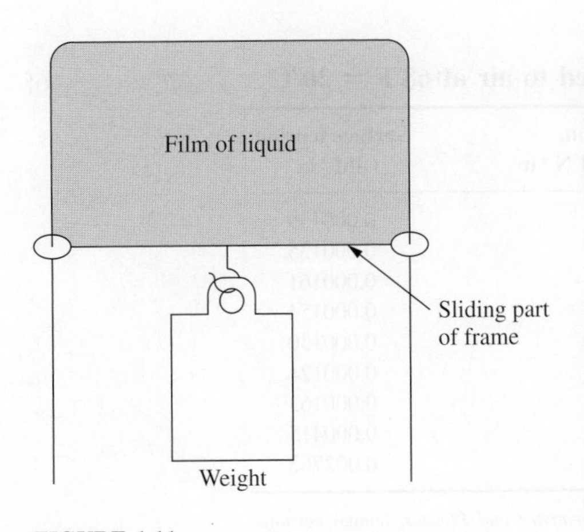

FIGURE 1.11
A very simple way to measure surface tension; see Example 1.3.

shown in Fig. 1.11. A wire frame with one movable side is dipped into a liquid and carefully removed with a film of liquid in the space formed by the frame. The film tries to assume a spherical shape, but since it adheres to the wire, it draws the movable part of the frame inward. The force necessary to resist this motion is measured by a weight. It is found experimentally that the ratio of the force to the length of the sliding part of the wire is always the same for a given liquid at a given temperature, regardless of the size of the apparatus. The liquid film in the frame has two surfaces (front and back), so the force-to-length ratio of *one* of the surfaces is exactly one-half of the total measurement. The surface tension of the liquid is then defined as

$$\text{Surface tension} = \frac{\text{force of one film}}{\text{length}} \quad \text{or} \quad \sigma = \frac{F}{l} \tag{1.7}$$

Example 1.3. The device in Fig. 1.10 has a sliding part 10 cm long. The mass needed to resist the inward pull of the fluid is 0.6 g, which exerts a force of 0.00589 N. What is the surface tension of the fluid?

From Eq. 1.7,

$$\sigma = \frac{F(\text{one film})}{l} = \frac{0.00589\ \text{N} / 2}{0.1\ \text{m}} = 0.0294\ \frac{\text{N}}{\text{m}} = 0.000168\ \frac{\text{lbf}}{\text{in}} \tag{1.K}$$

■

The device shown in Fig. 1.11 is easy to understand but not very practical as a measuring device; more practical ones are discussed in Chap. 14.

Surface tension is very slightly influenced by what the surrounding gas is—air or water vapor or some other gas. Typical values of the surface tension of liquids exposed to air are shown in Table 1.1. The traditional unit of surface tension is the dyne / cm = 0.001 N / m. At 68°F = 20°C, most organic liquids have about the same surface tension (≈25 dyne / cm) whereas that of water is about 3 times higher, and that of mercury is 20 times higher.

We indicated that the liquid adheres to the solid in the apparatus shown in Fig. 1.11. Liquids adhere strongly to some solids and not to others. For example, water adheres strongly to glass but very weakly to polyethylene. This greatly complicates the whole subject of surface tension; the phenomenon shown in Fig. 1.9 occurs much more often with glass, ceramic, or metal cups than with polyethylene or Teflon cups.

TABLE 1.1
Surface tensions of pure fluids exposed to air at 68°F = 20°C

Fluid	Surface tension, dyne / cm = 0.001 N / m	Surface tension, lbf / in
Acetic acid	27.8	0.000159
Acetone	23.7	0.000135
Benzene	28.25	0.000161
Carbon tetrachloride	26.95	0.000154
Ethyl alcohol	22.75	0.000130
n-Octane	21.8	0.000124
Toluene	28.5	0.000163
Water	72.74	0.000415
Mercury	484	0.002763

Extensive tables are available in the *Handbook of Chemistry and Physics,* annual editions, published by CRC Press, Boca Raton, Florida, and various other handbooks.

Two important effects attributable to surface tension are the capillary rise of liquids in small tubes and porous wicks (without which candles, kerosene lanterns or copper sweat-solder fittings would not work at all) and the tendency of jets of liquid to break up into drops (as from a garden hose or gasoline or diesel fuel injector or in an ink-jet printer). Surface tension effects are very important in systems involving large surface areas, such as emulsions (mayonnaise, cold cream, water-based paints) and multiphase flow through porous media (oil fields). We will discuss the effects in Chap. 14; see also references [5, 6].

1.6 PRESSURE

Pressure is defined as a compressive stress, or compressive force per unit area. In a stationary fluid (liquid or gas) the compressive force per unit area is the same in all directions. In a solid or in a moving fluid, the compressive force per unit area at some point is not necessarily the same in all directions. We can visualize why by squeezing a rubber eraser between our fingers; see Fig. 1.12. As we squeeze the eraser, it becomes thinner and longer, as shown. If we analyze the stresses in the eraser, we find that in the y direction the eraser is in compression, whereas in the x direction it is in tension. (This seems strange, but the eraser has been stretched in the x direction, and its elastic forces will pull it back when we let go; hence the tension.) The contraction in one direction and expansion in another in an elastic solid is described in terms of Poisson's ratio, discussed in any text on strength of materials. Because the tensile and compressive forces are at right angles to each other, there is also a strong shear stress at 45° to the x axis.

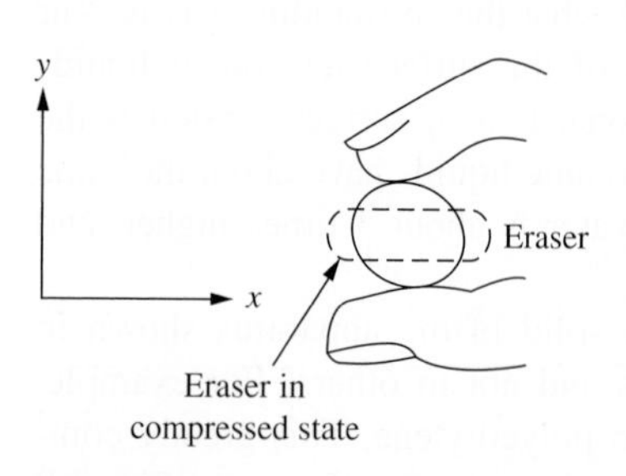

FIGURE 1.12
The response of an elastic solid to compression in one direction.

What would happen if we held our fingers in a cup of water and tried to squeeze the water between our fingers? Obviously, the water would run out from between our fingers, and our fingers would come together. Why? When we start to squeeze the water, it behaves like the eraser, setting up internal shear and tensile forces in the same directions as the eraser. However, ordinary fluids cannot permanently resist shear forces, so the water begins to flow and finally flows away. The eraser also flowed, until it had taken up a new shape, in which its internal tensile and shear resistance were enough to hold our fingers apart. Water cannot set up such resistance and so it simply flows away.

If we really wanted to squeeze the water, we would put it in some container that would prevent its flowing out to the side. If we did this with the eraser, then as we compressed it from the top, it would press out on the sides of the container. So also does water.

The foregoing is a description of why the pressure at a point in a fluid at rest is the same in all directions. It is not a proof of that fact; for a proof see App. B.1.

What we mean by pressure is not so clear for a solid as it is for a liquid or a gas. The compressive stresses at a given point in a solid are not the same in all directions. The usual definition of pressure in a solid is as follows: Pressure at a point is the average of the compressive stresses measured in three perpendicular directions. Since, as we have seen, these three stresses are all the same in a fluid at rest, the two definitions are the same. For a fluid in motion, the three perpendicular compressive stresses may not be the same. However, for this difference to be significant, the shear stresses must be very large, well outside the range of normal problems in fluid mechanics. Therefore, we normally extend the notion that pressure in a fluid at rest is the same in all directions to fluids in motion, with the reservation that at very high shear stresses (such as in the flow of metals or polymer melts through forming dies) this is not necessarily true. For polymer solutions and polymer melts the differences between the compressive stresses in directions at right angles to one another can be very significant and can lead to behavior quite different from the behavior of simple fluids; see [7].

In the solution of many problems, particularly those involving gases, it is most convenient to deal with pressures in an absolute sense, i.e., pressures relative to a compressive stress of zero; these are called *absolute pressures.* In the solution of many other problems, particularly those involving liquids with free surfaces, such as are encountered in rivers, lakes, and open or vented tanks, it is more convenient to deal with pressures above an arbitrary datum, the local atmospheric pressure. Pressures relative to the local atmospheric pressure are called *gauge pressures.*

Because both systems of measurement are in common use, it is necessary to make clear which kind of pressure we mean when we write "a pressure of 15 lb / in^2" [This unit is also called psi (pounds per square inch)]. It is usual to say "15 psi absolute" or "15 psia" for absolute pressure and "15 psi gauge" or "15 psig" for gauge pressure. The SI unit of pressure is the pascal, $Pa = N / m^2$. There does not seem to be a common set of abbreviations for Pascal absolute and Pascal gauge, so these must be written out.

Another two-datum situation familiar to the reader is found in the measurement of elevation. Mountain tops, road routes, and rivers are normally surveyed relative to

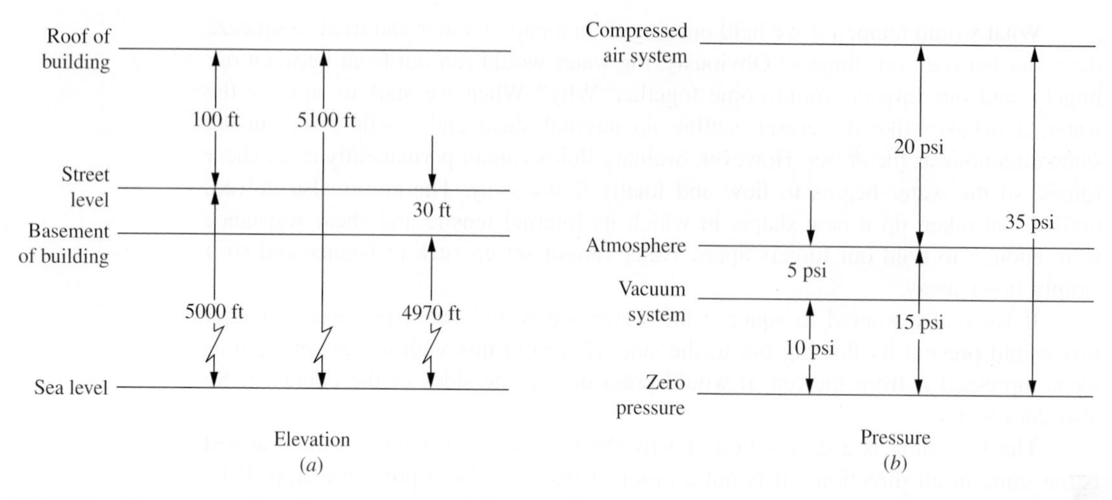

FIGURE 1.13
The relation between gauge and absolute pressure, and a comparison with elevation measurements.

mean sea level, which serves as an "absolute" datum, but most buildings are designed and constructed relative to some local elevation (usually a marker in the street); see Fig. 1.13. In both cases the most common measuring method gives answers in terms of the local datum. Most pressure gauges read the difference between the measured pressure and the local atmospheric pressure. For instance, the pressure gauge on the compressed air system in the figure would read 20 psig = 137.9 kPa gauge; the building height (by tape measure or transits) might be given as 100 ft = 30.5 m elevation. Both such measurements usually involve negative values, based on the local datum; the basement has a negative elevation relative to the street, −30 ft = −9.15 m, and the vacuum system has a negative pressure relative to the atmosphere, −5 psig = −34.5 kPa gauge.

Negative elevations relative to sea level can exist; the Dead Sea, for instance, is about 1200 feet (366 m) below sea level. Can negative absolute pressures exist? Certainly; a negative absolute pressure is a negative compressive stress, i.e., a tensile stress. These occur often in solids, very rarely in liquids, never in gases. They are rare in liquids because all liquids possess a finite vapor pressure. If the pressure of a liquid is reduced below its vapor pressure, the liquid boils and thus replaces the low pressure with the equilibrium vapor pressure of the liquid. However, this boiling never takes place spontaneously in an absolutely pure liquid [8], but rather occurs around small particles of impurities or at the wall of the container. (Most people have observed this phenomenon when they pour a cold carbonated drink into a glass; the bubbles form mostly at the edge of the glass, not in the bulk of the liquid. It can be shown dramatically by dropping some sugar into a cold, fresh glass of soft drink; do this over a sink!) Thus, if a liquid is very pure and the surfaces of its container are very smooth, the liquid can exist in tension at a negative absolute pressure. This situation is unstable, and a slight disturbance can cause the liquid to boil [9].

1.7 FORCE, MASS, AND WEIGHT

In fluid mechanics we are often concerned with forces, masses, and weights. The problem of units of force and mass is discussed in the next section. An unbalanced force makes things change speed or direction. Most forces in the world are balanced by opposite forces (a building exerts a force on the ground; the ground exerts an equal and opposite force on the building; neither moves). To make anything start moving or stop moving, we must exert an unbalanced force.

Mass is an indication of how much matter is present. The more matter, the more mass. (We may think of matter in any size, as bricks, molecules, atoms, nucleons, quarks, etc.). Mass is also an indicator of how hard it is to get some amount of matter moving or how hard it is to stop it once it is moving. We can all stop a baseball moving 50 ft / s (15.2 m / s) with little more damage than a possible sore hand. If we step in front of an automobile moving at the same speed, we will certainly be killed. The auto has much more mass; it is much harder to stop.

Weight is a force—the force that a body exerts due to the acceleration of gravity. When there is no gravity, there is no weight (e.g., in earth satellites there is no apparent gravity; this state is referred to as *weightlessness*).

1.8 UNITS AND CONVERSION FACTORS

Engineering is about real physical things, which can be measured and described in terms of those units of measure. Most engineering calculations involve these units of measure. It would be simple if there were only one set of such units that the whole world agreed upon and used; but that is not the case today. In the United States most measurements use the English system of units, based on the foot, the pound and the °F, but most of the world uses the metric (or SI) system of units based on the meter, the kg and the °C. The metric system has been legally accepted in the United States since 1866, and it has been the declared policy of the U.S. government to convert to metric since 1975 [10]. Progress has been disappointingly slow.

The situation is similar with languages; it would be easier if we all spoke one language. But we do not; the world has many languages. Educated Europeans all speak at least two languages well and generally can read one or two more. Similarly, U.S. engineers must be fluent in English and in metric units, be able to understand older literature written in the centimeter-gram-second (cgs) system, and in variant English systems that use the poundal or the slug and in specialized industrial units, like the 42-gal barrel for petroleum products or pressure differences expressed in inches of water. U.S. engineers must even deal with mixed systems, like automotive air pollutant emissions expressed in grams per mile. Furthermore, they must understand the differences between the common-use version of metric and SI, discussed below; they will be better able to deal with those differences if they understand why the differences arise.

In fluid mechanics we most often deal with dimensioned quantities, such as 12 ft / s (= 3.66 m / s), rather than with pure numbers such as 12 or 3.66. We often drop the units, for example, "I was driving 60," which in the United States normally means 60 mi / h, but in the rest of the world means 60 km / h. This is poor practice, but common. In 1999 [11] a $125 million NASA Mars probe was destroyed because someone failed to check their units. In technical work we always make clear the units in which any value is expressed! To become competent at solving fluid mechanics problems we must become virtually infallible in the handling of such units and their conversion factors. For most engineers the major sources of difficulties with units and conversion factors are carelessness and the simultaneous appearance of force and mass in the same equation.

A useful "system" for avoiding carelessness and consistently converting the dimensions of engineering quantities from one set of units to another has two rules:

1. *Always* (repeat, *always*) include the dimensions with any engineering quantity you write down.
2. Convert the dimensions you have written down to the dimensions you want in your answer by multiplying or dividing by 1.

Example 1.4. We are required to convert a speed of 327 mi / h to a speed in ft / s. The first step is to write the equation

$$\text{Speed} = 327\ \text{mi / h} \tag{1.L}$$

This is not the same as 327 km / h or 327. If we omit the dimensions, our equation is meaningless. We now write, as an equation, the definition of a mile:

$$1\ \text{mi} = 5280\ \text{ft} \tag{1.M}$$

Dividing both sides of this equation by 1 mi, we find

$$\frac{1\ \text{mi}}{1\ \text{mi}} = 1 = \frac{5280\ \text{ft}}{\text{mi}} \tag{1.N}$$

You may not be used to thinking of 5280 ft / mi as being the same thing as 1, but Eq. 1.N shows that they are the same. Similarly, we write the definition of an hour as an equation,

$$1\ \text{h} = 3600\ \text{s} \tag{1.O}$$

and divide both sides by 3600 s to find

$$\frac{3600\ \text{s}}{3600\ \text{s}} = 1 = \frac{\text{h}}{3600\ \text{s}} \tag{1.P}$$

Again, you may not be used to thinking of 1 h / 3600 s as the same thing as 1, but it is. Now let us return to Eq. 1.L and multiply both sides by 1 twice, choosing our equivalents of 1 from Eqs. 1.N and 1.P:

$$\text{Speed} \cdot 1 \cdot 1 = \frac{327\ \text{mi}}{\text{h}} \cdot \frac{5280\ \text{ft}}{\text{mi}} \cdot \frac{\text{h}}{3600\ \text{s}} \tag{1.Q}$$

We can now cancel the two 1's on the left side, because they do not change the value of "Speed," and we can cancel the units that appear both above and below the line on the right side to find

$$\text{Speed} = \frac{327\ \cancel{\text{mi}}}{\cancel{\text{h}}} \cdot \frac{5280\ \text{ft}}{\cancel{\text{mi}}} \cdot \frac{\cancel{\text{h}}}{3600\ \text{s}} = \frac{327 \cdot 5280\ \text{ft}}{3600\ \text{s}} = 480\,\frac{\text{ft}}{\text{s}} = 146\,\frac{\text{m}}{\text{s}} \tag{1.R}$$

■

This was an easy example, one you could certainly solve without going into as much detail as shown here, but it illustrates the procedure to be used in more complicated problems.

Example 1.5. Suppose Time equals 2.6 h. How many seconds is this? Again we begin by writing Time with its dimension as an equation:

$$\text{Time} = 2.6\text{ h} \tag{1.S}$$

We want to know its value in seconds, so we divide by 1,

$$\text{Time} = 2.6\text{ h} \cdot \frac{3600\text{ s}}{\text{h}} = 2.6 \cdot 3600\text{ s} = 9380\text{ s} \tag{1.T}$$

■

How did we know to multiply by 1 h / 3600 s in Example 1.4 and to divide by 1 h / 3600 s in Example 1.5? In each case we chose the value of 1 that allowed us to cancel the unwanted dimension. Three ideas are involved here:

1. Dimensions are treated as algebraic quantities and multiplied or divided accordingly.
2. Multiplying or dividing any quantity by 1 does not change its value.
3. Any dimensioned equation can be converted to 1 = 1 by dividing through by either side.

Using the last procedure, we can write

$$1 = \frac{60\text{ s}}{\text{min}} = \frac{12\text{ in}}{\text{ft}} = \frac{7000\text{ gr}}{\text{lbm}} = \frac{\text{mi}^2}{640\text{ acres}} = \frac{\text{Btu}}{252\text{ cal}} = \frac{\text{W}}{\text{VA}} = \text{etc.} \tag{1.8}$$

and as many other values of 1 as we like.

The previous examples did not involve the unit conversions that cause difficulties, the ones involving force and mass or thermal and mechanical energies. If everyone always used SI, we would never have those difficulties. In SI there is no difficulty with the units of force and mass; force is measured in newtons (N) and mass in kilograms (kg), and the only unit of energy is the mechanical-energy unit, the joule, where $\text{J} = \text{N} \cdot \text{m}$.

Unfortunately, in the English system (and in the traditional metric system as it is used by the public in Europe) there is difficulty with force-mass unit conversion. If we ask a typical European male what he weighs, he might well respond "80 kilos," meaning 80 kg. If he were speaking in SI he would not use kg as a unit of weight, because weight is a force and the SI unit of force is the newton. He should respond, "784.6 newtons" because that is the weight of an 80 kg mass in a standard gravitational field of $9.807\text{ m / s}^2 = 32.17\text{ ft / s}^2$. It is hard enough to teach novice engineers the difference between weight and mass; it is probably impossible to get the general public to take the view that a mass of 80 kg does not exert a force of 80 kg. To make this come out right, we need to decide that there are really *two* kilogram units, the kilogram-mass (kgm) and the kilogram-force (kgf). We can define these so that one kgm exerts a force of one kgf at standard gravity. That is what most of the people in the world actually do. Similarly, in the English system of units we need two kinds of pounds; pound-mass (lbm) and pound-force (lbf). Again we have defined these so that one lbm has a weight of (exerts a force of) one lbf at standard gravity.

Why does this cause problems? Because the kgm and kgf look like the same thing, so we are tempted to believe they are the same thing, and the lbm and the lbf look like the same thing, so we are tempted to believe they are the same thing. That is wrong. It is a trap for the unwary. They are not the same. This leads to serious errors in engineering calculations.

Newton's second law of motion is

$$F = ma \tag{1.9}$$

where F is force, m is mass, and a is acceleration. The pound-force (lbf) is defined as that force which, acting on a mass of 1 lbm, produces an acceleration of 32.2 ft / s^2. Substituting this definition into the last equation, we find

$$1 \text{ lbf} = \text{lbm} \cdot 32.2 \frac{\text{ft}}{\text{s}^2} \tag{1.U}$$

Dividing both sides of this by 1 lbf, we find

$$1 = \frac{\text{lbf}}{\text{lbf}} = 32.2 \frac{\text{lbm} \cdot \text{ft}}{\text{lbf} \cdot \text{s}^2} \tag{1.V}$$

If we then make the mistake of canceling the lbm on the top and the lbf on the bottom right-hand side, we will conclude that $1 = 32.2$ ft / s^2. This is clearly wrong, and if we do it in a problem we will find that the dimensions do not check and the numerical value of the answer will be wrong by a factor of 32.2 (if we use English units) or 9.8 (if we use metric units). Similarly, in the traditional metric system we have

$$1 \text{ kfg} = \text{kgm} \cdot 9.8 \text{ m} / \text{s}^2 \tag{1.W}$$

and if we divide both sides by kgf, we find

$$1 = \frac{\text{kgf}}{\text{kgf}} = 9.8 \frac{\text{kgm} \cdot \text{m}}{\text{kgf} \cdot \text{s}^2} \tag{1.X}$$

If we then cancel kgm and kgf on the right side we will conclude that $1 = 9.8$ m / s^2, which is equally absurd.

How can we get out of this difficulty? One way is to always work exclusively in SI. In that case kg will always mean kgm, and kgf will never appear. Instead the unit of force will always be the N = (1 / 9.8) kgf. However, then we will be unable to deal with the public, who speak (unintentionally) in kgf and lbf, or to deal with those parts of the engineering literature that use kgf and lbf. The other way is to decide we must live with the kgf and lbf, and so we will regularly have to use the force-mass conversion factor whenever units of force and of mass occur in the same equation. This conversion factor has the following values:

$$1 = 9.8 \frac{\text{kgm} \cdot \text{m}}{\text{kgf} \cdot \text{s}^2} = 1.0 \frac{\text{kgm} \cdot \text{m}}{\text{N} \cdot \text{s}^2} = 32.2 \frac{\text{lbm} \cdot \text{ft}}{\text{lbf} \cdot \text{s}^2} \tag{1.10}$$

Furthermore, we must know some history to understand the older literature. First, we must know that many older textbooks and articles used the symbol g_c to stand for this force-mass conversion factor. So whenever we see a g_c written into an equation, we must recognize it as a reminder that we must use the force-mass conversion factor. We must not confuse g_c, the force-mass conversion factor, with g, the acceleration of gravity; they are *not* the same.

Second, we should recognize that engineers using English units have tried to evade this difficulty by inventing two new units, the slug (1 slug = 32.2 lbm = 14.6 kg) and the poundal (pdl) (1 pdl = lbf / 32.2 = 0.138 N = 0.014 kgf). Using these, we have the following force-mass conversion factors:

$$1 = 9.8\,\frac{\text{kgm}\cdot\text{m}}{\text{kgf}\cdot\text{s}^2} = 1.0\,\frac{\text{kgm}\cdot\text{m}}{\text{N}\cdot\text{s}^2} = 32.2\,\frac{\text{lbm}\cdot\text{ft}}{\text{lbf}\cdot\text{s}^2} = 1.0\,\frac{\text{slug}\cdot\text{ft}}{\text{lbf}\cdot\text{s}^2} = 1.0\,\frac{\text{lbm}\cdot\text{ft}}{\text{pdl}\cdot\text{s}^2} \qquad (1.11)$$

The poundal finds little current use, but aeronautical engineers use the slug.

The kgf and the lbf have been around a long time, in spite of the efforts of scientists and engineers to replace them with the newton or the poundal. They survive because they seem natural to nonscientific users. Probably they will continue to be widely used, in spite of the efforts of the scientific community to replace them. Prudent engineers will learn to live with this fact, to use them when it seems appropriate, and to understand why they came about.

The second difficulty with units concerns mechanical and thermal units of energy. In SI the only unit of energy is the joule, 1 J = 1 N · m. This is clearly a mechanical unit, the product of a force and a distance. If we are transferring thermal energy (e.g., heating our houses or our soup), it seems natural to base the measurements on the quantity of thermal energy required to raise the temperature of some reference substance by some finite temperature interval. In the English system this quantity is the British thermal unit (Btu), which is the quantity of thermal energy required to raise the temperature of 1 lbm of water by 1°F. In the metric system the unit is the calorie (cal), which is the quantity of thermal energy required to raise the temperature of 1 g of water 1°C, or the kcal (kcal = 1000 cal; this is the "calorie" used in describing the energy content of foods). If we want to use the calorie or the Btu, then we need to convert from joules to calories or ft-lbf to Btu:

$$1 = \frac{\text{Btu}}{778\ \text{ft}\cdot\text{lbf}} = \frac{\text{Btu}}{1055\ \text{J}} = \frac{\text{cal}}{4.18\ \text{J}} = \frac{\text{kcal}}{4180\ \text{J}} = \frac{\text{kcal}}{4.18\ \text{kJ}} \qquad (1.12)$$

The Btu and the cal (or kcal) seem likely to continue in common usage; the Btu appears on almost all U.S. heating appliance and fuel bills (sometimes natural gas bills use the *therm* = 10^5 Btu), and kcal appears on numerous food products.

In summary, if we can do all our work in SI, we need never be concerned about force-mass conversions (N = kg · m / s) or energy conversions (J = N · m = W · s). If we are confronted with problems (or literature, or current U.S. legal definitions) involving the kgf, lbf, cal, kcal, or Btu, we must follow the rules outlined above. Always write down the dimensions, treat the dimensions as algebraic quantities, and multiply by 1 as often as needed to get the quantities into the desired set of units, using the appropriate values of the force-mass conversion factor and the thermal-mechanical energy conversion factor. Even in SI, if we stray from the basic units (m, kg, s, A, K, mol, and cd), we will need conversion factors such as

$$1 = \frac{1000 \text{ g}}{\text{kg}} = \frac{100 \text{ cm}}{\text{m}} = \frac{1000 \text{ mV}}{\text{V}} \tag{1.13}$$

Example 1.6. A mass of 10 lbm (4.54 kgm) is acted on by a force of 3.5 lbf (15.56 N or 1.59 kgf). What is the acceleration in ft / min^2?

Rearranging Eq. 1.9, we find

$$a = F / m \tag{1.14}$$

Substituting, we find

$$a = \frac{3.5 \text{ lbf}}{10 \text{ lbm}} \tag{1.Y}$$

Here we want the acceleration in ft / min^2, so we must multiply or divide by those equivalents of 1 that will convert the units:

$$a = \frac{3.5 \text{ lbf}}{10 \text{ lbm}} \cdot \frac{32.2 \text{ lbm} \cdot \text{ft}}{\text{lbf} \cdot \text{s}^2} \cdot \left(\frac{60 \text{ s}}{\text{min}}\right)^2 = \frac{3.5 \cdot 32.2 \cdot 60^2}{10} \frac{\text{ft}}{\text{min}^2} = 40{,}570 \frac{\text{ft}}{\text{min}^2} \tag{1.Z}$$

or

$$a = \frac{15.56 \text{ N}}{4.54 \text{ kg}} \cdot \frac{\text{kg} \cdot \text{m}}{\text{N} \cdot \text{s}^2} \cdot \left(\frac{60 \text{ s}}{\text{min}}\right)^2 = 123.4 \frac{\text{m}}{\text{min}^2} = 40{,}480 \frac{\text{ft}}{\text{min}^2} \tag{1.AA}$$

or

$$a = \frac{1.59 \text{ kgf}}{4.54 \text{ kgm}} \cdot \frac{9.8 \text{ kgm} \cdot \text{m}}{\text{kfg} \cdot \text{s}^2} \cdot \left(\frac{60 \text{ s}}{\text{min}}\right)^2 = 123.6 \frac{\text{m}}{\text{min}^2} = 40{,}540 \frac{\text{ft}}{\text{min}^2} \tag{1.AB}$$

The difference between these three answers is due to round-off error in the conversion factors used. If more figures had been carried (e.g., kgf = 9.80650 N), the answers would have agreed exactly, but since we know the input data to only two significant figures, our best answer, in all three cases, should be 40,500 ft / min^2. ■

Example 1.6 will be the last example in this book to use the kgf. Clearly the method of dealing with kgm and kgf is just the same as the method of dealing with lbm and lbf. For the rest of this book, we will use either lbm and lbf, or SI.

Example 1.7. An aluminum cell (Hall-Héroult process) has a current of 50,000 amp. If we assume it is 100% efficient, how much metallic aluminum does it produce per hour?

We first convert the current to gram equivalents per hour, using the necessary values of 1, one of which we take out of Prob. 1.16:

$$I = 50{,}000 \text{ A} \cdot \frac{\text{C}}{\text{A} \cdot \text{s}} \cdot \frac{3600 \text{ s}}{\text{h}} \cdot \frac{\text{g equiv}}{96{,}500 \text{ C}} = 1870 \frac{\text{g equiv}}{\text{h}} \tag{1.AC}$$

For aluminum,

$$27\text{ g} = 1\text{ mol} \tag{1.AD}$$

and

$$1\text{ mol} = 3\text{ g equiv} \tag{1.AE}$$

therefore,

$$I = 1870\,\frac{\text{g equiv}}{\text{h}} \cdot \frac{\text{mol}}{3\text{ g equiv}} \cdot \frac{27\text{ g}}{\text{mol}} \cdot \frac{\text{lbm}}{454\text{ g}} = 37.1\,\frac{\text{lbm}}{\text{h}} = 16.8\,\frac{\text{kg}}{\text{h}} \tag{1.AF}$$

■

In solving Example 1.7 we multiplied by 1 six times. Nonetheless, the procedure is simple and straightforward. Each multiplication by 1 gets rid of an undesired dimension and brings us closer to an answer in the desired units. We saw that an apparently complex problem was really a simple conversion-of-units problem. In the course of our studies and our professional careers we will have to convert units as quickly and as easily as we now add and subtract. It will be easiest if we develop the habit of following the two rules given at the start of Sec. 1.8, namely:

1. Always include the dimensions with any engineering quantity you write.
2. Convert the dimensions you have written to the dimensions you want in your answer by multiplying or dividing by 1.

A short table of these conversion factors can be found inside the front cover of this text. The American Society for Testing and Materials (ASTM) [12] has prepared a much longer and more complete table, which reveals some additional complexity. For example, there are five different calorie definitions in common usage. The largest is 1.002 times the smallest. Only in the most careful work is this small a difference relevant. But if we are doing that kind of work, it is worthwhile to find, study, and use the ASTM tables.

1.9 PRINCIPLES AND TECHNIQUES

As discussed in Sec. 1.3, there are very few underlying ideas in fluid mechanics. With these few ideas we can solve a great variety of problems. In so doing, we can focus our attention either on the application of principles or on the techniques of solving problems. The author recommends attention to the principles. In the 10 years following his graduation from college, the engineering business was revolutionized by the digital computer, the transistor, and the space industry, among other things. None of these amounted to much in 1954, and they were not part of undergraduate courses.

All these technologies rigidly obey Newton's laws and the laws of thermodynamics. Students who learned "cookbook" techniques for solving problems on 1954 were not well prepared for the technologies that appeared during the next 10 years, but those who learned the basic principles and how to apply them could adapt to any one of them. There seems to be little reason to believe that the pace of technological change will become slower in the future. If we concentrate on learning techniques, we may be faced in a few years with "technical obsolescence," but if we learn principles and their applications, we should have no such problem. The author believes that there will never be a surplus of people who *really* understand Newton's laws and the laws of thermodynamics.

1.10 ENGINEERING PROBLEMS

Although this book may fall into the hands of a practicing engineer, most of its readers will be college juniors; the following is addressed to them.

Engineering students start out in their freshman and sophomore years by doing "plug-in" problems. Given a problem statement, they select the appropriate formula either from the textbook or from their memory, and "plug in" the data in the problem to find the final answer. In their junior year they begin to find problems that can be readily reduced to plug-ins or to problems involving two or more equations that require some manipulations to be put in plug-in form. Furthermore, they may be exposed to problems that cannot be reduced to plug-ins and must be solved by trial and error. It is assumed that they can do simple plug-ins (such as gas-law calculations) without hesitation.

Instructors of third-year students would like to assign more complicated or difficult problems but generally cannot because:

1. The time required for them is too great—they cannot be done in the time that most students will devote to one homework problem.
2. The students would probably get intellectual indigestion on them. Therefore, at the third-year level most of the problems and examples in texts like this one are plug-ins or can be readily reduced to plug-ins.

When students start a senior laboratory or design course, they find their first real engineering problems. One of these may require 10 or 20 h of work and consist of 15 or 20 parts, each comparable to the problems and examples in this book. To deal with these problems, students break them into pieces small enough to handle as plug-ins. The interesting and exciting part of engineering is often the task of deciding how to divide a problem into reasonable pieces and how then to reassemble these pieces into a recognizable whole so that they fit together properly.

In the examples and problems in this book there are numerous simple plug-in problems. They are included because their solutions give the reader some feel for the numerical values involved in fluid mechanics. There are also more complex problems, in which two or more basic principles are involved (such as the mass balance and the energy balance). In these some manipulation is required to get the equations into plug-in form. The recommended procedure for solving such problems is this:

1. Make sure you understand precisely what the problem is; in particular, make sure you know precisely what is being asked for.
2. Decide which physical laws relate what you know to what you want to find.
3. Write the working form of these laws (as discussed later), and rearrange them to get the symbol for the quantity you seek standing alone to the left of the equal

sign. In so doing you will probably have to discard several terms in the physical-law equations. Discarding a term corresponds to making an assumption about the physical nature of the system (e.g., that a certain velocity is negligible). Thus, a list of such terms dropped is a list of assumptions made in solving the problem.

4. When step 3 is finished, the problem is reduced to a plug-in. Insert the given data, check the units, and find the numerical value of the answer.
5. Check the answer for plausibility: Does it indicate negative masses, velocities greater than the speed of light, or efficiencies greater than 100%? Does it pass the test of common sense, that is, do the results match your intuitive idea of what they should be? If not, is the difficulty with the calculations? or with your intuition? If neither is incorrect, perhaps you have made a new technical discovery! Also, re-examine the assumptions listed in step 3 to see whether they are consistent with the answer. If these checks are met, the answer probably is satisfactory.
6. If the problem is one that you may have to repeat with different data (such as the calculation of a fluid-flow rate from a measured pressure difference), then it might be worthwhile to see whether the answer can be put in a more convenient form, for example, some general plot or diagram. Perhaps the problem will occur often enough to justify programming its solution on a personal computer or entering it in a spreadsheet program.

In all engineering we must consider the degree of precision needed. Voltaire's famous dictum "The perfect is the enemy of the good!" describes the situation of the engineer. We could always spend more engineering effort, and do more testing, and thereby refine our design or our calculation a little more. But in any real problem the engineer's time is one of the limiting resources. We would all like the conditions that the famous architect Kobori Enshu demanded and received from the Japanese dictator Hideyoshi for the Katsura Villa: no limit on expense, no limit on time, and no client visits until the job is done. Many believe the result to be the greatest achievement of Japanese architecture and garden planning [13]. (If you are ever in Kyoto, visit it and decide for yourself.) But most engineers (and other professionals) are always working with limited time and limited budgets as well as clients who want intermediate progress reports. For us the goal is always to do the best possible, within the time, budget, and other constraints imposed by the client (or codes and regulations). So engineers must allocate their time well, handling routine things swiftly, and concentrating on those that are not routine and that may be a source of trouble. Much of what you learn in this book is routine to practicing engineers. The goal of this book is that students not only learn to do those routine things but also learn the scientific basis of the solution of those routine problems. In so doing, you will learn how engineers and scientists have turned yesterday's difficult problems into today's routine ones. That will help you to develop the habits of mind that will turn today's difficult problems into tomorrow's routine problems.

You should consider your degree of confidence in the answer to a problem. If the calculation used physical property data that is accurate to no more than $\pm5\%$, then it makes no sense to report the answer to 3 or more significant figures. If the solution presented required really speculative calculating approaches, or questionable input data, the reader should be alerted to that fact.

In the problems at the end of each chapter, one or two need to be broken down into simpler ones before they can be solved. The practice gained in doing these is well worth the effort.

1.11 WHY THIS BOOK IS DIFFERENT FROM OTHER FLUID MECHANICS BOOKS

Most undergraduate fluid mechanics books are written by mechanical or civil engineers. Please look at one; your impression will be that those books and this one are about totally different subjects. The reasons they look so different are these:

1. The fluid mechanics problems of greatest interest to mechanical and civil engineers (aerodynamics, flow around structures) are inherently two- or three-dimensional. They cannot be understood as or easily reduced to one-dimensional form. Most of the fluid mechanics problems of greatest interest to chemical engineers are inherently one-dimensional or can be understood and easily reduced to one-dimensional form. For this reason, civil and mechanical engineers start fluid mechanics as a three-dimensional study, and then derive the one-dimensional forms of greatest interest to chemical engineers from those three-dimensional forms.
2. Mechanical and civil engineers base most of their work on force and momentum. Those are the basic tools of the mechanical and civil engineer. Chemical engineers base most of their work on the conservation of mass and energy; the first course in chemical engineering is about mass and energy balances. Chemical engineers learn about force and momentum in physics but use them much less in their professional careers than they use mass and energy. The single most useful equation in fluid mechanics, Bernoulli's equation, can be found by starting with force and momentum, or with energy. Mechanical and civil engineers start with momentum. This book starts with energy. The energy approach makes much more sense to chemical engineers than does the momentum approach.
3. Momentum and force are vectors. For mechanical and civil engineers, fluid mechanics is inherently an exercise in vector calculus. Their books are full of vector equations. Many take the view that one of the main purposes of a fluid mechanics course is to immerse their students in the vector calculus, and make them exercise it. Mass and energy are scalars. Most of the quantities in chemical engineering are also scalars. Thus, chemical engineers have much less use of the vector calculus than do mechanical and civil engineers. Our graduate students are normally expected to become good at the vector calculus, but our undergraduates rarely use it.

For these reasons, this book uses scalars as much as possible and vectors only when necessary. It begins with the conservation of mass and energy and shows the vast range of practical fluid mechanical problems that can be solved with them, before it shows the momentum balance (which is inherently a vector balance) and shows the problems for which we need it. As a consequence, this book has far simpler mathematics than other fluids books. That does not mean that it sacrifices rigor; complexity is not rigor, or simplicity carelessness. In many cases the complete derivations are shown in appendices, with only the practical result shown in the main text.

In Parts II and III of this book, we cover the wide range of fluid mechanical problems of interest to chemical engineers that are best approached in a one-dimensional, energy-first approach. Then in Part IV we introduce the two- or three-dimensional, momentum-first approach, and discuss some of the chemical engineering problems that are best approached that way.

Figure 1.14 shows a chemical processing plant, in which lower-price chemicals are converted to higher-price (more useful) chemicals for profit and social benefit (and jobs for chemical engineers!). Many readers of this book will participate in the design, construction, and/or operation of similar plants. In such a plant the fluid flows are almost entirely inside pipes, pumps, vessels, fractionators, reactors, etc. We keep them inside because they are too valuable to waste and/or because their release would be dangerous or polluting. Almost all the flows in such a plant are most easily studied, predicted, and managed by the one-dimensional, mass-and-energy balance approach that forms Parts II and III of this book.

Figure 1.15 shows a schematic of a "cabin-type" industrial furnace. These are widely used for pyrolysis and reforming reactions in chemical engineering. Fifty years ago these were designed by hand using the one-dimensional methods presented in Parts II and III. With the recent spectacular advances in computer power, such furnaces are now designed using the two- and three-dimensional fluid mechanics methods presented in Part IV. Those methods and their computer implementation were largely developed by aeronautical engineers, to deal with the inherently

FIGURE 1.14
The mercaptan manufacturing unit at the Borger, Texas, complex of the Chevron Phillips Chemical Company. This plant is full of fluid flows, almost all of which are inside pipes, pumps, distillation columns, and associated vessels. (Courtesy of the Phillips Petroleum Company.)

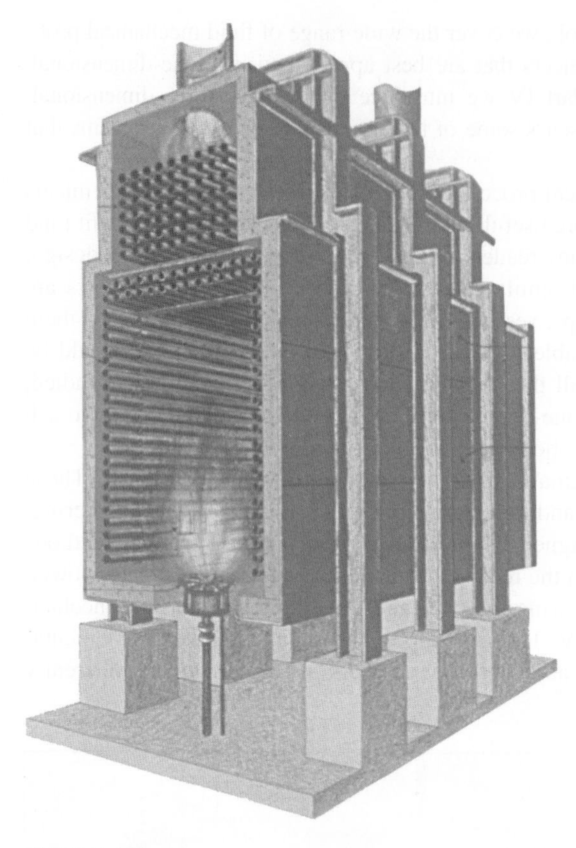

FIGURE 1.15
Cutaway drawing of a modern industrial furnace. The external steel frame supports the high-temperature refractory ceramic walls. There are multiple burners at the bottom, of which only one is shown. The flame heats the walls and the pipes through which the fluid being heated flows. Above the combustion chamber the hot gases pass over another bank of tubes, in which cooler fluid is warmed by the hot gases before they pass up the exhaust stacks seen at the top. (Courtesy of John Zinc Co. LLC.)

three-dimensional flow around airplanes. Furnace designers and other chemical engineers now use large computer codes to model the simultaneous three-dimensional fluid flow, heat transfer, and chemical reactions in such furnaces. The improvement in computational accuracy more than repays the additional cost and complexity. Part IV only introduces the basic ideas underlying such computations, and gives a bit of their history.

1.12 SUMMARY

1. Fluid mechanics is the study of forces and motions in fluids.
2. Fluids are substances that move continually when subjected to a shear force as long as the force is applied. Solids are substances that deform slightly when subjected to a shear force and then stop moving and permanently resist the force. There are, however, intermediate types of substance; the distinction between solid and liquid is one of degree rather than of kind.
3. Fluid mechanics is based on the principle of the conservation of matter, the first two laws of thermodynamics, Newton's laws of motion, and careful experiments.
4. Gases have weak intermolecular attractions and expand without limit. Liquids have much stronger intermolecular attractions and can expand very little. With increasing temperature and pressure, the differences between liquids and gases gradually disappear.
5. Density is mass per unit volume. Specific gravity of liquids is density / (density of water at 4°C). Specific gravity of gases is density / (density of air at the same T and P).

6. Viscosity is a measure of a fluid's resistance to flow. Most simple fluids are represented well by Newton's law of viscosity. The exceptions (non-Newtonian fluids) are generally complex mixtures, some of which are of great practical significance. Kinematic viscosity is viscosity divided by density.
7. Surface tension is a measure of a liquid's tendency to take a spherical shape, caused by the mutual attraction of the liquid's molecules.
8. Pressure is compressive force divided by area. It is the same in all directions for a fluid at rest and practically the same in all directions for most moving fluids.
9. In handling the units (dimensions) in this text, one should always write down the units of any dimensioned quantity and then multiply or divide by 1 to find the desired units in the answer.
10. Much of fluid mechanics can be based either on force and momentum, or on energy. This book, for chemical engineers, bases most of fluid mechanics on energy, thus dealing mostly with scalars instead of vectors. Momentum and vectors are used where they are needed.

PROBLEMS

See the Common Units and Values for Problems and Examples inside the back cover of this text. An asterisk (*) on the problem number indicates that the answer is in App. D.

1.1. In Sec. 1.3 the basic laws on which fluid mechanics rests are listed. How many of the basic laws of nature are not included in the list? To answer this question, make a list of what you consider to be the basic laws of nature. By basic laws, we mean laws that cannot be derived from other more basic ones; for example, Galileo's "laws of falling bodies" can be derived form Newton's laws and are not basic.

1.2. At low pressures there is a significant difference between the densities of liquids and of gases. For example, at 1 atm the densest gas known to the author is uranium hexafluoride, which has $M = 352$ g / mol; its normal boiling point is 56.2°C. Calculate its density in the gas phase at 1 atm and 56.2°C, assuming that it obeys the ideal gas law. The least dense liquid known to the author is liquid hydrogen, which at its normal boiling point, 20 K, has a density of 0.071 g / cm^3. Liquid helium also has a very low density, about 0.125 g / cm^3 (at 4 K). Excluding these remarkable materials, make a list of liquids which at 1 atm can exist at densities of less than 0.5 g / cm^3. A good source of data is *The Handbook of Chemistry and Physics,* CRC Press, Boca Raton, Florida, annual editions.

1.3.*For some oil and gas drilling operations we need a high-density drilling fluid (called "drilling mud"). Repeat Example 1.1 for a mud that is 50 wt. % water, 50 wt. % $BaSO_4$ (barite), $SG_{barite} = 4.49$.

1.4. Why are specific gravities most often referred to the density of water at 4°C instead of 0°C?

1.5.*A special-purpose piece of laboratory glassware, called a *pyncnometer,* is used to measure liquid densities. It has a volume of 25 cc, and a mass of 17.24 g when it is full of air. When filled with a liquid of unknown density, its mass = 45.00 g. What is the density of this liquid? How large an error do we make if we ignore the mass of the air that was in it when we weighed it and found $m = 17.24$ g?

1.6. The American Petroleum Institute (API) gravity (used extensively in the petroleum industry) is defined, in "degrees," by

$$\text{Deg API} = \frac{141.5}{\text{specific gravity}} - 131.5 \tag{1.AG}$$

Here the specific gravity is the ratio of the density of the liquid to that of water, both at 60°F. Sketch the relation between density in g / cm^3 and degrees API. What advantages of this scale might have led the petroleum industry to invent and adopt it?

1.7. Estimate the specific gravities (gas) for methane and propane. Their molecular weights are shown inside the back cover. (Commercial natural gas and commercial propane are mostly methane and propane, with small amounts of other substances, which may be ignored for this problem.) Which is more dangerous, a natural gas leak or a propane leak? Why?

1.8. What are the dimensions of dV / dy? What are the dimensions of shear stress? Shear stress in liquids is often called "momentum flux" [2]. Show that shear stress has the same dimensions as momentum / (area · time). What are the dimensions of viscosity?

1.9. List as many applications as you can of industrial, domestic, or other materials in which non-Newtonian viscosity behavior is desirable. In each case specify why this behavior is desirable.

1.10. In Example 1.2 we replaced a cylindrical problem with a linear approximation. The velocity distribution for this flow, taking the cylindrical character into account (see Prob. 15.22 and also [2], p. 91) is

$$V_\theta = \omega\left(\frac{k^2}{1 - k^2}\right)\cdot\left(\frac{R^2}{r} - r\right) \tag{1.AH}$$

where R is the radius of the outer cylinder, r is the local radius, $k = r_{\text{inner cylinder}} / R$, and ω is the angular velocity of the inner cylinder.

(*a*) Verify that this distribution shows a zero velocity at the radius of the outer, non-moving cylinder and shows $V_\theta = \omega kR$ at the surface of the inner, rotating cylinder.

(*b*) The shear rate in cylindrical coordinates, for a fluid whose velocity depends only on r (equivalent to dV / dy in rectangular coordinates) is given by

$$\sigma = \begin{pmatrix}\text{shear rate cylindrical} \\ \text{coordinates}\end{pmatrix} = r\frac{d}{dr}\left(\frac{V_\theta}{r}\right) \tag{1.AI}$$

Show that for the above velocity distribution, the shear rate at the surface of the inner cylinder is given by

$$\sigma = \omega\left(\frac{2}{1 - k^2}\right) \tag{1.AJ}$$

(*c*) Show that the shear rate computed by Eq. 1.AJ using the values in Example 1.2 is 12.26 / s, which is 1.15 times the value for the flat approximation in Example 1.2. The manual for the viscometer shown in Fig. 1.5 provides formulae equivalent to those in this problem.

1.11.*Calculate the surface / volume of a sphere, a cube, and a right cylinder of height equal to diameter. Which has the least surface / volume?

1.12. A liquid under tensile stress is unstable [9]; a small disturbance can cause it to boil and thereby change to a stable state. Make a list of other unstable situations demonstrable in a chemistry or physics laboratory. The working criterion of instability is that a very small disturbance can cause a large effect.

1.13. Earth may be considered a sphere with a diameter of ≈8000 mi and an average SG of ≈5.5. What is its mass? What is its weight? Explain your answer.

1.14. A cubic foot of water at 68°F = 20°C weighs 62.3 lbf on earth.
(*a*) What is its density?
(*b*) What does it weigh on the moon ($g \approx 6 \text{ ft / s}^2$)?
(*c*) What is its density on the moon?

1.15.*How many U.S. gallons are there in a cubic mile? The total proven oil reserves of the U.S. are roughly 30×10^9 bbl. How many cubic miles is this?

1.16. In electrochemical equations it is common to write in the symbol $\mathcal{F}$ (called *Faraday's constant*) to remind the user to convert from moles of electrons to coulombs. This is just like the force-mass and thermal energy-mechanical energy conversion factor, namely,

$$\mathcal{F} = 1 = \frac{96{,}500 \text{ C}}{\text{g equiv of electrons}} \tag{1.AK}$$

1 g equiv of electrons = $6.02 \cdot 10^{23}$ electrons. How many electrons are there in 1 C?

1.17. Older thermodynamics and fluids textbooks not only put the symbol g_c into equations to remind us to make the force-mass conversion but also put a J in equations to remind us to make the conversion from mechanical units of energy (e.g., ft · lbf) to thermal units of energy (e.g., Btu). Equation 1.11 shows the values of $g_c = 1$ for a variety of systems of units. Show the corresponding equation for J. (The use of the symbol g_c caused confusion because it is similar to g. Is there a symbol with which the J discussed in this problem can be confused?)

1.18.*As discussed in the text, the slug and the poundal were invented to make the conversion factor (mass length) / (force time^2) have a coefficient of 1. A new unit of length or a new unit of time could just as logically have been invented for this. Let us name those units the *toof* and the *dnoces*. What are the values of the toof and the dnoces in terms of the foot and the second?

1.19. In U.S. irrigation practice water is measured in acre-feet, which is the volume of water that covers an acre of land, one foot deep. What is the mass of an acre-foot of water (1 mi^2 = 640 acres)? What is the mass of a hectare-meter (ha · m) of water (km^2 = 100 ha)? Why would the acre-foot be a practical measure of irrigation water?

1.20. Einstein's equation $E = mc^2$ indicates that the speed of light squared must be expressible in units of energy per unit mass. What is the value of the square of the speed of light in Btu / lbm? In J / kg? The speed of light $c \approx 186{,}000$ mi / s = $2.998 \cdot 10^8$ m / s.

1.21.*A common basis for comparing rocket fuel systems is the *specific impulse,* defined as lbf of thrust produced divided by lbm / s of fuel and oxidizer consumed (see Chap. 7). The common values are 250 to 400 lbf · s / lbm. We frequently see the specific impulse referred to simply as "300 s." Is 300 s the same thing as 300 lbf s / lbm? European engineers regularly express the same quantity in terms of the equivalent exhaust velocity of the rocket. If a rocket has a specific impulse of 300 lbf s / lbm, what is its equivalent exhaust velocity?

1.22. Most U.S. engineers work with heat fluxes with the unit Btu / (h ft^2). In the rocket business the common unit is cal / (s cm^2). How many Btu / (h ft^2) is 1 cal / (s cm^2)? The proper SI unit is J / (m^2 s). How many Btu / (h ft^2) = 1 J / (m^2 s)?

1.23.*The Reynolds number, discussed in Chap. 6, is defined for a pipe as (velocity · diameter · density) / viscosity. What is the Reynolds number for water flowing at 10 ft / s in a pipe with a diameter of 6 in? What are its dimensions?

1.24. The flow of fluids through porous media (such as oil sands) is often described by *Darcy's equation* (see Chap. 11):

$$\frac{\text{Flow}}{\text{Area}} = \frac{\text{permeability}}{\text{viscosity}} \cdot \text{pressure gradient} \tag{1.AL}$$

The unit of permeability is the *darcy,* which is defined as that permeability for which a pressure gradient of 1 atm / cm for a fluid of 1 cP viscosity produces a flow of 1 cm^3 / s through an area of 1 cm^2. What are the dimensions of the darcy? What is its numerical value in the dimension? Give the answer both in English units and in SI units.

1.25.*What mass (weight?) would be needed in Example 1.3 if the liquid had been water?

1.26. Determine the value of X in the equation,

$$1.0 \frac{\text{Btu}}{\text{lbm} \cdot {}^\circ\text{F}} = X \frac{\text{cal}}{\text{g} \cdot {}^\circ\text{C}} \tag{1.AM}$$

1.27. In strict SI, the only unit of pressure is the Pascal (Pa). The most widely used derived unit is the bar (bar = 10^5 Pa = 0.1 MPa). What is the relation between the bar and the pressure of the atmosphere at sea level? Why is the bar a popular choice for a working SI derived unit?

1.28.*Air pollutant emissions from autos and trucks in the United States are reported in a mixed metric-English unit, g / mi. Suggest reasons why this might be a practical unit.

1.29. Many European pressure gauges give the pressure in kg / cm^2. Is this kgm or kgf? Why would this be a convenient unit of pressure?

1.30. In the third part of Example 1.6, what would have happened if we had taken the force-mass conversion factor as 32.2 lbm · ft / (lbf · s^2) instead of 9.8 kgm · m / (kgf · s^2)?

REFERENCES FOR CHAPTER 1

1. Reiner, M. "The Flow of Matter." *Scientific American 201(6),* (1959), pp. 122–138.
2. Bird, R. B., E. N. Stewart, and W. E. Lightfoot, *Transport Phenomena,* 2nd ed. New York: Wiley, 2002.
3. Poling, B. E., J. M. Prausnitz, and J. P. O'Connell, *The Properties of Gases and Liquids,* 5th ed., New York: McGraw-Hill, 2001.
4. Schneirla, T. C., and G. Piel, "The Army Ant." *Scientific American 178(6),* (1948), pp. 16–23.
5. Boys, C. V. *Soap Bubbles and the Forces Which Mould Them,* paperback ed., Garden City, New York: Doubleday, 1959. First published 1902. This interesting, informative, semitechnical book is highly recommended. Reading time, about 3 h.
6. Davies, J. T., and E. K. Rideal, *Interfacial Phenomena,* 2nd ed. New York: Academic Press, 1963.
7. Lodge, A. S. *Elastic Liquids; An Introductory Vector Treatment of Finite-Strain Polymer Rheology.* London: Academic Press, 1964.
8. Kaschiev, D. *Nucleation; Basic Theory with Applications.* Oxford: Butterworth Heinemann, 2000.
9. Zimmerman, M. H. "How Sap Moves in Trees." *Scientific American 208(3),* (1963), pp. 133–142.
10. de Nevers, N. "The Poundal per Square Foot, the Pascal and SI Units." *Engineering Education 78(2),* (1987), p. 137.
11. Pollack, A. "Missing What Didn't Add Up, NASA Subtracted an Orbiter." *New York Times* (Oct. 1, 1999), Section A, p. 1.
12. ASTM, *Metric Practice Guide,* ASTM Publication # E 380-97[e], Philadelphia, PA: ASTM, 1997.
13. Leavitt, R. "Kyoto" in *Fodor's Japan and Korea, 1982,* edited by A. Tucker. New York: Fodor's Modern Guides, Inc., 1982.

PART I

PRELIMINARIES

This part of the book (which includes the previous chapter) covers topics we need before we begin the study of flowing fluids. Chapter 1 introduced the basic idea of a fluid and its most important properties, as well as discussing units. Chapter 2 discusses fluid statics, the behavior of nonmoving fluids. We will need some of its results in subsequent chapters, but some are useful directly.

Chapters 3 and 4 discuss the conservation of mass and of energy. These are basic tools of the chemical engineer. Most readers of this book have been exposed to these in their previous course work. The treatment here is partly a review of that material and partly a recasting of these ideas in the form that is useful for the study of flowing fluids.

CHAPTER 2

FLUID STATICS

In this chapter we apply Newton's law of motion, $F = ma$, to fluids at rest. We will see that this leads to a remarkably simple equation:

$$\frac{dP}{dz} = -\rho g \tag{2.1}$$

This equation and its applications are almost the whole of fluid statics.

In Chap. 7 we will apply Newton's law of motion to moving fluids. This chapter is really only part of the more general application made in Chap. 7. In Chaps. 5, and 6, however, we will need some of the results from this chapter, and the kinds of problem we deal with here are different from (and simpler than) those in Chap. 7; for these reasons a separate chapter on fluid statics is practical at this point. Remember that all we do in this chapter is apply $F = ma$ to a static fluid; the more general applications, covering both moving and static fluids, are discussed in Chap. 7.

In the most careful work, we would write Eq. 2.1 as a vector equation, because the acceleration of gravity has both magnitude and direction, as does the *gradient* of the pressure. We will see that this chapter can be developed, to give all the correct and useful results, without using vector calculus. But remember than any application of Newton's law of motion is a vector application. We will say more about that in the introduction to Chap. 7, where we consider the classes of problems in which we must use the vector nature of forces and of momentum, and again in Chap. 15, where we reintroduce Newton's law in three-dimensional vector calculus form.

2.1 THE BASIC EQUATION OF FLUID STATICS

For a simple fluid at rest the pressure is the same in all directions. This idea seems hard for students. Suppose you compress a small coil spring between your thumb and forefinger. The spring exerts the same force on thumb and forefinger, but in opposite directions. If you rotate your hand any way, it still exerts equal forces in two opposite directions. So also with pressures in fluids at rest. There are no shear stresses in a fluid at rest. These facts lead to the basic equation of fluid statics. Consider a small block of fluid that is part of a large mass of fluid at rest in a gravity field; see Fig. 2.1. Since the fluid is at rest, there are no accelerations, and the sum of the forces on any part of the fluid in any direction is zero. Let us consider the z direction, opposite to the direction of gravity. The forces that act on the small block of fluid in the z direction are the pressure forces on the top and bottom and the force of gravity acting on the mass of the element. Their sum (positive upward) is

$$(P_{z=0})\,\Delta x\,\Delta y - (P_{z=\Delta z})\,\Delta x\,\Delta y - \rho g\,\Delta x\,\Delta y\,\Delta z = 0 \tag{2.2}$$

Dividing by $\Delta x\,\Delta y\,\Delta z$ and rearranging, we find

$$\frac{P_{z=\Delta z} - P_{z=0}}{\Delta z} = -\rho g \tag{2.3}$$

If we now let Δz approach zero, then

$$\lim_{\Delta z \to 0} \frac{\Delta P}{\Delta z} = \frac{dP}{dz} = -\rho g \tag{2.1}$$

This is the basic equation of fluid statics, also called the *barometric equation*. It is correct only if there are no shear stresses on the vertical faces of the cube in Fig. 2.1. If there are such shear stresses, then they may have a component in the vertical direction, which must be added into the sum of forces in Eq. 2.2. For simple Newtonian fluids, shear stresses in the vertical directions can exist only if the fluid has a different vertical velocity on one side of the cube from that on the other (see Eq. 1.4). Thus, Eq. 2.1 is correct if the fluid is not moving at all, which is the case in fluid statics; or if it is moving but only in the x and y directions; or if it has a uniform velocity in the z direction. In this chapter we will apply it only when a fluid has no motion relative to its container or to some set of fixed coordinates. In later chapters we will

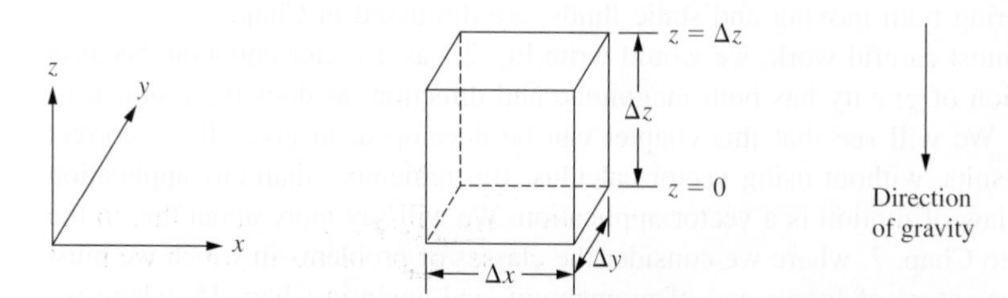

FIGURE 2.1
A small cube of fluid at rest.

apply it to flows in which there is no motion in the z direction or there is a motion with a uniform z component. We will also describe nonmoving fluids in accelerated motion in this chapter.

For complicated fluids, such as toothpaste, paints, and jellies, Eq. 2.1 is not correct, because these fluids can sustain small but finite shear stresses without any motion. The equation simply is not applicable. To find its equivalent, it is necessary to make up a sum of forces that includes shear forces on the vertical sides of the cube.

The barometric equation describes the change in pressure with distance upward, where "upward" is opposite to the direction of gravity, called z. (The minus sign appears in Eq. 2.1 because gravity points in the minus z direction.) If we want to know the change of pressure with distance in some other, nonvertical direction, call it direction a, then we can write

$$\frac{dP}{da} = \frac{dz}{da} \cdot \frac{dP}{dz} = -\rho g \frac{dz}{da} \tag{2.4}$$

But, as shown in Fig. 2.2,

$$\frac{dz}{da} = \frac{\Delta z}{\Delta a} = \cos\theta \tag{2.5}$$

where θ is the angle between the direction a and the z axis. Substituting this equation into Eq. 2.4, we have

$$\frac{dz}{da} = \frac{\Delta z}{\Delta a} = \cos\theta \quad \text{or} \quad \frac{dP}{da} = -\rho g \cos\theta \tag{2.6}$$

A particularly interesting direction a is the one at right angles to z, that is, any direction parallel to the x-y plane. For that direction θ is 90°, $\cos\theta$ is 0, and the pressure does not change with distance. Thus, from Eq. 2.6 we see that for a fluid at rest any surface that is perpendicular to the direction of gravity is a surface of constant pressure. The most interesting constant-pressure surface of a body of fluid at rest is the one with zero gauge pressure, that is, the surface in contact with the atmosphere. Since this is a constant-pressure surface, it must be everywhere perpendicular to the direction of gravity. On a global scale this makes the free surface of the oceans practically a sphere. (The earth is not quite spherical, being slightly flattened at the poles.) In typical engineering operations it means that the free surface of a liquid exposed to the atmosphere is practically a horizontal plane (Prob. 2.1).

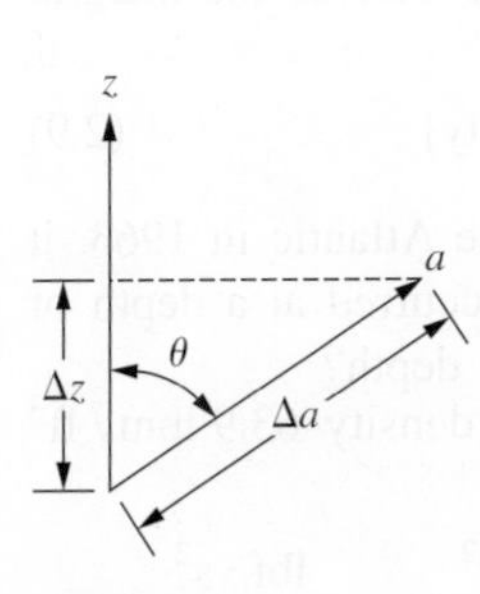

FIGURE 2.2
Relations between the a and z directions.

The product of density and gravity, which appears in Eq. 2.1, is often called the *specific weight,* and is given the symbol γ:

$$\rho g = \gamma = \text{specific weight} \tag{2.7}$$

At any place where the acceleration of gravity is equal to $32.2 \text{ ft/s}^2 = 9.81 \text{ m/s}^2$ (practically any place on the surface of the earth), the specific weight expressed in lbf/ft^3 (or kgf/m^3) is numerically equal to the density expressed in lbf/ft^3 (or kgf/m^3). The value in N/m^3 is numerically different.

Example 2.1. Calculate the specific weight of water at a place where the acceleration of gravity is 32.2 ft / s.

$$\gamma = \rho g = 62.3 \frac{\text{lbm}}{\text{ft}^3} \cdot 32.2 \frac{\text{ft}}{\text{s}^2} \cdot \frac{\text{lbf} \cdot \text{s}^2}{32.2 \text{ lbm} \cdot \text{ft}} = 62.3 \frac{\text{lbf}}{\text{ft}^3}$$

$$= 998.2 \frac{\text{kgf}}{\text{m}^3} = 9792 \frac{\text{N}}{\text{m}^3} \tag{2.A}$$

■

If one deals principally with fluid flows in which the forces of gravity are dominant, then one often can simplify the calculations by replacing the density in all equations by γ / g. In civil engineering hydraulics this is normally the case, and this is common practice. On the other hand, if one deals mostly with flows whose gravity terms are small compared with the other terms, then it is more convenient to work with ρ than with γ / g. In chemical engineering problems the gravity terms are normally small, so the specific weight is seldom used.

2.2 PRESSURE-DEPTH RELATIONSHIPS

Equation 2.1 is a separable, first-order differential equation that can be separated and integrated as follows:

$$\int dP = -\int \rho g \, dz \tag{2.8}$$

However, to perform the integration, it is necessary to have some relation between ρ, g, and z. In situations on the surface of the earth g is practically constant (see Sec. 2.8), so we may take it outside the integral sign. Several possible relations between ρ and z lead to simple integrations of the equation, as shown in the following material.

2.2.1 Constant-Density Fluids

No real substances have constant density; the density of all substances increases as the pressure increases. However, for most liquids at temperatures far below their critical temperatures, the effect of pressure on density is very small. For example, raising the pressure of water at 100°F from 1 to 1000 psia while holding the temperature constant causes the density to increase by 0.3%. In most engineering calculations we can neglect such small changes in density. Then we can take ρ outside the integral sign in Eq. 2.8 and find that the pressure change is

$$P_2 - P_1 = -\rho g(z_2 - z_1) \qquad \text{[constant density]} \tag{2.9}$$

Example 2.2. When the submarine *Thresher* sank in the Atlantic in 1963, it was estimated in the newspapers that the accident had occurred at a depth of 1000 ft (304.9 m). What is the pressure of the sea at that depth?

Seawater may be considered incompressible, with density 63.9 lbm / ft³ (1024 kg / m³). Thus

$$P_{1000 \text{ ft}} = 14.7 \frac{\text{lbf}}{\text{in}^2} + 63.9 \frac{\text{lbm}}{\text{ft}^3} \cdot 32.2 \frac{\text{ft}}{\text{s}^2} \cdot 1000 \text{ ft} \cdot \frac{\text{ft}^2}{144 \text{ in}^2} \cdot \frac{\text{lbf} \cdot \text{s}^2}{32.2 \text{ lbm} \cdot \text{ft}}$$

$$= 14.7 \frac{\text{lbf}}{\text{in}^2} + 444 \frac{\text{lbf}}{\text{in}^2} = 459 \frac{\text{lbf}}{\text{in}^2} \tag{2.B}$$

or

$$P_{304.9\text{ m}} = 101.3\text{ kPa} + 1024\frac{\text{kg}}{\text{m}^3}\cdot 9.81\frac{\text{m}}{\text{s}^2}\cdot 304.9\text{ m}\cdot\frac{\text{Pa}}{\text{N / m}^2}\cdot\frac{\text{N}\cdot\text{s}^2}{\text{kg}\cdot\text{m}}$$

$$= (101.3 + 3062.8)\text{ kPa} = 3.164\text{ MPa} \tag{2.C}$$

■

In hydraulics problems and in all problems involving a free surface exposed to the atmosphere, we can further simplify Eq. 2.9 by working in gauge pressure. The gauge pressure is zero at the free surface: $P_{1\text{ gauge}} = 0$. We now define the depth as the distance measured downward from the free surface and give it the symbol h,

$$h = z_{\text{free surface}} - z \tag{2.10}$$

in which case Eq. 2.9 simplifies to

$$P = \rho g h \qquad \text{[gauge pressure, constant density]} \tag{2.11}$$

In Example 2.2, at $h = 1000$ ft the gauge pressure is $P = 444$ psig $= 3062.8$ kPa, gauge.

Example 2.3. A cylindrical oil-storage tank is 60 ft deep and contains an oil of density 55 lbm / ft³. Its top is open to the atmosphere. What is the gauge-pressure-depth relation in this tank?

The gauge pressure is zero at the free surface. At the bottom it is

$$P_{\text{bottom}} = 55\frac{\text{lbm}}{\text{ft}^3}\cdot 32.2\frac{\text{ft}}{\text{s}^2}\cdot 60\text{ ft}\cdot\frac{\text{ft}^2}{144\text{ in}^2}\cdot\frac{\text{lbf}\cdot\text{s}^2}{32.2\text{ lbm}\cdot\text{ft}} = 22.9\frac{\text{lbf}}{\text{in}^2} = 158\text{ kPa} \tag{2.D}$$

From Eq. 2.11 we know that the pressure-depth relation is linear; see Fig. 2.3. ■

2.2.2 Ideal Gases

The density of gases changes significantly with pressure changes, so we must be cautious about taking the density outside the integral sign in Eq. 2.8. At low pressure the densities of most gases are well approximated by the ideal gas law,

$$\rho = \frac{PM}{RT} \qquad \text{[ideal gas]} \tag{2.12}$$

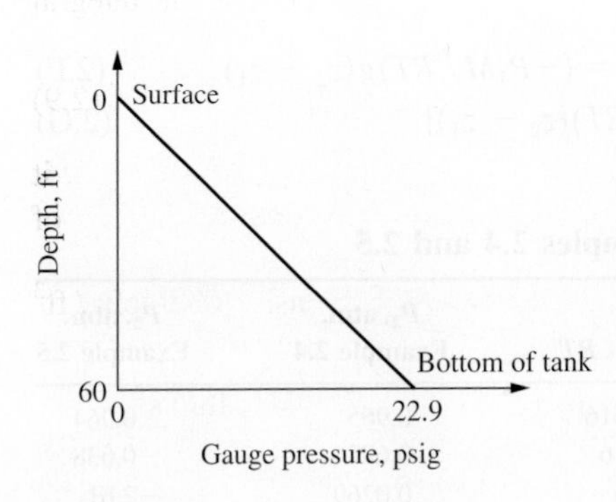

FIGURE 2.3
Pressure-depth relation in Example 2.3.

Here T is the absolute temperature, in ° Rankine or in Kelvins ($T\,°\text{R} = T\,°\text{F} + 459.69$, or $T\,\text{K} = T\,°\text{C} + 273.15$); R is the universal gas constant, whose value in various systems of units is shown on the inside back cover; M is the molecular weight, normally expressed in g / mol or lbm / lbmol. [This formulation of the ideal gas law gives the density in units of lbm / ft³. In chemistry one often sees the ideal gas law written as $\rho = P / RT$, which gives the density in lbmol / ft^3 or mol / m^3. Multiplying the latter density by the molecular weight (in lbm / lbmol or g / mol) gives the density, shown here.]

Substituting Eq. 2.12 for the density in Eq. 2.1, we find

$$\frac{dP}{dz} = -\frac{PM}{RT}g \quad \text{[ideal gas]} \tag{2.13}$$

If the temperature is constant, this can be separated and integrated as follows:

$$\int_1^2 \frac{dP}{P} = \frac{-gM}{RT}\int_1^2 dz \tag{2.14}$$

$$\ln\frac{P_2}{P_1} = \frac{-gM}{RT}(z_2 - z_1) \tag{2.15}$$

$$P_2 = P_1 \exp\left(\frac{-gM\Delta z}{RT}\right) \quad \text{[isothermal, ideal gas]} \tag{2.16}$$

Example 2.4. At sea level the atmospheric pressure is 14.7 psia and the temperature is 59°F = 15°C = 519°R. Assuming that the temperature does not change with elevation (a poor assumption, but one that simplifies the mathematics and that will be reexamined in a few pages), calculate the pressure at 1000, 10,000, and 100,000 ft. For $z = 1000$ ft, we find

$$P_2 = P_1 \exp\left(\frac{-32.2\ \text{ft}/\text{s}^2 \cdot 29\ \text{lbm}/\text{lbmol} \cdot 1000\ \text{ft}}{(10.73\ \text{lbf}/\text{in}^2 \cdot \text{ft}^3/\text{lbmol}\cdot{}^\circ\text{R}) \cdot 519^\circ\text{R}} \cdot \frac{\text{ft}^2}{144\ \text{in}^2} \cdot \frac{\text{lbf}\cdot\text{s}^2}{32.2\ \text{lbm}\cdot\text{ft}}\right)$$

$$= P_1 \exp(-0.03616) = \frac{P_1}{\exp 0.03616} = \frac{P_1}{1.0368} = 0.965\ \text{atm} \tag{2.E}$$

We can calculate the pressures at the other two elevations and show them, along with the results from the next example, in Table 2.1. ■

How much error would we have made if we had used the constant-density formulae instead of taking the change in density into account?

Example 2.5. Rework Example 2.4, assuming that air is a constant-density fluid, which has the same density at all elevations as it has at 14.7 psia and 59°F.

Here we use Eq. 2.9:

$$P_2 - P_1 = -\rho_1 g(z_2 - z_1) = (-P_1 M / RT) g (z_2 - z_1) \tag{2.F}$$

$$P_2 = P_1[1 - (gM/RT)(z_2 - z_1)] \tag{2.G}$$

TABLE 2.1

Calculated atmospheric pressures for Examples 2.4 and 2.5

Elevation, z, ft	Elevation, z, m	$gM\,\Delta z / RT$	P_2, atm, Example 2.4	P_2, atm, Example 2.5
1,000	304.8	0.03616	0.965	0.964
10,000	3,048	0.3616	0.697	0.638
100,000	30,480	3.616	0.0269	−2.61

For 1000 feet we find

$$P_2 = P_1 \cdot \left(1 - \frac{32.2 \text{ ft/s}^2 \cdot 29 \text{ lbm/lbmol} \cdot 1000 \text{ ft}}{(10.73 \text{ lbf/in}^2 \cdot \text{ft}^3\text{/lbmol} \cdot °R) \cdot 519°\text{R}} \cdot \frac{\text{ft}^2}{144 \text{ in}^2} \cdot \frac{\text{lbf} \cdot \text{s}^2}{32.2 \text{ lbm} \cdot \text{ft}}\right)$$

$$= P_1 \cdot (1 - 0.03616) = 0.964 \text{ atm} \tag{2.H}$$

This value, plus the corresponding ones for 10,000 and 100,000 ft, are shown in Table 2.1. ■

From Table 2.1 we see that up to 1000 ft the assumption of constant density creates a negligible error, at 10,000 ft it makes a 9% error, and at 100,000 ft it gives absurd results (negative absolute pressure in a gas??). Thus, for ordinary industrial-sized equipment (generally less than 1000 ft high) one can accurately calculate changes in gas pressure with elevation as if the gas had a constant density. On the other hand, in aeronautics and meteorological problems, in which the elevations are often from 10,000 to 100,000 ft, this simplification leads to disastrous errors.

In Examples 2.4 and 2.5 we made the simplifying assumption that the atmosphere was isothermal. Anyone who has gone to the mountains in the summer to get out of the heat did so because the atmosphere is not isothermal. To understand why the air temperature decreases with elevation, consider a mass of air being lifted from one elevation to a higher one (by a wind, for example, blowing it over a mountain range). The air mass expands because the pressure of the surrounding air decreases as it rises. The air mass is cooled because as it expands it does expansion work on the surrounding air. Air is a fairly poor conductor of heat, so during this process the rising air undergoes an expansion that is close to adiabatic and close to reversible. If it were exactly reversible and adiabatic, then the temperature-pressure-elevation relation would be exactly the isentropic one. For an isentropic atmosphere one can work out the following elevation-temperature and elevation-pressure relationships (Prob. 2.16):

$$P_2 = P_1\left(1 - \frac{k-1}{k} \cdot \frac{gM\Delta z}{RT_1}\right)^{k/(k-1)} \quad \text{[isentropic, ideal gas]} \tag{2.17}$$

$$T_2 = T_1\left(1 - \frac{k-1}{k} \cdot \frac{gM\Delta z}{RT}\right) \quad \text{[isentropic, ideal gas]} \tag{2.18}$$

Here k is the ratio of specific heats (discussed in Chap. 8); for air its value is practically constant at 1.4.

The isothermal atmosphere in Examples 2.4 and 2.5 would be observed if air were a perfect conductor of heat, evening out all temperature differences instantly. The isentropic atmosphere in Eqs. 2.17 and 2.18 would be observed if air were a perfect insulator against heat conduction, transferring no heat at all. Experimental measurements show that the real behavior of the atmosphere is intermediate between these two extremes. Heat is conducted outward from the earth not only by simple conduction in the air (which is fairly slow) but also by winds, which mix cold and warm air layers, and by condensation of water vapor and by infrared radiation. For calculation purposes meteorologists and aeronautical engineers have defined a "standard atmosphere," which agrees well with the *average* of many observations over the whole planet and all seasons of the year. As shown in Fig. 2.4, this standard atmosphere is

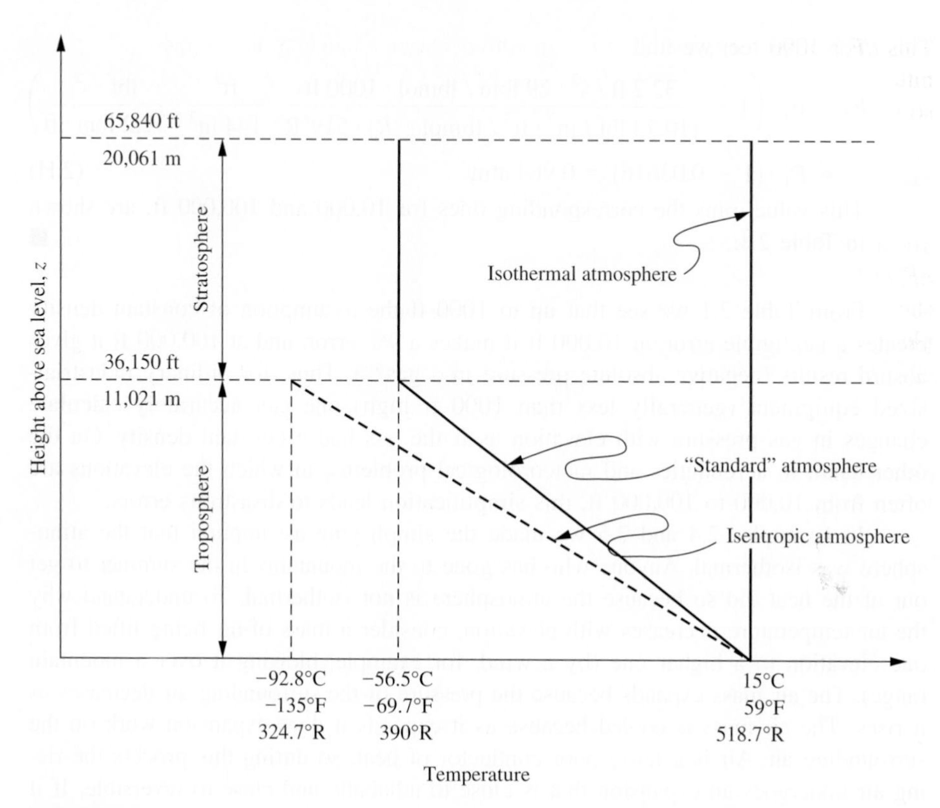

FIGURE 2.4
Comparison of standard atmosphere, isentropic atmosphere, and isothermal atmosphere.

indeed intermediate between the isothermal and isentropic atmospheres. It is an average; most interesting weather phenomena are caused by deviations from it. For a simple discussion of this see ([1], Chap. 5). From the standard-atmosphere temperature one may calculate a "standard" pressure-height curve (Prob. 2.17). Tables showing all the properties of the standard atmosphere are found in handbooks [2].

2.3 PRESSURE FORCES ON SURFACES

Static, simple fluids can exert only pressure forces on surfaces adjacent them. Since pressure is the normal (perpendicular) force per unit area, the pressure forces must act normal to the surface. Moving fluids can exert not only pressure forces but also shear forces, so the combined force exerted by a moving fluid on a surface is generally not normal to the surface. However, in problems involving moving fluids it is often convenient to treat the pressure and shear forces as separate and thus calculate the pressure force exactly as we do here, but use the pressure distribution on the surface corresponding to the flow situation rather than to the static-fluid one discussed in this chapter.

For an infinitesimal surface area the force exerted is

$$dF = P\,dA \tag{2.19}$$

This dF is a vector quantity: it has direction (perpendicular to the surface) and magnitude. For a plane surface all the differential dF vectors point in the same direction, so that we can find the total force simply by integrating this equation:

$$F = \int P \, dA \tag{2.20}$$

To calculate pressure forces on curved surfaces, we normally resolve the infinitesimal dF in Eq. 2.19 into its x and y components and integrate those. That calculation is shown in most civil and mechanical engineering fluid mechanics textbooks, for example, Reference 3 (Chap. 2).

If the pressure over an entire plane surface is constant, then Eq. 2.20 becomes

$$F = PA \qquad \text{[constant pressure, plane surface]} \tag{2.21}$$

Because the static pressure in gases changes very slowly with elevation, this is practically true for all moderate-sized flat surfaces exposed to gases, independent of the orientation of the surface. The same is not true for liquids.

For *horizontal* plane surfaces exposed to static liquids the pressure is constant over the entire surface, so Eq. 2.21 gives the required force.

Example 2.6. An oil-storage tank has a flat, horizontal, circular roof 120 ft in diameter. What force does the atmosphere exert on the roof?

$$F = PA = 14.7 \frac{\text{lbf}}{\text{in}^2} \cdot \frac{\pi}{4} (120 \text{ ft})^2 \cdot 144 \frac{\text{in}^2}{\text{ft}^2} = 2.39 \cdot 10^7 \text{ lbf} = 106.5 \text{ MN} \tag{2.I}$$

■

The roof of the storage tank can withstand this startlingly large downward force because the gas or air inside it exerts an equal upward force, so the *net* force due to the pressure of the atmosphere and the pressure of the gas inside the tank is zero. Since these forces ordinarily cancel out of force calculations, it is customary to make such calculation in gauge pressure, whenever both sides of the surface are subjected to the pressure of the atmosphere in addition to the gauge pressure of the liquid. Such tanks normally are vented to the atmosphere, to prevent having a gauge pressure or a vacuum in the tank (which it is not designed to withstand). Often the vent will have a *vapor conservation valve,* which prevents the tank from "breathing" in and out with small changes of atmospheric pressure or of the temperature of the tank contents; such valves are normally set to open for an internal pressure of vacuum of less than ± 0.1 psig. If for some reason that vent is closed (e.g., blocked by ice in a storm), then pumping liquid into the tank can bulge the tank outward, and pumping liquid out can collapse the tank inward. For fluids like gasoline this *tank breathing* can be a significant air pollution emission [1], which must be controlled.

Example 2.7. A layer of rain water 8 in deep collects on the roof of the oil-storage tank of Example 2.6. What net pressure force does it exert on the roof of the tank?

Here

$$P_{\text{gauge}} = \rho g h = 62.3 \frac{\text{lbm}}{\text{ft}^3} \cdot 32.2 \frac{\text{ft}}{\text{s}^2} \cdot \frac{8}{12} \text{ft} \cdot \frac{\text{lbf} \cdot \text{s}^2}{32.2 \text{ lbm} \cdot \text{ft}}$$

$$= 41.5 \frac{\text{lbf}}{\text{ft}^2} = 0.288 \frac{\text{lbf}}{\text{in}^2} = 1.989 \text{ kPa} \tag{2.J}$$

$$F = PA = 41.5 \frac{\text{lbf}}{\text{ft}^2} \cdot \frac{\pi}{4} (120 \text{ ft})^2 = 4.70 \cdot 10^5 \text{ lbf} = 3.24 \text{ MN} \tag{2.K}$$

■

We could have found exactly the same answer by asking what the weight W of the liquid on the roof was that is,

$$W = mg = V\rho g = \frac{8}{12} \text{ft} \cdot \frac{\pi}{4} (120 \text{ ft})^2 \cdot 62.3 \frac{\text{lbm}}{\text{ft}^3} \cdot 32.2 \frac{\text{ft}}{\text{s}^2} \cdot \frac{\text{lbf} \cdot \text{s}^2}{32.2 \text{ lbm} \cdot \text{ft}}$$

$$= 4.70 \cdot 10^5 \text{ lbf} = 3.24 \text{ MN} \tag{2.L}$$

where V is volume of the liquid. This is typical of fluid-statics problems involving horizontal surfaces. Since we found the basic equation of fluid statics by considering the weight of the fluid, we could work this kind of problem just as well by simply considering the weight of the fluid involved. This large a weight would collapse the roof of an ordinary tank and of some other light-duty structures; proper rainfall drainage is important.

For *vertical* plane surfaces the pressure is not constant over the whole surface. Therefore, Eq. 2.20 must be used to find the force, and in general we cannot take the pressure outside the integral sign.

Example 2.8. The lock gate of a canal (Fig. 2.5) is rectangular, 20 m wide and 10 m high. One side is exposed to the atmosphere, the other side to water whose top surface is level with the top of the lock gate. What is the net force on the lock gate?

The net force is the force exerted by the water on the front of the gate minus the force exerted by the atmosphere on the back of the gate. Over the short vertical distance involved, the pressure of the atmosphere may be considered constant $= P_{\text{atm}}$. Thus, the force exerted on the back of the gate by the atmosphere is

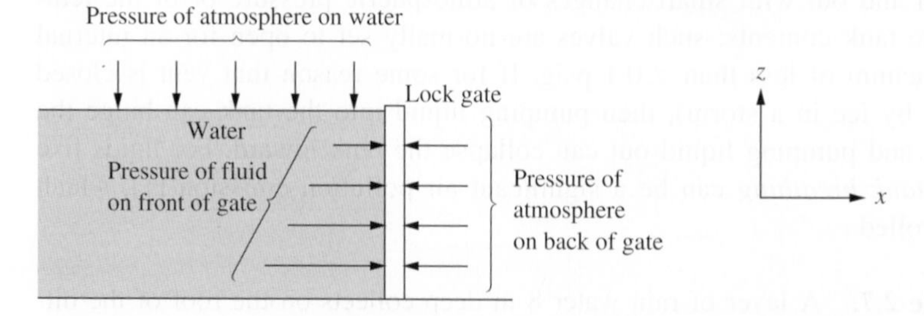

FIGURE 2.5
Horizontal pressure forces on a vertical surface.

$P_{atm}A$, where A is the area of the gate. The pressure at any point in the water is given by Eq. 2.11. Here we define W as the width of the gate and h as the depth below the free surface. Then, substituting Eq. 2.11 in Eq. 2.20, we find

$$F_{water} = \int P\,dA = \int (P_{atm} + \rho g h)\,dA = P_{atm}A + \rho g \int hW\,dh$$

$$= P_{atm}A + \rho g W \frac{h^2}{2}\bigg]_0^{10\text{ m}} \tag{2.M}$$

The net force in the x direction is

$$F_{net} = F_{water} - F_{air} = P_{atm}A + \rho g W \frac{h^2}{2}\bigg]_0^{10\text{ m}} - P_{atm}A \tag{2.N}$$

The two atmospheric pressure terms cancel each other, and

$$F_{net} = 998.2\,\frac{\text{kg}}{\text{m}^3}\cdot 9.81\,\frac{\text{m}}{\text{s}^2}\cdot 20\text{ m}\cdot \frac{h^2}{2}\bigg]_{h=0}^{h=10\text{ m}}\cdot\frac{\text{N}\cdot\text{s}^2}{\text{kg}\cdot\text{m}}$$

$$= 9.80\text{ MN} = 2.20\cdot 10^6\text{ lbf} \tag{2.O}$$

■

In this problem—and in all others in which a liquid, open to the atmosphere, acts on one side of a surface and the atmosphere acts on the opposite side—the effect of the atmospheric pressure cancels. Thus, such problems can be worked most easily by using gauge pressure. If it had been used in this problem, it would have given exactly the result just shown.

In the next section (and some other problems of practical interest) we want to know the x or y components of the pressure force on some surface that is curved, or that is flat but not perpendicular to the x or y axis. The basic procedure is to write

$$dF_x = \begin{pmatrix} x\text{-component} \\ \text{of } dF \end{pmatrix} = \sin\theta\cdot dF = \sin\theta\cdot P\,dA \tag{2.22}$$

and

$$dF_y = \begin{pmatrix} y\text{-component} \\ \text{of } dF \end{pmatrix} = -\cos\theta\cdot dF = -\cos\theta\cdot P\,dA \tag{2.23}$$

where θ is the angle between the normal to the surface and the vertical. If P is constant, then the x or y components of the pressure force are equal to P times the projected area of the surface in the x or y direction.

2.4 PRESSURE VESSELS AND PIPING

Figure 2.6 shows part of an oil refinery "tank farm." Three different types of storage vessels are shown. The largest are cylindrical with vertical axes and flat bottoms; these are used for liquids stored at approximately atmospheric pressure. The spherical tanks are used to store liquids (and rarely gases) at pressures substantially above atmospheric. The sausage-shaped tanks (horizontal cylinders with hemispherical ends) are also used to store liquids (and rarely gases) at high pressures. The choice between

FIGURE 2.6
Part of an oil refinery "tank farm" showing 24 flat-bottomed atmospheric pressure storage tanks, 2 high-pressure spherical storage tanks, and 24 high-pressure sausage-shaped storage tanks. [Courtesy of Chicago Bridge and Iron Company (CB&I).]

these three types of tank is based on economics, which is mostly driven by the necessity to make them strong enough to resist the pressure of the fluid they contain.

Returning to the oil-storage tank in Example 2.3 and Fig. 2.3, we can ask, how thick do the walls of the tank have to be to contain the fluid inside? Figure 2.7 shows an atmospheric-pressure, flat-bottomed tank like those shown in Fig. 2.6. Part (*a*) shows the whole tank, which is a cylindrical shell with a flat bottom and which rests on a concrete or gravel foundation. The tank has a lightweight roof (either flat or domed in Fig. 2.6).

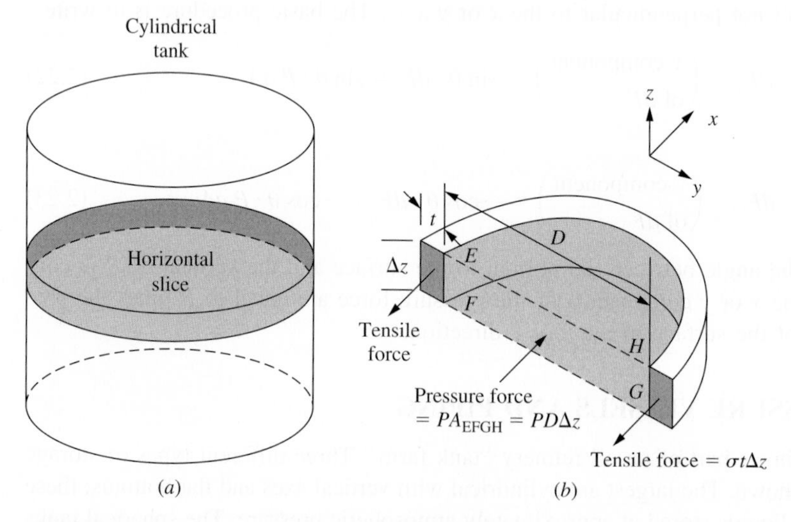

FIGURE 2.7
(*a*) A cylindrical flat-bottomed tank, showing a horizontal slice used in part (*b*). (*b*) A force balance on half of the horizontal slice, showing the pressure force in the x direction, and the tensile force in the two pieces of the tank's shell, which resists this pressure force.

Part (*a*) shows a horizontal slice (like one pancake in a stack) and part (*b*) shows that slice cut vertically (like a half pancake). Part (*b*) shows both the steel shell of the tank, whose thickness is exaggerated for clarity, and the liquid that is within the "half pancake."

Making a force balance in the x direction on the piece shown on Fig. 2.7(*b*), we see that the pressure force on the liquid surface E-F-G-H acts in the positive x direction, while the tensile forces in the two cut pieces of the shell of the tank act in the negative x direction. The liquid shown in the tank section in part (*b*) exerts a force radially outward over the part of the tank it contacts, but we are only interested in the x component of that force, which is equal to the x component of the force that the rest of the fluid exerts on this segment of fluid. We do not bother with forces in the y or z directions, because they do not concern us here. If the tank is not in the act of rupturing, then the sum of the forces in the x direction (or any other) must be zero, so we may write

$$P \cdot D \cdot \Delta z = 2\sigma_{\text{tensile}} \cdot \Delta z \cdot t \tag{2.24}$$

where σ_{tensile} is the tensile stress in the shell, P is the gauge pressure, also assumed to be uniform, and t is the thickness of the metal shell. Now we make the *thin-walled assumption* that σ_{tensile} is uniform over the wall thickness (see below). Solving for the required thickness of the shell, we find

$$t = \frac{PD}{2\sigma_{\text{tensile}}} \quad \text{[cylindrical, thin-walled assumption]} \tag{2.25}$$

The tensile stress in Eq. 2.25 is resisted by the external metal hoops in barrels and in wooden water tanks; it is normally called the *hoop stress.*

Example 2.9. If the design tensile stress (normally $\frac{1}{4}$ of the stress at rupture) of the tank wall is 20,000 psia, how thick must the shell of the tank in Example 2.3 be at the bottom of the tank? The diameter of the tank is 120 ft.

Substituting directly into Eq. 2.25, we write

$$t = \frac{(22.9 \text{ lbf / in}^2) \cdot 120 \text{ ft}}{2 \cdot 20{,}000 \text{ lbf / in}^2} = 0.0688 \text{ ft} = 0.825 \text{ in} = 2.10 \text{ cm} \tag{2.P}$$

■

From this example, we see that this large a tank requires a fairly thick wall. We also see from Fig. 2.3 that the gauge pressure falls to zero at the top, so the thickness required to resist the internal pressure has a triangular shape, thick at the bottom, zero at the top. As a practical matter we cannot have a zero thickness at the top—that would be impossible to build, would not support the roof of the tank, and would not resist wind and seismic forces. But such tanks are actually made up by welding (or bolting) together prefabricated curved plates, the thickness of which decreases from bottom to top (Prob. 2.28).

The actual design of this type of tank [4] adds a corrosion allowance to the thickness calculated in Eq. 2.25 and uses practically the thickness computed in Example 2.9 (adequate to resist pressure forces) at the bottom, but a thickness based on wind and seismic forces at the top. Equation 2.25 (and the corresponding equation for spherical containers shown below) makes the *thin-walled vessel* assumption. For pressures above about 3000 psia the required wall thicknesses become large enough that this uniform-stress assumption becomes inaccurate, and one must use *thick-walled vessel* equations, not shown here [5]. If $t/D > 0.25$ one should use such formulae. The barrels of firearms are normally thick-walled vessels, as are extremely high-pressure chemical reactors (see Prob. 2.33 and Table 2.2).

The other two types of tank shown in Fig. 2.6 (spherical and sausage-shaped) are *pressure vessels* designed to contain fluids with pressures much higher than that due to gravity (Examples 2.3 and 2.9). They are not vented to the atmosphere. The sausage-shaped tanks to the right in Fig. 2.6 are the standard storage tanks for propane, with a design pressure of 250 psig. The analysis of the required shell thickness for them is the same as in Fig. 2.7. In this case the cylinder is horizontal, so the pressure at the bottom is greater than the pressure at the top. However, these tanks are typically about 10 ft in diameter, and liquid propane has a SG of ≈0.5, so the difference in pressure from the top to the bottom of the liquid in the tank is

$$\Delta P = \rho g h = 0.5 \cdot 62.3 \frac{\text{lbm}}{\text{ft}^3} \cdot 32.2 \frac{\text{ft}}{\text{s}^2} \cdot 10 \text{ ft} \cdot \frac{\text{lbf} \cdot \text{s}^2}{32.2 \text{ lbm} \cdot \text{ft}} \cdot \frac{\text{ft}^2}{144 \text{ in}^2}$$
$$= 2.16 \text{ psi} = 14.9 \text{ kPa} \tag{2.Q}$$

which is less than 1% of the 250 psig design pressure, and is normally ignored. Thus, we may design such vessels by using Eq. 2.25.

Example 2.10. Estimate the necessary wall thickness for a horizontal cylindrical pressure vessel with a diameter of 10 ft, a working pressure of 250 psig, and a design tensile stress of 20,000 psig. This is similar to Example 2.9, in which the pressure was due to gravity,

$$t = \frac{(250 \text{ lbf}/\text{in}^2) \cdot 10 \text{ ft}}{2 \cdot (20{,}000 \text{ lbf}/\text{in}^2)} = 0.0625 \text{ ft} = 0.75 \text{ in} = 1.90 \text{ cm} \tag{2.R}$$

■

For spherical pressure vessels, the same calculation procedure leads to a different equation. Figure 2.8 shows a spherical pressure vessel cut in half, with the wall

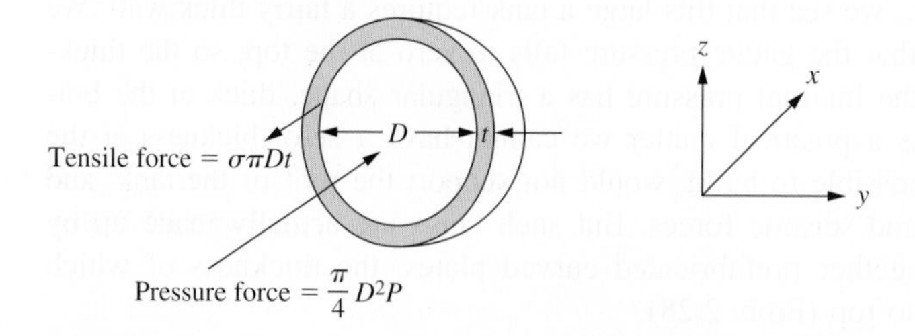

FIGURE 2.8
Half of a spherical pressure vessel, cut along its mid-plane, showing the pressure force in the x direction and the tensile force in the shell, which resists this pressure force.

thickness and diameter shown. The internal pressure force, acting in the plus x direction along the plane of the cut, is equal and opposite to the tensile stress in the wall in the plane of the cut, which acts in the minus x direction. Setting these equal and opposite, we find

$$P \cdot \frac{\pi}{4} D^2 = \sigma_{\text{tensile}} \cdot \pi D \cdot t \tag{2.26}$$

or

$$t = \frac{PD}{4\sigma_{\text{tensile}}} \quad \text{[spherical vessel, thin-walled assumption]} \tag{2.27}$$

This shows that for equal pressure and diameter, the required wall thickness for a spherical pressure vessel is exactly one-half that for a cylindrical pressure vessel. This suggests that one can store a given volume of high-pressure fluid in a container with much less metal in the walls if the container is spherical than if it is cylindrical. Figure 2.6 shows two such spherical containers in an oil refinery, holding high-pressure fluids. For space travel applications, where weight is critical, high-pressure fluids are always stored in spherical containers, to minimize container weight. The hemispherical ends of the sausage containers shown in Fig. 2.6 normally have thicknesses about one-half the thickness of the cylindrical section as suggested by Eqs. 2.25 and 2.27.

However, economics often dictate the use of the sausage-shaped containers of which 24 are shown in Fig. 2.6. These can be mass-produced in factories and shipped complete, whereas the spherical containers shown in Fig. 2.6 are too large to ship, so they are prefabricated in factories and then assembled in place. The supports for the spherical containers are more complex than those for the sausages, and the number of pieces is greater (look at the number of pieces on a soccer ball!). If a special steel with a high price per pound is needed (e.g., liquid natural gas shipping or storage), then the spherical container is often more economical. For most high-pressure liquid storage applications, the sausage container, in spite of its extra weight of metal, is often the most economical (see Prob. 2.36).

Ordinary pipes and tubes are thin-walled pressure vessels. The relation between their dimensions and their safe working pressure is given by Eq. 2.25 (with an added thickness as a corrosion allowance, and a joint efficiency term for welded pipe). The distillation towers, reflux drums, and other vessels in Fig. 1.14 are also pressure vessels, whose external shells are designed by the same formulae as are the sausage-shaped storage tanks in Fig. 2.6. For significant internal pressures (like the 250 psig in the previous examples) the required wall thickness to contain the pressure is enough that the resulting vessels are self-supporting and need no external structural support (other than foundations). Vessels with lower working pressures often have walls thin enough that they need external or internal bracing to resist gravity, wind forces, or seismic forces, [4]. The large trucks that deliver gasoline to local service stations have a standard truck chassis, on which the gasoline tank is mounted; the low-pressure gasoline tank is not very strong. The large trucks that deliver propane to regional distribution plants do not have a truck chassis; the high-pressure propane tank is strong enough that the wheels and axles are attached directly to it.

2.5 BUOYANCY

We can calculate the force exerted by static fluids on floating and immersed bodies by integrating the vertical component of the pressure force over the entire surface of the body. This leads to a very simple generalization, called *Archimedes' principle,* which is much easier to apply than the integration over the whole surface.

Consider the floating block of wood shown in Fig. 2.9. The block is at rest, so the sum of forces in any direction on it is zero. The only forces acting on it are the gravity force and the total pressure force around its entire surface; these must be equal and opposite. The vertical component of the pressure force integrated around the entire surface of a floating or submerged body is called a *buoyant force.* The buoyant force over the entire surface is then given by

$$F_{\text{vertical}} = F_z = \int -P \cos \theta \, dA \tag{2.28}$$

The $\cos \theta$ appears in this equation because the pressure forces act directly inward normal to all the surfaces they contact, whereas the vertical component is that pressure force times the cosine of the angle between the direction of the pressure force and the vertical direction, θ. For the block shown in Fig. 2.9, $\cos \theta$ is zero for the sides, -1 for the bottom, and $+1$ for the top; so

$$F_z = (P_{\text{bottom}} - P_{\text{top}}) \, \Delta x \, \Delta y \tag{2.29}$$

Here

$$(P_{\text{bottom}} - P_{\text{top}}) = \rho_{\text{liquid}} g h + \rho_{\text{air}} g (l - h) \tag{2.30}$$

Multiplying by $\Delta x \, \Delta y$ gives

$$F_z = \rho_{\text{liq}} g V_{\text{liq}} + \rho_{\text{air}} g V_{\text{air}} \tag{2.31}$$

where V_{liq} is the volume of liquid displaced, and V_{air} is the volume of air displaced. Thus the buoyant force is exactly equal to the weight of both fluids displaced. This is *Archimedes' principle.* In most cases the term for the weight of air in Eq. 2.31 is negligible compared with that for the weight of the water involved (density of water $\approx$800 times that of air!). For floating bodies Archimedes' principle is often restated: "A floating body displaces a volume of fluid whose weight is exactly equal to its own." If a body is completely immersed in a fluid, then there is only one term on the right in Eq. 2.31 and the statement becomes, "The buoyant force on a completely submerged body is equal to the weight of fluid displaced."

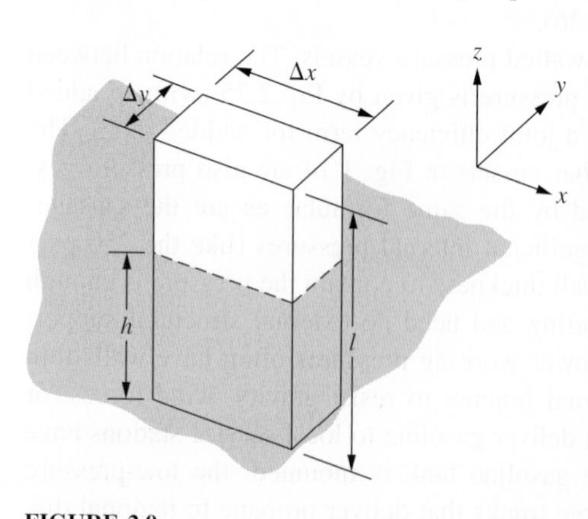

FIGURE 2.9
A floating block of wood, used to show Archimedes' principle.

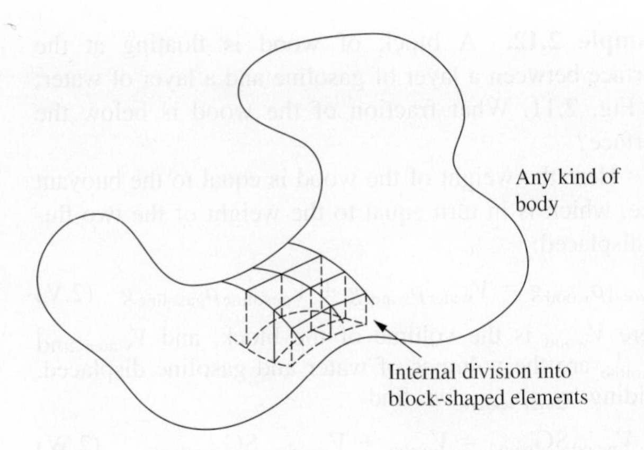

FIGURE 2.10
Any arbitrary-shaped object can be thought of as being made up of many blocks with vertical sides, like the one in Fig. 2.9.

The preceding statements were worked out for a block with the axis vertical. This was convenient, because the pressure on the vertical sides did not contribute to the buoyant force. However, the result is true for any kind of body because, as shown in Fig. 2.10, any shape at all can be visualized as made up of such blocks.

If the blocks are large, as shown in the figure, then their combined volume will be a rough approximation to the volume of the body. However, as the x and y dimensions of the blocks decrease, the blocks form a steadily improving approximation to the body, becoming identical with it as the x and y dimensions approach zero. Nonetheless, for each block, no matter how small, the foregoing argument holds, and therefore Archimedes' principle holds for any shape of body. Thus, although it would be very difficult to perform the indicated pressure integration over a body shaped like an octopus, if we know its volume (and hence the volume of fluid it displaces), then we can easily calculate the buoyant force by means of Archimedes' principle.

Example 2.11. A helium balloon is at the same pressure and temperature as the surrounding air (1 atm, 20°C) and has a diameter of 3 m. The weight of the plastic skin of the balloon is negligible. How much payload can the balloon lift?

The buoyant force is the weight of air displaced:

$$F_{\text{buoyant}} = \rho_{\text{air}} g V_{\text{balloon}} \tag{2.S}$$

The weight of helium is

$$W_{\text{helium}} = \rho_{\text{helium}} g V_{\text{balloon}} \tag{2.T}$$

Therefore, the payload is

$$\begin{aligned}
\text{Payload} &= F_{\text{buoyant}} - W_{\text{helium}} = V_{\text{balloon}} g(\rho_{\text{air}} - \rho_{\text{helium}}) \\
&= Vg\frac{P}{RT}(M_{\text{air}} - M_{\text{helium}}) \\
&= \frac{\pi}{6}\cdot(3\text{m})^3 \cdot \frac{9.81\text{ m}}{\text{s}^2}\cdot\frac{1\text{ atm}}{[8.2\cdot 10^{-5}\text{ m}^3\cdot\text{atm}/(\text{mol}\cdot\text{K})]\cdot 293.15\text{K}} \\
&\quad \cdot\left(29\frac{\text{g}}{\text{mole}} - 4\frac{\text{g}}{\text{mole}}\right)\cdot\frac{\text{kg}}{1000\text{ g}}\cdot\frac{\text{N}\cdot\text{s}^2}{\text{kg}\cdot\text{m}} = 144.2\text{ N} = 32.4\text{ lbf}
\end{aligned} \tag{2.U}$$

■

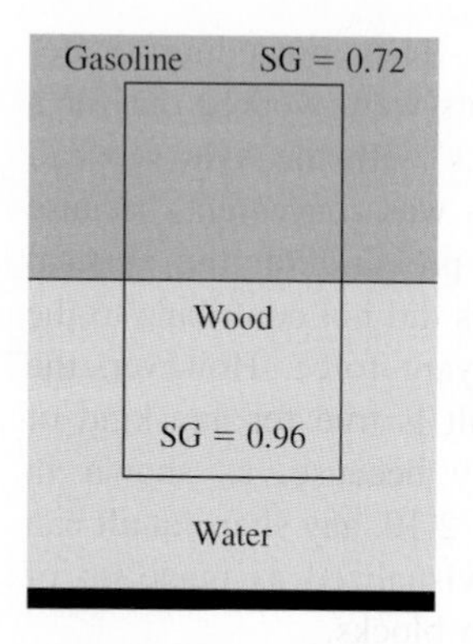

FIGURE 2.11
A block of wood floating at a gasoline-water interface.

Example 2.12. A block of wood is floating at the interface between a layer of gasoline and a layer of water; see Fig. 2.11. What fraction of the wood is below the interface?

Here the weight of the wood is equal to the buoyant force, which is in turn equal to the weight of the two fluids displaced:

$$V_{\text{wood}}\rho_{\text{wood}}g = V_{\text{water}}\rho_{\text{water}}g + V_{\text{gasoline}}\rho_{\text{gasoline}}g \quad (2.V)$$

where V_{wood} is the volume of the block, and V_{water} and V_{gasoline} are the volumes of water and gasoline displaced. Dividing by $g\rho_{\text{water}}$, we find

$$V_{\text{wood}}\text{SG}_{\text{wood}} = V_{\text{water}} + V_{\text{gasoline}}\text{SG}_{\text{gasoline}} \quad (2.W)$$

where SG is the specific gravity. But since

$$V_{\text{wood}} = V_{\text{water}} + V_{\text{gasoline}} \quad (2.X)$$

we may eliminate V_{gasoline}

$$V_{\text{wood}}\text{SG}_{\text{wood}} = V_{\text{water}} + (V_{\text{wood}} - V_{\text{water}})\,\text{SG}_{\text{gasoline}} \quad (2.Y)$$

and we then find

$$\frac{V_{\text{water}}}{V_{\text{wood}}} = \frac{\text{SG}_{\text{wood}} - \text{SG}_{\text{gasoline}}}{1 - \text{SG}_{\text{gasoline}}} = \frac{0.96 - 0.72}{1 - 0.72} = 0.857 \quad (2.Z)$$

■

This result appears paradoxical. The gasoline pushes down on the top of the block, not up on it at any point, yet the volume of gasoline displaced enters the buoyant force calculation. However, if we examine the pressure integral around the surface, we see that the pressure difference from top to bottom of the block does indeed involve the gasoline in the way shown.

Remember that the basic operation is the integration of the vertical components of the pressure force over the entire surface of the body. The convenient result of this integration is Archimedes' principle: the buoyant force is equal to the weight of the fluid displaced.

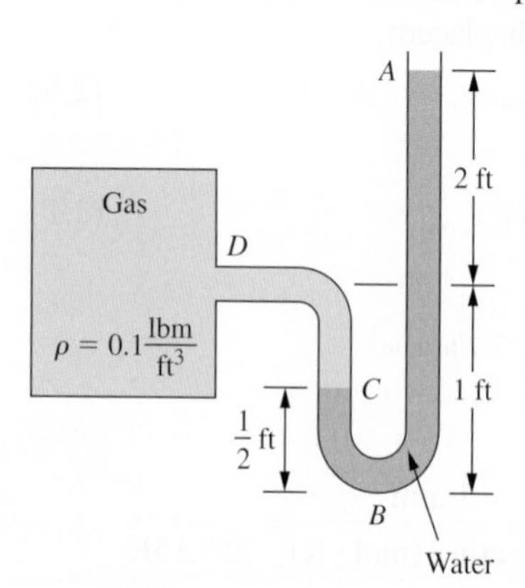

FIGURE 2.12
A simple manometer, filled with colored water.

2.6 PRESSURE MEASUREMENT

Pressures usually are measured by letting them act across some area and opposing them with either a gravity force or the force of a compressed spring. The gravity-force method uses a device called a *manometer*, described in the following example,

Example 2.13. Figure 2.12 shows a tank of gas connected to a manometer. The manometer is a U-shaped glass or transparent plastic tube open to the atmosphere at one end and containing colored water. From the elevations shown calculate the gauge pressure in the vessel.

We want to know the pressure at D. The simple way to work all manometer problems is to start with some pressure we know and work step by step to the pressure we want to find. In this case, we know the gauge pressure at A is zero, because the manometer is open to the atmosphere at A. The water is practically a constant-density fluid: therefore, we can use Eq. 2.11 to find the pressure at B:

$$P_B = P_A + \rho_{\text{water}} g h_B = 0 + \rho_{\text{water}} g \cdot 3 \text{ ft} \tag{2.AA}$$

To find the pressure at C we need Eq. 2.9:

$$P_C = P_B - (\rho_{\text{water}} g \cdot \tfrac{1}{2} \text{ ft}) \tag{2.AB}$$

To find the pressure at D we use the same equation (i.e., we assume that the density *change* of the gas is negligible):

$$P_D = P_C - (\rho_{\text{gas}} g \cdot \tfrac{1}{2} \text{ ft}) \tag{2.AC}$$

Adding these three equations and canceling like terms, we find

$$\begin{aligned} P_D &= (\rho_{\text{water}} g) \cdot \left(3 \text{ ft} - \frac{1}{2} \text{ft}\right) - \left(\rho_{\text{gas}} g \cdot \frac{1}{2} \text{ft}\right) \\ &= 32.2 \frac{\text{ft}}{\text{s}^2} \left[\left(62.3 \frac{\text{lbm}}{\text{ft}^3} \cdot 2.5 \text{ ft}\right) - \left(0.075 \frac{\text{lbm}}{\text{ft}^3} \cdot 0.5 \text{ ft}\right)\right] \cdot \frac{\text{lbm} \cdot \text{s}^2}{32.2 \text{ lbm} \cdot \text{ft}} \cdot \frac{\text{ft}^2}{144 \text{ in}^2} \\ &= 1.08 \frac{\text{lbf}}{\text{in}^2} - 0.0003 \frac{\text{lbf}}{\text{in}^2} = 1.08 \frac{\text{lbf}}{\text{in}^2} \text{gauge} = 7.46 \text{ kPa} \end{aligned} \tag{2.AD}$$

■

The example illustrates several points.

1. The contribution of the section of the manometer full of gas is only 0.03% of the answer. It is generally neglected in manometer problems.
2. Manometers that are open to the atmosphere are gauge-pressure devices and should be calculated in gauge pressure.
3. In reading such a device, we normally read an elevation; the actual operational reading in Example 2.13 was 2.5 ft. For many purposes it is convenient to think of pressures and to report them in terms of such manometer readings as heights; the U.S. air conditioning industry, for example, commonly refers to all pressure in air conditioning ducts as "inches of water," and most U.S. vacuum-equipment manufacturers refer to vacuums as "inches of mercury."
4. At no place in our last calculation did the cross-sectional area of the manometer tube enter. Therefore, this tube can be of any convenient size and need not be made of constant-diameter tubing. The only measurements necessary are the fluid densities, which can be looked up in handbooks, and the differences in elevation, which can be read directly with tape measures and rulers. Thus, manometers require neither calibration nor testing with standards; one simply connects them and takes the reading. There is no requirement that the tubes be vertical, only that we can read the vertical distance between the horizontal liquid surfaces.

5. It may seem that we went to a lot of trouble for such a simple problem. This is true; working engineers use shorter calculation methods than the one shown here. However, with complicated systems, such as two-fluid manometers, the shortcuts are confusing, and the step-by-step method just shown is always reliable.

Example 2.14. A two-fluid manometer is often used to make it unnecessary to read small differences in liquid level. The one shown in Fig. 2.13 is measuring the pressure difference between two tanks. What is that pressure difference?

We want to know $P_A - P_E$. All the fluids have practically constant density, so we can use Eq. 2.9. We begin by calling P_E known. Then

$$P_D = P_E + (\rho_{\text{water}} g \cdot 1 \text{ ft}) \tag{2.AE}$$

$$P_C = P_D + (\rho_{\text{oil}} g \cdot 2 \text{ ft}) \tag{2.AF}$$

$$P_B = P_C - (\rho_{\text{oil}} g \cdot 1 \text{ ft}) \tag{2.AG}$$

$$P_A = P_B - (\rho_{\text{water}} g \cdot 2 \text{ ft}) \tag{2.AH}$$

Adding these and canceling like terms, we find

$$\begin{aligned} P_A - P_E &= \rho_{\text{water}} g(1 \text{ ft} - 2 \text{ ft}) + \rho_{\text{oil}} g(2 \text{ ft} - 1 \text{ ft}) = 1 \text{ ft} \cdot g(\rho_{\text{oil}} - \rho_{\text{water}}) \\ &= 1 \text{ ft} \cdot 32.2 \frac{\text{ft}}{\text{s}^2} \cdot (1.1 - 1.0) \cdot 62.3 \frac{\text{lbm}}{\text{ft}^2} \cdot \frac{\text{lbf} \cdot \text{s}^2}{32.2 \text{ lbm} \cdot \text{ft}} \cdot \frac{\text{ft}^2}{144 \text{ in}^2} \\ &= 0.043 \frac{\text{lbf}}{\text{in}^2} = 298 \text{ Pa} \end{aligned} \tag{2.AI}$$

■

This reading corresponds to a pressure difference of 0.1 ft of water. The actual reading of this two-fluid manometer is 1 ft. If we assume that we can read liquid level differences with an accuracy of ±0.005 ft (= ±0.06 in), then a simple water manometer would have an uncertainty of 5% for this difference; the two-fluid manometer shown has an uncertainty of 0.5%.

Because a manometer is a device for measuring pressure differences, to use one to measure absolute pressure we must measure the difference between the pressure in question and a perfect vacuum. In principle this is impossible, because there is no such thing as a perfect vacuum, but in practice we may produce vacuums of sufficient quality that the error introduced by calling them perfect is negligible. This idea is used in the mercury barometer shown in Fig. 2.14. This common device for measuring the pressure of the atmosphere is found in most laboratories.

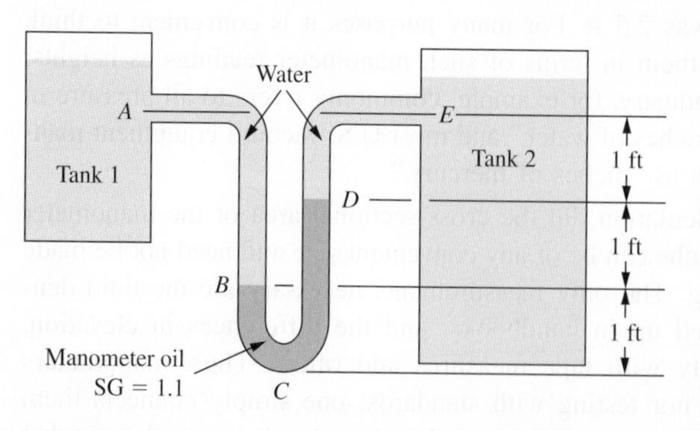

FIGURE 2.13
A two-fluid manometer, with water and a manometer oil.

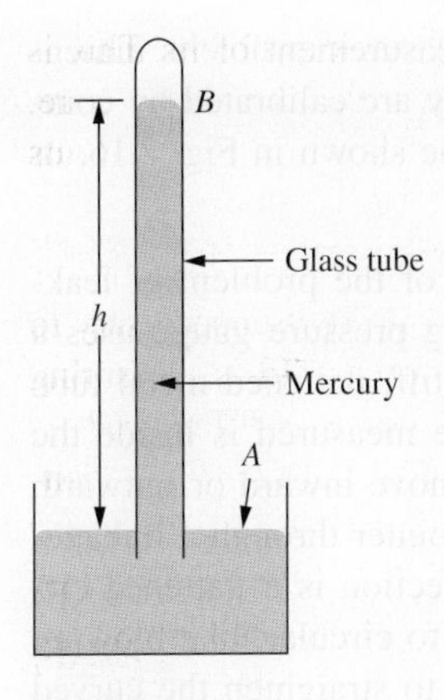

FIGURE 2.14
Mercury barometer.

The atmosphere acts on the mercury in the cup at the bottom, and the weight of the column of mercury opposes it. Calculating this, we find

$$P_A - P_B = \rho_{\text{Hg}} g h \tag{2.AJ}$$

where P_B is the pressure in the vapor space above the liquid mercury. In well-built manometers this will be simply the vapor pressure of mercury, which at 68°F = 20°C is about 10^{-6} atm; this is so small compared with 1 atm that it can be neglected. Thus, although the barometer, like all manometers, measures pressure differences, it can be used with satisfactory accuracy as an absolute-pressure device in this case; see Prob. 2.56.

The second way to measure pressure is to let the pressure act on some piston, which compresses a spring, and to measure the displacement. Figure 2.15 shows an *impractical but illustrative* way of doing this. The fluid whose pressure is to be measured presses on the piston, compressing the spring and moving the pointer along the scale. If we know the area of the piston, the spring constant, and the pointer reading for zero pressure, we can calculate the pressure on the piston from the pointer position.

Example 2.15. The piston in Fig. 2.15 has an area of 100 cm², and the spring constant k is 100 N / cm. We set the pointer so that there is a zero reading when both sides of the piston are exposed to the atmosphere. Now we attach the gauge to a tank with an unknown pressure, and the pointer moves to 2.5 cm. What is the pressure in the tank?

Here the net force acting on the piston is

$$F_{\text{net}} = (P_{\text{tank}} - P_{\text{atm}}) \cdot A = AP_{\text{tank, gauge}} \tag{2.AK}$$

This must be equal to the force on the spring, which is $k\,\Delta x$, and therefore

$$P_{\text{tank, gauge}} = \frac{k\,\Delta x}{A} = \frac{(100\text{ N / cm}) \cdot 2.5\text{ cm}}{100\text{ cm}^2} \cdot \left(\frac{100\text{ cm}}{\text{m}}\right)^2 = 25\text{ kPa} = 3.62\,\frac{\text{lbf}}{\text{in}^2} \tag{2.AL}$$

■

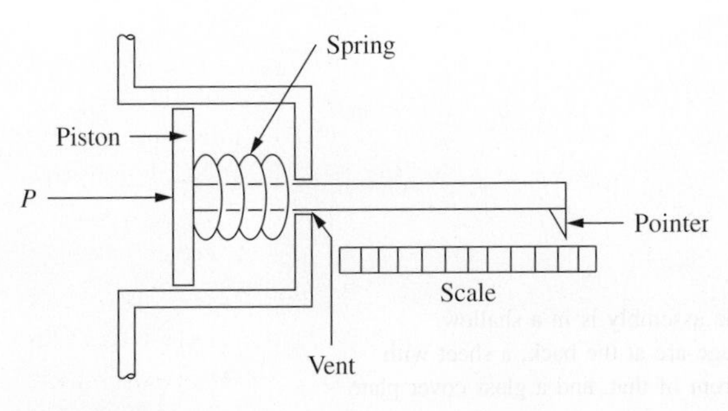

FIGURE 2.15
Piston-and-spring pressure gauge.

From this example we observe the following:

1. This device, like the manometer, measures pressure differences. To use it as an absolute-pressure device, we must make it compare pressure with a vacuum. We can do this by placing it in an evacuated chamber.

2. This device, unlike the manometer, requires a precise measurement of its dimensions or a calibration. Spring-type pressure gauges usually are calibrated by comparing their reading with those of manometers like the one shown in Fig. 2.16, or other equivalent devices (see Prob. 2.65).

The gauge shown in Fig. 2.15 is impractical because of the problem of leakage around the piston. The most widely used type of spring pressure gauge uses a *bourdon tube,* as shown in Fig. 2.16. A bourdon tube is a stiff, flattened metal tube bent into a circular shape; the fluid whose pressure is to be measured is inside the tube. One end of the tube is fixed, and the other is free to move inward or outward. The inward or outward movement of the free end moves a pointer through a linkage-and-gear arrangement. As Fig. 2.16 shows, the tube cross section is a flattened circle. Internal pressure makes its cross section become closer to circular, like blowing up a balloon, which stresses the outer surface, thus tending to straighten the curved tube. With a high enough pressure the tube would become straight with a circular cross section [6]. The tube itself serves as the spring; it is made of metal, which is stiff and has a reasonable spring constant. The internal pressures are low enough that the tube returns to the cross section shown in Fig. 2.16 when the pressure is removed. With such a tube the calculation of the movement as a function of the inside and outside pressures is more difficult than with the linear piston-and-spring gauge of Fig. 2.15. However, the bourdon tube is a very convenient shape and causes no

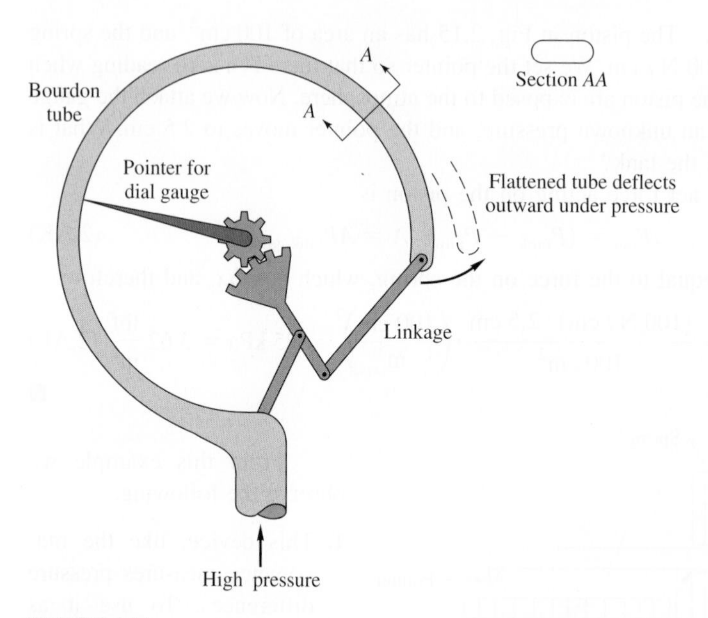

FIGURE 2.16
Bourdon-tube pressure gauge. The whole assembly is in a shallow, cylindrical container. The tube and linkage are at the back, a sheet with numbers comes next, the pointer is in front of that, and a glass cover plate protects the whole assembly.

leakage problems, as does the piston-and-spring gauge. Since both are calibrated devices, the difficulty in calculating the performance of the bourdon tube is not a real disadvantage. Bourdon-tube pressure gauges are simple, rugged, leak-free, reasonably reliable, and cheap; they are the most widely used type of industrial pressure gauge.

Neither the manometer nor the bourdon-tube gauge is suited to measuring rapidly changing pressures. Both are unsatisfactory for this purpose because of their high inertial mass; this mass makes them move slowly to accommodate a change in pressure, and so their readings lag behind a rapidly changing pressure. For rapidly changing pressures (such as pressure fluctuations in rocket motors, or the rapid oscillations in air pressure that we call sound, which are measured with a microphone), two other types of pressure gauges respond much more quickly. One is the diaphragm gauge, which is similar to that of Fig. 2.15 but has, instead of the piston and spring, a thin metal diaphragm, which acts as both. When the pressure increases, the diaphragm stretches very slightly; the stretch is detected by an electric strain gauge (or other electronic means) and recorded electrically. The advantage of the diaphragm over the bourdon tube is its very low mass, which allows it to move quickly in response to a change in pressure. The other type of rapid-response pressure gauge is the quartz-crystal piezometer, which uses the change in electrical properties of quartz crystals with change in pressure. Other electronic pressure gauges are available that take advantage of the response of fixed or oscillating microstructures to changes in external pressure or other electronic phenomena.

2.7 MANOMETER-LIKE SITUATIONS

In Sec. 2.6 we discussed manometers as pressure-measuring devices. There are many other fluid mechanical situations that are most easily understood if we analyze them just as we analyze manometers. Several examples are shown here.

Figure 2.17 shows a schematic cross section of a percolator-type coffee maker. In it, the pot is filled to a height z_1 with water. The basket above the water is filled with ground coffee. The whole assembly is placed on a stove and heated from below. When the water has been warmed, it begins to flow in irregular spurts up the central tube; it is diverted by the cap on the top, falls on the coffee grounds, and percolates through them, extracting the water-soluble constituents of the ground coffee, to make the hot drink many of us enjoy.

How can the fluid do this? Here we have a fluid flowing from a low elevation to a high one, with no mechanical device lifting it. How can that be? To answer the question, we compute the pressures at B and C. It will be easiest if we do this all in gauge pressure. In that case, the pressure at B will be

$$P_B = \rho g z_1 \tag{2.AM}$$

and if the fluid in the tube is up to the level where it spills out at D, then the pressure at C will be

$$P_C = \rho g z_2 \tag{2.AN}$$

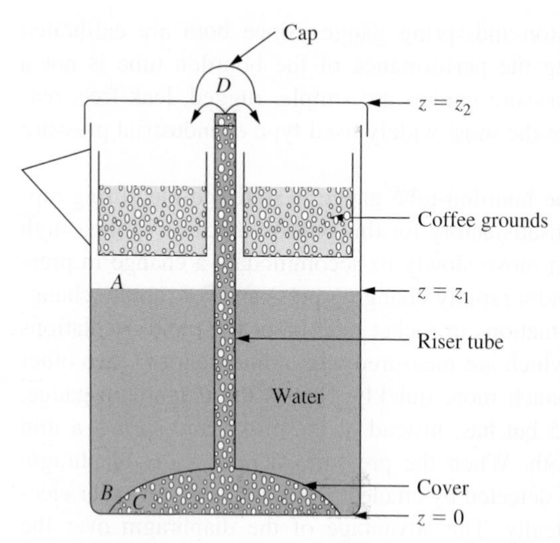

FIGURE 2.17
Coffee percolator, showing fluid flow driven by boiling.

and

$$P_B - P_C = g[(\rho z)_1 - (\rho z)_2] \tag{2.AO}$$

When we put the pot on the stove, the density inside and outside the riser tube will be the same (that of water), and the liquid in the tube will stand at the same level as the liquid outside, z_1. There will be no flow.

As the water at the bottom is heated by the stove, the loose-fitting cover prevents it from mixing with the rest of the fluid in the pot, so that a small amount of liquid is heated to its boiling point. When it boils, the bubbles of steam flow by buoyancy up through the riser tube. While they do so, the average density of the gas-liquid mixture in the riser tube decreases. If there is no net flow under the loose-fitting cover, then the pressure difference from one side of it to the other, $(P_B - P_C)$, must be zero, and the level in the riser tube, z_2, must increase to keep this pressure difference equal to zero.

When the generation rate of bubbles becomes high enough that z_2 becomes greater than the height of the top of the riser tube, then a mixture of steam and water will flow out the top of the tube. If the rate of generation of steam bubbles increases even more, then the average density of the steam-water mixture in the riser tube will fall low enough that $(P_B - P_C)$ can no longer be zero but must become a positive number. Then the pressure force due to gravity will force water from the pot under the loose-fitting cover, and the circulation will be established, with flow downward under the cover, up the riser tube, and down through the coffee grounds. For that flow, we can no longer use the simple equations of fluid statics, which we have used so far in this discussion; the methods of Chaps. 5, 6, 7, and 12 must be used. But this simple discussion shows how the pressure forces that move the fluids in coffee percolators arise. The exact same discussion applies to geysers (where the flow is intermittent instead of the steady flow in the boiling coffee pot) and to the circulation system in most steam boilers, in gas- and propane-fired refrigerators, and in the reboilers of many distillation columns. In all of these, the formation of bubbles of steam (or the vapor of some other liquid being boiled) lowers the average density in one leg of the "equivalent manometer," producing the pressure difference that drives the flow.

Such pressure differences can also arise in systems that do not involve boiling liquids, as is illustrated in Example 2.16

Example 2.16. Figure 2.18 shows a schematic of a home fireplace, with part of the house that surrounds it. The burning logs in the fireplace heat the gases

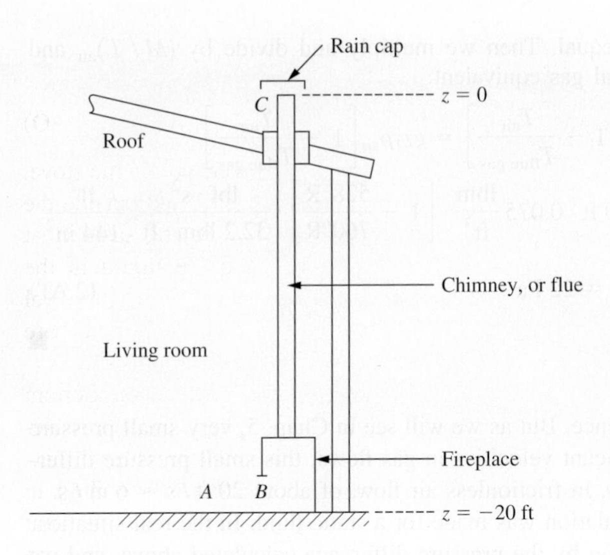

FIGURE 2.18
Home fireplace, showing fluid flow driven by temperature differences.

in the chimney to 300°F. If we treat this as a static situation, what will be the difference in pressure between the air in the room adjacent to the fireplace and the air inside the fireplace at the same level?

Here we assume that the house is leaky enough, or has an open window, so that the pressure inside the house is the same as the pressure in the atmosphere outside. (This is true for older houses, but not necessarily true for modern "tight" energy-conserving houses, which have much less air exchange with the surroundings!) Here, as shown in the figure, we have taken the elevation datum, $z = 0$, at the top of the chimney; this choice makes the solution simple. Taking the pressure at $z = 0$ to be atmospheric pressure, and working in gauge pressures, we can compute that

$$P_A = \rho_{\text{air}} g z_1 \tag{2.AP}$$

and that

$$P_B = \rho_{\text{flue gas}} g z_1 \tag{2.AQ}$$

The chimney is also called a *flue,* and the gas in it is normally called *flue gas.* Then

$$P_A - P_B = (\rho_{\text{air}} - \rho_{\text{flue gas}})\, g z_1 \tag{2.AR}$$

Assuming that both the air and the flue gas are ideal gases, we can express the density of each in terms of the ideal gas law and write

$$P_A - P_B = g z_1 \frac{P}{R}\left[\left(\frac{M}{T}\right)_{\text{air}} - \left(\frac{M}{T}\right)_{\text{flue gas}}\right] \tag{2.AS}$$

In the most careful work we need to take into account the small difference in molecular weights of air and flue gas, but here we can assume that the molecular weights are practically equal. Then we multiply and divide by $(M/T)_{\text{air}}$ and substitute ρ_{air} for its ideal gas equivalent:

$$P_A - P_B = gz_1 \frac{PM_{\text{air}}}{RT_{\text{air}}}\left[1 - \frac{T_{\text{air}}}{T_{\text{flue gas}}}\right] = gz_1\rho_{\text{air}}\left[1 - \frac{T_{\text{air}}}{T_{\text{flue gas}}}\right]$$

$$= 32.2\frac{\text{ft}}{\text{s}^2}\cdot 20\text{ ft}\cdot 0.075\frac{\text{lbm}}{\text{ft}^3}\cdot\left[1 - \frac{528°\text{R}}{760°\text{R}}\right]\cdot\frac{\text{lbf}\cdot\text{s}^2}{32.2\text{ lbm}\cdot\text{ft}}\cdot\frac{\text{ft}^2}{144\text{ in}^2}$$

$$= 0.0032\frac{\text{lbf}}{\text{in}^2} = 22\text{ Pa} \tag{2.AT}$$

■

This is a small pressure difference. But as we will see in Chap. 5, very small pressure differences can produce significant velocities in gas flows; this small pressure difference would produce a velocity, in frictionless air flow, of about 20 ft / s ≈ 6 m / s.

In this example the calculation was made for a static fluid. In the real situation, the fluid would be set in motion by the pressure difference calculated above, and we would need the methods developed in later chapters to compute the velocity. But this calculation shows how pressure differences can arise not only in boiling liquids but also in gases if one side of an "equivalent manometer" is heated to a temperature higher than the other. This explains how chimneys work. For all but the largest furnaces the air flow is driven through the furnace by the pressure difference computed here (called "natural draft"). That explains why large furnaces have tall stacks; the available pressure difference as just shown is proportional to the height of the stack. Many large furnaces now use powered fans to drive the gases through them (called "forced draft"). The choice between natural and forced draft is based on economics; the tall stacks are expensive, but once installed they require no power to run the forced-draft fans.

This calculation also explains many meteorological phenomena. Oceans and lakes are heated and cooled slowly by the sun because currents and waves mix their upper layers; the ground surface on the shore heats more rapidly in the daytime and cools more rapidly at night because it can transfer heat up and down only by conduction, which is slow compared to the convective mixing in bodies of water. Thus, during the day the hot ground surface heats the air above it, and hot air above the ground plays the same role as the hot flue gas in Example 2.16. The winds blow from the ocean or lake onto the shore. At night the ground cools, and cools the air above it, and the direction of the pressure gradient reverses, causing the wind to blow from the shore out over the body of water. The monsoon rains of India and parts of tropical Africa (and in weaker form, the southwestern United States) are the same phenomenon on a much larger scale. In the summer the heated air over the continents rises, and the cooler moist air from the adjacent oceans flows in and brings rains. The same description applies to firestorms, in which the rising, heated air above a large forest or urban fire induces strong winds blowing inward toward its center. Volcanoes are also manometer-like situations, but more complex than these meteorological examples.

2.8 VARIABLE GRAVITY

So far we have assumed that the acceleration of gravity is constant, $32.2 \text{ ft/s}^2 = 9.81 \text{ m/s}^2$ in the minus z direction. This is not exactly true in any problem involving two different elevations. However, the change in gravity with change in elevation is quite small. Near the surface of the earth, the acceleration of gravity is proportional to the reciprocal of the square of the distance from the center of earth. The radius of the earth is about 4000 mi = 6440 km, so the acceleration of gravity 1 mi = 1.609 km above the surface is $4000^2 / 4001^2 = 0.9995$ times the acceleration of gravity at the surface. Few engineering problems include data precise enough to justify making such corrections. (Deposits of metal ores perturb the acceleration of gravity above them; sensitive gravitometers flown in airplanes can detect the small gravity perbations caused by such deposits. Such gravitometers are widely used in mineral exploration.)

In two types of problem, however, nonconstant gravity is important:

1. Space travel and rocket problems; in these the distances from the earth become significant compared with 4000 mi, so the changing value of gravity must be taken into account.
2. Acceleration and centrifugal force problems.

Since this is a chapter on fluid statics, it seems a strange place to consider acceleration or centrifugal force problems, in which the fluid is certainly moving. We do so because, in these problems the fluid is not moving relative to its container or relative to other parts of the fluid. Really, all problems in terrestrial fluid statics involve moving fluids, because the fluids are on the earth, and the earth is rotating about its axis and revolving around the sun, and the sun is moving through space. As long as the individual particles of fluid are not moving relative to each other, we can treat such moving problems by the methods of fluid statics. Such motions of fluids are called *rigid-body motions*.

2.9 PRESSURE IN ACCELERATED RIGID-BODY MOTIONS

We now repeat the derivation of Eq. 2.1 for the case in which an entire mass of fluid is in some kind of accelerated rigid-body motion. Again we will use the small, cubical element of fluid shown in Fig. 2.1 and consider it to be part of a larger mass of fluid. In Sec. 2.1 we showed that if the fluid was not being accelerated, then the sum of the forces on it must be zero. If the fluid was being accelerated, then the sum of the forces acting on it, in the direction of the acceleration, must equal the mass times the acceleration. For the cubical element of fluid being accelerated in the vertical $(+z)$ direction, we rewrite Eq. 2.2 as

$$(P_{z=0})\,\Delta x\,\Delta y - (P_{z=\Delta z})\,\Delta x\,\Delta y - \rho g\,\Delta x\,\Delta y\,\Delta z = \rho\,\Delta x\,\Delta y\,\Delta z\,\frac{d^2 z}{dt^2} \quad \text{(2.AU)}$$

Dividing by $\Delta x \, \Delta y \, \Delta z$ and taking the limit as Δz approaches zero, we find

$$\frac{dP}{dz} = -\rho\left(g + \frac{d^2 z}{dt^2}\right) \tag{2.32}$$

which for constant-density fluids can be integrated to

$$P_2 - P_1 = -\rho\left(g + \frac{d^2 z}{dt^2}\right)(z_2 - z_1) \qquad \text{[constant density]} \tag{2.33}$$

and which for gauge pressure simplifies further to

$$P = -\rho h\left(g + \frac{d^2 z}{dt^2}\right) \qquad \text{[constant density, gauge pressure]} \tag{2.34}$$

Example 2.17. An open tank containing water 5 m deep is sitting on an elevator. Calculate the gauge pressure at the bottom of the tank

(*a*) when the elevator is standing still,
(*b*) when the elevator is accelerating upward at the rate of 5 m / s^2, and
(*c*) when the elevator is accelerating downward at the rate of 5 m / s^2.

From Eq. 2.11, part (*a*) is simply

$$P_{\text{bottom}} = \rho g h = 998.2 \frac{\text{kg}}{\text{m}^3} \cdot 9.81 \frac{\text{m}}{\text{s}^2} \cdot 5 \text{ m} \cdot \frac{\text{N} \cdot \text{s}^2}{\text{kg} \cdot \text{m}} \cdot \frac{\text{Pa}}{\text{N / m}^2}$$

$$= 49.0 \text{ kPa} = 7.11 \frac{\text{lbf}}{\text{in}^2} \tag{2.AV}$$

For parts (*b*) and (*c*) we use Eq. 2.34:

$$P_{\text{bottom}} = \rho h\left(g + \frac{d^2 z}{dt^2}\right) = 998.2 \frac{\text{kg}}{\text{m}^3} \cdot 5 \text{ m} \cdot \left(9.81 \frac{\text{m}}{\text{s}^2} + 5 \frac{\text{m}}{\text{s}^2}\right) \cdot \frac{\text{N} \cdot \text{s}^2}{\text{kg} \cdot \text{m}} \cdot \frac{\text{Pa}}{\text{N / m}^2}$$

$$= 74.0 \text{ kPa} = 10.73 \frac{\text{lbf}}{\text{in}^2} \tag{2.AW}$$

and

$$P_{\text{bottom}} = \rho h\left(g + \frac{d^2 z}{dt^2}\right) = 998.2 \frac{\text{kg}}{\text{m}^3} \cdot 5 \text{ m} \cdot \left(9.81 \frac{\text{m}}{\text{s}^2} - 5 \frac{\text{m}}{\text{s}^2}\right) \cdot \frac{\text{N} \cdot \text{s}^2}{\text{kg} \cdot \text{m}} \cdot \frac{\text{Pa}}{\text{N / m}^2}$$

$$= 24.0 \text{ kPa} = 3.48 \frac{\text{lbf}}{\text{in}^2} \tag{2.AX}$$

■

Most of us can tell which way an elevator starts to move, by sensing these small changes in our weight.

If the acceleration is not in the same direction as gravity or in the direction opposite to it, it must be in some direction a, as in Fig. 2.2. Then we can take the summation of forces in the a direction and substitute $g \cos \theta$ for g in Eq. 2.34:

$$\frac{dP}{da} = -\rho\left(g \cos \theta + \frac{d^2 a}{dt^2}\right) \tag{2.35}$$

Example 2.18. A rectangular tank of orange juice on a cart is moving in the x direction with a steady acceleration of 1 ft / s^2; see Fig. 2.19. What angle does its free surface make with the horizontal?

Here we assume that the tank has been under acceleration so long that the initial sloshing back and forth of the liquid at the start of acceleration has died out and that the fluid is truly in rigid-body motion. In the figure the points A and B are both on the free surface; neglecting the very slight change in atmospheric pressure over this change in elevation, we may say that the gauge pressure is zero at both points. Then we can calculate the pressure at C from the pressure at A by using Eq. 2.35. Here we are applying it in the y direction ($a = y$), so we have $\cos \theta = 1$ and $d^2y / dt^2 = 0$. Hence, the result is the same as Eq. 2.11:

$$P_C = -\rho g \, \Delta y \tag{2.AY}$$

(Here Δx and Δy are both negative moving from the free surface, so P_C is positive.) We may also calculate the gauge pressure at C by using Eq. 2.35 for the horizontal direction, in which case we have $a = x$, $\cos \theta = 0$, and

$$P_C = -\rho \, \Delta x \frac{d^2 x}{dt^2} \tag{2.AZ}$$

But the pressure at C is the same no matter how we calculate it, so we may eliminate P_C between these two equations and rearrange to find

$$\frac{\Delta y}{\Delta x} = \frac{d^2x / dt^2}{g} = \tan \theta \tag{2.BA}$$

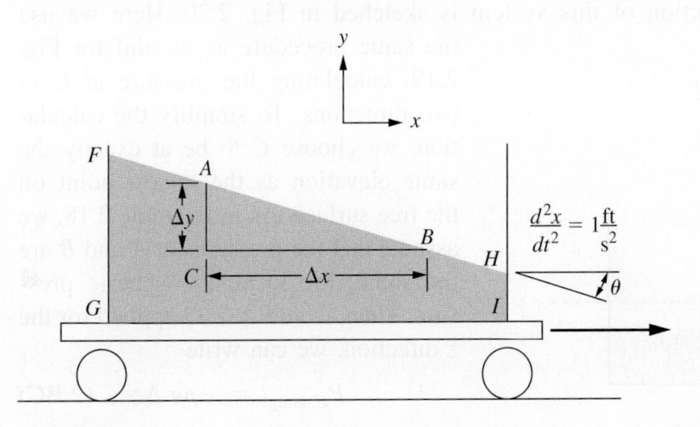

FIGURE 2.19
A system in linear acceleration.

where θ is the angle shown in Fig. 2.19. For this problem

$$\theta = \text{arc tan} \frac{d^2x / dt^2}{g} = \text{arc tan} \frac{1 \text{ ft} / \text{s}^2}{32.2 \text{ ft} / \text{s}^2} = 1.76° \quad (2.BB)$$

■

To calculate the pressure at any point in the tank, we may now use Eq. 2.11, being careful to measure the depth from the free surface vertically above the point in question. The force on the wall *FG*, for example, is exactly the same at every point as it would be if the cart were standing still and filled with liquid up to point *F*. The force on wall *HI* is exactly the same as it would be if the cart were standing still and filled with liquid up to the level of *H*.

Example 2.18, a case of uniform, rectilinear acceleration holds little practical interest, because such an acceleration acting for a reasonable period of time (e.g., long enough for the sloshing to die out) would produce enormous velocities. However, it serves as an introduction to the more interesting case of *rigid-body rotation*. Consider an open-topped cylindrical tank of water with a vertical axis. The system is initially at rest; then the tank is set in steady motion, rotating about its vertical axis. At first the fluid in the center will not be affected by the rotation of the walls but will stand still, and only the fluid near the walls will rotate. This sets up motions of parts of the fluid relative to each other, so that this is not a fluid-statics problem. Eventually, however, the shear forces due to this relative motion will bring the fluid at the center to the same angular velocity as the tank, and thereafter there is no relative motion within the fluid. Once the fluid in the center reaches the same angular velocity as the wall of the container, the whole of the fluid moves as if it were a rigid body; hence the name, "rigid-body rotation." Pressures in rigid-body rotation can be calculated by the method of fluid statics.

Example 2.19. An open-topped can of water 30 cm in inside diameter is rotating at 78 rpm. It has been rotating a long time and is in rigid-body rotation. What is the shape of the free surface?

A cross section of this system is sketched in Fig. 2.20. Here we use the same procedure as we did for Fig. 2.19, calculating the pressure at *C* in two directions. To simplify the calculation, we choose *C* to be at exactly the same elevation as the lowest point on the free surface. As in Example 2.18, we assume that the pressures at *A* and *B* are the same, the local atmospheric pressure. Then, from Eq. 2.35 applied for the *z* direction, we can write

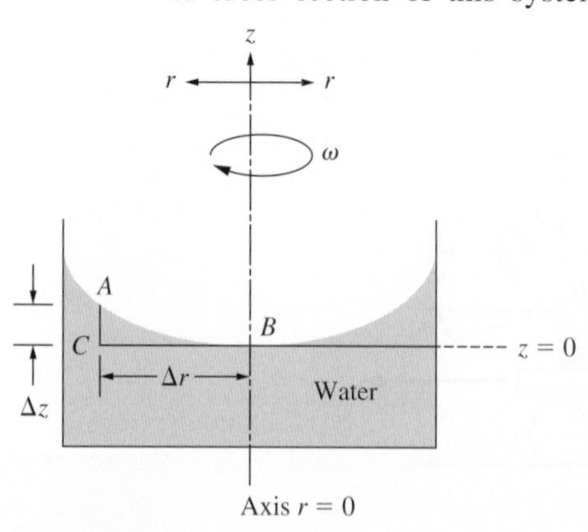

FIGURE 2.20
A system with rotation, leading to centrifugal acceleration.

$$P_{C,\text{ gauge}} = -\rho g \, \Delta z \quad (2.BC)$$

because the rotational acceleration is perpendicular to the *z* axis. In the radial direction, which is the *r* direction in Fig. 2.20, the only forces acting on the element of fluid are the pressure forces and the centripetal acceleration, whose magnitude is given by

$$\text{Centripetal acceleration} = -(\text{angular velocity})^2 \cdot \text{radius}$$

$$a_c = -\omega^2 r \tag{2.BD}$$

Substituting this for d^2a / dt^2 in Eq. 2.35 and noting that $\cos\theta$ is zero for the radial direction, we find

$$\frac{dP}{dr} = \rho\omega^2 r \tag{2.36}$$

(No minus sign appears here because the centrifugal force points in the $+r$ direction, whereas the gravity force points in the $-z$ direction.) We then find the gauge pressure at C:

$$P_C = \int_{r=0}^{r=\Delta r} \rho\omega^2 r\, dr = \rho\omega^2 \left.\frac{r^2}{2}\right]_0^{\Delta r} = \rho\omega^2 \frac{(\Delta r)^2}{2} \tag{2.37}$$

The pressure at C is the same no matter how we calculate it, so we may eliminate P_C between these two equations and divide ρg to find

$$-\Delta z = \frac{\omega^2}{2g}(\Delta r)^2 \tag{2.BE}$$

If we now let the elevation of point B (the lowest point on the free surface) be $z = 0$, then the length Δz is minus the value of z at point A, and Δr is the value of r at point A; so those points on the free surface are described by

$$z = \frac{\omega^2}{2g} r^2 \tag{2.38}$$

The free surface is a parabola with vertex at the center of the can. The height of the free surface at the wall of the can is

$$z = \frac{(2\pi \cdot 78\ \text{rpm})^2 \cdot (15\ \text{cm})^2}{2 \cdot 9.81\ \text{m/s}^2} \cdot \left(\frac{\text{min}}{60\ \text{s}}\right)^2 \cdot \frac{\text{m}}{100\ \text{cm}} = 7.65\ \text{cm} = 3.01\ \text{in} \tag{2.BF}$$

■

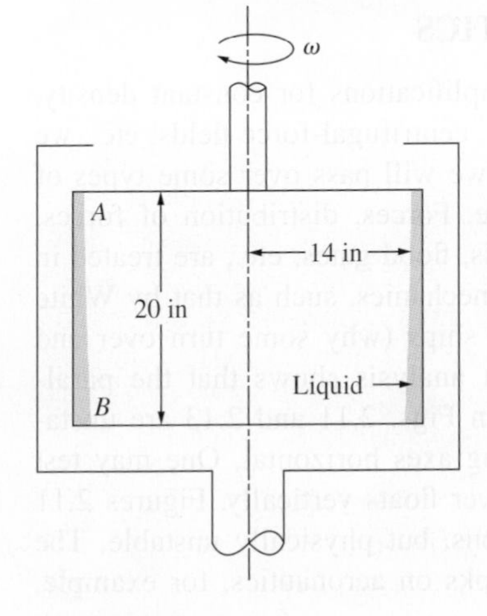

FIGURE 2.21
The basket of an industrial centrifuge.

To find the pressure at any point in the rotating system (with the axis of rotation vertical) we use Eq. 2.11 and measure the distance down from the free surface directly upward from the point in question. The pressure at any point on the wall of the can in Fig. 2.20 is exactly the same as if the can were not rotating and were filled to the level of the rotating free surface at the edge of the can.

Example 2.20. An industrial centrifuge has a basket with a 30 in diameter and is 20 in high. Its speed is 1000 rpm; see Fig. 2.21. If the liquid layer against the wall of the centrifuge is 1 in thick at the top, how thick is it at the bottom?

This is really the same problem as Ex. 2.19, except that only part of the parabolic free surface is present. To solve it, we write Eq. 2.38 twice, for points A and B on the figure, and we then subtract one from the other;

$$z_A - z_B = \frac{\omega^2}{2g}(r_A^2 - r_B^2) \tag{2.39}$$

The only unknown here is r_B. Solving for it, we find

$$\begin{aligned} r_B &= \left[r_A^2 - (z_A - z_B)\frac{2g}{\omega^2} \right]^{1/2} \\ &= \left[(14 \text{ in})^2 - \frac{20 \text{ in} \cdot 2 \cdot 32.2 \text{ ft / s}^2}{(2\pi \cdot 1000 \text{ / min})^2} \cdot \left(\frac{60 \text{ s}}{\text{min}}\right)^2 \cdot \left(\frac{12 \text{ in}}{\text{ft}}\right) \right]^{1/2} \\ &= (196 \text{ in}^2 - 1.4 \text{ in}^2)^{1/2} = 13.95 \text{ in} = 0.354 \text{ m} \end{aligned} \tag{2.BG}$$

■

Thus, the liquid film is 1.05 in (2.67 cm) thick at the bottom. One may readily calculate that at the outer side of the centrifuge, the ratio of the centrifugal acceleration to that of gravity is

$$\left(\frac{\text{Centrifugal acceleration}}{\text{Gravity acceleration}}\right) = \frac{\omega^2 r}{g} = \frac{(2\pi \cdot 1000 \text{ / } 60 \text{ s})^2 \cdot 1.25 \text{ ft}}{32.2 \text{ ft / s}^2} = 426 \tag{2.BH}$$

This ratio explains why liquids can be separated from solids much more effectively in such a centrifuge than by simple gravity draining, both in industrial usage and in home clothes washing machines. We will see in Chap. 10 that a centrifugal pump is a modified centrifuge; and in studying air pollution control, the widely used cyclone separators that collect particles are also modified centrifuges.

2.10 MORE PROBLEMS IN FLUID STATICS

Having worked out the basic equation and its simplifications for constant density, gauge pressure, isothermal and isentropic ideal gas, centrifugal-force fields, etc., we can attack a wide range of problems. In this text we will pass over some types of problems that have been widely treated elsewhere. Forces, distribution of forces, overturning moments, etc., on dams, retaining walls, flood gates, etc., are treated in all texts on civil or mechanical engineering fluid mechanics, such as that by White [3]. The subject of the buoyancy and stability of ships (why some turn over and others do not) is treated in the same texts. That analysis shows that the parallelepiped blocks floating with their axes vertical in Figs. 2.11 and 2.13 are unstable; they will instantly turn over to have their long axes horizontal. One may test that with a short piece of lumber in a pond; it never floats vertically. Figures 2.11 and 2.13 are convenient for showing the derivations, but physically unstable. The behavior of lighter-than-air craft is covered in books on aeronautics, for example, Prandtl and Tietjens [7].

2.11 SUMMARY

1. For simple fluids at rest, the pressure-depth relationship is given by the basic equation of fluid statics, $dP / dz = -\rho g$, found by considering the weight of a small element of fluid and the pressure change with depth necessary to support that weight.
2. For constant-density fluids the basic equation can be integrated to $P_2 - P_1 = -\rho g(z_2 - z_1)$. This equation is an excellent approximation for liquids and a good approximation for gases when the change in elevation is small.
3. For changes in elevation measured in thousands of feet, gases cannot be treated as constant-density fluids. For isothermal, isentropic, or constant-temperature-gradient behavior, the basic equation can be easily integrated for ideal gases.
4. Problems involving liquids with free surfaces are generally easiest to work in gauge pressure, in which case the basic equation simplifies further to $P_{\text{gauge}} = \rho gh$.
5. The force exerted by a static fluid on any infinitesimal surface is given by $dF = P\,dA$ and points normal to the surface. This equation can be integrated to find total forces and the x and y components of pressure forces for a wide variety of situations.
6. The necessary wall thickness of pressure vessels and pipes for modest internal pressures are calculated using the thin-walled approximation.
7. The buoyant force exerted by a fluid on a floating or submerged body is equal to the weight of the fluid displaced.
8. Most pressure-measuring devices either balance the pressure against the weight of a column of fluid, in which case the height of the fluid column is the reading, or let the pressure act on some area, compressing a spring, in which case the deflection of the spring is the reading.
9. The manometer, used for measuring fluid pressure differences, is a logical model for *manometer-like* flows, including most chimneys and flues, and circulating fluid flows, driven by heating one side of the *equivalent manometer.*
10. Problems involving accelerated motion can be handled by the methods of fluid statics if the particles of fluid do not move relative to each other, for example, in rigid-body rotation.

PROBLEMS

See the Common Units and Values for Problems and Examples inside the back cover of this text. An asterisk (*) on the problem number indicates that the answer is in App. D.

2.1.* A large petroleum storage tank is 100 ft in diameter. The free surface is really a very small part of a sphere with radius $\approx$4000 mi (the radius of the earth). If one drew an absolutely straight line from the liquid surface at one side of the tank to the liquid surface directly across the diameter on the other side, how deep into the fluid would that line go? In most fluid mechanics problems we ignore the curvature of the earth. Does this calculation support that simplification?

2.2. Calculate the specific weight of water at a place where the acceleration of gravity is $32.2\ \text{ft}/\text{s}^2 = 9.81\ \text{m}/\text{s}^2$. Express your answer in lbf/ft^3 and in kgf/m^3. Calculate its specific weight on the moon, where $g \approx 6\ \text{ft}/\text{s}^2 \approx 2\ \text{m}/\text{s}^2$.

2.3.* Calculate the specific weight of water in the SI system of units.

2.4. Calculate the pressure gradient due to gravity in water, in psi / ft. Most experienced engineers round this to 0.5 psi / ft, and use it for making routine calculations in their heads.

2.5. Most swimmers find the pressure at a depth of about 10 ft painful to ears. What is the gauge pressure at this depth?

2.6. A new submarine can safely resist an external pressure of 1000 psig. How deep in the ocean can it safely dive?

2.7. The tallest buildings in the world (excluding TV towers, which are not buildings in the common sense) are the Petronas Twin Towers in Kuala Lumpur, 1483 ft tall. If the pressure in the supply line to the drinking fountain on the top floor (perhaps 1450 ft high) is 15 psig, what is the required pressure in the supply line at street level? Assume zero flow in the water line.

2.8.* The deepest point in the oceans of the world is believed to be in the Marianas Trench, southeast of Japan; there the depth is about 11,000 m. What is the pressure at that point?

2.9. In the deep oil fields of Louisiana one occasionally encounters a fluid pressure of 10,000 psig at a depth of 15,000 ft. If this pressure is greater than the hydrostatic pressure of the drilling fluid in the well from the surface, the result may be a blowout, which is dangerous to life and property. Assuming that you are responsible for selecting the drilling fluid for an area where such pressures are expected, what is the minimum density drilling fluid you can use, assuming a surface pressure 0 psig for the drilling fluid?

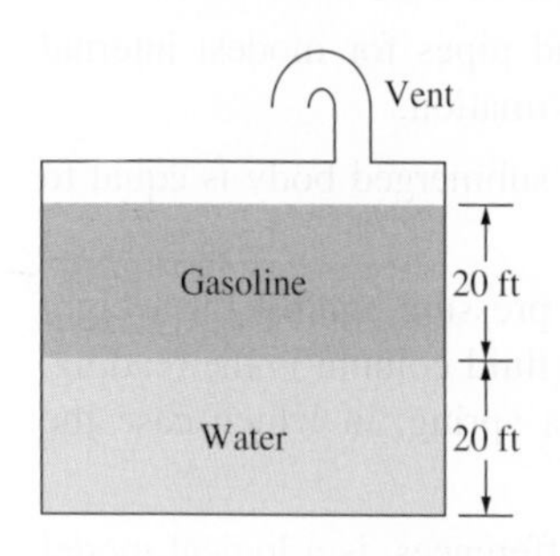

FIGURE 2.22
A storage tank holding two fluids.

2.10. The tank in Fig. 2.22 contains gasoline and water. What is the absolute pressure at the bottom? Sketch the curve of gauge pressure versus depth for this tank.

2.11. Large hydrocarbon storage tanks normally have a valve on their vents that allows free flow of air in or out when liquid is being pumped in or out but that prevents air flow for small pressure differences caused by wind, solar heating of the tank, and changes in atmospheric pressure. These reduce the *breathing losses* of valuable hydrocarbons as well as the amount of atmospheric pollution. These valves are typically set to open at an internal pressure of 4 in of water and an external pressure of 2 in of water.

(*a*) Estimate the force on the roof of the tank in Example 2.6 at the opening pressures for internal pressure and for vacuum.

(*b*) Why is the setting larger for internal pressure than external pressure?

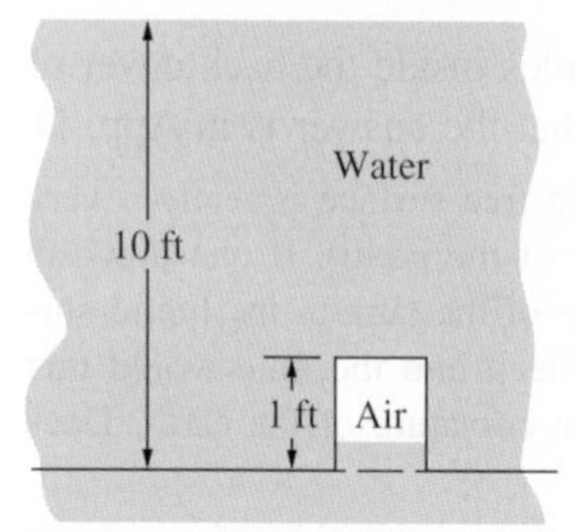

FIGURE 2.23
Figure for Prob. 2.12.

2.12.* An open-ended can 1 ft long is originally full of air at 70°F. The can is now immersed in water, as shown in Fig. 2.23. Assuming that the air stays at 70°F and behaves as a ideal gas, how high will the water rise in the can?

2.13. Normally we assume that liquids are constant-density fluids. To find out how large an error we make that way, compute the pressure at the deepest point in the oceans (about 11,000 m) two ways;

(*a*) Assume seawater is a constant-density fluid with properties shown inside the back cover of this book.

(*b*) Assume that the density of water is given by $\rho = \rho_0[1 + \beta(P - P_0)]$. The definitions of the symbols in this equation and the value of β for water are given in App. A.6 and A.9.

2.14. On a very cold day in Antarctica the temperature of the air is $-60°F$. Assuming that the air remains isothermal up to a 10,000-ft elevation and that the pressure at sea level is 1 atm, estimate the pressure at 10,000 ft.

2.15. An airplane takes off from sea level and is climbing at 2000 ft / min. The plane is not pressurized, so that the pressure of the cabin is falling as the plane rises. At sea level (just after takeoff), how fast is the pressure falling (psi / min or kPa / min)?

2.16. Derive Eqs. 2.17 and 2.18, starting with P / ρ^k = constant and $\rho = PM / (RT)$.

2.17. For the "standard atmosphere" shown in Fig. 2.4,
(*a*) derive the pressure-height relation for the troposphere,
(*b*) calculate the pressure at the troposphere-stratosphere interface, and
(*c*) derive the pressure-height relation for the stratosphere.

2.18.*(*a*) At what height does the equation for an isentropic atmosphere, Eq. 2.17, indicate that the temperature of the air is 0 K? Assume that the surface temperature is 59°F = 15°C.
(*b*) What is the physical significance of this prediction?
(*c*) What is the predicted pressure (Eq. 2.17) for this elevation?

2.19. For most problems we assume that P_{atm} = 14.7 psia. This is a reasonable approximation for sea level but not for other elevations. What is the average atmospheric pressure at
(*a*) Salt Lake City, whose elevation is 4300 ft;
(*b*) 10,000 ft, the elevation to which the cabins on commercial airliners are pressurized;
(*c*) on the top of Mt. Everest (29,028 ft)?
(For simplicity, use the isothermal atmosphere; but see also Prob. 2.21.)

2.20.*What is the sea-level temperature gradient in °F / ft in
(*a*) the standard atmosphere (Fig. 2.4) and
(*b*) the isentropic atmosphere (Eq. 2.17) with a surface temperature of 59°F = 15°C?
(The negative of this gradient, called the *lapse rate,* is widely used in meteorology.)

2.21. The conditions at sea level are 14.7 psia and 59°F = 15°C. Calculate the pressure and temperature at 10,000 ft according to
(*a*) the isothermal atmosphere,
(*b*) the isentropic atmosphere, and
(*c*) the standard atmosphere.

2.22.*What is the mass of the entire atmosphere of the earth? The earth may be considered a sphere of radius ≈4000 mi. All of the atmosphere is so close to the surface of the earth that all of it may be considered to be subjected to the same acceleration due to gravity.

2.23. The oil-storage tank in Examples 2.6 and 2.7 has a vent to the atmosphere to allow air to move in or out as the tank is filled or emptied. This vent is plugged by snow in a blizzard while the oil is being pumped out of the tank, and the gauge pressure in the tank falls to -1 psig. What is the *net* force on the roof of the tank?

2.24.*In the hydraulic lift in Fig. 2.24 the total mass of car, rack, and piston is 1800 kg. The piston has a cross-sectional area of 0.2 m^2. What is the pressure in the hydraulic fluid in the cylinder if the car is not moving?

2.25. Hoover Dam is approximately 230 m high and 76 m wide at the top. Consider it to be a rectangle (only approximately true). When the water is up to the top, what is the pressure at the bottom? What is the net force tending to move the dam?

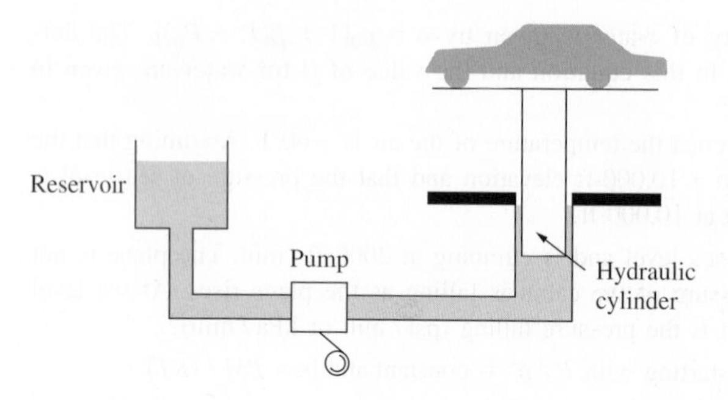

FIGURE 2.24
Hydraulic lift.

2.26. Example 2.8, leading to Eq. 2.M, was simple because the width of the surface was constant. Suppose that instead of being a rectangle, the lock was an isosceles triangle, apex down, 20 m wide at the top and 10 m deep.

(*a*) Show that the width, instead of being constant is given by

$$W = 20\text{ m} \cdot \left(1 - \frac{h}{10\text{ m}}\right) \quad \text{(2.BI)}$$

(*b*) Show that replacing the W in Eq. 2.M with this value of W and carrying out the integration leads to

$$F = \rho g \cdot 20\text{ m} \int h\left(1 - \frac{h}{10\text{ m}}\right) dh = \rho g \cdot 20\text{ m}\left[\frac{h^2}{2} - \frac{h^3}{3 \cdot 10\text{ m}}\right]_0^{10\text{ m}} \quad \text{(2.BJ)}$$

(*c*) Calculate the net force on the triangular lock gate, and compare it to that on the rectangular lock gate in Example 2.8.

(*d*) Show that the right-hand integral in Eq. 2.M can be written as

$$F = \rho g A h_c \quad \text{(2.BK)}$$

where h_c is the depth of the centroid of the surface exposed to the fluid, defined as

$$\begin{pmatrix}\text{Depth of}\\ \text{centroid}\end{pmatrix} = h_c = \frac{\int h dA}{A} \quad \text{(2.BL)}$$

for any shape.

(*e*) For a triangle, $h_c = h_{\text{maximum}} / 3$ measured from the base toward the apex. Repeat part (*c*) using this simplification. Are the answers the same?

2.27.* A dam has an upstream face that is vertical and has the shape of a semicircle with a diameter of 100 m at the top. Water is up to the top of the dam. The atmosphere presses on the rear of the dam. What is the net horizontal force on the dam? Work this problem two ways:

(*a*) by direct integration of the pressure force as shown in part (*b*) of the preceding problem.

(*b*) by using the centroid of depth (as in parts *d* and *e* of the preceding problem). The centroid of a semicircle about its diameter is $2D / 3\pi$. Centroids of all common geometric figures are shown in books on strength of materials.

2.28. Example 2.9 shows the calculated thickness at the bottom needed for steel plate making up the shell of a vertical-axis, flat-bottomed, atmospheric-pressure storage tank. In the United States such tanks are normally made from steel sheets either 8 or 10 ft wide, so their heights are normally multiples of 8 or 10 ft. The 60-ft-high tank in that example would be made of 6 bands, one atop the other, each made of steel sheets 10 ft wide, each with its own uniform thickness. The lowest band would have the thickness calculated in Example 2.9.

(*a*) What would the required thickness be for the remaining bands above the bottom band?

(*b*) If the calculation in part (*a*) leads to a thickness of less than about 0.25 in, then a plate with 0.25 in thickness will be used. Is that the case here? Suggest reasons why a plate with less than 0.25 in thickness will not be used.

(*c*) The thicknesses calculated above are wasteful, because they use a uniform thickness, instead of one that tapers from bottom to top, which would keep the stress constant in the whole wall of the tank. If we could buy such tapered plates, in which the thickness at the top of one band exactly matched the thickness at the bottom of the next upper band, how many pounds of steel (SG = 7.9) would we save on that tank? For really big tanks, the steel mills will roll tapered such plates for you, to make this savings, but that is uncommon.

2.29. The largest vertical-axis, flat-bottomed, atmospheric-pressure storage tanks have heights of about 70 ft and diameters of about 400 ft. The largest spherical storage tanks are about 80 ft in diameter. The largest sausage-shaped tanks are about 11 ft in diameter and 90 ft long. Suggest reasons for these maximum dimensions for each shape of container.

2.30.*We want to select a pipe with an inside diameter of 1 ft that will withstand an internal pressure of 1000 psig. The steel to be used has a maximum allowable tensile stress of 40,000 psi but, to allow for a safety factor of 4, we design for a maximum stress of 10,000 psi. How thick must the pipe walls be?

2.31. From the data given in App. A.2 on the diameter and wall-thickness of schedule 40 pipes, sizes 2.5 in and larger, show that these correspond almost exactly to the formula

$$t = A + BD \quad (2.BM)$$

where t is the wall thickness, D is the diameter, A is an arbitrary constant known as the "corrosion allowance," and B is the value one would compute from Eq. 2.25. Calculate the values of A and B in this equation from the best straight line through a plot of thickness versus diameter data for schedule 40 pipes.

2.32. The thin-walled formulae are based on the assumption that the stress is practically uniform across the cross section of the vessel wall. It is generally used when $D_o / D_i < 1.5$. Sketch what the stress distribution would be for a vessel if internal pressure caused both the inside and outside diameter to increase by the same amount. Estimate how big the difference in pressure from inside to outside would be for a pipe or vessel with $D_o / D_i < 1.5$.

2.33. The thin-walled formulae in the text, Eqs. 2.25 and 2.27, are simplifications of the formulae in the piping codes [5], which are reproduced in Table 2.2.

(*a*) Repeat Example 2.10, using the thin-walled formula for a cylinder from Table 2.2, with $E_J = 1.00$ and $C_C = 0.0$. How much difference does it make?

(*b*) Same as part (*a*), but use the thick-walled formula from Table 2.2.

(*c*) For a cylindrical vessel with $r_i = 5.00$ ft, $S = 20{,}000$ psi, $E_J = 1.00$, and $C_C = 0.0$, prepare a plot of t versus P showing the calculated t from Eq. 2.25 and from the thin- and thick-walled equations from Table 2.2. Cover the pressure range from 0 to 10,000 psi.

2.34. An ordinary rifle has a maximum pressure of about 50,000 psig during firing. This peak value occurs a short way down the barrel from the chamber and lasts about 0.001 s. Farther down the barrel, the peak pressure is substantially less than this [8].

(*a*) If the inside diameter of the barrel is 0.22 in, estimate the required thickness of the barrel wall and the barrel diameter by the thin-walled and thick-walled formulae

TABLE 2.2
Vessel wall thickness formulae, Ref. [5]

Thickness equations	Limiting conditions		
Cylindrical shells			
Thin-walled: $t = \dfrac{P \cdot r_i}{SE_J - 0.6P} + C_C$	$t \le r_i/2$	or	$P \le 0.385SE_J$
Thick-walled: $t = r_i\left(\dfrac{SE_J + P}{SE_J - P}\right)^{1/2} - r_i + C_C$	$t > r_i/2$	or	$P > 0.385SE_J$
Spherical shells			
Thin-walled: $t = \dfrac{P \cdot r_i}{2SE_J - 0.2P} + C_C$	$t \le 0.356r_i$	or	$P \le 0.665SE_J$
Thick-walled: $t = r_i\left(\dfrac{2SE_J + 2P}{2SE_J - P}\right)^{1/3} - r_i + C_C$	$t > 0.356r_i$	or	$P > 0.665SE_J$

Here t = wall thickness, r_i = internal radius, S = design stress, E_J = joint efficiency, and C_C = corrosion allowance.

in Table 2.2 and by Eq. 2.25. Use $S \approx 80{,}000$ psig. Rifles are seamless, so that they have no joint, making $E_J = 1.00$. Their owners take good care of them, so that $C_C = 0.0$.

(*b*) The common practice in design of ordinary pressure vessels is to use a value of S (in Table 2.2) of 20,000 psig instead of 80,000 psig, thus providing a safety factor of 4. Does this work for the formulae in Table 2.2? For Eq. 2.27? Could one design rifles using $S = 80{,}000$ psi and then apply a suitable safety factor to the calculated wall thickness? Would that work in the formulae in Table 2.2? For Eq. 2.31?

(*c*) A simple, single-shot target rifle has an inside barrel diameter of 0.22 in. The barrel tapers from the chamber to the outlet, have a diameter of 0.74 in near the chamber and 0.605 in at the muzzle. Using all three of the values calculated in part (*a*), estimate the safety factor in the diameter at the thick end of the barrel. (Much thinner-walled barrels can be used on guns. They work, but they have very small safety factors and are risky [8]. They are also inaccurate, because the thin-walled barrel is flexible, and flexes during firing. Target rifles often have barrels thicker than the values shown here, not for safety but for stiffness and accuracy.)

2.35. Estimate the required wall thickness for the lowest ring of a flat-bottomed atmospheric pressure tank that will store water, with height 64 ft, diameter 200 ft, and $\sigma = 30{,}000$ psig, using Eq. 2.25. The American Petroleum Institute (API) Standard [9] uses a calculation method somewhat different from the equations in Table 2.2 and Eq. 2.25. Using it, they compute a wall thickness (page K-7) of 1.092 in. How does that compare with the value you compute in this example?

2.36. We wish to purchase a steel tank to store 20,000 gallons of propane; design pressure is 250 psig, and design stress is 20,000 psi. What will be the weight of metal in the shell (excluding foundations, valves, manholes, etc.) if the tank is

(*a*) cylindrical with hemispherical ends, like those in Fig. 2.6. The cylindrical section will have length = 6 times diameter.

(*b*) spherical.

Assume that the simple thin-walled formulae (Eqs. 2.25 and 2.27) may be used. For steel, SG = 7.9.

2.37. Assuming that steel pipes have an allowable wall stress of 10,000 psi, calculate the maximum internal pressure allowable for a 5 in, schedule 40 pipe that has an outside diameter of 5.563 in and an inside diameter of 5.047 in.

2.38. For a cylindrical vessel with spherical ends, what is the relation between the circumferential (hoop) stress in the cylindrical section and the axial stress in the same section?

2.39.* Archimedes is said to have discovered the buoyancy rules, which are called Archimedes' principle, when he was asked by the King of Syracuse in Sicily to determine whether a crown was pure gold, as the goldsmith said, or was an alloy. At that time no chemical means were known for settling the question without destroying the crown. Archimedes was struck with the idea of how to do so while taking a bath, and he jumped out of his tub and ran through the streets yelling "Eureka" ("I have found it"). The story goes that he was so excited that he did not bother to get dressed before doing this.

Suppose that in testing the crown Archimedes found that in air it had a weight of 5.0 N and a weight of 4.725 N in water. Assuming that the crown was made of gold or of silver or of an alloy of both, what percentage by volume was the gold? Assume that the density of gold-silver alloys is $\rho_{\text{alloy}} = \rho_{\text{silver}} + (\text{vol.\% gold}) \cdot (\rho_{\text{gold}} - \rho_{\text{silver}}) / 100$. The densities of gold and silver are 19.3 and 10.5 g / cm^3, respectively.

2.40.* A helium balloon has a flexible skin of negligible weight and infinite capacity for expansion, so that the helium is always at the same pressure as the surrounding air. If the balloon moves up and down slowly, then the temperature of the gas in the balloon will be practically the same as that of the surrounding air. If the mass of helium in the balloon is 10 lbm, how much payload can it lift under the following conditions:

(*a*) 1 atm and 70°F,

(*b*) 0.01 atm and 0°F, and

(*c*) 0.001 atm and −100°F? Assume that helium behaves as an ideal gas.

2.41. Helium is preferred to hydrogen in balloons because it is nonflammable. However, hydrogen has only half the weight of helium. By how much would the payload of the balloon in Example 2.11 have been increased if hydrogen had been used to fill it instead of helium?

2.42. Currently, recreational balloons are not filled with hydrogen or helium but with hot air; the pilot has a small propane burner to heat the air in the balloon. If the balloon is a sphere 20 m in diameter, and if the total weight of balloon, pilot, passenger compartment, propane burner, propane tank, ropes, etc. is 200 kg, what average temperature must the air in the balloon have to just barely lift the balloon? Assume that the air inside and outside the balloon both have atmospheric pressure and have equal molecular weights, 29 gm / mol. (The latter is slightly inaccurate because of the products of combustion inside the balloon; this inaccuracy is small.)

2.43.* A sample of lead is weighed on a pan balance by means of brass weights. It weighs 2.500 lbf.

(*a*) With the same set of brass weights, what would the lead weigh if the entire scale with weights and lead were at the bottom of a tank of water?

(*b*) If they were in a vacuum chamber? Here $\text{SG}_{\text{brass}} = 8.5$ and $\text{SG}_{\text{lead}} = 11.3$.

2.44. Rework Example 2.12, not by Archimedes' principle, but by assuming the block has only vertical and horizontal faces and calculating the difference in pressure between the top and bottom faces.

2.45. A 150-lb drunkard falls in a vat of whiskey. Whiskey has SG = 0.92, whereas the drunkard has SG = 0.99. The drunkard, who wants to stay alive long enough to drink his fill of the whiskey, "treads water," keeping his head above the whiskey. If his head up to his mouth is 15% of his body volume, how much upward force must he exert by "treading water" to keep his head out of the whiskey?

2.46.*It is proposed to build a raft of pine logs for carrying a cargo on a river. The cargo will weigh 500 kg, and it must be kept entirely above the water level. How many kilograms of pine logs must we use to make the raft, if the logs may be entirely submerged, and they have SG = 0.80?

2.47. A sunken battleship weighs 40,000 tons. It may be considered to be all steel, SG = 7.9.

(*a*) We now propose to raise the battleship by sinking steel tanks adjacent to it, attaching them to the battleship, and then blowing the water out of them with compressed air, making them buoyant. Assuming that the compressed-air tanks will have negligible mass, what volume must they have to raise the battleship? Assume that the battleship is in seawater and that the insides of the battleship are completely filled with water.

(*b*) It has been suggested that we could raise the ship by attaching a cable to it and hauling it up. If the cable has a working tensile stress of 20,000 psi, how thick would it have to be?

(*c*) If the battleship is 1000 ft deep and the cable is of uniform thickness, what is the stress in the cable at the top due to the weight of the cable alone?

2.48. A swimming pool is emptied for cleaning. The dimensions of the pool are 20 ft · 30 ft · 6 ft (average depth). A rainstorm causes the water table in the ground around the pool to rise so that the water level in the ground is up to 1 ft below the surface of the ground and thus up to within one foot of the top of the pool. The liquid pressure exerted by this groundwater on the pool is the same as if the pool were immersed to a depth of 5 ft in pure water. What is the upward (buoyant) force exerted by the groundwater on the pool?

2.49.*On July 2, 1982, Larry Walters attached 42 helium-filled weather balloons to a lawn chair, sat in it, and took a balloon ride high over Los Angeles [10]. People who saw him were amazed; air traffic controllers were dumbfounded. Estimate the diameter of the individual weather balloons, which are assumed to be spherical and to all have the same diameter. Assume that Mr. Walters, plus the lawn chair, plus the ropes, the empty balloons, and the miscellaneous things he took along had a weight of 200 lbf.

2.50. A rowboat is in a circular swimming pool with diameter 10 ft. The person in the rowboat throws overboard a 100 lbm block of steel, SG = 7.9, which sinks to the bottom. Does this action cause the level in the swimming pool to rise, stay the same, or decline? How much?

2.51. The fluid shown shaded in the manometer of Fig. 2.25 is ethyl iodide, SG = 1.93. The heights are h_1 = 44 in and h_2 = 8 in

(*a*) What is the gauge pressure in the tank?

(*b*) What is the absolute pressure in the tank?

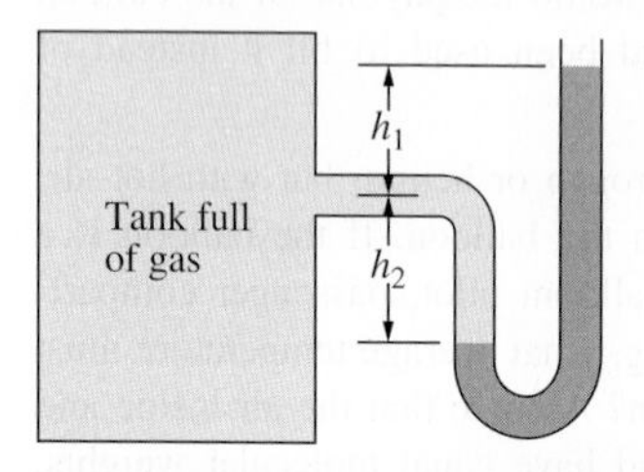

FIGURE 2.25
Simple manometer.

2.52.*The two tanks in Fig. 2.26 are connected through a mercury manometer. What is the relation between Δz and Δh?

2.53. Figure 2.27 is a schematic diagram of a general two-fluid manometer. What is $P_A - P_B$ in terms of h, g, ρ_1, and ρ_2? If we want maximal sensitivity—that is, $\Delta h / (P_A - P_B)$ as large as possible—what relation of ρ_1 to ρ_2 should we choose?

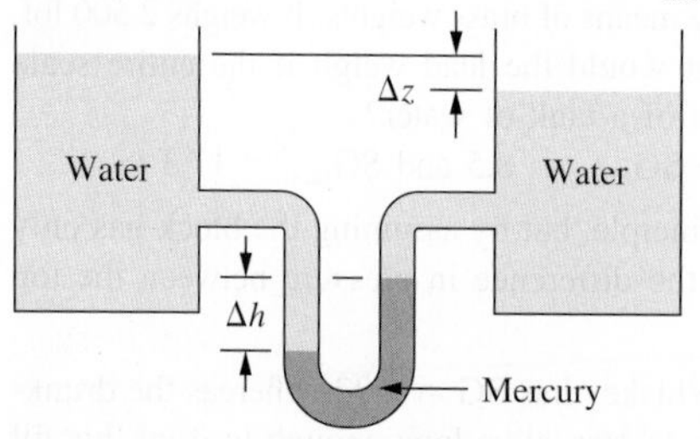

FIGURE 2.26
Mercury-water two-fluid manometer.

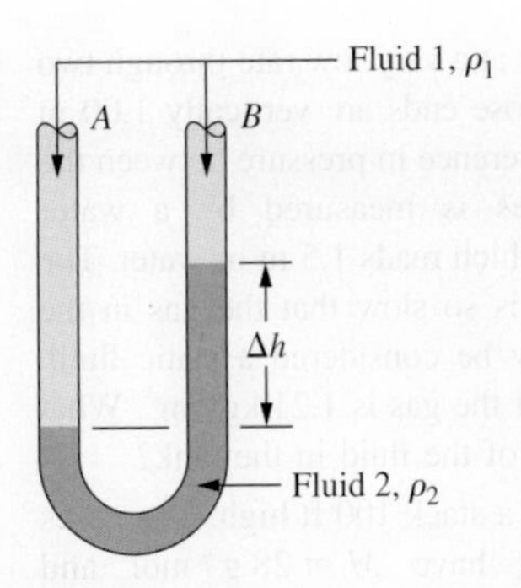

FIGURE 2.27
Two-fluid manometer.

2.54.* For low pressure differences the inclined manometer shown in Fig. 2.28 is often used (this device is so often used to measure the "draft" of a furnace that its common name is a *draft tube*).
(*a*) If the scale is set to read zero length at $P_A = P_B$ and the manometer fluid is colored water, what will the reading be at $P_A - P_B = 0.1$ lbf / in^2?
(*b*) What would the reading of an ordinary manometer with vertical legs be for this pressure difference?

2.55. The manometer in Prob. 2.54 has as its reservoir a cylinder with a diameter of 2 in. The tube has a diameter of $\frac{1}{8}$ in. The scale is set to read zero at $P_A = P_B$. When the level is at the 10 in mark, how much has the level in the reservoir fallen?

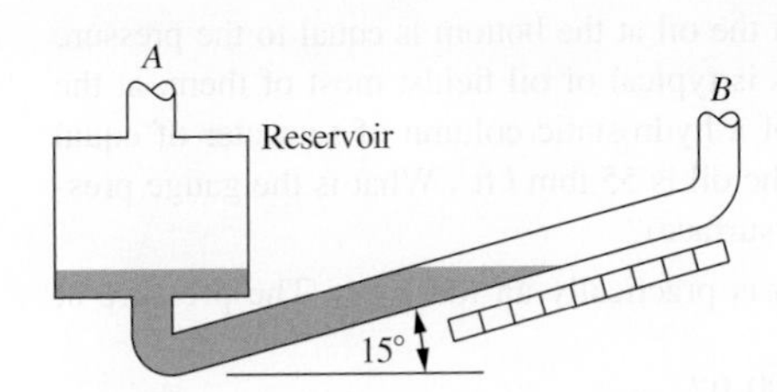

FIGURE 2.28
Draft tube.

2.56. The conventional barometer shown in Fig. 2.14 is filled with mercury.
(*a*) How high must it be to record a pressure of 1 atm?
(*b*) How high would it have to be if we used water instead of mercury as the barometer fluid.
(*c*) How large an error in pressure would we make with a water barometer by ignoring the pressure of the water in space above the liquid?

2.57. Television and newspaper meteorologists regularly show atmospheric high and low pressure regions. Typically, a high will have a sea level pressure of about 1025 millibar and a low will have a sea level pressure of 995 millibar. (Engineers would state these as 1.025 and 0.995 bars, but meteorologists always use the millibar.) Assuming a static atmosphere (impossible, but useful for this problem), estimate the average temperature difference between ground level and the top of the troposphere between the high and the low needed to cause this pressure difference. Could these pressure difference be caused by differences in moisture content?

2.58.* A common scheme for measuring the liquid depth in tanks is shown in Fig. 2.29. Compressed air or nitrogen flows slowly through a "dip tube" into the liquid. The gas-flow rate is so low that the gas may be considered a static fluid. The pressure gauge is 6 ft above the end of the dip tube.

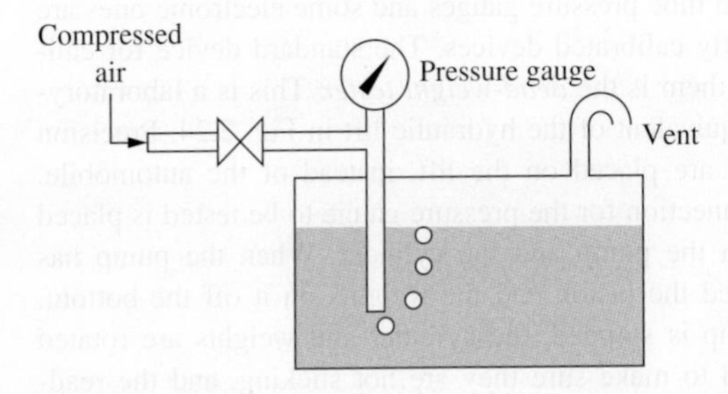

FIGURE 2.29
Dip-tube depth gauge.

(*a*) If the pressure gauge reads 2 psig and the dip tube is 6 in from the bottom of the tank, what is the depth of the liquid in the tank? Here $\rho_{\text{liquid}} = 60$ lbm / ft^3 and $\rho_{\text{gas}} = 0.075$ lbm / ft^3.
(*b*) Customarily, engineers read these gauges as if ρ_{gas} were zero. How much error is made by such a simplification?

2.59. The system shown in Fig. 2.30 is used to measure the density of a liquid in a tank. Compressed air or

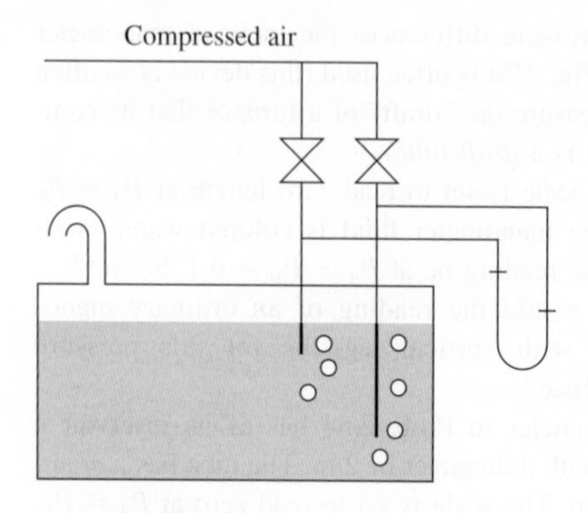

FIGURE 2.30
Two dip-tube density gauge.

nitrogen flows at a very low rate through two dip tubes, whose ends are vertically 1.00 m apart. The difference in pressure between the two dip tubes is measured by a water manometer, which reads 1.5 m of water. The gas-flow rate is so slow that the gas in the dip tubes may be considered a static fluid. The density of the gas is 1.21 kg / m^3. What is the density of the fluid in the tank?

2.60.*A furnace has a stack 100 ft high. The gases in the stack have $M = 28$ g / mol and $T = 300$°F. If the pressures of the air and the gas in the stack are equal at the top of the stack, what is the pressure difference at the bottom of the stack?

2.61. An oil well is 10,000 ft deep. The pressure of the oil at the bottom is equal to the pressure of a column of seawater 10,000 ft deep. (This is typical of oil fields; most of them, at the time of discovery, have about the pressure of a hydrostatic column of seawater of equal depth; there are exceptions.) The density of the oil is 55 lbm / ft^3. What is the gauge pressure of the oil at the top of the well (at the surface)?

2.62.*A natural-gas well contains methane, which is practically an ideal gas. The pressure at the surface is 1000 psig.

(*a*) What is the pressure at a depth of 10,000 ft?

(*b*) How much error would be made by assuming that methane was a constant-density fluid? Assume the temperature is constant at 70°F.

2.63. An oil pipeline was constructed to transport an oil with SG = 0.8 for a distance of 10 mi. The country was hilly, so that the line made many ups and downs. These may be considered equivalent to 10 rises of 200 ft, followed by descents of 200 ft. When the pipe was completed it was tested by pumping water through it. The water flowed satisfactorily with an inlet pressure of 150 psi. Then the oil was slowly fed into the pipe. As the oil flowed the pressure at the inlet end began to rise, and the flow rate began to fall. Finally, the flow stopped altogether, while the pressure at the inlet side remained at 150 psi. Explain what caused this. (*Hint:* This is a manometer problem.)

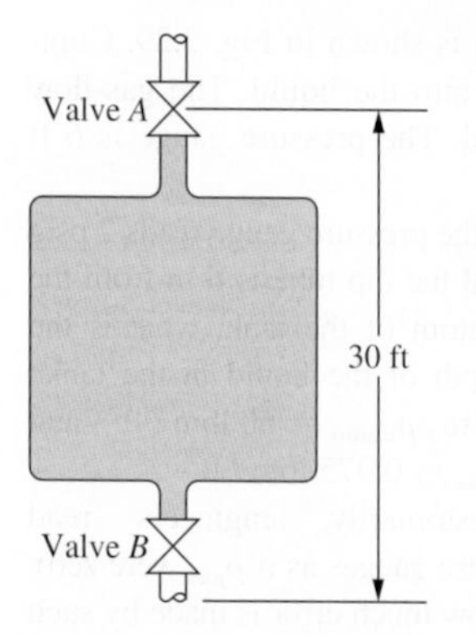

FIGURE 2.31
A tank that can be collapsed by draining.

2.64. The tank in Fig. 2.31 is completely full of water; there is no air. Both valves are closed; now we open valve *B* and allow the water to drain out, without opening valve *A*. What is the minimum pressure that will be reached in the tank?

2.65.*Bourdon tube pressure gauges and some electronic ones are inherently calibrated devices. The standard device for calibrating them is the *dead-weight tester.* This is a laboratory-sized equivalent of the hydraulic lift in Fig. 2.24. Precision weights are placed on the lift, instead of the automobile. The connection for the pressure gauge to be tested is placed between the pump and the cylinder. When the pump has just lifted the piston and the weights on it off the bottom, the pump is stopped, the cylinder and weights are rotated by hand to make sure they are not sticking, and the reading of the pressure gauge is recorded. More weights are

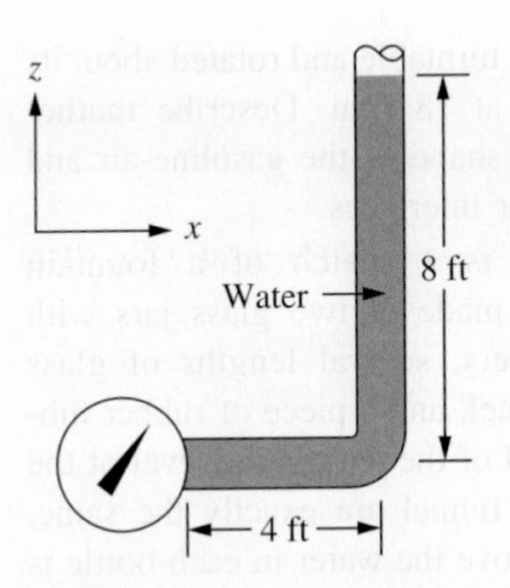

FIGURE 2.32
A very simple accelerometer.

added and the process repeated. If the diameter of the piston is 0.500 in, and the weight of the piston plus weights is 25.00 lbf, what is the pressure exerted on the pressure gauge? Manufacturers claim that these testers are accurate to ±0.5% of the pressures calculated this way.

2.66. A popular, low cost way of marking the tops of fence posts level uses a transparent plastic garden hose. One partly fills the hose with water, holds the two ends to two different posts, and adjusts the liquid level in one end of the hose (by raising or lowering it) until it is level with the top of the post that is being used for reference. Then one marks the level of the fluid in the hose on the next post; and so on. Will this system work if there are trapped air bubbles in the hose?

2.67. Rework Example 2.17 for the elevator falling freely, i.e., for downward acceleration of 9.81 m / s^2.

2.68.*The device in Fig. 2.32 consists of two pieces of pipe of 1 in inside diameter that is connected to a pressure gauge. The whole apparatus is on a elevator, which moves in the *z* direction. The pressure gauge reads 5 psig.
(*a*) How fast is the elevator accelerating?
(*b*) Which way?

2.69. The rectangular tank in Fig. 2.33 is sitting on a cart. We now slowly accelerate the cart. What is the maximum acceleration we can give the cart without having the fluid spill over the edge of the tank?

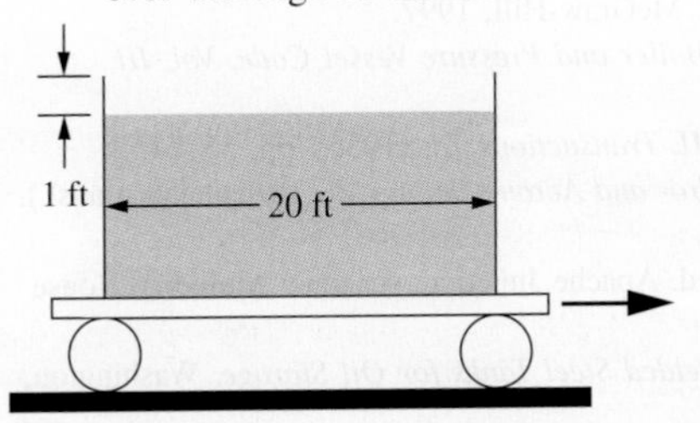

FIGURE 2.33
Figure for Prob. 2.69.

2.70.*A closed tank contains water and heating oil, SG = 0.96, and is completely full of the two liquids, with no air space at the top. The tank is being steadily accelerated in the *x* direction at 1 ft / s^2. What angle does the water-oil interface make with the vertical?

2.71. If the fluid in the centrifuge in Example 2.20 is water, what is the gauge pressure at the outer wall of the centrifuge (under the layer of water 1 in thick)?

2.72.*In the centrifuge in Example 2.20 a solid particle of volume 0.01 in^3 is settling through the fluid.
(*a*) When it is almost at the wall, where the radius is 15 in, what is the buoyant force acting on it?
(*b*) Which way does the buoyant force act?

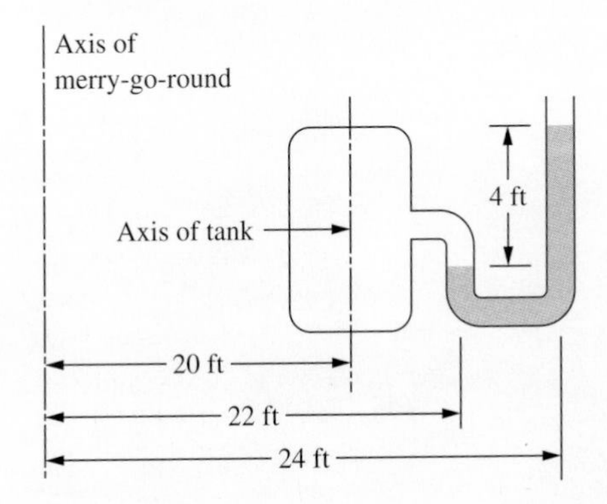

FIGURE 2.34
Manometer on merry-go-round.

2.73. The tank and manometer shown in Fig. 2.34 are mounted on a merry-go-round that is revolving at 10 rpm. The vessel is filled with a gas of negligible density; the manometer fluid is water. What is the pressure in the vessel?

2.74. A cylindrical, open-topped can contains a layer of gasoline 4 in deep on top of a layer of water 4 in deep. The can is now set on a phonograph turntable and rotated about its vertical axis at 78 rpm. Describe mathematically the shape of the gasoline-air and gasoline-water interfaces.

2.75. Figure 2.35 is a sketch of a fountain arrangement made of two glass jars with rubber stoppers, several lengths of glass tubing, a funnel, and a piece of rubber tubing. The level of the jet and the level of the water in the funnel are exactly the same. The space above the water in each bottle is full of air, as is the rubber tube connecting the two bottles. An inventor has come to us, telling us that with this arrangement the water will squirt high in the air, much higher than the water level in the funnel. Is she right? Explain your answer.

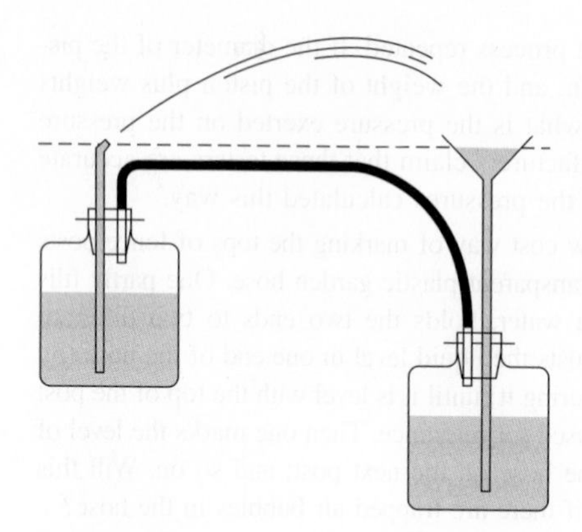

FIGURE 2.35
Gravity fountain.

REFERENCES FOR CHAPTER 2

1. de Nevers, N. *Air Pollution Control Engineering,* 2nd ed. New York: McGraw-Hill, 2000, p. 120.
2. *Handbook of Chemistry and Physics,* Boca Raton, Florida: CRC Press, annual editions.
3. White, F. M. *Fluid Mechanics,* 4th ed. New York: McGraw-Hill, 1999.
4. Meyers, P. E. *Aboveground Storage Tanks.* New York: McGraw-Hill, 1997.
5. "Rules for Construction of Pressure Vessels." *ASME Boiler and Pressure Vessel Code, Vol. III, Div. 1.* New York: ASME, 1998 (updated regularly).
6. Jennings, F. B. "Theories of the Bourdon Tube." *ASME Transactions 78,* (1956), pp. 55–64.
7. Prandtl, L., and O. G. Tietjens. *Fundamentals of Hydro- and Aeromechanics* (L. Rosenhead, transl.). New York: Dover, 1957.
8. Rinker, R. A. *Understanding Firearm Ballistics.* 4th ed. Apache Junction, Arizona: Mulberry House Publishers, 2000.
9. American Petroleum Institute (API). *Standard 650, Welded Steel Tanks for Oil Storage.* Washington, D.C.: API, 1997.
10. Plimpton, G. "The Man in the Flying Lawn Chair." *New Yorker* (June 1, 1998), pp. 62–67.

CHAPTER 3

THE BALANCE EQUATION AND THE MASS BALANCE

Much of engineering is simply careful accounting of things other than money. The accountings are called mass balances, energy balances, component balances, momentum balances, etc. In this chapter we examine the basic idea of a balance and then apply it to mass. The result is the *mass balance,* one of the four basic ideas listed in Sec. 1.3. Chemical engineers use some form of the balance equation in almost every problem they encounter.

3.1 THE GENERAL BALANCE EQUATION

Let us illustrate the general balance idea by making a *population balance* around the State of Utah. The population of Utah can change by:

1. Births
2. Deaths
3. Immigration
4. Emigration

Adding these with the correct algebraic signs and equating them to the increase in population, we get

$$\text{Increase in population} = \text{births} - \text{deaths} + \text{immigration} - \text{emigration} \qquad (3.1)$$

This equation is a special case of the general balance equation:

$$\text{Accumulation} = \text{creation} - \text{destruction} + \text{flow in} - \text{flow out} \qquad (3.2)$$

We can now make four comments:

1. These equations must apply to some period of time. If we simultaneously talk about the births in one year and the deaths in one month, the balance will be queer indeed. If in the population balance we are talking about one year, then we can divide Eq. 3.1 by one year to find

$$\begin{pmatrix}\text{Annual increase}\\ \text{in population}\end{pmatrix} = \begin{pmatrix}\text{annual}\\ \text{birth rate}\end{pmatrix} - \begin{pmatrix}\text{annual}\\ \text{death rate}\end{pmatrix} + \begin{pmatrix}\text{annual}\\ \text{immigration rate}\end{pmatrix} - \begin{pmatrix}\text{annual}\\ \text{emigration rate}\end{pmatrix} \qquad (3.3)$$

   This is a *rate equation.* If someone promises you a million dollars, you will be happy. If he pays you at the rate of $0.01 per year, you will be unhappy; we are all normally interested in rates.

2. If we apply the population balance to the State of Utah for a one-day period, we will find misleading rates. The number of births per day fluctuates; the annual rate is practically constant. To get meaningful rates, the period over which measurements are made must be long enough to average out fluctuations. (There are some situations in which we want to study the short-time fluctuations, e.g., the statistical study of turbulence. For such studies it is worthwhile to make balances over time periods short enough for these fluctuations not to "average out.")

3. In the example above, the balance was made over an identifiable set of boundaries (the legal boundaries of the state of Utah; see Prob. 3.1). A general principle of engineering balances is that there can be no meaningful balance without a carefully *defined and stated set of boundaries.* The set of boundaries need not be fixed, but they must be identifiable. Suppose a group of people was shipwrecked in the Antarctic and took refuge on a floating iceberg. We could make a population balance around the iceberg, and it would have the same terms in it as did our population balance around the state of Utah. The boundaries of the iceberg are perfectly well-defined, but the iceberg is not fixed in place, and its size is constantly changing.

   Whatever is inside a set of boundaries is often called *system.* Everything that is outside the boundaries we will call the *surroundings.* Thus, the boundaries divide the whole universe into two parts, the system and the surroundings.

   For some problems it is convenient to choose as our system the contents of some closed container, which does not allow flow into or out of it. For such a system the balance equation reduces to

$$\text{Accumulation} = \text{creation} - \text{destruction} \qquad [\text{closed system}] \qquad (3.4)$$

Such a system is called a *closed system.* An example might be the population of a sealed space capsule traveling through space, for which the population balance equation would be

$$\text{Increase in population} = \text{births} - \text{deaths} \tag{3.A}$$

The closed system is widely used in chemistry. It is very convenient when a chemical reaction is taking place in a closed container, in which new species may be created by chemical reaction and old ones destroyed but none flow into or out of the container.

An *open system* is usually some kind of container or vessel that has flow in and out across its boundaries at some small number of places. This is used much more commonly in engineering than is the closed system and will be used extensively in this book.

We consider flows in and out of most open systems only at some small number of places: for example, a household water heater that has one cold water inlet pipe, one hot water outlet pipe, one drain pipe, and one connection for a pressure relief valve. If we choose as a system some arbitrary region of space that can have flow in or out over its entire boundary, then this system is called a *control volume.* In this book we will treat any control volume as a special kind of open system.

4. The balance equation deals only with *changes* in the thing being accounted for, not with the total amount present. The population balance given above tells the change in the population of Utah but not the numerical value of the population. If we want to know the numerical value of the population of that state, we may conduct a census. Alternatively, if we could find birth, death, immigration, and emigration data from the time that the first person entered the state to the present, we could compute the change in population starting with population zero. Mathematically, this is

$$\text{Current population} = \int_{\text{time at population}=0}^{\text{present}} (\text{rate of change of population})\, d(\text{time}) \tag{3.5}$$

Beginners are often tempted to find a place in their balances for the total amount contained, such as a numerical value of population; resist this temptation!

To what can the balance equation be applied? It can be applied to any countable set of units or to any *extensive property.* An extensive property is one that doubles when the amount of matter present doubles. Some examples are mass, energy, entropy, mass of any chemical species, momentum, and electric charge. Some examples of countable units are people, apples, pennies, molecules, home runs, electrons, and bacteria.

The balance equation cannot be applied to uncountable individuals (units) or to *intensive properties.* Intensive properties are independent of the amount of matter present. Some examples are temperature, pressure, viscosity, hardness, color, honesty, electric voltage, beauty, and density. An example of uncountable individuals is all the decimal fractions between 0 and 1.

Because the balance equation is so important, we consider one more nonengineering illustration.

Example 3.1. Write out the appropriate terms to apply Eq. 3.2 to your bank account.

By inspection we can write that

$$\begin{pmatrix}\text{Increase in bank}\\ \text{balance}\end{pmatrix} = \begin{pmatrix}\text{interest}\\ \text{payments}\end{pmatrix} - \text{charges} + \text{deposits} - \text{withdrawals} \tag{3.B}$$

■

As in the population example, the value of the current bank balance does not enter into this equation. (It is unfortunate that the current value of your account is called a *bank balance,* which conflicts with our use of the term *balance.*) In both the population example and this example, there are terms that are roughly proportional to the current value of what is accounted for. The birth and death rates in any state are roughly proportional to the total population, and the interest payment in your bank account is proportional to the current amount in the account. Thus the current values do often enter such balance equations indirectly. But they have no direct entry into the accountings.

3.2 THE MASS BALANCE

Our example of a balance equation in the preceding section would be of interest to demographers but not necessarily to engineers. The most important chemical engineering balance is the mass balance. Mass obeys the general balance equation: creation and destruction terms are zero. Thus, the mass balance is

$$\begin{pmatrix}\text{Increase in mass within}\\ \text{the chosen boundaries}\end{pmatrix} = \begin{pmatrix}\text{flow of}\\ \text{mass in}\end{pmatrix} - \begin{pmatrix}\text{flow of}\\ \text{mass out}\end{pmatrix} \tag{3.6}$$

The careful application of this equation is necessary to most fluid-mechanics problems. We can divide by time and find

$$\begin{pmatrix}\text{Rate of increase of mass}\\ \text{within the chosen boundaries}\end{pmatrix} = \begin{pmatrix}\text{flow rate}\\ \text{of mass in}\end{pmatrix} - \begin{pmatrix}\text{flow rate}\\ \text{of mass out}\end{pmatrix} \tag{3.7}$$

The mass balance cannot be derived from any prior principle. Like all the other basic “laws of nature,” it rests on its ability to explain observed facts. Every careful experimental test indicates that it is correct. Mass can exist in a variety of forms, for example, solid, liquid, gas, and some other bizarre forms, and can convert from one to the other. When liquid water evaporates we see the liquid disappear, but we have no visual evidence that the mass of the surrounding air increased by exactly the mass of the water vapor thus produced. Lavoisier made the first clear statement of the law [1] and demonstrated that if processes similar to the evaporation of water were carried out in a closed glass jar resting on a balance, there was no loss of mass; the visible water had changed to invisible water vapor, but the mass of the contents of the jar did not change. The idea that mass is conserved seems quite obvious to us now, but it was not known nor believed by the human race before about 1780. The key discovery was that gases had mass, which was not intuitively obvious to scientists or the public before then.

We will see in Chap. 4 that mass and energy can be converted from one to the other. In most engineering problems we can neglect this fact and use the simple formulation in Eq. 3.5 (but we cannot neglect it in dealing with atomic bombs or the energy source of the sun). There is no experimental evidence on earth that matter is created except by conversion of energy to mass, as just described. A more interesting prospect is the idea of the "steady-state universe," put forward by the British astronomer Fred Hoyle. According to his theories, matter is being created all the time, everywhere; however, the rate is very slow, about one hydrogen atom per hour per cubic mile of space [2]. No instruments now exist that could detect such an event, so a confirmation of this theory on an earthbound scale seems impossible at present. Hoyle claimed that the experimental observations of the behavior of the farthest galaxies support his theories; most other astronomers disagree. Although these theories have no foreseeable application to engineering problems, it is well to keep an open mind on the subject of the *absolute* nature of the mass balance or the other laws of nature as we currently understand them.

Example 3.2. Consider the simple pot-bellied stove, burning natural gas, shown in Fig. 3.1. Applying Eq. 3.7 to this stove, we choose as our system boundaries the walls of the stove. Then Eq. 3.7 becomes

$$\begin{pmatrix}\text{Rate of increase of}\\ \text{mass within the chosen}\\ \text{boundaries}\end{pmatrix} = \begin{pmatrix}\text{mass flow rate}\\ \text{of gas in}\end{pmatrix} + \begin{pmatrix}\text{mass flow}\\ \text{rate of air in}\end{pmatrix} - \begin{pmatrix}\text{mass flow rate}\\ \text{of exhaust}\\ \text{gas out}\end{pmatrix} \tag{3.C}$$

■

Here we have two mass-flow-in terms. There is no limit to the number of such terms. Recall our population balance around the state of Utah; we could have immigration by airplane, car, boat, train, etc. We would have a term for each and add the terms to get the total immigration term. Similarly here we add the individual mass-flow-in terms to get the total mass-flow-in term.

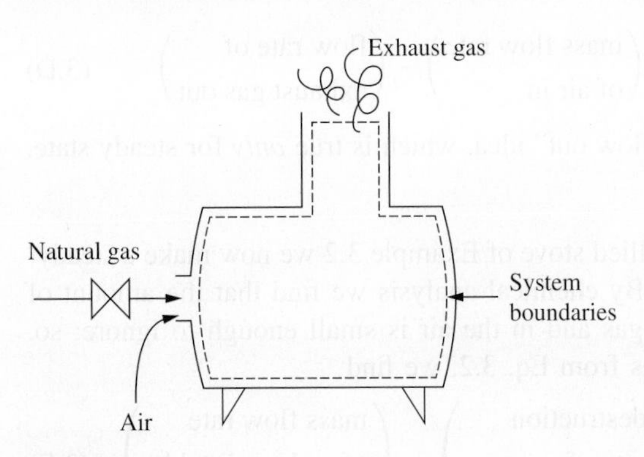

FIGURE 3.1
Pot-bellied stove.

The mass balance has several other names that are in wide use. These are the *principle of conservation of mass,* the *continuity equation,* the *continuity principle,* and the *material balance.* They all mean exactly the same thing as mass balance, namely, that mass obeys the general balance equation, with no creation or destruction.

3.3 STEADY-STATE BALANCES

When the pot-bellied stove in Fig. 3.1 is first lighted after being turned off for a long time, the temperature of its various parts will change rapidly. After a certain time it will be warmed up, and thereafter the temperature of the various parts will not change with time. During the warm-up period, the velocities and temperatures of the gases passing through it at some fixed point will be changing with time. A thermometer at some fixed point in the flue will register a continually increasing temperature. After the stove has warmed up, this thermometer will register a constant temperature. When the stove has warmed up and is running steadily we speak of it as being at steady state.

A steady state does not mean that nothing is changing; it means that nothing is changing with respect to time. Consider a waterfall with a steady flow over it. From the viewpoint of a particle of water there is a rapid increase in velocity as it falls and a sudden decrease in velocity at the bottom. From the viewpoint of an observer watching one specific point in space the waterfall is always the same: there is always water going by at a fixed velocity. Mathematically, if the velocity V is some function of time and position,

$$V = f(t, x, y, z) \tag{3.8}$$

then at steady state

$$\left(\frac{\partial V}{\partial t}\right)_{x,\,y,\,z} = 0 \qquad \text{[steady state]} \tag{3.9}$$

We may similarly write for steady state that $(\partial / \partial t)_{x,\,y,\,z}$ of *any* measurable property of the system at *any* point is zero. Thus, if we write the balance equation for some measurable quantity such as mass and divide by dt to find the rate form, then we see that the left-hand side (the time rate of mass increase within the system) must be zero, because at every point in the system the mass contained is not changing with time. Entirely analogous arguments indicate that at steady state the accumulation term must be zero for all possible balances, including the energy and momentum balances, which we will discuss in Chaps. 4 and 7.

Returning now to the pot-bellied stove of Example 3.2, we see that, at steady state, the mass balance simplifies to

$$0 = \begin{pmatrix}\text{Mass flow rate}\\ \text{of gas in}\end{pmatrix} + \begin{pmatrix}\text{mass flow rate}\\ \text{of air in}\end{pmatrix} - \begin{pmatrix}\text{flow rate of}\\ \text{exhaust gas out}\end{pmatrix} \tag{3.D}$$

This is the familiar "flow in equals flow out" idea, which is true *only* for steady state, with no creation or destruction.

Example 3.3. For the pot-bellied stove of Example 3.2 we now make a steady-state carbon dioxide balance. By chemical analysis we find that the amount of carbon dioxide in the natural gas and in the air is small enough to ignore; so, omitting the unnecessary terms from Eq. 3.2, we find

$$0 = \begin{pmatrix}\text{Creation rate}\\ \text{of carbon}\\ \text{dioxide}\end{pmatrix} + \begin{pmatrix}\text{destruction}\\ \text{rate of}\\ \text{carbon dioxide}\end{pmatrix} - \begin{pmatrix}\text{mass flow rate}\\ \text{of carbon dioxide}\\ \text{out in exhaust gas}\end{pmatrix} \tag{3.E}$$

Chemical analysis of the exhaust gas indicates that it contains 8% to 12% carbon dioxide, so the mass flow rate out is not negligible. Thus, for this equation to be satisfied, there must be significant creation minus destruction of carbon dioxide in the stove; i.e., carbon dioxide is formed by combustion in the stove. In this case, the destruction term is negligible. ■

If we made a similar balance for natural gas we would see that the destruction term would be approximately equal to the mass-flow-in term. In the field of chemical reactions, the creation and destruction terms are very important and cannot be ignored. The momentum balance (Chap. 7) includes creation and destruction terms, as does the entropy balance or the second law of thermodynamics (see any elementary textbook on thermodynamics). Thus, although the two most common balances, the mass and energy balances, have no creation or destruction terms, one should remember that these terms are very important in some other balances.

3.4 THE STEADY-STATE FLOW, ONE-DIMENSIONAL MASS BALANCE

Consider the steady-state flow of some fluid in a pipe of varying cross section, Fig. 3.2. If we apply the steady-state mass balance equation to the system shown, we find

$$\text{Mass flow rate in at point 1} = \text{mass flow rate out at point 2} \tag{3.F}$$

In general, velocity is not the same at every point in a cross section of a pipe; it is faster near the center than at the walls. (One may verify this for the analogous open-channel flow by dropping bits of wood or leaves on a flow of water in a ditch or gutter and noting that those in the center go faster than those at the side.) Therefore, to calculate the total flow rate in across the system boundaries at point 1, we should break the area across which the flow is coming in into small subareas (A), over each of which the flow is practically uniform:

$$\text{Mass flow rate in at point 1} = \sum_{\text{many subareas}} \rho A V \tag{3.10}$$

Here the individual elements of area must be taken perpendicular to the local flow velocity. For flow in a straight pipe or channel this is no problem, because the flow is all in one direction, and the area we normally consider is one perpendicular to the flow. If we then take the limit, as each subarea becomes infinitely small, the term on the right becomes the integral, over the entire system boundary at point 1, of $\rho V\,dA$. Therefore, the steady-state mass balance for the system shown in Fig. 3.2 is

$$0 = \int_{\text{area 1}} \rho V\,dA - \int_{\text{area 2}} \rho V\,dA \tag{3.G}$$

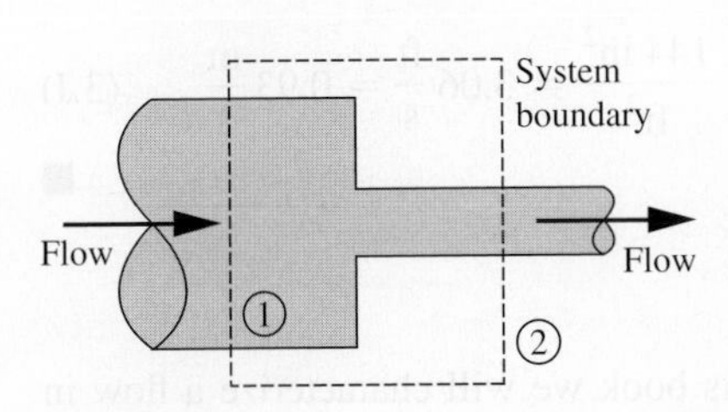

FIGURE 3.2
A system with one flow in and one flow out.

But we could choose points 1 and 2 to be any locations in the pipe, so for steady-state flow in a pipe or channel this equation becomes

$$\int_{\substack{\text{area at any boundary}\\ \text{perpendicular to the flow}}} \rho V\, dA = \text{constant} \qquad \text{[steady flow in a pipe or channel]} \tag{3.11}$$

3.4.1 Average Velocity

No real flow has a completely uniform velocity over the whole cross section. But for many problems we use an appropriate average velocity as if it were uniform across the whole cross section. The constant in Eq. 3.11 is the total mass per unit time passing down the pipe or channel, called the *mass flow rate.* It is normally measured in kg / s or lbm / s and given the symbol $\dot{m}$. If the density is uniform across the cross section of the pipe or channel (almost always practically true) then we may further define

$$\begin{pmatrix}\text{Volumetric}\\ \text{flow rate}\end{pmatrix} = Q = \frac{\text{mass flow rate}}{\text{density}} = \frac{\dot{m}}{\rho} \tag{3.12}$$

(In civil engineering books this quantity is called the *discharge.*) If we now divide the volumetric flow rate by the cross-sectional area of the pipe or channel, we find

$$\begin{pmatrix}\text{Average}\\ \text{velocity}\end{pmatrix} = V_{\text{average}} = \frac{Q}{A} \tag{3.13}$$

Example 3.4. A typical self-service gasoline pump puts 15 gal of fuel into our tank in 2 min. The inside diameter of the nozzle is 1.0 in. What are the volumetric flow rate, mass flow rate, and average velocity?

The volumetric flow rate is

$$Q = \frac{V}{t} = \frac{15\text{ gal}}{2\text{ min}} = 7.5\,\frac{\text{gal}}{\text{min}} = 0.0167\,\frac{\text{ft}^3}{\text{s}} = 0.00047\,\frac{\text{m}^3}{\text{s}} \tag{3.H}$$

The density of gasoline varies from refiner to refiner and with time of the year. On average, it has an SG of about 0.72, so that

$$\dot{m} = Q\rho \approx 0.0167\,\frac{\text{ft}^3}{\text{s}} \cdot 0.72 \cdot 62.3\,\frac{\text{lbm}}{\text{ft}^3} = 0.75\,\frac{\text{lbm}}{\text{s}} = 0.34\,\frac{\text{kg}}{\text{s}} \tag{3.I}$$

and

$$V_{\text{average}} = \frac{Q}{A} = \frac{0.0167\text{ ft}^3/\text{s}}{(\pi/4)\cdot(1\text{ in})^2} \cdot \frac{144\text{ in}^2}{\text{ft}^2} = 3.06\,\frac{\text{ft}}{\text{s}} = 0.93\,\frac{\text{m}}{\text{s}} \tag{3.J}$$

■

3.4.2 Velocity Distributions

For most of the rest of Parts I, II, and III of this book we will characterize a flow in a pipe or channel as having one velocity (the *block flow* or *plug flow* assumption), the average velocity calculated above. How good an approximation is that? How big a price in accuracy do we pay for the huge calculational simplification we get that way?

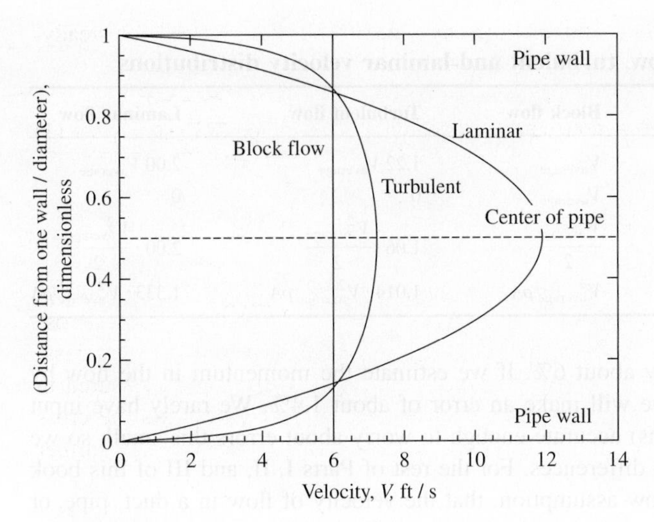

FIGURE 3.3
Velocity distributions in a circular pipe with an average velocity of 6 ft / s. The Block flow curve (vertical line) corresponds to the simplification that the whole flow may be represented by its average velocity. The Laminar flow curve (discussed in detail in Chap. 6) shows that for this type of flow the velocity at the center of the pipe is twice the average velocity, and the velocity distribution is parabolic. The Turbulent curve is an approximation of experimental measurements, represented by "Prandtl's 1 / 7 power rule" (see Prob. 3.10 and Chap. 17).

Figure 3.3 compares the velocity distributions in a pipe computed by various assumptions. In all three cases the average velocity, $V_{\text{average}} = 6$ ft / s, a common velocity in industrial pipe flow. Figure 3.3 shows that for the *block flow* assumption, the velocity is constant at 6 ft / s over the entire cross section of the pipe. *Turbulent flow,* discussed in Chap. 6, is the most common type of flow in industrial pipes, tubes, and channels. The curve shown for turbulent flow is an approximation; see Prob. 3.8. It shows that the velocity goes to zero at each of the pipe walls, as you can observe in flows in a river or rain gutter. For the average velocity to be 6 ft / s, the maximum velocity at the center of the channel must be 7.35 ft / s. *Laminar flow,* also discussed in Chap. 6, occurs in very small pipes and channels (e.g., almost all the blood flows in your body) and for high-viscosity fluids (pouring syrup on your pancakes) but not very frequently in common industrial flows. We will speak more about it in subsequent chapters. For laminar flow, as for turbulent flow, the velocity at the pipe walls is zero. To have the average velocity be 6 ft / s, the maximum velocity at the center must be 12 ft / s.

The average kinetic energy (KE) and average momentum in a pipe flow (discussed in Chap. 7) are always somewhat larger than those corresponding to the average velocity, because in calculating them (Probs. 3.10 and 3.11) we see that the velocity appears to the second or third power. Table 3.1 shows how much difference it makes if we assume either of these three distributions. For example, laminar flow is quite different from block or turbulent flows; we will speak more about that in Chap. 6. But turbulent flow is not much different from block flow. If we estimate the kinetic energy in a turbulent flow by the block flow (average velocity) assumption we

TABLE 3.1
Comparisons of block flow, turbulent and laminar velocity distributions

	Block flow	Turbulent flow	Laminar flow
Maximum velocity	V_{average}	$1.22\ V_{\text{average}}$	$2.00\ V_{\text{average}}$
Minimum velocity	V_{average}	0	0
Kinetic energy per unit mass	$\dfrac{V^2_{\text{average}}}{2}$	$1.06 \cdot \dfrac{V^2_{\text{average}}}{2}$	$2.00 \cdot \dfrac{V^2_{\text{average}}}{2}$
Total momentum in the flow	$V^2_{\text{average}}\,\rho A$	$1.014 \cdot V^2_{\text{average}}\,\rho A$	$1.333 \cdot V^2_{\text{average}}\,\rho A$

will make an error of only about 6%. If we estimate the momentum in the flow by the same simplification, we will make an error of about 1.4%. We rarely have input data (in industrial situations) accurate enough to worry about errors this small, so we will normally ignore these differences. For the rest of Parts I, II, and III of this book we will make the block-flow assumption, that the velocity of flow in a duct, pipe, or channel is adequately represented by its average value. Where we do not make that assumption, we will make that clear in the text. For the most careful work, reconsider that assumption, looking again at Table 3.1. Table 3.1 applies only to flow in a circular pipe or duct; for other geometries (the atmosphere, the oceans, all two- and three-dimensional flows), the results are more complex.

With this simplification, and the additional (very good) assumption that the density of the fluid is constant across the cross section, the integration in Eq. 3.11 can be easily performed, giving

$$\rho_1 A_1 V_1 = \rho_2 A_2 V_2 = \dot{m} = \text{constant} \qquad \text{[steady flow in a pipe or channel]} \quad (3.14)$$

Example 3.5. In a natural-gas pipeline at station 1 the pipe diameter is 2 ft and the flow conditions are 800 psia, 60°F, and 50 ft / s velocity. At station 2 the pipe diameter is 3 ft and the flow conditions are 500 psia, 60°F. What is the velocity at station 2? What is the mass flow rate?

Solving Eq. 3.14 for V_2, we find

$$V_2 = V_1 \frac{\rho_1}{\rho_2} \frac{A_1}{A_2} = 50 \frac{\text{ft}}{\text{s}} \cdot \frac{\rho_1}{\rho_2} \cdot \frac{\left(\dfrac{\pi}{4}\right)(2\text{ ft})^2}{\left(\dfrac{\pi}{4}\right)(3\text{ ft})^2} \quad (3.K)$$

The density of natural gas (principally methane) at 800 psia and 60°F is approximately 2.58 lbm / ft^3, and at 500 psia and 60°F it is approximately 1.54 lbm / ft^3 [3]. Therefore,

$$V_2 = V_1 \frac{\rho_1}{\rho_2} \frac{A_1}{A_2} = 50 \frac{\text{ft}}{\text{s}} \cdot \frac{2.58(\text{lbm / ft}^3)}{1.54(\text{lbm / ft}^3)} \cdot \frac{\left(\dfrac{\pi}{4}\right)(2\text{ ft})^2}{\left(\dfrac{\pi}{4}\right)(3\text{ ft})^2} = 37.2 \frac{\text{ft}}{\text{s}} = 11.3 \frac{\text{m}}{\text{s}} \quad (3.L)$$

$$\dot{m} = \rho_1 V_1 A_1 = 2.58 \frac{\text{lbm}}{\text{ft}^3} \cdot 50 \frac{\text{ft}}{\text{s}} \cdot \frac{\pi}{4} (2\text{ ft})^2 = 405 \frac{\text{lbm}}{\text{s}} = 184 \frac{\text{kg}}{\text{s}} \quad (3.M)$$

■

For liquids at temperatures well below their critical temperature the changes in density with moderate temperature and pressure changes are small. Therefore, for liquids we can divide the density out of Eq. 3.14, finding

$$A_1 V_1 = A_2 V_2 = \frac{\dot{m}}{\rho} = \text{constant} \quad \left[\begin{array}{l}\text{constant density}\\ \text{steady flow in a}\\ \text{pipe or channel}\end{array}\right] \tag{3.15}$$

Mass divided by density equals volume; therefore, the constant in this equation (the mass flow rate divided by the density) is the *volumetric flow rate, Q,* discussed above.

Example 3.6. Water is flowing in a pipe. At point 1 the inside diameter is 0.25 m and the velocity is 2 m / s. What are the mass flow rate and the volumetric flow rate? What is the velocity at point 2 where the inside diameter is 0.125 m?

$$\dot{m} = \rho_1 V_1 A_1 = 998.2\,\frac{\text{kg}}{\text{m}^3} \cdot 2\,\frac{\text{m}}{\text{s}} \cdot \frac{\pi}{4}(0.25\ \text{m})^2 = 98.0\,\frac{\text{kg}}{\text{s}} = 216\,\frac{\text{lbm}}{\text{s}} \tag{3.N}$$

$$Q = \frac{\dot{m}}{\rho} = V_1 A_1 = 2\,\frac{\text{m}}{\text{s}} \cdot \frac{\pi}{4}(0.25\ \text{m})^2 = 0.09817\,\frac{\text{m}^3}{\text{s}} = 3.46\,\frac{\text{ft}^3}{\text{s}} \tag{3.O}$$

$$V_2 = V_1 \frac{A_1}{A_2} = 2\,\frac{\text{m}}{\text{s}}\,\frac{\left(\dfrac{\pi}{4}\right)(0.25\ \text{m})^2}{\left(\dfrac{\pi}{4}\right)(0.125\ \text{m})^2} = 8\,\frac{\text{m}}{\text{s}} = 24.2\,\frac{\text{ft}}{\text{s}} \tag{3.P}$$

■

3.5 UNSTEADY-STATE MASS BALANCES

The steady-state behavior of systems, shown in the preceding examples, is very important. Most of the examples and problems shown in elementary textbooks concern steady-state behavior. However, unsteady-state behavior is probably more important. The characteristics of the two are compared in Table 3.2.

A power plant burns fuel and produces electricity by means of a boiler, turbine, condenser, generator, etc.; its steady-state behavior is fairly easy to calculate. However, its behavior when the power demand on the generator is suddenly increased or decreased is much more difficult to calculate. The power company would prefer

TABLE 3.2
Comparison of steady-state and unsteady-state processes

Property	Steady state	Unsteady state
Calculations	Generally easy	More difficult
Normally requires calculus?	No	Yes
Setup in laboratory	Difficult	Easy
Large-scale industrial use	Desirable	Undesirable
Efficiency	Generally high	Generally lower
Capital cost per unit of production		
Large-volume product (e.g., gasoline)	Low	High
Small-volume product (e.g., pharmaceuticals)	High	Low

to have a steady load, because then they could always operate the plant at its maximum efficiency. Nonetheless, they must plan for and be equipped for sudden load disturbances (e.g., a lightning strike shuts down a major consumer, thus quickly reducing the power demand). It has also been observed that most industrial disasters, such as explosions and fires, do not occur during periods of steady-state operation but during startup or shutdown of some processing unit, for example, the Chernobyl nuclear disaster in 1986 near Kiev. It occurred during an unusual shutdown. Thus we see that the unsteady-state behavior is very important and worthy of our attention.

Unsteady-state mass balances do not introduce any new ideas beyond those seen so far. However, as shown by the following examples, they generally lead to more complicated mathematics.

Example 3.7. The microchip diffusion furnace in Fig. 3.4 contains air, which may be considered an ideal gas. The vacuum pump is pumping air out prior to beginning the thermal diffusion step. During the pumpout process the heating coils in the tank hold the temperature in the tank constant at 68°F. The volumetric flow rate at the inlet of the pump, independent of pressure, is 1.0 ft^3 / min. How long does it take the pressure to fall from 1 atm to 0.0001 atm?

We choose as our system the tank up to the pump inlet. For this system the mass balance gives

$$\left(\frac{dm}{dt}\right)_{\text{system}} = -\dot{m}_{\text{out}} \tag{3.Q}$$

But we know that

$$m_{\text{system}} = V_{\text{system}}\,\rho_{\text{system}} \tag{3.R}$$

where V_{system} is the volume of the system, which does not change. Thus,

$$\left(\frac{dm}{dt}\right)_{\text{system}} = V_{\text{system}}\frac{d\rho_{\text{system}}}{dt} \tag{3.S}$$

Furthermore,

$$\dot{m}_{\text{out}} = Q_{\text{out}}\,\rho_{\text{out}} \tag{3.T}$$

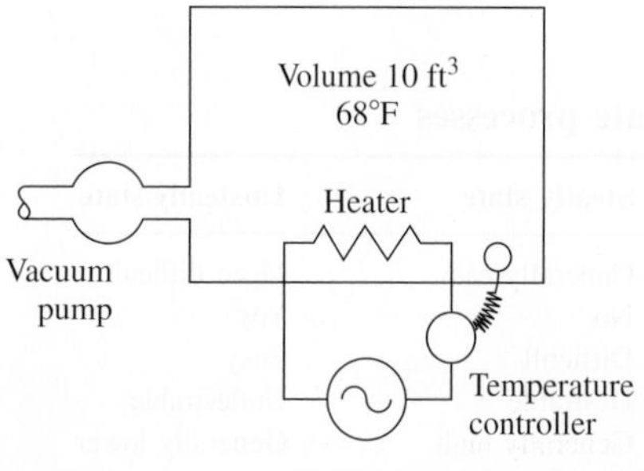

But Q_{out} is constant and

$$\rho_{\text{out}} = \rho_{\text{system}} \tag{3.U}$$

so that

$$V_{\text{sys}}\frac{d\rho_{\text{sys}}}{dt} = -Q_{\text{out}}\,\rho_{\text{sys}} \tag{3.V}$$

This is separable, first-order differential equation, which can be rearranged to

$$\frac{d\rho_{\text{sys}}}{\rho_{\text{sys}}} = -\frac{Q_{\text{out}}}{V_{\text{sys}}}\,dt \tag{3.W}$$

FIGURE 3.4
Evacuation of a microchip diffusion furnace.

and integrated from initial to final states, yielding

$$\ln \frac{\rho_{\text{sys, final}}}{\rho_{\text{sys, initial}}} = -\frac{Q_{\text{out}}}{V_{\text{sys}}} \Delta t \tag{3.16}$$

For low-pressure gases at constant temperature the densities are proportional to the pressures, so we can solve for the required time:

$$\Delta t = \frac{V_{\text{sys}}}{Q_{\text{out}}} \ln \frac{P_{\text{initial}}}{P_{\text{final}}} = \frac{10\,\text{ft}^3}{1\,\text{ft}^3\,/\,\text{min}} \ln \frac{1\ \text{atm}}{0.0001\,\text{atm}} = 92.1\,\text{min} \tag{3.X}$$

■

Example 3.8. In Example 3.7 we considered a vacuum system with zero leaks. No real vacuum systems are totally leak-free; engineers work very hard to keep the leakage rate as low as possible. If the tank in Example 3.7 has a leak of 0.0001 lbm / min of air, what will the pressure-time plot look like, and what will be the final pressure?

Equation 3.Q becomes

$$\left(\frac{dm}{dt}\right)_{\text{system}} = -\dot{m}_{\text{out}} + \dot{m}_{\text{in}} \tag{3.Y}$$

we then follow the preceding problem, retaining $\dot{m}_{\text{in}}$ as a constant in our equations. We find

$$V_{\text{sys}} \frac{d\rho_{\text{sys}}}{dt} = -Q_{\text{out}}\,\rho_{\text{sys}} + \dot{m}_{\text{in}} \tag{3.Z}$$

$$\frac{d\rho_{\text{sys}}}{[\rho_{\text{sys}} - (\dot{m}_{\text{in}}\,/\,Q_{\text{out}})]} = -\frac{Q_{\text{out}}}{V_{\text{sys}}}\,dt \tag{3.AA}$$

$$\ln \frac{\rho_{\text{sys, final}} - (\dot{m}_{\text{in}}\,/\,Q_{\text{out}})}{\rho_{\text{sys, initial}} - (\dot{m}_{\text{in}}\,/\,Q_{\text{out}})} = -\frac{Q_{\text{out}}}{V_{\text{sys}}} \Delta t \tag{3.17}$$

If we ask how long it takes this system to reach 0.0001 atm, we will find that it can never get there. To see why, we ask what its steady state pressure is by setting $\Delta t = \infty$. That can only be possible if the numerator of the fraction on the left becomes zero, or

$$\rho_{\text{sys, final}} = \frac{\dot{m}_{\text{in}}}{Q_{\text{out}}} = \frac{0.0001\ \text{lbm / min}}{1\ \text{ft}^3\ /\ \text{min}} = 0.0001\,\frac{\text{lbm}}{\text{ft}^3} \tag{3.AB}$$

At 68°F = 20°C the density of air is 0.075 lbm / ft^3, and for ideal gases densities are proportional to pressures, so

$$P_{\text{steady state}} = 1\ \text{atm} \cdot \frac{0.0001\ \text{lbm / ft}^3}{0.075\ \text{lbm / ft}^3} = 0.00133\ \text{atm} \tag{3.AC}$$

Solving Eq. 3.17 for the density at any time, we find

$$\rho_{\text{sys, any time}} = \left(\rho_{\text{sys, initial}} - \frac{\dot{m}_{\text{in}}}{Q_{\text{out}}}\right)\exp\left(-\frac{Q_{\text{out}}}{V_{\text{sys}}}\Delta t\right) + \frac{\dot{m}_{\text{in}}}{Q_{\text{out}}}$$

$$= (0.075 - 0.0001)\frac{\text{lbm}}{\text{ft}^3}\exp\left(-\frac{0.1}{\text{min}}\Delta t\right) + 0.0001\,\frac{\text{lbm}}{\text{ft}^3} \tag{3.AD}$$

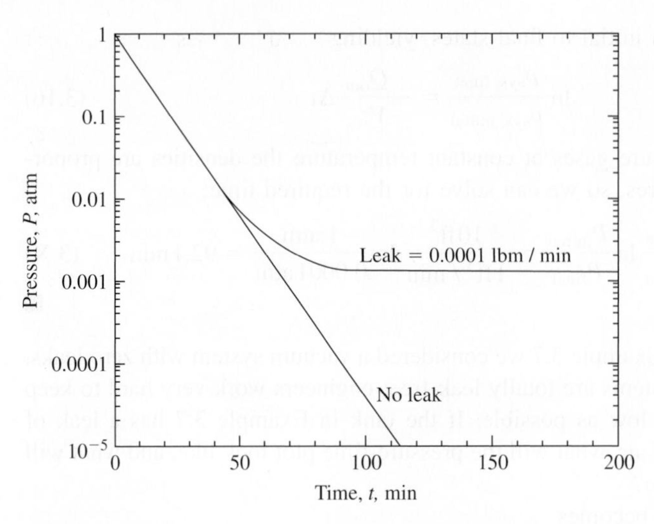

FIGURE 3.5
Calculated pressure-time behavior of the microchip diffusion furnace with zero leakage (Example 3.7) and with a constant 0.0001 lbm / min leakage (Example 3.8).

For $\Delta t = 50$ min, we have

$$\rho_{\text{sys, 50 min}} = 0.0749 \frac{\text{lbm}}{\text{ft}^3} \exp\left(-\frac{0.1}{\text{min}} \cdot 50 \text{ min}\right) + 0.0001 \frac{\text{lbm}}{\text{ft}^3}$$

$$= 0.000505 + 0.0001 = 0.000605 \frac{\text{lbm}}{\text{ft}^3} \tag{3.AE}$$

and

$$P_{50 \text{ min}} = 1 \text{ atm} \frac{0.000605 \text{ lbm / ft}^3}{0.075 \text{ lbm / ft}^3} = 0.00806 \text{ atm} \tag{3.AF}$$

The same calculation is repeated for other times, on a spreadsheet. The resulting pressure-time curves for this example and the previous one are shown in Fig. 3.5. ■

In many unsteady-state mass-balance problems it is convenient to take as the system the fluid within some container. Thus, as the mass of fluid increases or decreases, the volume of the system will change.

Example 3.9. A cylindrical tank 3 m in diameter, with axis vertical, has an inflow line of 0.1 m inside diameter and an outflow line of 0.2 m inside diameter. Water is flowing in the inflow line at a velocity of 2 m / s and leaving by the outflow line at a velocity of 1 m / s. Is the level in the tank rising or falling? How fast?

Here we take as our system the instantaneous mass of water in the tank. For this system

$$\left(\frac{dm}{dt}\right)_{\text{system}} = \dot{m}_{\text{in}} - \dot{m}_{\text{out}} \tag{3.AG}$$

For any fluid we have $m = \rho V$ and $\dot{m} = \rho Q$. Substituting these into the last equation and canceling the constant density, we find

$$\left(\frac{dV}{dt}\right)_{\text{system}} = Q_{\text{in}} - Q_{\text{out}} \tag{3.18}$$

The volumetric flow in or out is equal to *VA*, so

$$\left(\frac{dV}{dt}\right)_{\text{sys}} = 2\,\frac{\text{m}}{\text{s}} \cdot \frac{\pi}{4}(0.1\text{ m})^2 - 1\,\frac{\text{m}}{\text{s}} \cdot \frac{\pi}{4}(0.2\text{ m})^2$$

$$= 0.0157 - 0.0314 = -0.0157\,\frac{\text{m}^3}{\text{s}} \tag{3.AH}$$

The volume of liquid in the tank is decreasing, and the level is falling. The rate of decrease of volume is equal to the cross-sectional area times the rate of fall of the level:

$$\left(\frac{dV}{dt}\right)_{\text{system}} = A\,\frac{dz_{\text{surface}}}{dt}$$

$$\frac{dz_{\text{surf}}}{dt} = \frac{1}{A}\frac{dV}{dt} = \frac{1}{(\pi/4)(3\text{ m})^2}\left(-0.0157\,\frac{\text{m}^3}{\text{s}}\right)$$

$$= -0.0022\,\frac{\text{m}}{\text{s}} = -0.00673\,\frac{\text{ft}}{\text{s}} \tag{3.AI}$$

■

3.6 MASS BALANCES FOR MIXTURES

In the preceding examples, the flowing materials have been uniform single species, such as air or water. In most of the rest of this book we will deal with such uniform single species. However, there are many problems of great interest in which two or more components mix inside the system we are considering. If we make the simplest possible mixing assumption—perfect mixing of all components—then we can apply the simple balance equation as we have done before and find useful answers. The perfect mixing assumption is obviously a great simplification of what must occur in nature, but it is often used because the results are so simple and useful. Several examples illustrate the idea.

Example 3.10. Figure 3.6 is a sketch of a rectangular city with length L and width W. The wind blows over the city in the x direction with velocity V. Atmospheric turbulence mixes the air over the city up to the height H, so we may assume that the air in the "box" with dimensions L times W times H is well mixed and has the same pollutant concentration c everywhere. The air flowing into the upwind side of the city has pollutant concentration b (which stands for *background concentration*). The city emits pollutants into the atmosphere uniformly over its surface with an emission rate q. {Here q would have dimensions like $\text{kg}/(\text{m}^2 \cdot \text{s})$. This uniform-emission assumption is a fair one for emissions from autos or small industry, which are more or less uniformly spread over the city, but a very poor one for emissions from a single large factory or power

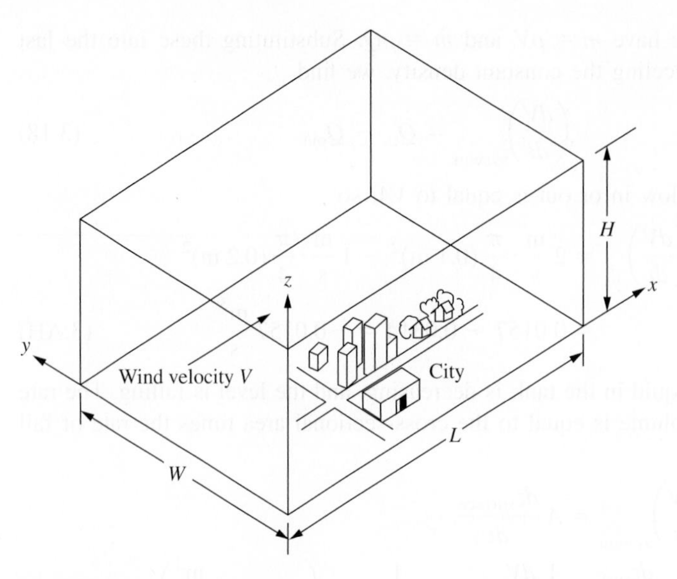

FIGURE 3.6
Idealized city used for Example 3.10.

plant; such emissions are treated a very different way in air pollutant modeling and regulation. [4], Chap. 6.} What is the concentration of pollutant in the air over the city in terms of q, V, W, L, and H?

Here we make the steady-state assumption, that the concentration is not changing with time, so that the algebraic sum of the flows of pollutant in and out must be zero. Writing that sum, we see

$$0 = \begin{pmatrix} \text{Flow rate of} \\ \text{pollutant into} \\ \text{city from upwind} \end{pmatrix} + \begin{pmatrix} \text{flow rate of} \\ \text{pollutant into} \\ \text{city air from city} \end{pmatrix} - \begin{pmatrix} \text{flow rate of pollutant} \\ \text{out of downwind} \\ \text{edge of city} \end{pmatrix} \tag{3.AJ}$$

The pollutant flow rates are expressed as concentrations (e.g., kg / m^3) times volumetric flow rates (e.g., m^3 / s), so

$$0 = bVWH + qLW - cVWH$$

$$c = b + \frac{qL}{VH} \tag{3.19}$$

Equation 3.19 says that the pollutant concentration in the city is equal to that in the air entering the city (the background concentration) plus a term (qL / VH) that indicates how much the pollutant concentration has been increased by the emissions from the city itself. This is the "box model" or "proportional" or "rollback" equation, which has played a very important role in the formulation of air pollution regulations in the USA [4]. ■

Example 3.11. Our paint shop will use a special paint that contains benzene as a solvent. In the course of an 8-h day the paint will evaporate 5 kg (22 lb) of benzene ($q = 5\text{ kg} / 8\text{ hr}$). The shop dimensions are $10\text{ m} \cdot 4\text{ m} \cdot 4\text{ m}$. To protect the health of our workers, we must limit the concentration of benzene in the shop air to less than or equal to the industrial hygiene standard for benzene [5], which was $1.3\text{ mg} / \text{m}^3$ in 2003. If we want to keep the concentration c of benzene in the shop at or below this permitted concentration, how large a flow of ventilating air must we supply?

This problem is very similar to Example 3.10. Here we assume that the benzene is well mixed into the air in the shop and that the air leaving the shop will have the permitted benzene concentration. Making a steady-state benzene balance on the shop, taking the inlet air flow as Q, we write

$$0 = \dot{m}_{\substack{\text{benzene in}\\ \text{inlet air}}} + \dot{m}_{\substack{\text{benzene evaporated}\\ \text{from paint}}} - \dot{m}_{\substack{\text{benzene in}\\ \text{outlet air}}} = Qb + q - Qc \qquad \text{(3.AK)}$$

Now we observe that there is negligible benzene in the incoming air ($b = 0$), so we can solve for Q, finding

$$Q = \frac{q}{c} = \frac{5\text{ kg} / 8\text{ h}}{1.3\text{ mg} / \text{m}^3} \cdot 10^6 \frac{\text{mg}}{\text{kg}} = 481{,}000 \frac{\text{m}^3}{\text{h}}$$

$$= 8000 \frac{\text{m}^3}{\text{min}} = 283{,}000 \frac{\text{ft}^3}{\text{min}} \qquad \text{(3.AL)}$$

■

This example (called the well-mixed model in industrial hygiene [6]) shows that, for the assumption of perfect mixing of the benzene into the shop air, it is quite straightforward to compute the required dilution air to meet the industrial hygiene standard. We can also see that this is an impossibly large airflow rate. If we divide the above flow rate by the cross-sectional area of the shop ($4\text{ m} \cdot 4\text{ m}$), we find

$$\text{Velocity} = \frac{Q}{A} = \frac{8000\text{ m}^3 / \text{min}}{16\text{ m}} = 500 \frac{\text{m}}{\text{min}} = 8.33 \frac{\text{m}}{\text{s}} = 27 \frac{\text{ft}}{\text{s}} = 18 \frac{\text{mi}}{\text{h}} \qquad \text{(3.AM)}$$

This very high velocity could hardly be used inside a paint shop. Our practical alternatives are to choose a less toxic solvent, for which the permitted concentration is higher, or to devise some kind of ventilation system, like a laboratory fume hood, that will prevent the mixing of the benzene with the air the workers breathe or to provide the workers with personal protective devices. We also need to consider the air pollution consequences of emitting 5 kg / day of benzene to the atmosphere; in most U.S. cities that would require a permit and probably some form of capture or destruction of the benzene.

These two examples appear here because every chemical engineer is both an environmental engineer and a safety engineer. We are responsible for protecting the public and the workers under our supervision from harm due to our activities. The completely mixed model used here is also useful for more traditional chemical engineering problems.

TABLE 3.3
Comparison of Examples 3.7 and 3.12

Type of variable	Example 3.7, vacuum pump down	Example 3.12, tank washout
Capacity variable	Tank volume, V_{system}	Liquid volume, V_{liquid}
Flow variable	Pump-out rate, Q_{out}	Flow-through rate, Q_{liquid}
Concentration variable	Gas density, ρ_{system}	Salt concentration, c_{salt}
Starting variable	Initial gas density, $\rho_{\text{sys, init}}$	Initial concentration, $c_{\text{salt, init}}$
Resulting ratio	$\dfrac{\rho}{\rho_{\text{init}}} = \dfrac{P}{1 \text{ atm}}$	$\dfrac{c_{\text{salt}}}{c_{\text{salt, initial}}}$
Time variable	Time, t	Time, t

Example 3.12. A tank contains 1000 m^3 of salt solution, with salt concentration $= 10 \text{ kg} / \text{m}^3$. At time zero, salt-free water starts to flow into the tank at a rate of 10 m^3 / min. Simultaneously, salt solution flows out of the tank at 10 m^3 / min, so that the volume of solution in the tank is always 1000 m^3. A mixer in the tank keeps the concentration of salt in the entire tank uniform so that the concentration in the effluent is the same as the concentration in the tank. What is the concentration in the effluent as a function of time?

This example is exactly the same as Example 3.7, with the variables renamed, as shown in Table 3.3. The reader may make the substitutions shown there and have the resulting solution to this problem. See also Prob. 3.21. This same problem appears in heat transfer and mass transfer, with the variables renamed. ■

3.7 SUMMARY

1. Balances are important in engineering.
2. All balances can be made from the general balance equation (accumulation = creation − destruction + flow in − flow out) by dropping the unnecessary terms.
3. All balances can be divided by time to make rate equations.
4. In any balance it is necessary to choose and state the boundaries over which the balance is made. Whatever is inside the boundaries over which the balance is made is called the "system." Whatever is outside is called the "surroundings."
5. The most important engineering balance is the mass balance, in which the creation and destruction terms are zero. This is also called "the continuity equation" or "the principle of conservation of mass."
6. In applying the mass balance to flowing fluids, we normally speak of the mass flow rate and the volumetric flow rate. We also normally assume that the average velocity adequately represents the fluid behavior, although we know that the most precise work must take into account the fact that the velocity is not uniform across the flow.
7. The completely mixed model, used in several examples in this chapter, is immensely useful.

PROBLEMS

See the Common Units and Values for Problems and Examples inside the back cover of this text. An asterisk (*) on the problem number indicates that the answer is in App. D.

3.1. In our balance equation for the population of the state of Utah, we must resolve several questions of definition. For example, are out-of-state students to be counted in the population of Utah? Are tourists driving through the state to be counted while they are here? List several other ambiguous groups for which we must make a definition.

3.2. Write out the balance for the number of one-dollar bills in circulation in the United States.

3.3. Write out the balance for the mass of refined sugar in the state of Idaho (which is a sugar-producing state).

3.4. A child's toy balloon has just had its neck released. It is zipping through the air and shooting air out its neck. Write out the balance for the mass of air involved. Are your system boundaries fixed in space? Are they fixed in size? Are they identifiable?

3.5. Write a carbon-atom balance for an automobile that is driving at a constant speed; include the carbon atoms that are bound in chemical compounds as well as the free carbon atoms. Consider other flows of carbon atoms than those in the exhaust gas. There are more terms in this balance than most students expect.

3.6. Write a mass balance for an exploding firecracker.

3.7.*A river has a cross section that is approximately a rectangle 10 ft deep and 50 ft wide. The average velocity is 1 ft / s. How many gallons per minute pass a given point? What is the average velocity (assuming steady flow) at a point downstream, where the channel shape has changed to 7 ft in depth and 150 ft in width?

3.8. The annual flow of the Colorado River below Glen Canyon Dam is approximately 10^7 acre-ft / yr, where an acre-ft is the volume needed to cover one acre, one foot deep $\approx 4.35 \cdot 10^4$ ft^3. This flow is not steady over the year, but varies from season to season. At a point where the river is 200 ft wide and 10 ft deep (assume for this problem that the river has a rectangular cross section), what is the velocity of the river, averaged over the whole year?

3.9. In March 1996 a special release of water, $Q = 45{,}000$ ft^3 / s, was made from the Glen Canyon Dam, to create an "artificial flood" in the Grand Canyon.

(*a*) The flow was through 8 pipes, each with an internal diameter of 8 ft. Estimate the velocity through those pipes.

(*b*) Estimate the average velocity of the river at some point downstream of the dam, where the width of the river was 200 ft and its average depth was 10 ft.

3.10. There is steady flow in a circular pipe. The average velocity is given by

$$V_{\text{average}} = \frac{Q}{A} = \int_{r=0}^{r=r_{\text{wall}}} V \cdot 2\pi r \, dr \,/\, \pi r_{\text{wall}}^2 \tag{3.20}$$

Calculate the ratio of the average velocity to the maximum velocity for each of the following cases. Compare your results to those shown in Table 3.1.

(*a*) The flow is laminar (to be discussed in Chap. 6), and the local velocity at any point in the pipe is given by

$$V = V_{\max}\left(\frac{r_{\text{wall}}^2 - r^2}{r_{\text{wall}}^2}\right) \tag{3.21}$$

where r is the radial distance from the center of the pipe and r_{wall} is the radius at the wall of the pipe.

(*b*) The flow is turbulent (to be discussed in Chap. 6), and the local velocity is given by

$$V = V_{\max}\left(\frac{r_{\text{wall}} - r}{r_{\text{wall}}}\right)^{1/7} \tag{3.22}$$

(*c*) This is *Prandtl's 1 / 7 power rule,* which is a good but not outstanding approximation of the velocity distribution in turbulent flow in a circular pipe. It is the best *simple* mathematical description of that distribution; see Table 17.1 and Fig. 17.7. At higher average velocities the 1 / 7 is replaced by 1 / 10. Repeat part (*b*) using 1 / 10 instead of 1 / 7.

3.11. See the preceding problem.

(*a*) The average kinetic energy per unit mass in any flow in any circular conduit is given by

$$\begin{pmatrix}\text{Average kinetic}\\ \text{energy, per}\\ \text{unit mass}\end{pmatrix} = \frac{\displaystyle\int_{r=0}^{r=r_{\text{wall}}} (V^2/2)\cdot V\cdot 2\pi r\,dr}{\displaystyle\int_{r=0}^{r=r_{\text{wall}}} V\cdot 2\pi r\,dr} \tag{3.23}$$

but the denominator of this fraction equals

$$\int_{r=0}^{r=r_{\text{wall}}} V\cdot 2\pi r\,dr = Q = V_{\text{average}}\,A = \pi r_{\text{wall}}^2\,V_{\text{average}} \tag{3.24}$$

so that Eq. 3.23 simplifies to

$$\begin{pmatrix}\text{Average kinetic}\\ \text{energy, per}\\ \text{unit mass}\end{pmatrix} = \frac{\displaystyle\int_{r=0}^{r=r_{\text{wall}}} V^3\cdot r\,dr}{r_{\text{wall}}^2\,V_{\text{average}}} \tag{3.25}$$

Show that substituting Eqs. 3.21 and 3.22 for V in this equation, integrating, and simplifying leads to the values shown in Table 3.1.

(*b*) The total momentum flow in a pipe is given by

$$\begin{pmatrix}\text{Total momentum}\\ \text{flow}\end{pmatrix} = \int_{\text{whole flow area}} V\,dm = \int_{r=0}^{r=r_{\text{wall}}} V\cdot\rho V\cdot 2\pi r\,dr \tag{3.26}$$

For block flow this simplifies to

$$\begin{pmatrix}\text{Total momentum}\\ \text{flow}\end{pmatrix} = V_{\text{average}}^2\,\rho A \qquad [\text{block flow}] \tag{3.27}$$

Show that substituting Eqs. 3.21 and 3.22 for V in Eq. 3.26, integrating, and simplifying leads to the values shown in Table 3.1. Show the corresponding values for 1 / 10 instead of 1 / 7 in Eq. 3.22.

3.12. An ideal gas is flowing in a constant-diameter pipe at a constant temperature. What is the relation of average velocity to pressure?

3.13.* A column of soldiers is marching 12 abreast at a speed of 4 mi / h. To get through a narrow pass they must crowd in to form a column 10 men abreast. Assuming steady flow, how fast are the soldiers moving when they are 10 abreast?

3.14.* A water tank has an inflow line 1 ft in diameter and two outflow lines of 0.5 ft diameter. The velocity in the inflow line is 5 ft / s. The velocity out one of the outflow lines is 7 ft / s. The mass of water in the tank is not changing with time. What are the volumetric flow rate, mass flow rate, and velocity in the other outflow line?

3.15. A compressed-air vessel has a volume of 10 ft^3. Cooling coils hold its temperature constant at 68°F. The pressure now in the vessel is 100 psia. Air is flowing in at the rate of 10 lbm / h. How fast is the pressure increasing?

3.16.*Repeat Example 3.8 for a leak rate of 0.001 lbm / min.

3.17. The tank in Example 3.8 has a leak that admits air at an *unknown* but constant rate. We find that it takes 72 min to reach a pressure of 0.001 atm.
(*a*) What is the leakage rate, in lbm / min or some equivalent units?
(*b*) What will the steady-state pressure be?

3.18. The tank in Example 3.8 has a leak that admits air at the rate

$$\dot{m} = \frac{0.0005 \text{ lbm}}{\text{min} \cdot \text{atm}} (P_{\text{outside}} - P_{\text{tank}}) \tag{3.AN}$$

How long does it take the pump to reduce the pressure in the tank from 1 atm to 0.01 atm? What is the steady-state pressure in the tank?

3.19.*A lake has a surface area of 100 km^2. One river is bringing water into the lake at the rate of 10,000 m^3 / s, while another is taking water out at 8000 m^3 / s. Evaporation and seepage are negligible. How fast is the level of the lake rising or falling?

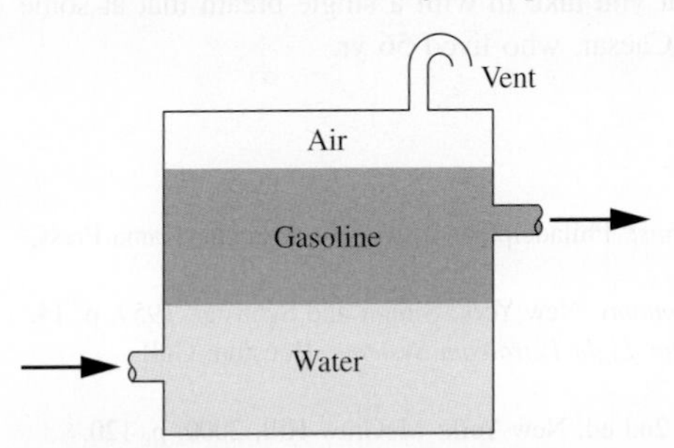

FIGURE 3.7
Tank with two fluids for Prob. 3.20.

3.20. The tank in Fig. 3.7 has an inflow line with a cross-sectional area of 0.5 ft^2 and an outflow line with a cross-sectional area of 0.3 ft^2. Water is flowing in the inflow line at a velocity of 12 ft / s, and gasoline is flowing out the outflow line at a velocity of 16 ft / s. How many lbm / s of air are flowing through the vent? Which way?

3.21.*A vacuum chamber has a volume of 10 ft^3. When the vacuum pump is running, the steady-state pressure in the chamber is 0.1 psia. The pump is shut off, and the following pressure-time data are observed:

Time after shutoff, min	Pressure, psia
0	0.1
10	1.1
20	2.1
30	3.1

Calculate the rate of air leakage into the vacuum chamber when the pump is running. Air may be assumed to be an ideal gas. The air temperature may be assumed constant at 68°F.

3.22. Finish Example 3.12, showing the resulting numerical values and their dimensions. Can you use Fig. 3.5 in this problem?

3.23. Repeat Prob. 3.22, except now there is a layer of solid salt on the bottom of the tank, which is steadily dissolving into the solution at a rate of 5 kg / min, and the inflowing water contains no salt.

3.24. Repeat Prob. 3.22, except that the outflow is only 9 m^3 / min, so that the total volume of liquid contained in the tank is increasing by 1 m^3 / min.

3.25. Rework Example 3.7 with the following change. The tank is now somewhat flexible, so that it is being slowly crushed by the surrounding pressure. If it is crushed at such a rate that its volume decreases steadily by 0.1 ft^3 / min, and this rate of volume decrease begins as soon as the vacuum pump starts, how long does it take the pressure to fall from 1 atm to 0.0001 atm?

3.26.*While Moses was crossing the Red Sea, he took up a liter of water, examined it, and then threw it back. The tides, currents, evaporation, and rainfall, have been steadily mixing the waters of the world's oceans since, so we may assume (for this problem only!) that the molecules in that liter of water have now been uniformly distributed over the waters of all the oceans of the world. If you pick up a liter of water from the ocean and examine it, how many molecules will it contain that were in the liter which Moses examined? State clearly your assumptions and simplifications.

3.27. The typical human being breathes about 10 times / min and takes in about 1 liter per breath. Assuming that the atmosphere has been perfectly mixed since Julius Ceasar's time, estimate the number of air molecules that you take in with a single breath that at some time were breathed in and out by Julius Caesar, who lived 56 yr.

REFERENCES FOR CHAPTER 3

1. Poirier, J. P. *Lavoisier, Chemist, Biologist, Economist.* Philadelphia: University of Pennsylvania Press, 1996.
2. Gamow, G., "Modern Cosmology." *The New Astronomy,* New York: Simon and Schuster, 1959, p. 14.
3. Starling, K. E. *Fluid Thermodynamic Properties for Light Petroleum Systems,* Houston: Gulf Publishing, 1973, p. 12.
4. de Nevers, N. *Air Pollution Control Engineering.* 2nd ed. New York: McGraw-Hill, 2000, p. 120.
5. *Threshold Limit Values for Chemical Substances and Physical Agents.* Cincinnati: American Conference of Governmental Industrial Hygienists. Annual editions. The value shown is from the 2001 edition.
6. Keil, C. B. *Mathematical Models for Estimating Occupational Exposure to Chemicals.* Fairfax, Virginia: American Industrial Hygienists Association, 2000.

CHAPTER 4

THE FIRST LAW OF THERMODYNAMICS

We have seen how the general balance equation applies to people and mass. We now apply it to the abstract quantity energy to find the *energy balance,* which is also called the *first law of thermodynamics* or the law of *conservation of energy.* This is one of the four basic ideas listed in Sec. 1.3 and one of the few truly fundamental laws of nature. Like the others, it cannot be derived from any more basic principle; rather, it rests on its ability to explain all the pertinent observations of nature ever made and on the fact that all experiments designed to prove or disprove it have indicated that it is true.

4.1 ENERGY

The idea of energy and the first law of thermodynamics arose out of observations of friction heating. By the early 1800s scientists knew that a moving body possessed what we would call kinetic energy. They also knew that if it was allowed to come to rest by sliding across a rough surface, it lost this kinetic energy, but it and the surface became hotter. Various explanations of this phenomenon were tried, but, principally through the work of Rumford, Joule, and Mayer [1], the idea was introduced that in any such process there is a quantity called *energy* that is conserved. This quantity could appear in the form of kinetic energy or of heat. We will see that it can also appear in other forms.

Like the law of conservation of mass, the law of conservation of energy seems intuitively obvious to us today. But it was far from obvious to scientists or the public before about 1800 that the various forms of energy were all manifestations of the same quantity. Furthermore, there is no satisfactory, simple definition of energy. The definition can be simple or accurate, but not both. The technically accurate definition is that energy is an abstract quantity, which can appear in various forms, which can be converted from one form to another subject to some restrictions, and which appears to be conserved in all energy transactions.

Some quantities in engineering have absolute values: temperature, entropy, length, and mass (excluding relativistic effects). For each of these there is defined a standard unit of measurement, and the meaning of a zero amount of the quantity is clear.

Other quantities in engineering have only relative values. The simplest example is elevation. We can speak of elevation relative to mean sea level, relative to some convenient bench mark, relative to ground level, or relative to the center of the earth. Any of these is useful. However, an "elevation of 23 ft" without mention of the datum is meaningless. Another example of a relative quantity is velocity. Normally we consider velocity relative to the local surface of the earth, and a "velocity of 23 km / hr" is perfectly clear. However, this statement has a built-in assumption of a datum, namely, the surface of the earth. If we speak of a star moving at 23 km / s, we probably mean relative to the sun, but we could mean relative to the earth or to the center of our galaxy. We must state our datum, unless the datum is understood by all.

All energy quantities are relative to some arbitrary datum. This statement is made simply so that you will remember it; a more precise statement is that no one has yet found a way to measure or to calculate absolute values of energies. From this it follows that all energy calculations will be based on *changes* in energy or on energies relative to some arbitrary datum. This need not trouble us; all the buildings in the world were designed relative to some arbitrary elevation datum without any particular trouble about the datum.

4.2 FORMS OF ENERGY

As already implied, energy has many forms, and they are interconvertible, subject to some restrictions (which apply only to the direction of conversion). Consider a 1 kg ball of steel. What forms of energy can it possess?

4.2.1 Internal Energy

If the steel ball is at a temperature of 20°C, you can hold it in your hand. If it is at a temperature of 200°C, you cannot hold it in your hand very long. Clearly, the ball at 200°C produces effects that the ball at 20°C cannot. Yet, if we measure the mass of the ball, it is the same at 20°C as it is at 200°C (within the precision of current measuring techniques). If we could label the atoms when the ball was at 20°C and take a census of them when the ball was at 200°C, we would find exactly the same atoms present. Therefore, the difference between what the ball will do at 20°C and what it will do at 200°C is not dependent on changing the mass or identity of the *matter* present. Something else obviously is involved. We will say for now that a body that is hot possesses more *internal energy* than the same body does when cold.

Now suppose that instead of an iron ball we had a balloon that contains a mixture of gasoline and oxygen with a total mass of 1 kg at 20°C. Now we can introduce a small spark, and the contents of the balloon will become very hot (explosively). After a moment the contents will be much hotter than at the start, and they will have a different chemical composition; instead of being oxygen and gasoline, they will be carbon dioxide and water vapor. Clearly, the oxygen-gasoline mixture at 20°C can produce effects that the mixture of carbon dioxide and water (when cooled to 20°C) cannot. Therefore, there must be a difference in energy. This we classify as a change of internal energy.

Thus, an *approximate* rule (with exceptions to be seen later) is that internal energy is a measure of hotness plus the ability to cause heat-releasing chemical reactions. A more complete definition will be given in Sec. 4.6. We will denote internal energy of some mass of matter by U and internal energy per unit-mass by u, ($U = mu$). The use of uppercase letters for the total quantity and lowercase letters for the quantity per unit mass is very common in thermodynamics and will be used extensively in this chapter. The quantities per unit mass are often called specific quantities; for example, u would be called *specific internal energy.* This use of two symbols makes sense only for extensive properties—those that double when the mass doubles. It is never used for intensive properties—those that do not depend on the amount of mass present. (What would one mean by "the temperature per unit mass?")

4.2.2 Kinetic Energy

Let us return to our 1 kg ball of steel. If I hand it to you gently, you can easily hold it. If I deliver it to you at a velocity of 100 m / s and you are foolish enough to get in the way, it will certainly kill you. The fast-moving ball can produce effects different from those produced by the slow-moving ball. The difference we call the difference in *kinetic energy*. Kinetic energy is the energy that a moving body possesses because of its motion. We will call the kinetic energy of some mass of matter "KE" and kinetic energy per unit-mass "ke."

4.2.3 Potential Energy

If our 1 kg ball of steel is resting on the floor, it is not likely to damage the floor. If it is resting on a shelf 100 m above the floor and is then gently pushed off, it will probably go right through an ordinary wood floor. When it reaches the floor, its kinetic energy is very large. However, when it was sitting on a shelf a 100 m above the floor, it did not have that kinetic energy, but it obviously had the potentiality of acquiring it by falling. This potentiality to do work, or to acquire kinetic energy, we call *potential energy*. In this instance it is the energy which the body possesses by being some distance above "bottom" in a gravity field. We will call potential energy of some mass of matter "PE" and potential energy per unit-mass "pe."

If we fabricate our iron ball into a coil spring, it will have a certain length when relaxed. When we compress it, it will have another length but, given the opportunity, it will return to its original length and in doing so can give some other body kinetic energy. Toy guns work exactly this way. If we compare the compressed spring with the relaxed one, we see that the compressed spring has the potentiality of speeding up a projectile or doing some other useful work. This is due to a difference in energy; one might logically call this "spring energy," but most writers have agreed to call it potential energy. This is particularly true in physical chemistry, in which the "springs" are the repulsive force fields between atoms or subatomic particles. Throughout the remainder of this chapter we will consider potential energy to mean the energy that a body possesses due to its position in a gravitational field; however, we must remember this other meaning.

4.2.4 Electrostatic Energy

Suppose we take our ball of iron and some convenient dielectric and fabricate a huge electric condenser out of it. If it is not charged, we can touch the terminals with our fingers without effect. If it is charged, placing our fingers over the terminals will be a shocking experience. The charged condenser can do things that an uncharged one cannot. This difference is due to what we call *electrostatic energy*.

4.2.5 Magnetic Energy

If our ball of iron is shaped into a rod or horseshoe and annealed, it will not attract iron filings. If we subject it to a properly designed magnetic field and then remove the field, it will attract the filings; it has become a magnet. A magnet can do things that a bar of unmagnetized steel cannot. The difference is due to what we call *magnetic energy*.

4.2.6 Surface Energy

Salad oil, egg yolks, and vinegar do not form a homogeneous mixture. If they are gently shaken together and then allowed to settle, they will separate cleanly. If, however, they are beaten very vigorously, to break the oil into small droplets, then they will form a stable system called mayonnaise. Under normal conditions mayonnaise will not separate back into salad oil, egg yolks, and vinegar; it is an emulsion. Emulsions possess properties their unmixed constituents do not. These are due to the *surface energy* of all the microscopic droplets that make up an emulsion.

4.2.7 Nuclear Energy

Einstein showed that matter and energy are interconvertible; their conversion is the basis of nuclear explosives and nuclear power plants and the source of energy in the sun and the stars. We will discuss this conversion in Sec. 4.11; for the moment, we restrict ourselves to saying that it is *convenient* to talk of some materials *as if* they possessed another kind of energy, called *nuclear energy*.

Of the kinds of energy that matter can possess, only internal, kinetic, and potential appear in common fluid mechanics problems. In most of the rest of this book we will consider only these three. The other kinds of energy are important in other fields. Electrostatic energy is the basis of xerographic copying machines, TV tubes, lightning, and electrostatic air cleaners. Magnetic energy is the basis of computer disk storage devices and videotapes and all electromagnetic devices, such as electric motors and generators. Nuclear energy is the source of all life; solar energy is based on nuclear energy. But these three play little role in simple fluid mechanics and will not be discussed further. Surface energy does play a role in fluid mechanics. Chapter 14 examines that role. For the rest of this book, we will ignore it as well, so for fluid mechanics we normally consider that a mass of fluid has three types of energy; internal, kinetic, and potential.

4.3 ENERGY TRANSFER

If a kilogram of matter can possess energy, how can that energy be transferred from one body to another?

One way to transfer energy from one body to another is to place two bodies at different temperatures in contact with each other. It is our universal experience that in such circumstance the internal energy of the hotter body will decrease and the internal energy of the colder body will increase. Therefore, energy must have flowed from one to the other. The energy that flows directly between two bodies in contact because of a temperature difference we call *heat*.

Our definition of heat is different from the one in common use. We say, "Heat is energy in transit from one body to another because of a temperature difference." English speakers commonly use "heat" interchangeably with "temperature." This leads to expressions such as "It's not the heat; it's the humidity" or "Beat the heat with a Brand X air conditioner." Clearly, these rest on the human experience that, when the temperature of the air is high, energy will flow into our bodies, uncomfortably. While it is flowing, it is "heat."

The idea of energy "in transit" is also contradicted by common usage. Many people refer to a hot body as containing a large amount of heat rather than a large amount of internal energy. Rain is water in transit from clouds to ground under the influence of gravity. We would scarcely look at a cloud and say, "Look at all that rain," or look at the ocean and refer to "all that rain." Another useful analogy is to electric current. While electrons are flowing from a battery to a condenser, we speak of them as an electric current, but when the flow has stopped, we speak of them as "charge" and speak of a "charged battery" or a "charged condenser." We would scarcely speak of a "currented" condenser or a "currented" battery. These ideas are summarized in Table 4.1.

TABLE 4.1
Comparison of three kinds of flows

Name of species at rest	Potential difference causing species to flow	Name of species flowing
Water	Elevation difference	Rain
Electrons or charge	Voltage difference	Current
Energy	Temperature difference	Heat

The second way two bodies can exchange energy is by doing *work* upon each other. Again we must distinguish between an engineer's idea of work and common English usage. For an engineer

$$\text{Work} = \int \text{force} \cdot d(\text{distance}) = \int F\,dx \tag{4.1}$$

or its equivalent. A hod carrier lifting plaster up a ladder is doing work in the engineering sense of the word. However, if he is required to stand and hold a load of plaster on his shoulder for an hour, he is not doing work in the engineering sense, although he will certainly feel just as tired as if he had kept moving. Similarly, a baby-sitter is "working" in common usage but not in the engineering sense. If we rub two articles together, such as two pencil erasers, and they resist the rubbing, then we must exert a force and move them. Thus, rubbing is work in the engineering sense.

We have considered work only as a force times a distance, $F\,dx$. There can also be the work of rotating shafts and electrical and magnetic work, considered in Sec. 4.11.

The third way two bodies can exchange energy is by *radiation.* The sun heats the earth by radiation, and x-rays and gamma rays change the energy of bodies by radiation. Radiation does not fit perfectly into either of the categories *work* and *heat.* However, with a little adjustment of the definitions, it can be made to appear as heat. When a radiation is due to a difference in temperature, such as that between the sun and the earth or between the glowing wires in the toaster and the slice of bread, then that radiation fits our definition of heat well, except that the bodies are not in contact. However, if gamma radiation is flowing from a cold piece of radium to a warmer piece of lead, heat appears to flow from a cold body to a hot body. If we focus our attention, not on the average temperature of the bulk of the radium, but on the individual atom emitting the gamma ray, we see that at the instant of emitting the ray it undergoes a nuclear event, which raises its instantaneous "temperature" to a very high value. If one considers this "temperature" instead of the temperature of the large mass of radium, then radiation fits the definition of heat fairly well.

4.4 THE ENERGY BALANCE

Now we are ready to write the general energy balance equation. To write any balance equation, we need a well-defined system of boundaries. Let us choose as our system the tank shown in Fig. 4.1.

We begin our balance by excluding from consideration magnetic, electrostatic, surface, and nuclear energies. Thus, the only kinds of energy that a pound of matter can contain will be internal, kinetic, and potential, $u + \text{ke} + \text{pe}$. The general balance says that

$$\text{Accumulation} = \text{flow in} - \text{flow out} + \text{creation} - \text{destruction} \tag{3.2}$$

However, observations of nature have led to the conclusion that energy can be neither created nor destroyed (excluding nuclear effects, to be discussed later); so, for energy the balance equation is

$$\text{Accumulation} = \text{flow in} - \text{flow out} \tag{4.A}$$

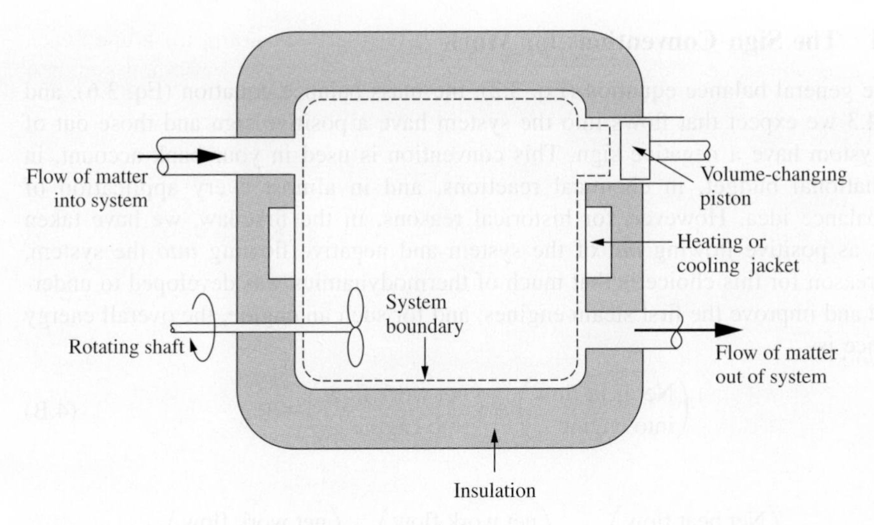

FIGURE 4.1
Tank used as system for energy balance.

Accumulation is the differential of the energy contained within the system boundary. The only such energy is that associated with the matter within the boundary. If the matter is uniform (if all has the same u, ke, and pe), then accumulation is $d[m \cdot (u + \text{pe} + \text{ke})]$, where m is the mass in the system.

Energy can enter in three ways. One way is by matter coming in the inlet pipe. For every infinitesimal amount of matter that flows in, the amount of energy that flows in with it is $(u + \text{pe} + \text{ke})_{\text{in}}\, dm_{\text{in}}$. Obviously, for matter flowing out the outlet line, the amount of energy flowing out is $(u + \text{pe} + \text{ke})_{\text{out}}\, dm_{\text{out}}$. The two other ways energy can flow in or out are via heat through the heating or cooling jacket, which we will call dQ, and via mechanical work of various forms, which we will call dW. As discussed in most thermodynamics texts dQ and dW are inexact differentials. This means that the value of $\int_a^b dQ$ or of $\int_a^b dW$ depends not only on the initial and final states of the system but also on the path followed. Quantities such as dz (an infinitesimal change in elevation) are exact differentials. The elevation change going from New York to Chicago is the same, independent of the route taken; it depends only on the initial and final elevations. However, the fact that dQ and dW are inexact differentials has no effect on their role in most fluid-mechanics problems, so we shall not dwell further on this distinction.

Substituting these terms into Eq. 4.A, we find

$$d[m(u + \text{pe} + \text{ke})]_{\text{sys}} = [(u + \text{pe} + \text{ke})_{\text{in}}\, dm_{\text{in}} + dQ_{\text{in}} + dW_{\text{in}}] - [(u + \text{pe} + \text{ke})_{\text{out}}\, dm_{\text{out}} + dQ_{\text{out}} + dW_{\text{out}}] \quad (4.2)$$

By letting dQ be the algebraic sum of the heat flows in and out, and dW be the algebraic sum of the work flows in and out, this becomes

$$d[m(u + \text{pe} + \text{ke})]_{\text{sys}} = (u + \text{pe} + \text{ke})_{\text{in}}\, dm_{\text{in}} - (u + \text{pe} + \text{ke})_{\text{out}}\, dm_{\text{out}} + dQ + dW \quad (4.3)$$

4.4.1 The Sign Conventions for Work

In the general balance equation (Eq. 3.2), the mass balance equation (Eq. 3.6), and Eq. 4.3 we expect that flows into the system have a positive sign and those out of the system have a negative sign. This convention is used in your bank account, in the national budget, in chemical reactions, and in almost every application of the balance idea. However, for historical reasons, in the first law, we have taken work as positive flowing *out* of the system and negative flowing *into* the system. The reason for this choice is that much of thermodynamics was developed to understand and improve the first steam engines, and for such an engine, the overall energy balance is

$$\begin{pmatrix}\text{Net heat flow}\\ \text{into engine}\end{pmatrix} + \begin{pmatrix}\text{net work flow}\\ \text{into engine}\end{pmatrix} = 0 \tag{4.B}$$

or

$$\begin{pmatrix}\text{Net heat flow}\\ \text{into engine}\end{pmatrix} = -\begin{pmatrix}\text{net work flow}\\ \textit{into}\text{ engine}\end{pmatrix} = \begin{pmatrix}\text{net work flow}\\ \textit{out}\text{ of engine}\end{pmatrix} \tag{4.C}$$

If one defines work as positive *when flowing outward,* then this becomes

$$dQ = dW \quad \text{or} \quad dQ - dW = 0 \quad \text{[classical sign convention]} \tag{4.D}$$

This is the classical definition, appearing in all thermodynamics books before about 1990, and in the first two editions of this book.

That definition has been the source of unending confusion for students, and textbooks have begun to abandon it, and take all flows of anything into the system as positive. That changes Eq. 4.D to

$$dQ = -dW \quad \text{or} \quad dQ + dW = 0 \quad \text{[“modern” sign convention]} \tag{4.E}$$

This sign convention has the advantage that all flows in are positive and all flows out are negative, but the drawback that for a power plant like a steam engine the product (work delivered) is negative. This edition of this book uses this convention; work, like anything else, is positive flowing into the system and negative flowing out. Students will certainly encounter texts (all older texts, some modern ones) that make an exception to this rule and take work as positive when it flows out of the system (Eq. 4.D).

4.5 KINETIC AND POTENTIAL ENERGIES

Equation 4.3 can be used only if we can find a way to assign numerical values to the various symbols in it. We already have an expression for work, Eq. 4.1. It has the dimension of force times distance; in the SI system its unit is the Joule = Newton · meter, (J = N · m). In the English engineering system of units, its unit is the foot · pound force, (ft · lbf).

We deduced Eq. 4.3 for the system shown in Fig. 4.1, but it applies equally well to many other systems. Let us again choose as our system a 1 kg steel ball. We lift it slowly by a distance of dz. We insulate it so that during this lifting process no heat is transferred to or from the surroundings: $dQ = 0$. Moreover, no matter flows into or out of the system: $dm_{\text{in}} = dm_{\text{out}} = 0$. Since no matter flows in or out, we have $d[m(u + \text{pe} + \text{ke})]_{\text{sys}} = m[d(u + \text{pe} + \text{ke})]$. Substituting this in Eq. 4.3, we find

$$md[(u + \text{pe} + \text{ke})]_{\text{sys}} = dW \tag{4.F}$$

If we have proceeded without friction heating, the final temperature is the same as the initial temperature; so we conclude that du_{sys} is zero. The final and initial velocities are also zero; so $d(\text{ke})_{\text{sys}}$ is zero. Furthermore, according to Eq. 4.1, dW equals $F\,dz$. Here the sign of the work is positive because work is done on the system. The force needed to lift the ball is the same as the weight of the ball, i.e., its mass times the acceleration of gravity; so

$$md(\text{pe}) = m_{\text{sys}}g\,dz \tag{4.G}$$

$$d(\text{pe}) = g\,dz \tag{4.4}$$

Here, then, is a convenient equation for the change in potential energy. If the acceleration of gravity is constant (practically true in all earth-bound problems but certainly not true in interplanetary space problems), we may integrate both sides of Eq. 4.4, taking the g outside the integral sign, and find

$$\text{pe} = gz + \text{constant} \tag{4.H}$$

Here the constant is chosen to make the potential energy zero when the elevation above some arbitrary datum (such as sea level or local ground level) is zero. If z is measured above this datum, then the constant is zero, and

$$\text{pe} = gz \tag{4.5}$$

Example 4.1. Determine the change in potential energy of a 10 kg bag of feathers that is raised a vertical distance of 23 m.

Since we are dealing with a *change* in potential energy (as we do in all practical problems), we need not concern ourselves with the datum. We can see this by applying Eq. 4.H to the initial and final states:

$$\Delta\text{pe} = \text{pe}_{\text{fin}} - \text{pe}_{\text{init}} = gz_{\text{fin}} + \text{constant} - (gz_{\text{init}} + \text{constant}) = g(z_{\text{fin}} - z_{\text{init}})$$

$$= 9.81\,\frac{\text{m}}{\text{s}^2}\cdot 23\text{ m} = 225.6\,\frac{\text{m}^2}{\text{s}^2} = 740.0\,\frac{\text{ft}^2}{\text{s}^2} \tag{4.I}$$

This is the change in the potential energy per unit mass. We calculate the total change in potential energy by multiplying the potential energy change per unit mass by the mass present:

$$\Delta\text{PE} = m\,\Delta\text{pe} = 10\text{ kg}\cdot 225.6\,\frac{\text{m}^2}{\text{s}^2} = 2256\,\frac{\text{kg}\cdot\text{m}^2}{\text{s}^2} = 53{,}400\,\frac{\text{lbm}\cdot\text{ft}^2}{\text{s}^2} \tag{4.J}$$

To find the answer in joules or ft · lbf, we use the force-mass conversion factor:

$$\Delta\text{PE} = 2256\,\frac{\text{kg}\cdot\text{m}^2}{\text{s}^2}\cdot\frac{\text{J}}{\text{N}\cdot\text{m}}\cdot\frac{\text{N}\cdot\text{s}^2}{\text{kg}\cdot\text{m}} = 2256\text{ J} = 1664\text{ ft}\cdot\text{lbf} \tag{4.K}$$

■

Now we take our 1 kg ball of steel and throw it horizontally. Again we take the ball as the system. As before, we can insulate it so that no heat flows in or out ($dQ = 0$). During the throwing process no matter flows into or out of the ball; so $dm_{in} = dm_{out} = 0$. Furthermore, if we proceed without friction heating, the temperature will not change; so $du_{sys} = 0$. If we throw it perfectly horizontally, then there is no change in elevation during the throwing; so $d(gz)_{sys} = 0$. As before, $dW = F\,dx$. This F is the force exerted *on* the system. Substituting all of these in Eq. 4.3 yields

$$m_{sys}\,d(\text{ke}) = F\,dx \tag{4.L}$$

We may replace F with $m_{sys}a_{sys}$ (according to Newton's law) to get

$$d(\text{ke}) = a_{sys}\,dx \tag{4.M}$$

But $a = dV/dt$, where V is the velocity; so $a\,dx = dV\,dx/dt$. Furthermore, $dx/dt = V$; so $a\,dx = V\,dV$. We can now integrate both sides of Eq. 4.M to get

$$\text{ke} = \frac{V^2}{2} + \text{constant} \tag{4.N}$$

Here, too, we may choose any value we like for the constant. The logical choice is zero, which makes the kinetic energy zero for a body at rest

$$\text{ke} = \frac{V^2}{2} \tag{4.6}$$

Example 4.2. What is the kinetic energy of a 0.01 lbm bullet traveling 2000 ft / s relative to the barrel of the gun it has just left?

Here the velocity is measured relative to the same datum as we want the kinetic energy to be relative to, so we have no problem with the datum.

$$\text{KE} = m \cdot \text{ke} = m\frac{V^2}{2}$$
$$= \frac{0.01\ \text{lbm} \cdot (2000\ \text{ft/s})^2}{2} \cdot \frac{\text{lbf} \cdot \text{s}^2}{32.2\ \text{lbm} \cdot \text{ft}} = 621\ \text{ft} \cdot \text{lbf} = 842\ \text{J} \tag{4.O}$$

■

Example 4.3. Suppose the gun of Example 4.2 were mounted, facing backward, on an airplane that just flew past us at a velocity of 1990 ft / s. What then would the kinetic energy of the bullet be (*a*) relative to the airplane and (*b*) relative to us?

Obviously, the bullet is moving 2000 ft / s relative to the airplane, so the kinetic energy relative to the airplane is the same as in Example 4.2. However, relative to us, the bullet is moving 10 ft / s, and its kinetic energy is

$$\text{KE} = m\frac{V^2}{2} = \frac{0.01\ \text{lbm} \cdot (10\ \text{ft/s})^2}{2} \cdot \frac{\text{lbf} \cdot \text{s}^2}{32.2\ \text{lbm} \cdot \text{ft}}$$
$$= 0.016\ \text{ft} \cdot \text{lbf} = 0.021\ \text{J} \tag{4.P}$$

■

This result appears startling but is correct. It shows us that all energy measurements are relative to some datum. As a practical matter, if you were in front of the airplane in Example 4.2, the bullet would be lethal. If you are behind, the bullet fired backward would reach you at 10 ft / s, and you could easily catch it in your hand. From the viewpoint of someone riding in the airplane, the bullets are the same: they leave at a velocity of 2000 ft / s.

4.6 INTERNAL ENERGY

Now that we have the numerical forms of kinetic energy per unit mass and potential energy per unit mass, we can rewrite Eq. 4.3 as

$$d\left[m\left(u + gz + \frac{V^2}{2}\right)\right]_{\text{sys}} = \left(u + gz + \frac{V^2}{2}\right)_{\text{in}} dm_{\text{in}} - \left(u + gz + \frac{V^2}{2}\right)_{\text{out}} dm_{\text{out}} + dQ + dW \tag{4.7}$$

This is the semifinal form of the energy-balance equation.

At this point we must reconsider our idea of the internal energy. Since the potential energy has to do with elevation and gravity, we should expect its numerical formulation (Eq. 4.5) to involve g and z, as it does. Likewise, the kinetic energy depends on velocity, and its formulation (Eq. 4.6) indicates this. The internal energy, as we suggested before, is related to the "hotness" of a body, so we should expect is formulation to be related to heat in some way. The common unit of energy used for kinetic and potential energies is the ft · lbf or J, but this is an inconvenient unit for heat flows or for the internal energy. Instead, we use a "heat" unit, the *British thermal unit* (Btu) or the *calorie*. The Btu, defined as the amount of energy that must be transferred into 1 lbm of water to raise its temperature 1°F starting at 59.5°F, is the unit used in English-speaking countries to measure most heat flows. (Inquisitive readers can find ratings in Btu per hour on the name plates of most U.S. household furnaces and water heaters, and they can find the heating value of natural gas in Btu per cubic foot on their natural gas bill.) The calorie (cal) is defined as the amount of energy that must be transferred into 1 g of water to raise its temperatures 1°C starting at 0°C. The calorie is an impractically small engineering unit; we normally use *kcal* = 1000 cal. (The "calorie" in diet books is the kcal. A normal adult doing moderate work needs to eat about 2500 kcal of food per day.) The Btu is also impractically small for industrial-sized equipment; we normally use 10^6 Btu as the working unit. (In 2004 the world wholesale price of natural gas was about \$3 / 10^6 Btu, that of coal about \$1 / 10^6 Btu. Natural gas bills often use the *therm* = 10^5 Btu.)

Now suppose we take as our system the tank shown in Fig. 4.1. We close the valves in the inlet and outlet lines, so that $dm_{\text{in}} = dm_{\text{out}} = 0$. We also stop the rotating shaft and do not move the volume-changing piston, so there will be no work done ($dW = 0$). Now we transfer 100 Btu of energy into the tank from the heating jacket:

$$d\left[m\left(u + gz + \frac{V^2}{2}\right)\right]_{\text{sys}} = dQ = 100 \text{ Btu} \tag{4.Q}$$

However, in this operation the elevation and velocity of the material in the tank did not change, so that $d(gz)_{sys} = d(V^2/2)_{sys} = 0$ and, since m_{sys} remained constant, we may take it out of the differential.

This leads to

$$du_{sys} = \frac{dQ}{m_{sys}} = \frac{100 \text{ Btu}}{m_{sys}} = \frac{25{,}200 \text{ cal}}{m_{sys}} = \frac{25.2 \text{ kcal}}{m_{sys}} \qquad \text{(4.R)}$$

The potential and kinetic energies per unit mass are expressed in units of ft · lbf / lbm or J / kg. Here we have a change in internal energy expressed in Btu / lbm or cal / kg. In our balance equation we obviously need some way to interconvert these units so that the sum $(u + gz + V^2/2)$ is in a consistent set of units. All efforts to calculate this conversion factor from some more basic principle have failed; however, it can be determined experimentally.

Suppose we cool our system in Fig. 4.1 to its initial state by removing 100 Btu of energy via the cooling jacket. Now we start the stirrer and measure the work input required to produce the same temperature rise as was caused by the addition of 100 Btu as heat. In this case Eq. 4.3 simplifies to

$$du_{sys} = \frac{dW}{m_{sys}} \qquad \text{(4.S)}$$

By carefully measuring the temperature changes, we can find the exact number of ft · lbf or J of work that produces the same heating effects as 1 Btu or 1 cal of energy added as heat. This experiment was made by Joule [2] in 1849: it formed the keystone in constructing the first law of thermodynamics. His experimental result (as corrected by later workers with better equipment) is

$$1 \text{ Btu} = 778 \text{ ft} \cdot \text{lbf}; \qquad 1 \text{ cal} = 4.184 \text{ J} \qquad \text{(4.8)}$$

This is an experimental fact, reproducible in any well-equipped laboratory. Using the conversion factors in Eq. 4.8, we can easily convert all the terms in the energy balance to a common basis. In the SI system the use of the calorie is discouraged; thermal energy quantities are to be expressed only in Joules. However, the use of the calorie (or kcal) in metric-using countries is quite common; today's student will have to be familiar with its use.

We said before that internal energy might be thought of as hotness plus chemical energy. However, there can also be internal-energy changes at constant temperature. Suppose we have some mass of some substance in an absolutely rigid vessel. Now we transfer heat into the vessel. For this process, Eq. 4.7 yields

$$m_{sys}\, du = dQ \qquad \text{(4.T)}$$

Thus, we see that *for a simple constant-volume heating* we have $du = dQ / m$. What are the possible external signs of such an increase in internal energy?

1. The substance may increase in temperature.
2. The substance may undergo an energy-consuming chemical reaction, such as

$$2NH_3 \rightarrow N_2 + 3H_2 \qquad \text{(4.U)}$$

3. The substance may undergo a phase change such as ice → water or water → steam.
4. The substance may undergo a crystal-structure change, such as $\alpha_{\text{iron}} \rightarrow \gamma_{\text{iron}}$. This is really a phase change but is not so obvious as those shown above.
5. Any combination of the four items listed above may occur simultaneously.

We see that any exact definition of an internal-energy change must be based on a consideration of all the terms in Eq. 4.7. Thus, Eq. 4.7 *is* the exact definition of the change in internal energy. If we restrict ourselves to a closed system of constant mass, with no changes in kinetic or potential energy, Eq. 4.7 simplifies to

$$m\,du = dQ + dW \tag{4.V}$$

Integrating this, we find

$$U = mu = Q + W + \text{constant} \tag{4.W}$$

Here there is no obvious choice for the constant, as there was in the case of kinetic or potential energy. In making up tabulations of thermodynamic properties we must arbitrarily select a value for this constant. For the common steam tables the constant is chosen to make $u = 0$ for liquid water at the triple point. This choice is made on the basis of convenience alone.

In sum, internal energy may be thought of approximately as hotness plus chemical energy. Its exact formulation is Eq. 4.7, which allows us to calculate changes of internal energy. Using this equation, and an arbitrarily selected value in some datum state, we can make up a table showing the numerical value of the internal energy per unit mass for any state of any substance, e.g., the steam table.

4.7 THE WORK TERM

So far, we have said little about the work term in Eq. 4.7. Suppose our system is the 1 kg steel ball described previously. The system is practically rigid, and the work done on it generally consists of something, e.g., our hand, pushing it. This work is shown by Eq. 4.1.

Now consider the system shown in Fig. 4.1. Let us assume that the material in the tank is something easily compressed, such as air or steam. In this case we can do work on the system by moving the volume-changing piston; the magnitude of this work is shown by Eq. 4.1. As shown in Fig. 4.1, to compress the system we move the piston in the negative x direction. The force required to move the piston is equal to the piston's cross-sectional area times the pressure in the tank. Further, the product of the piston's cross-sectional area and the distance traveled is equal to the decrease in volume of the tank; so the work done *on* the system is

$$dW = F\,dx = PA\,dx = -P\,dV \tag{4.X}$$

where V is the volume of the tank. The minus sign appears because the volume of the tank decreases as work is done *on* the system. This result is correct for any work done by a moving boundary. However, for boundaries moving at supersonic speeds the pressure at the boundary may be different from the pressure in the nearby fluid. As long as the P in Eq. 4.X is the P experienced by the boundary, this result is correct. If we close the inlet and outlet lines of the tank, turn off the heating and cooling coils, and then move the piston inward, Eq. 4.7 shows the following:

$$(m\,du)_{\text{sys}} = dW = -P\,dV \tag{4.Y}$$

When we move the piston inward, dV is negative, so du_{sys} is positive. When du_{sys} is positive, the temperature of the system will rise unless a phase change or chemical reaction occurs. This phenomenon can be observed easily in an ordinary bicycle pump. Driving the piston inward causes the air in the pump to become hot. So we see that one form of work we must consider is the work of moving the boundaries of the system. This work is equal to $-P\,dV$ and is often simply referred to as "$P\,dV$ work."

4.8 INJECTION WORK

If we considered only such systems as a cannon ball or a tank with no flow in or out, we would never need to introduce the idea of *injection work.* However, it is often advantageous to choose an *open system,* that is, a certain set of boundaries through which mass flows. If, for example, we wanted to analyze the power plant at Hoover Dam, we would find it easier to choose as our system the power plant from water inlet to water outlet than to choose 1 lbm or 1 kg of water passing through the plant as our system. By choosing the open system with mass flow through it we will have a much simpler analysis, because we do not need to consider the many changes in pressure, elevation, and velocity along the complex flow path taken by the water through valves, turbines, wicket gates, and so forth. However, we do have to consider the injection work.

Suppose our system is the tank shown in Fig. 4.1. We now bring into the tank a mass dm_{in} from the inlet line; nothing flows out, there is no heat transfer, and there is no work due to moving the volume-changing piston or to turning the shaft. What will be the energy balance for this operation? This is easiest to see when we do it by a two-step process; see Fig. 4.2.

In the first step we let the mass dm flow in and simultaneously move the volume-changing piston out. We move the piston at a rate such that the fluid originally in the tank is not compressed. This means that all the fluid pushed aside by the fluid coming in is pushed into the space vacated by the volume-changing piston. Thus, there is no net work done on the system because, for all the fluid involved, there is no volume change. Therefore, the compression work, $-P\,dV,$ is zero. The energy balance for step 1 is

$$d\left[m\left(u + gz + \frac{V^2}{2}\right)\right]_{\text{sys}} = \left(u + gz + \frac{V^2}{2}\right)_{\text{in}} dm_{\text{in}} \tag{4.Z}$$

Now, to get to the desired final state, we must move the volume-changing piston back to its original position. It must move back by a volume exactly equal to the volume of the fluid that moved in, which is $v_{\text{in}}\,dm_{\text{in}}$; then the work to move it back

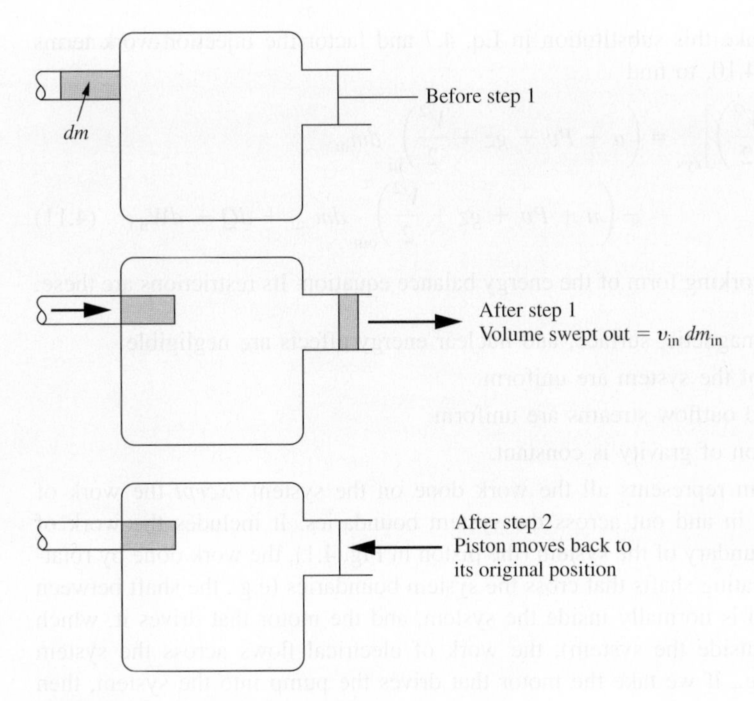

FIGURE 4.2
Two-step process to illustrate injection work.

is $dW = Pv_{in}\, dm_{in}$. The energy balance for this second step is

$$d\left[m\left(u + gz + \frac{V^2}{2}\right)\right]_{sys} = dW = (Pv)_{in}\, dm_{in} \tag{4.AA}$$

The energy balance for the entire injection process, which is the equivalent of the two steps given above, must be the sum of the energy balances for the two separate steps:

$$d\left[m\left(u + gz + \frac{V^2}{2}\right)\right]_{sys} = \left(u + gz + \frac{V^2}{2}\right)_{in} dm_{in} + (Pv)_{in}\, dm_{in}$$

$$= \left(u + Pv + gz + \frac{V^2}{2}\right)_{in} dm_{in} \tag{4.9}$$

What does this $(Pv)_{in}\, dm_{in}$ term represent? We call it injection work, because it is exactly the work that is needed to inject the mass dm_{in} across the system boundaries. It is also sometimes called *intrusion work, flow work,* and *flow energy*.

Obviously, we could repeat the calculation for fluid flowing out the outlet line. How, then, will we reconcile this injection-work idea with Eq. 4.7? Equation 4.7 is correct as it stands; for the process described above the $(Pv)_{in}\, dm_{in}$ term is included in the dW term. However, we now break up the dW term as follows:

$$dW = dW_{inj} + dW_{n.f.} = (Pv)_{in}\, dm_{in} + dW_{n.f.} \tag{4.10}$$

where the subscripts "inj" and "n.f." denote "injection" and "non-flow," respectively.

We now make this substitution in Eq. 4.7 and factor the injection-work terms as shown in Eq. 4.10, to find

$$d\left[m\left(u + gz + \frac{V^2}{2}\right)\right]_{\text{sys}} = \left(u + Pv + gz + \frac{V^2}{2}\right)_{\text{in}} dm_{\text{in}} - \left(u + Pv + gz + \frac{V^2}{2}\right)_{\text{out}} dm_{\text{out}} + dQ + dW_{\text{n.f.}} \quad (4.11)$$

This is the final working form of the energy balance equation. Its restrictions are these:

1. Electrostatic, magnetic, surface, and nuclear energy effects are negligible.
2. The contents of the system are uniform.
3. The inflow and outflow streams are uniform.
4. The acceleration of gravity is constant.
5. The $dW_{\text{n.f.}}$ term represents all the work done on the system *except* the work of driving matter in and out across the system boundaries. It includes the work of moving the boundary of the system (the piston in Fig. 4.1), the work done by rotating or reciprocating shafts that cross the system boundaries (e.g., the shaft between a pump, which is normally inside the system, and the motor that drives it, which is normally outside the system), the work of electrical flows across the system boundaries (e.g., if we take the motor that drives the pump into the system, then the wires that bring electricity to the motor must cross the system boundary), and some others.

Using this equation, we can solve an immense array of problems of great variety. Furthermore, by making slight changes we can relax these four restrictions so that the equation will apply to *any* problem.

4.9 ENTHALPY

In Eq. 4.11 the combination $u + Pv$ occurs in the flow-in and flow-out terms. This combination occurs so often in thermodynamics that it has been given a name and a symbol:

$$u + Pv = h = \text{enthalpy per unit mass, or specific enthalpy} \quad (4.12)$$

Enthalpy is also called *total heat, inherent heat,* and several other names in older thermodynamics texts. Obviously, it is the combination of the internal energy per unit mass and the injection work per unit mass. Its use is practically universal in classical thermodynamics, and most tables of thermodynamic properties show h but not u, because users of those tables prefer that. Substituting Eq. 4.12 in Eq. 4.11, we find its *exact equivalent:*

$$d\left[m\left(u + gz + \frac{V^2}{2}\right)\right]_{\text{sys}} = \left(h + gz + \frac{V^2}{2}\right)_{\text{in}} dm_{\text{in}} - \left(h + gz + \frac{V^2}{2}\right)_{\text{out}} dm_{\text{out}} + dQ + dW_{\text{n.f.}} \quad (4.13)$$

Let us summarize how we found this equation:

1. We discussed the general idea of the balance equation.
2. We then introduced the abstract quantity energy.
3. We then asserted, *without proof,* that this abstract quantity, energy, obeys the balance equation, the creation and destruction terms being set equal to zero. This assertion is unprovable; it rests on its ability to explain all the careful experiments ever run to test it.
4. We then chose a fairly general system and listed a set of restrictions that would apply to that system.
5. We wrote out in detail the balance equation for that system, subject to the restrictions and subject to the sign conventions for heat and work, finding Eq. 4.7.
6. We then introduced the idea of injection work, split up the work term in Eq. 4.7, and regrouped terms to find Eq. 4.11.
7. Finally, we introduced the definition of enthalpy to find Eq. 4.13.

Recognize that this is not a *derivation* of the first law of thermodynamics; that law is underivable. Rather, this is a set of algebraic manipulations and definitions that converts the statement "energy obeys the balance equation without creation or destruction" into a very convenient and useful working equation.

4.10 RESTRICTED FORMS

Equation 4.13 is powerful because it is so general. However, whenever we write it, we have, in effect, written the four restrictions listed previously. The procedure recommended for solving all thermodynamics problems is to write Eq. 4.11 or Eq. 4.13, select a system of boundaries, and cancel the terms that appear negligible. Each cancellation represents an assumption. For example, if we assume no heat exchange with the surroundings, then dQ is zero. By crossing out dQ, we are making this assumption. When all of the unnecessary terms have been canceled, we have not only a working equation but also a list of the assumptions on which that equation is based.

Several restricted forms of Eq. 4.13 are in common use.

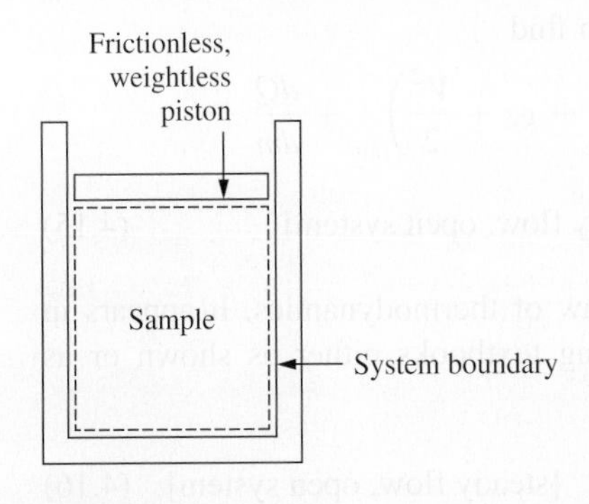

FIGURE 4.3
Simple piston and cylinder.

Example 4.4. Air and coal are contained in the constant-pressure cylinder shown in Fig. 4.3. This cylinder has a frictionless, weightless piston, so the pressure inside the cylinder is always exactly the same as the pressure of the atmosphere. A small spark is now introduced, causing the coal to burn. When the burning is over, the piston has moved so that the volume of the contents has increased by 1 ft^3. The heat transferred to the surroundings was 42 Btu. What is the internal-energy change for this reaction?

We choose as our system the contents of the cylinder. In this system there is no flow in or out, so that $dm_{in} = dm_{out} = 0$. Furthermore, there is negligible change in the kinetic or potential energies of the material contained in the system: $d(gz)_{sys} = d(V^2 / 2)_{sys} = 0$. Because there is no mass flow in or out, we have $d(mu)_{sys} = m\, du_{sys} = dU_{sys}$. Making these substitutions in Eq. 4.13, we find

$$dU_{sys} = dQ + dW_{n.f.} \qquad \text{[closed system]} \tag{4.14}$$

This formula appears in most chemistry books as the basic statement of the first law of thermodynamics. (In this statement the subscript for "non-flow" on the work term is unnecessary, since there can be no injection work into or out of a closed system. Thus, this term is usually written simply dW.) From the foregoing we can see that Eq. 4.14 is a much more restricted form than that we have chosen (Eq. 4.13). We can now substitute for dW from Eq. 4.X and find

$$\begin{aligned} dU_{sys} &= dQ + dW_{n.f.} = dQ - P\,dV \\ &= -42\ \text{Btu} - 14.7\frac{\text{lbf}}{\text{in}^2} \cdot 1\ \text{ft}^3 \cdot \frac{144\ \text{in}^2}{\text{ft}^2} \cdot \frac{\text{Btu}}{778\ \text{ft} \cdot \text{lbf}} \\ &= -44.7\ \text{Btu} = -11{,}270\ \text{cal} = -47{,}153\ \text{J} \end{aligned} \tag{4.AB}$$

■

Here dQ and dW are both negative according to our sign convention; heat flowed out and the system did work by expanding against the surroundings. The device in this example is a *calorimeter*. Refined versions of it are regularly used to determine the heating values of fuels like coal and natural gas, whose prices are adjusted up or down based on changes in heating value; see Prob. 4.18.

Example 4.5. A steady-flow water power plant has its water inlet 15 m above its water outlet. The water enters the plant with a velocity of 3 m / s and leaves with a velocity of 10 m / s. What is the work done by the plant per kilogram of water passing through it?

We choose as our system the plant from inlet to outlet. If the flow is steady, then, as discussed in Sec. 3.3, we have $d[m(u + gz + V^2 / 2)]_{sys} = 0$. Furthermore, for the assumption of only one inlet and one outlet stream dm_{in} equals dm_{out}. We can then divide by dm to find

$$0 = \left(h + gz + \frac{V^2}{2}\right)_{in} - \left(h + gz + \frac{V^2}{2}\right)_{out} + \frac{dQ}{dm} + \frac{dW_{n.f.}}{dm} \qquad \text{[steady flow, open system]} \tag{4.15}$$

This is the steady-flow form of the first law of thermodynamics. It appears in most chemical and mechanical engineering textbooks either as shown or as rearranged to

$$dh + g\,dz + d\left(\frac{V^2}{2}\right) = \frac{dQ}{dm} + \frac{dW_{n.f.}}{dm} \qquad \text{[steady flow, open system]} \tag{4.16}$$

In Eq. 4.16, we have applied Eq. 4.15 to two points negligibly far apart in a steadily flowing stream. The energy balance form shown in Eq. 4.16 will be used for high-velocity gas flows in Chap. 8; it is the convenient form for those flows. For low-velocity gas flows and all liquid flows, we find the convenient form by replacing dh by

$$dh = d(u + Pv) = d\left(u + \frac{P}{\rho}\right) = du + d\left(\frac{P}{\rho}\right) \tag{4.AC}$$

Substituting this in Eq 4.16, we find

$$du + d\left(\frac{P}{\rho}\right) + g\,dz + d\left(\frac{V^2}{2}\right) = \frac{dQ}{dm} + \frac{dW_{\text{n.f.}}}{dm} \quad \text{[steady flow, open system]} \tag{4.17}$$

which will reappear at the start of Chap. 5.

Returning to the water power plant, we assume that there is no heat transfer to the plant ($dQ = 0$) and that the enthalpy of the outlet water is the same as the enthalpy of the inlet water (this is equivalent to the assumption that the inlet and outlet water streams are at the same temperature and pressure). Then

$$\frac{dW_{\text{n.f.}}}{dm} = g(z_{\text{in}} - z_{\text{out}}) + \frac{V_{\text{in}}^2 - V_{\text{out}}^2}{2} = \left(9.81\,\frac{\text{m}}{\text{s}^2} \cdot 15\text{ m}\right) + \frac{(3\text{ m/s})^2 - (10\text{ m/s})^2}{2}$$

$$= 147.15\,\frac{\text{m}^2}{\text{s}^2} - 43.50\,\frac{\text{m}^2}{\text{s}^2} = 101.65\,\frac{\text{m}^2}{\text{s}^2} \cdot \frac{\text{J}}{\text{N}\cdot\text{m}} \cdot \frac{\text{N}\cdot\text{s}^2}{\text{kg}\cdot\text{m}} = 101.65\,\frac{\text{J}}{\text{kg}}$$

$$= 34.01\,\frac{\text{ft}\cdot\text{lbf}}{\text{lbm}} = 0.044\,\frac{\text{Btu}}{\text{lbm}} \tag{4.AD}$$

■

In Sec. 3.2 we divided the mass balance by dt to find the rate form, which showed the mass flow rate, $\dot{m}$. If we divide Eq. 4.13 by dt, we find

$$\frac{d\left[m\left(u + gz + \frac{V^2}{2}\right)\right]_{\text{sys}}}{dt} = \left(h + gz + \frac{V^2}{2}\right)_{\text{in}} \dot{m}_{\text{in}} - \left(h + gz + \frac{V^2}{2}\right)_{\text{out}} \dot{m}_{\text{out}} + \dot{Q} + \dot{W}_{\text{n.f.}} \tag{4.18}$$

The term on the left is the time rate of change of the energy contained in the system. We see two $\dot{m}$ terms on the right. One is the mass flow rate in (lbm / s or kg / s) and the other the mass flow rate out. For steady-state flow these are equal. We also see a $\dot{Q}$ term, which represents the net *heat flow rate* into the system. This has no common name other than heat flow rate. The same is not true for the $\dot{W}_{\text{n.f.}}$ term on the right, which logically should be called the work flow rate into the system. But before we had modern terminology, the work flow rate had a common name, the *power,* Po, which it retains. Both the heat flow rate and the power have dimensions (ft · lbf / s or J / s or Btu / s or cal / s or horsepower or watts; these terms are interconvertible by simple conversion factors.)

In most of simple fluid mechanics the $\dot{Q}$ term is negligible, or, as we will see in Chap. 5, it is practically offset by $(h_{in} - h_{out})$. However, the $\dot{W}_{n.f.}$ term will be important if the system we are considering contains a pump or a turbine. Most of the electric power in the world is produced by turbines, driven by water or steam or hot combustion gases or windmills, so this term is very important in the fluid mechanics of those devices; see Chap. 10.

4.11 OTHER FORMS OF WORK AND ENERGY

So far we have discussed only $F\,dx$ work and kinetic, potential, and internal energies. In this section we consider some of the other kinds of work and energy.

In most modern machinery work is done by rotating shafts. In a simple crank arrangement, Fig. 4.4, a force is being exerted on a crank and being resisted by the shaft to which the crank is attached; the torque Γ in the shaft is given by

$$\Gamma = FL \tag{4.19}$$

If we now allow the shaft to rotate about its axis, the force always being applied at right angles to the crank, the distance through which the force has moved will be

$$dx = L\,d\theta \tag{4.AE}$$

where $d\theta$ is the angular displacement about the center, in radian measure. Solving Eq. 4.19 for F and Eq. 4.AE for dx, we find

$$dW = F\,dx = \left(\frac{\Gamma}{L}\right)L\,d\theta = \Gamma\,d\theta \tag{4.AF}$$

The power is given by

$$\text{Po} = \frac{dW}{dt} = \Gamma\frac{d\theta}{dt} = \Gamma\omega \tag{4.20}$$

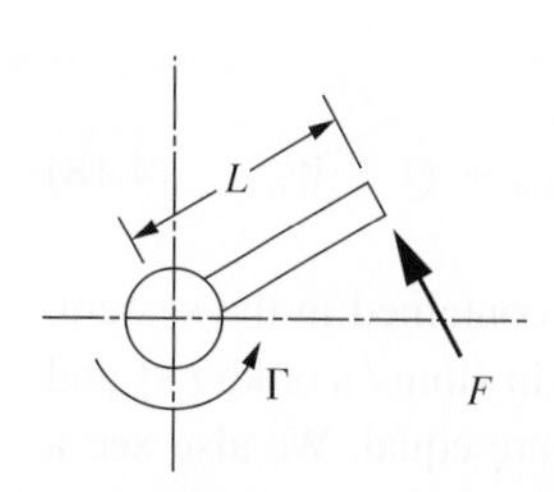

FIGURE 4.4
Relation between force, lever arm, and torque in a simple crank.

Here we see that the power output of a rotating device is the product of its torque Γ and its rotating speed ($\omega = 2\pi$ · revolutions per minute or per second). Generally, the torque a device can develop is roughly proportional to its size, so that to have a given amount of power we can use a large low-speed device or a small high-speed device. In speedboat, automobile, and airplane power plants, where low weight is important, the trend of the past 150 years has been to higher and higher rotational speeds, to get lighter and lighter weight engines (up to 6000 rpm in normal autos, 12,000 rpm in racing autos and motorcycles). In large stationary engines like those in diesel electric power plants or large ship engines, where weight is of no serious concern, the economic choice is a large, low-speed engine (generally less than 1000 rpm), to minimize frictional resistance. Dental drill turbines, which are very small, rotate at 350,000 rpm! Auto superchargers are also small, and rotate up to 200,000 rpm.

Much of the work of modern societies is done electrically. It is shown in texts on electricity that the force required to move a charge Q in an electric field is

$$F = Q\frac{dE}{dx} \tag{4.21}$$

Here dE / dx is the potential (or voltage) gradient. Substituting this in Eq. 4.1, we find

$$W = \int F\,dx = \int Q\frac{dE}{dx}\,dx \tag{4.AG}$$

For any fixed amount of charge we can integrate this to find

$$W = Q\Delta E \tag{4.AH}$$

If we now consider that both the voltage difference and the charge can vary, we can differentiate this equation and find

$$dW = Q\,d(\Delta E) + \Delta E\,dQ \tag{4.22}$$

Ordinarily, we take either of two views of electric flows. One view is to consider some fixed piece of equipment with a steady flow of electrons through it and a fixed voltage difference across it; this corresponds to a steady-flow open system with respect to electrons (or to "charge," which by the usual conventions is the negative of electrons). In this case, since nothing is changing with time, $d(\Delta E)$ is zero and

$$dW = \Delta E\,dQ \tag{4.AI}$$

Dividing this equation by dt, we have

$$\text{Po} = \frac{dW}{dt} = \frac{dQ}{dt}\Delta E = I\,\Delta E \tag{4.23}$$

Here $(dQ/dt) = I$ is the current, so this is the familiar statement that the electric power is the product of the voltage difference and the current. This is our usual way of looking at motors, generators, electric cells, etc.

The other way of looking at electric flows is to consider some fixed amount of charge (or a fixed number of electrons). This corresponds to a closed system for electrons. In this case dQ is zero, and Eq. 4.22 becomes

$$dW = Q\,d(\Delta E) \tag{4.24}$$

We normally consider one of the voltages to be a fixed ground voltage, to which we assign the arbitrary value of zero, so that $d(\Delta E)$ becomes simply dE. Equation 4.24 is the usual way of regarding capacitors, television tubes, xerographic printers, and electron ballistics in general.

So far we have treated energy and mass as two completely distinct entities. In most engineering problems this is a satisfactory approximation, but it is impossible to understand nuclear reactors or atomic explosions or the behavior of the sun without taking into account the conversion of matter into energy. Einstein showed that this conversion may occur and that, when it does, it obeys the rule

$$E = mc^2 \tag{4.25}$$

where c is the speed of light in a vacuum. This equation indicates that c^2 must have the dimensions of energy divided by mass. Simple calculations show that

$$c^2 = 3.85 \cdot 10^{13} \frac{\text{Btu}}{\text{lbm}} = 8.95 \cdot 10^{16} \frac{\text{J}}{\text{kg}} \tag{4.26}$$

In the statement of Eq. 4.13, we specifically excluded such nuclear effects. We have also indicated that mass is conserved. What is really true is that mass plus energy together obey a conservation law, with neither creation or destruction. In this case we could write a mass balance and add it to the energy balance:

$$dm_{\text{sys}} = dm_{\text{in}} - dm_{\text{out}} \tag{4.AJ}$$

We now multiply both sides of this equation by c^2 and add it to Eq. 4.13 to find

$$d\left[m\left(u + gz + \frac{V^2}{2}\right)\right]_{\text{sys}} + c^2\,dm_{\text{sys}} = \left(h + gz + \frac{V^2}{2} + c^2\right)_{\text{in}} dm_{\text{in}}$$

$$- \left(h + gz + \frac{V^2}{2} + c^2\right)_{\text{out}} dm_{\text{out}} + dQ + dW_{\text{n.f.}} \tag{4.27}$$

Obviously, if there is no conversion of mass to energy, the c^2 terms added here cancel, and we find the same result as that from Eq. 4.13.

Example 4.6. A nuclear power plant, running steadily, produces $7 \cdot 10^8$ watts of electric power and rejects $13 \cdot 10^8$ watts of heat to the cooling water taken from a nearby river. How much matter is being converted into energy each hour?

We choose as our system the complete power plant, excluding the cooling water passing through it. Then, in the period between fuel refuelings there is no mass flow into or out of the system (we will overlook such things as boiler feedwater makeup due to leaks or water-treating chemicals, since they are insignificant). Furthermore, at steady state the internal, potential, and kinetic energies of the various parts of the plant are not changing (again, this is slightly inaccurate, because chemical changes accompany nuclear fission, but the error is insignificant). Thus, the only terms remaining in Eq. 4.27 are

$$c^2\,dm_{\text{sys}} = dQ + dW_{\text{n.f.}} \tag{4.AK}$$

We divide by dt and solve for $dm\,/\,dt$:

$$\frac{dm}{dt} = \frac{(dQ\,/\,dt) + (dW_{\text{n.f.}}\,/\,dt)}{c^2}$$

$$= \frac{(-13 \cdot 10^8\ \text{W}) - (7 \cdot 10^8\ \text{W})}{8.95 \cdot 10^{16}\ \text{J}\,/\,\text{kg}} \cdot \frac{\text{J}}{\text{W} \cdot \text{s}} = -2.24 \cdot 10^{-8} \frac{\text{kg}}{\text{s}}$$

$$= -17.8 \cdot 10^{-5} \frac{\text{lbm}}{\text{hr}} \tag{4.AL}$$

This is a medium-sized power plant; it converts less than 2 / 10,000 of a pound-mass per hour into energy. ■

Einstein's relativity theory shows not only that mass and energy can be converted one to another but also that radiant energy (such as light and infrared radiation) has mass. This was demonstrated in classic experiments that showed that light beams passing close to the sun are deflected by the sun's gravity field. The theory shows that the mass of such radiation is proportional to its energy, and the proportionality is given by Eq. 4.26. Based on this and other arguments physics teaches that in *all* energy transactions there is a change in mass, with the proportionality given by Eq. 4.26.

Example 4.7. A sample of water has a mass of exactly 1.0 lbm. It is heated over a hot plate from 59.5 to 60.5°F. How much, if any, does its mass increase?

We have so chosen the problem that Eq. 4.13 indicates that

$$m\,du = dQ = 1 \text{ Btu} \tag{4.AM}$$

Dropping the unnecessary terms from Eq. 4.27, we have

$$\Delta m = \frac{\Delta Q}{c^2} = \frac{1 \text{ Btu}}{3.85 \cdot 10^{13} \text{ Btu / lbm}} = 2.6 \cdot 10^{-14} \text{ lbm} = 5.7 \cdot 10^{-14} \text{ kg} \tag{4.AN}$$

Thus, we conclude that the final mass of the water is 1.000,000,000,000,026 lbm. ■

Detecting the difference between this mass and a mass of 1 lbm would be a serious measurement challenge. However, because Einstein's theory works whenever tests are possible, physicist are confident that, if such measurements were made, they would show this result. Obviously, for most engineering problems we may neglect this effect.

Other kinds of energy generally considered are surface, electrostatic, and magnetic energies. Although they are easily described, their mathematical formulation is more difficult than that of kinetic or potential energy. The reason is that a change in surface or electrostatic field or magnetic field usually is accompanied by an absorption or rejection of heat or by a change in internal energy, whereas we can represent kinetic energy by a single term, because we can increase a body's kinetic energy without any exchange of heat with the surroundings or any change in internal energy. The usual treatment of the other energies requires simultaneous applications of the first and second laws of thermodynamics and, hence, is beyond the scope of this chapter. An intuitive introduction to surface energy is given in Chap. 14.

4.12 LIMITATIONS OF THE FIRST LAW

The first law of thermodynamics is a conservation law. It *accounts for* the quantity called energy. It says nothing about the direction in which changes of energy occur. It is equally well satisfied with water flowing downhill and with water flowing uphill. It makes no distinction between gas flowing out of a high-pressure vessel into the atmosphere and gas flowing from the atmosphere into a high-pressure vessel. To see whether we can make such an interchange, we must rely on another basic principle of nature, the second law of thermodynamics.

4.13 SUMMARY

1. The first law of thermodynamics resulted from a study of friction heating.
2. This study led to the definition of an abstract quantity called energy and to the statement that (excluding nuclear reactions) energy follows the balance principle, with neither creation or destruction.
3. All energy quantities are measured relative to some arbitrary datum.
4. The first law of thermodynamics is not derivable or provable; it rests solely on its ability to predict the outcome of all the experiments ever run to test it.
5. The first law of thermodynamics can be used to solve an enormous number of problems.
6. The first law of thermodynamics says nothing about the direction of an energy change; that is covered by the second law of thermodynamics.

PROBLEMS

Please see the Common Units and Values for Problems and Examples inside the back cover! An asterisk (*) on the problem number indicates that the answer is in App. D.

Several problems in this section deal with ideal gases. It may be shown that for an ideal gas the enthalpy and internal energy depend on temperature alone. If an ideal gas has a constant heat capacity (which may be assumed in all the ideal-gas problems in this chapter), it is very convenient to choose an enthalpy datum that leads to $h = C_P T$ and $u = C_V T$ where T is the absolute temperature; these values may be used in the ideal-gas problems in this chapter.

4.1. The groups $u + gz + V^2/2$ and $h + gz + V^2/2$ occur in most thermodynamics problems. To evaluate the relative magnitude of the individual terms, calculate gz and $V^2/2$ in Btu / lbm and J / kg for the following: z = 10, 100, 1000, 10,000 ft; V = 10, 100, 1000, 10,000 ft / s. Show these results on a log-log plot.

4.2. A gun fires a bullet vertically upward. The bullet has a mass of 0.02 lbm and leaves the gun at a velocity of 2000 ft / s. Air resistance is negligible.

(*a*) How much kinetic energy does the bullet possess when it leaves the gun?

(*b*) How high will the bullet go?

(*c*) At its highest point, how much potential energy will it have relative to the gun barrel?

4.3.*On the moon $g \approx 6$ ft / s^2. How much work is required to raise a 2.0 lbm ball of steel 10 ft on the moon?

4.4. A steel ball with mass of 3.0 kg is dropped from an airplane and goes 500 m in free fall. Air resistance may be neglected. How much work is done on the ball? What does the work?

4.5. A hydraulic lift is shown in Fig. 2.24. The combined mass of the piston, rack, and car is 4000 lbm. The working fluid is water. There is no heat transfer to or from the water, and the internal energy of the water per unit mass is constant. The water may be considered incompressible.

(*a*) Taking all the water in the reservoir, line, and hydraulic cylinder as the system (i.e., taking the closed-system approach), calculate the work necessary to raise the rack and car 1 ft (neglect the change in potential energy of the water in the system).

(*b*) Repeat part (*a*), taking all the water plus the car and the rack as the system.

(*c*) Repeat part (*a*), taking an open-system approach; choose as your system the volume of the hydraulic cylinder, excluding the piston, rack, and car. If the absolute pressure in the system is 1000 lbf / in^2, calculate the volume that must flow in to raise the car 1 ft.

4.6.*At Hoover Dam the difference in elevation from lake level to the stream below the generators ≈ 750 ft. With 100 percent efficient machines, how many kilowatt-hours can be recovered per kilogram of water passing through the system?

4.7.*A boiler feed pump takes water from a tank at 95°C and 100 kPa absolute (1 bar) and delivers it at 190°C and 2000 kPa absolute (20 bar). The flow is steady, and the heat loss from the pump is 2 kJ / kg of water passing through it. What is the work input to the pump in kJ / kg of water passing through it? Here use the density of water at 1 atm and 20°C. If you have access to a steam table, rework the problem using the values from it, and comment on how much different the answer is.

4.8. In a power plant, the inlet water has a velocity of 100 ft / s and an elevation of 80 ft above the outlet of the plant. The outlet water has a velocity of 5 ft / s. How much work can be extracted by the power plant per unit-mass of water passing through the system? The enthalpy of the outlet water may be assumed identical with the enthalpy of the inlet water.

4.9.*Water flows steadily through a power plant. The enthalpy of the outlet water is the same as the enthalpy of the inlet water. The water inlet is 40 m above the water outlet. Both inlet and outlet are at atmospheric pressure. The inlet velocity is 9 m / s, and the outlet velocity is 15 m / s. The flow rate is 5000 kg / s. What is the power output of the plant?

4.10. An ideal gas is flowing steadily in a horizontal, adiabatic nozzle. The inlet conditions are $T_1 = 600°\text{F}$ and $V_1 = 300$ ft / s. The outlet velocity is 2000 ft / s. The heat capacity C_P is 0.3 Btu / lbm · °F. What is the temperature of the gas leaving the nozzle?

4.11.*The rigid vessel shown in Fig. 4.5 has a volume of 1 ft^3 and contains an ideal gas. The temperature is 100°F. The input to the heater is regulated to hold the temperature constant at 100°F. Originally it is at a pressure of 100 lbf / in^2. The valve is partly opened, and the gas is allowed to escape slowly. When the pressure in the vessel has fallen to 20 lbf / in^2, the valve is closed. How much heat was added by the heater during this period?

4.12. A rigid container is initially evacuated. Its valve is then opened and air flows in until the pressure is 1 atm. Assume that air is an ideal gas and that heat transfer is negligible. During the filling process the kinetic energy of the inflowing gas is substantial, but at the end of the process the gas in the container has all come to rest. Write the energy balance for this process, and show the relation between the final temperature and the temperature of the surrounding air.

4.13. The preceding problem is the simplest *adiabatic bottle-filling problem*. The general adiabatic bottle-filling problem allows for the container not to be initially empty, but instead to have some gas at some T and P. The final pressure may be less than the pressure of the inflowing gas (if we shut off the valve before the process stops on its own). Repeat the preceding problem for that case, thus producing the general solution of ideal-gas adiabatic bottle-filling problems. Present your answer as T_f = function of (T_{in}, etc.) Here the initial and final mass may not appear in the solution, but the final and initial pressures may appear.

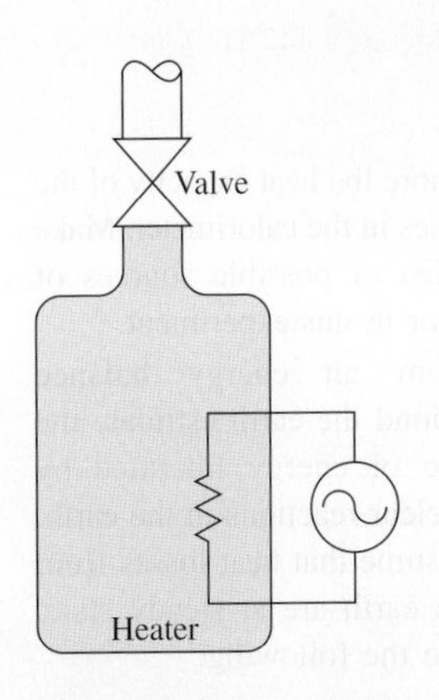

FIGURE 4.5
Rigid vessel with internal temperature held constant by an electric heater.

4.14. A rigid, adiabatic container contains air at 0.5 atm and 20°C. The surroundings air is at 1.0 atm and 20°C. We now open a valve between the container and the surroundings and allow air to flow in until the inside and outside pressures are equal. What is the final temperature of the air in the container? (The "general solution" to ideal gas bottle-filling problems, worked out in Prob. 4.13, may be useful here.)

4.15.*In rating the energy release of nuclear explosives the Department of Energy uses the energy unit "kiloton," where 1 kton $= 10^{12}$ cal. This is roughly the energy release involved in detonating 10^3 tons of TNT (trinitrotoluene). The Hiroshima bomb was reported to be about 14 ktons. How much matter was converted to energy in detonating it?

4.16. Typical high explosives liberate about 1800 Btu / lbm of thermal energy on exploding. It has been suggested that a high-velocity projectile might liberate as much thermal energy on being stopped, by conversion of its kinetic energy to thermal energy. How fast must such a projectile be going in order that its kinetic energy, if all turned to internal energy, would be the same as that of the typical explosive described above? Under what circumstances could a projectile have this kind of velocity?

4.17. Simple diet books show that fats yield 9 kcal / g and carbohydrates yield 4 kcal / g. Present a simple description, based on organic chemistry, for this difference. Based on this difference in energy content, comment on where one would expect fats to occur in plants and animals.

4.18. Figure 4.6 shows a simple combustion calorimeter. The sample is ignited electrically. After a few minutes the temperature of the water and calorimeter is constant at ΔT higher than the starting temperature. The heat of combustion is defined as

$$\Delta u_{\text{combustion}} = \frac{U_{\text{final products of combustion}} - U_{\text{initial fuel + oxygen}}}{m_{\text{sample}}} \tag{4.AO}$$

Determine the heat of combustion of a sample from the following data:

Sample mass	4 g
Calorimeter mass	500 g
Water mass	5000 g
$C_{V\text{ calorimeter}}$	0.12 cal / g · C°
$C_{V\text{ water}}$	1.0 cal / g · C°
ΔT	5 C°

Ignore the heat capacity of the gases in the calorimeter. Make a list of possible sources of error in this experiment.

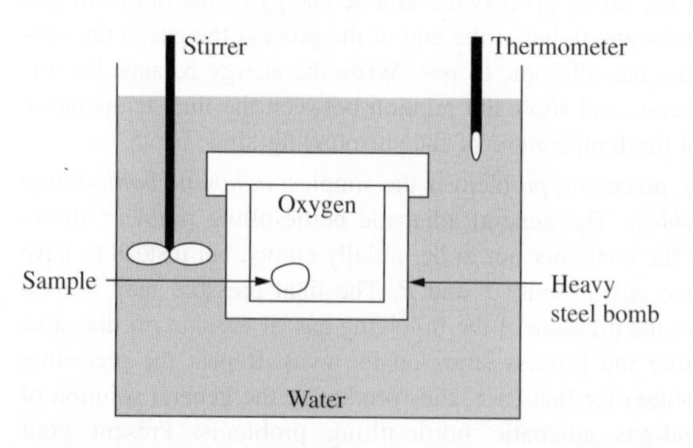

FIGURE 4.6
Simple combustion calorimeter.

4.19. From an energy balance around the earth estimate the rate of energy liberated by nuclear reactions in the earth. Assume that heat losses from the earth are at steady state. Use the following:

Earth is roughly a sphere 8000 mi in diameter.

Geothermal gradient of temperature, dT/dx, is approximately 0.02 F° / ft.
Thermal conductivity k of earth (near the surface) is about 1 Btu / hr · F° · ft.
Heat flow is estimated from $dQ/dt = kA(dT/dx)$, where A is area.

Also calculate the rate at which matter is being converted to energy in the earth.

4.20. Write an energy balance for the sun, indicating which terms are probably important and which are probably negligible.

4.21.*A steady-flow water power plant has the following inlet and outlet conditions:

	Inlet	Outlet
Pressure, P, psig	0	0
Elevation, z, ft	75	0
Velocity, V, ft / s	400	50
Temperature, T, °F	70.0	70.1

The plant is adiabatic. How much work does it deliver per lbm of fluid flowing through?

4.22. An *adiabatic throttle* or *throttling valve* is a valve or orifice in a pipe, with an opening much smaller than the pipe diameter, such that although the fluid velocity through it may be quite high, the velocities upstream and downstream are negligible. Work out the steady-flow energy balance for such a throttle two ways, assuming negligible heat transfer:

(*a*) Using the steady-flow form of the first law, taking the throttle as the system.

(*b*) Using the closed system form, taking 1 kg of matter passing through as your system.

Are the resulting equations the same? Should they be?

4.23. Do any of the following processes violate the first law of thermodynamics? Do these processes violate common sense? Do they violate the second law of thermodynamics?

(*a*) A baseball lying on a table spontaneously jumps to another table that is 10 ft higher. When the process is over, the temperature of the ball has fallen sufficiently for du to be -0.01284 Btu / lbm.

(*b*) In a rigid, insulated container 1 lbm of dry, saturated steam at 30 psia spontaneously converts to 0.105 lbm of ice at 32°F and to 0.895 lbm of superheated steam at 640°F and 42.8 psia. The properties of these materials are:

	Saturated steam	Ice	Superheated steam
T, °F	250.34	32	640
P, psia	30.0	0.0886	42.8
v, ft^3 / lbm	13.748	0.01747	15.23
u, Btu / lbm	1088.0	−143.35	1232.2

(*c*) Freon 12 flows through a throttling valve. The velocities on both sides of the valve are negligible. The conditions before the valve (upstream) are 150°F and 14.7 lbf / in^2. The conditions after (downstream) are 160°F and 66 lbf / in^2. The enthalpies upstream and downstream are both ≈1230 Btu / lbm.

REFERENCES FOR CHAPTER 4

1. Von Baeyer, H. C. *Maxwell's Demon: Why Warmth Disperses and Time Passes*. New York: Random House, 1998.
2. Joule, J. P. "On the Mechanical Equivalent of Heat." Originally presented to the Royal Society, June 21, 1849. See *The Scientific Papers of James Prescott Joule,* Physical Society of London (1884). This short paper is highly recommended as an example of the care and thought that must go into even a relatively simple experiment if the results are to be accurate to plus or minus 1%. Note also that in Joule's time the distinction between the ideas "temperature" and "heat" had not become clear, so he occasionally interchanges them.

PART II

FLOWS OF FLUIDS THAT ARE ONE-DIMENSIONAL, OR THAT CAN BE TREATED AS IF THEY WERE

This part of the book covers most of the flows of greatest interest to chemical engineers. No real fluid flow is completely one-dimensional. But many flows are practically one-dimensional, or can be studied and predicted with considerable accuracy using one-dimensional approximations. For most of these flows one can have a deeper and richer understanding if one reconsiders them from the two- or three-dimensional viewpoint. But learning about them first as one-dimensional is quicker and easier and does not prevent that later reconsideration in Part IV.

Chapters 5, 6, 7, and 8 cover most of the flows of greatest technical interest and introduce most of the basic ideas and terminology used for these flows.

CHAPTER 5

BERNOULLI'S EQUATION

The energy balance for steady, incompressible flow, called Bernoulli's equation, is probably the most useful single equation in fluid mechanics.

5.1 THE ENERGY BALANCE FOR A STEADY, INCOMPRESSIBLE FLOW

We begin with Eq. 4.17,

$$du + d\left(\frac{P}{\rho}\right) + g\,dz + d\left(\frac{V^2}{2}\right) = \frac{dQ}{dm} + \frac{dW_{\text{n.f.}}}{dm} \qquad \text{[steady flow, open system]} \quad (4.17)$$

which applies to the changes from one point to the next along the direction of flow in any steady flow of a homogeneous fluid. Electrostatic, magnetic, and surface energies are assumed to be negligible.

Multiplying by minus 1 and regrouping produce

$$\Delta\left(\frac{P}{\rho} + gz + \frac{V^2}{2}\right) = \frac{dW_{\text{n.f.}}}{dm} - \left(\Delta u - \frac{dQ}{dm}\right) \qquad (5.1)$$

Here ΔP stands for $P_{\text{out}} - P_{\text{in}}$, etc. This equation is the preliminary form of Bernoulli's equation. To save paper, in the rest of this chapter we will speak of Bernoulli's equation as B.E. The original form of B.E. was developed by Daniel Bernoulli (1700–1782) in an entirely different way. By considering momentum balance (Chap. 7) for a frictionless fluid he found $\Delta(P / \rho + gz + V^2 / 2) = 0$, the same as Eq. 5.1 but without the two terms to the right of the equal sign. The original equation was not applicable to flows containing pumps or turbines or to flows in which fluid friction was important. Equation 5.1, based on the energy balance, is applicable to all the flows to which the original, momentum-based, frictionless B.E. applies as well as to those that have significant friction and / or pumps. Some writers refer to Eq. 5.1 as the *extended form of B.E.* or the *engineering form of B.E.*

Before converting it to the final form, let us see what each of the terms represents physically. The P/ρ terms are injection-work terms, representing the work required to inject a unit mass of fluid into or out of the system, or both. The gz terms are potential-energy terms, representing the potential energy of a unit mass of fluid above some arbitrary datum plane. Since they appear only as Δgz, it is unnecessary in most problems to know or to state what that datum is. The $V^2/2$ terms show the kinetic energy per unit mass of fluid. The $dW_{\text{n.f.}}/dm$ term represents the amount of work done on the fluid per unit mass of fluid passing through the system (this does not include injection work, which was specifically excluded). Most often this represents work input from a pump or compressor, or work output in a turbine or expansion engine.

5.2 THE FRICTION-HEATING TERM

We are all familiar with friction heating, as seen in the smoking brakes and tires of an auto that has stopped suddenly and in the high temperature of a saw that is cutting wood. We are less familiar with the idea of friction heating in fluids, because the temperature increases produced by friction heating in fluids are generally much less than those produced by rubbing two solids together. These temperature increases are less for the following reasons:

1. The amount of frictional work per unit mass in typical fluid-flow problems is generally less than in the examples cited above. In these examples the friction-heating energy is concentrated in a small volume; in fluid flows it is spread over a larger volume of fluid.
2. The heat capacity of liquids is generally greater than that of solids. For example, the amount of heat required to raise the temperature of 1 lbm of water by 1°F will raise the temperature of 1 lbm of steel by about 8°F.

Example 5.1. One kilogram of water falls over a 100 m waterfall and lands in the pool at the bottom. This converts the potential energy it had at the top of the fall to internal energy. How much does the temperature of the water increase?

In real waterfalls we must consider evaporation of part of the falling water, which cools the remaining water. But ignoring that for this example, we solve Eq. 5.1 for the change in internal energy,

$$\Delta u = -g(\Delta z) = -9.81\,\frac{\text{m}}{\text{s}^2}(-100\text{ m})\cdot\frac{\text{N}\cdot\text{s}^2}{\text{kg}\cdot\text{m}}\cdot\frac{\text{J}}{\text{N}\cdot\text{m}}$$

$$= 981\,\frac{\text{J}}{\text{kg}} = 328.1\,\frac{\text{ft}\cdot\text{lbf}}{\text{lbm}} \tag{5.A}$$

and the temperature increase is

$$\Delta T = \frac{\Delta u}{C_V} = \frac{981 \text{ J/kg}}{4184 \text{ J/(kg} \cdot \text{°C)}} = 0.23\text{°C} = 0.42\text{°F} \tag{5.B}$$

■

This example shows why we rarely think about friction heating in liquids; the calculated temperature increase, even for this large change in potential energy, is below our ability to sense by sticking our finger in the water.

Friction heating involves the conversion of some other kind of energy (kinetic or potential) or of external work (injection, shaft, or expansion) into internal energy. For constant-density materials (gas, liquid, or solid) the only other way (excluding magnetic, electrostatic, etc.) the internal energy per unit mass can change is through external heating or cooling. Thus,

$$\Delta u = \frac{d(\text{friction heating})}{dm} + \frac{dQ}{dm} \quad \begin{bmatrix}\text{constant-density} \\ \text{materials only}\end{bmatrix} \tag{5.2}$$

Solving this equation for the friction heating per unit mass, we see that it is given by the $\Delta u - dQ/dm$ term on the right of Eq. 5.1.

This friction heating is not connected with any heating or cooling of the fluid through heat transfer with the surroundings and has the same meaning whether the fluid is being heated or cooled. This may be seen by considering the simple, frictionless heater for a constant-density fluid shown in Fig. 5.1. For such a heater there is no change in elevation or velocity and, because there is no friction, there is no change in pressure. Similarly, there is no pump or compressor work, so B.E. simplifies to

$$0 = -\left(\Delta u - \frac{dQ}{dm}\right) \quad \text{[frictionless heater]} \tag{5.3}$$

If, however, there were friction in the heater, then $\Delta u - dQ/dm$ would be a positive number, whose value would be exactly equal to the amount of friction heating per unit mass.

The increased internal energy produced by friction heating is generally useless for industrial purposes, so friction heating is often referred to as *friction loss.* Energy does not disappear in this case. Rather, energy of a valuable form is converted to energy of a normally useless form; hence the "loss" of energy (really, of useful energy).

As discussed in Sec. 2.2, there is no such thing as an absolutely incompressible fluid. Furthermore, there are some situations in which even a fluid with a very small compressibility, such as water, behaves in a compressible way. Thus, we speak of an *incompressible flow,* by which we mean a flow in which the changes

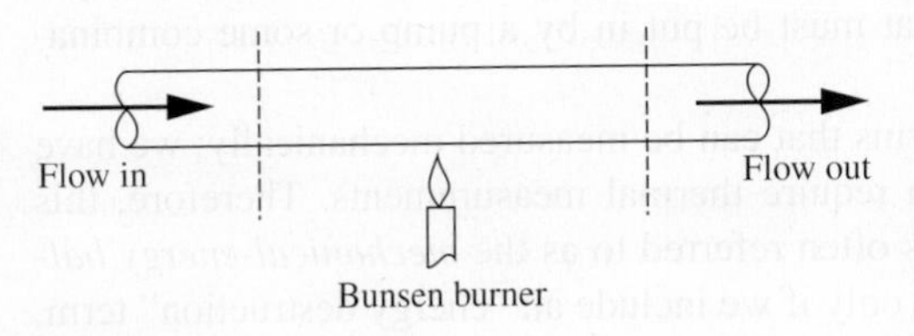

FIGURE 5.1
A simple frictionless heater.

in density are unimportant, rather than of an incompressible fluid. As a general rule, all steady flows of liquids and most steady flows of gases at low velocities (see Sec. 5.6) may be considered incompressible, whereas some unsteady flows of liquids (see Sec. 7.4) and all steady flows of gases at high velocities may not be considered incompressible. We will consider the flow of gases at high velocities in Chap. 8, where we will see that the same terms that appear in B.E. will reappear in different combinations. Therefore, we will apply B.E. only to incompressible flows and use only the incompressible-flow meaning of $\Delta u - dQ/dm$, that is, friction heating per unit mass.

To save writing, we now introduce a new symbol for the friction heating per unit mass,

$$\Delta u - \frac{dQ}{dm} = \mathscr{F} = \begin{pmatrix} \text{friction heating} \\ \text{per unit mass} \end{pmatrix} \quad \begin{bmatrix} \text{constant-density} \\ \text{flow} \end{bmatrix} \tag{5.4}$$

Here we use $\mathscr{F}$ to avoid confusion with F for force. Most civil engineering texts call this quantity gh_f or gh_L, where g is the acceleration of gravity and h_f or h_L stands for *friction head loss* (Sec. 5.4). Some thermodynamics textbooks introduce the idea of the lost work in explaining the second law of thermodynamics. It can be shown that for a constant-density fluid at the heat reservoir temperature the friction heating per unit mass is exactly equal to the lost work per unit mass, so some texts call this term *LW.* Other texts call it $(-\Delta P/\rho)_{\text{friction}}$, since for the most common pipe friction problem, steady flow in horizontal, constant-area pipes, $\mathscr{F} = (-\Delta P/\rho)_{\text{friction}}$.

Substituting the definition of $\mathscr{F}$ into Eq. 5.1 changes it to the final working form of B.E.,

$$\Delta\left(\frac{P}{\rho} + gz + \frac{V^2}{2}\right) = \frac{dW_{\text{n.f.}}}{dm} - \mathscr{F} \tag{5.5}$$

One may show as a consequence of the second law of thermodynamics that $\mathscr{F}$ is zero for frictionless flows and positive for all real flows. One sometimes calculates flows in which $\mathscr{F}$ is negative. This indicates that the assumed direction of the flow is incorrect; for the assumed conditions at the inlet and outlet locations the flow is thermodynamically possible only in the opposite direction. On the other hand, frictionless flows are reversible; any flow described by B.E. in which $\mathscr{F}$ is zero could be reversed in direction without any change in magnitude of the velocities, pressures, elevations, etc.

Since for all real flows $\mathscr{F}$ is positive, the effect in Eq. 5.5 with a minus sign before $\mathscr{F}$ is to indicate that friction causes a decrease in pressure or a decrease in elevation or a decrease in velocity or a decrease in the work that can be extracted by a turbine or an increase in the work that must be put in by a pump or some combination of these effects.

In Eq. 5.5 we now have only terms that can be measured mechanically; we have eliminated the Q and u terms, which require thermal measurements. Therefore, this equation, the working form of B.E., is often referred to as the *mechanical-energy balance.* Mechanical energy is conserved only if we include an "energy destruction" term, $\mathscr{F}$. This equation has the same restrictions as Eq. 5.1 and, in addition, the restriction that the effects of changes in density are negligible.

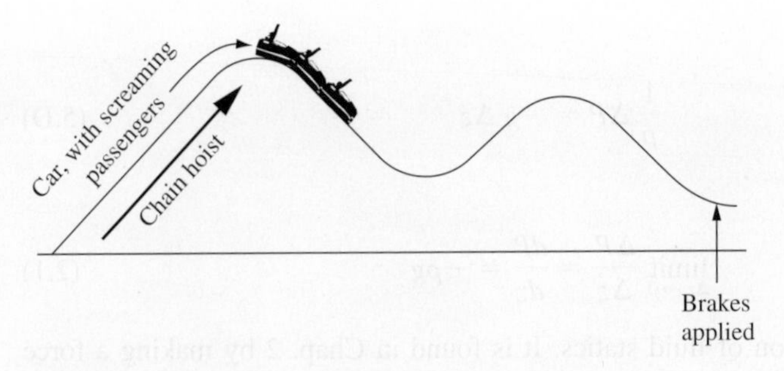

FIGURE 5.2
A simple roller coaster, which illustrates four of the five terms in Bernoulli's equation.

In most applications we will be dealing with flow in a pipe or channel and will assume that the fluid velocity is constant across a given cross section perpendicular to the flow. This approximation is excellent for most engineering problems (see Table 3.1); one interesting exception is discussed in Sec. 5.11.

B.E. deals with the conversion of one kind of energy to another. These changes are illustrated in a common roller coaster, Fig. 5.2. At the left, the car with passengers is lifted from ground level to the top of the first hill by a chain hoist driven by an electric motor, which engages teeth on the bottom of the car. For this part of the trip the change in potential energy, Δgz, is equal to the work input ($\Delta W_{\text{n.f.}} / m$). At the top of the first hill, the car disengages from the chain, pauses a moment for the passengers to anticipate what comes next, and then descends to the first valley. In this part of the trip the decrease in potential energy is practically equal to the increase in kinetic energy; at the first valley the car is going very fast. From the first valley to the top of the second hill, the car's kinetic energy decreases as its potential energy increases. The top of the second hill is always somewhat lower than the top of the first hill, because there has been some friction slowing the car, both due to air resistance and due to rolling friction on the track. If there were no friction, the car could go up and down to the same original height forever; with friction, the top of each succeeding high point must be lower than the preceding one. At the end of the ride (which has more than the two hills shown here), brakes on the track slow the car (convert kinetic energy to friction heating), bringing it to a safe stop at the end of the ride. Four of the five terms in B.E. appear in this description of a roller coaster. The fifth, involving pressure, is discussed in Sec. 5.5.

5.3 ZERO FLOW

The basic equation of fluid statics is a limited form of Eq. 5.5. If we apply Eq. 5.1 between any two points in a fluid flow in which the velocities are slowly becoming zero, then there will be no work or friction and the kinetic energy terms will approach zero so that

$$\Delta\left(\frac{P}{\rho} + gz\right) = 0 \qquad \text{[zero flow]} \tag{5.C}$$

Rearranging, we find

$$\frac{1}{\rho}\Delta P = -g\,\Delta z \tag{5.D}$$

or

$$\underset{\Delta z \to 0}{\text{limit}}\frac{\Delta P}{\Delta z} = \frac{dP}{dz} = -\rho g \tag{2.1}$$

which is the basic equation of fluid statics. It is found in Chap. 2 by making a force balance around an elemental particle of fluid. The derivation shown here points out only that Eq. 5.5 is general enough to include cases of zero flow.

5.4 THE HEAD FORM OF BERNOULLI'S EQUATION

In many problems, particularly those involving flow of water in dams, canals, and open channels, it is convenient to divide both sides of Eq. 5.5 by g to find

$$\Delta\left(\frac{P}{\rho g} + z + \frac{V^2}{2g}\right) = \frac{dW_{\text{n.f.}}}{g\,dm} - \frac{\mathscr{F}}{g} \tag{5.6}$$

which is called the *head form* of B.E.

Every term in Eq. 5.6 has the dimension of length. The lengths are at least conceptually convertible into elevation Δz above some datum plane. These elevations are commonly referred to as "heads." ("Head" is apparently a variant spelling and pronunciation of "height.") Thus, we would refer to the various terms in Eq. 5.6 as the *pressure head, gravity head, velocity head, pump or turbine head,* and *friction head loss.* One occasionally sees the terms *static head,* which is the sum of the pressure and gravity heads, and *dynamic head,* which is the sum of the static head and the velocity head.

There is no simple, universal rule for deciding when to use the head form of B.E. and when to use the energy form, Eq. 5.5; if correctly applied, both give the same result. Through practice engineers learn which is the most convenient for a given problem. Civil engineers use the head form much more than do chemical engineers; but the terms velocity head and pump head occur often in chemical engineering.

5.5 DIFFUSERS AND SUDDEN EXPANSIONS

In the following sections we will see several examples of flow in which a moving fluid is slowed to a stop. Here we consider two ways of slowing down a fluid: a diffuser and a sudden expansion. A diffuser is a gradually expanding pipe or duct, as sketched in Fig. 5.3.

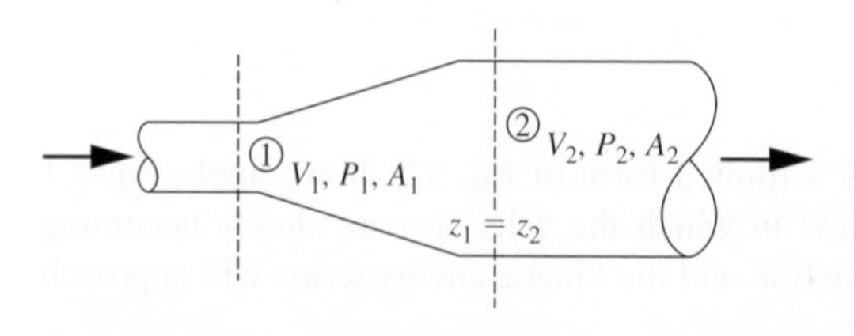

FIGURE 5.3
A simple diffuser in which a fluid flow is slowed in an orderly fashion.

Writing B.E. for the pipe between locations 1 and 2, both at the same elevation, we find

$$\frac{P_2 - P_1}{\rho} + \frac{V_2^2 - V_1^2}{2} = -\mathcal{F} \tag{5.E}$$

From the mass balance for a constant-density fluid we have

$$V_2 = \frac{V_1 A_1}{A_2} \tag{5.F}$$

and, substituting for V_2 in Eq. 5.E, we find

$$P_2 - P_1 = \rho \frac{V_1^2}{2}\left(1 - \frac{A_1^2}{A_2^2}\right) - \rho\mathcal{F} \tag{5.7}$$

This increase in pressure that accompanies the decrease in velocity is often called *pressure recovery.* In such a device kinetic energy is converted partly into injection work (shown by an increase in pressure) and partly into friction heating.

Students find it hard to visualize why the pressure increases as the fluid slows down in steady flow. First, observe that in a constant-density steady flow the velocity can only change from one point to another if the cross-sectional area of the flow changes. In a constant-cross-section area pipe or duct, in steady flow, the velocity is the same at each downstream location. Figure 5.4 shows two types of flow channel, one of which contracts in the flow direction, the other of which expands in the flow direction.

The left part of Fig. 5.4 is the common garden-hose nozzle with which the reader is familiar. In it the cross-sectional area decreases in the flow direction, and the velocity increases. Most students have observed that behavior; the slow-moving flow in the garden hose is converted to the much-faster moving jet of water by the nozzle. If we consider the small section marked Δx, we see that the fluid in it must be accelerating, From Newton's second law we know that $F = ma$; and if the acceleration is in the flow direction, then there must be a net force acting on this slice of fluid, in the flow direction. The only forces acting are the pressure forces, which act on the slice from behind and from in front (we ignore the small shear forces at the walls of the duct). For the algebraic sum of these forces to point in the flow direction (which must occur if the flow is accelerating), the downstream pressure must be less than the upstream pressure. The right part of Fig. 5.4 is obviously the mirror

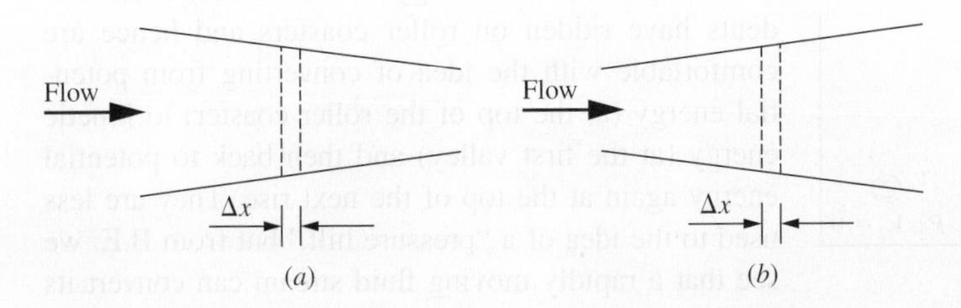

FIGURE 5.4
Two flow channels, one of which, (*a*), contracts in the flow direction, the other of which, (*b*), expands in the flow direction.

image of the left. Students have seldom experienced this flow—it has no common example like a garden hose nozzle. It is called a *diffuser* and appears in various industrial devices. In it the flow is slowing down, because the cross-sectional area perpendicular to the flow is increasing. If we apply the $F = ma$ discussion to it, we see that it is the same, with the words upstream and downstream and increasing and decreasing interchanged. The pressure must *increase* in the flow direction across the section marked Δx for there to be a net force that decelerates the fluid.

Students have experienced the increase in velocity in rolling down a hill on bicycles or skates and the decrease in velocity in rolling up a hill. The gz and $V^2/2$ terms in B.E. show that interaction. The preceding paragraph shows that there is an analogous behavior as fluid flows up or down a *pressure hill;* down the pressure hill, the fluid speeds up (in frictionless flow) and up the hill, the fluid slows down. The P/ρ and $V^2/2$ terms in B.E. show that behavior.

It is possible to build diffusers in which friction heating is only about one-tenth of the decrease in kinetic energy; or, as is commonly stated, the *pressure recovery* is about 90 percent of the maximum possible from a frictionless diffuser.

Now consider a fluid flowing through a duct into a large tank of fluid with no net velocity, as shown in Fig. 5.5. This is called a *sudden expansion.* Here point 2 is chosen far away from the fluid inlet, so that the velocity at point 2 is negligible. Writing B.E. between points 1 and 2, we find

$$P_2 - P_1 = \frac{\rho V_1^2}{2} - \rho \mathcal{F} \tag{5.G}$$

which is quite similar to Eq. 5.7. Here, however, the friction term is much larger than that for the diffuser, because instead of the fluid being brought to rest in an orderly fashion it is stopped by a chaotic mass of eddies, which convert all its kinetic energy into internal energy. Thus, it is an experimental observation that for such sudden expansions the friction heating per unit mass is almost exactly equal to the decease in kinetic energy per unit mass, and there is no pressure recovery at all. Therefore, the pressure of a fluid flowing into such a sudden expansion is the same as the pressure of the fluid into which it flows. This conclusion is limited to flows with velocities less than the speed of sound; it does not apply to sonic or supersonic flows, which we will discuss in Chap. 8.

These two ways of stopping a fluid are analogous to stopping a fast-moving auto by letting it run up a hill and thereby converting its kinetic energy into useful potential energy and to stopping it with its brakes and thereby converting its kinetic energy into useless internal energy in the brakes. Most students have ridden on roller coasters and hence are comfortable with the idea of converting from potential energy (at the top of the roller coaster) to kinetic energy (at the first valley) and then back to potential energy again at the top of the next rise. They are less used to the idea of a "pressure hill," but from B.E. we see that a rapidly moving fluid stream can convert its kinetic energy to potential energy by climbing a gravity hill, or into injection work by climbing a "pressure hill," or into internal energy by friction heating.

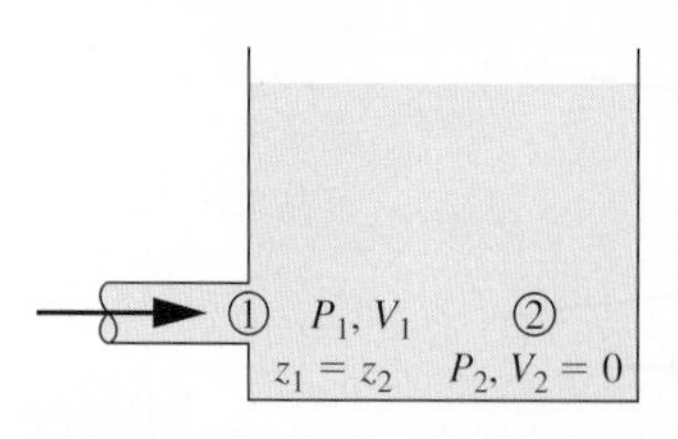

FIGURE 5.5
A sudden expansion in which a fluid flow is slowed in a chaotic fashion.

5.6 B.E. FOR GASES

B.E., as we have written it, is exactly correct for constant-density fluids and practically correct for all flows in which the density changes are unimportant. For liquids this includes almost all steady flows. We show here that it also is practically correct for low-velocity gas flows.

Example 5.2. The tank in Fig. 5.6 is full of air at 68°F = 20°C. The air is flowing out at a steady rate through a smooth, frictionless nozzle to the atmosphere. What is the flow velocity for various tank pressures?

At point 1 the velocity is negligible and, as discussed in Sec. 5.5, the pressure at point 2 is equal to the local atmospheric pressure, if the flow is subsonic. Making these insertions in B.E., without friction, taking 1 to be in the tank away from the nozzle, and 2 to be in the jet, just outside the nozzle, we find

$$V_2 = \left[\frac{2(P_1 - P_{\text{atm}})}{\rho} \right]^{1/2} \tag{5.8}$$

Which value of the density should we use here? It is obviously different at the two states, because the pressure is not the same at the two states. However, if the pressure change is small, the two densities will be practically the same. Let us use the upstream density (but see Prob. 5.5). This will be given by substituting the ideal gas law, $\rho = MP_1 / RT_1$ in Eq. 5.8.

$$V_2 = \left[\frac{2RT_1}{P_1 M} (P_1 - P_{\text{atm}}) \right]^{1/2} \tag{5.9}$$

Using this equation, we can calculate V_2 for various values of P_1. For example, if P_1 is (P_{atm} + 0.01 psig), then

$$V_2 = \left[\frac{2 \cdot (10.73 \text{ psi} \cdot \text{ft}^3 / {}^\circ\text{R} \cdot \text{lbmol}) \cdot 528^\circ\text{R}}{(14.71 \text{ psi})(29 \text{ lbm / lbmol})} \cdot 0.01 \frac{\text{lbf}}{\text{in}^2} \cdot \frac{144 \text{ in}^2}{\text{ft}^2} \cdot \frac{32.2 \text{ lbm} \cdot \text{ft}}{\text{lbf} \cdot \text{s}^2} \right]^{1/2}$$

$$= \left[1231 \frac{\text{ft}^2}{\text{s}^2} \right]^{1/2} = 35 \frac{\text{ft}}{\text{s}} = 10.7 \frac{\text{m}}{\text{s}} \tag{5.H}$$

Equation 5.9 is based on the assumption of a constant-density fluid, which is not exactly correct here; the exactly correct result for this system, taking gas expansion into account, is developed in Chap. 8. The velocities calculated from Eq. 5.9 and the correct solution from Chap. 8 are compared in Table 5.1. ■

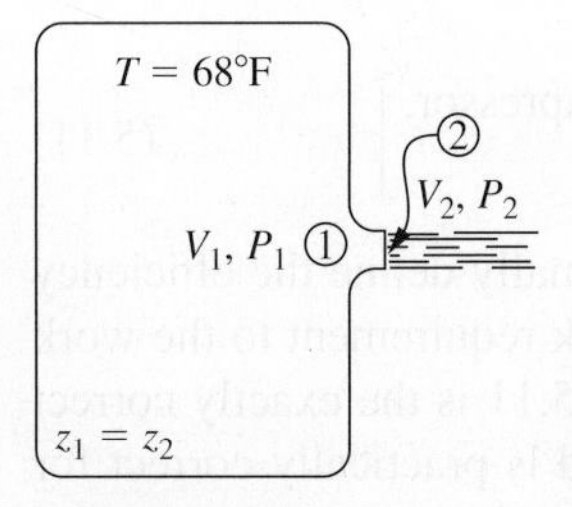

FIGURE 5.6
A gas flowing through a nozzle driven by a modest pressure difference.

From the values in Table 5.1, it is clear that to assume that gas flows are incompressible and are described by B.E. causes a very small error at gas velocities below about 200 ft / s. Even at a velocity of 700 ft / s (213 m / s) the error caused by assuming incompressible flow is only about 5%. The rightmost column in Table 5.1 shows why the answers from simple B.E. and the high-velocity calculations in Chap. 8 differ. Using the methods in Chap. 8, we

TABLE 5.1
Comparison of Bernoulli's equation and high-velocity gas flow equations for the flow in Fig. 5.6

$(P_1 - P_2)$, psia	V_2 from Eq. 5.9, ft / s	V_2 from Chap. 8, ft / s	T_2 from Chap. 8, °F
0.01	35	35	67.9
0.1	110	111	67.0
0.3	190	191	65.0
0.6	267	268	62.0
1	340	343	58.2
2	466	476	49.1
3	554	572	40.7
5	678	713	25.6

see that as the velocity increases, the gas temperature falls; for the lowest row in the table the temperature is 42°F less than the starting temperature. In a high-velocity gas flow the gas can convert some of its internal energy to kinetic energy, so the velocity will be higher and the temperature lower than those we calculate using the constant density assumption in B.E.

Most air-conditioning and low-speed aircraft problems involve velocities below 200 ft / s (61 m / s), so these problems can be solved with engineering accuracy by B.E. On the other hand, where there are significant pressure changes for gases in flow, which lead to high velocities, the density changes must be taken into account, as shown in Chap. 8. Observe also the very high velocities caused by very small pressure differences acting on gases. The inverse of this observation is that for ordinary flow velocities the pressure differences in gases are at least an order of magnitude smaller than for the corresponding flow velocities in liquids. The reason is that the pressure appears in this type of calculation only as $(\Delta P / \rho)$ and for gases ρ is typically about 1 / 800 the value for liquids.

Application of B.E. to a simple, horizontal pump or compressor with equal-sized inlet and outlet pipes (so that there is no velocity change) leads to

$$\frac{dW_{\text{n.f.}}}{dm} = \frac{\Delta P}{\rho} + \mathcal{F} \tag{5.10}$$

If we ignore friction, this equation becomes

$$\frac{dW_{\text{n.f.}}}{dm} = \frac{\Delta P}{\rho} \quad \begin{bmatrix}\text{frictionless pump or compressor,} \\ \text{constant-density fluid}\end{bmatrix} \tag{5.11}$$

Real pumps and compressors are never frictionless. We normally define the efficiency of a pump or compressor as the ratio of this frictionless work requirement to the work actually needed to drive the pump or compressor. Equation 5.11 is the exactly correct frictionless work requirement for constant-density fluids and is practically correct for most pumps that are pumping real liquids. For gases, whose density will change in a compressor, it is not exactly correct. The correct result, taking the change of gas density into account, is developed in Chap. 10. However, if the pressure change ΔP is small compared with the inlet pressure P_{in}, then Eq. 5.11 gives a very good estimate of the required frictionless work. For example, if $\Delta P / P$ is 0.1 or less, then the result from Eq. 5.11 is certain to be within 10 percent of the calculated result which takes density changes into account. This pressure range includes most fans, blowers, air-conditioning systems, vacuum cleaners, etc., but does not include air compressors that inflate the tires of our vehicles or that drive paint sprayers or pneumatic tools (see Prob. 5.58).

5.7 TORRICELLI'S EQUATION AND ITS VARIANTS

The most interesting applications of B.E. include the effects of friction. Before we can solve these, we must learn how to evaluate the $\mathcal{F}$ term, which we will do in Chap. 6. However, there are many flow problems in which the friction heating terms are small compared with the other terms and can be neglected. We can solve these by means of B.E. without the friction heating term. A good example of this type of problem is the tank-draining problem, which leads to Torricelli's equation.

Example 5.3. The tank in Fig. 5.7 is full of water and open at the top. There is a frictionless nozzle near the bottom, the diameter of which is small compared with the diameter of the tank. What is the velocity of the flow out of the nozzle?

To solve this problem, we apply Eq. 5.5 between the free surface at the top of the tank, location 1, and the jet of fluid as it leaves the tank, location 2. In addition to the assumptions built into B.E. we make the following:

1. The diameter of the tank is so large that the velocity at the free surface is practically zero, $V_1 \approx 0$.
2. The pressures at locations 1 and 2 are the local atmospheric pressures. The pressure of the atmosphere is not exactly the same at both points, but it is practically the same; so we assume $\Delta P = 0$.
3. There is no friction or external work.
4. Flow is steady; that is, the level at the top of the tank is not falling. This means that fluid must be flowing into the tank somewhere exactly as fast as it flows out at location 2.

Subject to these restrictions, we may write

$$g(z_2 - z_1) + \frac{V_2^2}{2} = 0 \tag{5.I}$$

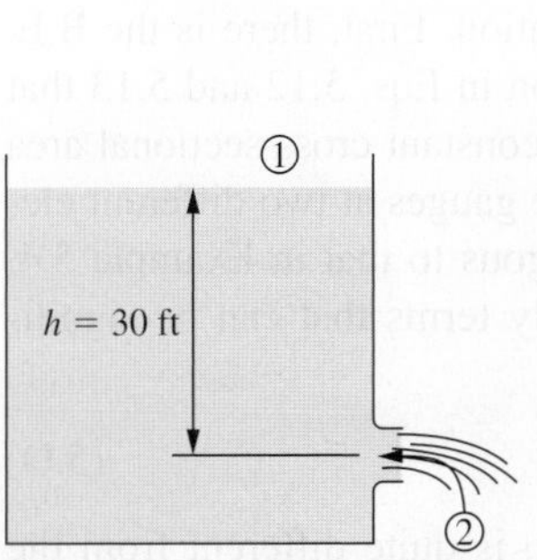

Here $z_2 - z_1 = -h$, so

$$V_2 = (2gh)^{1/2} \qquad \text{[Torricelli's equation]} \tag{5.12}$$

Torricelli's equation says that the fluid velocity is exactly the same as the velocity the fluid would attain by falling freely from rest a distance h. Substituting the numerical values, we find

$$V_2 = \left(2 \cdot 32.2 \frac{\text{ft}}{\text{s}^2} \cdot 30 \text{ ft}\right)^{1/2} = 43.9 \frac{\text{ft}}{\text{s}} = 13.4 \frac{\text{m}}{\text{s}} \tag{5.J}$$

■

FIGURE 5.7
The flow described by Torricelli's equation.

This is the classic tank-draining solution. It is correct only for situations in which the assumptions made in finding Torricelli's equation apply; in Examples 5.4 and 5.5 we examine some situations in which they may not apply.

Example 5.4. Repeat Example 5.3, making the area of the outlet nozzle 1 ft^2 and the cross-sectional area of the tank 4 ft^2.

In this case we cannot assume that the velocity at the free surface is zero, as we did in Example 5.3, so

$$g(-h) + \frac{V_2^2 - V_1^2}{2} = 0 \tag{5.K}$$

Using the mass balance for a constant-density fluid, we can solve for V_1 in terms of V_2, A_1, and A_2 and substitute for V_1, finding

$$g(-h) + \frac{1}{2}\left[V_2^2 - \left(\frac{V_2 A_2}{A_1}\right)^2\right] = 0 \qquad -gh + \frac{V_2^2}{2}\left[1 - \left(\frac{A_2}{A_1}\right)^2\right] = 0 \tag{5.L}$$

$$V_2 = \left[\frac{2gh}{1 - (A_2 / A_1)^2}\right]^{1/2} \tag{5.13}$$

Inserting the numerical values, we find

$$V_2 = \left(\frac{2 \cdot 32.2 \text{ ft / s}^2 \cdot 30 \text{ ft}}{1 - (1/4)^2}\right)^{1/2} \tag{5.M}$$

This is the answer from Example 5.3, divided by $(15/16)^{1/2}$:

$$V_2 = \frac{43.9 \text{ ft / s}}{(15/16)^{1/2}} = 45.3 \frac{\text{ft}}{\text{s}} = 13.8 \frac{\text{m}}{\text{s}} \tag{5.N}$$

■

Why does the water flow faster in this case? All the water in the tank has measurable kinetic energy; it is flowing down at a velocity of 11.4 ft / s. In Example 5.3 the water in the tank has immeasurably small kinetic energy.

What happens in Example 5.4 if the cross-sectional area of the tank is equal to the cross-sectional area of the outlet, that is, if $A_2 = A_1$? If we substitute this in Eq. 5.13, it predicts an infinite velocity! Therefore, Eq. 5.13 does not describe this situation. Recall the assumptions that went into that equation. First, there is the B.E. assumption of steady flow. Second, there is the assumption in Eqs. 5.12 and 5.13 that friction is negligible. Suppose we have a vertical pipe of constant cross-sectional area and steady flow downward. Suppose also that the pressure gauges at two different elevations read the same value. Then this situation is analogous to that in Example 5.4, with $A_1 = A_2$. Returning to Eq. 5.5, we see that the only terms that can be significant are

$$g(z_2 - z_1) = -\mathcal{F} \tag{5.O}$$

This situation, in which the friction forces are dominant, is quite different from the situation shown in Fig. 5.7, from which we found Torricelli's equation, and is not covered by the frictionless assumption of Torricelli's equation.

Example 5.5. Repeat Example 5.3, making the tank contain carbon dioxide gas at the same temperature and pressure as the surrounding atmosphere.

This looks strange: an open tank full of gas! However, it is easy to demonstrate in the laboratory or kitchen by mixing a little bicarbonate of soda and vinegar in a cup. When the bubbling has stopped, the cup will be filled with carbon dioxide gas. This gas is heavier than air and can be poured from cup to cup, visibly. However, it mixes slowly with the air and ultimately will disperse by diffusion.

Returning to the problem, it appears at first to be the same as Example 5.3. However, there is a big difference, namely, we cannot ignore the difference in atmospheric pressure between locations 1 and 2. The other assumptions for Example 5.3 appear sound, so B.E. becomes

$$\frac{P_2 - P_1}{\rho} + g(z_2 - z_1) + \frac{V_2^2}{2} = 0 \tag{5.P}$$

From the basic equation of fluid statics we can calculate

$$P_2 - P_1 = -\rho_{\text{air}} g(z_2 - z_1) \tag{5.Q}$$

Now we must be careful, because there are two densities in our problem: the ρ_{air} shown here and the ρ in B.E. If we follow the derivation of B.E. back to its source, we see that the ρ in it is the ρ of the fluid that is flowing; we label it ρ_{CO_2}. Combining these two equations, we find

$$\frac{-\rho_{\text{air}} g(z_2 - z_1)}{\rho_{CO_2}} + g(z_2 - z_1) + \frac{V_2^2}{2} = 0;$$

$$0 = \frac{V_2^2}{2} + g(z_2 - z_1)\left(1 - \frac{\rho_{\text{air}}}{\rho_{CO_2}}\right) \tag{5.R}$$

Solving for V_2, we find

$$V_2 = \left[2gh\left(1 - \frac{\rho_{\text{air}}}{\rho_{CO_2}}\right)\right]^{1/2} \tag{5.14}$$

If we assume the air and carbon dioxide behave as ideal gases and are at the same temperature and pressure, their densities are proportional to their molecular weights, 29 and 44 g / mol, respectively, so

$$V_2 = \left(2 \cdot 32.2 \frac{\text{ft}}{\text{s}^2} \cdot 30 \text{ ft}\right)^{1/2} \cdot \left(1 - \frac{29}{44}\right)^{1/2} = 43.9 \frac{\text{ft}}{\text{s}} \cdot 0.34^{1/2}$$

$$= 25.6 \frac{\text{ft}}{\text{s}} = 7.8 \frac{\text{m}}{\text{s}} \tag{5.S}$$

■

If the difference in atmospheric pressure is important in Example 5.5, is it also important in Example 5.3? Equation 5.14 applies as well to Example 5.3 as it does to Example 5.5. Therefore, if we want to take the effect of the difference in atmospheric pressure into account in Example 5.3, we should use Eq. 5.14. This is equivalent to multiplying the answer in Example 5.3 by $(1 - \rho_{\text{air}} / \rho_{\text{water}})^{1/2}$. For water and air at normal temperature and pressure, this is about

$$\left(1 - \frac{\rho_{\text{air}}}{\rho_{\text{water}}}\right)^{1/2} = \left(1 - \frac{0.075 \text{ lbm / ft}^3}{62.3 \text{ lbm / ft}^3}\right)^{1/2} = (0.9988)^{1/2} = 0.9994 \tag{5.T}$$

Ignoring the change in atmospheric pressure in Torricelli's equation for air and water makes an error of ≈0.06 percent (much less than the error introduced by some of the other assumptions). We are justified in leaving this term out if the ratio of the density of the surrounding fluid to that of the flowing fluid, $\rho_{\text{surrounding fluid}}/\rho_{\text{moving fluid}}$, is much less than 1. This is true in most hydraulics problems but not in two-liquids problems (Probs. 5.15 and 5.16).

We will discuss one more variant of Torricelli's equation in Sec. 5.10.

5.8 B.E. FOR FLUID-FLOW MEASUREMENT

Several important types of fluid-flow measuring devices are based on the frictionless form of B.E. Where the friction effects in these devices become significant, they are normally accounted for by introducing empirical coefficients and retaining the frictionless form of B.E., rather than by introducing the friction term into B.E. Thus, we consider these devices before we discuss the friction term in B.E., even though these devices obviously involve some friction. These devices have been in common use for at least 100 years. Modern electronics and computers have made possible other types of flow-measuring devices not based on B.E. The devices described here are still more widely used than the computer-electronic ones, because the B.E. devices are simple, reliable, and cheap.

5.8.1 Pitot Tube

The simplest *pitot tube* (H. Pitot, 1695–1771) is sketched in Fig. 5.8. This is sometimes called an *impact tube* or *stagnation tube.* It consists of a bent, transparent tube with one vertical leg projecting out of the flow and another leg pointing directly upstream in the flow.

At location 1 the flow is practically undisturbed by the presence of the tube and hence has the velocity that would exist at location 2 if the tube were not present. At location 2 the flow has been completely stopped by the tube that has been inserted, so $V_2 = 0$. Writing B.E. between locations 1 and 2, we find

$$\frac{P_2 - P_1}{\rho} - \frac{V_1^2}{2} = -\mathscr{F} \tag{5.U}$$

But inside the pitot tube the fluid is not moving, so the pressure at location 2 is given by

$$P_2 = P_{\text{atm}} + \rho g(h_1 + h_2) \tag{5.V}$$

If all the fluid flow is in the horizontal direction, then the basic equation of fluid statics can be used to find the vertical change in pressure with depth inside the flow, so that

$$P_1 = P_{\text{atm}} + \rho g h_2 \tag{5.W}$$

Substituting Eqs. 5.V and 5.W in Eq. 5.U and rearranging, we find

$$V_1 = (2gh_1 + 2\mathscr{F})^{1/2} \tag{5.X}$$

It has been found experimentally that the friction heating term in Eq. 5.X is normally less than

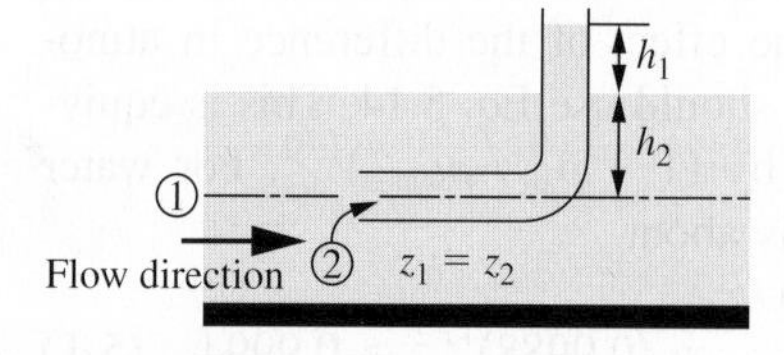

FIGURE 5.8
Pitot tube for fluid velocity measurement.

1 percent of the total; it may be ignored, giving

$$V_1 = (2gh_1)^{1/2} \qquad \text{[pitot tube]} \tag{5.15}$$

The pitot tube allows one to measure a liquid height (a very easy thing to measure) and calculate a velocity from it by means of B.E. The device, exactly as shown in Fig. 5.8, is used for finding velocities at various points in open-channel flow and for determining the velocities of boats.

Example 5.6. A pitot tube exactly as shown in Fig. 5.8 is used for measuring the velocity of a sailboat. When the water level in the tube is 1 m above the water surface, how fast is the boat going?

$$V_1 = \left(2 \cdot 9.81 \frac{\text{m}}{\text{s}^2} \cdot 1\ \text{m}\right)^{1/2} = \left(19.62 \frac{\text{m}^2}{\text{s}^2}\right)^{1/2} = 4.43 \frac{\text{m}}{\text{s}} = 14.5 \frac{\text{ft}}{\text{s}} \tag{5.Y}$$

■

5.8.2 Pitot-Static Tube

The pitot tube shown in Fig. 5.8 is suitable for liquid open-channel flow but not for flow of the atmosphere or flow in pipes. For the latter two uses, it is combined with a second tube, called a *static tube,* shown in Fig. 5.9. This is the most common type, with the pitot (or impact) tube inside the surrounding static tube. This combination is often simply called a pitot tube.

As the figure shows, the tube that faces the flow is the high-pressure side, whereas the surrounding tube that has openings perpendicular to the flow is the low-pressure side. These two are connected to opposite sides of some appropriate pressure-difference measuring device. Experimental tests have shown that for a well-designed

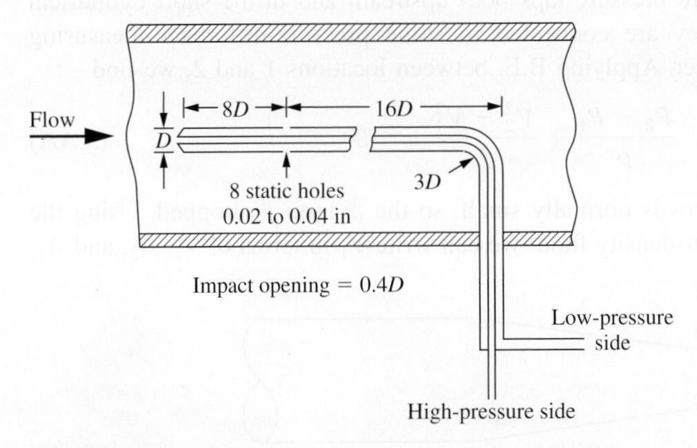

FIGURE 5.9
Pitot-static tube for fluid velocity measurement. The "low pressure side" and "high pressure side" are connected to some appropriate pressure-difference measuring device. The dimensions shown are typical of those on the devices used to measure stack velocities in air pollution sampling. The pitot-static tubes used in aircraft are similar in concept but somewhat different in dimensions. They are often heated to prevent ice-plugging.

pitot-static tube the friction effect is negligible, so we may read the pressure difference from the meter and calculate the velocity from Eq. 5.U rearranged:

$$V_1 = \left(\frac{2\Delta P}{\rho}\right)^{1/2} \quad \text{[pitot-static tube]} \tag{5.16}$$

Example 5.7. Air is flowing in the duct in Fig. 5.9. The pressure-difference gauge attached to the pitot-static tube indicates a difference of 0.05 psi. What is the air velocity?

$$V = \left(\frac{2 \cdot 0.05 \text{ lbf / in}^2}{0.075 \text{ lbm / ft}^3} \cdot \frac{144 \text{ in}^2}{\text{ft}^2} \cdot \frac{32.2 \text{ lbm} \cdot \text{ft}}{\text{lbf} \cdot \text{s}^2}\right)^{1/2} = 78.6 \frac{\text{ft}}{\text{s}} = 23.9 \frac{\text{m}}{\text{s}} \tag{5.Z}$$

■

The pitot-static tube is the standard device for measuring the air speed of airplanes and is often used for measuring the local velocity in pipes or ducts, particularly in air pollution sampling procedures. One can easily identify the pitot-static probes on airplanes. Multiengine planes have them near the nose, at the side below the pilot's window. Single-engine propeller planes place the probe below the wing, far enough out from the center not to be influenced by the propeller. Look for these the next time you are at the airport! For measuring flow in enclosed ducts or channels, the venturi meter and orifice meters discussed below are more convenient and more frequently used.

5.8.3 Venturi Meter

Figure 5.10 shows a horizontal *venturi meter* (G. Venturi, 1746–1822). It consists of a truncated cone in which the cross-sectional area perpendicular to flow decreases, a short cylindrical section, and a truncated cone in which the cross-sectional area increases to its original value. There are pressure taps both upstream and in the short cylindrical section (the "throat"); they are connected to some pressure-difference-measuring device, usually a manometer. Applying B.E. between locations 1 and 2, we find

$$\frac{P_2 - P_1}{\rho} + \frac{V_2^2 - V_1^2}{2} = -\mathcal{F} \tag{5.AA}$$

The friction in these devices is normally small, so the $\mathcal{F}$ term is dropped. Using the mass balance for a constant-density fluid, we can write V_1 in terms of V_2, A_2, and A_1.

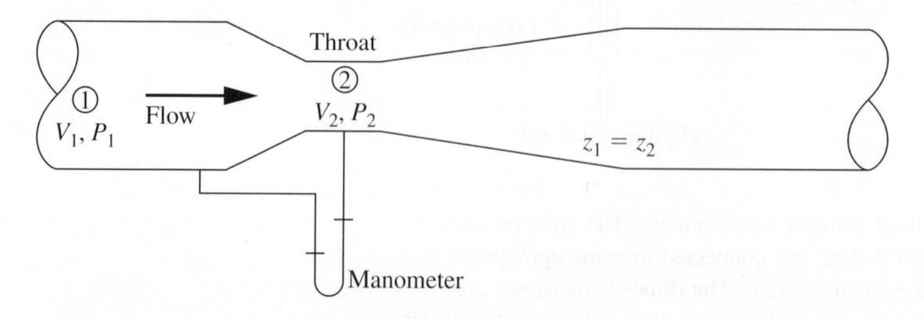

FIGURE 5.10
Venturi meter for fluid velocity measurement.

Substituting in Eq. 5.AA and rearranging, we find

$$V_2 = \left[\frac{2(P_1 - P_2)/\rho}{1 - (A_2^2/A_1^2)}\right]^{1/2} \qquad \text{[venturi meter]} \qquad (5.17)$$

Example 5.8. The venturi meter in Fig. 5.10 has water flowing through it. The pressure difference $P_1 - P_2$ is 1 psi. The diameter at point 1 is 1 ft, and that at point 2 is 0.5 ft. What is the volumetric flow rate through this meter?

From Eq. 5.17,

$$V_2 = \frac{\left(\frac{2 \cdot 1 \text{ lbf/in}^2}{62.3 \text{ lbm/ft}^3} \cdot \frac{144 \text{ in}^2}{\text{ft}^2} \cdot \frac{32.2 \text{ lbm} \cdot \text{ft}}{\text{lbf} \cdot \text{s}^2}\right)^{1/2}}{\{1 - [(\pi/4)(0.5 \text{ ft})^2]^2 / [(\pi/4)(1 \text{ ft})^2]^2\}^{1/2}} = 12.7 \frac{\text{ft}}{\text{s}} = 3.9 \frac{\text{m}}{\text{s}} \qquad (5.AB)$$

The volumetric flow rate is

$$Q = V_2 A_2 = 12.7 \frac{\text{ft}}{\text{s}} \cdot \frac{\pi}{4}(0.5 \text{ ft})^2 = 2.49 \frac{\text{ft}^3}{\text{s}} = 0.070 \frac{\text{m}^3}{\text{s}} \qquad (5.AC)$$

■

It is found experimentally that the flow rate calculated from Eq. 5.17 is slightly higher than that actually observed. This is due partly to friction heating in the meter, which we have assumed to be zero, partly to the fact that the flow is not entirely uniform across any cross section of the pipe, and partly to the fact that the flow is not perfectly one-dimensional, as we have also tacitly assumed. One could attempt to account for these differences by using a more complicated formula than Eq. 5.17; however, the more common approach is to introduce an empirical coefficient into Eq. 5.17, called *the coefficient of discharge, C_v:*

$$V_2 = C_v \left[\frac{2(P_1 - P_2)/\rho}{1 - (A_2^2/A_1^2)}\right]^{1/2} \qquad (5.18)$$

A large number of experimental tests have shown that C_v depends only on the Reynolds number, a dimensionless group whose significance will be discussed in Chaps. 6 and 9; these results are summarized in Fig. 5.11.

Example 5.9. Rework Example 5.8, taking into account the experimental results summarized in Fig. 5.11.

This requires a trial-and-error solution because, to calculate V, we need to know C_v, which is a function of V. The procedure is as follows.

1. Assume $V = V_{\text{Ex. 5.7}} = 12.7$ ft / s.
2. Compute the Reynolds number, $\mathcal{R}$ at point 1.

$$\mathcal{R}_1 = \frac{V_1 D_1 \rho}{\mu} = \frac{V_2(A_2/A_1)D_1\rho}{\mu}$$

$$= \frac{(12.7 \text{ ft/s}/4) \cdot 1 \text{ ft} \cdot 62.3 \text{ lbm/ft}^3}{1 \text{ cP} \cdot 6.72 \cdot 10^{-4} \text{ lbm/ft} \cdot \text{s} \cdot \text{cP}} = 2.9 \cdot 10^5 \qquad (5.AD)$$

3. On Fig. 5.11 we read $C_v = 0.984$.

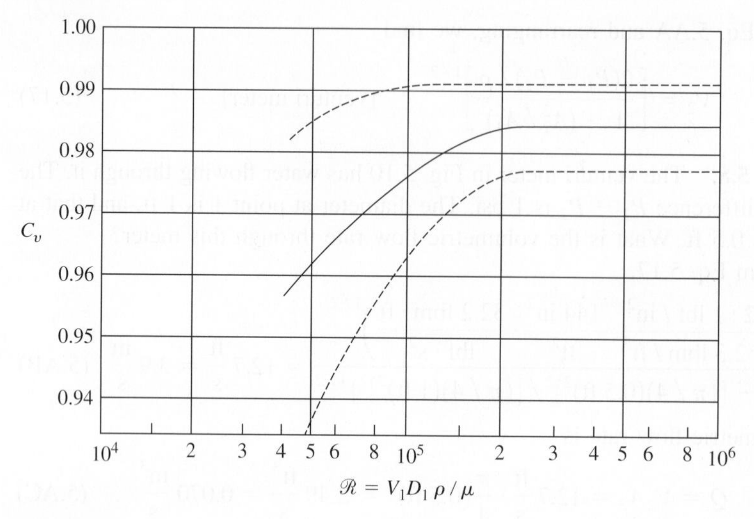

FIGURE 5.11
Discharge coefficients for venturi meters. Here velocities and diameters, V_1 and D_1 are measured at point 1 in Fig. 5.10. The solid line represents the best average of the available data; the dotted lines represent the range of the scatter in the experimental data. (From *Fluid Meters, Their Theory and Practice,* 5th ed., ASME, New York, 1959. Reproduced with permission of the publisher.)

4. $V_{\text{revised}} = 0.984 \cdot 12.7 \frac{\text{ft}}{\text{s}} = 12.5 \frac{\text{ft}}{\text{s}}.$ (5.AE)

5. We should now repeat steps 2 and 3, using this revised value of V. However, in comparing these, we ask, "How much would C_v be changed by using $V_{\text{revised}} = 12.5$ ft / s in calculating the Reynolds number (step 2) and then using a new value of C_v?" Clearly, because of the shape of Fig. 5.11 this would cause a negligible change; so a revised C_v would be the same, and we accept $V = 12.5$ ft / s as a satisfactory estimate of the velocity. Then

$$Q = 12.5 \frac{\text{ft}}{\text{s}} \cdot \frac{\pi}{4} (0.5 \text{ ft})^2 = 2.45 \frac{\text{ft}^3}{\text{s}} = 0.069 \frac{\text{m}^3}{\text{s}} \quad (5.\text{AF})$$

If the velocity had been much lower, not corresponding to the horizontal part of the curve in Fig. 5.11, this trial-and-error solution probably would have taken several steps; normally these meters are designed to operate at high velocities, on the right-hand side of Fig. 5.11, so that this trial and error is very simple. ■

The foregoing is all based on a horizontal venturi meter. If we use the setup shown in Fig. 5.12 and take the manometer reading as a pressure difference to get our value of $(P_1 - P_2)$ in Eq. 5.18, then the result is quite independent of the angle to the vertical of the venturi meter. The reason is that the elevation change in the meter is compensated by the elevation change in the manometer legs. Consider the venturi meter in Fig. 5.12.

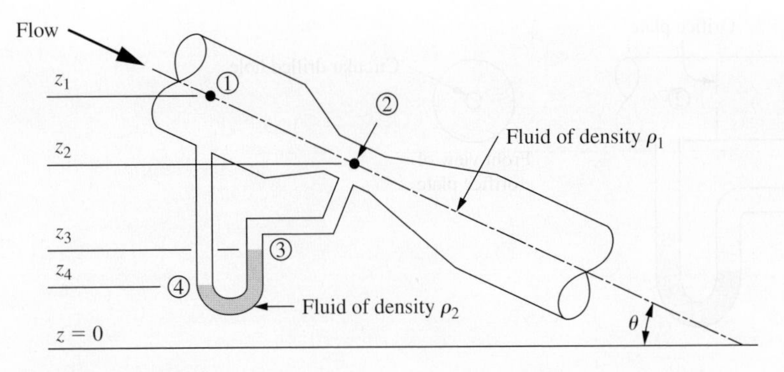

FIGURE 5.12
An inclined venturi meter, with the pressure difference measured by a manometer.

Applying B.E. between points 1 and 2 on this figure and solving for $V_2(1 - A_2^2/A_1^2)^{1/2}$ gives

$$V_2\left(1 - \frac{A_2^2}{A_1^2}\right)^{1/2} = \left[\frac{2(P_1 - P_2)}{\rho} + 2g(z_1 - z_2)\right]^{1/2} \tag{5.AG}$$

To solve for $(P_1 - P_2)$, let us call P_2 known and work our way through the manometer step by step:

$$P_3 = P_2 + \rho_1 g(z_2 - z_3) \tag{5.AH}$$

$$P_4 = P_3 + \rho_2 g(z_3 - z_4) \tag{5.AI}$$

$$P_1 = P_4 - \rho_1 g(z_1 - z_4) \tag{5.AJ}$$

Adding these equations and canceling like terms, we find

$$P_1 = P_2 + \rho_1 g[(z_2 - z_1) - (z_3 - z_4)] + \rho_2 g(z_3 - z_4) \tag{5.AK}$$

$$P_1 - P_2 = -\rho_1 g(z_1 - z_2) + g(z_3 - z_4)(\rho_2 - \rho_1) \tag{5.AL}$$

Substituting this in Eq. 5.18, we see that the elevation $(z_1 - z_2)$ does indeed cancel, and we find

$$V_2 = \left[\frac{2g(z_3 - z_4)(\rho_2 - \rho_1)}{\rho_1(1 - A_2^2/A_1^2)}\right]^{1/2} \qquad \left[\begin{array}{l}\text{inclined venturi}\\ \text{meter with manometer}\end{array}\right] \tag{5.19}$$

But $g(z_3 - z_4)(\rho_2 - \rho_1)$ is precisely the pressure difference we would have calculated for the manometer reading if we had not taken the difference in length of the manometer legs into account. The result found above is true for any angle θ; so we conclude that, if the venturi meter is connected as shown in Fig. 5.12, we can neglect the angle to the vertical and simply use Eq. 5.19 (but see Prob. 5.34).

5.8.4 Orifice Meter

The venturi meter described above is a reliable flow-measuring device. Furthermore, it causes little pressure loss (that is, the actual value of $\mathcal{F}$ is small). For these reasons it is widely used, particularly for large-volume liquid and gas flows. However, the meter

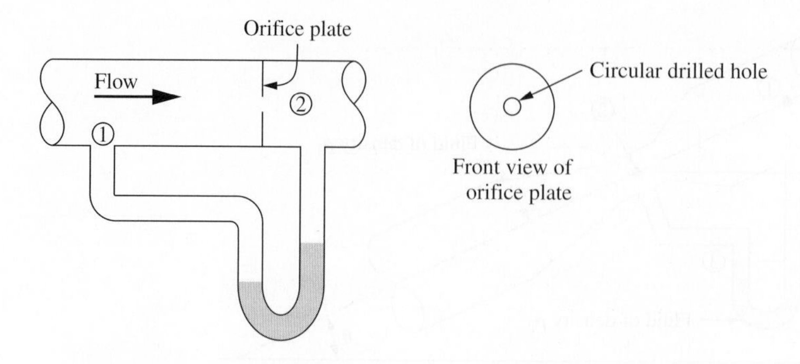

FIGURE 5.13
Orifice meter for fluid velocity measurement.

is relatively complex to construct and hence expensive. For small pipelines, its cost seems prohibitive, so simpler devices have been invented, such as the *orifice meter*.

As shown in Fig. 5.13, the orifice meter consists of a flat orifice plate with a circular hole drilled in it. There is a pressure tap upstream from the orifice plate and another just downstream. If the flow direction is horizontal and we apply B.E., ignoring friction, from point 1 to point 2 in the figure, we find Eq. 5.17, exactly the same equation we found for a venturi meter. However, in this case we cannot so easily assume frictionless flow and uniform flow across any cross section of the pipe as we can in the case of the venturi meter.

As in the case of the venturi meter, experiments indicate that, if we introduce a discharge coefficient and thus form Eq. 5.18, then that coefficient is a fairly simple function of the ratio of the diameter of the orifice hole to the diameter of the pipe, D_2 / D_1, and the Reynolds number; the relation is shown in Fig. 5.14.

Example 5.10. Water is flowing at a velocity of 1 m / s in a pipe 0.4 m in diameter. In the pipe is an orifice with a hole diameter of 0.2 m. What is the measured pressure drop across the orifice?

Rearranging Eq. 5.18, we find

$$\Delta P = \frac{\rho V_2^2}{2C_v^2} \cdot \left(1 - \frac{A_2^2}{A_1^2}\right) = \frac{\rho V_2^2}{2C_v^2}\left(1 - \frac{D_2^4}{D_1^4}\right) \tag{5.AM}$$

From the mass balance for steady flow, we know that

$$V_2 = V_1 \frac{A_1}{A_2} = 1 \frac{\text{m}}{\text{s}} \cdot \frac{(\pi / 4) \cdot (0.4 \text{ m})^2}{(\pi / 4) \cdot (0.2 \text{ m})^2} = 4 \frac{\text{m}}{\text{s}} = 13.1 \frac{\text{ft}}{\text{s}} \tag{5.AN}$$

The Reynolds number $\mathscr{R}_2$ based on D_2 is calculable and will be found to be about $1.6 \cdot 10^6$; so, from Fig. 5.14, we have $C_v = 0.62$. Hence

$$P_1 - P_2 = \frac{(998.2 \text{ kg / m}^3) \cdot (4 \text{ m / s})^2}{2 \cdot 0.62^2} \cdot (1 - 0.5^4) \cdot \frac{\text{N} \cdot \text{s}^2}{\text{kg} \cdot \text{m}} \cdot \frac{\text{Pa}}{\text{N / m}^2}$$
$$= 19.5 \text{ kPa} = 2.83 \text{ psi} \tag{5.AO}$$

■

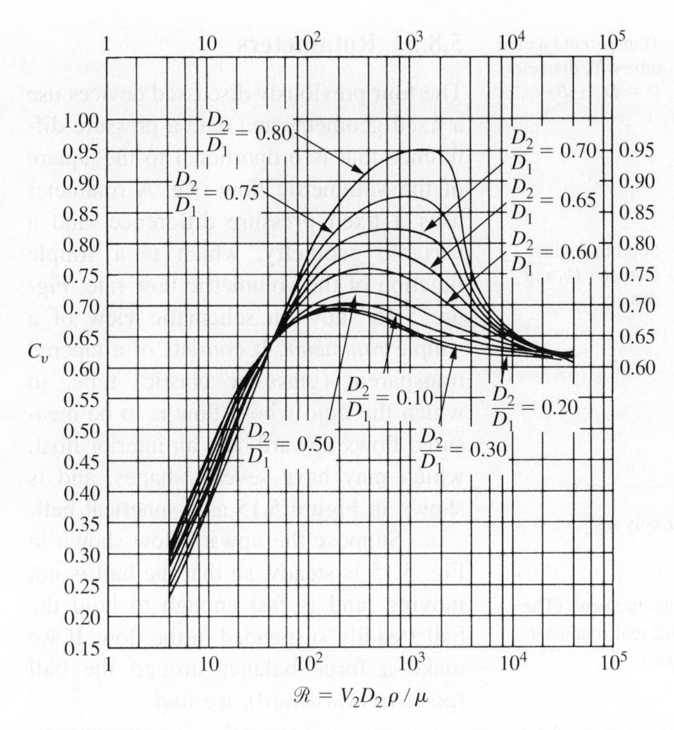

FIGURE 5.14
Discharge coefficients for drilled-plate orifices. (From G. L. Tuve and R. E. Sprenkle, "Orifice discharge coefficients for viscous liquids," *Instruments* 6:201 (1933). Reproduced by permission of the publisher.)

From Fig. 5.14 we see that for small orifice holes ($D_2 / D_1 \leq 0.4$) and high flow rates ($\mathscr{R}_2 > \sim 1000$), C_v is approximately equal to 0.6. These conditions occur in most typical industrial orifice applications, so many practicing engineers automatically write down $C_v = 0.6$ for orifice meters, or for the flow through any simple orifice. In new applications it is best to check Fig. 5.14 to see whether this simplification applies. By the mathematical methods of potential flow (Chap. 16), one may show that an ideal orifice should have $C_v = \pi / (\pi + 2) = 0.611$ [1].

Figure 5.14 is based on a standard location of the upstream and downstream pressure taps. When the taps are in some other location, the value of C_v will be different [2]. In comparison with venturi meters, orifice meters have high pressure losses—high $\mathscr{F}$—and correspondingly high pumping costs, but because they are mechanically simple they are cheap and easy to install. For flows in small-sized pipes, orifice meters are much more common than venturi meters.

The values of C_v in Fig. 5.14 are applicable only to drilled-plate orifices (sometimes called *square-edge orifices,* because the edges of the hole are not rounded). Some other standard types also are used, and sets of C_v curves for these have been published [3].

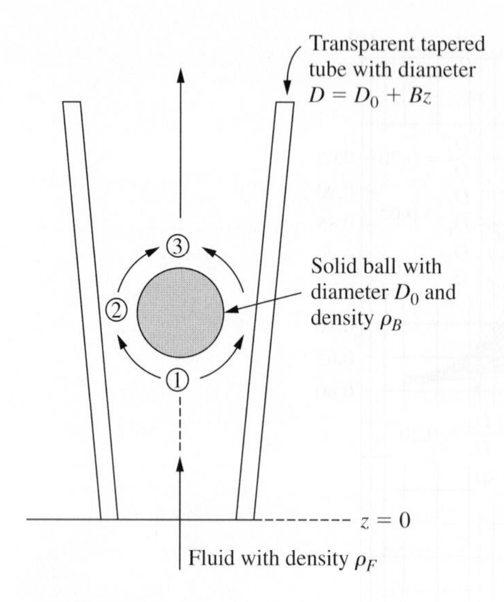

FIGURE 5.15
Rotameter for fluid velocity measurement. (The taper of the tube is exaggerated; real rotameter tubes have a much smaller taper.)

5.8.5 Rotameters

The four previously discussed devices use a fixed geometry and read a pressure difference that is proportional to the square of the volumetric flow rate. A rotameter uses a fixed pressure difference, and a variable geometry, which is a simple function of the volumetric flow rate. Figure 5.15 shows a schematic view of a simple *rotameter*. It consists of a tapered transparent (glass or plastic) tube, in which the fluid whose flow is to be measured flows upward, and an interior float, which may have several shapes, and is shown in Figure 5.15 as a spherical ball.

Suppose the upward flow shown in Fig. 5.15 is steady, so that the ball is not moving, and is fast enough to hold the ball steadily suspended in the flow. If we make a force balance around the ball (positive downward), we find

$$0 = F_{\text{gravity}} + F_{\text{pressure from above}} - F_{\text{buoyancy}} - F_{\text{pressure from below}} \tag{5.AP}$$

If we assume that the pressure below the ball is practically uniform across the ball's lower surface, and similarly for the pressure across the ball's upper surface, and remember from Chap. 2 that the z component of that pressure force will be simply the pressure times the projected area of the ball, we find

$$0 = \frac{\pi}{6} D_0^3 \rho_{\text{ball}} g + P_3 \frac{\pi}{4} D_0^2 - \frac{\pi}{6} D_0^3 \rho_{\text{fluid}} g - P_1 \frac{\pi}{4} D_0^2 \tag{5.AQ}$$

$$\frac{\pi}{6} D_0^3 (\rho_{\text{ball}} - \rho_{\text{fluid}})\, g = \frac{\pi}{4} D_0^2 (P_1 - P_3) \tag{5.AR}$$

From B.E., we can find that

$$P_1 - P_2 = \rho_{\text{fluid}} \cdot \left(\frac{V_2^2}{2} - \frac{V_1^2}{2} \right) = \rho_{\text{fluid}} \cdot \frac{V_2^2}{2} \cdot \left(1 - \frac{A_2^2}{A_1^2} \right) \tag{5.AS}$$

But $(A_2 / A_1)^2$ is generally much less than 1, so we can drop it in the last term above. And as discussed previously, the flow from 2 to 3 is a sudden expansion, so that P_3 is very nearly the same as P_2. Making these substitutions in Eq. 5.AS and solving for V_2, we find

$$V_2 = \left[\frac{4 D_0 g}{3} \cdot \frac{\rho_{\text{ball}} - \rho_{\text{fluid}}}{\rho_{\text{fluid}}} \right]^{1/2} \quad \text{[rotameter]} \tag{5.20}$$

Thus, applying B.E. (and some judicious assumptions), we find that for a given diameter of the ball and of the densities of ball and fluid, there is only one possible value of V_2 that will keep the ball steadily suspended! That means that for any flow rate, Q, the ball must move to that elevation in the tapered tube where $V_2 = Q / A_2$. But

$$A_2 = \frac{\pi}{4}[(D_0 + Bz)^2 - D_0^2] = \frac{\pi}{4}[2Bz + (Bz)^2] \tag{5.AT}$$

The taper of the tube, B, is generally small enough that the $(Bz)^2$ term in Eq. 5.AT is small compared to $2Bz$ and can be dropped. Then it follows that the height, z, at which the ball stands is linearly proportional to the volumetric flow rate Q.

This treatment is simple; more complex treatments [4] lead to similar conclusions. Because of some of the assumptions that went into finding Eq. 5.20, we should not assume that we can compute the true velocities from it; an empirical coefficient like the orifice coefficient would enter. However, most rotameters are treated as calibrated devices; for a given tube, float, and fluid, the Q-z curve is measured and thereafter one simply reads the float position and looks up the flow rate from that calibration curve.

Example 5.11. Our rotameter has been calibrated for nitrogen at room temperature and atmospheric pressure; the calibration shows that for a reading (float position) of 50 percent of the height of the rotameter tube the volumetric flow rate is 100 cm³ / min. We now need to measure the flow of helium at room temperature and atmospheric pressure, using the same rotameter. When the reading is 50 percent of full scale, estimate the helium volumetric flow rate.

From Eq. 5.20 we know that the velocity V, and hence the volumetric flow rate, for a given float position

$$V \propto \left(\frac{\rho_{\text{ball}} - \rho_{\text{fluid}}}{\rho_{\text{fluid}}}\right)^{1/2} \tag{5.AU}$$

Here the density of the ball, if it is made of almost any solid material, is at least 1000 times the density of nitrogen at atmospheric pressure, so we can safely drop the ρ_{fluid} in the numerator, from which it follows that the velocity is proportional to $1 / (\text{fluid density})^{1/2}$. Thus

$$Q_{\text{helium}} = Q_{\text{nitrogen}}\left(\frac{\rho_{\text{nitrogen}}}{\rho_{\text{helium}}}\right)^{1/2} = Q_{\text{nitrogen}}\left(\frac{M_{\text{nitrogen}}}{M_{\text{helium}}}\right)^{1/2}$$

$$= 100\,\frac{\text{cm}^3}{\text{min}}\left(\frac{28}{4}\right)^{1/2} = 265\,\frac{\text{cm}^3}{\text{min}} \tag{5.AV}$$

■

Rotameters are very widely used for measuring low flow rates. The simple spherical ball float is used for the smallest flows, and more complex float designs are used for larger flow rates.

5.9 NEGATIVE ABSOLUTE PRESSURES: CAVITATION

In certain flows B.E. can predict negative absolute pressures, as shown by the following two examples. In gases, negative absolute pressures have no physical meaning at all. When B.E. predicts a negative absolute pressure for a gas flow, then the flow probably contains velocities much too high for the assumptions of B.E.; the equations developed in Chap. 8 must then be used.

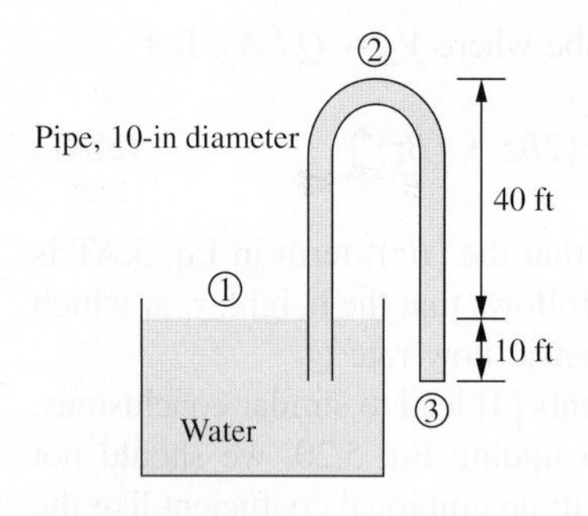

FIGURE 5.16
A siphon that will not work; see Example 5.12.

In liquids, negative absolute pressures can exist under very rare conditions, but they are unstable. Normally, when the absolute pressure on a liquid is reduced to the vapor pressure of the liquid, the liquid boils. This converts the flow to a two-phase flow, which has a much higher value of $\mathcal{F}$ than does the corresponding one-phase flow. Thus, when B.E. predicts a pressure less than the vapor pressure of the liquid, the flow as calculated is physically impossible; the actual flow will have a much higher friction effect, and the flow velocity will be less than that assumed in the original calculation.

Example 5.12. Figure 5.16 shows a siphon that is draining a tank of water. What is the absolute pressure at point 2?

Applying B.E. without friction from the free surface, point 1, to the outlet, point 3, we find

$$V_3 = [2g(h_1 - h_3)]^{1/2} = [2 \cdot (32.2 \text{ ft / s}) \cdot 10 \text{ ft}]^{1/2}$$
$$= 25.3 \text{ ft / s} = 7.71 \text{ m / s} \tag{5.AW}$$

Then applying B.E. between points 1 and 2, we find

$$P_2 = P_1 - \rho\left[\frac{V_2^2}{2} + g(z_2 - z_1)\right]$$
$$= 14.7\frac{\text{lbf}}{\text{in}^2} - 62.3\frac{\text{lbm}}{\text{ft}^3}\left[\frac{(25.3 \text{ ft / s})^2}{2} + 32.2\frac{\text{ft}}{\text{s}^2} \cdot 40 \text{ ft}\right] \cdot \frac{\text{lbf} \cdot \text{s}^2}{32.2 \text{ lbm} \cdot \text{ft}} \cdot \frac{\text{ft}^2}{144 \text{ in}^2}$$
$$= 14.7\frac{\text{lbf}}{\text{in}^2} - 21.6\frac{\text{lbf}}{\text{in}^2} = -6.91\frac{\text{lbf}}{\text{in}^2} = -47.6 \text{ kPa} \quad ??? \tag{5.AX}$$

■

This flow is physically impossible. One may show that, when water is open to the atmosphere, such siphons can never lift water more than about 34 ft (10.4 m) above the water surface, even with zero velocity; the siphon shown in Fig. 5.16 will not flow at all. In this example the physically unreal, negative pressure was mostly a result of the gravity term in B.E. Negative absolute pressures can also be predicted by B.E. for horizontal flows in which gravity plays no role.

Example 5.13. Water flows from a pressure vessel through a venturi meter to the atmosphere; see Fig. 5.17. P_1 = 10 psig and $A_2 / A_3 = 0.50$. What is the pressure at location 2?

Applying B.E. without friction between locations 1 and 3, we find

$$V_3 = \left[2\frac{(P_1 - P_3)}{\rho}\right]^{1/2} = \left[2 \cdot \frac{10 \text{ lbf / in}^2}{62.3 \text{ lbm / ft}^3} \cdot \frac{32.2 \text{ lbm} \cdot \text{ft}}{\text{lbf} \cdot \text{s}^2} \cdot \frac{144 \text{ in}^2}{\text{ft}^2}\right]^{1/2}$$
$$= 38.6\frac{\text{ft}}{\text{s}} = 11.8\frac{\text{m}}{\text{s}} \tag{5.AY}$$

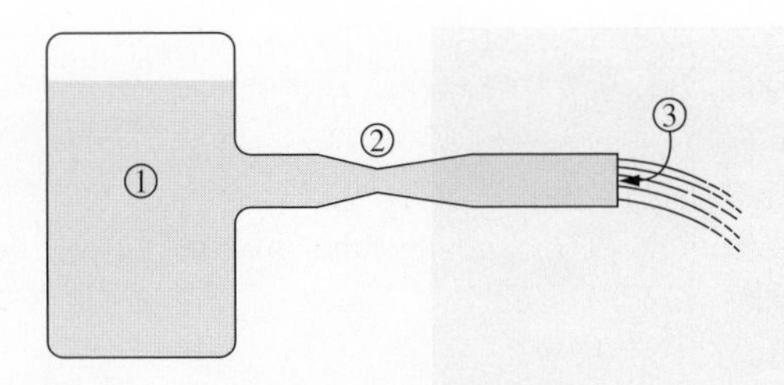

FIGURE 5.17
A horizontal venturi; see Example 5.13.

The mass balance gives us

$$V_2 = V_3 \frac{A_3}{A_2} = 38.6 \frac{\text{ft}}{\text{s}} \cdot \frac{1}{0.50}$$

$$= 77.2 \frac{\text{ft}}{\text{s}} = 23.5 \frac{\text{m}}{\text{s}} \qquad (5.AZ)$$

Applying B.E. without friction between locations 1 and 2, we find

$$P_2 = P_1 - \rho \frac{V_2^2}{2} = 24.7 \text{ psia} - \frac{62.3 \text{ lbm / ft}^3 (77.2 \text{ ft / s})^2}{2} \cdot \frac{\text{lbf} \cdot \text{s}^2}{32.2 \text{ lbm} \cdot \text{ft}} \cdot \frac{\text{ft}^2}{144 \text{ in}^2}$$

$$= 24.7 \text{ psia} - 40.1 \text{ psi} = -15.4 \text{ psia} = -106 \text{ kPa, abs} \qquad ??? \qquad (5.BA)$$

■

This flow also is physically unreal. At such high velocities the frictional effects become large; so the frictionless assumption above is a poor one. If the frictional effect were negligible, then the fluid would boil in the venturi and thus convert to a two-phase flow with a much lower flow rate. The device sketched in Fig. 5.17 is widely used as a vacuum pump. An opening in the side of the tube at point 2 will suck in air; such devices, attached to faucets, are widely used as a laboratory source of modest vacuum. As the above example shows, with a high flow one produces a negative absolute pressure. But with modest flows one does not produce an impossible flow, and does produce a useful vacuum.

Example 5.13 shows that as the velocity increases in horizontal flow, the pressure falls. The pressure decrease can cause boiling of the liquid. Dramatic examples of this phenomenon occur in pumps, turbines, and ship's propellers, as shown in Fig. 5.18. In these devices the fluid is often speeded up to a velocity at which it forms a vapor bubble. Then the bubble flows to a region of higher pressure and collapses. The collapse can cause a sudden pressure pulse, and the pulses, occurring at high frequencies, can damage the pump, turbine, etc. The phenomenon of local boiling due to velocity increase is called *cavitation,* the study of which is an important part of modern research in fluid machines [5]. We will say a little more about this in Chap. 10.

5.10 B.E. FOR UNSTEADY FLOWS

B.E. is a steady-flow equation; however, it can be successfully applied to some unsteady flows if the changes in flow rate are slow enough to be ignored. To decide how slow the change must be to be ignored, we reason as follows. For a steady flow $(\partial V / \partial t)_{x,y,z}$ is zero. This means that, although an observer riding with the fluid would observe a changing velocity, an observer watching a specific point in the system would observe no change in velocity with respect to time. We are safe in ignoring unsteady-flow effects if $(\partial V / \partial t)_{x,y,z}$ for all points in the system is small compared with the acceleration we are considering, that is, the acceleration of gravity or the acceleration due to pressure forces, $(dP / dL) / \rho$. If, on the other hand, $(\partial V / \partial t)_{x,y,z}$ at any point in the

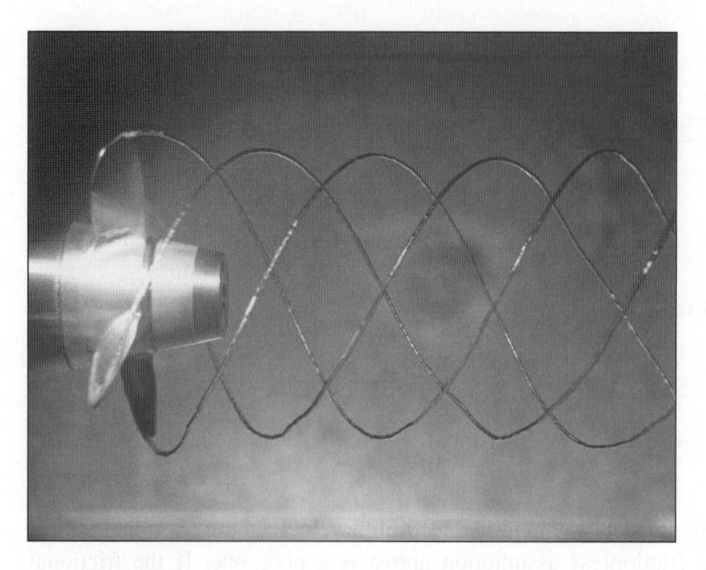

FIGURE 5.18
Cavitation bubbles formed at the tips of a propeller. The propeller is rotating in a channel with flow from left to right. At the tips of the blades the pressure is low enough to cause the water to boil. The bubbles have short lifetimes; the collapse of the unstable bubbles produces shock waves in the water, which can be destructive. (Photograph from the Garfield Thomas Water Tunnel Building, Applied Research Laboratory, The Pennsylvania State University.)

system is comparable to the largest of the other acceleration terms in the system, then we cannot safely apply B.E. to the system.

Example 5.14. If the tank in Fig. 5.7 is cylindrical with a diameter of 10 m, and if the outlet nozzle is 1 m in diameter, how long does it take the fluid level to drop from 30 m above the tank outlet to 1 m above the tank outlet?

Here we assume that the $(\partial V / \partial t)_{x,y,z}$ is small; we will check that assumption later. The instantaneous flow rate is assumed to be given by B.E., which here takes the form of Torricelli's equation:

$$V_2 = (2gh)^{1/2} \tag{5.12}$$

But, by the mass balance for an incompressible fluid,

$$V_2 = V_1 \frac{A_1}{A_2} \tag{5.BB}$$

where V_1 is the rate at which the free surface of the tank is moving downward, which is equal to $-dh / dt$; so

$$V_2 = \frac{-dh}{dt} \cdot \frac{A_1}{A_2} = (2gh)^{1/2}: \qquad \frac{-dh}{h^{1/2}} = \frac{A_2}{A_1}(2g)^{1/2}\,dt \tag{5.BC}$$

$$-\int_{h_1}^{h_2} \frac{dh}{h^{1/2}} = -2h^{1/2}\Big]_{h_1}^{h_2} = \frac{A_2}{A_1}(2g)^{1/2}\int_{t_1}^{t_2} dt = \frac{A_2}{A_1}(2g)^{1/2}\,t\Big]_{t_1}^{t_2} \tag{5.BD}$$

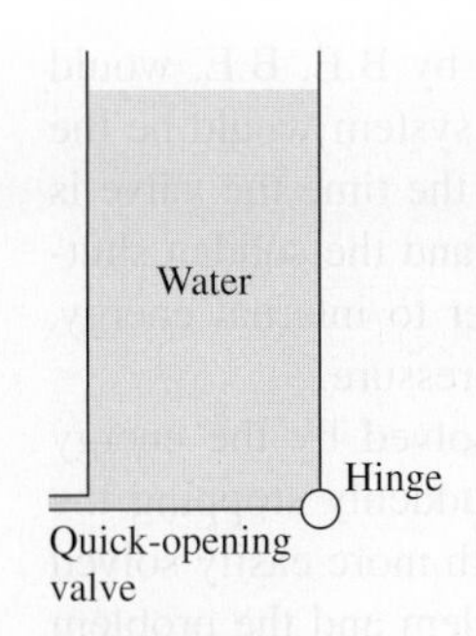

FIGURE 5.19
Opening the valve produces a flow not described by Bernoulli's equation.

Therefore,

$$\Delta t = t_2 - t_1 = \frac{-2(h_2^{1/2} - h_1^{1/2})}{(A_2 / A_1)(2g)^{1/2}} \quad \begin{bmatrix}\text{frictionless} \\ \text{tank draining}\end{bmatrix} \tag{5.21}$$

Inserting numbers, we find

$$\Delta t = \frac{-2[(1\text{ m})^{1/2} - (30\text{ m})^{1/2}]}{[(\pi/4)\cdot(1\text{ m})^2/(\pi/4)\cdot(10\text{ m})^2]\cdot(2\cdot 9.81\text{ m/s}^2)^{1/2}}$$

$$= 2.02 \cdot 10^2\text{ s} = 3.37\text{ min} \tag{5.BE}$$

The maximum velocity in this tank is at the outlet, and all other velocities are proportional to it; therefore, the maximum value of $(\partial V/\partial t)_{x,y,z}$ must occur at the outlet. Differentiating Torricelli's equation with respect to time, we find

$$\frac{\partial V_2}{\partial t} = \frac{(2g)^{1/2}}{2h^{1/2}}\frac{dh}{dt} \tag{5.BF}$$

Substituting for dh/dt, we find

$$\frac{\partial V_2}{\partial t} = \left(\frac{g}{2h}\right)^{1/2} \cdot V_2\frac{A_2}{A_1} = \left(\frac{g}{2h}\right)^{1/2}(2gh)^{1/2}\frac{A_2}{A_1} = g\frac{A_2}{A_1} \tag{5.BG}$$

Thus, in this example the maximum value of $(\partial V/\partial t)_{x,y,z}$ is $\frac{1}{100}$ the acceleration of gravity, and the unsteady-flow aspect of the problem can safely be neglected. ■

Most of the flow problems in which the unsteady flow cannot be neglected and hence in which B.E. cannot be applied involve starting the flow from rest or sudden stopping of the flow. Consider the pipe and valve shown in Fig. 5.19. Initially the pipe is practically full of fluid and the valve is closed. Then the valve is suddenly opened. If the friction effect is negligible, then the fluid will fall freely, maintaining its cylindrical shape, just as a solid rod would. In this case the entire outflow process takes place during the flow-starting period; the whole fluid is still accelerating when the last particle of fluid leaves the pipe. Friction and surface tension complicate the picture, but for low-viscosity fluids in short, large-diameter pipes the result described above is experimentally observed.

In this case all the fluid has the same velocity, and thus $(\partial V/\partial t)_{x,y,z}$ is the same at all points where there is fluid. Here it is equal to g, so it is of the same size as the largest accelerations in B.E., and the test indicates that we cannot safely apply B.E. to this problem.

The other general type of unsteady-flow problem that cannot be solved by B.E. is the problem with sudden valve closing, which leads to a phenomenon called *water hammer.*

Figure 5.20 shows a tank from which a liquid flows through a pipe, at the end of which is a quick-closing valve. If the liquid is flowing steadily and the valve is suddenly

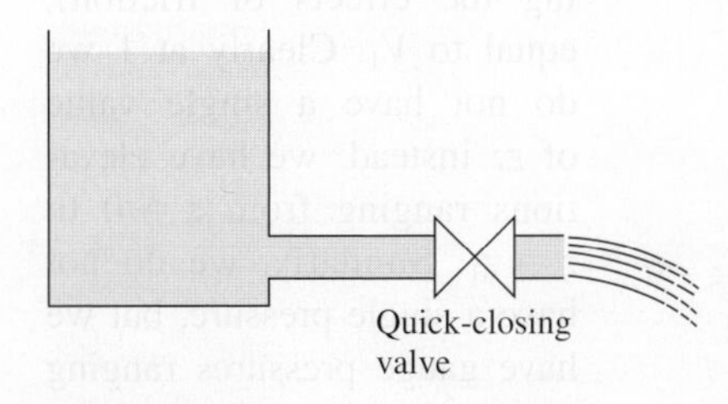

FIGURE 5.20
Closing the valve quickly produces a flow not described by Bernoulli's equation.

closed, the flow during the closing process cannot be described by B.E. B.E. would indicate that, once the valve closed, the pressure throughout the system would be the pressure given by the basic equation of fluid statics. Actually, at the time the valve is being closed, the fluid in the pipe has significant kinetic energy, and the sudden shutting of the valve requires that kinetic energy be converted either to internal energy, with a rise in temperature, or to injection work, with a rise in pressure.

This chapter has concentrated on problems most easily solved by the energy balance (of which B.E. is a restricted form). The problem of suddenly stopping the fluid, in Fig. 5.20, and the problem of starting it from rest are both more easily solved by the momentum balance (Chap. 7). We will return to this problem and the problem of what happens when the valve in Fig. 5.20 is suddenly opened in Chap. 7. For now we simply note that although B.E. is immensely useful, there are some problems for which it is not useful; the starting and stopping of the flow in Fig. 5.20 is one of those problems.

5.11 NONUNIFORM FLOWS

So far in this chapter, and in the vast majority of problems in pipes, channels, ducts, etc., we assume that the velocity is practically uniform across the pipe, duct, or channel, so that we may associate one velocity with the entire flow at any one downstream location perpendicular to the flow. In most flows of practical interest to chemical engineers this simplification introduces negligible errors (see Table 3.1). However there are some very simple and common flows for which this is not the case. The simplest and most illustrative example of this type is the flow over a sharp-edged weir.

Figure 5.21 shows schematically the flow in an open channel that passes over a sharp-edged weir. One may study a very similar flow in a kitchen sink by pouring water out of a rectangular baking dish, at a high enough velocity that the flow does not dribble down the side of the dish but rather flows freely away from the edge, as shown in Fig. 5.21. The flow over the weir in Fig. 5.21 is simpler than the flow out of baking dish, because the weir is assumed to extend a long way into and out of the page, so that the complications where it meets the walls of the channel, equivalent to the effects of the corners of the dish, can be ignored.

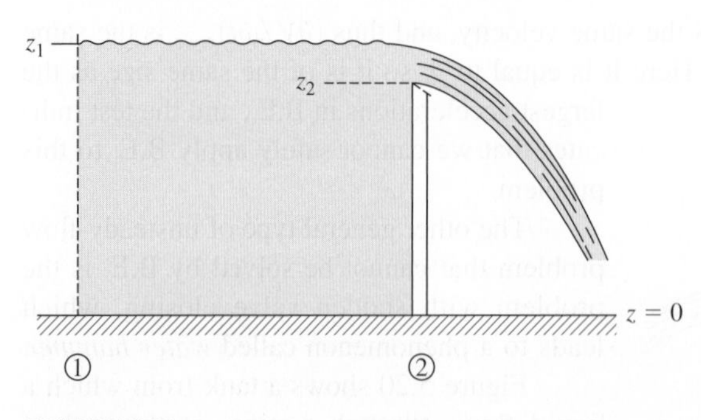

FIGURE 5.21
Flow over a weir.

Here at 1, far upstream from the weir, the velocity is presumably uniform (ignoring the effects of friction), equal to V_1. Clearly at 1 we do not have a single value of z; instead, we have elevations ranging from $z = 0$ to $z = z_1$. Similarly, we do not have a single pressure, but we have gauge pressures ranging from zero at the free surface to $P = \rho g z_1$ at the bottom.

Fortunately, this causes little difficulty because the sum $P/\rho + gz + V^2/2$ is constant at point 1, independent of z, because as z declines the pressure rises, according to the basic equation of fluid statics, to keep the sum of the $P/\rho + gz$ terms constant.

The same is not true at 2. There the flow is open to the air at both sides, so that the gauge pressure at 2 must be zero, at all elevations above the weir. We can thus write B.E. between an arbitrary upstream point at 1 and some elevation z (above z_2) at point 2. Because the sum $P/\rho + gz$ is independent of z at 1, we choose $z = z_1$, for which $P_1 = 0$, and write

$$gz_1 + \frac{V_1^2}{2} = gz + \frac{V_2^2}{2} \qquad V_2 = \left[2g(z_1 - z) + \frac{V_1^2}{2}\right]^{1/2} \tag{5.BH}$$

This says that at the free surface the velocity should be the same as the velocity at 1, which in most cases is negligibly small, and that at the bottom of the overflowing stream, the velocity should be a maximum, with the value given by the above equation, with $z = z_2$. One can try this by pouring water from a baking dish in the sink and seeing that this is not exactly the case. The internal friction in the flow does not allow the fluid at the surface to go that slowly, adjacent to the much faster-flowing fluid just below it. Instead the faster-flowing fluid drags the surface fluid along, faster than Eq. 5.BH predicts. One may also see that as the surface fluid speeds up, its elevation falls, so that instead of being completely level up to the weir, as Fig. 5.21 shows, the free surface actually falls slightly just before the weir, to satisfy the energy balance.

Ignoring this disagreement between Eq. 5.BH and what we can see in the sink, we can compute the expected flow rate by considering a length W into the page in Fig. 5.21. To simplify the integration, we now measure elevations downward from the surface, defining $h = z_1 - z$, and write

$$Q = \int V\,dA = W\int_0^{h_1}\left(2gh + \frac{V_1^2}{2}\right)^{1/2} dh \tag{5.BI}$$

Normally V_1 is small enough that we can drop it from the right side of the above equation, and integrate to find

$$Q = W\sqrt{2g}\,\frac{h_1^{3/2}}{3/2} \tag{5.22}$$

Experimental results show that the $3/2$ power dependence of Q on h_1 is correct but that the flow rate is less than predicted by Eq. 5.22, typically about 67 percent of what that equation predicts [6].

The same differences in velocity from top to bottom of the flow that we calculate here are certainly present in all the horizontal-flow examples in this chapter. However, in a flow like that shown in Fig. 5.7, the difference in elevation from the top to the bottom of the exit flow is so small compared to the elevation change from the free surface in the tank to the centerline of the exit that we make a negligible error in ignoring the minor differences in velocity from top to bottom of the exit flow. The same is true of most of the flows of practical interest to chemical engineers. But for shallow gravity-driven flows, for example, the flow over weirs in distillation columns, clarifiers, etc., one must take them into account.

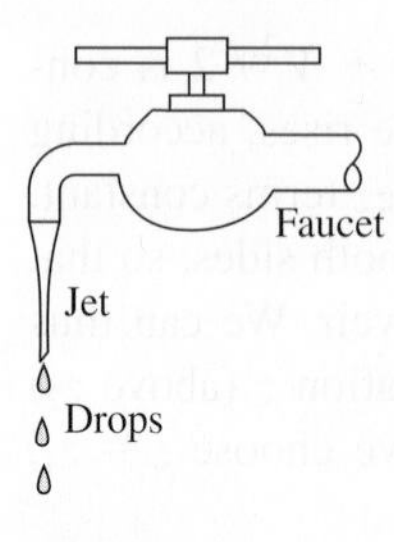

FIGURE 5.22
Slow flow from a partly opened faucet, producing a jet that contracts as it falls and finally breaks up into drops.

Figure 5.21 shows the jet that flows over the weir becoming thinner as it flows. This is also explained by B.E., as shown in the next example.

Example 5.15. A faucet in an ordinary sink with a partially opened valve is shown in Fig. 5.22. The flow tapers from the faucet, forming a thinner and thinner stream as it flows downward. Eventually it breaks up into drops, due to surface tension, discussed in Chap. 14. If the diameter of the falling column of fluid is 2 cm and its velocity 0.5 m / s where it leaves the faucet, what is the expected diameter at 0.5 m below the faucet?

Taking 1 to be the exit of the faucet and 2 to be 0.5 m lower, we find from B.E. that

$$V_2^2 - V_1^2 = 2gh \qquad V_2 = (V_1^2 + 2gh)^{1/2} \tag{5.BJ}$$

$$V_2 = \left[\left(0.5\,\frac{\text{m}}{\text{s}}\right)^2 + 2 \cdot 9.81\,\frac{\text{m}}{\text{s}^2} \cdot 0.5\text{ m}\right]^{1/2}$$

$$= \left(10.06\,\frac{\text{m}^2}{\text{s}^2}\right)^{1/2} = 3.17\,\frac{\text{m}}{\text{s}} = 10.4\,\frac{\text{ft}}{\text{s}} \tag{5.BK}$$

By material balance for a constant-density fluid,

$$A_2 = A_1 \frac{V_1}{V_2} = \frac{0.5\text{ m / s}}{3.17\text{ m / s}} = 0.158 \tag{5.BL}$$

and correspondingly

$$D_2 = D_1\left(\frac{A_2}{A_1}\right)^{1/2} = 2\text{ cm} \cdot 0.158^{1/2} = 0.79\text{ cm} = 0.31\text{ in} \tag{5.BM}$$

As the velocity of the fluid increases according to B.E., the column shrinks laterally according to the material balance. ■

For inherently two- or three-dimensional flows, like the flow around an airplane, the simple application of B.E. from one point to another in the flow, which we have used here, is only applicable if the two points chosen are on a single streamline, as discussed in Chap. 16.

5.12 SUMMARY

1. B.E. is the energy-balance equation for steady flow of constant-density fluids.
2. For constant-density fluids the term $(\Delta u - dQ / dm)$ in B.E. represents the friction heating per unit mass, $\mathscr{F}$.
3. Although no fluid has exactly constant density, B.E. can be applied with negligible error to almost all steady flows of liquids and to steady flow of gases at low velocities, because in those flows the effect of changes in density is negligible.

4. A large number of fluid-measuring devices are based on the frictionless form of B.E. Friction is present in these devices. The commonly used equations for these devices retain the frictionless B.E. form and add empirical correction factors to deal with the effects of friction.
5. B.E. can predict negative absolute pressures for some impossible flows. For gas flows, this prediction normally means that the velocities are too high for B.E. to apply. For liquids, it normally means that the fluid will boil, leading to a two-phase flow and a much lower velocity than predicted.
6. Although B.E. is a steady-flow equation, it can be used for unsteady flows if the time rate of change of velocity at every point in the system is small compared with the accelerating forces (e.g., the acceleration of gravity). It is not useful for problems involving sudden starting and stopping of flows; those are best solved with the momentum balance.
7. Normally we ignore the differences in velocity perpendicular to the flow in applying B.E. to the flow in pipes and channels. This causes negligible errors, except in shallow, gravity-driven flows, like the flows over weirs.

PROBLEMS

See the Common Units and Values for Problems and Examples inside the back cover. An asterisk (*) on the problem number indicates that the answer is in App. D.

5.1.*If a body falls 1000 ft in free fall and then is stopped by friction in such a way that all its kinetic energy is converted into internal energy, how much will the temperature of the body increase if

(*a*) It is steel, $C_V = du / dT = 0.12$ Btu / lbm · °F.

(*b*) If it is water, $C_V = du / dT = 1.0$ Btu / lbm · °F? Here C_V is the heat capacity at constant volume.

5.2. Show that in the head form of B.E. each term has the dimension of a length.

5.3.*Water is flowing in a pipe at a velocity of 8 m / s. Calculate the pressure increase and the increase in internal energy per unit mass for each of the following ways of bringing it to rest:

(*a*) A completely frictionless diffuser with infinitely large A_2.

(*b*) A diffuser that has 90 percent of the pressure recovery of a frictionless diffuser, with infinitely large A_2.

(*c*) A sudden expansion.

5.4. A fluid is flowing in a frictionless diffuser in which $A_2 / A_1 = 3$ and $V_1 = 10$ ft / s. Calculate the pressure recovery $(P_2 - P_1)$

(*a*) For the fluid being water.

(*b*) For the fluid being air.

5.5. Rework Example 5.2, calculating the density from the formula $\rho = MP_{avg} / RT$ where P_{avg} is $0.5(P_1 + P_{atm})$. Compare the results with those shown in Table 5.1.

5.6. Torricelli's equation can be reduced to a simple plot of V as a function of h. Prepare such a plot for heights up to 1000 ft.

5.7.*The tank shown in Fig. 5.7 is modified to have an outflow area of 2 ft^2. The diameter of the tank is so large that it may be considered infinite. The height h is now 12 ft. How many cubic feet per second are flowing out? Assume frictionless flow.

5.8. Repeat Example 5.3 when the fluid is gasoline.

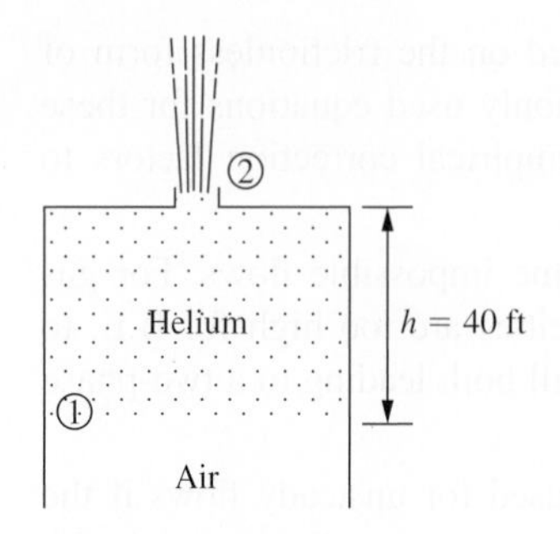

FIGURE 5.23
Buoyancy-driven gas flow.

5.9.*Repeat Prob. 5.7, except that now the horizontal cross-sectional area of the tank is 5 ft^2.

5.10. Hoover Dam has a height of 726 ft. Assume the water is up to the top of the dam on the upstream side. If one were to drill a hole through its base and let the water squirt out, and if friction were negligible, what velocity of water jet would we expect?

5.11. An ocean liner strikes an iceberg, which tears a 5 m^2 hole in its side. The center of the hole is 10 meters below the ocean surface. Estimate the volumetric flow rate of water into the ship.

5.12. Rework Example 5.5, making the fluid in the tank air be at the same temperature and pressure as the air of the atmosphere. Is the answer from Eq. 5.12 plausible? Would the answer from Eq. 5.14 be plausible?

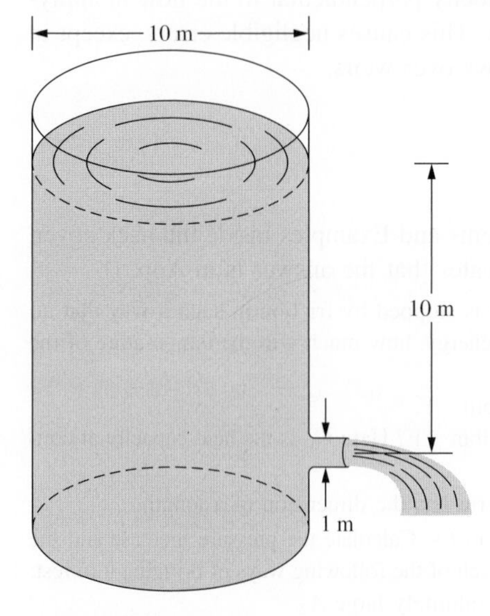

FIGURE 5.24
Tank-draining flow.

5.13.*The tank in Fig. 5.23 is full of helium, at the same temperature and pressure as the surrounding atmosphere. Assuming steady, frictionless flow, what is the velocity of helium through the hole?

5.14. The tank in Figure 5.24 is cylindrical with a diameter of 10 m. The outlet is a cylindrical frictionless nozzle, with diameter 1 m. The top of the tank is open to the atmosphere. When the level in the tank is 10 m above the centerline of the outlet, how fast is the level in the tank falling?

5.15. In Fig. 5.25 a tank of water is immersed in a larger tank of gasoline, and the water is flowing out through a hole in the bottom. What is the velocity of this flow?

5.16.*In the vessel in Fig. 5.26 water is flowing steadily in frictionless flow under the barrier. What is the velocity of the water flow under the barrier?

5.17. In the tank and standpipe in Fig. 5.27, which way is the fluid flowing? *Hint:* Write B.E., taking the two free surfaces as points 1 and 2. Compute the magnitude and sign of $\mathcal{F}$ for flow in each direction.

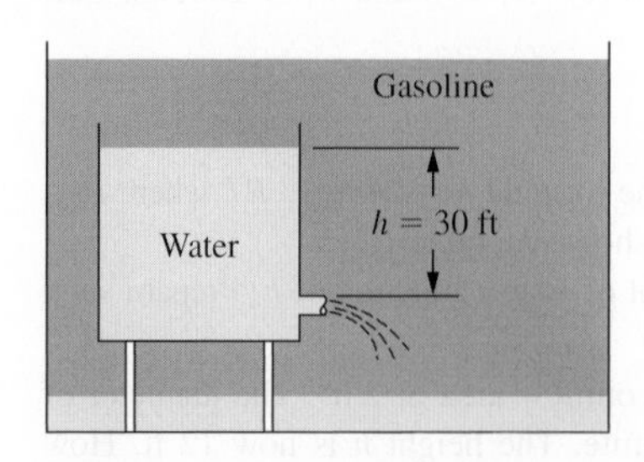

FIGURE 5.25
Two-fluid, gravity-driven flow.

5.18.*In the tank in Fig. 5.28, water is under a layer of compressed air that is at a pressure of 20 psig. The water is flowing out through a frictionless nozzle that is 5 ft below the water surface. What is the velocity of the water?

5.19. In the preceding problem, if the liquid level remains constant, and we slowly lower the air pressure, at some air pressure the velocity will be 10 ft / s. What pressure will that be?

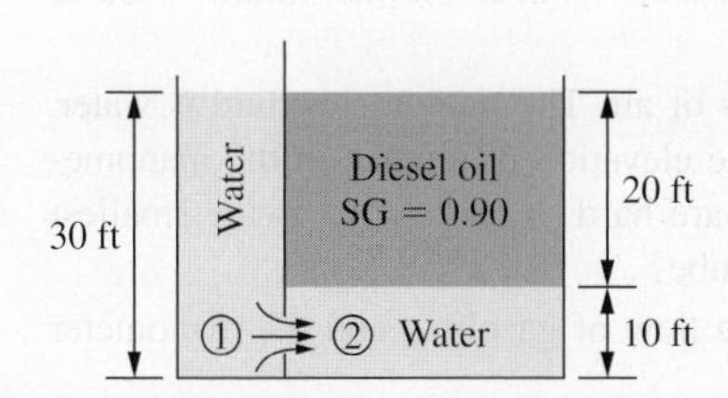

FIGURE 5.26
Two-fluid, gravity-driven flow.

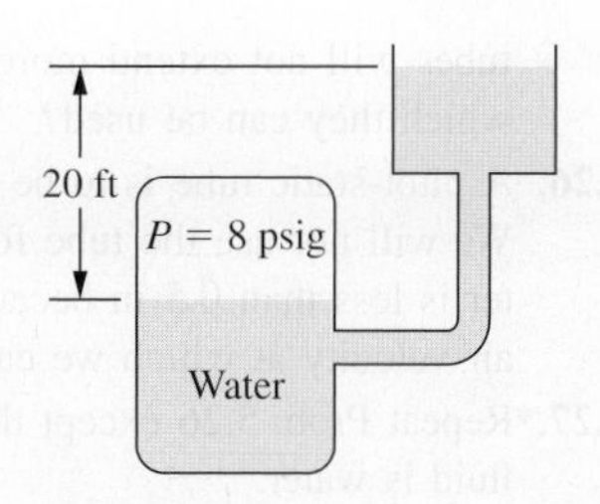

FIGURE 5.27
Which way does this system flow?

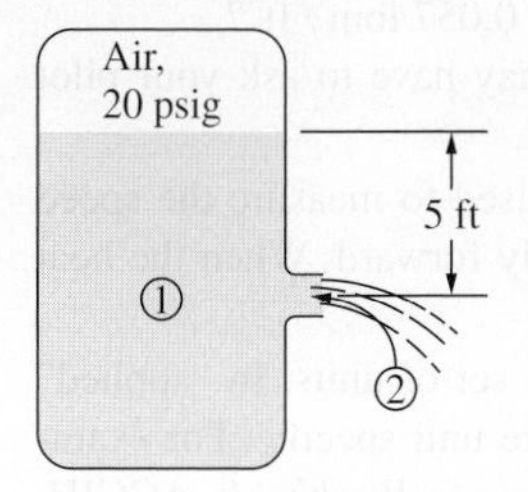

FIGURE 5.28
Flow driven by gravity and pressure.

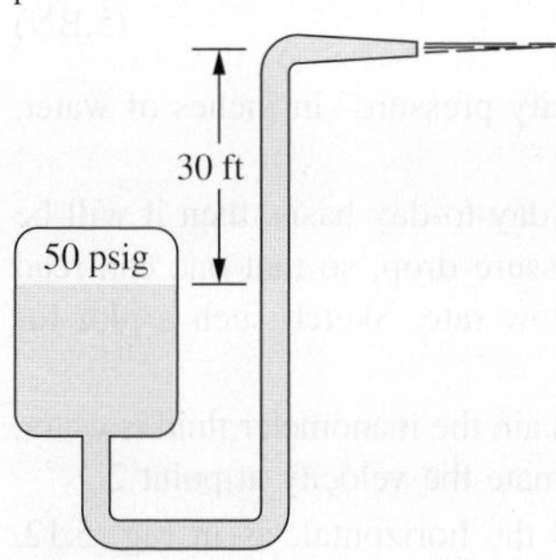

FIGURE 5.29
Flow driven by pressure against gravity.

5.20. The system in Fig. 5.29 consists of a water reservoir with a layer of compressed air above the water and a large pipe and nozzle. The pressure of the air is 50 psig, and the effects of friction can be neglected. What is the velocity of the water flowing out through the nozzle?

5.21.*The tank in Fig. 5.30 has a layer of mercury under a layer of water. The mercury is flowing out through a frictionless nozzle. What is the velocity of the fluid leaving the nozzle?

5.22. The compressed-air-driven water rocket shown in Fig. 5.31 is ejecting water vertically downward through a frictionless nozzle. When the pressure and elevation are as shown, what is the velocity of the fluid leaving the nozzle?

5.23.*An industrial centrifuge is sketched in Fig. 5.32. The fluid in the basket is water. The radii are $r_1 = 21$ in and $r_2 = 20$ in. The basket is revolving at 2000 rpm. There is a small hole in the outer wall of the centrifuge, through which the fluid is flowing in frictionless flow. What is the velocity of flow through this hole?

5.24. Flow-recorder charts frequently have a square scale rather than a linear one. Why?

5.25.*A pitot tube is being designed for use as a speedometer on power boats. For ease of construction the

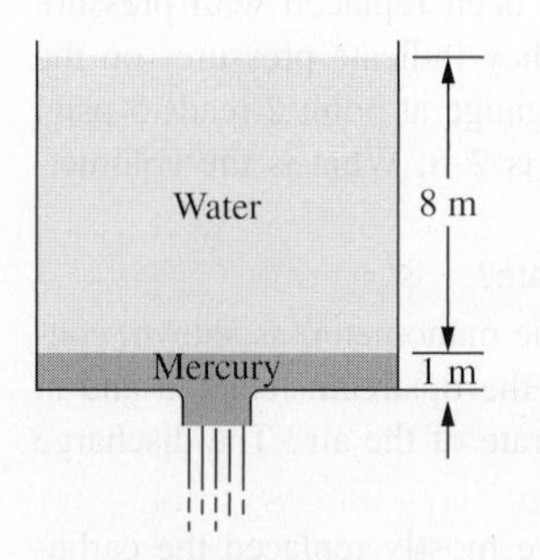

FIGURE 5.30
Two-fluid, unsteady tank draining.

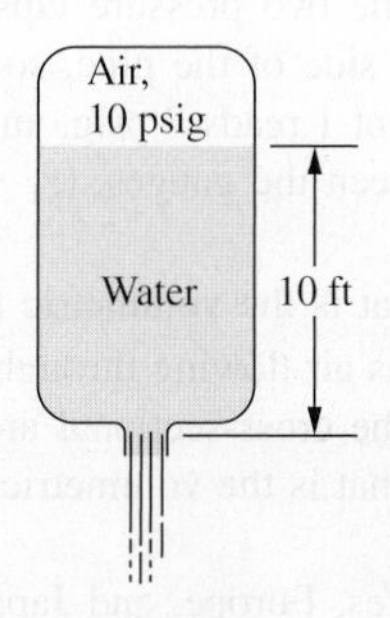

FIGURE 5.31
Compressed air-water rocket.

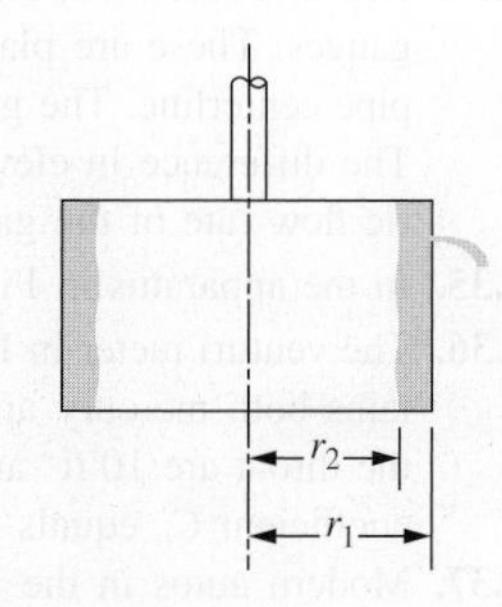

FIGURE 5.32
Centrifuge basket with leak.

tubes will not extend more than 10 ft above the water. What is the maximum speed at which they can be used?

5.26. A pitot-static tube is to be used to measure a flow of air. The manometer fluid is water. We will not use the tube for flows so slow that the elevation difference in the manometer is less than 0.5 in because smaller differences are hard to read. What is the smallest air velocity at which we can use this pitot-static tube?

5.27.*Repeat Prob. 5.26 except that now we measure the flow of gasoline, and the manometer fluid is water.

5.28. A pitot-static tube is used to measure an airplane's air speed. When the pressure-difference gauge reads 0.3 psig, how fast is the plane going?

(*a*) At sea level where the air density is about 0.075 lbm / ft^3?

(*b*) At an altitude of 10,000 ft, where the air density is about 0.057 lbm / ft^3?

(*c*) Does this difference cause problems for the pilot? (You may have to ask your pilot or aeronautical engineering friends for help on part *c*.)

5.29. A pitot tube, connected to a bourdon-tube pressure gauge, is used to measure the speed of a boat. The tube is just below the waterline and faces directly forward. When the boat is going 60 km / h, what is the reading of the pressure gauge?

5.30. In this book and most textbooks, equations are correct for any set of units. In "applied" or "practical" publications, one regularly sees equations that are unit specific. For example, Jack Caravanos (*Quantitative Industrial Hygiene; A Formula Workbook,* ACGIH, Cincinnati, 1991, page 66) gives the following equation for the air-flow velocity in a duct, based on measurements with a pitot-static tube:

$$V = 4005\sqrt{VP} \tag{5.BN}$$

where V is the velocity in ft / min, and VP is the "velocity pressure" in inches of water. Is this consistent with Eq. 5.16?

5.31. If the venturi meter in Example 5.10 is to be used on a day-to-day basis, then it will be useful to have a plot of volumetric flow rate versus pressure drop, so that one can read the pressure drop and simply look up the volumetric flow rate. Sketch such a plot for flow rates of 1 to 10 ft^3 / s.

5.32. In the venturi meter shown in Fig. 5.10, the flowing fluid is air, the manometer fluid is water, $(D_2 / D_1) = 0.5$, and the manometer reading is 1 ft. Estimate the velocity at point 2.

5.33. The venturi meter in Example 5.8 is now set at 30° to the horizontal, as in Fig. 5.12. The flowing fluid is gasoline. The fluid in the bottom of the manometer is colored water. The reading of the manometer is $z_3 - z_4 = 1$ ft. What is the volumetric flow rate of the gasoline?

5.34.*Repeat Prob. 5.33, except that the two pressure taps have been replaced with pressure gauges. These are placed on the side of the pipe, so that they indicate pressures on the pipe centerline. The gauge at point 1 reads 7 psig, and the gauge at point 2 reads 5 psig. The difference in elevation between the gauges, $(z_1 - z_2)$, is 2 ft. What is the volumetric flow rate of the gasoline?

5.35. In the apparatus in Fig. 5.33 what is the volumetric flow rate?

5.36.*The venturi meter in Fig. 5.34 has air flowing through it. The manometer, as shown, contains both mercury and water. The cross-sectional areas at the upstream location and at the throat are 10 ft^2 and 1 ft^2. What is the volumetric flow rate of the air? The discharge coefficient C_v equals 1.0.

5.37. Modern autos in the United States, Europe, and Japan have mostly replaced the carburetors of older autos with fuel injectors. But autos produced for developing countries still have carburetors, as do power tools such as lawnmowers. The carburetors in automobiles and power tools are much more complicated versions of the carburetor shown in

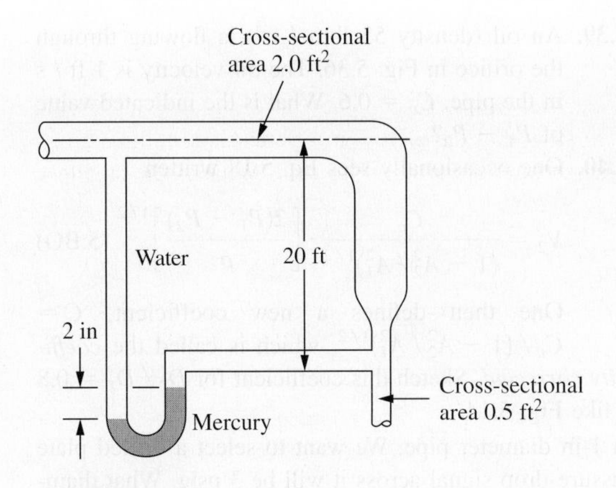

FIGURE 5.33
Device for Prob. 5.35.

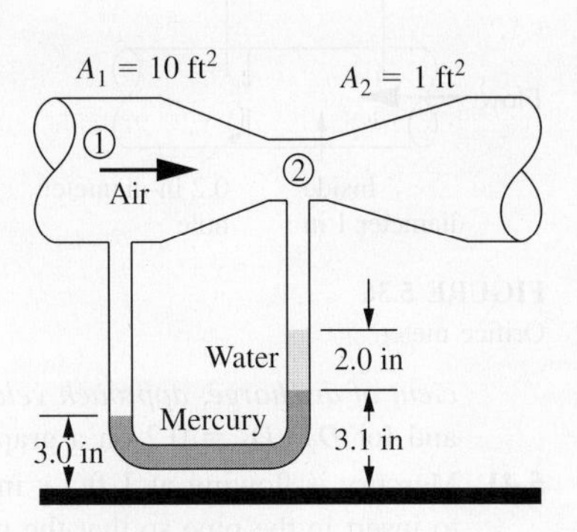

FIGURE 5.34
Venturi meter with two-fluid manometer.

Fig. 5.35, but they operate the same way as that simple one. The cross-sectional areas at points 1 and 3 are large enough that the velocities there can be considered negligible compared to the velocity at point 2, and the pressures at points 1 and 3 are both approximately equal to atmospheric pressure. The gasoline enters from a constant-liquid-level, atmospheric-pressure reservoir through a large-diameter tube and a small jet, which may be considered a frictionless nozzle with diameter D_j. The diameter at the throat of the venturi, point 2, is D_2.

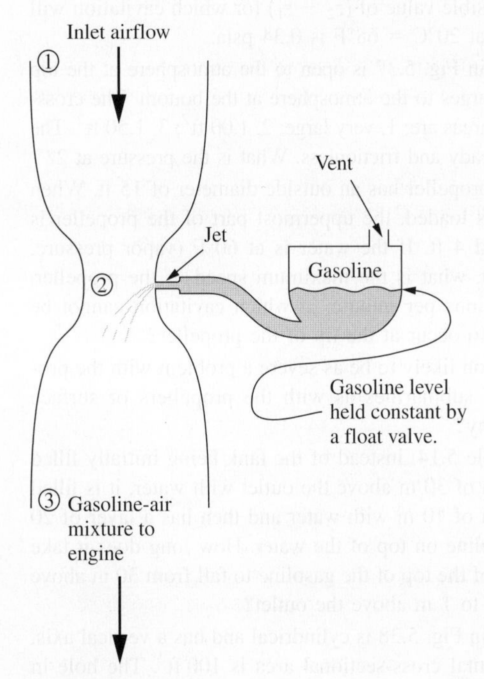

FIGURE 5.35
Elementary carburetor. The liquid level in the reservoir is held constant by a float valve.

(*a*) Write the equation for the air-fuel ratio, which is the (mass flow rate of air) / (mass flow rate of fuel), in terms of the diameters of the throat, the jet, etc.

(*b*) How does this air-fuel ratio change with changes in air flow rate to the engine? (The air flow rate to the engine is governed by the setting of the throttle plate, which is connected to the driver's accelerator pedal, and located between the part of the carburetor shown here and the engine.)

(*c*) If we want an air-fuel ratio of 15 lbm / lbm (typical of gasoline engines), what ratio of D_1 / D_2 should we choose?

(*d*) If the carburetor shown here gives an air-fuel ratio of 15 at sea level, will it give the same, a higher, or a lower fuel-air ratio in Denver, elevation 5280 ft above sea level?

5.38. In the United States natural gas is normally piped inside buildings at a pressure of 4 in of water, whereas propane is piped inside buildings at a pressure of 11 in of water. Why?

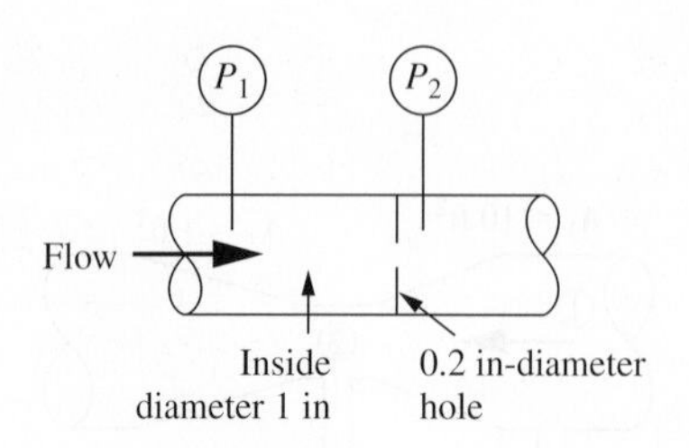

FIGURE 5.36
Orifice meter.

5.39. An oil (density 55 lbm / ft^3) is flowing through the orifice in Fig. 5.36. The oil velocity is 1 ft / s in the pipe. $C_v = 0.6$. What is the indicated value of $P_1 - P_2$?

5.40. One occasionally sees Eq. 5.18 written

$$V_2 = \frac{C_v}{(1 - A_2^2 / A_1^2)^{1/2}} \cdot \left[\frac{2(P_1 - P_2)}{\rho}\right]^{1/2} \tag{5.BO}$$

One then defines a new coefficient, $C = C_v / (1 - A_2^2 / A_1^2)^{1/2}$, which is called the *coefficient of discharge, approach velocity corrected.* Sketch this coefficient for $D_2 / D_1 = 0.8$ and for $D_2 / D_1 = 0.2$ on a graph like Fig. 5.14.

5.41. Mercury is flowing at 1 ft / s in a 1-in diameter pipe. We want to select a drilled plate to insert in the pipe so that the pressure-drop signal across it will be 3 psig. What diameter should we select for the orifice hole?

5.42.*A venturi meter, Fig. 5.10, has $A_2 / A_1 = 0.5$. The fluid flowing is water. The pressure at point 1 is 20 psia.

(*a*) What is the velocity at point 2 that corresponds to a pressure at point 2 of 0.0 psia?

(*b*) If the water is at 200°F, its vapor pressure is 11.5 psia. What is the highest velocity possible at point 2 at which water at 200°F will not boil?

5.43. For a siphon similar to that sketched in Fig. 5.16, we want the fluid to have a velocity of 10 ft / s in the siphon pipe. If we assume that flow is frictionless and that the minimum pressure allowable is 1 psia, what is the maximum height that the top of the siphon may have above the liquid surface level?

5.44. In Example 5.12, what is the highest possible value of $(z_2 - z_1)$ for which cavitation will not occur? The vapor pressure of water at 20°C = 68°F is 0.34 psia.

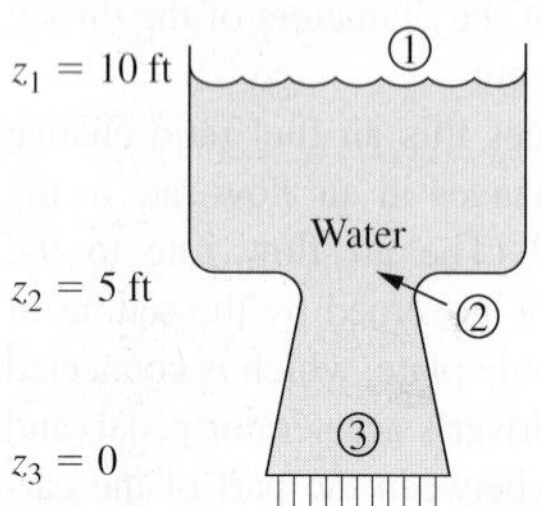

FIGURE 5.37
Vertical, gravity-driven venturi.

5.45. The tank in Fig. 5.37 is open to the atmosphere at the top and discharges to the atmosphere at the bottom. The cross-sectional areas are; 1, very large; 2, 1.00 ft^2; 3, 1.50 ft^2. The flow is steady and frictionless. What is the pressure at 2?

5.46.*A ship's propeller has an outside diameter of 15 ft. When the ship is loaded, the uppermost part of the propeller is submerged 4 ft. If the water is at 60°F (vapor pressure, 0.26 psia), what is the maximum speed of the propeller, in revolutions per minute, at which cavitation cannot be expected to occur at the tip of the propeller?

5.47. Is cavitation likely to be as severe a problem with the propellers of submarines as with the propellers of surface ships? Why?

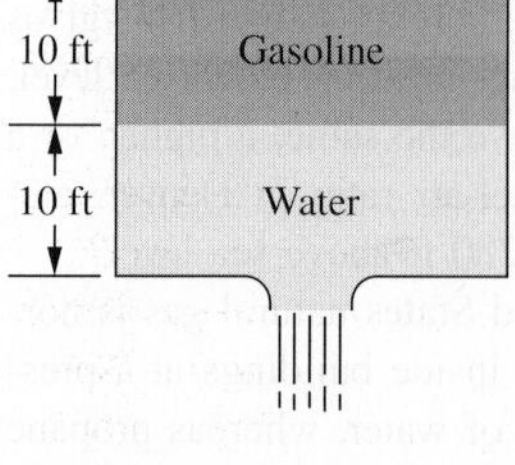

5.48. In Example 5.14, instead of the tank being initially filled to a depth of 30 m above the outlet with water, it is filled to a depth of 10 m with water and then has a layer of 20 m of gasoline on top of the water. How long does it take the level of the top of the gasoline to fall from 30 m above the outlet to 1 m above the outlet?

10 ft Gasoline
10 ft Water

FIGURE 5.38
Two-fluid tank draining.

5.49.*The tank in Fig. 5.38 is cylindrical and has a vertical axis. Its horizontal cross-sectional area is 100 ft^2. The hole in the bottom has a cross-sectional area of 1 ft^2. The interface between the gasoline and the water remains perfectly horizontal at all times. That interface is now 10 ft above

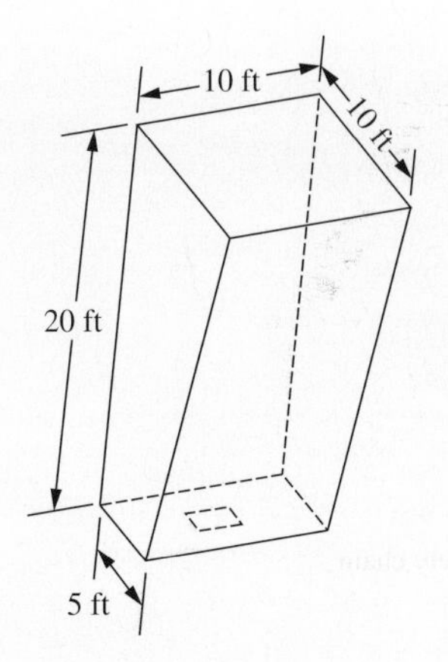

FIGURE 5.39
Tank draining with non-constant cross section.

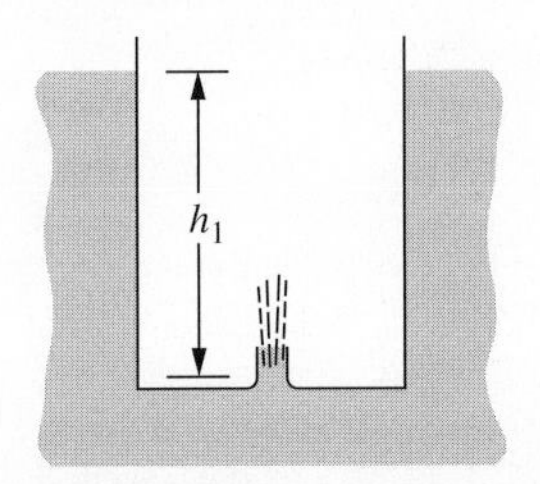

FIGURE 5.40
Gravity inflow.

the bottom. How soon will gasoline start to flow out the bottom? Assume frictionless flow.

5.50. (*a*) In Figure 5.28, if the tank diameter is 10 ft, the outlet diameter is 1 ft and a pressure regulator supplies compressed air as needed to keep the pressure in the gas space constant at 20 psig, how long will it take for the water level to fall from 5 ft to 1 ft?

(*b*) Repeat part (*a*) except that the compressed air system is turned off so that the pressure above the water falls as the liquid flows out and the gas expands. Assume that the gas is an ideal gas and that its temperature remains constant at 20°C = 68°F. Initially the gas space is 1 ft high, and at P = 20 psig.

5.51. The open-topped tank shown in Fig. 5.39 is full to the top with water. The bottom opening is uncovered, so that the water runs out into the air. The cross-sectional area of the bottom opening is 1 ft^2. How long does it take the tank to empty? Assume frictionless flow.

5.52. An open-ended tin can, Fig. 5.40, has a hole punched in its bottom. The can is empty and is suddenly immersed in water to the depth h_1 shown and then held steady. The area of the hole is 0.5 in^2, and the horizontal cross-sectional area of the can is 20 in^2. If we assume that the flow through the hole in the bottom of the can is frictionless, how long does it take the can to fill up to the level of the surrounding water?

5.53. A fluid mechanics demonstration device has the same flow diagram as Fig. 5.6. The tank is rectangular, 6 in by 5.5 in. The outlet opening is circular, D = 0.30 in. The tank is filled with water and allowed to drain. How long will it take the level to fall from 11 in above the centerline of the opening to 1 in above it?

5.54. A 1-gal paint can, diameter 6.5 in and height 7.5 in, is filled with methane. A 0.25 in. diameter hole in the top is covered with masking tape, as is a 0.5 in. diameter hole at the bottom; see Fig. 5.41. At time zero, the two masking tapes are removed, and a stopwatch is started. The methane flows upward out the hole in the lid by gravity and is lighted, producing a yellowish flame. For all the calculations below, assume that the flow resistance through the hole in the bottom is negligible.

(*a*) What is the initial velocity of the methane through the hole in the top?

(*b*) As the methane in the can is replaced by the inflowing air, that velocity falls and the flame becomes smaller and bluer. What is the relation between that velocity and elapsed time? Here use the two classic mixing models of chemical engineering, totally unmixed flow, in which the air and methane form separate layers, one above the other, and totally mixed flow, in which the concentration of methane in the mixture inside the container is always uniform throughout the container.

(*c*) When the flow rate through the opening at the top becomes less than the laminar flame speed for methane-air mixtures, 1.1 ft / s, the flame burns back into the container, where the velocity is less than 1.1 ft / s and spreads rapidly, producing a bang and a flash, and propelling the container's lid into the air. The internal safety chain prevents the lid from hurting anyone, and pulls the can up off its bottom. How long does it take for this to occur, according to the totally unmixed and totally mixed models in part (*b*)? The observed time is about 325 s. This demonstration is described in detail in [7].

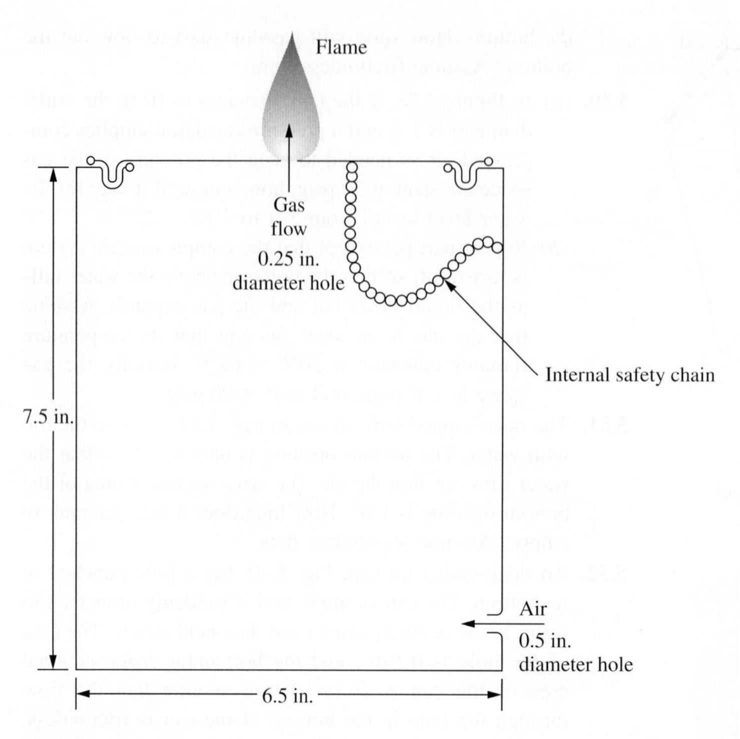

FIGURE 5.41
Simple time bomb; see Prob. 5.54.

5.55. Repeat Example 5.14 with the water in the tank being replaced with propane. Assume zero mixing between the propane and the air above it.

5.56. Figure 5.42 shows a toy fluid-mechanics demonstrator, which consists of a wooden (or plastic) spool, a piece of cardboard, and a thumbtack. When one blows hard enough downward into the spool, the cardboard is held firmly against the spool; when one stops blowing, the cardboard falls away by gravity. Sketch a pressure-radius plot for pressure along line A–A in the sketch while air is flowing. Use the axes shown in the lower part of the figure.

The function of the thumbtack is to prevent the cardboard from moving sideways; otherwise it plays no role in the device. Assume that the cardboard is stiff enough that the distance between the cardboard and the spool is constant, independent of radius. The device works with a piece of ordinary flexible paper, but the mathematics are more complex because the distance between paper and spool is not constant. A piece of adhesive tape holding the thumbtack in place helps.

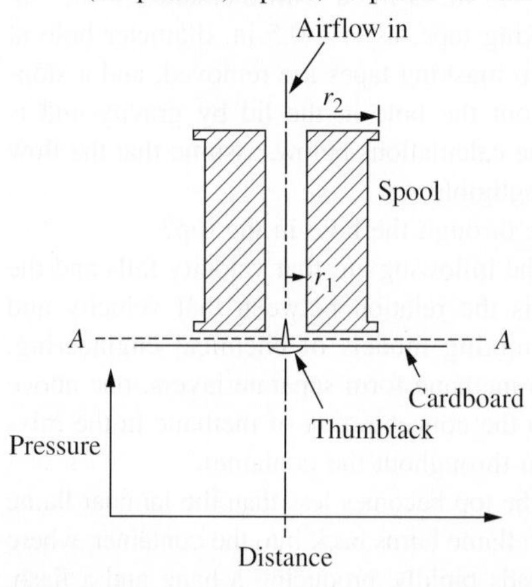

FIGURE 5.42
Spool and cardboard fluid mechanics demonstrator.

5.57. In the spool-and-cardboard demonstrator in the preceding problem, the hole in the spool has a diameter of 7.1 mm, the outside diameter of the spool is 35 mm, and the space between the spool and the cardboard disk is estimated to be 0.2 mm. One lung full of air is about 1 L and is blown out in about 2 s. Based on these values, estimate the lowest pressure likely to occur in the space between the spool and the cardboard.

5.58. For frictionless pumps and compressors pumping constant-density fluids, the required work is given by Eq. 5.11. If the fluid is an ideal gas, then that equation becomes

$$\frac{dW_{\text{n.f.}}}{dm} = \frac{RT}{M}\int \frac{dP}{P} \qquad \text{(5.BP)}$$

For very small pressure changes this is practically

$$\frac{dW_{\text{n.f.}}}{dm} = \frac{RT}{M}\frac{\Delta P}{P_1} \qquad [\Delta P << P_1] \qquad \text{(5.BQ)}$$

Almost all real compressors are intermediate between adiabatic (no heat transfer to the surroundings) and isothermal (complete thermal equilibrium with the surroundings). For those two cases the required work for ideal gases is shown in Chap. 10 to be

$$\frac{dW_{\text{n.f.}}}{dm} = \frac{RT}{M}\ln\frac{P_2}{P_1} \qquad \text{[isothermal, frictionless]} \qquad \text{(5.BR)}$$

and

$$\frac{dW_{\text{n.f.}}}{dm} = \frac{RT_1}{M}\cdot\frac{k}{k-1}\left[\left(\frac{P_2}{P_1}\right)^{(k-1)/k} - 1\right] \qquad \text{[adiabatic, frictionless]} \qquad \text{(5.BS)}$$

Here T_1 is the inlet temperature, k is the ratio of specific heats (to be discussed in Chap. 8), which is practically constant for any gas ($\approx$1.40 for air), and P_1 and P_2 are the inlet and outlet pressures, respectively. To show how these formulae compare, prepare a plot of $[M/RT_1]\cdot(dW_{\text{n.f.}}/dm)$ versus P_2/P_1 for air, showing curves for each of the three equations, for the range $1.0 < P_2/P_1 < 1.3$. Here the calculated work is that done to drive the pump or compressor, which is work done on the system and has a positive sign.

5.59. Figure 5.43 shows an air-cushion car, of the type widely used to slide heavy loads over relatively smooth surfaces. In it, a fan or blower forces air under pressure into the confined space under the car. This air supports the car and its load. Some of the air continually leaks out through the gap between the skirt of the car and the ground; the fan must supply enough air to make up for this leakage. Assuming that the car and its payload have a total mass of 5000 lbm, that the car is circular with a diameter of 10 ft, and that the clearance between the skirt of the car and the floor is 0.01 in, calculate the air flow rate. Then, assuming that the blower is 100 percent efficient and isothermal (Prob. 5.58), calculate the required blower horsepower.

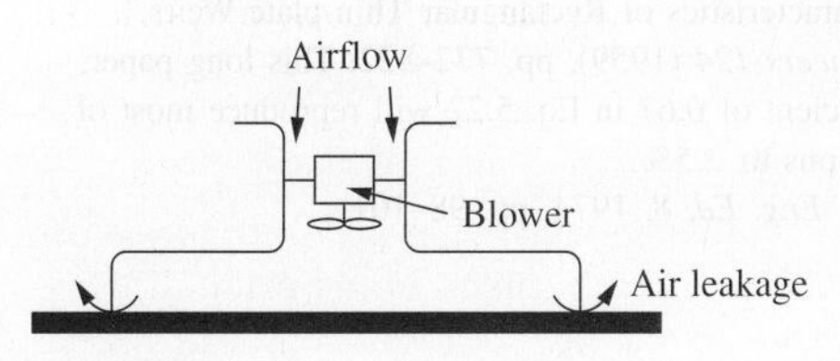

FIGURE 5.43
Air-cushion car.

5.60.*An air-inflated "bubble" structure has a skin that weighs 1.0 lbf/ft^2. Real ones are cylindrical domes, but for this problem consider it to be a flat roof, held up by the pressure inside it. The floor area is 20,000 ft^2. All such structures have some leakage, which must be supplied by a fan that runs constantly. The true leakage area consists of many small pinholes in the

fabric, leaky seams, etc. For this problem assume that the leakage is equivalent to the frictionless flow through 5 ft^2 of opening.
(*a*) Estimate the gauge pressure inside the structure.
(*b*) Estimate the leakage flow rate that must be made up for by the fan.
(*c*) Estimate the power requirement for the fan, which is assumed to be 100 percent efficient.

5.61. Water flows (100 m^3 / s) in a channel 50 m wide and spills over a sharp-edged weir.
(*a*) Estimate the difference in elevation between the upstream flow and the top of the weir.
(*b*) If the upstream channel is 5 m deep, what is the upstream velocity?
(*c*) How large a percentage error are we likely to have made in neglecting this velocity in formulating Eq. 5.22?

5.62. In Example 5.3 we computed the exit velocity by Torricelli's equation, which does not take into account the fact that at the bottom of the jet the velocity will be higher than at the top, as discussed in Sec. 5.11. How large an error are we likely to have made? If the jet is passing thorough a perfectly rounded entrance with an outlet diameter of 0.5 ft, and the centerline of the jet is 30 ft below the fluid surface, how much difference should there be between the velocities at the top and the bottom of the jet?

5.63. A slow-moving stream of water flows from a faucet into a sink. It is observed that the width of the stream decreases with distance from the faucet. If the flow leaves the faucet, vertically downward, in the form of a cylindrical jet with diameter 0.25 in. and a velocity of 1 ft / s, what will be its diameter one ft below the faucet?

5.64. In Example 5.15, if the column is expected to break into drops when its diameter is 0.1 in, how far below the faucet should this occur?

5.65. Equation 5.BJ shows V as $f(h)$ for flow from a faucet. Show the corresponding equation for D as $f(h)$.

5.66. A meteorologist, discussing a record-breaking hurricane said, "It had a pressure of 850 millibars in the center, so it had winds of 250 miles an hour!" Explain this statement in terms of B.E.

REFERENCES FOR CHAPTER 5

1. Lamb, H. *Hydrodynamics,* New York: Dover, 1945, p. 99.
2. Boyce, M. P. "Transport and Storage of Fluids." In *Perry's Chemical Engineers' Handbook, 7,* ed. R. H. Perry, D. W. Green, and J. O. Maloney. New York: McGraw-Hill, 1997, pp. 10–14.
3. "Flows of Fluids Through Valves, Fittings and Pipes," *Technical Paper No. 410,* Crane Company, 475 N. Gary Ave., Carol Stream, IL, 1957, p. A-19.
4. Fischer, K. "How to Predict Calibration of Variable-Area Flow Meters." *Chemical Engineering 59,* no. 6 (1952), pp. 180–184. One must read this paper carefully to see that it is discussing corrections of only a few percent to the simple equations presented in this text.
5. Li, S. C., ed. *Cavitation of Hydraulic Machinery,* London: Imperial College Press, 2000.
6. Kindsvater, C. E. and R. W. Carter. "Discharge Characteristics of Rectangular Thin-plate Weirs." *Transactions of the American Society of Civil Engineers 124* (1959), pp. 772–822. This long paper, with comments by others, shows that using a coefficient of 0.67 in Eq. 5.22 will reproduce most of the published test data over a wide range of conditions to $\pm 5\%$.
7. de Nevers, N. "An Inexpensive Time Bomb" *Chem. Eng. Ed. 8,* 1974, pp. 98–101.

CHAPTER 6

FLUID FRICTION IN STEADY, ONE-DIMENSIONAL FLOW

In Chap. 5 we found the working form of Bernoulli's equation (B.E.)

$$\Delta\left(\frac{P}{\rho} + gz + \frac{V^2}{2}\right) = \frac{dW_{\text{n.f.}}}{dm} - \mathscr{F} \tag{5.5}$$

and applied it to problems in which we could set the friction term, $\mathscr{F}$, equal to zero. In this chapter we show how to evaluate the $\mathscr{F}$ term for the very important and practical case of steady flow in one dimension, as in a pipe, duct, or channel. Using the $\mathscr{F}$ terms we evaluate here, we can use Eq. 5.5 for a much wider range of problems than those we have considered so far, including many problems of great practical interest to chemical engineers. Keep in mind that our main reason for evaluating $\mathscr{F}$ is to put the proper relation for $\mathscr{F}$ into Eq. 5.5, and then solve the resulting equation for the appropriate pressures, velocities, elevations, pipe diameters, etc.

The form of the friction-loss term is strongly dependent on the geometry of the system. The problem is much simpler if the flow is all in one direction, as in a pipe, rather than in two or three dimensions, as around an airplane. Therefore, we will first consider fluid friction in long, constant-diameter pipes in steady flow. This case is of great practical significance and is the easiest case to treat mathematically. Starting and stopping of flow in pipes are discussed in Sec. 7.4. In Sec. 6.13 we will consider the frictional drag on particles in steady, rectilinear motion, which, although it is two-dimensional, gives results quite similar to those found in long, straight pipes.

In Part IV we will investigate two- and three-dimensional flows by using some of the ideas from this chapter and introducing several others.

6.1 THE PRESSURE-DROP EXPERIMENT

The classic pressure-drop experiment to determine $\mathcal{F}$ is performed on an apparatus like that shown in Fig. 6.1. In this experiment we set the volumetric flow rate of the fluid with the flow-regulating valve. We measure the volumetric flow rate with the tank or bucket on the scale and a stop watch. At steady state we read pressure gauges P_1 and P_2 and record their difference. Usually we are interested in pressure drop per unit length, so we divide the pressure drop by distance Δx (the length of the test section) and plot $[(P_1 - P_2) / \Delta x]$ against volumetric flow rate Q.

Regardless of what Newtonian liquid is flowing or what kind of pipe we use, the result is always of the form shown in Fig. 6.2, and for all gases *at low velocities* the result is the same as that shown.

The salient features of Fig. 6.2 are that *for one specific fluid flowing in one specific pipe:*

1. At low flow rates the pressure drop per unit length is proportional to the volumetric flow rate to the 1.0 power.
2. At high flow rates the pressure drop per unit length is proportional to the volumetric flow rate raised to a power that varies from 1.8 (for very smooth pipe) to 2.0 (for very rough pipes).
3. At intermediate flow rates there is a region where the experimental results are not easily reproduced. The two curves for the other two regions are shown dotted, extrapolated into this region. The flow can oscillate back and forth between these two curves, and take up values between them. If the experimental apparatus is like that in Fig. 6.1, with a more or less constant value of dP / dx, then the volumetric flow rate will oscillate horizontally between the two curves, producing an irregular pulsing flow.

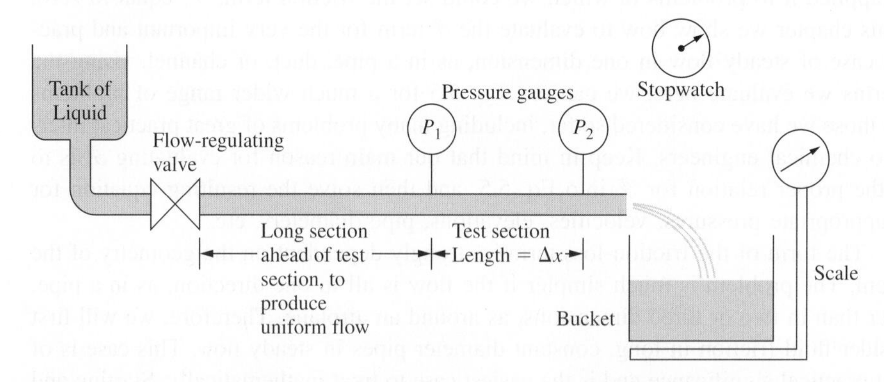

FIGURE 6.1
Apparatus for the pressure-drop experiment. One may read either the increase in weight on the scale or the increase in volume in the bucket based on calibration marks on its sides. The flow-rate-measuring method here, called "bucket and stopwatch," is the most accurate method known and is used to calibrate other methods, such as those shown in Chap. 5.

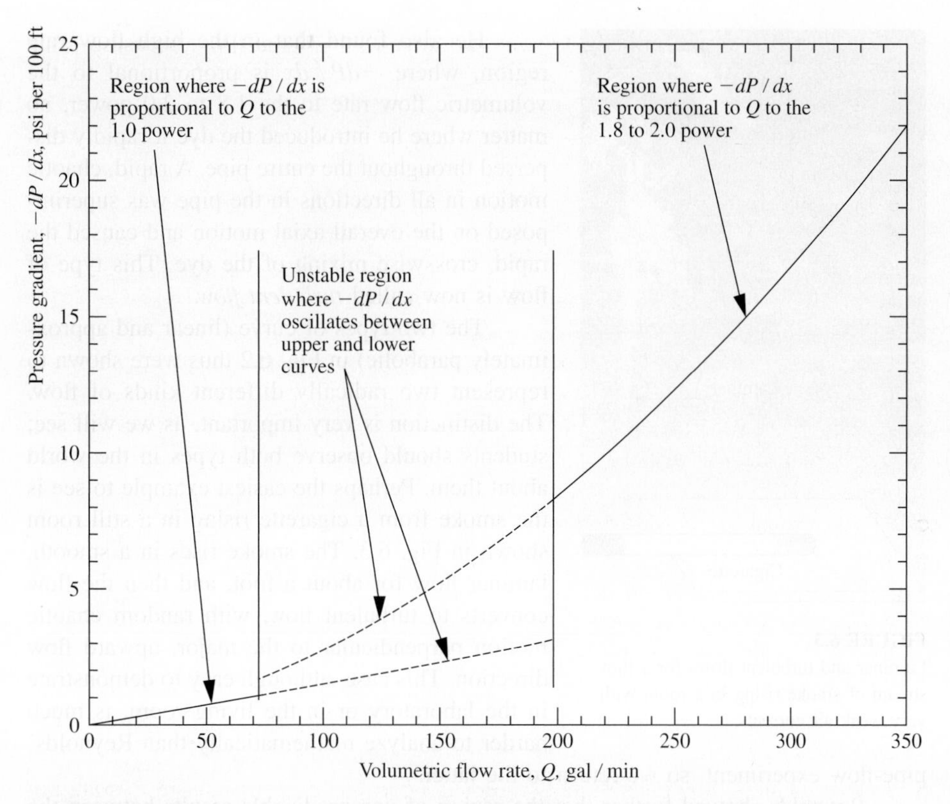

FIGURE 6.2
Typical pressure-drop curve for a specific fluid in a specific pipe. These are calculated values for an oil with SG = 1.0 and $\mu = 50$ cP, flowing in a 3 in schedule 40 pipe. For other fluids and other pipes the plot looks the same, but the numerical values are different. If the volumetric flow rate, Q, is constant, then in the unstable region the flow will oscillate vertically between the two curves of $-dP/dx$. If, instead, $-dP/dx$ is fixed (e.g., a flow by gravity from a reservoir), then in the unstable region the flow will oscillate horizontally between two values of Q.

This experiment is relatively easy to run, and the curves have been found for many combinations of pipe and fluid. However, since all possible combinations have not been tested, it would be convenient to have some way of calculating the results of a new combination without having to test it. Furthermore, no inquisitive mind will be satisfied with Fig. 6.2 without asking why it has three regions so different from each other.

6.2 REYNOLDS' EXPERIMENT

Osborne Reynolds [1] explained the strange shape of Fig. 6.2. In an apparatus similar to that of Fig. 6.1 but made of glass, he arranged to introduce a liquid dye into the flowing stream at various points. He found that, in the low flow rate region (in which $-dP/dx$ is proportional to the flow rate), the dye he introduced formed a smooth, thin, straight streak down the pipe; there was no mixing perpendicular to the axis of the pipe. This type of flow, in which all the motion is in the axial direction, is now called *laminar flow* (the fluid appears to move in thin shells or layers, or *laminae*).

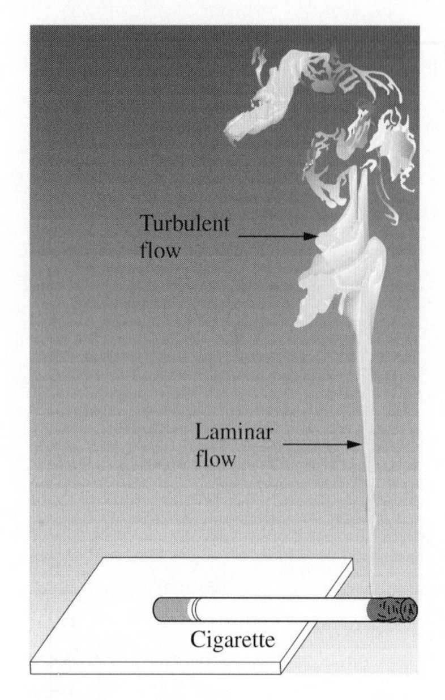

FIGURE 6.3
Laminar and turbulent flows for a thin stream of smoke rising in a room with very weak air currents.

He also found that in the high flow rate region, where $-dP/dx$ is proportional to the volumetric flow rate to the 1.8 to 2.0 power, no matter where he introduced the dye it rapidly dispersed throughout the entire pipe. A rapid, chaotic motion in all directions in the pipe was superimposed on the overall axial motion and caused the rapid, crosswise mixing of the dye. This type of flow is now called *turbulent flow.*

The two types of curve (linear and approximately parabolic) in Fig. 6.2 thus were shown to represent two radically different kinds of flow. The distinction is very important, as we will see; students should observe both types in the world about them. Perhaps the easiest example to see is the smoke from a cigarette rising in a still room shown in Fig. 6.3. The smoke rises in a smooth, laminar flow for about a foot, and then the flow converts to turbulent flow, with random chaotic motion perpendicular to the major, upward flow direction. This case, although easy to demonstrate in the laboratory or in the living room, is much harder to analyze mathematically than Reynolds' pipe-flow experiment, so we return to the latter.

Reynolds showed further that the region of unreproducible results between the regions of laminar and of turbulent flow is the region of transition from the one type of flow to the other, called the *transition region.* The reason for the poor reproducibility here is that laminar flow can exist in conditions in which it is not the stable flow form, but it fails to switch to turbulent flow unless some outside disturbance such as microscopic roughness on the pipe wall or very small vibrations in the equipment triggers the transition. Thus, in the transition region the flow can be laminar or turbulent, and the pressure drop or flow rate can suddenly change by a factor of 2. Under some circumstances the flow can alternate back and forth between being laminar and turbulent, causing the pressure drop to oscillate between a higher and a lower value; or for a constant pressure drop, as in Fig. 6.1, the velocity can oscillate between a higher and a lower value.

Besides clarifying the strange shape of Fig. 6.2, Reynolds made the most celebrated application of dimensional analysis (Chap. 9) in the history of fluid mechanics. He showed that for smooth, circular pipes, for all Newtonian fluids, and for all pipe diameters the transition from laminar to turbulent flow occurs when the dimensionless group $DV\rho/\mu$ has a value of about 2000. Here D is the pipe diameter, V is the average fluid velocity in the pipe, ρ is the fluid density, and μ is the fluid viscosity. This dimensionless group is now called the Reynolds number, $\mathcal{R}$:

$$\begin{pmatrix}\text{Reynolds number for}\\ \text{flow in a circular pipe}\end{pmatrix} = \mathcal{R} = \frac{DV\rho}{\mu} = \frac{DV}{\nu} = \frac{4Q}{\pi D\nu} \tag{6.1}$$

TABLE 6.1
Comparison of laminar, transition, and turbulent flows

	Type of flow		
	Laminar	**Transition**	**Turbulent**
Behavior of dye streak	Dye in, Flow	Oscillates between laminar and turbulent	Dye in, Flow
Pressure drop proportional to	$Q^{1.0}$	Oscillates from one value to another; very difficult to measure	$Q^{1.8}$ (very smooth pipes) to $Q^{2.0}$ (very rough pipes)
Reynolds number	<2000	≈2000 to 4000	>4000

The transition region on Fig. 6.2 corresponds to Reynolds numbers between about 2000 and 4000. For Reynolds numbers above about 4000, the flow is stably turbulent. For flows other than pipe flow, some other appropriate length is substituted for the pipe diameter in the Reynolds number, producing a different Reynolds number, as will be discussed later. All Reynolds numbers are (some length · velocity · density / viscosity).

The difference between laminar and turbulent flows is one of the most important differences in fluid mechanics. The equations in this book for laminar flow do not describe turbulent flow, nor do the turbulent flow equations describe laminar flow. If you learn nothing else in this chapter, learn that. In pipe flow, the boundary between laminar and turbulent flow is the region from Reynolds number ≈2000 to ≈4000. This means that almost all flows of gases and liquids like water in ordinary-sized pipes are turbulent. The only exceptions to that statement are flows of fluids much more viscous than water, such as asphalt, maple syrup, or polymer solutions. (The fluid used as an example to make up Fig. 6.2 is 50 times as viscous as water; if that figure had been made for water, the laminar region would have practically disappeared into the left axis!) However, in very small tubes or other flow passages the flow is normally laminar. The flow in the heart and the major arteries near it in our bodies and those of most animals our size are turbulent. The rest of the blood flow in our bodies is laminar, as is the flow of fluids in filters, in groundwater, and in oil fields. (These latter are not exactly pipe flow, but as shown in Chap. 11, the flow passages between the solid particles in filters and in the ground behave as irregular-shaped pipes.) River flows are mostly turbulent, and the main flows of the atmosphere are turbulent, but in low-wind situations and in the stratosphere the atmosphere can be laminar. Both laminar and turbulent flows are important; you could not read this statement without the turbulent flow near your heart or the laminar flow of blood to your brain and eyes.

The results of Reynolds' experiments are summarized on Table 6.1.

6.3 LAMINAR FLOW

Laminar flow is the simplest flow, so we discuss it first. Consider a steady laminar flow of an incompressible Newtonian fluid in a horizontal circular tube or pipe. A section of the tube Δx long with inside radius r_0 is shown in Fig. 6.4. We arbitrarily select a rod-shaped

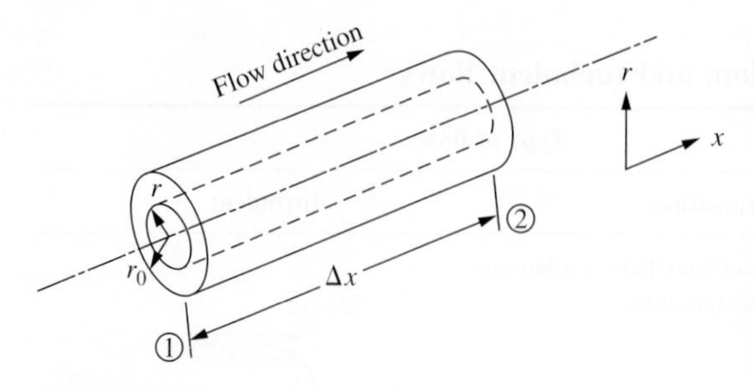

FIGURE 6.4
Force-balance system in pipe flow. The balance is made around the cylindrical rod-shaped volume, symmetrical about the centerline.

element of the fluid, symmetrical about the center, with radius *r*, and compute the forces acting on it. Here it is assumed that location 1 is well downstream from the place where the fluid enters the tube. This analysis is not correct for the tube entrance (see Part IV). The flow is steady and all in the axial direction. There is no acceleration in the x direction, so the sum of the forces acting in the x direction on the rod-shaped element we have chosen must be zero. There is a pressure force acting on each end, equal to the pressure times the cross-sectional area of the end. These act in opposite directions; their sum *in the positive x direction* is

$$\text{Pressure force} = P_1(\pi r^2) - P_2(\pi r^2) = \pi r^2 (P_1 - P_2) \tag{6.2}$$

Along the cylindrical surface of our rod-like element the pressure forces have no component in the x direction and can be ignored, but there is a shear force resisting the flow. The shear force acts in the direction opposite to the pressure gradient, which is in the flow direction, and its magnitude is

$$\text{Shear force} = 2\pi r\,\Delta x \cdot (\text{shear stress at } r) = 2\pi r\,\Delta x \cdot \tau \tag{6.3}$$

Since the pressure force and shear force are the only forces acting in the x direction, and since the sum of the forces is zero, these must be equal and opposite. Equating their sum to zero and solving for the shear stress at *r*, we find

$$\tau = \begin{pmatrix} \text{Shear stress acting} \\ \text{on the central rod} \\ \text{at radius } r \end{pmatrix} = \frac{-r(P_1 - P_2)}{2\Delta x} \tag{6.4}$$

The minus sign shows that our intuition is correct, and τ acts in the minus x direction. (See the discussion of the sign of the shear stress in Sec. 1.5). Equation 6.4 applies to steady laminar or turbulent flow of any kind of fluid in any circular pipe or tube.

Here we have applied Newton's law, $F = ma$, to the particularly simple case in which there is no acceleration and sum of the forces is therefore zero; in Chap. 7 and Part IV we will see how to apply it to more complicated cases.

We saw in Chap. 1 that for Newtonian fluids in laminar motion the shear stress is equal to the product of the viscosity and the velocity gradient. Substituting in Eq. 6.4, we find

$$\mu \frac{dV}{dr} = -r \frac{P_1 - P_2}{2\Delta x} \tag{6.5}$$

For steady laminar flow the pressure gradient $(P_1 - P_2)/\Delta x$ does not depend on radial position in the pipe, so we may integrate this to

$$V = \frac{-r^2}{4\mu} \cdot \frac{P_1 - P_2}{\Delta x} + \text{constant} \tag{6.6}$$

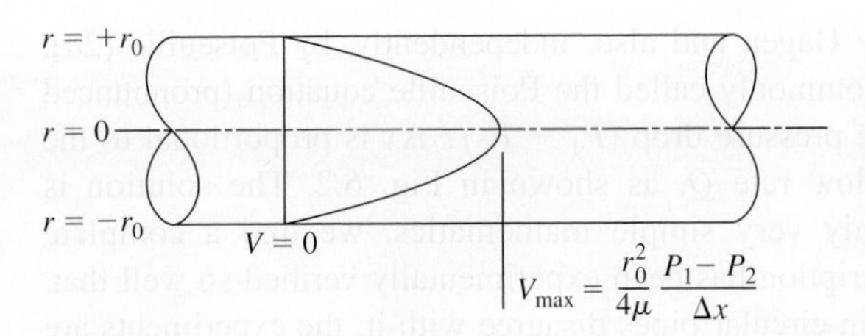

FIGURE 6.5
Velocity distribution in steady, laminar flow of a Newtonian fluid in a circular pipe.

To solve for the value of the constant, we need one observational fact: that for the flow of everything except rarefied gases, the fluid at the solid surface clings to the surface. This is not an intuitively obvious fact, and we cannot derive it from some prior principle. The behavior is quite different from that of solids, whose sliding surfaces do slip over one another so that there is a sharp discontinuity in velocity at the sliding boundary. However, one may observe that it is so by watching the behavior of bits of wood or leaves on the surface of a stream: those at the center move rapidly, those near the bank slowly, and those right at the bank not at all (this condition is often referred to as the *no-slip condition;* one kind of rarefied gas flow is called, logically enough, *slip flow*). From this observational fact it follows that at $r = r_0$ (at the pipe wall), $V = 0$; so

$$0 = \frac{-r_0^2}{4\mu} \cdot \frac{P_1 - P_2}{\Delta x} + \text{constant} \tag{6.7}$$

Substituting this value of the constant in Eq. 6.6 and factoring, we find

$$V = \frac{r_0^2 - r^2}{4\mu} \cdot \frac{P_1 - P_2}{\Delta x} \tag{6.8}$$

This equation says that for steady, laminar flow of Newtonian fluids in circular pipes:

1. The velocity is zero at the tube wall ($r = r_0$).
2. The velocity is a maximum at the center of the pipe ($r = 0$).
3. The magnitude of this maximum velocity is

$$V_{max} = \frac{r_0^2}{4\mu} \cdot \frac{P_1 - P_2}{\Delta x} \tag{6.A}$$

4. The pressure drop per unit length is independent of fluid density and is proportional to the first power of the local velocity and the first power of the viscosity.
5. The velocity-radius plot is a parabola; see Fig. 6.5.

In engineering we are generally more interested in the volumetric flow rate Q than in the local velocity V. To find the Q of a uniform-velocity flow we multiply the velocity by the cross-sectional area perpendicular to flow (see Table 3.1). The velocity of the laminar flow described above is not uniform, so we must integrate velocity times area over the whole pipe cross section.

$$Q = \int_{\text{tube}} V\,dA = \int_{r=0}^{r=r_0} \frac{r_0^2 - r^2}{4\mu} \cdot \frac{P_1 - P_2}{\Delta x} \cdot 2\pi r\,dr$$

$$= \frac{P_1 - P_2}{\Delta x} \cdot \frac{\pi}{2\mu}\left[\frac{r_0^2 r^2}{2} - \frac{r^4}{4}\right]_{r=0}^{r=r_0} = \frac{P_1 - P_2}{\Delta x} \cdot \frac{\pi}{\mu} \cdot \frac{r_0^4}{8} = \frac{P_1 - P_2}{\Delta x} \cdot \frac{\pi}{\mu} \cdot \frac{D_0^4}{128} \tag{6.9}$$

This equation was developed by Hagen and also, independently, by Poiseuille [2*a*]. In the United States it is most commonly called the Poiseuille equation (pronounced "pwah-zoo-y"). It shows that the pressure drop $(P_1 - P_2) / \Delta x$ is proportional to the first power of the volumetric flow rate Q, as shown in Fig. 6.2. The solution is immensely satisfying; using only very simple mathematics, we find a complete description of the flow. The description has been experimentally verified so well that, when laminar-flow experiments in circular pipes disagree with it, the experiments are in error.

From Eq. 6.9 it can also be shown (Prob. 6.4) that

$$V_{\text{avg}} = \frac{Q}{\pi r^2} = \frac{V_{\text{max}}}{2} \tag{6.10}$$

$$\text{Average ke} = \frac{\int (V^2/2)\, dQ}{\int dQ} = \frac{\int (V^2/2) V \cdot 2\pi r\, dr}{\int V \cdot 2\pi r\, dr} = (V_{\text{avg}})^2 \tag{6.11}$$

Both of these values are shown in Table 3.1.

To see how this fits in with B.E. (Eq. 5.5), we apply B.E. from point 1 to point 2 in Fig. 6.4 and substitute for $-\Delta P$ from Eq. 6.9 to find

$$\mathcal{F} = \frac{-\Delta P}{\rho} = Q\, \Delta x \frac{\mu}{\rho} \cdot \frac{128}{\pi D_0^4} \tag{6.12}$$

Equation 6.12 relates the $\mathcal{F}$ in B.E. to the flow rate, diameter, length, and (viscosity / density) of a horizontal flow in which gravity plays no role. If we repeat the entire derivation for a vertical flow in a pipe in which the pressure is constant throughout (Prob. 6.1), we find that the ΔP term is replaced with a $\rho g\, \Delta z$ term. When we substitute this in Eq. 6.12 and then calculate $\mathcal{F}$ from B.E., we find

$$\mathcal{F} = -g\Delta z = Q\, \Delta x \frac{\mu}{\rho} \cdot \frac{128}{\pi D_0^4} \tag{6.13}$$

Thus, for either horizontal or vertical laminar flow we find

$$\mathcal{F} = Q\, \Delta x \frac{\mu}{\rho} \cdot \frac{128}{\pi D_0^4} \qquad \begin{bmatrix} \text{laminar flow} \\ \text{only!} \end{bmatrix} \tag{6.14}$$

We may readily extend the argument to show that this equation applies also to flows at any angle to the vertical, so that Eq. 6.14 is the general description of friction heating in laminar flow of Newtonian fluids in circular pipes.

Example 6.1. Oil at a rate of 50 gal / min is flowing steadily from tank A to tank B through 3000 ft of 3-in schedule 40 pipe; see Fig. 6.6. (Appendix A.2 shows the dimensions of standard U.S. schedule 40 pipe sizes. The inside diameter (ID) of 3-in schedule 40 pipe is 3.068 in.) The oil has a density of 62.3 lbm / ft^3 and a viscosity of 50 cP. The levels of the free surfaces are the same in both tanks. Tank B is vented to the atmosphere. What is the gauge

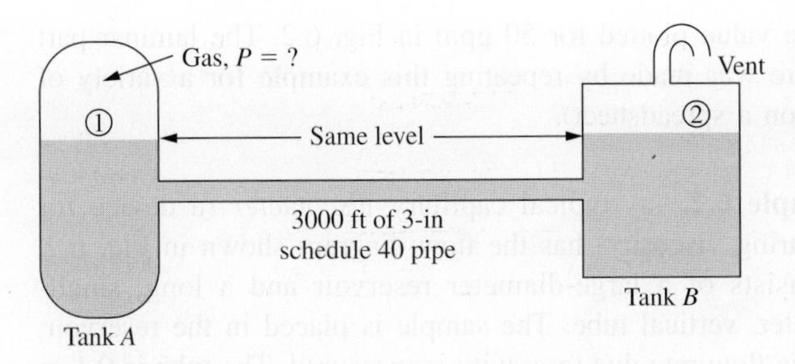

FIGURE 6.6
A pressure-driven fluid transfer from one tank to another, used in Examples 6.1 and 6.4.

pressure in tank A required to produce this flow rate? (Pushing liquids from one container to another with gas pressure is common in industry, particularly when a leak of the fluid would be dangerous.)

Applying B.E. between the free surface in tank A (point 1) and the free surface in tank B (point 2), we see that the velocities are negligible. Since there is no change in elevation or pump or compressor work, we have

$$\Delta \frac{P}{\rho} = -\mathcal{F} \tag{6.B}$$

The density is constant, so in tank A the gauge pressure $(P_1 - P_2)$ is $-(-\rho\mathcal{F})$. *If the flow in the pipe is laminar,* then we can solve for this pressure from Eq. 6.14. The average velocity is

$$V_{\text{avg}} = \frac{Q}{A} = \frac{50 \text{ gal / min}}{(\pi / 4)(3.068 \text{ in})^2} \cdot \frac{144 \text{ in}^2}{\text{ft}^2} \cdot \frac{\text{min}}{60 \text{ s}} \cdot \frac{\text{ft}^3}{7.48 \text{ gal}}$$

$$= 2.17 \frac{\text{ft}}{\text{s}} = 0.66 \frac{\text{m}}{\text{s}} \tag{6.C}$$

Therefore, the Reynolds number is

$$\mathcal{R} = \frac{(3.068 / 12) \text{ ft} \cdot 2.17 \text{ ft / s} \cdot 62.3 \text{ lbm / ft}^3}{50 \text{ cP} \cdot 6.72 \cdot 10^{-4} \text{ lbm / (ft} \cdot \text{s} \cdot \text{cP)}} = 1028 \tag{6.D}$$

As shown before, steady pipe flow is laminar if $\mathcal{R} < 2000$; so we have laminar flow here and are safe in substituting for $\mathcal{F}$ from Eq. 6.14. Multiplying through by $-\rho$, we find

$$-\Delta P = P_1 - P_2 = Q \frac{128}{\pi} \cdot \frac{\mu}{D_0^4} \Delta x$$

$$= 50 \frac{\text{gal}}{\text{min}} \cdot \frac{128}{\pi} \cdot \frac{50 \text{ cP}}{(3.068 \text{ in})^4} \cdot 3000 \text{ ft} \cdot 231 \frac{\text{in}^3}{\text{gal}}$$

$$\cdot 2.09 \cdot 10^{-5} \frac{\text{lbf} \cdot \text{s}}{\text{cP} \cdot \text{ft}^2} \cdot \frac{\text{min}}{60 \text{ s}} \cdot \frac{\text{ft}}{12 \text{ in}} = 23.1 \frac{\text{lbf}}{\text{in}^2} = 159 \text{ kPa} \tag{6.E}$$

This is the gauge pressure in tank A required to produce a flow of 50 gpm. ■

Readers may check that this corresponds to

$$-\frac{\Delta P}{\Delta x} = \frac{23.1 \text{ psi}}{3000 \text{ ft}} = 0.77 \frac{\text{psi}}{100 \text{ ft}} \tag{6.F}$$

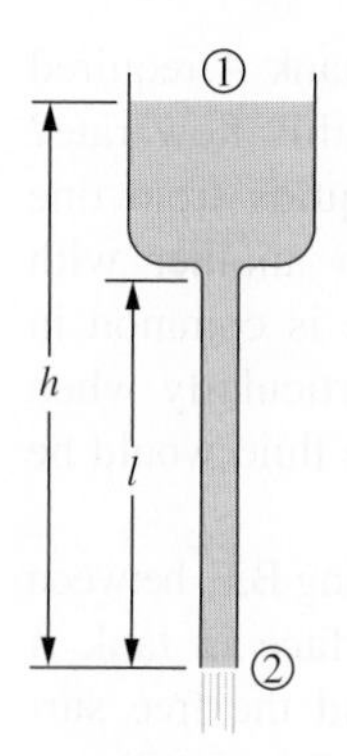

FIGURE 6.7
Typical capillary viscometer; see Example 6.2.

which is the value plotted for 50 gpm in Fig. 6.2. The laminar part of that figure was made by repeating this example for a variety of flow rates (on a spreadsheet).

Example 6.2. A typical capillary *viscometer* (a device for measuring viscosity) has the flow diagram shown in Fig. 6.7. It consists of a large-diameter reservoir and a long, small-diameter, vertical tube. The sample is placed in the reservoir, and the flow rate due to gravity is measured. The tube is 0.1 m long and has a 1 mm ID. The height of the fluid in the reservoir above the inlet to the tube is 0.02 m. The fluid being tested has a density of 1050 kg / m^3. The flow rate is 10^{-8} m^3 / s. What is the viscosity of the fluid?

Applying B.E. between the free surface in the reservoir (point 1) and the fluid leaving the bottom of the viscometer (point 2), we see that the pressure at each point is atmospheric and that there is no pump or compressor work. We can neglect the velocity in the reservoir, so B.E. becomes

$$g(z_2 - z_1) + \frac{V_2^2}{2} = -\mathcal{F} \tag{6.G}$$

The kinetic-energy term here is negligible compared with the other two terms. This is found in most laminar-flow problems, so we drop the kinetic-energy term and find

$$\mathcal{F} = -g\,\Delta z \tag{6.H}$$

Substituting for $\mathcal{F}$ from Eq. 6.14 and solving for μ, we find

$$\begin{aligned}\mu &= \frac{\rho g(-\Delta z)\,\pi D_0^4}{128 Q\,\Delta x}\\ &= \frac{1050\ \text{kg / m}^3 \cdot 9.81\ \text{m / s}^2 \cdot 0.12\ \text{m} \cdot \pi \cdot (0.001\ \text{m})^4}{128 \cdot (10^{-8}\ \text{m}^3\text{ / s}) \cdot 0.1\ \text{m}} \cdot \frac{10^3\ \text{cP} \cdot \text{m} \cdot \text{s}}{\text{kg}}\\ &= 30.3\ \text{cP} = 0.0303\ \text{Pa} \cdot \text{s}\end{aligned} \tag{6.I}$$

■

Viscometers of the same type, but slightly more complicated than the one described above, are very widely used. In using them we recognize that:

1. They must not be used at flow rates so high that flow is turbulent (Prob. 6.8).
2. They must be long enough that the error introduced by applying the Poiseuille equation (which only applies well downstream from the entrance) to the whole tube is small. For a brief introduction to the problem of *entrance flow,* to which Poiseuille's equation does not apply, see Part IV.
3. The Poiseuille equation applies only to Newtonian fluids, so this type of apparatus can be used simply only for such fluids; see Chap. 13.

4. Because the viscosity measured by such a device is proportional to D_0^4, a small error in the diameter measurement leads to a large error in the viscosity measurement. For this reason these devices ordinarily are calibrated by using a fluid of known viscosity and by determining the appropriate average diameter from the calibration.
5. As discussed in Chap. 5, the $\mathscr{F}$ term represents the conversion of mechanical energy into internal energy. Normally this conversion results in an increase of temperature. It can be shown (Prob. 6.6) that in this case the temperature change is negligible. However, in more viscous liquids, which are pumped through capillary viscometers, there can be a significant temperature rise. In most fluids a small temperature rise can cause a large viscosity change, so the temperature rise must be minimized.
6. The commercial versions of this device normally require the user to use a stopwatch and measure the time for the liquid level to pass from one mark on the glass tube to another. The resulting measurement is a time. For that reason viscosities are often reported as times, e.g., *Saybolt Seconds Universal* (SSU), the standard viscosity measurement for fuel oils in the United States. Formulae for converting from SSU or other time-for-the-interface-to-pass-the-marks in various standard capillary-tube-gravity viscometers are given in handbooks.

6.4 TURBULENT FLOW

Why does the preceding analysis not work for turbulent flow? Equation 6.4 is correct for steady laminar or turbulent flow of any kind of fluid, but the substitution of $\mu(dV/dy)$ for the shear stress is correct only for *laminar flow* of *Newtonian fluids*. In laminar flow in a tube there is no motion perpendicular to the tube axis. In turbulent flow there is no *net* motion perpendicular to the tube axis, but there does exist an intense, local, oscillating motion perpendicular to the tube axis. The transfer of fluid perpendicular to the net axial motion causes an increase in shear stress over the value given above for laminar flow of Newtonian fluids. This is most easily seen in an analogy. Consider two students playing catch with baseballs. One is standing on the ground, the other on a railroad car; see Fig. 6.8.

In Fig. 6.8(a) the railroad car is not moving, and both students throw the ball back and forth in the plus and minus y direction. Each time one catches the ball, the student experiences a force and, if the student throws it back at the same speed, the student exerts an equal force in the opposite direction. Therefore, the net effect of their throwing the ball back and forth is that a force is exerted on each one, tending to move them apart in the plus or minus y direction. There is no force in the x direction.

In Fig. 6.8(b) the train is moving at constant speed in the x direction. Each student still throws the ball in the plus or minus y direction. However, because of their relative motion, each one receives the ball moving, not in the y direction, but at an angle between the x and y directions; the directions of the balls (relative to the two students) are shown by the arrows. Since each one receives the balls in these directions, the force exerted by a student in stopping the ball consists, not only of the

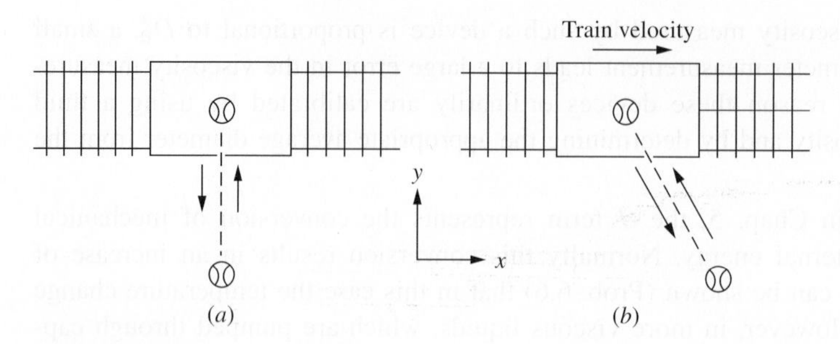

FIGURE 6.8
Illustration by analogy of shear forces due to turbulence: (*a*) top view of students playing catch, neither moving; (*b*) top view of students playing catch, one moving perpendicular to the direction of throwing the ball.

y component, which the other student put into the ball by throwing it, but also of the x component due to their relative motion. When the train is moving, in addition to the y-directed force there is a force tending to retard the train and to drag the stationary student along in the x direction.

Exactly the same thing happens in turbulent fluid flow. The exchange of fluid between the faster-moving fluid in the center of the tube and the slower-moving fluid near the wall increases the shear stress over that which would exist in laminar flow. This extra stress is called a *Reynolds stress* after Reynolds, who first explained it. Thus, the actual $\mathcal{F}$ in turbulent flow is greater than that predicted by Poiseuille's equation.

In the case of the students throwing the balls, the extra stress is proportional to the velocity of the train and the number of times per second which they throw the balls back and forth. In the case of the corresponding stress in a fluid in turbulent flow the stress is proportional to the velocity gradient dV/dy times the average mass of fluid passing back and forth across a surface of constant y (across which there is no *net* flow). Since the velocity goes from zero at the pipe wall to the average velocity near the center, the velocity gradient should be some function of V_{avg}/D. If we now assume that this is a linear proportion and that the magnitude of the flow of mass back and forth across a surface of constant y is proportional to the average velocity, then it follows that

$$\mathcal{F} \propto \frac{V_{avg}^2}{D} \tag{6.15}$$

Furthermore, the friction heating should be proportional to the length of the pipe. Including this idea, we find

$$\mathcal{F} \propto \frac{\Delta x V_{avg}^2}{D} \tag{6.16}$$

This equation rests on the plausible assumption that the friction heating is proportional to the length of the pipe and on the more questionable assumption that it is proportional to V_{avg}^2/D. Do these assumptions agree with the experimental data? The answer

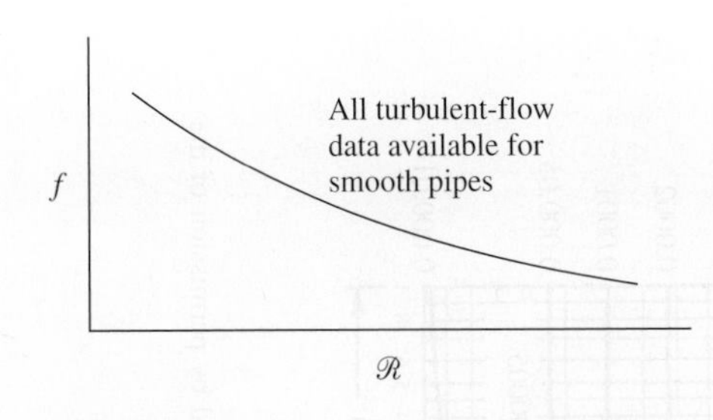

FIGURE 6.9
Blasius and Stanton's friction factor plot.

is, Yes and no. To save writing, we now define a new term, the *friction factor f*, which is equal to half the proportionality constant in Eq. 6.16, and drop the avg subscript on the velocity, so that

$$\mathcal{F} = 2f\frac{\Delta x V^2}{D} = 4f\frac{\Delta x}{D}\cdot\frac{V^2}{2} \tag{6.17}$$

$$f \equiv \text{friction factor}$$

$$\equiv \frac{\mathcal{F}}{4(\Delta x / D)(V^2/2)} \tag{6.18}$$

To test these assumptions, Blasius and Stanton [2*b*] calculated the friction factor for a large variety of pipe-flow experiments with smooth pipes. They found that the friction factor was not a constant, as predicted by the above simple theory, but decreased slowly with increasing Reynolds number. However, all the data for smooth pipes of various diameters at all velocities for a large range of fluids formed a single curve on a plot of friction factor versus Reynolds number; see Fig. 6.9.

Once plots like this came into common use, it became apparent that they were very good for smooth pipes, such as glass pipes or drawn metal tubing, but that the pressure drops they predicted were too low for rough pipes, such as those made from cast iron or concrete. It appeared that the roughness of the pipe surface influenced f. To resolve the question Nikuradse [3] measured the pressure drop in various smooth pipes to the inside of which he had glued sand grains. He found that for a given value of ε / D, where ε is the size of the sand particle and D is the pipe diameter, he could plot all his results on one curve on Fig. 6.9, but that there were different curves for different values of ε / D. This ratio, ε / D, is called the *relative roughness*. Figure 6.10 is currently the most commonly used friction factor plot (which chemical engineers normally call a *friction factor plot,* and some other disciplines refer to as a *Moody diagram*), prepared by Moody [4*a*], who based it on Nikuradse's data and on all the other available data on flow in pipes.* Moody also suggested the working values for the absolute roughness, shown in Table 6.2.

Figure 6.10 shows that, as the relative roughness becomes greater and greater, the assumptions that went into Eq. 6.16 become better and better; f becomes a constant that is independent of diameter, velocity, density, and fluid viscosity.

To make life hard for the working engineer, there are two values of the friction factor in common use The one shown in Eq. 6.18 appears in most chemical engineering books, but in mechanical engineering and civil engineering books there appears

$$f_{\text{civ, mech}} = \frac{\mathcal{F}}{(\Delta x / D)(V^2/2)} = 4f_{\text{chem}} \tag{6.19}$$

*The friction behavior of a pipe with sand grains glued to the wall is somewhat different from that of a commercial pipe. This is believed to be due to the wide range of sizes and shapes of the rough spots in a commercial pipe compared with the uniform size and shape of the sand grains used by Nikuradse. Moody [4] made Fig. 6.9 according to the Colebrook equation [5], which agrees with the data on commercial pipes. The differences between the two kinds of roughness have been discussed ([6], p. 529).

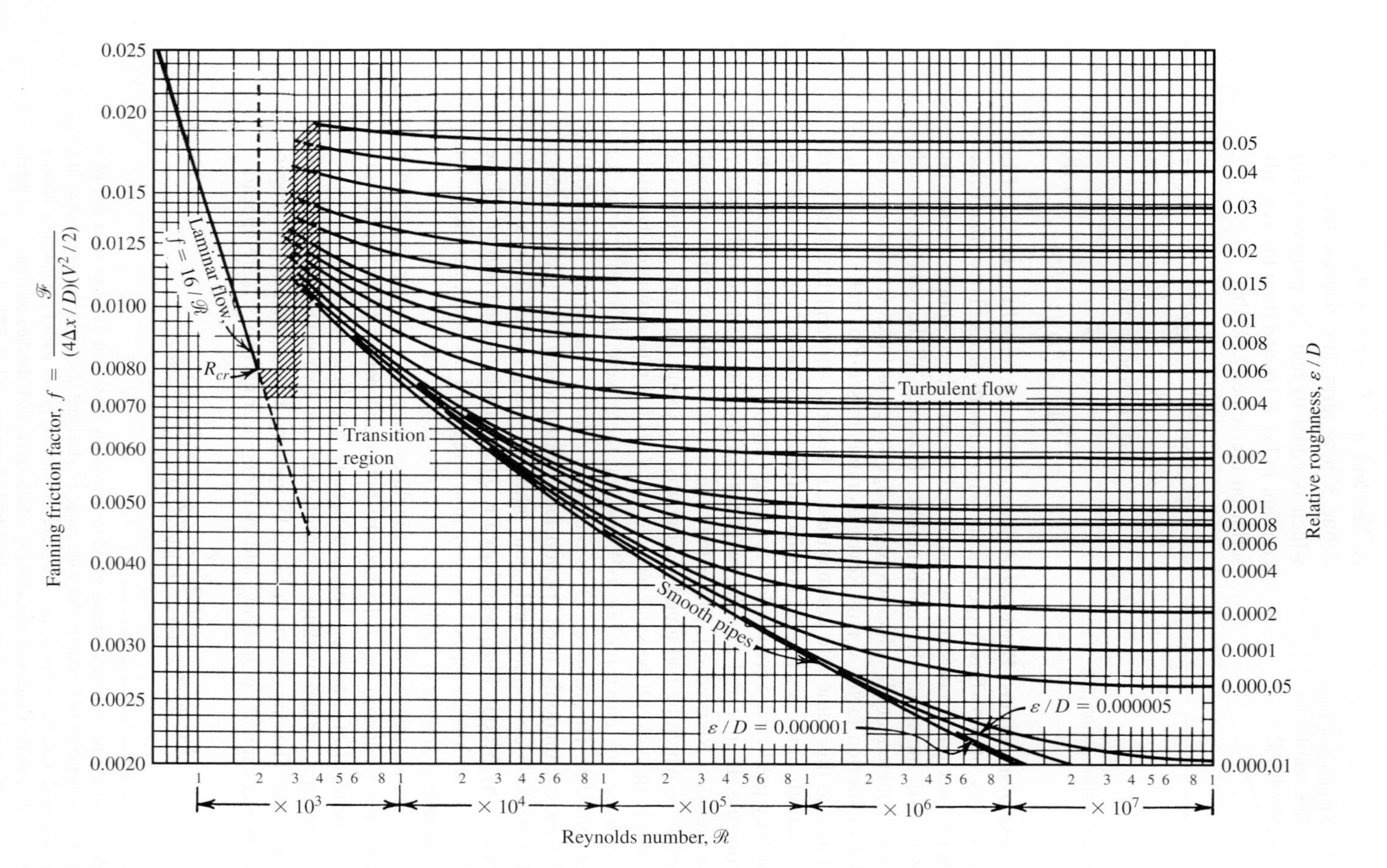

FIGURE 6.10
Friction factor plot for circular pipes. (From L. W. Moody, "Friction factors for pipe flow," *Trans. ASME 66*:672 (1944). Reproduced by permission of the publisher.)

TABLE 6.2
Values of surface roughnesses for various materials,* to be used with Fig. 6.10

	Surface roughness	
	ε, ft	ε, in
Drawn tubing (brass, lead, glass, etc.)	0.000005	0.00006
Commercial steel or wrought iron	0.00015	0.0018
Asphalted cast iron	0.0004	0.0048
Galvanized iron	0.0005	0.006
Cast iron	0.00085	0.010
Wood stave	0.0006–0.003	0.0072–0.036
Concrete	0.001–0.01	0.012–0.12
Riveted steel	0.003–0.03	0.036–0.36

*From Moody [4].

The existence of the two values means that, whenever engineers plan to use a chart like Fig. 6.10 or an equation with f in it, they must check to see on which of the two f values the chart or equation is based. Throughout this book we will use the value of f_{chem} defined by Eq. 6.18. It is often called the *Fanning friction factor,* while the one 4 times as large is called the *Darcy* or *Darcy-Weisbach friction factor:* $f_{\text{Fanning}} = \tau / (\rho V^2 / 2)$; $f_{\text{Darcy-Weisbach}} = 4\tau / (\rho V^2 / 2)$.

There is really not much point in having a curve for laminar flow on a friction factor plot, since laminar flow in a pipe can be solved analytically. Poiseuille's equation (Eq. 6.8) can be rewritten (Prob. 6.12) as

$$f = \frac{16}{\mathcal{R}} \tag{6.20}$$

Plotting any equation this simple is unnecessary. However, the laminar-flow line usually is included in friction factor plots, as it is in Fig. 6.10. Furthermore, the turbulent and transition region curves on Fig. 6.10 can be represented with very good accuracy by [7]

$$f = 0.001375 \cdot \left[1 + \left(20{,}000 \frac{\varepsilon}{D} + \frac{10^6}{\mathcal{R}} \right)^{1/3} \right] \tag{6.21}$$

which has no theoretical basis, but reproduces the turbulent region in Fig. 6.10 well (see Prob. 6.39).

Example 6.3. Read the value of the friction factor from Fig. 6.10 for $\mathcal{R} = 10^5$ and $(\varepsilon / D) = 0.0002$, and compare that value to the value from Eq. 6.21.

From Fig. 6.10, as closely as I can read it, $f = 0.00475$. From Eq. 6.21,

$$f = 0.001375 \cdot \left[1 + \left(20{,}000 \cdot 0.0002 + \frac{10^6}{10^5} \right)^{1/3} \right] = 0.0047 \tag{6.J}$$

The difference between the two values is less than our ability to read Fig. 6.10. ■

As seen here, Fig. 6.10 can be reduced to two equations; so why bother with it? It has great historic significance and considerable intuitive content. Most modern engineers have quick computer programs that solve the type of problems presented in most of the rest of this chapter. So the hand solutions, using Fig. 6.10, are presented to help the student understand and develop an intuitive feel for what is going on in those programs. The student is advised to program Eq. 6.21 into a spreadsheet and, after making a few chart lookups on Fig. 6.10, only to glance at that figure and use the spreadsheet, which is its equivalent, to find working values for problems (and exams, if you can bring your spreadsheet with you!).

6.5 THE THREE FRICTION FACTOR PROBLEMS

The friction factor plot, Fig. 6.10, relates six parameters of the flow:

1. Pipe diameter, D.
2. Average velocity, V_{avg}.
3. Fluid density, ρ.
4. Fluid viscosity, μ.
5. Pipe roughness, ε.
6. The friction heating per unit mass, $\mathcal{F}$.

Therefore, given any five of these, we can use Fig. 6.10 to find the sixth.

Often, instead of being interested in the average velocity V_{avg}, we are interested in the volumetric flow rate,

$$Q = \frac{\pi}{4} D^2 V_{\text{avg}} \tag{6.K}$$

The three most common types of problem are shown in Table 6.3. For all of these problems the equations to be solved are shown in Table 6.4. This appears to be a formidable list of equations, but as the following examples show, their solution, while tedious by hand, is straightforward, and they are readily solved by computer. We will begin with a type 1 problem by hand and then do types 2 and 3 by spreadsheet.

Example 6.4. In Example 6.1, Fig. 6.6, we have decided that we wish to transport 300 gal / min, instead of the 50 gal / min in that example. Now what is the required pressure in Tank *A*?

This is 6 times the required volumetric flow rate in Example 6.1; the average velocity is $V = 2.17 \text{ ft/s} \cdot 6 = 13.0 \text{ ft/s}$. If the flow here were laminar, as it was in that example, then we could simply multiply the required pressure by 6. But in that example the Reynolds number $\mathcal{R} = 1028$. Here we have the same pipe diameter, viscosity, and density, but 6 times the velocity, so we must have $\mathcal{R} = 1032 \cdot 6 = 6164$. This is >4000, so the flow is sure to be turbulent.

TABLE 6.3
The three friction factor problems

Type	Given	To find
1	$D, \varepsilon, \rho, \mu, Q$	$\mathcal{F}$
2	$D, \varepsilon, \rho, \mu, \mathcal{F}$	Q
3	$\varepsilon, \rho, \mu, \mathcal{F}, Q$	D

TABLE 6.4
The equations to be solved in all pipe-flow-with-friction problems

Bernoulli's equation	$\Delta\left(\frac{P}{\rho} + gz + \frac{V^2}{2}\right) = \frac{dW_{\text{n.f.}}}{dm} - \mathcal{F}$
Friction heating term in B.E.	$\mathcal{F} = 4f\frac{\Delta x}{D} \cdot \frac{V^2}{2}$
Reynolds number	$\mathcal{R} = \frac{DV\rho}{\mu} = \frac{DV}{\nu} = \frac{4Q}{\pi D \nu}$
Friction factor, laminar flow, if $\mathcal{R} < 2000$ or	$f = \frac{16}{\mathcal{R}}$
Friction factor, turbulent flow, if $\mathcal{R} > 4000$	$f = 0.001375 \cdot \left[1 + \left(20{,}000\frac{\varepsilon}{D} + \frac{10^6}{\mathcal{R}}\right)^{1/3}\right]$
Volumetric flow rate as function of velocity, some problems	$Q = \frac{\pi}{4}D^2 V_{\text{avg}}$

To use Fig. 6.10 or Eq. 6.21, we need a value of ε / D. Reading the value for commercial steel pipe from Table 6.2, we have

$$\frac{\varepsilon}{D} = \frac{0.0018 \text{ in}}{3.068 \text{ in}} = 0.0006 \qquad (6.\text{L})$$

Then we enter Fig. 6.10 at the right at $\varepsilon / D = 0.0006$ and follow that curve to the left to $\mathcal{R} = 6192$, finding (as best we can read that crowded part of the chart) $f = 0.009$. We may check that value from Eq. 6.21, finding 0.00905. We will discuss later the uncertainties in friction factor values, so for now we accept 0.0091 as a good estimate of f.

The B.E. analysis is the same as in Example 6.1, leading to Eq. 6.B. Combining that with Eqs. 6.17 and 6.21, we find

$$\Delta P = 4f\frac{\Delta x}{D}\rho\frac{V^2}{2}$$

$$= 4 \cdot 0.0091 \frac{3000 \text{ ft}}{(3.068 / 12) \text{ ft}} \cdot 62.3 \frac{\text{lbm}}{\text{ft}^3} \cdot \frac{(13.0 \text{ ft / s})^2}{2} \cdot \frac{\text{lbf} \cdot \text{s}^2}{32.2 \text{ lbm} \cdot \text{ft}} \cdot \frac{\text{ft}^2}{144 \text{ in}^2}$$

$$= 484 \text{ psi} = 3340 \text{ kPa} \qquad (6.\text{M})$$

This corresponds to 16.1 psi / 100 ft, which is the value shown on the turbulent flow line in Fig. 6.2, which in turn was made by repeating this calculation in a spreadsheet for various values of Q. It is $(484 / 23.1) \approx 21$ times the value in Example 6.1. If the flow had remained laminar, it would be 6 times the value in Example 6.1. ■

In all such problems (and the ones that follow) it is necessary to convert the volumetric flow rate (gal / min or ft^3 / s or m^3 / s) into linear velocity (in this case, 300 gal / min in a 3-in pipe = 13.0 ft / s). This routine calculation can be simplified by the use of App. A.2, which shows the volumetric flow rate in gal / min corresponding to a velocity of 1 ft / s for all schedule 40 standard U.S. pipe sizes. In the foregoing example we could have looked up the value of 23.00 (gal / min) / (ft / s) for

3-in pipe and computed

$$V = \frac{300 \text{ gal / min}}{23.0(\text{gal / min}) / (\text{ft / s})} = 13.0 \frac{\text{ft}}{\text{s}} \tag{6.N}$$

Like all Type 1 problems, this was quite straightforward. We used all the equations in Table 6.4 except the laminar friction factor equation. The sequence of operations was

$$\begin{pmatrix}\text{Volumetric} \\ \text{flow rate, } Q\end{pmatrix} \to V \to \mathscr{R} \to f \to \mathscr{F} \to \Delta P \tag{6.O}$$

Please review the calculation to see that all these steps were used. For Types 2 and 3, we cannot proceed as easily as this but must resort to a trial-and-error solution; that is easy with a spreadsheet.

Example 6.5. A gasoline storage tank drains by gravity to a tank truck; see Fig. 6.11. The pipeline between the tank and the truck is 100 m of 0.1 m diameter commercial steel pipe. The properties of gasoline are given in the Common Units and Values for Problems and Examples. Both tank and truck are open to the atmosphere, and the level in the tank is 10 m above the level in the truck. What is the volumetric flow rate of the gasoline?

Applying B.E. between the free surface in the tank, point 1, and the free surface in the truck, point 2, we see that all terms cancel except

$$\Delta(gz) = -\mathscr{F} = -4f \frac{\Delta x}{D} \cdot \frac{V^2}{2} \tag{6.22}$$

This is a type 2 problem. The equation contains two unknowns, V and f; therefore, to solve it we need an additional equation or relationship among the variables listed. The second relationship is provided by Fig. 6.10, which relates f and V. We could use Eq. 6.21 to replace either f or V in terms of the other; some computer programs do that. Others follow the trial-and-error procedure here, replacing the chart lookups with applications of Eq. 6.21.

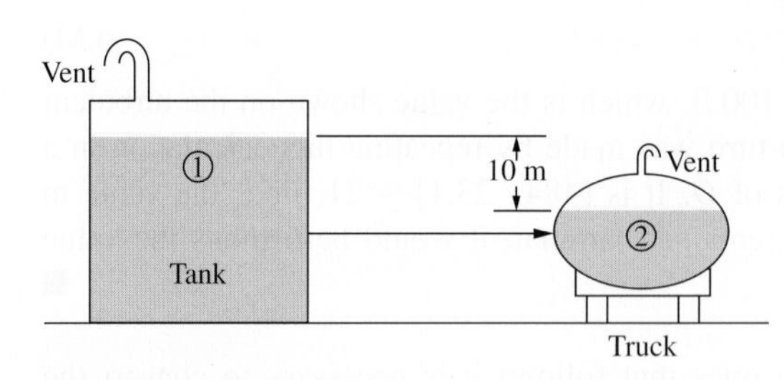

FIGURE 6.11
A gravity-driven fluid transfer from one tank to another, used in Example 6.5. The air from the headspace displaced by the liquid flow into the truck exits by the vent on the tank roof. In the United States such tank vent emissions are controlled to prevent air pollutant emissions.

Here we know the fluid properties and the pipe diameter (0.1 m ID). From Table 6.2 we have $\varepsilon = 0.0018$ in; so

$$\frac{\varepsilon}{D} = \frac{0.0018 \text{ in}}{0.1 \text{ m}} \cdot \frac{\text{m}}{39.37 \text{ in}} = 0.00046 \tag{6.P}$$

From Fig. 6.10 we see that for this value of the relative roughness the possible range of f for turbulent flow is 0.0042 to about 0.008. As our first guess, let us try $f_{\text{first guess}} = 0.005$. Then from Eq. 6.22, rearranged to solve for

$V_{\text{first guess}}$, we have

$$V_{\text{first guess}} = \left[\frac{2g(-\Delta z)}{4f}\cdot\frac{D}{\Delta x}\right]^{1/2} = \left(\frac{2\cdot 9.81\ \text{m/s}^2\cdot 10\ \text{m}}{4\cdot 0.005}\cdot\frac{0.1\cdot\text{m}}{100\ \text{m}}\right)^{1/2}$$

$$= 3.13\ \frac{\text{m}}{\text{s}} \tag{6.Q}$$

If this is a good guess, then the computed f should match our first $f_{\text{first guess}}$. Using $V_{\text{first guess}}$, we compute

$$\mathcal{R}_{\text{first guess}} = \frac{0.1\ \text{m}\cdot(3.13\ \text{m/s})\cdot 720(\text{kg/m}^3)}{0.0006\ \text{kg/m}\cdot\text{s}} = 3.76\cdot 10^5 \tag{6.R}$$

From Fig. 6.10 for this value of $\mathcal{R}$ and ε/D we read $f \approx 0.0045$, and from Eq. 6.21 we compute $f \approx 0.00450$. We could repeat the process by hand, using $f_{\text{second guess}} = 0.00450$, and continue until we had satisfactory agreement between the successive values of f. But computers do this very easily for us, so we proceed on a spreadsheet as shown in Table 6.5.

The first column of Table 6.5 shows the names of the variables, the second shows the nature of each variable, and the third shows the values shown above, based on $f_{\text{first guess}} = 0.005$. We see at the bottom of the second column that the ratio of $f_{\text{computed}}/f_{\text{guessed}} = 0.901$. We next ask the spreadsheet's numerical solution package ("goal seek" on Excel spreadsheets) to make the value of $f_{\text{computed}}/f_{\text{guessed}}$ become equal to 1.00 by changing the value of f_{guessed}. We see that for an f_{guessed} of 0.00449, that ratio becomes $1.0001 \approx 1.00$. We could get more significant figures of agreement, but the input data do not justify that so we accept the values in the column at the right as correct. Then

$$Q = \frac{\pi}{4}\cdot(0.1\ \text{m})^2\cdot 3.304\ \frac{\text{m}}{\text{s}} = 0.0260\ \frac{\text{m}^3}{\text{s}} = 0.916\ \frac{\text{ft}^3}{\text{s}} = 411\ \frac{\text{gal}}{\text{min}} \tag{6.S}$$

■

TABLE 6.5
Numerical solution to Example 6.5

Variable	Type	First guess	Solution
D, m	Given	0.1	0.1
L, m	Given	100	100
Δz, m	Given	−10	−10
ε, in	Given	0.0018	0.0018
ρ, kg/m^3	Given	720	720
μ, cP	Given	0.6	0.6
f, guessed	**Guessed**	**0.005**	**0.00449**
V, m/s	Calculated	3.13	3.304
$\mathcal{R}$	Calculated	375,900	396,500
ε/D	Calculated	0.000457	0.000457
f_{computed}	Calculated	0.004505	0.00449
$f_{\text{computed}}/f_{\text{guessed}}$	**Check value**	0.901	**1.0001**

Example 6.6. We want to transport 500 ft^3 / min of air horizontally from our air conditioner to an outbuilding 800 ft away. The air is at 40°F and a pressure of 0.1 psig. At the outbuilding the pressure is to be 0.0 psig. We will use a circular sheet metal duct, which has a roughness of 0.00006 in. Find the required duct diameter.

Here we are applying B.E. to a compressible fluid. However, as discussed in Sec. 5.6, for low fluid velocities B.E. gives the same result as the analysis that takes compressibility into account. Applying B.E. from the inlet of the duct, point 1, to its outlet, point 2, we find

$$\frac{\Delta P}{\rho} = -\mathscr{F} = -4f\frac{\Delta x}{D}\cdot\frac{V^2}{2} \tag{6.23}$$

This is the Type 3 problem, in which we know everything but the pipe diameter. The equation contains the three unknowns f, D, and V; therefore, we need two additional relations. One is supplied by Fig. 6.10; the other, by the continuity equation, which shows that Q (which is given in the problem statement) is equal to $V(\pi / 4)\,D^2$. We could use this relation to eliminate V or D from Eq. 6.23, but this is not particularly convenient. Rather, we proceed by trial and error. First we guess a pipe diameter and calculate the pressure drop from Eq. 6.23. Then we compare the calculated pressure drop with the known value, 0.1 psi, and readjust the guessed pipe diameter until we find the diameter for which the pressure drop is 0.1 psi.

For the density of air we use the average pressure between inlet and outlet and the ideal gas law:

$$\rho_{\text{air}} = \frac{PM}{RT} = \frac{14.75\ \text{lbf / in}^2 \cdot 29\ \text{lbm / lbmol}}{10.73[\text{lbf}\cdot\text{ft}^3 / (\text{in}^2\cdot\text{lbmol}\cdot{}^\circ\text{R})]\cdot 500^\circ\text{R}} = 0.080\,\frac{\text{lbm}}{\text{ft}^3} \tag{6.T}$$

For air at 40°F we have $\mu = 0.017$ cP (see App. A.1). For our first trial we select $D_{\text{first guess}} = 1$ ft. Then

$$\left(\frac{\varepsilon}{D}\right)_{\text{first guess}} = \frac{0.00006\ \text{in}}{12\ \text{in}} = 0.000005 \tag{6.U}$$

$$V_{\text{first guess}} = \frac{(500\ \text{ft}^3 / \text{min})\cdot(\text{min} / 60\ \text{s})}{(\pi / 4)(1\ \text{ft})^2} = 10.6\,\frac{\text{ft}}{\text{s}} = 3.23\,\frac{\text{m}}{\text{s}} \tag{6.V}$$

$$\mathscr{R}_{\text{first guess}} = \frac{1\ \text{ft}\cdot 10.6\ \text{ft / s}\cdot 0.080\ \text{lbm / ft}^3}{0.017\ \text{cP}\cdot 6.72\cdot 10^{-4}\,\text{lbm / ft}\cdot\text{s}\cdot\text{cP}} = 7.43\cdot 10^4 \tag{6.W}$$

From Fig. 6.10 we read $f \approx 0.0049$ and, from Eq. 6.21, $f = 0.00465$. Then

$$\begin{aligned}\Delta P &= \rho_{\text{air}}\left(-4f\frac{\Delta x}{D}\cdot\frac{V^2}{2}\right)\\ &= -0.080\,\frac{\text{lbm}}{\text{ft}^3}\cdot\frac{4}{2}\cdot 0.00465\cdot\frac{800\ \text{ft}}{1\ \text{ft}}\cdot\left(10.6\,\frac{\text{ft}}{\text{s}}\right)^2\cdot\frac{\text{lbf}\cdot\text{s}^2}{32.2\ \text{lbm}\cdot\text{ft}}\cdot\frac{\text{ft}^2}{144\ \text{in}^2}\\ &= -0.0146\,\frac{\text{lbf}}{\text{in}^2} = -101\ \text{Pa}\end{aligned} \tag{6.X}$$

TABLE 6.6
Numerical solution to Example 6.6

Variable	Type	First guess	Solution
Q, cfm	Given	500	500
D, ft	**Guessed**	**1**	**0.667**
L, ft	Given	800	800
ε, in	Given	0.00006	0.00006
ρ, lbm / ft^3	Given	0.08	0.08
μ, cP	Given	0.017	0.017
V, ft / s	Calculated	10.6	23.8
$\mathcal{R}$	Calculated	74301	111396
ε / D	Calculated	0.000005	7.496E-06
f_{computed}	Calculated	0.00465	0.00425
$\Delta P_{\text{computed}}$, psi	Calculated	0.01446	0.1000
Allowed ΔP, psi	Given	0.1	0.1
$\Delta P_{\text{computed}} / \Delta P_{\text{allowed}}$	**Check value**	0.1446	**1.000**

Our first guess of the diameter is too large, because it would result in a pressure drop due to friction that is only about one-seventh of that available; i.e., a duct 1 ft in diameter will do, but we can use a smaller one and still get the required flow with the available pressure difference. We could make a second guess of the diameter and repeat the calculation, but it is much easier to let our computer do this. Table 6.6 shows the spreadsheet solution. As in Table 6.5, the first column lists the variables, the second describes the variables, and the third corresponds to $D_{\text{first guess}} = 1$ ft, with the values shown above in this example. To find the solution in the fourth row we let the spreadsheet's numerical solution routine vary the value of D to make the check value in the lower right corner become 1.0000. This shows that the required diameter is 0.667 ft = 8.00 in = 0.203 m. ■

6.6 SOME COMMENTS ABOUT THE FRICTION FACTOR METHOD AND TURBULENT FLOW

1. The three preceding examples show how we use the friction factor plot, Fig. 6.10, or its numerical equivalent, Eqs. 6.20 and 6.21. However, the calculations for turbulent and transition flow are not reliable to better than ±10 percent, because the exact values of the roughnesses are seldom known to better than that accuracy. Furthermore roughnesses of pipes change over time as they corrode or collect deposits. It is common practice in the long-distance oil and gas pipeline industry to regularly force a scraper (called a "pig") through their pipelines. This cleans the inner pipe surface, thus greatly lowering the roughness and lowering the required pressure drop. The savings in pumping cost more than repay the cost of this regular cleaning.

2. The plot is made up for sections of pipe that contain no valves, elbows, sudden contractions, sudden expansions, etc. These are probably present in all the actual situations described in Examples 6.3, 6.4, 6.5, and 6.6. We will discuss how to account for them in Secs. 6.8 and 6.9.

3. The data on which the plot is based are all taken well downstream of the entrance to the pipe. We will discuss the entrance region briefly in Part IV.
4. The friction factor plot is a generalization of experimental data. One should not attach much theoretical significance to it. So far no one has been able to calculate friction factors for turbulent flow without starting with experimental data.
5. It can be easily shown that for turbulent flows the heat-transfer and mass-transfer coefficients are related fairly simply to the friction factor f. This is so because the eddy that transports momentum (and thus increases the shear stress) also transports heat and mass and, thus, increases the heat and mass transfer. This subject is discussed in heat-transfer and mass-transfer texts as the *Reynolds analogy.*
6. In Part IV we will discuss briefly the measured velocity distributions in turbulent pipe flow. Now we simply note that the velocity profile of turbulent pipe flow is much flatter than that of laminar pipe flows (see Fig. 3.3). Most of the fluid flows in a central core, which moves almost as a unit at nearly the same velocity throughout. There is a thin layer near the pipe wall in which the velocity drops rapidly from the high velocity of the central core to the zero velocity at the wall. Thus, it is quite reasonable to treat the average velocity of a turbulent pipe flow as the velocity representing the entire flow.
7. Now that we all have computers, and many of us have access to programs that solve the above examples quickly and easily, the hand solution of these problems and the reading of friction factors from Fig. 6.10 are part of an chemical engineer's cultural background, but most likely not part of her day-to-day tool kit. However, knowing how these methods work helps the engineer know what those convenient computer programs are doing, and whether or not they are applicable in some unusual situation.

6.7 MORE CONVENIENT METHODS

The friction factor plot, Fig. 6.10, is a very great generalization; all the pressure-drop data for all Newtonian fluids, pipe diameters, and flow rates are put on a single graph. However, as shown in Examples 6.4, 6.5, and 6.6, the plot is tedious to use by hand. Therefore, before the computer age, working engineers rearranged the same experimental data in numerous forms that are more convenient. The resulting methods are more convenient but less compact than the friction factor methods; for instance, one might be using 20 charts instead of only one. These are now mostly of historical interest, because our computer programs are so good. However, the student is likely to encounter some of them and wonder how they are organized. Furthermore, many of them allow more intuitive insight into these flows than the computer programs do. Some of these methods are shown in this section.

Suppose we decide to build an oil refinery, a city water supply system, or an aircraft carrier. We will have to deal with a very large number of fluid flows. We could calculate the friction effect for each from Fig. 6.10. However, in any of these projects we would probably use U.S. standard pipe sizes for practically all of the flows. From App. A.2 we see that they constitute a fairly small number of sizes. For pipes of a given size and of the same material, the diameter and relative roughness are constant. Therefore, for a given size of pipe there are only four variables: $\mathcal{F}$ per foot, Q, ρ, and μ. These can be plotted (for one pipe size) in a way that makes calculations of friction loss very easy. Thus, if we spend the time to make about 10 such plots for the common U.S. pipe sizes, we can save considerable work in designing our refinery, water system, or aircraft carrier. Naturally, oil companies, water-supply companies, and the Navy have done just that.

In making such plots and tables it is customary to set them up for the most common problem, which is the long, horizontal, constant-diameter pipe. For such a pipe, B.E., rearranged, is

$$\left(\frac{-\Delta P}{\Delta x}\right)_{\text{friction}} = \frac{\rho}{\Delta x}\mathcal{F} = \rho\left(\frac{4f}{D}\cdot\frac{V^2}{2}\right) \quad \begin{pmatrix}\text{steady flow, horizontal}\\ \text{pipe, with no pumps or}\\ \text{compressors}\end{pmatrix} \tag{6.24}$$

so the charts customarily can be read directly in $-\Delta P / \Delta x$, dropping the "friction" subscript. If we must use such a chart for some other type of problem, we may read the appropriate $-\Delta P / \Delta x$ and then use Eq. 6.24 to find $\mathcal{F}$.

Figure 6.12 is an example of such a plot. This figure shows, for a 3-in pipe, the pressure drop per 1000 ft as a function of volumetric flow rate, kinematic viscosity (viscosity / density), and specific gravity. The plot is logarithmic on both axes, but the log scale is different for each. A plot like this can be made directly from Fig. 6.10 (Prob. 6.29).

Example 6.7. Rework Example 6.1 by using Fig. 6.12.

Here the B.E. analysis is the same as in Example 6.1. We start on the chart at the right at 50 gal / min and read horizontally to the left to 50.0 cSt line, and then vertically downward. The bottom section allows for fluids of various specific gravities; here the specific gravity is ≈1.00, so we read to the bottom of the plot, finding 7.7 psi / 1000 ft. In this example the pipe is 3000 ft long, so the pressure drop is 3 times 7.7 = 23.1 psi. The perfect agreement with Example 6.1 should not surprise us; the laminar part of Fig. 6.12 was made up from the same equations we used there. ■

Figure 6.12 is a "convenience" chart made up from Fig. 6.10. It is well suited to the needs of an oil company, which spends large sums of money in pumping fluids with a wide range of viscosities, sometimes in laminar flow, sometimes in turbulent flow. But it is poorly suited to the needs of a city water-supply company, which deals almost exclusively with water. When Fig. 6.12 was made from Fig. 6.10, the pipe diameter and roughness were held constant. If we are dealing with water, we can assume that the temperature is constant (which is approximately true in city water systems) and that the absolute roughness of the pipe wall is constant (also approximately true in city water systems). Then the pressure drop as a function of pipe diameter and flow rate can be tabulated for all flows of water at the chosen temperature. Appendix A.3 is such a table, made up for the flow of water at

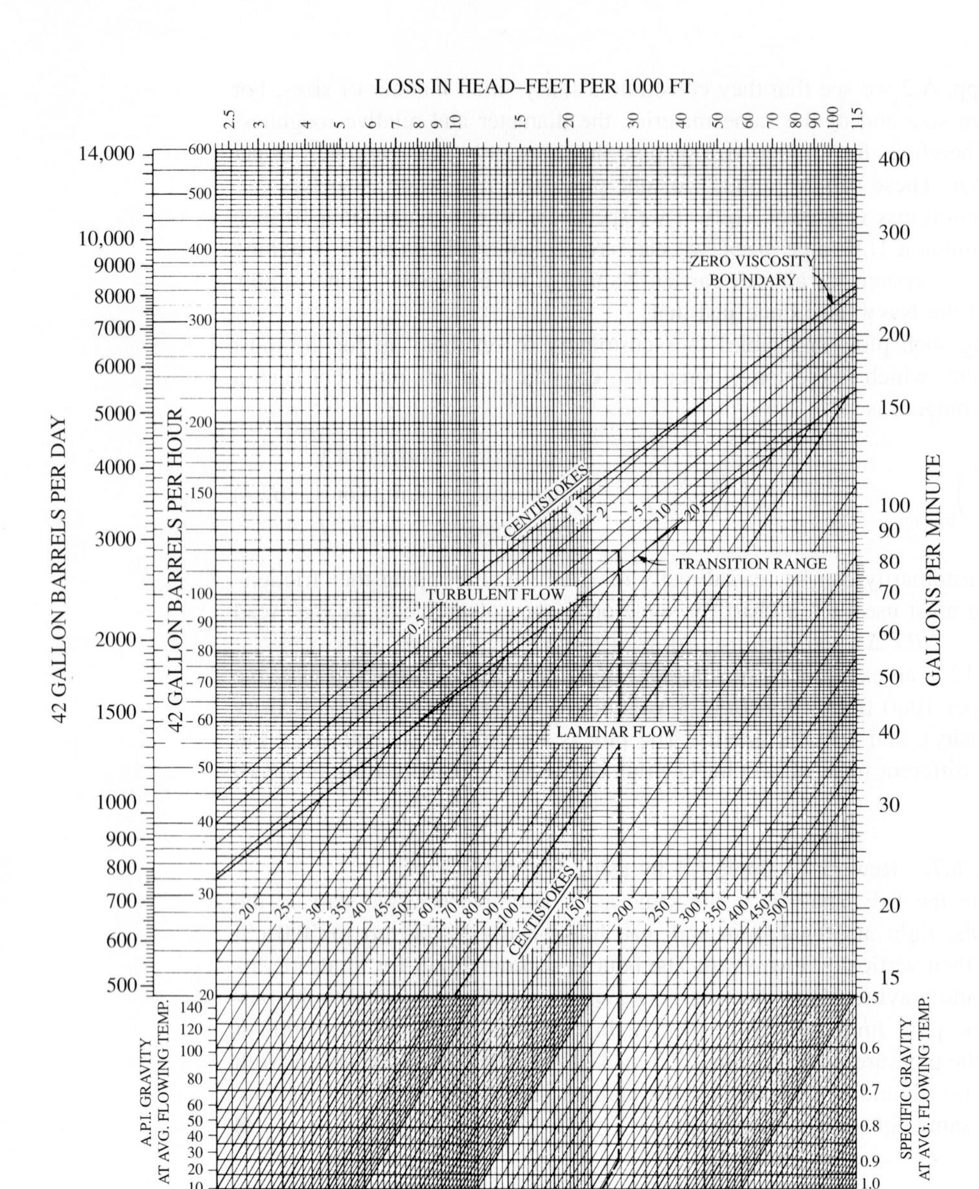

FIGURE 6.12
Pressure drop in a 3-in schedule 40 pipe, 3.068 in inside diameter. Example shown; flow rate = 120 barrels per hour (BPH); kinematic viscosity = 10 cSt; specific gravity = 0.9; pressure loss (follow dashed line) = 10.7 psi / 1000 ft. (Courtesy of the Board of Engineers, Standard Oil Company of California.)

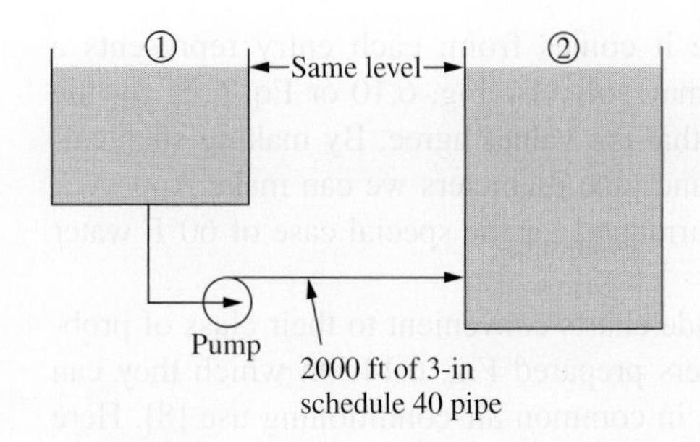

FIGURE 6.13
A pump-driven fluid transfer from one tank to another, used in Example 6.8.

60°F through schedule 40 pipe (the most common size in United States industrial practice).

Example 6.8. Two reservoirs are connected by 2000 ft of 3-in pipe. We want to pump 200 gal / min of water from one to the other. The levels in the reservoirs are the same, and both are open to the atmosphere; see Fig. 6.13. What are the pump work per unit mass, the required pump pressure rise, and the required pump power?

Applying B.E. from the free surface of the first reservoir, point 1, to the free surface of the second, point 2, we see that all terms are zero except

$$0 = \frac{dW_{\text{n.f.}}}{dm} - \mathscr{F} \tag{6.Y}$$

The pump work (positive because of the thermodynamic sign convention) is equal to the friction loss. We could solve this problem using Fig. 6.10 or the spreadsheets in Tables 6.5 and 6.6. But it is faster and easier using App. A.3.

We start at the left of App. A.3 at 200 gal / min and read horizontally to the column for 3-in pipe, where the pressure drop is indicated as (3.87 psi) / (100 ft). The pipe is 2000 ft long, so the pressure drop due to friction is

$$\mathscr{F} = \left(\frac{dW_{\text{n.f.}}}{dm}\right)_{\text{pump}} = \left(\frac{-\Delta P}{\rho}\right)_{\text{friction}}$$

$$= \frac{1}{\rho} \cdot 3.87 \frac{\text{psi}}{100\text{ ft}} \cdot 2000\text{ ft} = \frac{1}{\rho} \cdot 77.4 \frac{\text{lbf}}{\text{in}^2} \tag{6.Z}$$

The pump must increase the pressure of the fluid flowing through it by 77.4 psi to overcome the friction in the 2000 ft of pipe.

The pump power required is

$$\text{Po} = \frac{dW_{\text{n.f.}}}{dt} = \frac{dW_{\text{n.f.}}}{dm}\dot{m} = \dot{m}\frac{1}{\rho} \cdot 77.4 \frac{\text{lbf}}{\text{in}^2} \tag{6.AA}$$

Here

$$\dot{m} = 200 \frac{\text{gal}}{\text{min}} \cdot \frac{\text{min}}{60\text{ s}} \cdot 62.3 \frac{\text{lbm}}{\text{ft}^3} \cdot \frac{\text{ft}^3}{7.48\text{ gal}} = 27.8 \frac{\text{lbm}}{\text{s}} \tag{6.AB}$$

$$\text{Po} = \frac{77.4\text{ lbf / in}^2}{62.3\text{ lbm / ft}^3} \cdot 27.8 \frac{\text{lbm}}{\text{s}} \cdot \frac{32.2\text{ lbm}\cdot\text{ft}}{\text{lbf}\cdot\text{s}^2} \cdot \frac{144\text{ in}^2}{\text{ft}^2} \cdot \frac{\text{hp}\cdot\text{s}}{550\text{ ft}\cdot\text{lbf}}$$

$$= 8.90\text{ hp} = 6.63\text{ kW} \tag{6.AC}$$

The pump power computed here is the amount of mechanical power delivered to the fluid. For a 100 percent efficient pump this would also be the power input to the pump. Real pumps are never 100 percent efficient (nor are the electric motors which drive most of them); their actual behavior is discussed in Chap. 10. ■

In using App. A.3 remember where it comes from; each entry represents a calculation like that in Example 6.4. One may solve by Fig. 6.10 or Eq. 6.21 for the value we read from the table here and see that the values agree. By making such calculations for a large number of flow rates and pipe diameters we can make App. A.3. Thus, that appendix is simply Fig. 6.10, rearranged for the special case of 60°F water flowing in schedule 40 pipes.

Just as oil refinery engineers have made charts convenient to their class of problems, so also have air-conditioning engineers prepared Fig. 6.14, on which they can quickly solve most pressure drop problems in common air-conditioning use [8]. Here they have recognized that almost all flow in air-conditioning ducts is of air at about 70°F and 14.7 psia, for which the viscosity and density are known, and that most of the ducts are made of galvanized steel or aluminum, for which ε is known. Thus, they have one fewer variable than the oil refinery engineers (who treat fluids with a variety of viscosities), so that instead of having to have a separate plot for each size pipe, like Fig. 6.13, they can have one plot that covers all sizes.

Example 6.9. Repeat Example 6.6 by using Fig. 6.14. Here the figure was made for air at 68°F, which we can tell from the assumed density of 0.075 lbm / ft^3, while our problem is for 40°F and a density of 0.080 lbm / ft^3. We know that the viscosities do not match perfectly either. However, we ignore these differences for the moment and simply use Fig. 6.14. We know the flow rate (500 ft^3 / min), the pressure drop (0.1 psi), and the pipe length (800 ft). We need to convert the pressure drop to pressure drop per unit length, which we can do either as

$$\frac{\Delta P}{\Delta x} = \frac{0.1 \text{ psi}}{800 \text{ ft}} \cdot \frac{27.69 \text{ in H}_2\text{O}}{\text{psi}} = 0.00346 \frac{\text{in H}_2\text{O}}{\text{ft}} = 0.346 \frac{\text{in H}_2\text{O}}{100 \text{ ft}} \tag{6.AD}$$

or as

$$\frac{\Delta P}{\Delta x} = \frac{0.1 \text{ psi}}{800 \text{ ft}} \cdot \frac{6.895 \cdot 10^3 \text{ Pa}}{\text{psi}} \cdot \frac{3.28 \text{ ft}}{\text{m}} = 28.3 \frac{\text{Pa}}{\text{m}} \tag{6.AE}$$

These give the same entry point on the abscissa. Reading at the intersection of this pressure gradient and 500 cfm, we find that the required pipe diameter is about 8.2 in and the velocity is about 1350 ft / min = 22.5 ft / s. The close agreement with Example 6.5 (8.0 in, 23.5 ft / s) simply shows that Fig. 6.14 was made using the standard friction factor plot. The small differences between the density and viscosity of air at 70°F and 40°F are the probable cause of the differences shown. ■

Such convenient charts as Figs. 6.13 and 6.14 and App. A.3 are widely used in industry for routine calculations, even in the computer age. When engineers leave the university and join industrial firms, they find that their colleagues have a large supply of them. It is worth the young engineer's while to trace them back to their sources. Not only will they discover how the convenient methods fit in with the ideas learned in the university, but also they will see more clearly the limitations of the convenient methods. Then they can use them for the routine parts of complex jobs, saving their creative efforts for the non-routine parts that will test their talents and education.

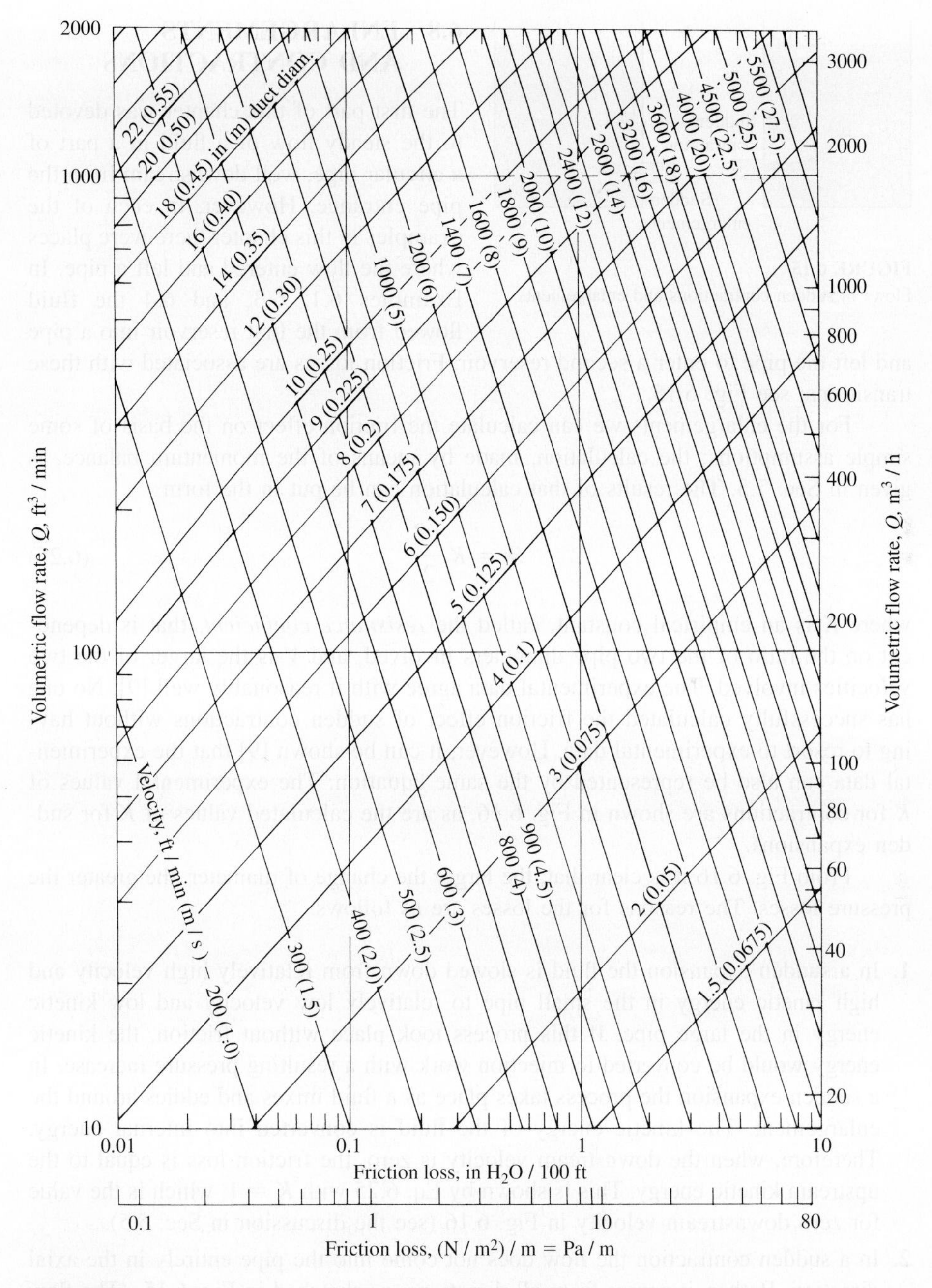

FIGURE 6.14
Friction of air in straight ducts for volumetric flow rates of 10 to 2000 ft^3 / min (20 to 3000 m^3 / h). Based on standard air of 0.075 lb / ft^3 (1.2 kg / m^3) density, flowing through average, clean, round, galvanized metal ducts having approximating 40 joints per 100 ft (30 m). Do not extrapolate below the chart. (Reprinted from the *1972 ASHRAE Handbook—Fundamentals,* with permission.)

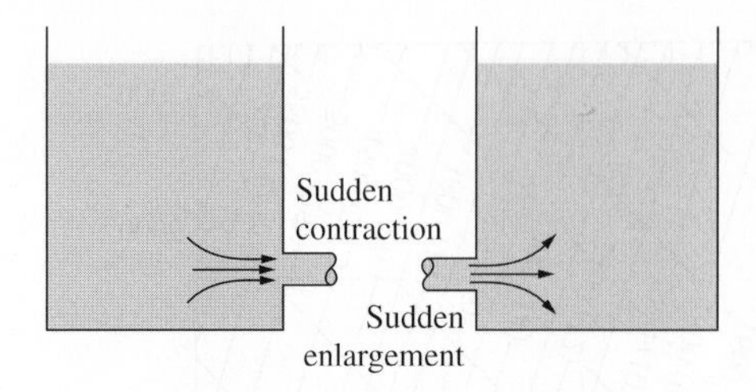

FIGURE 6.15
Flows in sudden contractions and enlargements.

6.8 ENLARGEMENTS AND CONTRACTIONS

The first part of this chapter was devoted to the steady flow of a fluid in a part of a circular pipe, well downstream from the pipe entrance. However, in each of the examples in this chapter there were places where the flow entered and left a pipe. In Examples 6.1, 6.3, and 6.4 the fluid flowed from the first reservoir into a pipe and left the pipe to enter a second reservoir. Friction losses are associated with these transitions; see Fig. 6.15.

For the enlargements we can calculate the friction effect on the basis of some simple assumptions; the calculation, made by means of the momentum balance, is given in Sec. 7.3. The results of that calculation can be put in the form

$$\mathcal{F} = K\frac{V^2}{2} \tag{6.25}$$

where K is an empirical constant, called the *resistance coefficient,* that is dependent on the ratio of the two pipe diameters involved, and V is the larger of the two velocities involved. The experimental data agree with it reasonably well [9]. No one has successfully calculated the friction effect of sudden contractions without having to resort to experimental data. However, it can be shown [9] that the experimental data can also be represented by the same equation. The experimental values of K for contractions are shown in Fig. 6.16, as are the calculated values of K for sudden expansions.

From Fig. 6.16 it is clear that, the larger the change of diameter, the greater the pressure losses. The reasons for the losses are as follows.

1. In a sudden expansion the fluid is slowed down from relatively high velocity and high kinetic energy in the small pipe to relatively low velocity and low kinetic energy in the large pipe. If this process took place without friction, the kinetic energy would be converted to injection work with a resulting pressure increase. In a sudden expansion the process takes place as a fluid mixes and eddies around the enlargement. The kinetic energy of the fluid is converted into internal energy. Therefore, when the downstream velocity is zero, the friction loss is equal to the upstream kinetic energy. This is shown by Eq. 6.25 with $K = 1$, which is the value for zero downstream velocity in Fig. 6.16 (see the discussion in Sec. 5.5).
2. In a sudden contraction the flow does not come into the pipe entirely in the axial direction. Rather, it comes from all directions, as sketched in Fig. 6.15. (The flow is not completely one-dimensional, but rather two- or three-dimensional.) On entering the pipe the flow follows the pattern shown in Fig. 6.17.

The fluid forms a neck, called the *vena contracta,* just downstream of the tube entrance. The flow into the neck is caused by the radial inward velocity of the fluid

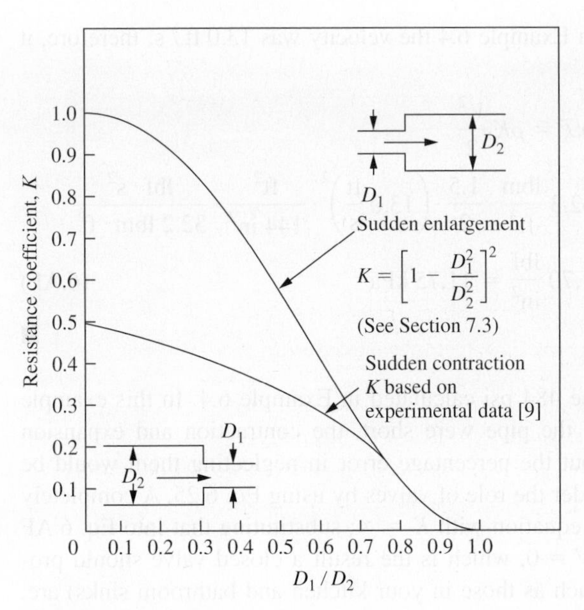

FIGURE 6.16
Resistance due to sudden enlargements and contractions. The resistance coefficient K is defined in Eq. 6.25. (From Crane Technical Paper No. 410, reproduced by permission of the Crane Company.)

approaching the tube. Because it is coming radially inward, the fluid overshoots the tube wall and goes into the neck. This neck is surrounded by a collar of stagnant fluid. In the neck the velocity is greater than the velocity farther downstream. Thus, the kinetic energy decreases from the neck to some point downstream, where the velocity is practically uniform over the cross section of the pipe. This kinetic energy is not all recovered as increased pressure but leads to the friction loss shown in Eq. 6.25 with the values of K from Fig. 6.16.

Our discussion of entrance and exit losses has concerned turbulent flow only. In laminar flow these effects generally are negligible, because the kinetic energies generally are negligible compared with the viscous effects.

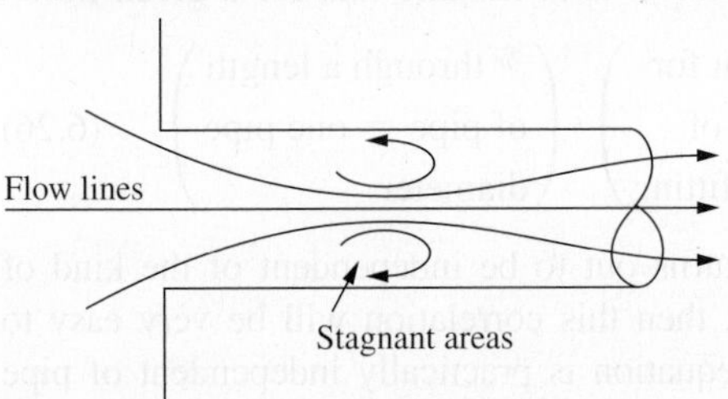

FIGURE 6.17
Flow pattern (turbulent flow) in a sudden contraction, showing that the flow is not truly one-dimensional, but comes in with a radial component, causing it to form a narrow neck before it straightens out into one-dimensional pipe flow.

Example 6.10. Calculate the error made in Example 6.4 by neglecting the expansion and contraction losses.

From Fig. 6.16 we see that for flow from a tank to a pipe the coefficient K is 0.5 and for flow from a pipe to a tank it is 1.0. Thus, the friction loss due to the expansion and contraction should be $1.0 + 0.5 = 1.5$ (kinetic energy of the fluid in the pipe). In Example 6.4 the velocity was 13.0 ft / s; therefore, it can be shown that

$$-\Delta P_{\substack{\text{expansion}\\ \text{and contraction}}} = \rho\mathcal{F} = \rho K \frac{V^2}{2}$$

$$= 62.3 \frac{\text{lbm}}{\text{ft}^3} \cdot \frac{1.5}{2} \cdot \left(13.0 \frac{\text{ft}}{\text{s}}\right)^2 \cdot \frac{\text{ft}^2}{144 \text{ in}^2} \cdot \frac{\text{lbf} \cdot \text{s}^2}{32.2 \text{ lbm} \cdot \text{ft}}$$

$$= 1.70 \frac{\text{lbf}}{\text{in}^2} = 11.75 \text{ kPa} \tag{6.AF}$$

■

This is ≈0.4 percent of the 484 psi calculated in Example 6.4. In this example the pipe was long (3000 ft). If the pipe were short, the contraction and expansion losses would be just as large, but the percentage error in neglecting them would be much greater. We can also consider the role of valves by using Eq. 6.25. A completely closed valve is the same as that equation with $K = \infty$; substituting that into Eq. 6.AF and then into Eq. 5.5 leads to $V = 0$, which is the result a closed valve should produce. Flow-regulating valves (such as those in your kitchen and bathroom sinks) are, in effect variable Ks; wide open, they have a small value of K, fully closed they have an infinite value of K, and by handle adjustment they can take up any value in between, thus controlling the flow.

6.9 FITTING LOSSES

In addition to expansions and contractions, in most fluid systems we must take into account the effect of valves, elbows, etc. They are much more complex to analyze than the one-dimensional flows we have considered so far (take apart an ordinary household faucet and study its flow path; it is much more complicated than that of a straight pipe). Efforts at calculating the friction losses in such fittings have been made, and the results have been correlated in two convenient ways which allow us to treat them as if they were one-dimensional problems.

The first way to correlate these test results is to assume that for a given flow

$$\begin{pmatrix} \mathcal{F} \text{ through} \\ \text{a valve or fitting} \end{pmatrix} = \begin{pmatrix} \text{a constant for} \\ \text{that kind of} \\ \text{valve or fitting} \end{pmatrix} \cdot \begin{pmatrix} \mathcal{F} \text{ through a length} \\ \text{of pipe} = \text{one pipe} \\ \text{diameter} \end{pmatrix} \tag{6.26}$$

If for a given kind of fitting this constant turns out to be independent of the kind of flow, and independent of the pipe diameter, then this correlation will be very easy to use. In turbulent flow, the constant in this equation is practically independent of pipe size, flow rate, and nature of the fluid flowing. If we know the value of the constant for a particular kind of fitting, we can calculate an "equivalent length of pipe" that would have the same friction effect as the fitting, and we can add this length to the actual length of the pipe to find an *adjusted length,* which gives practically the same friction effect as does the actual pipe including fittings. The constant in the equation is dimensionless. Typical values are shown in Table 6.7; they are referred to as *equivalent lengths.*

TABLE 6.7

Equivalent lengths and *K* values for various kinds of fitting*

Type of fitting	Equivalent length, L/D, dimensionless	Constant, K, in Eq. 6.25, dimensionless
Globe valve, wide open	350	6.3
Angle valve, wide open	170	3.0
Gate valve, wide open	7	0.13
Check valve, swing type	110	2.0
90° standard elbow	32	0.74
45° standard elbow	15	0.3
90° long-radius elbow	20	0.46
Standard tee, flow-through run	20	0.4
Standard tee, flow-through branch	60	1.3
Coupling	2	0.04
Union	2	0.04

**Source:* Reference 10.

Example 6.11. Rework Example 6.4 on the assumption that, in addition to the 3000 ft of 3-in pipe, the line contains two globe valves, a swing check valve, and nine 90° standard elbows.

Using the constants in Table 6.7, we can calculate the equivalent length of 3-in pipe that would have the same friction effect as these fittings. This is:

$$\sum L/D = 2 \cdot 350 + 1 \cdot 110 + 9 \cdot 32 = 1098 \qquad \text{(6.AG)}$$

From Eq. 6.26 we see that this is the number of pipe diameters needed to have the same friction loss as the fittings. Thus, the equivalent length is $1098 \cdot [(3.068/12)\text{ ft}] = 281$ ft. Therefore, the adjusted length of the pipe is

$$\begin{pmatrix}\text{Adjusted}\\\text{length}\end{pmatrix} = \begin{pmatrix}\text{actual pipe}\\\text{length}\end{pmatrix} + \begin{pmatrix}\text{equivalent length}\\\text{for fittings}\end{pmatrix}$$
$$= 3000 + 281 = 3281\text{ ft} \qquad \text{(6.AH)}$$

The total pressure drop is

$$-\Delta P_{\text{total}} = 484\text{ psi} \cdot \frac{3281\text{ ft}}{3000\text{ ft}} = 529\text{ psi} \qquad \text{(6.AI)}$$

and that due to the valves and fittings

$$-\Delta P_{\text{valves and fittings}} = 529\text{ psi} - 484\text{ psi} = 45\text{ psi} = 310\text{ kPa} \qquad \text{(6.AJ)}$$

■

The second way to represent the same experimental data for the friction losses in valves or fittings is to assign a value of K in Eq. 6.25 to each kind of fitting. Those values, based on friction-loss experiments, are also shown in Table 6.7.

Example 6.12. Repeat Example 6.11, using the K values in Table 6.7.

Using those values, we compute that

$$\sum K_{\text{valves and fittings}} = 3 \cdot 6.3 + 1 \cdot 2.0 + 9 \cdot 0.74 = 27.56 \qquad \text{(6.AK)}$$

and

$$-\Delta P_{\text{valves and fittings}} = 27.56 \cdot 62.3 \frac{\text{lbm}}{\text{ft}^3} \cdot \frac{(13.0 \text{ ft/s})^2}{2} \cdot \frac{\text{lbf} \cdot \text{s}^2}{32.2 \text{ lbm} \cdot \text{ft}} \cdot \frac{\text{ft}^2}{144 \text{ in}^2}$$

$$= 31 \text{ psi} = 216 \text{ kPa} \tag{6.AL}$$

■

The fact that the second method of making this estimate gives an answer only 69% of the first reminds us that these procedures give only a fair estimate of the pressure drop, not as reliable an estimate as we can make for flow in a straight pipe. These two methods appear to be quite different, but are not. If we write the friction heating for each method,

$$\mathcal{F} = 4f\left(\frac{L}{D}\right)_{\text{equivalent}} \frac{V^2}{2} = K_{\text{fitting}} \frac{V^2}{2} \tag{6.27}$$

we see that they are the same if $4f(L/D)_{\text{equivalent}} = K_{\text{fitting}}$. The equivalent length method lets $\mathcal{F}$ of a fitting vary with the size of the pipe and the Reynolds number, whereas the K method makes it independent of those. Both seem to match the experimental data about as well as each other. Lapple [10] suggests that the equivalent length method matches experimental results better when $\mathcal{R} < 10^5$ and the K method matches experimental results better when $\mathcal{R} > 10^5$.

Laminar flow has yielded little experimental data on which to base pressure-drop correlations for valves and fittings. Generally, the adjusted length calculated by the method given above will be correct for turbulent flow but will be too large for laminar flow. Empirical guides to estimating the adjusted length for laminar flow have been published [11].

Do not attach theoretical significance to these empirical relations for fitting losses. They are simply the results of careful tests of specific cases, arranged in a way that is useful in predicting the behavior of new systems. Fitting losses and expansion and contraction losses are often lumped as *minor losses* even though for a short piping system they may be larger than the straight pipe loss.

6.10 FLUID FRICTION IN ONE-DIMENSIONAL FLOW IN NONCIRCULAR CHANNELS

6.10.1 Laminar Flow in Noncircular Channels

Steady laminar flow in a circular pipe is one of the simplest flow problems. A somewhat harder problem is steady flow of an incompressible Newtonian fluid in some constant-cross-section duct or pipe that is not circular, such as a rectangular duct or an open channel. For laminar flow of a Newtonian fluid the problem can be solved analytically for several shapes. Generally, the velocity depends on two dimensions. In several cases of interest the problems can be solved by the same method we used to find Eq. 6.9; i.e., setting up a force balance around some properly chosen section of the flow, solving for the shear stress, introducing the Newtonian law of viscosity for the shear stress, and integrating to find the velocity distribution. From the velocity distribution the volumetric flow rate-pressure drop relation is found.

That these are all similar to the solution for laminar flow in a horizontal circular tube may be seen by comparing the horizontal, steady-flow solutions with that for a circular tube. For a circular tube,

$$Q = \left(\frac{P_1 - P_2}{\Delta x} \cdot \frac{1}{\mu}\right) \cdot \frac{\pi}{128} D_0^4 \tag{6.9}$$

For a slit between two parallel plates (Prob. 6.48),

$$Q = \left(\frac{P_1 - P_2}{\Delta x} \cdot \frac{1}{\mu}\right) \cdot \frac{1}{12} lh^3 \tag{6.28}$$

where h is the distance between plates and l is the width of the slit. If both sides are divided by l, the left-hand side becomes the volumetric flow rate per unit width. For an annulus (Prob. 6.49),

$$Q = \left(\frac{P_1 - P_2}{\Delta x} \cdot \frac{1}{\mu}\right) \cdot \frac{\pi}{128} (D_o^2 - D_i^2) \left[D_o^2 + D_i^2 - \frac{D_o^2 - D_i^2}{\ln(D_o / D_i)} \right] \tag{6.29}$$

where D_o is the outer diameter and D_i is the inner diameter. These equations differ only by the terms at the far right, which account for the different geometries. Most of the cases that can be worked out by simple mathematics have been summarized by Bird [12, Chaps. 2–4] and Sakiadis [13]. One may show (Prob. 6.52) that as the spacing in the annulus become small ($D_i \rightarrow D_o$), Eq. 6.29 reduces to

$$Q = \left(\frac{P_1 - P_2}{\Delta x} \cdot \frac{1}{\mu}\right) \cdot \frac{1}{12} \cdot \pi D \left(\frac{D_o - D_i}{2}\right)^3 \tag{6.30}$$

which is the same as Eq. 6.28 with the slit length and slit width renamed. This is one of many circular or annular problems that can be simplified by converting them to equivalent straight or planar problems.

6.10.2 Seal Leaks

An extremely important chemical engineering application of Eq. 6.30 is the problem of seal leakage. Figure 6.18 shows three kinds of seals. Figure 6.18(*a*) shows a static seal, as exists between the bottle cap and the top of a soft-drink or beer bottle. A thin washer of elastomeric material is compressed between the metal cap and the glass bottle top. This compressed material forms a seal that prevents the escape of CO_2 (carbonation). The leakage rate is not exactly zero, but it is small enough to hold the carbonation for many years. Leaks through this kind of seal are generally unimportant. Sealing is more difficult when one of the sealed surfaces moves relative to the other.

Figure 6.18(*b*) shows a simple compression seal between a housing and a shaft. The example shown is a water faucet, in which a nut screws down over the body of

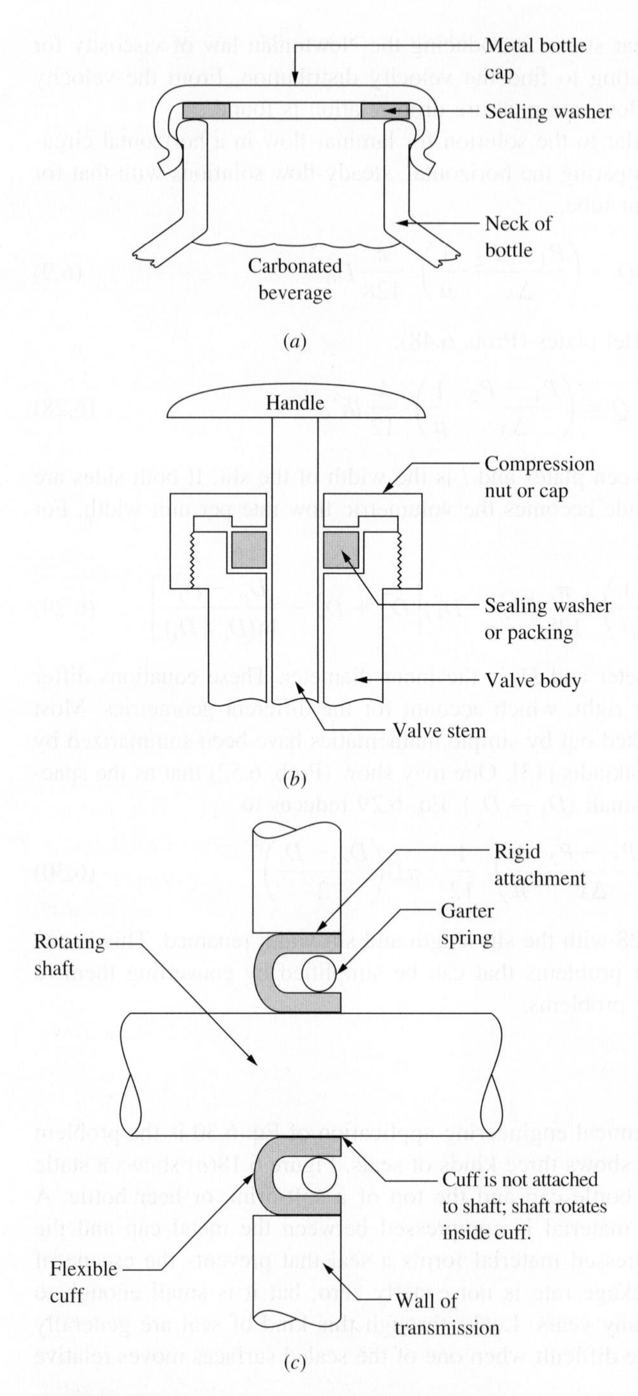

FIGURE 6.18
Three kinds of seals: (*a*) a static seal, as exists between a soft-drink or beer bottle and its bottle cap, (*b*) a packed seal, as exists between the valve stem and valve body of simple faucets, and as also exists on many simple pumps, (*c*) a rotary seal of the type common on the drive shafts of automobiles and some pumps.

the faucet to compress an elastomeric seal, which is trapped between the body of the faucet and the stem of the valve. The compressed seal must be tight enough to prevent leakage of the high-pressure water inside the valve out along the edge of the stem, but not so tight that the valve cannot be easily rotated by hand. Students are probably aware from personal experience that this type of seal often leaks. If the leak is a small amount of water into the bathroom sink, that causes little problem; tightening the nut normally reduces the leak to a rate low enough that it becomes invisible (but does not become zero!).

Example 6.13. A valve has a seal of the type shown in Fig. 6.18(*b*). Inside the valve is gasoline at a pressure of 100 psig. The space between the seal and the valve stem is assumed to have an average thickness of 0.0001 in. The length of the seal, in the direction of leakage, is 1 in. The diameter of the valve stem is 0.25 in. Estimate the gasoline leakage rate.

This is the case described by Eq. 6.30, Inserting values, we have

$$Q = \frac{100 \text{ lbf / in}^2}{1 \text{ in}} \cdot \frac{1}{12 \cdot 0.6 \text{ cP}} \cdot \pi \cdot 0.25 \text{ in} \cdot (10^{-4} \text{ in})^3 \cdot \frac{\text{cP} \cdot \text{ft}^2}{2.09 \cdot 10^{-5} \text{ lbf} \cdot \text{s}} \cdot \frac{144 \text{ in}^2}{\text{ft}^2}$$

$$= 7.5 \cdot 10^{-5} \frac{\text{in}^3}{\text{s}} = 0.27 \frac{\text{in}^3}{\text{h}} = 1.3 \cdot 10^{-9} \frac{\text{m}}{\text{s}} \tag{6.AM}$$

$$\dot{m} = Q\rho = 0.27 \frac{\text{in}^3}{\text{h}} \cdot 0.026 \frac{\text{lbm}}{\text{in}^3} = 0.007 \frac{\text{lbm}}{\text{h}} = 0.0032 \frac{\text{kg}}{\text{h}} \tag{6.AN}$$

Tests indicate that the average leak rate from many oil refinery valves, processing this kind of liquid, is about 0.024 lbm / h, 3.5 times the value calculated here [13]; see Prob. 6.49. There are enough of these valves in a typical oil refinery or chemical plant that they contribute significantly to the overall emissions of hydrocarbons and chemicals to the atmosphere [14, p. 347]. ■

Figure 6.18(*c*) shows in greatly simplified form the seal that surrounds the drive shaft of an automobile, where that shaft exits from the transmission. The inside of the transmission is filled with oil. The flexible seal is like a shirt cuff turned back on itself, with the outside held solidly to the wall of the transmission and the inside held loosely against the rotating shaft by a "garter spring." If we set that spring loosely, then there will be a great deal of leakage. If we set it very tight, then the friction and wear between the cuff and the shaft that rotates inside it will be excessive. The setting of the tension on that spring is a compromise between the desire for low leakage and the desire for low friction and wear (buyers expect these seals to last as long as the auto!). That compromise normally leads to a low, but not a zero leakage rate; a small amount of oil is always dripping out and accumulating on the floor of our garages. Valves and pumps also have shafts that must rotate and hence have the same kind of leakage problem. Pumps and valves in chemical engineering all have the same kind of leakage problem shown Example 6.13. The seals regularly used are more complex versions of the ones shown there.

6.10.3 Turbulent Flow in Noncircular Channels

We are no more able to calculate the pressure drop in steady, *turbulent* flow in a noncircular channel than we are in a circular one. However, it seems reasonable to expect that we could use the friction-loss results for circular pipes to estimate the results for other shapes. Let us *assume* that for a given fluid the shear stress at the wall of any conduit is the same for a given V_{average}, independent of the shape of the conduit. Then, from a force balance on a horizontal section like that leading to Eq. 6.3, we conclude that in steady flow

$$\Delta P \cdot \begin{pmatrix}\text{area perpendicular}\\ \text{to the flow}\end{pmatrix} = \begin{pmatrix}\text{wall shear}\\ \text{stress}\end{pmatrix} \cdot \begin{pmatrix}\text{wetted}\\ \text{perimeter}\end{pmatrix} \cdot \Delta x \tag{6.31}$$

Rearranging this, we find

$$\frac{\Delta P}{\Delta x} = \tau \cdot \begin{pmatrix}\text{wetted}\\ \text{perimeter}\end{pmatrix} \Big/ \begin{pmatrix}\text{area perpendicular}\\ \text{to the flow}\end{pmatrix} \tag{6.32}$$

We now define a new term:

$$\begin{pmatrix}\text{Hydraulic}\\ \text{radius}\end{pmatrix} = \text{HR} = \begin{pmatrix}\text{area perpendicular}\\ \text{to the flow}\end{pmatrix} \Big/ \begin{pmatrix}\text{wetted}\\ \text{perimeter}\end{pmatrix} \tag{6.33}$$

For a circular pipe this is

$$\text{HR} = \frac{\pi r^2}{2\pi r} = \frac{r}{2} = \frac{D}{4} \qquad [\text{circular pipe}] \tag{6.AO}$$

If the assumptions that went into Eq. 6.31 are correct, then we can construct the ratio of the pressure drop per unit length in a noncircular conduit to that in a circular one:

$$\frac{(\Delta P / \Delta x)_{\text{noncircular}}}{(\Delta P / \Delta x)_{\text{circular}}} = \frac{1 / \text{HR}}{4 / D} = \frac{D}{4\,\text{HR}} \tag{6.AP}$$

$$\left(\frac{\Delta P}{\Delta x}\right)_{\text{noncircular}} = \left(\frac{\Delta P}{\Delta x}\right)_{\text{circular}} \left(\frac{D}{4\text{ HR}}\right) \tag{6.AQ}$$

But for turbulent flow $(\Delta P / \Delta x)_{\text{circular}}$ is given by the friction-factor equation, Eq. 6.23. Substituting, we get

$$\left(\frac{\Delta P}{\Delta x}\right)_{\text{noncircular}} = \frac{-4f\rho V^2}{2D} \cdot \frac{D}{4\text{ HR}} = \frac{-f\rho V^2}{2\text{ HR}} \tag{6.34}$$

Alternatively, we may write

$$\mathscr{F}_{\text{noncircular}} = \frac{f\,\Delta x}{\text{HR}} \cdot \frac{V^2}{2} \tag{6.35}$$

What value of f should we use in Eqs. 6.34 and 6.35? Experimental results indicate that those equations work *fairly well* if one uses the ordinary friction factor plot (Fig. 6.10) but replaces the diameter in $\mathscr{R}$ and in ε / D with 4 HR. The equations do not work well for shapes that depart radically from circles, such as long, narrow slits.

Example 6.14. Air at 1 atm and 68°F is flowing in a long, rectangular duct whose cross section is 1 ft by 0.5 ft, with $V_{avg} = 40$ ft / s. The roughness of the duct is 0.00006 in. What is the pressure drop per unit length?

First we calculate the hydraulic radius:

$$\text{HR} = \frac{0.5\ \text{ft}^2}{(2 \cdot 1\ \text{ft}) + (2 \cdot 0.5\ \text{ft})} = 0.167\ \text{ft} \tag{6.AR}$$

$$\mathcal{R} = \frac{V\rho(4\ \text{HR})}{\mu} = \frac{40\ \text{ft / s} \cdot 0.075\ \text{lbm / ft}^3 \cdot (4 \cdot 0.1667)\ \text{ft}}{0.018\ \text{cP} \cdot (6.72 \cdot 10^{-4}\ \text{lbm / ft} \cdot \text{s} \cdot \text{cP})} = 1.65 \cdot 10^5 \tag{6.AS}$$

$$\frac{\varepsilon}{D} = \frac{\varepsilon}{4\ \text{HR}} = \frac{0.00006\ \text{in}}{(4 \cdot 0.1667)\ \text{ft} \cdot (12\ \text{in / ft})} = 7.5 \cdot 10^{-6} \tag{6.AT}$$

From Fig. 6.10 (for this low an ε / D we use the "smooth-tubes" curve) we find $f = 0.0039$ or, using Eq. 6.21, we find $f = 0.00390$. Using the latter value in Eq. 6.34, we find

$$-\frac{\Delta P}{\Delta x} = \frac{0.00390 \cdot 0.075\ \text{lbm / ft}^3 \cdot (40\ \text{ft / s})^2}{2 \cdot 0.1667\ \text{ft}} \cdot \frac{\text{ft}^2}{144\ \text{in}^2} \cdot \frac{\text{lbf} \cdot \text{s}^2}{32.2\ \text{lbm} \cdot \text{ft}}$$

$$= 3.0 \cdot 10^{-4}\ \frac{\text{lbf / in}^2}{\text{ft}} = 0.82\ \frac{\text{in H}_2\text{O}}{100\ \text{ft}} = 6.8\ \frac{\text{Pa}}{\text{m}} \tag{6.AU}$$

■

We may check this result by using Fig. 6.14. Here we assume that the pressure drop in this rectangular duct should be similar but not identical to that for the same volumetric flow rate in a circular duct with the same cross-sectional area. The diameter of such a duct would be

$$D = \sqrt{\frac{4}{\pi} A} = \sqrt{\frac{4}{\pi}\ 0.5\ \text{ft}^2} = 0.80\ \text{ft} = 9.6\ \text{in} \tag{6.AV}$$

Entering Fig. 6.14 at $V = 2400$ ft / min, and interpolating between the 9-in and 10-in diameter lines we find approximately the same pressure gradient values shown in Eq. 6.AU. Many air-conditioning ducts are rectangular or square, because for a given cross-sectional area they can fit in a smaller ceiling space than an equivalent circular duct. This comparison shows that the pressure gradient in them is very similar to that in a circular duct of equal cross-sectional area.

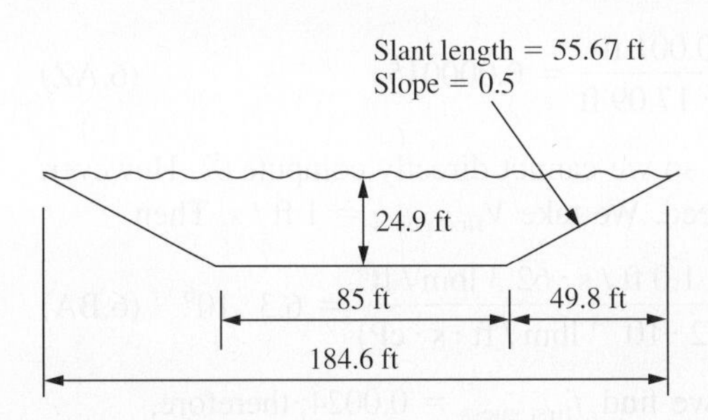

FIGURE 6.19
Cross-sectional view of one of the large irrigation water canals of California's Central Valley Project. The sloping sides have $dy / dx = 0.50$. The cross-sectional area (at the design depth of 24.9 ft) = 3356 ft^3, the wetted perimeter = 196.4 ft, and HR = 17.09 ft. See Example 6.15.

Example 6.15. Figure 6.19 shows the cross section of one of the canals in the Central Valley Project irrigation system in California. The slope is 0.00004 ft / ft = 0.21 ft / mile. What are the velocity and volumetric flow rate in this canal?

First we apply B.E. from some upstream point in the canal to some downstream point in the canal. Since both points are open to the atmosphere, the pressures are the same. For steady flow of a constant-density fluid in a canal of constant cross-sectional area the velocities at the two points are the same. There is no pump or turbine work. Therefore, the remaining terms are

$$g\,\Delta z = -\mathcal{F} \tag{6.AW}$$

This says that the decrease in potential energy is exactly equal to the energy "loss" due to friction, i.e., the mechanical energy converted to internal energy. Substituting for $\mathcal{F}$ from Eq. 6.23, we find

$$-g\,\Delta z = \frac{f\,\Delta x\,V^2}{2\,\text{HR}}; \qquad V = \left(\frac{2 \cdot \text{HR} \cdot g}{f} \cdot \frac{-\Delta z}{\Delta x}\right)^{1/2} \tag{6.AX}$$

In the previous example the wetted perimeter was the entire perimeter of the duct. Here we do not include the part of the perimeter facing the air, because the air exerts little resistance to the flow compared with the walls of the canal. The reader can verify this by watching the flow of leaves or bits of wood on any open stream or irrigation ditch; those at the center move much faster than those at the edges. If the air restrained the flow as much as the solid walls do, then the whole top surface of the flow would not move at all, just as the fluid right at the solid boundaries does not move. Therefore, the hydraulic radius is

$$\text{HR} = \frac{\text{flow area}}{\text{wetted perimeter}} = \frac{3356\ \text{ft}^2}{196.36\ \text{ft}} = 17.09\ \text{ft} \tag{6.AY}$$

The absolute roughness of a concrete-lined irrigation ditch is estimated from Table 6.2 at the low end of the range of values shown, 0.001 ft, so the estimated relative roughness is

$$\frac{\varepsilon}{4\,\text{HR}} = \frac{0.001\ \text{ft}}{4 \cdot 17.09\ \text{ft}} = 0.000015 \tag{6.AZ}$$

Here we do not know the velocity, so we cannot directly compute $\mathcal{R}$. However, we can guess a velocity, and proceed. We take $V_{\text{first guess}} = 1$ ft / s. Then

$$\mathcal{R}_{\text{first guess}} = \frac{4 \cdot 17.09\ \text{ft} \cdot 1.0\ \text{ft / s} \cdot 62.3\ \text{lbm / ft}^3}{1.0\ \text{cP} \cdot (6.72 \cdot 10^{-4}\ \text{lbm / ft} \cdot \text{s} \cdot \text{cP})} = 6.3 \cdot 10^6 \tag{6.BA}$$

Thus, from Fig. 6.10 or Eq. 6.21 we find $f_{\text{first guess}} = 0.0024$; therefore,

$$V_{\text{second guess}} = \left(\frac{2 \cdot 17.09\ \text{ft} \cdot 32.2\ \text{ft / s}^2}{0.0024} \cdot 0.00004\right)^{1/2} = 4.28\ \frac{\text{ft}}{\text{s}} \tag{6.BB}$$

The design value is 3.89 ft / s, indicating that we have chosen a smaller value for the roughness of the concrete canal walls than did the designer (who had experimental data on such canals). We could compute a second guess of $\mathcal{R}$ and f, finding $f = 0.0023$, but we are not justified in doing this, because of our uncertainty of the roughness.

Using the design value of the velocity, we find

$$Q = VA = 3.89 \frac{\text{ft}}{\text{s}} \cdot 3356 \text{ ft}^2 \approx 13{,}100 \frac{\text{ft}^3}{\text{s}} \tag{6.BC}$$

which is the design volumetric flow rate of this section of the canal system. ■

6.11 MORE COMPLEX PROBLEMS INVOLVING B.E.

Now that we can evaluate all the terms in B.E., we may consider some of the more interesting types of problem that this equation can be used to solve.

Example 6.16. A large, high-pressure chemical reactor contains water at a pressure of 2000 psi. A 3-in schedule 80 line connecting to it ruptures at a point 10 ft from the reactor. What is the flow rate through this break?

This is an unsteady-state problem; the reactor pressure will fall during the outflow. However, if the reactor is large, the unsteady-state contribution can be neglected, and we will do so here. Applying B.E. from the free liquid surface in the reactor to the exit of the pipe, we neglect the potential-energy terms, which are negligible, and the small velocity at the free surface. The remaining terms are

$$\frac{P_2 - P_1}{\rho} + \frac{V_2^2}{2} = -\mathcal{F} \tag{6.36}$$

The flow rate in this case will be much higher than is used in common industrial practice, so App. A.3 will be of no use to us. Here the friction loss consists of two parts: the entrance loss into the pipe and the loss due to the flow through the 10 ft of pipe. Substituting from Eqs. 6.25 and 6.17, we find

$$\mathcal{F} = \underbrace{K\frac{V^2}{2}}_{\textit{entrance}} + \underbrace{4f\frac{\Delta x}{D}\cdot\frac{V^2}{2}}_{\textit{straight pipe}} \tag{6.BD}$$

Substituting Eq. 6.BD into Eq. 6.36, we find

$$\frac{P_2 - P_1}{\rho} = \frac{-V_2^2}{2} - \frac{KV_2^2}{2} - 4f\frac{\Delta x}{D}\cdot\frac{V_2^2}{2} = \frac{-V_2^2}{2}\left(1 + K + 4f\frac{\Delta x}{D}\right) \tag{6.BE}$$

$$V_2 = \left(\frac{2(P_1 - P_2)/\rho}{1 + K + 4f(\Delta x/D)}\right)^{1/2} \tag{6.BF}$$

From Fig. 6.16 we can read $K = 0.5$ (the diameter of the line is much smaller than the tank diameter). From App. A.2 we find that for 3-in schedule 80 pipe the inside diameter is 2.900 in. Then, from Table 6.2,

$$\frac{\varepsilon}{D} = \frac{0.0018 \text{ in}}{2.900 \text{ in}} = 0.00062 \tag{6.BG}$$

It is safe to assume that the Reynolds number here will be very high; so on Fig. 6.10 we select as our first guess a friction factor at the far right of the diagram, which for an ε / D of 0.00062 gives us $f = 0.0043$. Then

$$V_2 = \left[\frac{2(2000 - 15) \text{ lbf / in}^2}{62.3 \dfrac{\text{lbm}}{\text{ft}^3} \cdot \left(1 + 0.5 + \dfrac{4 \cdot 0.0043 \cdot 10 \text{ ft}}{(2.900 / 12) \text{ ft}}\right)} \cdot \frac{32.2 \text{ lbm} \cdot \text{ft}}{\text{lbf} \cdot \text{s}^2} \cdot \frac{144 \text{ in}^2}{\text{ft}^2} \right]^{1/2} \tag{6.BH}$$

$$= \left[\frac{2 \cdot 1985 \cdot 32.2 \cdot 144}{62.3 \cdot (1 + 0.5 + 0.71)} \cdot \frac{\text{ft}^2}{\text{s}^2} \right]^{1/2} = 365 \frac{\text{ft}}{\text{s}} = 111 \frac{\text{m}}{\text{s}} \tag{6.BI}$$

We then check to see whether our assumed friction factor is correct:

$$\mathcal{R} = \frac{(2.9 / 12) \text{ ft} \cdot 62.3 \cdot \text{lbm / ft}^3 \cdot 365 \text{ ft / s}}{1.0 \text{ cP} \cdot (6.72 \cdot 10^{-4} \text{ lbm / ft} \cdot \text{s} \cdot \text{cP})} = 8.2 \cdot 10^6 \tag{6.BJ}$$

From Fig. 6.10 we see that our assumed f was correct. From App. A.2 we see that for 1 ft / s the flow rate is 20.55 gal / min; so the flow rate is

$$Q = 365 \frac{\text{ft}}{\text{s}} \cdot \frac{20.55 \text{ gal / min}}{\text{ft / s}} = 7500 \frac{\text{gal}}{\text{min}} = 0.47 \frac{\text{m}^3}{\text{s}} \tag{6.BK}$$

This is the instantaneous volumetric flow rate. As the flow continues, the pressure and flow rate will both decrease. ■

Appendix A.3 is of no use in this case, because the flow velocity is much larger than normal pipeline velocities. Ignoring the kinetic energy of the exit fluid, or the friction loss in the pipe, or the entrance loss would have given a significantly incorrect answer.

Example 6.17. A fire truck, Fig. 6.20, is sucking water from a river and delivering it through a long hose to a nozzle, from which it issues at a velocity of 100 ft / s. The total flow rate is 500 gal / min. The hoses have a diameter equivalent to that of a 4-in schedule 40 pipe and may be assumed to have the same relative roughness. The total length of hose, corrected for valves, fittings, entrance, etc., is 300 ft. What is the power required of the fire truck's pump?

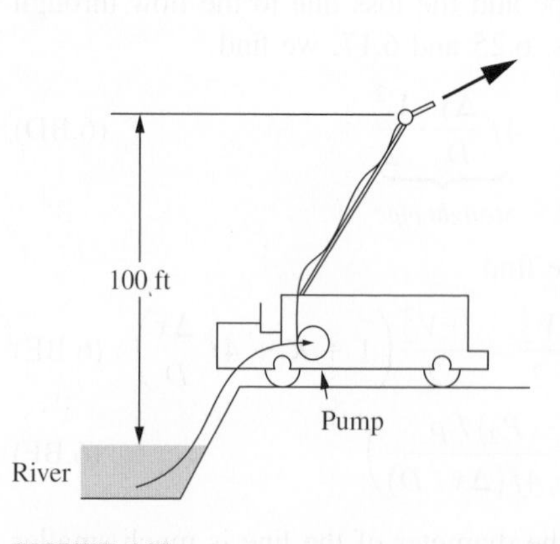

FIGURE 6.20
Fire truck pumping water from a river, used in Example 6.17.

Applying B.E. from the surface of the river, point 1, to the outlet of the nozzle, point 2, we find

$$g(z_2 - z_1) + \frac{V_2^2}{2} = \frac{dW_{\text{n.f.}}}{dm} - \mathcal{F} \tag{6.BL}$$

Here we may find the friction-loss term from App. A.3 and Eq. 6.24:

$$\mathscr{F} = \frac{-\Delta P}{\Delta x} \cdot \frac{\Delta x}{\rho} = \frac{5.65 \text{ lbf / in}^2}{100 \text{ ft}} \cdot \frac{300 \text{ ft}}{62.3 \text{ lbm / ft}^3} \cdot \frac{32.2 \text{ lbm} \cdot \text{ft}}{\text{lbf} \cdot \text{s}^2} \cdot \frac{144 \text{ in}^2}{\text{ft}^2}$$

$$= 1260 \frac{\text{ft}^2}{\text{s}^2} \tag{6.BM}$$

Then we have

$$\frac{dW_{\text{n.f.}}}{dm} = 32.2 \frac{\text{ft}}{\text{s}^2} \cdot 100 \text{ ft} + \frac{(100 \text{ ft / s})^2}{2} + 1260 \frac{\text{ft}^2}{\text{s}^2} = 9480 \frac{\text{ft}^2}{\text{s}^2} \tag{6.BN}$$

$$\dot{m} = 500 \frac{\text{gal}}{\text{min}} \cdot 8.33 \frac{\text{lbm}}{\text{gal}} \cdot \frac{\text{min}}{60 \text{ s}} = 69.5 \frac{\text{lbm}}{\text{s}} = 31.6 \frac{\text{kg}}{\text{s}} \tag{6.BO}$$

Therefore,

$$\text{Po} = \frac{dW_{\text{n.f.}}}{dt} = \frac{dW_{\text{n.f.}}}{dm} \dot{m}$$

$$= 9480 \frac{\text{ft}^2}{\text{s}^2} \cdot 69.5 \frac{\text{lbm}}{\text{s}} \cdot \frac{\text{lbf} \cdot \text{s}^2}{32.2 \text{ lbm} \cdot \text{ft}} \cdot \frac{\text{hp} \cdot \text{s}}{550 \text{ ft} \cdot \text{lbf}}$$

$$= 37 \text{ hp} = 27.6 \text{ kW} \tag{6.BP}$$

■

Figure 6.21 illustrates a class of problems that occurs very often in water supply networks in which multiple reservoirs are connected to multiple users. Water flows from one reservoir through a pipe to a division point (called a *node*), whence it flows to two other reservoirs via separate pipes. The elevations of the reservoirs are shown. The task is to compute the flow through each pipe branch, assuming steady flow. The three-reservoir example illustrates the idea, but not the complexity that can exist in municipal and industrial plant water supply networks that have grown and been modified over time.

For such long pipes the kinetic-energy terms in B.E. will be negligible, and the gauge pressures at the free surfaces are all zero. If these two simplifications are not appropriate, the problems can still be solved, but not as simply as is shown here. Writing B.E. for the three sections of the pipe in Fig. 6.21, we find

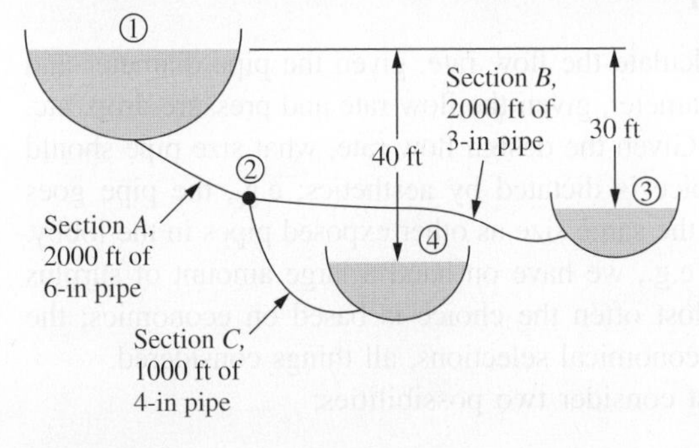

FIGURE 6.21
Multiple-reservoir system with branching pipes, typical of city water systems that have grown and been modified over time.

$$\frac{P_2}{\rho} + g(z_2 - z_1) = -\mathscr{F}_A$$

$$= -\left(4f \frac{\Delta x}{D} \cdot \frac{V^2}{2}\right)_A \tag{6.BQ}$$

$$\frac{-P_2}{\rho} + g(z_3 - z_2) = -\mathscr{F}_B$$

$$= -\left(4f \frac{\Delta x}{D} \cdot \frac{V^2}{2}\right)_B \tag{6.BR}$$

$$\frac{-P_2}{\rho} + g(z_4 - z_2) = -\mathscr{F}_C$$

$$= -\left(4f \frac{\Delta x}{D} \cdot \frac{V^2}{2}\right)_C \tag{6.BS}$$

From the mass balance we find

$$Q_A = Q_B + Q_C; \qquad V_A = V_B \frac{A_B}{A_A} + V_C \frac{A_C}{A_A} \tag{6.BT}$$

Here we have four equations relating four unknowns (the three Vs and P_2). However, to solve the problem, we must use the correct values of the three fs, which are related to the pipe diameters and the Vs by the friction-factor chart, Fig. 6.10, or by Eq. 6.21. Thus, we could also think of this as a system with seven unknowns and seven equations (taking the friction-factor chart or Eq. 6.21 three times). Because of the forms of Eq. 6.21 there is no possibility that one can solve these equations analytically. The solution must be by trial and error.

In the problem statement the elevation at point 2 is not given, and the pressure at point 2 is unknown. In practice, we will know the elevation at point 2, and the flows will determine the pressure there, which will be different from the pressure which exists there when no fluid is flowing. The problem is inherently a trial and error, easy on computers. We begin by defining

$$\alpha = \left(\frac{P_2 - P_1}{\rho g} + z_2 - z_1 \right) \tag{6.BU}$$

and guessing a value of α. Using it we can solve each of Equations 6.BQ, 6.BR and 6.BS, for the velocities in each of the pipe sections. From those velocities we compute the three volumetric flow rates and check to see if the algebraic sum of the volumetric flow rates into point 2 is zero. If not, we make a new guess of α and repeat the flow calculations, to find the unique value of α which makes that sum = zero.

For this three-branch, one-node example the trial and error is quite easy (Prob. 6.68). For more complex examples it is not. A widely used systematic procedure for solving this type of system was developed by Cross [15]. Computer programs to carry out that solution are available [16].

6.12 ECONOMIC PIPE DIAMETER, ECONOMIC VELOCITY

From the foregoing we can easily calculate the flow rate, given the pipe diameter and pressure drop, or calculate the pipe diameter, given the flow rate and pressure drop, etc. A much more interesting question is, Given the design flow rate, what size pipe should we select? It is possible that the choice is dictated by aesthetics; e.g., the pipe goes through a lobby, and we want it to be the same size as other exposed pipes in the lobby. Or it may be dictated by the supply; e.g., we have on hand a large amount of surplus 4-in pipe that we want to use up. Most often the choice is based on economics; the engineer is asked to make the most economical selections, all things considered.

For economic analysis we must consider two possibilities:

1. The fluid is available at a high pressure and will eventually be throttled to a low pressure, so the energy needed to overcome friction losses may come from the available pressure drop.

2. The fluid is not available at a high pressure, so a pump or compressor is needed to overcome the effects of fluid friction.

The first is simple: we select the smallest size of pipe that will carry the required flow with the available pressure drop. Example. 6.6 is that case.

If the effects of friction must be overcome by a pump or compressor, then the total annual costs of the pump-pipeline system are the following:

1. Power to run the pump.
2. Maintenance charges on pump and line.
3. Capital-cost charges for both line and pump.

How these change with increasing pipe diameter is sketched in Fig. 6.22. The figure indicates the following:

1. The larger the pipe diameter, the greater the capital charges. The cost of pipelines is roughly proportional to the pipe diameter; bigger pipes cost more to buy, require more expensive supports, take longer to install, etc. The cost of the pump is proportional to the cost of the pipe and is included in it.
2. The maintenance cost is practically independent of the pipe size.
3. The pumping cost goes down rapidly as the pipe size goes up. The pumping cost is proportional to the pressure drop (see Example 6.8), which for turbulent flow is proportional to the velocity to the 1.8 to 2.0 power divided by the diameter. The velocity (for constant flow rate) is proportional to the reciprocal of the square of the diameter, so the pumping cost is proportional to the reciprocal of the diameter to the 4.6 to 5 power.

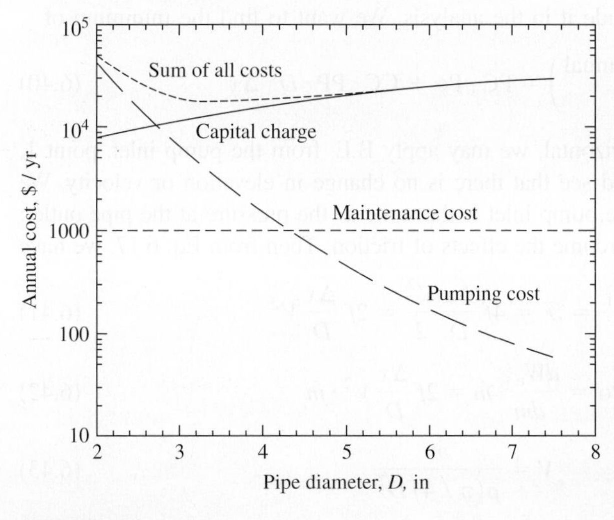

FIGURE 6.22
Relation between capital, operating, and maintenance costs for a pipeline with pumped flow. The numerical values are based on the economic data in Example 6.18.

As Fig. 6.22 shows, the sum of these has a rather broad minimum. This minimum occurs at the economic pipe diameter. Here we are taking the sum of a power cost during some finite period, e.g., a year, and the one-year charge for owning the pipeline and the pump, whose lifetime will be many years. Books on engineering economics or process design show various sophisticated ways to do that [17]. Here we use the simplest possible way, computing the annual capital charge, which is equivalent to what we would pay each year if we did not own the pipeline but instead financed it through a bank, making annual payments to "buy it on time," the same way most of us buy houses and cars. The charge we calculate here is equivalent to what the bank would charge us to pay for the pipeline in annual installments over the life of the pipeline.

We begin the economic analysis with

$$\begin{pmatrix}\text{Purchase}\\ \text{price}\end{pmatrix} = \text{PP}\cdot\begin{pmatrix}\text{pipe}\\ \text{diameter}\end{pmatrix}\cdot\begin{pmatrix}\text{pipe}\\ \text{length}\end{pmatrix} = \text{PP}\cdot D\cdot\Delta x \tag{6.37}$$

where the purchase price is what we would have to pay a contractor for both supplies and labor to build the complete pipeline and pump for us. PP is a constant with dimension \$ / inch · ft. Typically, the purchase price is about three times the cost of the materials used—pipe, fittings, supports, etc. The remainder is the labor cost to install the pipe and other associated construction costs. Then

$$\begin{pmatrix}\text{Annual capital}\\ \text{charge}\end{pmatrix} = \text{CC}\cdot\begin{pmatrix}\text{purchase}\\ \text{price}\end{pmatrix} \tag{6.38}$$

where CC is a constant with dimension 1 / yr. The annual capital charge is our "equivalent annual payment" to buy the pipeline over time, proportional to the purchase price. It includes the portion of the initial price paid, interest on the remaining debt, taxes, and insurance; these are the same components of a typical house mortgage payment (which is normally paid monthly rather than annually in this calculation). Then

$$\begin{pmatrix}\text{Annual pumping}\\ \text{charge}\end{pmatrix} = \text{PC}\cdot\begin{pmatrix}\text{power input to}\\ \text{drive the pump}\end{pmatrix} \tag{6.39}$$

where PC is a constant with dimension \$ / kWh.

As shown in Fig. 6.22, the maintenance cost is practically independent of the pipe diameter, so we will not include it in the analysis. We want to find the minimum of

$$\begin{pmatrix}\text{Total annual}\\ \text{cost}\end{pmatrix} = \text{PC}\cdot\text{Po} + \text{CC}\cdot\text{PP}\cdot D\cdot\Delta x \tag{6.40}$$

Assuming that the pipe is horizontal, we may apply B.E. from the pump inlet, point 1, to the pipe outlet, point 2, and see that there is no change in elevation or velocity. We assume that the pressure at the pump inlet is the same as the pressure at the pipe outlet; i.e., the pump only has to overcome the effects of friction. Then from Eq. 6.17, we have

$$\frac{dW_{\text{n.f.}}}{dm} = \mathcal{F} = 4f\,\frac{\Delta x}{D}\,\frac{V^2}{2} = 2f\,\frac{\Delta x}{D}\,V^2 \tag{6.41}$$

$$\text{Po} = \frac{dW_{\text{n.f.}}}{dm}\,\dot{m} = 2f\,\frac{\Delta x}{D}\,V^2\cdot\dot{m} \tag{6.42}$$

$$V = \frac{\dot{m}}{\rho(\pi / 4)\,D^2} \tag{6.43}$$

and therefore

$$\text{Po} = \frac{\dot{m}^3\,2f\,\Delta x(4 / \pi)^2}{\rho^2\,D^5} \tag{6.44}$$

Substituting Eq. 6.44 into Eq. 6.40, we find

$$\begin{pmatrix}\text{Total annual} \\ \text{cost}\end{pmatrix} = \text{PC} \cdot \frac{\dot{m}^3 2f\,\Delta x(4/\pi)^2}{\rho^2 D^5} + \text{CC} \cdot \text{PP} \cdot D \cdot \Delta x \tag{6.45}$$

We now differentiate the total annual cost with respect to diameter D and set the derivative equal to zero:

$$0 = \frac{d(\text{cost})}{dD} = \left[\text{PC} \cdot \dot{m}^3\, 2f\,\Delta x \left(\frac{4}{\pi}\right)^2 \frac{1}{\rho^2} \cdot \frac{-5}{D^6}\right] + (\text{CC} \cdot \Delta x \cdot \text{PP}) \tag{6.46}$$

Solving for the D that minimizes costs, D_{econ}, we find

$$D_{\text{econ}} = \left[\frac{10 \cdot \text{PC} \cdot \dot{m}^3 f(4/\pi)^2(1/\rho^2)}{\text{CC} \cdot \text{PP}}\right]^{1/6} \tag{6.47}$$

This equation shows that the economic pipe diameter is independent of how long the pipe is. This should be no surprise: Both the pumping and capital costs are proportional to the pipe length. The equation also shows that the economic diameter is proportional to the friction factor to the $\frac{1}{6}$ power; we can use a rough estimate of the friction factor with little error. Equation 6.47 is only correct for turbulent flow, because we have assumed f is a constant. The corresponding equation for laminar flow is shown in Prob. 6.73.

Example 6.18. We want to transport 200 gal / min of water 5000 ft in a horizontal carbon-steel pipe. We will install a pump to overcome the friction loss. Given the economic data shown below, what is the economic pipe diameter?

$$\text{PC} = \frac{\$0.04}{\text{kWh}}, \qquad \text{PP} = \frac{\$2}{\text{in of diameter} \cdot \text{ft of length}}, \qquad \text{CC} = \frac{0.40}{\text{yr}} \tag{6.BV}$$

First we guess that the pipe will have an inside diameter of 3 in. Then, from Table 6.2 we have $\varepsilon / D = 0.0018 / 3 = 0.0006$. The friction factor (Fig. 6.10) will probably be about 0.0042. The mass flow rate is $200 \text{ gal/min} \cdot 8.33 \text{ lbm/gal} = 1666 \text{ lbm/min}$. Substituting these and the values of PC, CC, and PP in Eq. 6.48 produces

$$D_{\text{econ}} = \left[\frac{\dfrac{\$0.04}{\text{kW} \cdot \text{h}} \cdot \left(\dfrac{1666 \text{ lbm}}{\text{min}}\right)^3 \cdot 10 \cdot 0.0042 \cdot \left(\dfrac{4}{\pi}\right)^2 \cdot \left(\dfrac{\text{ft}^3}{62.3 \text{ lbm}}\right)^2}{(0.4/\text{yr}) \cdot (\$2/\text{in} \cdot \text{ft})}\right]^{1/6}$$

$$\cdot \left[\frac{\text{kW} \cdot \text{h}}{3600 \text{ kJ}} \cdot \frac{\text{kJ}}{737.6 \text{ ft} \cdot \text{lbf}} \cdot \frac{\text{lbf} \cdot \text{s}^2}{32.2 \text{ lbm} \cdot \text{ft}} \cdot \frac{\text{min}^2}{3600 \text{ s}^2} \cdot \frac{5.256 \cdot 10^5 \text{ min}}{\text{yr}} \cdot \frac{\text{ft}}{12 \text{ in}}\right]^{1/6}$$

$$= (0.00577 \text{ ft}^6)^{1/6} = 0.289 \text{ ft} = 3.24 \text{ in} = 0.082 \text{ m} \tag{6.BW}$$

Because of the approximate nature of the economic data used, a 4-in pipe would probably be selected. It would be appropriate to check the assumed friction factor (Prob. 6.62). ■

Because calculations such as these are long and tedious, companies that install many pipelines have solved the problem for a large number of cases and have summarized the results in convenient form. The most popular method is to calculate the economic velocity:

$$\text{Economic velocity, } V_{\text{econ}} = \frac{Q}{(\pi / 4)(D_{\text{econ}})^2} \tag{6.48}$$

Substituting for the economic diameter from Eq. 6.47, we find

$$V_{\text{econ}} = \frac{\dot{m} / \rho}{\dot{m}(1 / \rho^{2/3}) f^{1/3} \cdot \text{constants}} = \text{constants} \cdot \frac{1}{f^{1/3} \rho^{1/3}} \tag{6.49}$$

In Example 6.18 the velocity corresponding to the economic diameter is 6.8 ft / s. Equation 6.49 says that for a given set of cost data the economic velocity is independent of the mass flow handled and dependent only on the fluid density and on the friction factor. More thorough analyses and far more complicated cost equations lead to substantially the same conclusion. For example, for schedule 40 carbon-steel pipe Boucher and Alves [18] give the economic velocities shown in Table 6.8.

Table 6.8 refers to turbulent flow only. For laminar flow the value of f goes up quite rapidly as the viscosity increases, making the economic velocity go down. Oil companies spend more money pumping viscous liquids (crude oils, asphalt, heating oils, etc.) than do other companies; they have made up the most convenient economic-velocity plots for laminar flow, such as Fig. 6.23. With it one can rapidly select the economic velocity and pipe diameter for laminar flow, subject to the restriction that the economic data on the line to be installed are the same as those shown on the plot. Figure 6.23 has nomenclature similar to that of Fig. 6.12, and the comments on the latter are applicable here. Figure 6.23 also shows the economic velocity and diameter for turbulent flow, which are practically the same as one would estimate from Table 6.8.

Why does App. A.3 show the velocity in feet per second for all the water flows given? From Table 6.8 and Fig. 6.23, we can see that for water (which is almost always in turbulent flow in industrial equipment) the economic velocity is almost always about 6 ft / s. Experienced engineers often simply select pipe sizes for water or similar fluids by looking on App. A.3 for the pipe size that gives a velocity of about 6 ft / s (2 m / s). Similarly, air-conditioning engineers normally select the duct size that gives a velocity of about 40 ft / s (12 m / s), without bothering with a detailed

TABLE 6.8
Economic velocity for flow in schedule 40 steel pipe*

Fluid density, lbm / ft³	Economic velocity, ft / s
100	5.1
50	6.2
10	10
1	19.5
0.1	39.0
0.01	78.0

*From Reference 18.

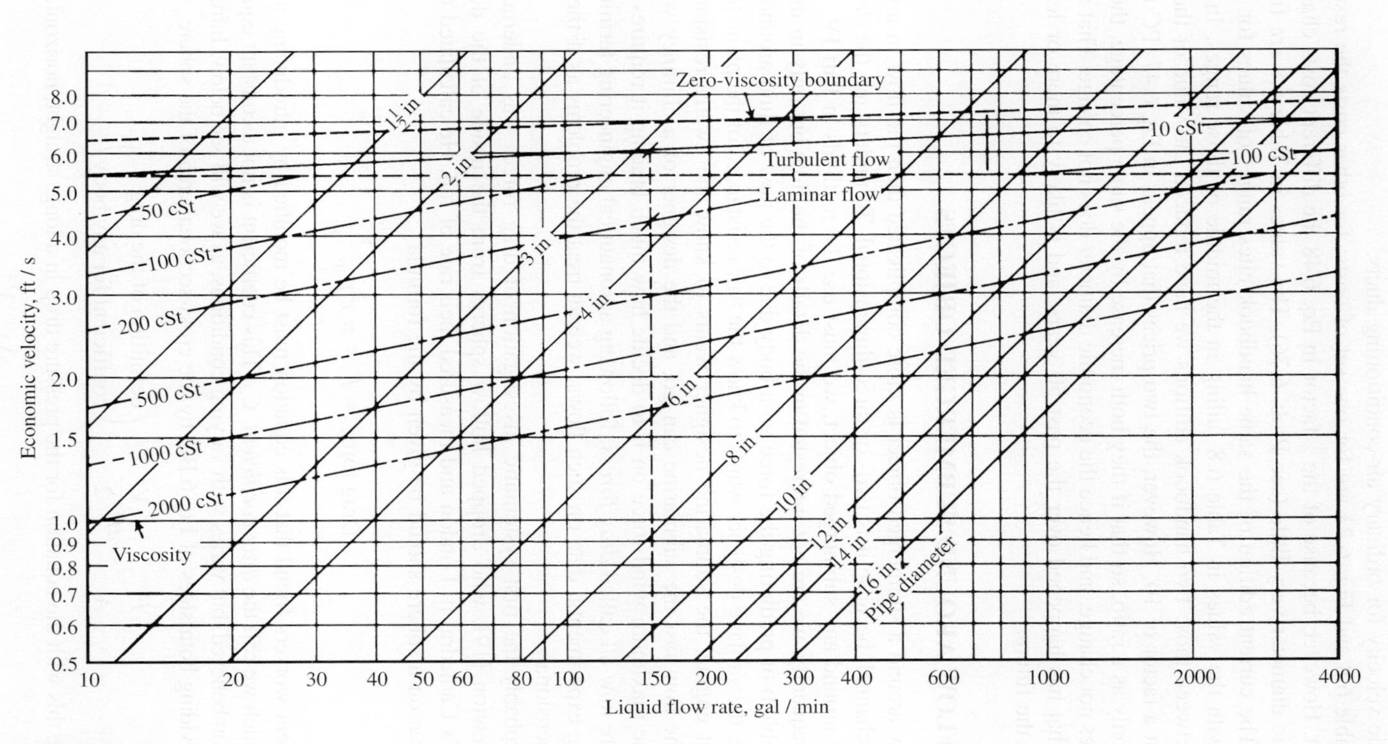

FIGURE 6.23
Economic pipe size for pumped liquids (carbon-steel pipe). Assumptions: pumping cost = \$135 per horsepower year (≈\$0.02 / kWh); line cost = \$1 per in of diameter per ft; fixed charges per year on line = 0.40 times line cost; liquid specific gravity = 0.80 (not very critical). Examples: for 150 gal / min and 200 cSt, use the 4-in line; for 150 gal / min and 10 cSt, use the 3-in line. (These prices are from the early 1960s; current prices are higher, but they have risen more or less together, so that the economic pipe sizes have not changed much.) (Courtesy of the Board of Engineers, Standard Oil Company of California.)

calculation. From their experience or Table 6.8 they know that this is close to the true economic velocity for ordinary air-conditioning ducts.

Table 6.8 and Fig. 6.23 are for one set of costs; for other costs the results are different. However, because of the $\frac{1}{6}$ factor in Eq. 6.48 the different costs change the economic diameter very little (see Prob. 6.76). The values in Table 6.8 are from the 1960s. The current edition of the same handbook cites ranges of values for various fluids, with the values in Table 6.8 falling in the middle of those ranges. In the 44 years between those two handbook editions we have reduced the value of the dollar by about a factor of 10. However, the two prices that appear in Eq. 6.47, PC and PP, appear only as a ratio, so that if they both increase by the same percentage, then their ratio does not change, and hence the economic diameter does not change. That appears to be what has happened over the past 50 years, and is likely to more or less continue in the future.

6.13 FLOW AROUND SUBMERGED OBJECTS

The flow around a submerged object is more complicated than the flow in a straight pipe or channel, because it is two- or three-dimensional. To understand the *details* of the flow around any submerged object, we must use the methods in Part IV.

Frequently, we are not interested in the details of the flow but only in the practical problem of predicting the force on a body due to the flow of fluid around it. For example, the airplane designer wants to know the "air resistance" of the plane to select the right engine, the submarine designer wants to know the "water resistance" to determine how fast the submarine can go, and the designer of a chimney wants to know the maximum wind force on it to decide how much bracing it requires. These forces are now all called *drag forces,* following aeronautical engineering terminology. By using experimental data on such flows, we can treat the problems as if they were one-dimensional.

Probably the first systematic investigation of drag forces was undertaken by Isaac Newton [19], who dropped hollow spheres from the inside of the dome of St. Paul's Cathedral in London and measured their rate of fall. He calculated that the drag force on a sphere should be given by the formula

$$\text{Drag force} = F = \pi r^2 \rho_{\text{air}} \frac{V^2}{2} \tag{6.BX}$$

Subsequent workers found that this equation must be modified by introducing a coefficient, which we call the *drag coefficient,* C_d. This coefficient is not a constant equal to 1, as Newton believed, but varies with varying conditions, as we will see below. Introducing it and dividing both sides of Eq. 6.BX by the cross-sectional area of the sphere, we find

$$\frac{F}{A} = C_d \rho \frac{V^2}{2} \quad \left(\begin{matrix}\text{definition of the drag} \\ \text{coefficient for any body}\end{matrix}\right) \tag{6.50}$$

Compare this with the equation for the pressure drop in a long, straight, horizontal pipe:

$$-\Delta P = 4f \frac{\Delta x}{D} \rho \frac{V^2}{2} \tag{6.51}$$

From these equations we see that C_d plays the same role as f. Equation 6.51 contains the factor $\Delta x / D$, which describes the geometry of the system (long, thin pipes have more pressure drop than short, thick ones for equal velocity and density), but since all spheres have the same shape, there is no need to include such a factor in Eq. 6.51.

In the case of steady pipe flow it was found experimentally that f depends only on the Reynolds number and the relative roughness. It has been found similarly that the drag coefficient for smooth spheres in steady motion depends only on the Reynolds number. Here we must redefine the Reynolds number, which previously included the pipe diameter. The common practice is to define a *particle Reynolds number,* in which the particle diameter takes the place of the pipe diameter:

$$\begin{pmatrix}\text{Particle Reynolds}\\ \text{number}\end{pmatrix} \equiv \mathcal{R}_p \equiv \frac{\text{particle diameter} \cdot \text{velocity} \cdot \text{fluid density}}{\text{fluid viscosity}} \tag{6.52}$$

With this definition, all the steady-state drag data on single, smooth spheres moving in infinite, quiescent, Newtonian fluids at moderate velocities can be represented by a single curve on Fig. 6.24. This figure shows also drag coefficients for disks and cylinders, to be discussed later. It is limited to steady velocities of less than about one half the local speed of sound; velocities higher than this are discussed elsewhere [19].

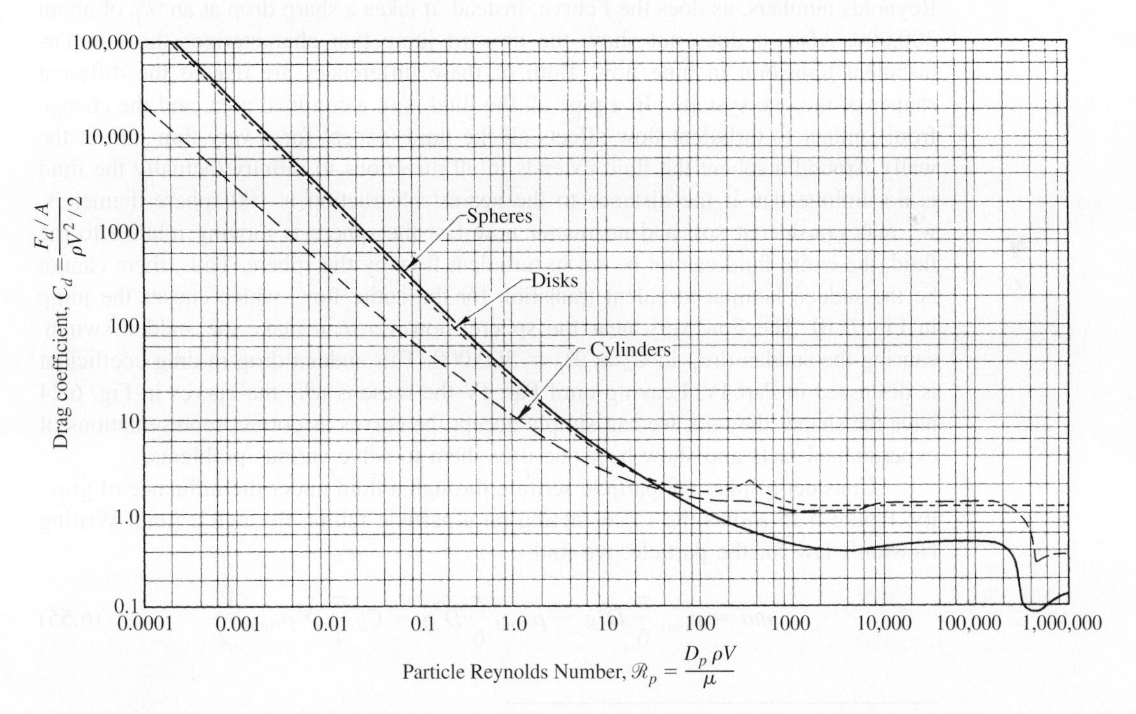

FIGURE 6.24
Drag coefficients for spheres, disks, and cylinders. R. Perry, D. Green, "Perry's Chemical Engineer's Handbook," McGraw-Hill 1997, reproduced with permission from the McGraw-Hill companies, based on C. E. Lapple and C. B. Shepherd, "Calculation of particle trajectories", *Ind. Eng. Chem, 32,* 605–617, (1940). *For spheres only,* the curve for $\mathcal{R}_p < 0.3$ is given exactly by $C_d = 24 / \mathcal{R}_p$ and for $0.3 < \mathcal{R}_p < 1000$, the curve is given to a satisfactory approximation by $C_d = (24 / \mathcal{R}_p) \cdot (1 + 0.14 \cdot \mathcal{R}_p^{0.7})$.

Figure 6.24 and the friction factor plot, Fig. 6.10, show marked similarities. In both, at low Reynolds numbers, there is a region where f or C_d is proportional to $1/\mathcal{R}$ or $1/\mathcal{R}_p$: i.e., a straight line of slope $-45°$ on log paper. For pipes this is Poiseuille's equation, which can be written $f = 16/\mathcal{R}$; for spheres (straight line on Fig. 6.23) it is *Stokes' law**, which can be written

$$C_d = \frac{24}{\mathcal{R}_p} \quad \begin{pmatrix} \text{Stokes' drag coefficient for} \\ \text{low } \mathcal{R}_p \text{ only} \end{pmatrix} \tag{6.53}$$

which can also be rewritten as

$$F_{\text{drag}} = 3\pi\mu DV \quad \begin{pmatrix} \text{Stokes' drag force for} \\ \text{low } \mathcal{R}_p \text{ only} \end{pmatrix} \tag{6.54}$$

which is sometimes the more convenient form.

Both figures have a horizontal section at high $1/\mathcal{R}$ or $1/\mathcal{R}_p$ in which f or C_d is practically independent of the Reynolds number. The value of f or C_d on this horizontal section cannot be calculated; it is found from experimental data.

However, the curve of the sphere drag coefficient has some marked differences from the friction factor plot. It does not continue smoothly to higher and higher Reynolds numbers, as does the f curve; instead, it takes a sharp drop at an $\mathcal{R}_p$ of about 300,000. Also, it does not show the upward jump that characterizes the laminar-turbulent transition in pipe flow. Both of these differences are due to the different shapes of the two systems. In a pipe all the fluid is in a confined area, and the change from laminar to turbulent flow affects all the fluid (except for a very thin film at the wall). Around a sphere the fluid extends in all directions to infinity (actually the fluid is not infinite but, if the distance to the nearest obstruction is 100 sphere diameters, we may consider it so), and no matter how fast the sphere is moving relative to the fluid, the entire fluid cannot be set in turbulent flow by the sphere. Thus, there cannot be the sudden laminar-turbulent transition for the entire flow, which causes the jump in Fig. 6.10. The flow very near the sphere, however, can make the sudden switch, causing the sudden drop in C_d at $\mathcal{R}_p = 300{,}000$. This sudden drop in drag coefficient is discussed in Part IV. Leaving until Part IV the reasons *why* the curves in Fig. 6.24 have the shapes they do, we can simply accept the curves as correct representations of experimental facts and show how one uses them to solve various problems.

Consider a spherical particle settling through a fluid under the influence of gravity. Figure 6.25 shows the forces acting on a particle falling through a fluid. Writing Newton's law for the particle, we find

$$ma = \rho_{\text{part}} \frac{\pi}{6} D^3 g - \rho_{\text{fluid}} \frac{\pi}{6} D^3 g - C_d \frac{\pi}{4} D^2 \rho_{\text{fluid}} \frac{V^2}{2} \tag{6.55}$$

---

*Stokes' law, like the Poiseuille equation, can be derived mathematically without the aid of experimental data. In doing so we must assume that the flow is laminar, that Newton's law of viscosity holds, and that the resulting terms in the equations involving velocities squared are negligible. The latter condition is called *creeping flow*. Even using these assumptions, the derivation takes several pages [20]. The greater complexity, compared with the Poiseuille equation, is due to the three-dimensionality of the flow around a sphere.

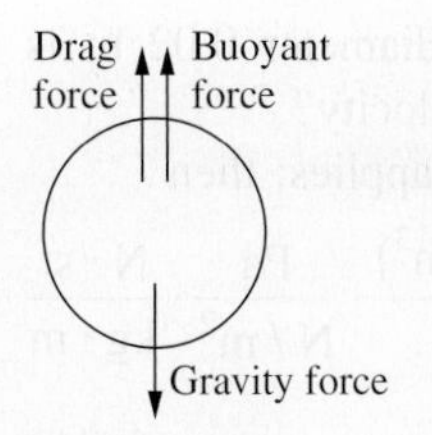

FIGURE 6.25
Gravity, buoyancy, and drag forces acting on a particle settling in a fluid.

If the particle starts from rest, its initial velocity is zero, so the drag force is initially zero. The particle accelerates rapidly; as it accelerates, the drag force increases as the square of the velocity increases, until it equals the gravity force minus the buoyant force. This is the state of *terminal velocity;* the sum of the forces acting is zero, so the particle continues to fall at a constant velocity. To find this velocity, we set the acceleration to zero in Eq. 6.55 and solve for V:

$$V^2 = \frac{4Dg(\rho_{\text{part}} - \rho_{\text{fluid}})}{3C_d\rho_{\text{fluid}}} \tag{6.56}$$

This equation is correct for any value of $\mathscr{R}_p$. If the particle is very small, it probably obeys Stokes' drag relation, Eq. 6.53, Substituting this in Eq. 6.56 and rearranging, we find

$$V = \frac{D^2 g(\rho_{\text{part}} - \rho_{\text{fluid}})}{18\,\mu} \quad \text{[Stokes' law]} \tag{6.57}$$

Example 6.19. Estimate the terminal settling velocity in air of a dust sphere with diameter 1 μ and SG = 2. (Fine particle calculations in the United States are almost always done in SI. $1\,\mu = 1$ micron $= 10^{-3}$ mm $= 10^{-6}$ m $=$ 0.000039 in. A typical human hair is about 50 microns in diameter.)

Substituting those values in Eq. 6.57, we find

$$V = \frac{(9.81\text{ m/s}^2)(10^{-6}\text{ m})^2[(2000\text{ kg/m}^3) - (1.20\text{ kg/m}^3)]}{(18)(1.8\cdot 10^{-5}\text{ Pa}\cdot\text{s})}$$

$$= 6.05\cdot 10^{-5}\,\frac{\text{m}}{\text{s}} = 0.00605\,\frac{\text{cm}}{\text{s}} = 1.99\cdot 10^{-4}\,\frac{\text{ft}}{\text{s}} \tag{6.BY}$$

We see that the air density in the $(\rho_{\text{particle}} - \rho_{\text{fluid}})$ term contributed little to the answer. If we had set it equal to zero, our answer would have been 1.0006 times the answer shown above. We will rarely know the actual particle diameters to this accuracy, so for most applications of particles moving in gases we drop the ρ_{fluid} term. For high-pressure gases this might lead to significant error; for gravitational settling in liquids we can seldom use this simplification.

Now we must check our assumption that Stokes' law applies here. The particle Reynolds number is

$$\mathscr{R}_p = \frac{10^{-6}\text{ m}\cdot(1.20\text{ kg/m}^3)\cdot(6.05\cdot 10^{-5}\text{ m/s})}{1.8\cdot 10^{-5}\text{ kg/(m}\cdot\text{s)}} = 4.03\cdot 10^{-6} \tag{6.BZ}$$

From Fig. 6.24 it appears that Stokes' law (the straight-line part of the figure) holds to an $\mathscr{R}_p$ of about 0.3; so the Stokes' law assumption was good here (see Prob. 6.89). ■

Stokes' law has been well verified for the range of conditions in which its assumptions hold good. However, both for very large and for very small particles these assumptions are not correct.

Example 6.20. A solid steel sphere of SG = 7.85 and diameter 0.02 m is falling at its terminal velocity through water. What is its velocity?

As a first trial we assume that Stokes' law, Eq. 6.57, applies; then

$$V_{\text{Stokes}} = \frac{(0.02\text{m})^2 \cdot (9.81 \text{ m/s}^2) \cdot (7.85 - 1) \cdot (998.2 \text{ kg/m}^3)}{18 \cdot 1.002 \cdot 10^{-3} \text{ Pa} \cdot \text{s}} \cdot \frac{\text{Pa}}{\text{N/m}^2} \cdot \frac{\text{N} \cdot \text{s}^2}{\text{kg} \cdot \text{m}}$$

$$= 1488 \frac{\text{m}}{\text{s}} = 4880 \frac{\text{ft}}{\text{s}} \quad ??? \qquad (6.CA)$$

The corresponding particle Reynolds number is

$$\mathscr{R}_p = \frac{0.02 \text{ m} \cdot (998.2 \text{ kg/m}^3) \cdot (1488 \text{ m/s})}{1.002 \cdot 10^{-3} \text{ Pa} \cdot \text{s}} \cdot \frac{\text{Pa}}{\text{N/m}^2} \cdot \frac{\text{N} \cdot \text{s}^2}{\text{kg} \cdot \text{m}}$$

$$= 2.96 \cdot 10^7 \quad ??? \qquad (6.CB)$$

Clearly, Stokes' law *does not apply* here. However, Eq. 6.56 applies for *any* value of the Reynolds number. To solve it, we must assume a value of the drag coefficient, calculate the corresponding velocity, and check our assumed C_d. This is a fairly simple trial and error. On our first trial we let $(C_d)_{\text{first guess}} = 0.4$; then

$$V_{\text{first guess}} = \left[\frac{8 \cdot 0.01\text{m} \cdot (9.81 \text{ m/s}^2) \cdot (7.85 - 1) \cdot (998.2 \text{ kg/m}^3)}{3 \cdot (998.2 \text{ kg/m}^3) \cdot 0.4}\right]^{1/2}$$

$$= 2.12 \frac{\text{m}}{\text{s}} = 6.94 \frac{\text{ft}}{\text{s}} \qquad (6.CC)$$

The corresponding particle Reynolds number is:

$$(\mathscr{R}_p)_{\text{first guess}} = \frac{0.02\text{m} \cdot (998.2 \text{ kg/m}^3) \cdot (2.12 \text{ m/s})}{1.002 \cdot 10^{-3} \text{ Pa} \cdot \text{s}} \cdot \frac{\text{Pa}}{\text{N/m}^2} \cdot \frac{\text{N} \cdot \text{s}^2}{\text{kg} \cdot \text{m}}$$

$$= 4.2 \cdot 10^4 \qquad (6.CD)$$

From Fig. 6.24 we see that the drag coefficient corresponding to this Reynolds number is about 0.5. On our second trial we use a $(C_d)_{\text{second guess}} = 0.5$; then

$$V_{\text{second guess}} = 2.12 \frac{\text{m}}{\text{s}} \cdot \left(\frac{0.4}{0.5}\right)^{1/2} = 1.90 \frac{\text{m}}{\text{s}} = 6.21 \frac{\text{ft}}{\text{s}} \qquad (6.CE)$$

$$(\mathscr{R}_p)_{\text{second guess}} = 4.2 \cdot 10^{-4} \cdot \frac{1.90}{2.12} = 3.8 \cdot 10^4 \qquad (6.CF)$$

From Fig. 6.24 we see that the assumed C_d and $\mathscr{R}_p$ agree; so this velocity is the desired solution. ■

These calculations are simple if tedious. They have been carried out for a variety of spherical particles in air and water, and are summarized on Fig. 6.26. From it we see the following:

1. Figure 6.26 has curves for spherical particles of various specific gravities, at their terminal velocities, both in air and in water.
2. Stokes' law (Eq. 6.57) shows the terminal settling velocity to be proportional to the square of the diameter. Thus, on this plot, Stokes' law plots as a straight line with slope 2.00.

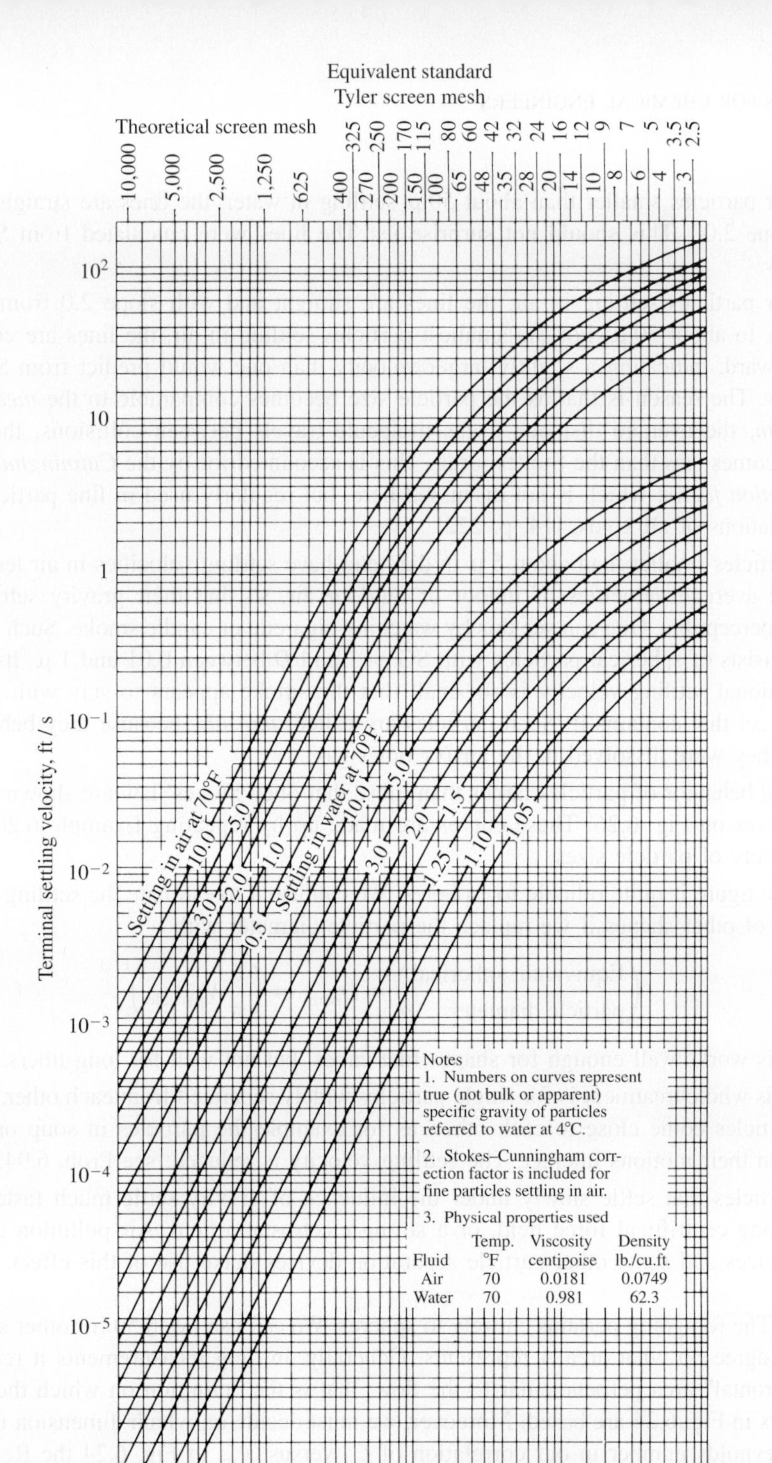

FIGURE 6.26
Terminal velocities of spherical particles of different densities settling in air and water at 70°F under the influence of gravity. [From C. E. Lapple et al., *Fluid and Particle Mechanics,* (Newark: University of Delaware, 1951), p. 292.] The screen-size values shown at the top allow one to convert from screen size to diameter in microns. A particle that will pass through a 325 screen has a diameter $\leq 40\mu$. (Observe that the scales are 1, 1.5, 2, 2.5, 3, 3.5, 4, 5. . .)

3. For particles smaller than about 50 μ settling in water, the lines are straight, with slope 2.00. This should not surprise us: The lines were calculated from Stokes' law.
4. For particles settling in air, the lines are straight and with slope 2.0 from about 5 μ to about 30 μ. For the smallest particles settling in air, the lines are concave upward, indicating a slightly larger velocity than one would predict from Stokes' law. The reason is that as the particle size becomes comparable to the *mean free path,* the average distance a gas molecule travels between collisions, the drag becomes less than the Stokes' drag. This is accounted for by the *Cunningham correction factor,* which is not discussed here, but regularly used in fine particle calculations of all kinds, [14, p. 222].
5. Particles smaller than about 5 μ in diameter have settling velocities in air less than the average velocities of indoor or outdoor air, so that their gravity settling is imperceptible. One can see this by watching cigarette or candle smoke. Such smoke consists of spherical particles with SG ≈ 1 and D between 0.01 and 1 μ. Its gravitational settling velocity is so small that the smoke appears to stay with the air parcel that contains it. Such particles are called *aerosols,* because they behave as if they were dissolved in the air (or other gas).
6. The behavior of particles larger than those that obey Stokes' law are shown by the curves on Fig. 6.26. These curves are made up by repeating Example 6.20 for a variety of particle sizes.
7. The figure is quite reliable for spheres. We can use it to estimate the settling velocity of other shapes if we replace the particle diameter with

$$\begin{pmatrix}\text{Equivalent spherical}\\ \text{particle diameter}\end{pmatrix} = D_{\text{equiv}} = \left(\frac{6}{\pi} V_{\text{particle}}\right)^{1/3} \tag{6.58}$$

   This works well enough for shapes like cubes, but not well for long fibers.
8. This whole treatment is for single particles widely separated from each other. When particles come close to each other, as for example the particles in soup or mud, then their motions interact. The settling velocity is reduced; see Prob. 6.94.
9. Particles that settle slowly under the influence of gravity settle much faster in a strong centrifugal force field, or a strong electrostatic field. Air pollution control devices and some other particle separation devices make use of this effect.

The foregoing pertains entirely to spheres. We can use Eq. 6.50 for other shapes, if we agree on what area A represents. Generally, in drag measurements it refers to the "frontal" area perpendicular to the flow; that is the definition on which the coefficients in Fig. 6.24 are based. Moreover, we must decide on which dimension to base the Reynolds number in our correlation of C_d versus $\mathscr{R}_p$: In Fig. 6.24 the Reynolds number for cylinders takes the cylinder diameter as D, and that for disks takes the disk diameter.

Example 6.21. A cylindrical chimney is 5 ft in diameter. The wind is blowing horizontally at a velocity of 20 mi / h. What is the wind force per foot of height on the chimney?

For this case the Reynolds number, based on the cylinder diameter, is

$$\mathscr{R}_{\text{cylinder}} = \frac{5\ \text{ft} \cdot (20\ \text{mi}/\text{h}) \cdot (0.075\ \text{lbm}/\text{ft}^3)}{0.018\ \text{cP}} \cdot \frac{5280\ \text{ft}}{\text{mi}} \cdot \frac{\text{h}}{3600\text{s}} \cdot \frac{\text{cP} \cdot \text{s} \cdot \text{ft}}{6.72 \cdot 10^{-4}\ \text{lbm}}$$

$$= 0.91 \cdot 10^6 \tag{6.CG}$$

From the curve for cylinders on Fig. 6.24 we can read $C_d \approx 0.35$; therefore,

$$F = AC_d\rho\frac{V^2}{2} = \frac{[5\ \text{ft}^2/(\text{ft of height})] \cdot 0.35 \cdot 0.075\ \text{lbm}/\text{ft}^3 \cdot (20\ \text{mi}/\text{h})^2}{2}$$

$$\left(\frac{5280\ \text{ft}}{\text{mi}}\right)^2 \cdot \left(\frac{\text{h}}{3600\ \text{s}}\right)^2 \cdot \frac{\text{lbf} \cdot \text{s}^2}{32.2\ \text{lbm} \cdot \text{ft}}$$

$$= 1.75\frac{\text{lbf}}{\text{ft of height}} = 2.61\frac{\text{N}}{\text{m}} \tag{6.CH}$$

■

This is not a very large force. However, we see from Eq. 6.50 that the force is proportional to the square of the wind velocity; so for a 100 mi / h wind the force is 25 times as great. To make matters worse, the wind force on long, thin objects can be oscillatory. The oscillatory motion is caused by the formation of vortices, which break away rhythmically. If the frequency of the shedding of these vortices is close to the natural frequency of oscillation of the system, then the wind force can drive that natural frequency disastrously. The most famous case of this was the Tacoma Narrows Bridge, which was destroyed by such oscillations by a wind of only ≈40 miles / h [21].

In aircraft calculations the drag coefficient of a wing usually is based on the wing's horizontal surface, rather than on the area perpendicular to the flow, and the length in $\mathscr{R}$ taken as the *chord* (the length of the wing from front to rear). In addition, aeronautical engineers define a lift coefficient, C_l, with exactly the same form as Eq. 6.50. In that equation F stands for the upward force of the air exerted on an airplane's wings, A stands for the horizontal wing surface, and C_d is replaced with C_l.

Example 6.22. An airplane has wings 15 m long (tip to tip) and 1.5 m wide (front to rear). The lift coefficient in level flight is $C_l = 0.8$, and the drag coefficient is $C_d = 0.04$. The drag and lift on parts other than the wing may be neglected. How much force must be exerted by the propeller to keep the plane moving 150 km / h? What is the maximum weight of the loaded airplane in this condition?

The drag force is

$$F_{\text{drag}} = AC_d\rho\frac{V^2}{2}$$

$$= \frac{(15\ \text{m} \cdot 1.5\ \text{m}) \cdot 0.04 \cdot (1.20\ \text{kg}/\text{m}^3) \cdot (150\ \text{km}/\text{h})^2}{2}$$

$$\cdot\left(\frac{1000\ \text{m}}{\text{km}} \cdot \frac{\text{h}}{3600\ \text{s}}\right)^2 \cdot \frac{\text{N} \cdot \text{s}^2}{\text{kg} \cdot \text{m}} = 945\ \text{N} = 212\ \text{lbf} \tag{6.CI}$$

This drag force is equal and opposite to the "thrust" force that must be supplied by the engine and propeller to keep the plane moving at this speed. The lift force is given by

$$F_{\text{lift}} = AC_l\rho\frac{V^2}{2} \tag{6.CJ}$$

which is the same as the drag force multiplied by C_l / C_d;

$$F_{\text{lift}} = 945\ \text{N}\cdot\frac{0.8}{0.04} = 18{,}900\ \text{N} = 4249\ \text{lbf} \tag{6.CK}$$

The lift is equal to the maximum gross, loaded weight of the aircraft. ■

The last example shows why the lift and drag coefficients are so useful to aeronautical engineers. Their ratios, C_l / C_d, are equal to the allowable ratio of total aircraft weight to thrust of the power plant. Normally both C_l and the C_l / C_d ratio are functions of aircraft speed and of the angle between the oncoming airstream and the wing surface [22]. This also shows why commercial aircraft fly as high as they can. To maintain level flight they need a lift equal to their weight and a thrust equal to their drag. The lift and drag are both proportional to ρV^2. For a given weight the required speed goes up as the square root of the air density goes down. The drag has the same relationship. So the higher they go, the lower the air density, and the faster they can go for a given hourly fuel input. Thus, their fuel cost per hour remains constant as they go up, but their fuel cost per mile goes down. (They also deliver the customers to their destination sooner, which the customers like, and pay for fewer hours of work to pilots and flight attendants.)

6.14 SUMMARY

1. The steady flow of fluids in constant-cross-section conduits can be of two radically different kinds: laminar, in which all the motion is all in the flow direction, and turbulent, in which there is a chaotic crosswise motion perpendicular to the net flow direction. The same is true for unconfined flows like the oceans or the atmosphere.
2. In laminar flow the pressure drop per unit length is proportional to the first power of the volumetric flow rate. The entire flow behavior can be calculated simply. The calculation requires the observational fact that fluid clings to solid surfaces, i.e., the velocity at the surface is zero.
3. In turbulent flow the pressure drop per unit length is proportional to the flow rate to the 1.8 to 2.0 power. The behavior cannot be calculated without experimental data.
4. All experimental data on the turbulent flow of Newtonian fluids in circular pipes can be represented on the friction factor plot.
5. The friction factor plot can be replaced by two fairly simple equations. The first, for laminar flow, is simply a rearrangement of Poisueille's equation and is restricted to laminar flow in a circular tube, for which it is rigorous. The second is simply a satisfactory fit of the experimental data. With these two equations we can completely replace the friction factor plot. But the plot has considerable intuitive content and is still useful for hand calculations.

6. All data on turbulent flow through valves and fittings can be correlated by assuming that each kind of fitting contributes as much friction as a certain number of pipe diameters of straight pipe. That number is about the same for one kind of fitting, independent of pipe size, fluid properties, etc. An alternative approach assigns a resistance coefficient, K, to each kind of fitting.

7. Laminar flow in a few kinds of noncircular conduits can be analyzed by the same technique used for circular pipes.

8. Turbulent flow friction losses in many kinds of noncircular conduits can be estimated by substituting 4 times the hydraulic radius for the diameter in the Reynolds number, ε / D, the friction factor plot, and B.E.

9. The economic size for a pipe is the size with the lowest sum of annual charges for the purchased cost of the pipe and pump and the annual power cost of running the pump or compressor needed to overcome friction. For turbulent flow this results in an economic velocity which is practically independent of everything but fluid density; it is about 6 ft / s for most low viscosity liquids and about 40 ft / s for air and other gases under normal conditions.

10. The forces of fluids flowing over bodies are ordinarily correlated by the drag equation, in which the drag coefficient plays the same role as does the friction factor in pipe flow.

PROBLEMS

See the Common Units and Values for Problems and Examples inside the back cover. An asterisk (*) on a problem number indicates that its answer is shown in App. C. In all problems in this chapter, unless a statement is made to the contrary, assume that all pipes are schedule 40, commercial steel (see App. A.2).

6.1. Derive the equivalents of Eqs. 6.5 and 6.9 for fluid flow in the vertical direction, taking gravity into account. Then generalize them for fluid flow at any angle, taking gravity into account.

6.2. Air is flowing through a horizontal tube with a 1.00 in inside diameter. What is the maximum average velocity at which laminar flow will be the stable flow pattern? What is the pressure drop per unit length at this velocity?

6.3.*Repeat Prob. 6.2, for water.

6.4. Show the derivations of Eqs. 6.10 and 6.11.

6.5. Show the effect on the calculated viscosity, in the viscometer in Example 6.2, of the 10 percent error in the measurement of

(*a*) Flow rate
(*b*) Fluid density
(*c*) Tube diameter.

6.6. In Example 6.2 how much does the internal energy per unit mass of the fluid increase as it passes through the viscometer? Assume that there is no heat transfer from the fluid to the wall of the viscometer. If the heat capacity of the fluid is 0.5 Btu / lbm · °F = 2.14 kJ / kg · °C how much does the fluid's temperature rise?

6.7.*A circular, horizontal tube contains asphalt, with μ = 100,000 cP (= 1000 Poise) and ρ = 70 lbm / ft^3. The tube radius is 1.00 in. Asphalt may be considered a Newtonian fluid for the purposes of this problem, although it is not always one. We now apply a pressure gradient of 1.0 (psi) / ft. What is the steady state flow rate?

6.8. What is the Reynolds number in Example 6.2? What is the lowest fluid viscosity for which one should use this viscometer?

6.9. The simple capillary viscometer in Example 6.2 is not the same as that actually used. In the practical versions, there are two marks on the glass, and the user reads the time for the fluid level to pass between the two marks. In Example 6.2, if the upper reservoir has a diameter of 10 mm, how many seconds does it take the level in the reservoir to fall from 0.02 to 0.01 m? (In actual practice the marks are placed in narrow parts of a tube above and below the reservoir, so that reading the passage of the interface is easy. To see how these work look at the viscometers in any laboratory glassware catalog.)

6.10.*If the students in Fig. 6.8(*a*) throw a standard 0.31 lbm baseball back and forth, if its velocity in flight is 40 mi / h, and if each one throws it an average of once every 10 s, what is the average force in the y direction tending to separate them?

6.11. Show that in Fig. 6.8(*b*) the force in the x direction is independent of how fast the balls move in the y directions.

6.12. Show that, if we define the shear stress at the pipe wall as $\tau = f\rho V^2 / 2$ and then calculate the pressure gradient for horizontal flow, we find Eq. 6.18.

6.13. Show that Poiseuille's equation may be rewritten as $f = 16 / \mathcal{R}$.

6.14.*A fluid is flowing in a pipe. The pressure drop is 10 psi per 1000 ft. We now double the volumetric flow rate, holding the diameter and fluid properties constant. What is the pressure drop if the new Reynolds number
(*a*) Is 10?
(*b*) Is 10^8?

6.15.*As discussed in the text, there are two friction factors in common use, which means that there are two versions of Fig. 6.10 in common use. When one encounters a friction factor plot and wants to know on which of the definitions it is based, the easiest way is to look at the label on the laminar flow line. For the Fanning friction factor used in this book, that is labeled $f = 16 / \mathcal{R}$. For a chart based on the Darcy-Weisbach friction factor, what is the label on the laminar flow line?

6.16. Water is flowing at an average velocity of 7 ft / s in a 6-in pipe. What is the pressure drop per unit length?

6.17. An oil with a kinematic viscosity of 5 cSt and SG = 0.80 is flowing in a 3-in pipe. The pressure drop is 30 psi per 1000 ft. What is the flow rate in gallons per minute? Show the solution two ways:
(*a*) Using Fig. 6.12.
(*b*) Using Fig. 6.10 and/or Eqs. 6.20 and 6.21.

6.18. We want to transport 200 gal / min of fluid through a 3-in pipe. The available pressure drop is 28 psi per 1000 ft. The fluid properties are SG = 0.75 and μ = 0.1 cP. Is a 3-in pipe big enough?

6.19.*Oil is flowing at a rate of 150 gal / min, with μ = 1.5 cP, and SG = 0.87 in a 3-in pipe 1000 ft long. What is the pressure drop? Calculate it two ways:
(*a*) By Fig. 6.10 and/or Eqs. 6.20 and 6.21.
(*b*) By Fig. 6.12.

6.20. In Example 6.5, how much does the temperature of the gasoline rise as it flows through the pipe? Assume that there is no heat transfer from the gasoline and that its heat capacity is 0.6 Btu / lbm · °F.

6.21. (*a*) Set up the spreadsheet program shown in Table 6.5 and verify the values.
(*b*) Using that spreadsheet, find the corresponding values for $\Delta z = -30$ m.

6.22. (*a*) Set up the spreadsheet program shown in Table 6.6 and verify the values.
(*b*) Using that spreadsheet, find the corresponding values for Q = 2000 cfm.

6.23.*Two large water reservoirs are connected by 5000 ft of 8-in pipe. The level in one reservoir is 200 ft above the level in the other, and water is flowing steadily through the pipe from one reservoir to the other. Both reservoirs are open to the atmosphere. How many gallons per minute are flowing? Show both the solution based on Table A.3 and that based on Fig. 6.10/Eq.6.21.

6.24. In Prob. 6.23 we want to replace the existing pipe with a new one, which will transmit 10,000 gal / min under the same conditions. What size of pipe should we choose?

6.25.*Two tanks are connected by 500 ft of 3-in pipe. The tanks contain an oil with μ = 100 cP and SG = 0.85. The level in the first tank is 20 ft above the level in the second, and the pressure in the second is 10 psi greater than the pressure in the first. How much oil is flowing through the pipe? Which way is it flowing?

6.26. We are offered some pipes made of a new kind of plastic. To test their roughness, we pump water through a 3-in pipe made of this material at an average velocity of 40 ft / s. The observed friction factor is 0.0032. Estimate the absolute roughness of this plastic.

6.27. As discussed in Sec. 6.5, the friction factor plot, Fig. 6.10, relates six variables and therefore can be used for finding any of the six if the other five are known. Examples 6.4, 6.5, and 6.6 show how to find three of these quantities, given all the others. Problem 6.26 shows how to find a fourth, given all the others. The remaining two are the density and viscosity of the flowing fluid. *Turbulent-flow* pressure drops are almost never used for determining fluid viscosities or densities. Discuss why this is so.

6.28. Equation 6.21 leads easily to quick trial-and-error solutions to all the problems in Sec. 6.5. One could also use it to eliminate a variable and thus reduce those trial-and-error solutions to a single equation. Show the algebra of that elimination and the resulting single-variable equations for Examples 6.5 and 6.6. Are those equations likely to be easier to solve analytically or numerically?

6.29. On a piece of log paper 2 cycles by 2 cycles, make up the equivalent of Fig. 6.12 for a 2-in pipe. Show the following:

(*a*) The zero-viscosity boundary.
(*b*) The laminar-flow region.
(*c*) The turbulent-flow region.
(*d*) The transition region.

6.30. Calculate the pressure drop per unit length for the flow of 100 gal / min of air in a 3-in pipe by using Fig. 6.12.

6.31.*Oil of a kinematic viscosity of 20 cSt is flowing in a 3-in pipe. According to Fig. 6.12,

(*a*) What is the highest volumetric flow rate at which the flow is certain to be laminar?
(*b*) What Reynolds number does this correspond to?
(*c*) What is the lowest volumetric flow rate at which the flow is certain to be turbulent?
(*d*) What Reynolds number does this correspond to?

6.32. Do any of the values in App. A.3 correspond to laminar flow?

6.33. Check the values in App. A.3 for 24-in pipe to see whether they correspond to a constant friction factor or whether the pipe size is so large that the friction factor corresponds to the "smooth tubes" curve in Fig. 6.10. The inside diameter of 24-in schedule 40 pipe is 22.624 in.

6.34. Rework Example 6.5 by using App. A.3. Show what corrections, if any, are needed in the solution if it is assumed that gasoline has the same flow properties as water. In that example the pipe ID was 0.1 m = 3.94 in. From App. A.2 we see that 4-in schedule 40 pipe has 4.03 in ID. Work the problem assuming that 3.94 $\approx$ 4.03, and then estimate how much difference this simplification makes.

6.35. Estimate the pressure loss for 1000 ft^3 / min of air flowing in a 12-in diameter air-conditioning duct 1000 ft long.

6.36. Estimate the required pipe diameter to transport 100 m^3 / h of air with a friction loss of 1 Pa / m.

6.37. Estimate the volumetric flow rate of air for a pressure drop of 5 Pa / m in a duct with diameter 0.125 m.

6.38. Estimate the pressure drop for 1000 ft^3 / min of hydrogen flowing in a 6-in diameter pipe 500 ft long in two ways:
(*a*) Using Fig. 6.10, or Eq. 6.21.
(*b*) Using Fig. 6.14, and suitable corrections for its much lower density than that of air.

6.39. The common friction factor plot (Fig. 6.10) is based on the Colebrook equation [5],

$$\frac{1}{\sqrt{f}} = -4 \log\left(\frac{\varepsilon / D}{3.7} + \frac{1.255}{\mathcal{R}\sqrt{f}}\right) \tag{6.59}$$

which itself is a data-fitting equation with no theoretical basis. It is difficult to use, because f appears on both sides, once as the argument of a logarithm. There are other data-fitting equations that attempt to reproduce Eq. 6.59 with a more easily used form, of which one of the most popular is that due to Haaland [23],

$$f = 0.25 \bigg/ \left(-1.8 \log\left[\frac{6.9}{\mathcal{R}} + \left(\frac{\varepsilon / D}{3.7}\right)^{1.11}\right]\right)^2 \tag{6.60}$$

another equation, [24], is

$$f = 0.0624 \bigg/ \left[\log\left(\frac{\varepsilon}{3.7 D_H} + \frac{5.74}{\mathcal{R}^{0.8}}\right)\right]^2 \tag{6.61}$$

where D_H is the hydraulic diameter, twice the hydraulic radius.
Find the friction factor for $\mathcal{R} = 2.00 \cdot 10^5$, and $\varepsilon / D = 0.0006$,
(*a*) From Fig. 6.10.
(*b*) From the Colebrook equation (Eq. 6.59).
(*c*) From the Haaland equation (Eq. 6.60).
(*d*) From Eq. 6.21.
(*e*) From Eq. 6.61.

6.40. Figure 6.14 is the "standard chart" for air-conditioning applications. It is based on Fig. 6.10 and the assumptions that the air flowing is at 1 atm and 68°F, i.e., the same assumption as given inside the back cover of the book. The plot is logarithmic on both axes, but the length of a decade on the vertical axis is greater than that on the horizontal axis.
(*a*) If a given duct diameter corresponded to a fixed value of f, what should the (line? curve?) for that diameter look like on this plot?
(*b*) Is that shape observed for the small ducts, i.e., those with $D < 5$ in?
(*c*) Is that shape observed for large ducts, i.e., those with $D > 10$ in?
(*d*) Why are these different?
(*e*) Figure 6.14 shows a pressure gradient of almost exactly 0.2 in of water / 100 ft for 1000 ft^3 / min in a 12-in duct. What value of the absolute roughness ε does that correspond to?
(*f*) How does the value just determined in part (*e*) compare to the value for steel in Table 6.2? Explain!

6.41. Examples 6.11 and 6.12 show significantly different values for the pressure drop due to the valves and fittings, calculated by the equivalent length method and the K method. Is that result a function of pipe diameter? To answer this question, prepare a plot of $\Delta P_{\text{valves and fittings}}$ versus pipe diameter for the same fluid as in Example 6.4 (SG = 1.00, μ = 50 cP) and the same average velocity, V_{avg} = 13.0 ft / s, covering the pipe diameter range of 3 in (that example) to 12 in, with the same fittings as in Examples 6.11 and 6.12, by the methods in Examples 6.11 and 6.12.

6.42.*Water is flowing at a rate of 1500 gal / min in a horizontal, 10-in pipe that is 50 ft long and contains two standard 90° elbows and a swing-type check valve. Estimate the pressure drop using both methods of accounting for the elbows and the valve.

6.43. A piping system consists of 100 ft of 2-in pipe, a sudden expansion to 3-in pipe, and then 50 ft of 3-in pipe. Water is flowing at 100 gal / min through the system. What is the pressure difference from one end of the pipe to the other?

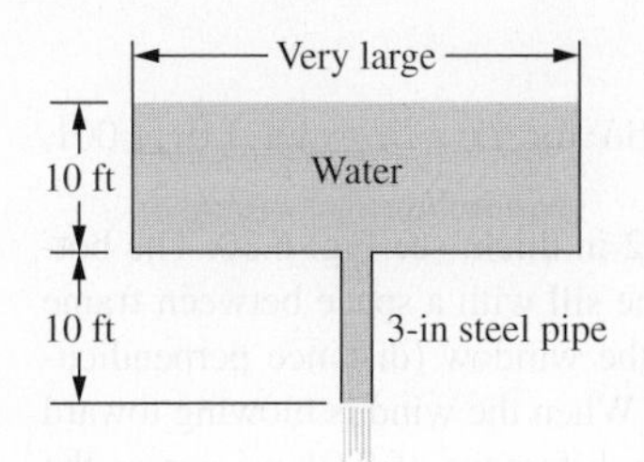

FIGURE 6.27
Tank draining by gravity, with pipe and entrance friction, Prob. 6.45.

6.44.*Two large water tanks are connected by a 10 ft piece of 3-in pipe. The levels in the tanks are equal. When the pressure difference between the tanks is 30 psi, what is the volumetric flow rate through the pipe?

6.45. The water in Fig. 6.27 is flowing steadily. What is the flow rate?

6.46.*We are going to lay a length of 6-in steel pipe for a long distance and allow water to flow through it by gravity. If we want a flow rate of 500 gal / min, how much must we slope the pipe (i.e., how many feet of drop per foot of pipe length or how many ft / mi)?

6.47. A 1-gal can full of water has the dimensions shown in Fig. 6.28. There is a horizontal piece of $\frac{1}{4}$-in galvanized pipe inserted in the bottom. The end of the pipe is unplugged, and the water is allowed to flow out of the tank.

(*a*) How long will it take the level in the tank to fall from 7 in above the centerline of the pipe to 1 in above the centerline of the pipe? Make whatever assumptions seem plausible.

(*b*) As the level falls, the flow slows down, until it finally converts from turbulent to laminar. How far will the level be above the centerline of the pipe when this transition occurs?

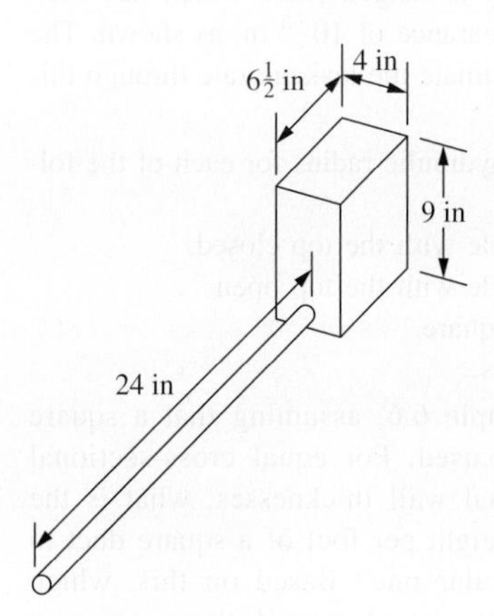

FIGURE 6.28
A portable demonstrator of tank draining with friction, Prob. 6.47.

6.48. Derive Eq. 6.28. It is suggested that you use the coordinates shown in Fig. 6.29. Here the flow is in the x direction from left to right, and the slit extends a distance l in the z direction. Choose as your element for the force balance a piece symmetrical about the y axis (other choices are possible but lead to more difficult mathematics). *Hint:* This is a repeat of the derivation of Eq. 6.8 in a different geometry. Simply follow that derivation, changing the geometry.

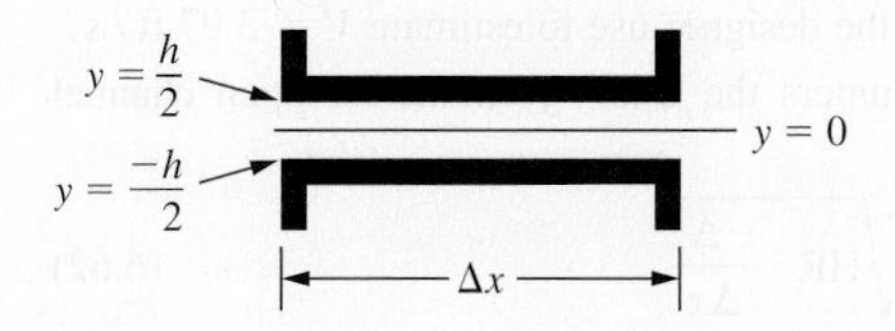

FIGURE 6.29
Suggested dimensions for Prob. 6.48.

6.49. Derive Eq. 6.29. This equation is derived in detail in Bird et al. [12, p. 54].

6.50. Example 6.13 shows that the calculated leakage rate is less than the average observed rate for typical valves in oil refineries by a factor of 3.5. In that example we assumed that the average thickness of the leakage path was 0.0001 in. If we held all the other values in that example constant except this thickness, what value of the thickness corresponds to the observed leakage rate?

6.51. In Example 6.13 we replaced Eq. 6.29 (flow in an annulus) with Eq. 6.30 (the linear simplification of Eq. 6.29). How much difference does it make in our answer? Check by repeating Example 6.13, using Eq. 6.29.

6.52. (*a*) Show how one obtains Eq. 6.30 from Eq. 6.28.
(*b*) Show the ratio of $Q_{\text{Eq. 6.30}} / Q_{\text{Eq. 6.28}}$.
(*c*) Using a spreadsheet, show the value of this ratio for $D_o / D_i = 1.1, 1.01, 1.001$, and 1.0001.

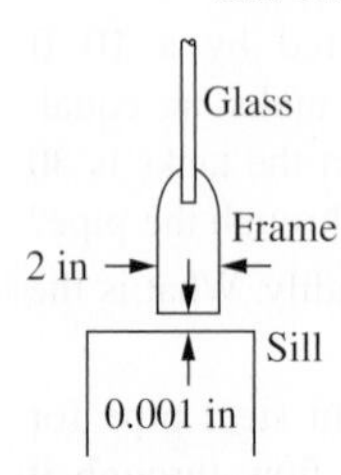

FIGURE 6.30
Leakage flow beneath a window, Prob. 6.53.

6.53.*The wooden frame of a window is 2 in thick; see Fig. 6.30. The bottom of the window closes against the sill with a space between frame and sill of 0.001 in. The width of the window (distance perpendicular to the paper in the figure) is 2 ft. When the wind is blowing toward the window and creating a pressure difference of 0.01 psi across the window, what is the volumetric flow rate of air through the space between frame and sill?

6.54. The cylindrical vessel in Fig. 6.31 is full of water at a pressure of 1000 psig. The top is held on by a flanged joint, which has been ground smooth and flat, with a clearance of 10^{-5} in, as shown. The diameter of the vessel is 10 ft. Estimate the leakage rate through this joint.

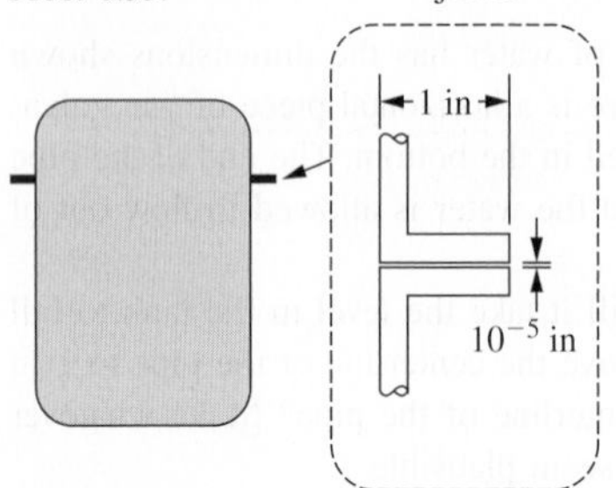

FIGURE 6.31
Leakage flow through a flanged joint, Prob. 6.54.

6.55. Calculate the hydraulic radius for each of the following shapes:
(*a*) A semicircle with the top closed.
(*b*) A semicircle with the top open.
(*c*) A closed square.
(*d*) An annulus.

6.56.*Rework Example 6.6, assuming that a square duct is to be used. For equal cross-sectional areas and equal wall thicknesses, what is the ratio of the weight per foot of a square duct to that of a circular one? Based on this, which would normally be chosen if there were no space constraints? Look around public buildings in which such ducts are visible, and examine where circular ducts are used and where square or rectangular ducts are used. Do your observations agree with your answer to this problem?

6.57.*In Example 6.15, what values of f and ε did the designer use to estimate $V = 3.97$ ft / s?

6.58. (*a*) In hydraulics books one regularly encounters the *Chézy formula* for open channel flow,

$$V = C\sqrt{\text{HR} \cdot \frac{-\Delta z}{\Delta x}} \tag{6.62}$$

in which we have changed from the notation normally shown in those books to the notation in this book. What value of C makes Eq. 6.62 the same as Eq. 6.AX?

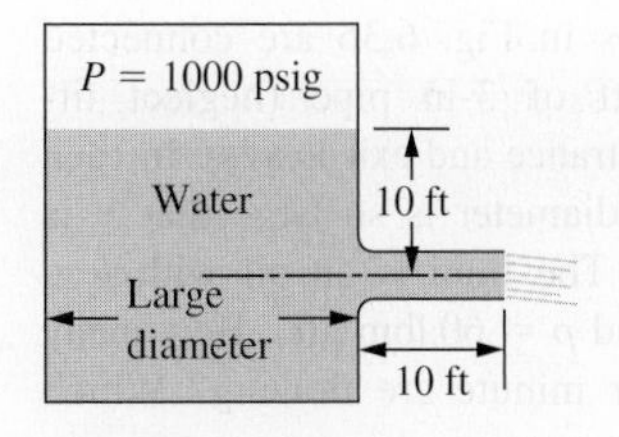

FIGURE 6.32
Pressure and gravity-driven flow with friction, Prob. 6.59.

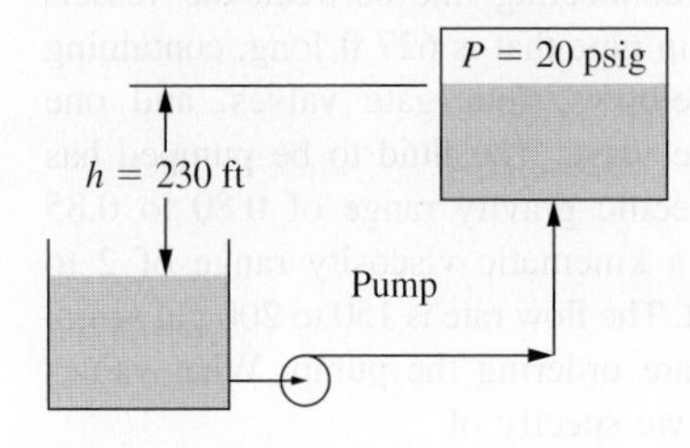

FIGURE 6.33
Pumped fluid transfer with both pressure and elevation change, Prob. 6.60.

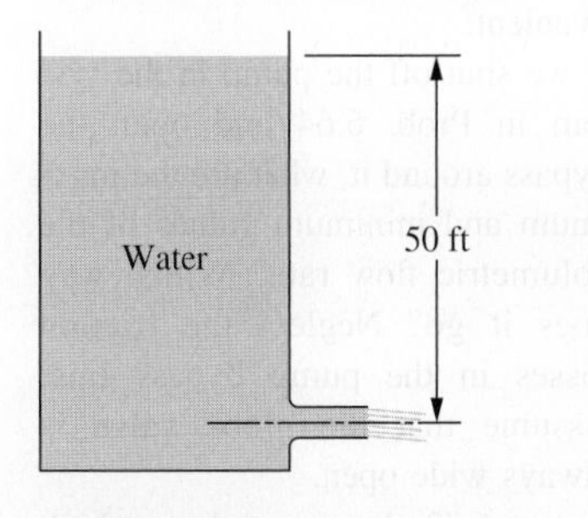

FIGURE 6.34
Gravity draining flow with friction, Prob. 6.61.

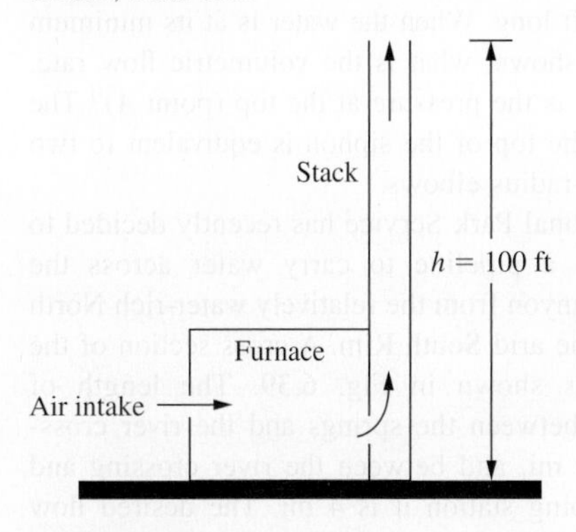

FIGURE 6.35
Flow in a furnace and chimney, Prob. 6.62.

(*b*) The same books often show the *Manning coefficients* to use in the Chézy formula, given by

$$C = \alpha \frac{\text{HR}^{1/6}}{n} \tag{6.63}$$

in which n is a roughness parameter, whose value is ≈ 0.012 for finished cement and α is a dimensional conversion factor. Determine the value α that corresponds to Example 6.15.

6.59.*In Fig. 6.32 the 3-in pipe is joined to the tank by a well-designed adapter, in which there is no entrance loss. What is the instantaneous velocity in the pipe?

6.60. In Fig. 6.33 water is being pumped through a 3-in pipe. The length of the pipe plus the equivalent length for fittings is 2300 ft. The design flow rate is 150 gal / min.

(*a*) At this flow rate, what pressure rise across the pump is required?

(*b*) If there are no losses in pump, motor, coupling, etc., how many horsepower must the pump's motor deliver?

6.61.*The tank in Fig. 6.34 is attached to 10 ft of 5-in pipe. The losses at the entrance from the reservoir to the pipe are negligible. What is the velocity at the exit of the pipe?

6.62. The flue gas in the stack in Fig. 6.35 is at 350°F and has $M = 28$ g / mol. The stack diameter is 5 ft, and the friction factor in the stack is 0.005. In passing through the furnace the air changes significantly in density, because it is heated by the combustion and then cooled in giving up heat to the working parts of the furnace. Thus, we cannot rigorously apply B.E. in the form we use in this problem (we could do so by integrating from point to point, over points so close together that the density change was negligible, but that would be very difficult in such a complex flow through a furnace). However, experimental data on the friction effects of furnaces indicate that if we treat them as constant-density devices with flowing fluids having the density and viscosity of the gas in the stack, then we can use Eq. 6.25. For this furnace, for those assumptions, $K \approx 3.0$. Thus, in applying B.E. to this furnace and stack, assume that the air changes at the inlet to the furnace to a gas with $M = 28$ g / mol and $T = 350$°F, and then maintains that M and T throughout the furnace and the stack. Estimate the velocity of the gases in the stack.

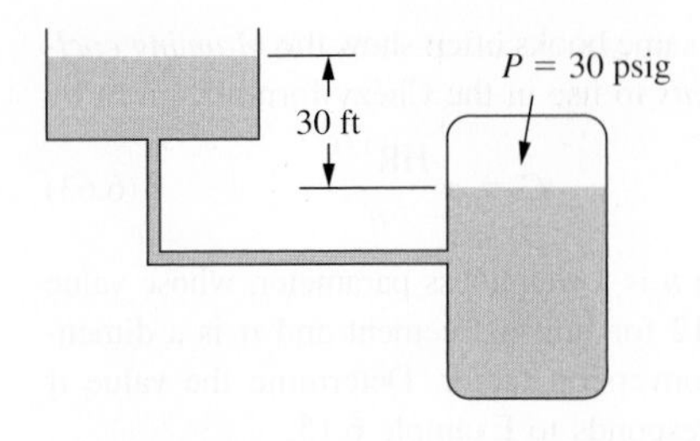

FIGURE 6.36
Flow driven one way by gravity and the opposite way by pressure difference, Prob. 6.63.

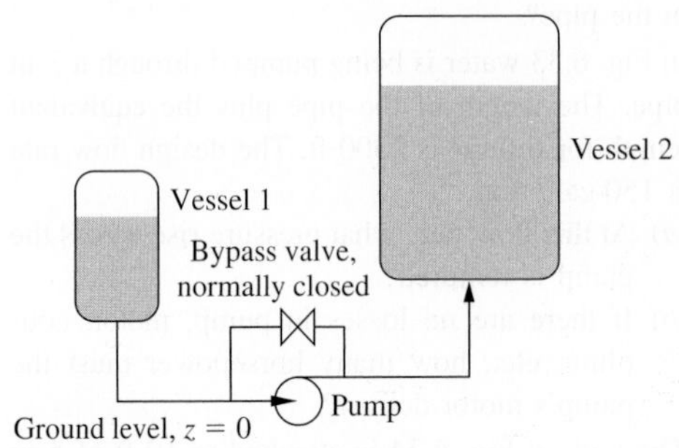

FIGURE 6.37
A somewhat more realistic pumping situation, Prob. 6.64.

TABLE 6.A
Values for Prob. 6.64

	Vessel 1	Vessel 2
P_{max}, psig	20	81
P_{min}, psig	8	47
Max liquid level, above $z = 0$, ft	43	127
Min liquid level, above $z = 0$, ft	21	100

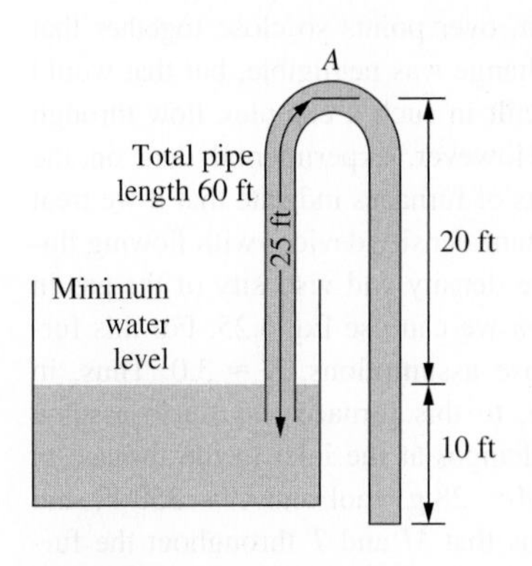

FIGURE 6.38
Siphon with friction, Prob. 6.66.

6.63.*The vessels in Fig. 6.36 are connected by 1000 ft of 3-in pipe (neglect fitting and entrance and exit losses). In each vessel the diameter is so large that V is negligible. The fluid is an oil with $\nu =$ 100 cSt, and $\rho = 60 \text{ lbm / ft}^3$. How many gallons per minute are flowing? Which way?

6.64. The two vessels shown in Fig. 6.37 have the design conditions shown in Table 6.A. The connecting line between the vessels is 3-in pipe that is 627 ft long, containing six elbows, four gate valves, and one globe valve. The fluid to be pumped has a specific gravity range of 0.80 to 0.85 and a kinematic viscosity range of 2 to 5 cSt. The flow rate is 150 to 200 gal / min. We are ordering the pump. What values will we specify of

(*a*) the flow rate?

(*b*) the pump head, $\Delta P / \rho g$, in feet? For this problem the head form of B.E. is convenient.

6.65.*If we shut off the pump in the system in Prob. 6.64 and open the bypass around it, what are the maximum and minimum values of the volumetric flow rate? Which way does it go? Neglect the friction losses in the pump bypass line. Assume that the globe valve is always wide open.

6.66. Figure 6.38 shows a siphon, which will be used to empty water out of a tank. The siphon is made of 10-in pipe, 60 ft long. When the water is at its minimum level, as shown, what is the volumetric flow rate, and what is the pressure at the top (point A)? The bend at the top of the siphon is equivalent to two 90° long-radius elbows.

6.67. The National Park Service has recently decided to construct a pipeline to carry water across the Grand Canyon from the relatively water-rich North Rim to the arid South Rim. A cross section of the system is shown in Fig. 6.39. The length of pipeline between the springs and the river crossing is 10 mi, and between the river crossing and the pumping station it is 4 mi. The desired flow rate is 1000 gal / min. The pressure at the springs and at the pumping station may be assumed

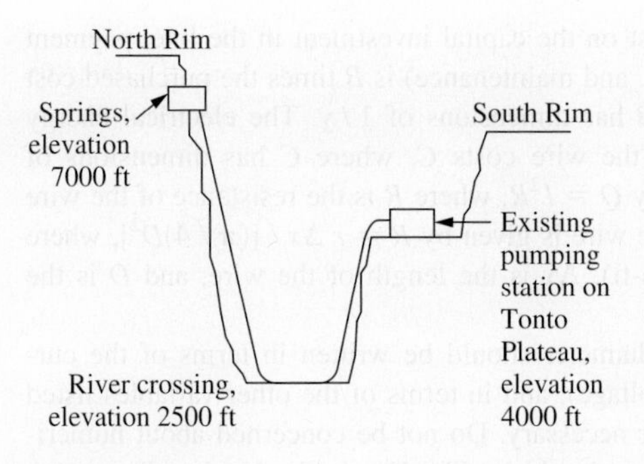

FIGURE 6.39
Elevations for the freshwater pipeline across the Grand Canyon, Prob. 6.67.

atmospheric. Because all the materials must be brought into place on muleback, which is costly, there is a considerable incentive to make the pipe as lightweight as possible. Recommend what material the pipe should be made of, what its inside diameter should be, and what its wall thickness should be.

6.68. Solve for the steady state flows in Fig. 6.21. As a first guess take $\alpha = 10$ ft, which is close to, but not equal the value of α which makes the sum of the volumetric flow rates into point 2 = 0.

6.69. Repeat Prob. 6.68 for the case where $(z_3 - z_1) = -5$ ft and $(z_4 - z_1) = -40$ ft. *Hint:* In this case, not all the flows will be in the same direction as in Prob. 6.68.

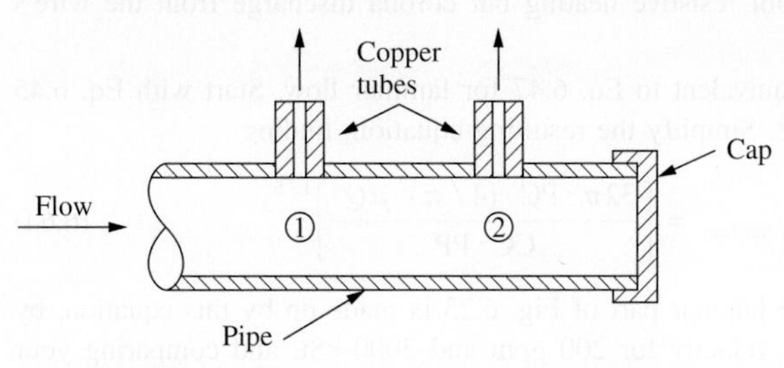

FIGURE 6.40
Part of a manifold, Prob. 6.70.

6.70.*Figure 6.40 shows the end of a *manifold* in which one major pipe feeds into a number of smaller pipes that branch from it. Many air-conditioning ducts are manifolds. This figure shows only the last two branches. The flow out through the branches depends almost entirely on the pressure in the main channel opposite them, and very little on the velocity in that channel. For the flow as shown, will the velocity out of tube 1 be greater or less than that out of tube 2;

(*a*) For zero friction in the pipe flow (because it is so much bigger than the tubes)?

(*b*) For substantial friction in the pipe flow (because a cylindrical rod has been inserted, thus making the cross-sectional area perpendicular to flow much less)?

(*c*) Dr. J. Q. Cope, vice president of Chevron Research when the author started there in 1958, used this device to teach humility to new Ph.D.s. He would describe the device, without mentioning the insertable rod. Then he would goad the new Ph.D. into betting him which jet would squirt higher when water was introduced. Then he would go and get the device, inserting or removing the rod as needed before returning to collect on his bet. What practical lessons might a new engineer learn from this story [25]?

6.71. Check the assumed friction factor in Example 6.18. For the value of relative roughness shown, the range of possible friction factors in turbulent flow is 0.004 to 0.01. How much would the economic diameter differ from Example 6.18 if $f = 0.01$?

6.72. The type of calculation of economic diameter of pipes shown in Sec. 6.12 was apparently first performed by Lord Kelvin [26] in connection with the problem of selecting the economic diameter for long-distance electric wires. To see how he obtained his result, derive the formula for the economic diameter of an electric conductor (analogous to Eq. 6.47) using the following information: The purchased cost of the whole transmission line (including poles, insulators, land, construction labor, etc.) is A times the mass of metal in the wire, where A has dimensions of \$ / lbm. The annual cost of owning the transmission line (which includes interest on the capital investment in the line, payment on the principal of that investment, taxes, and maintenance) is B times the purchased cost of the whole transmission line, where B has dimensions of 1 / y. The electrical energy that is lost due to resistive heating in the wire costs C, where C has dimensions of \$ / kWh. The resistive heating is given by $Q = I^2R$, where R is the resistance of the wire and I is the current. The resistance of the wire is given by $R = r\,\Delta x / [(\pi / 4)D^2]$, where r is the resistivity (with dimension ohm-ft), Δx is the length of the wire, and D is the wire diameter.

Your formula for the economic diameter should be written in terms of the current to be carried (*not* in terms of the voltage), and in terms of the other variables listed above, plus any others that you consider necessary. Do not be concerned about numerical units conversions; your final equation should be like Eq. 6.47, showing D_{econ} as a function of the appropriate variables to the appropriate powers. (Kelvin's solution is still correct for low-voltage transmission lines, but not for the high-voltage lines now used, in which the major loss is not resistive heating but corona discharge from the wire's surface.)

6.73. (*a*) Work out the equation equivalent to Eq. 6.47 for laminar flow. Start with Eq. 6.45 and substitute $f = 16 / \mathcal{R}$. Simplify the resulting equation, finding

$$D_{\text{econ, laminar}} = \left[\frac{32\pi \cdot \text{PC} \cdot (4 / \pi)^2 \, \mu Q^2}{\text{CC} \cdot \text{PP}} \right]^{1/5} \tag{6.64}$$

(*b*) Check to see whether the laminar part of Fig. 6.23 is made up by this equation, by calculating the economic velocity for 200 gpm and 2000 cSt, and comparing your answer to the value you read from that figure. Use the economic values that are shown with that figure.

(*c*) The lines of constant viscosity on Fig 6.23 have slope ≈ 0.2. Does this agree with Eq. 6.64?

(*d*) Figure 6.23 indicates that the economic velocity is practically independent of fluid density. Does this agree with Eq. 6.64?

6.74. It has been proposed to solve Los Angeles' air pollution problem by pumping out the contaminated air mass every day. The area of the L.A. Basin is 4083 mi^2. The contaminated air layer is roughly 2000 ft thick. Suppose we plan to pump it out every day, a distance of 50 mi to Palm Springs. (It is assumed that the residents of Palm Springs will not object, which is not a very good assumption.)

(*a*) Estimate the economic velocity in the pipe.
(*b*) Estimate the required pipe diameter.
(*c*) Estimate the pressure drop.
(*d*) Estimate the pumping power requirement.
(*e*) Comment on the feasibility of this proposal.

6.75. It has been proposed to solve the water problem in Los Angeles by importing water from the mouth of the Columbia River, where vast amounts flow into the sea. One way to do this would be with a pipeline and pumping station. Both ends of the pipe are at sea level, so the only pumping cost would be the cost of overcoming the friction loss. The pipe length would be about 1000 mi. Assuming that we want to move 10^7 acre-ft a year (1 acre-ft = 43,560 ft^3), estimate the horsepower of pumps required. State your assumptions.

6.76. If in Example 6.18 the fluid were water contaminated with hydrofluoric acid, we would have to use a special corrosion-resistant pipe. Suppose that this pipe had a value of PP exactly 10 times that of carbon-steel pipe. What would D_{econ} be?

6.77. If Table 6.8 were based on Eq. 6.49 and the friction factor were held constant, then the product of the economic velocity and the cube root of the density would be a constant. How much does it vary from being a constant? What is the cause of this variation?

6.78. Does the result in Example 6.18 agree exactly with the data in Table 6.8 and Fig. 6.23? If not, how much does it disagree, and what is the most probable cause of the difference?

6.79. You are selected to design the fuel line for a Mars-landing rocket. Money is unimportant; low mass is the main goal. Decide what the significant "economic" factors in this problem are, and write in general form the equivalent of Eq. 6.47 for this problem.

6.80. A 1 μ diameter spherical particle with SG = 2.0 is ejected from a gun into air at a velocity of 10 m / s. How far does it travel before it is stopped by viscous friction? (This distance is the *Stokes' stopping distance,* which appears often in the fine particle literature.) Ignore the effect of gravity.

(*a*) Work out the equation in general terms, by writing $F = ma$. The drag force is the only force acting on the particle, after it leaves the gun, operating in the direction opposite the direction of motion, and is given by Eq. 6.54

$$F = -3\pi\mu DV = ma = \frac{\pi}{6} D^3 \rho \frac{dV}{dt}; \qquad \frac{dV}{dt} = -\frac{18\mu V}{D^2 \rho} \tag{6.65}$$

Substitute $dt = dx / V$, separate the variables, cancel the two V terms, and integrate from $V = V_0$ at $x = 0$ to $V = 0$ at $x_{\text{Stokes' stopping distance}}$ to find

$$x_{\text{Stokes' stopping}} = \frac{V_0 D^2 \rho}{18\,\mu} \tag{6.66}$$

One often sees this equation with a C, for the Cunningham correction factor in the numerator.

(*b*) Insert the numerical values and find the value of $x_{\text{Stokes' stopping distance}}$.

(*c*) On the basis of the logic of this calculation, how long does it take the particle to come to exactly zero velocity? How long does it take it to come to 1 percent of V_0?

(*d*) How far does the particle fall by gravity (which we ignored in this derivation) in the time it takes to come to 1 percent of V_0?

6.81.*Rework Example 6.19 for the particle settling in water at 68°F instead of in air.

6.82. Rework Example 6.20 for the ball falling in glycerin instead of in water. $\mu_{\text{glyc}} = 800$ cP, and $\rho_{\text{glyc}} = 78.5$ lbm / ft^3.

6.83.*A spherical balloon is 10 ft in diameter and has a buoyant force 0.1 lbf greater than its weight. What is its terminal velocity rising through air?

6.84. A standard baseball has a diameter of 2.9 in and a mass of 0.31 lbm. Good fast-ball pitchers can throw one at about 100 mi / h.

(*a*) Neglecting the effect of the stitching on the ball and the spin of the ball, estimate the drag force of the air on the ball.

(*b*) The distance from the pitcher's mound to home plate is 60 ft. If the ball left the pitcher's hand at 100 mi / h, how fast will it be going when it reaches home plate, subject to the simplifications in part (*a*)?

6.85. Probably the most-studied kick in soccer history was David Beckham's free-kick goal in the England-Greece World Cup Qualifiers in 2001, [27]. This kick even made a movie title ("Bend it like Beckham," 2002).

The kick left his foot at 36 m / s, 27 m from the goal. It was high enough to pass over the screen of defenders, and spinning enough on a vertical axis to curve toward the corner of the goal. It appeared to be aimed above the goal, but suddenly slowed down dramatically in flight, and fell into the upper corner of the goal. Explain in terms of Fig. 6.24 how this was possible. A standard soccer ball weighs ≈425 g, and has a diameter of ≈22.3 cm.

6.86. You have certainly observed that you can walk much faster in air than you can when you are up to your neck in a swimming pool. Why? Is that mostly because the water is more viscous than air? Or because the water is more dense than air? Or some other answer?

6.87.*A 120-lbm parachutist jumps from a plane and falls in free fall for a while before opening her chute.

(*a*) If she falls head first, then her projected area perpendicular to the direction of fall is $\approx 1\ ft^2$ and $C_d \approx 0.7$. What is her terminal velocity? How many seconds must she fall to reach 99 percent of this terminal velocity, assuming that the drag coefficient is independent of velocity? How far does she fall in this number of seconds?

(*b*) If instead of falling head first, she spreads out her arms and legs and lies horizontally, then her projected area will be $\approx 6\ ft^2$ and $C_d \approx 1.5$. Repeat part (*a*) for this condition.

6.88. We want to design a parachute. The requirement is that at terminal velocity the rider will have a velocity equal to the maximum velocity the rider would reach jumping to the ground from a 10-ft-high roof. The rider weighs 150 lbf. The parachute will be circular and its drag coefficient $C_d = 1.5$. What diameter must the parachute have?

6.89. Occasionally, car companies advertise that their sports-model automobiles have very low values of C_d, typically about 0.3 for teardrop-shaped cars. That drag coefficient is based on the frontal area. If a car has that C_d and a width of 6 ft and a height of 5 ft and is going 70 mi / h,

(*a*) What is the air resistance of the car?

(*b*) How much power must be expended to overcome this air resistance?

6.90. In Examples 6.19 and 6.20, we assumed Stokes' law applies, calculated V, and then checked $\mathcal{R}_p$ to see whether the assumption of Stokes' law was a good one. If our assumption was not a good one, then the V calculated in the first step was a wrong velocity, and the calculated $\mathcal{R}_p$ was wrong, too. Is there any chance that this procedure can lead to a combination of V and $\mathcal{R}_p$ that indicates that Stokes' law should be obeyed when actually the $\mathcal{R}_p$ based on the correct solution is outside the range of Stokes' law?

6.91. Check the results of Examples 6.19 and 6.20 on Fig. 6.26.

6.92. A spherical raindrop with a diameter of 0.001 in is falling at its terminal velocity in still air. How fast is it falling?

6.93. In James Bond movies the hero is often swimming and has to dive deep into the water to escape the bullets from the enemy helicopter flying above him. How deep should he dive? Assume the bullet is a sphere of diameter 0.5 in and mass 0.027 lbm. It hits the surface of the water vertically at a velocity of 1000 ft / s and will not inflict serious injury if it is slowed down to a velocity of 100 ft / s or less. For the purposes of this problem only, assume that the drag coefficient is constant, independent of velocity, and equal to 0.1.

6.94. A bullet was fired straight up at 2700 ft / s. The bullet had a mass of 150 grains (the standard mass unit in U.S. gun lore, 1 lbm = 7000 grains), a more-or-less cylindrical shape with a sharp point, and a diameter of 0.30 in.

(*a*) If there were zero air resistance, how high would it go? How long would it take to reach that altitude? How long would it take it to come back to ground? What would its velocity be when it came back to ground?

(*b*) It actually went up 9000 ft in 18 s and returned to earth in 31 more seconds for total time of 49 s and arrived with a velocity of 300 ft / s [28]. Based on these values, estimate the drag coefficient going up, and the drag coefficient coming down. Assume in both cases that the bullet retained its vertical orientation due to its axial rotation (i.e., did not tumble end-over-end) and that the drag coefficient was a constant (only a fair assumption).

6.95. Figure 6.26 (and the calculations on which it is based) are widely used in chemical engineering to describe separations processes based on gravity. Large particles can be separated into distinct sizes by screens, but smaller particles cannot. Instead they are separated into size fractions or different density fractions by processes that take advantage of their different settling velocities in air or water. Many mineral separations function by making a very uniform particle size sample with screens, and then separating by specific gravity in differential settlers, mostly in water. In these devices the settling velocity is less than that shown in Fig. 6.26, because that figure (and the examples in the chapter) assume that each particle is far from any other particle. In these devices the particles are close to one another.

(*a*) Sketch the flow around a particle which is settling in a closed container, by itself, and then in a mud or slurry in which there are many other nearby particles. Indicate why one would expect the other particles (and the flow of fluid which they induce) to cause the particle to settle more slowly.

(*b*) This slowing is called *hindered settling*. Its effect can be estimated [29] by

$$V_{\text{terminal, hindered}} = V_{\text{terminal, isolated}}(1 - c)^n \tag{6.67}$$

in which c is the volume fraction of solids and $n = 4.65$ in the Stokes' law region, with value decreasing to 2.33 for $\mathcal{R}_p > 1000$. Using this value, estimate the settling velocity of a simple spherical particle of SG = 3 and $D = 20\ \mu$ in water all by itself, and in a mud that has $c = 0.4$.

REFERENCES FOR CHAPTER 6

1. Reynolds, O. "An Experimental Investigation of the Circumstances Which Determine Whether the Motion of Water Shall Be Direct or Sinuous and of the Law of Resistance in Parallel Channels." *Philosophy Transactions of the Royal Society 174* (1883). The history of this work and its relation to others are discussed in Rouse and Ince [2, p. 206].
2. Rouse, H., S. Ince. *History of Hydraulics.* Iowa City, Iowa: Iowa Institute of Hydraulic Research, 1957, (*a*) p. 160, (*b*) p. 323.
3. Nikuradse, J. "Stroemungsgesetze in rauhen Rohren (Flow Laws in Rough Tubes)." *VDI-Forsch.-Arb. Ing.-Wes. Heft 361* (1933).
4. Moody, L. W. "Friction Factors for Pipe Flow." *Transactions of the American Society of Mechanical Engineers 66* (1944), (*a*) pp. 671–684, (*b*) Hunter Rouse's comments at the end of the article.
5. Colebrook, C. F. "Turbulent Flow in Pipes, with Particular Reference to the Transition Region Between Smooth and Rough Pipe Laws," *Journal of the Institute of Civil Engineering 11* (1938–1939), pp. 133–155.
6. Schlichting, H., K. Gersten. *Boundary-Layer Theory.* 8th ed. Berlin: Springer, 2000.
7. Massey, B. S., "Mech. of Fluids", 1970, p. 180 attributes Eq. 6.21 to Moody. However it is not shown in Ref (4), so it must be somewhere else.
8. *ASHRAE Handbook of Fundamentals.* Atlanta, Georgia: ASHRAE. New editions every few years.
9. O'Brian, M. P., G. H. Hickox. *Applied Fluid Mechanics.* New York: McGraw-Hill. 1937, p. 211.
10. Lapple, C. E. "Velocity Head Simplifies Flow Computation." *Chemical Engineering 56* (1949), *56(5),* pp. 96–104.

11. "Flows of Fluids Through Valves, Fittings and Pipes." *Technical Paper No. 410.* Crane Company, 475 N. Gary Ave., Carol Stream, Illinois, 1957, p. A-19.
12. Bird, R. B., E. N. Stewart; W. E. Lightfoot. *Transport Phenomena.* 2nd ed. New York: Wiley, 2002.
13. "Emission Factors for Equipment Leaks of VOC and HAP." U.S. Environmental Protection Agency Report No. EPA-450/3-86-002. Washington, D.C.: U.S. Government Printing Office, 1986.
14. de Nevers, N. *Air Pollution Control Engineering,* 2nd ed. New York: McGraw-Hill, 2000.
15. Cross, H. "Analysis of Flow in Networks of Conduits or Conductors." *University of Illinois Bulletin 86,* 1946. This method is described in detail in most older hydraulics books.
16. Carnahan, B., H. A. Luther, J. O. Wilkes. *Applied Numerical Methods.* New York: John Wiley, 1969, p. 310.
17. Peters, M. S., K. D. Timmerhaus. *Plant Design and Economics for Chemical Engineers,* 4th ed. New York: McGraw-Hill, 1991, Chaps. 4–10.
18. Boucher, D. F., G. E. Alves. "Fluid and Particle Mechanics." In *Chemical Engineers' Handbook,* 4th ed. R. H. Perry; C. H. Chilton; S. D. Kirkpatrick. New York: McGraw-Hill, 1963, pp. 5–30.
19. Newton, I. *Philos. Nat. Primo. Math. 2* (1710), Sect. 2. Cited in L. B. Torobin; W. H. Gauvin, *Canadian Journal of Chemical Engineering 37* (1959), p. 129.
20. Lamb, H. *Hydrodynamics.* New York: Dover, 1945.
21. Steinman, D. B. "Suspension Bridges: The Aerodynamic Stability Problem and Its Solution." *American Scientist 42* (1954), p. 397.
22. Von Mises, R. *Theory of Flight.* New York: Dover, 1945, 139 *et seq.* This book from before the jet era gives a thorough, readable treatment of basic aerodynamics.
23. Haaland, S. E. "Simple and Explicit Formulas for the Friction Factor in Turbulent Pipe Flow." *Journal of Fluids Engineering 105* (March 1983), pp. 89–90.
24. Oosthuizen, P. H., W. E. Carscallen. *Compressible Fluid Flow.* New York: McGraw-Hill, 1997.
25. de Nevers, N. "Bernoulli's Equation with Friction." *Chemical Engineering Ed. 7* (1973), pp. 126–128.
26. Thompson, W. (Lord Kelvin). "On the Economy of Metal in Conductors of Electricity." *Report to the British Association for the Advancement of Science 1881.* London: John Murray, 1882, p. 526.
27. Fluent Company Press Release PR43. May 20, 2002. www.fluent.com/about/news/pr/pr43.htm.
28. Rinker, R. A. *Understanding Firearm Ballistics,* 4th ed. Apache Junction, Arizona: Mulberry House Publishers, 2000.
29. Tilton, J. N. "Fluid and Particle Dynamics." In *Perry's Chemical Engineers' Handbook;* 7th ed. R. H. Perry; D. W. Green; J. O. Maloney. New York: McGraw-Hill, 1997, pp. 6–52.

CHAPTER 7

THE MOMENTUM BALANCE

Newton's second law of motion, often called Newton's *equation of motion*, is commonly written

$$F = ma \tag{7.1}$$

It is easily applied in this form to the motion of rigid bodies like the falling bodies of elementary physics. It also can be easily applied in this form to the motion of fluids that are moving in rigid-body motion, as discussed in Chap. 2. However, for fluids that are moving in more complicated motions, e.g., in pipes or around airplanes, it is difficult to use Eq. 7.1 in the form shown. Therefore, in this chapter we will rewrite the equation in the form of a momentum balance. The momentum balances given in this chapter are rearrangements of Eq. 7.1, and they, too, are often referred to in the engineering literature as *equations of motion.*

The momentum balance form will prove very convenient for solving fluid-flow problems. In particular, it will allow us to find out something about complicated flows through a system without having to know in detail what goes on inside the system. In this way the momentum balance is similar to the mass and energy balances. For example, by using the mass and energy balances we can find out some things about a turbine or compressor from the inlet and outlet streams only, without knowing in detail what goes on inside. We also will frequently apply the momentum balance "from the outside."

Remember that all of this chapter is simply the manipulation and application of Eq. 7.1. In this chapter the applications are one-dimensional; in Part IV we apply the same ideas to two- and three-dimensional flows.

7.1 MOMENTUM

Momentum, like energy, is an abstract quantity. Unlike energy, it is defined in terms of simpler quantities, mass and velocity. The definition of momentum is given in terms of the momentum of a body:

$$\begin{pmatrix}\text{Momentum}\\ \text{of a body}\end{pmatrix} = \begin{pmatrix}\text{mass of}\\ \text{the body}\end{pmatrix} \cdot \begin{pmatrix}\text{velocity of}\\ \text{the body}\end{pmatrix} = mV \tag{7.2}$$

It makes no sense to speak of momentum as separate from bodies because, as we see from this equation, if there is no mass, there is no momentum. Only bodies—of solid, liquid, or gas—have mass, so momentum can exist only in connection with some body.*

Furthermore, momentum is a vector. We have applied the balance equations to mass and energy, which are scalar, here we will apply it to a vector and get similar results. Most often in dealing algebraically with vectors, one uses the scalar components of the vector rather than the vector itself. For example, Eq. 7.1 may be written in vector form

$$\mathbf{F} = m\mathbf{a} \tag{7.3}$$

(**boldface** indicates a vector). However, any vector can be resolved into the vector sum of three scalar components multiplied by unit vectors, in three mutually perpendicular directions.** For example,

$$\mathbf{F} = F_x\mathbf{i} + F_y\mathbf{j} + F_z\mathbf{k} \tag{7.4}$$

where F_x, F_y, and F_z are the scalar components of the vector **F** in the x, y, and z directions, and **i, j,** and **k** are unit vectors in the x, y, and z directions, respectively. Similarly, we can resolve the acceleration vector **a** and rewrite Eq. 7.3 in the following forms:

$$F_x\mathbf{i} + F_y\mathbf{j} + F_z\mathbf{k} = m(a_x\mathbf{i} + a_y\mathbf{j} + a_z\mathbf{k}) \tag{7.5}$$

$$(F_x - ma_x)\mathbf{i} + (F_y - ma_y)\mathbf{j} + (F_z - ma_z)\mathbf{k} = 0 \tag{7.6}$$

But this equation is the equation of a new vector, the ($\mathbf{F} - m\mathbf{a}$) vector, which is seen to be zero. For a vector to be zero each one of its scalar components must be zero, so this equation is exactly equivalent to

$$F_x - ma_x = 0; \quad F_y - ma_y = 0; \quad F_z - ma_z = 0 \tag{7.7}$$

---

*Neutrinos and light quanta (photons) apparently possess momentum but not rest mass and thus might be considered exceptions to this statement. However, they are observable only when moving at high velocities, at which time they have considerable energy and, hence, relativistic mass, so this statement is correct even for them.

**As far as we know, we live in a three-dimensional universe, so we speak of three mutually perpendicular directions. If we lived in an n-dimensional universe there would be n mutually perpendicular directions. In some problems it is convenient to consider n-dimensional "spaces," in which a vector is resolved into n "perpendicular" components.

This shows us that we may consider any vector equation as a shorthand way of writing three scalar equations. Vector calculus is a powerful tool for deriving equations describing multidimensional problems. In electromagnetic problems and problems involving moving coordinate axes (e.g., gyroscopes) it is easiest to work directly with the vector quantities. However, for solving practical fluid-mechanics problems it is almost always more convenient to use the three scalar (component) equations, which are the exact equivalent of the vector equation. In this chapter we will show the momentum balance both as a vector equation and as its more useful scalar equivalents. We will show the application of the vector calculus approach to fluid mechanics problems in Part IV.

One distinct complication with the momentum balance as compared with the mass and energy balances concerns the algebraic signs of the momentum terms. In the energy and mass balances we have little trouble with signs, because we seldom consider negative energies and never consider negative masses.† On the other hand, if we wish to represent a velocity in the minus x direction, we write it as

$$\mathbf{V} = V_x\mathbf{i} + V_y\mathbf{j} + V_z\mathbf{k} \tag{7.A}$$

where V_z and V_y are zero, and V_x is a negative number. Therefore, in our scalar equations we will have to be more careful of algebraic signs than we were with the mass and energy balances.

7.2 THE MOMENTUM BALANCE

In Chap. 3 we saw that the general balance equation (Eq. 3.2) can be applied to any extensive property—any property that is proportional to the amount of matter present. Since momentum is proportional to the amount of matter present, it is an extensive property and must obey a balance equation. Here, as in all other balance equations, we must be careful to choose and define our system.

Figure 7.1 shows the system used to state the momentum balance; it consists of some tank or vessel with flow of matter in or out and system boundaries as shown. The momentum contained within the system boundaries is

$$\left(\begin{array}{c}\textbf{Momentum inside}\\ \textbf{system boundaries}\end{array}\right) = \int_{\substack{\text{all mass}\\ \text{in system}}} \mathbf{V}\,dm \tag{7.8}$$

We simplify this by assuming that all the mass inside the system has the same velocity, so that this integral simplifies to $(m\mathbf{V})_{\text{sys}}$. The momentum-accumulation term becomes

$$\textbf{Momentum accumulation} = d(m\mathbf{V})_{\text{system}} \tag{7.9}$$

†Although in an absolute sense energies can never be negative, energies, relative to an arbitrary datum can be negative. In one-component systems, such as steam power plants or refrigeration systems, the datum usually is so chosen that none of the energy terms is negative. However, the common datum for combustion work and chemical-reaction problems results in negative energy terms.

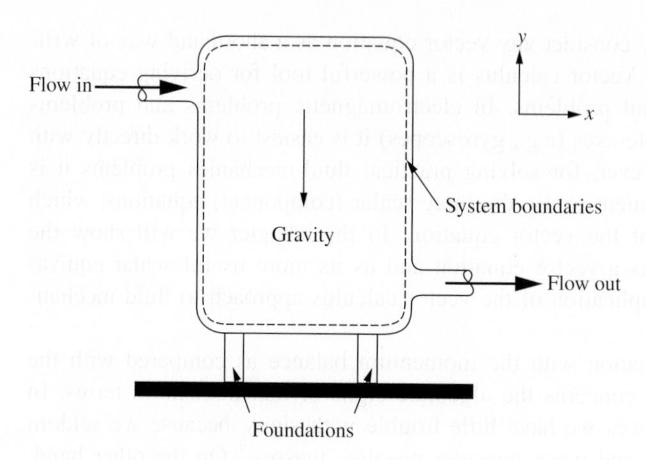

FIGURE 7.1
System used for stating the momentum balance.

For one flow in and one flow out, as in the figure, the momentum flow in minus momentum flow out is

$$\textbf{Momentum flow in} - \textbf{momentum flow out} = \mathbf{V}_{\text{in}}\, dm_{\text{in}} - \mathbf{V}_{\text{out}}\, dm_{\text{out}} \quad (7.10)$$

If there is more than one flow in or out, there will be summation terms for momentum flows in and out, just as there are summation terms for mass and energy flows into and out of a system in the mass and energy balances.

In Sec. 3.4 we discussed velocity distributions in flows in pipes. There we showed that simplifying the problems by replacing actual flows, which have a velocity gradient from centerline to edge, with flows all at one velocity, $V = V_{\text{avg}}$, changed the calculated kinetic energy in the flow by about 6 percent for most turbulent flows, which we considered negligible. Table 3.1 shows that a similar simplification changes the calculated momentum in such a flow by about 1 percent, which we will also consider negligible. Thus, for the rest of this chapter we will replace the real flow in pipes, channels, and jets, which has some nonuniform velocity distribution, with a flow with a uniform velocity distribution, $V = V_{\text{avg}}$. If there is any question whether this is permissible, please refer to Table 3.1 and the problems associated with it.

Now, to account for the creation or destruction of momentum, we invoke Eq. 7.1, which can be rewritten

$$\mathbf{F} = m\mathbf{a} = m\frac{d\mathbf{V}}{dt} \quad (7.11)$$

The possible changes of mass of the system are accounted for by the flow-in or flow-out terms, and the creation or destruction terms must apply equally well to constant-mass and variable-mass systems. For a constant-mass system, we can take the m inside the differential sign in the last equation and rearrange, to show that

$$d(m\mathbf{V})_{\text{sys}} = \mathbf{F}\,dt \tag{7.12}$$

so that the momentum creation or destruction term is $\mathbf{F}\,dt$.*

If more than one force is acting, the **F** in Eq. 7.12 must be replaced with a sum of forces. Usually there are several forces acting in fluid-flow problems, so we write the momentum balance with a $\sum \mathbf{F}$. The forces acting on the system shown in Fig. 7.1 are the external pressure on all parts of its exterior and the force of gravity. Other forces that we might consider are electrostatic or magnetic forces. If we had chosen our system such that the boundary passed through the foundations in Fig. 7.1, then there would be a compressive force in the structural members of the foundation, which would have to be taken into account.

Writing all the terms together, we find the vector form of the momentum balance:

$$d(m\mathbf{V})_{\text{sys}} = \mathbf{V}_{\text{in}}\,dm_{\text{in}} - \mathbf{V}_{\text{out}}\,dm_{\text{out}} + \sum \mathbf{F}\,dt \tag{7.13}$$

Here we have not included a destruction term, because the $\sum \mathbf{F}$ in the equation is the vector sum of all the forces acting on the system. If this sum is in the opposite direction of the velocities, then the $\sum \mathbf{F}\,dt$ term is a momentum destruction term; it will enter with a minus sign. Most often we divide Eq. 7.13 by dt to find the *rate form* of the momentum balance:

$$\frac{d(m\mathbf{V})_{\text{sys}}}{dt} = \mathbf{V}_{\text{in}}\,\dot{m}_{\text{in}} - \mathbf{V}_{\text{out}}\,\dot{m}_{\text{out}} + \sum \mathbf{F} \tag{7.14}$$

This is not a *derivation* of the momentum balance, but simply a restatement of Newton's second law in a convenient form. Furthermore, Newton's laws, like the laws of thermodynamics and the law of conservation of mass, are underivable; they cannot be demonstrated from any prior principle but rest solely on their ability to predict correctly the outcome of any experiment ever run to test them.

Equations 7.13 and 7.14 are balance equations entirely analogous to the mass and energy balances we discussed in Chaps. 3 and 4. They have the same basic restriction of those balances, namely, that they may be applied only to a carefully defined system.

---

*Equation 7.12 implies that momentum is creatable, which can be misleading. If a person standing on the earth throws a ball, the momentum of the ball is increased in one direction and the momentum of the earth in the opposite direction by an equal amount. Thus, the momentum of the earth-ball system is unchanged. Because the mass of the earth is so much larger than that of the ball, we do not perceive this change in the earth's velocity (Prob. 7.1). However, Eq. 7.12 is correct, because creating momentum in one system results in creating equal and opposite momentum in some other system (usually the earth); so the net change of momentum of the universe is zero.

Modern physicists prefer Eq. 7.12 to Eq. 7.1 as the basic statement of Newton's law of motion. The reason is that, as the velocity of a body approaches the speed of light, the force exerted on it results mostly in an increase in mass rather than an increase in velocity. Thus, Eq. 7.1 is limited to constant-mass systems and excludes any system that is being accelerated to a speed near that of light, whereas Eq. 7.12 applies not only to constant-mass systems but also to bodies being accelerated to speeds near that of light, such as the particles in linear accelerators.

With the mass balance we can choose any system in which we can account for all flows of matter across the boundaries. With the energy balance we must choose a system in which we can account not only for any flows of matter across the boundaries but also for heat flows across the boundaries and work due to electric current, changing magnetic fields, moving boundaries, and rotating and reciprocating shafts. In applying the momentum balance we must choose a system in which it is possible to account for all flows of matter across the boundaries and also for all external forces acting on the system. In most fluid-flow problems this means that it must be possible to calculate the pressure on every part of the system boundary. Notice, however, that there is no term in Eq. 7.13 or 7.14 for heat flows, rotating shafts, or electric current flows, so we may choose systems for the momentum balance without necessarily being able to calculate those quantities over the boundaries of the system (we must account for electrostatic or magnetic fields, if they are significant). As we do with mass and energy balances, we may consider closed systems, in which the $\dot{m}$ terms are zero, or steady-flow systems, in which the accumulation is zero. Skill in applying the momentum balance is largely a matter of skill in choosing a system in which one can conveniently calculate all the terms in the balance.

Equations 7.13 and 7.14 are vector equations; each of them can be represented by three scalar equations, showing the components of the vectors in the x, y, and z directions or the r, θ, and z directions or in spherical coordinates. The x-component scalar equations equivalent to them are

$$d(mV_x)_{\text{sys}} = V_{x_{\text{in}}}\, dm_{\text{in}} - V_{x_{\text{out}}}\, dm_{\text{out}} + \sum F_x\, dt \tag{7.15}$$

$$\frac{d(mV_x)_{\text{sys}}}{dt} = V_{x_{\text{in}}}\, \dot{m}_{\text{in}} - V_{x_{\text{out}}}\, \dot{m}_{\text{out}} + \sum F_x \tag{7.16}$$

The corresponding y and z equations can be found from these simply by replacing all the x subscripts with y or z subscripts. The r, θ, and z component equations for cylindrical coordinates are shown in App. C.

To illustrate the application of the momentum balance, we consider first two very simple examples not involving fluids.

Example 7.1. A baseball is thrown in a horizontal direction. What terms of the momentum balance apply?

Taking the ball as our system and using the x component of the momentum balance, we see that there is no flow of matter in or out; therefore,

$$d(mV_x)_{\text{sys}} = (m\, dV_x)_{\text{sys}} = F_x\, dt; \qquad F_x = m\frac{dV_x}{dt} = ma_x \tag{7.B}$$

This is a simple restatement of $F = ma$ for a constant-mass system. ■

Example 7.2. A duck has a mass of 3 lbm and is flying due west at 15 ft / s. The duck is struck by a bullet with a mass of 0.05 lbm, which is moving due east at 1000 ft / s. The bullet comes to rest in the duck's gizzard. What is the final velocity of the duck-bullet system?

Here the problem is one-dimensional, so we work with the x-directed scalar-component equation, Eq. 7.15, and choose east as the positive x direction. First we work the problem by taking as our system the combined bullet and duck. No matter is flowing in or out of this system, nor does any external force act on it (we ignore the wind force, if a wind is blowing); so Eq. 7.15 becomes

$$d(mV_x)_{\text{sys}} = 0 \tag{7.C}$$

$$(mV_x)_{\text{sys, fin}} = (mV_x)_{\text{duck, init}} + (mV_x)_{\text{bullet, init}} \tag{7.D}$$

When we solve for $V_{x_{\text{sys, fin}}}$, we find

$$V_{x_{\text{sys, fin}}} = \frac{(mV_x)_{\text{duck, init}} + (mV_x)_{\text{bullet, init}}}{m_{\text{duck-bullet, fin}}}$$

$$= \frac{[3\ \text{lbm} \cdot (-15\ \text{ft / s})] + (0.05\ \text{lbm} \cdot 1000\ \text{ft / s})}{3.05\ \text{lbm}}$$

$$= +1.6\,\frac{\text{ft}}{\text{s}} = +0.73\,\frac{\text{m}}{\text{s}} \tag{7.E}$$

Now we do the problem over, taking the duck as our system. In this case there is mass flow into the system, so we have

$$d(mV_x)_{\text{sys}} = V_{x_{\text{in}}}\, dm_{\text{in}} \tag{7.F}$$

Integrating, we find

$$(mV_x)_{\text{sys, fin}} - (mV_x)_{\text{sys, init}} = V_{x_{\text{in}}}\, m_{\text{in}} \tag{7.G}$$

When we solve for $V_{x_{\text{sys, fin}}}$, we find

$$V_{x_{\text{sys, final}}} = \frac{V_{x_{\text{in}}}\, m_{\text{in}} + (mV_x)_{\text{sys, init}}}{m_{\text{sys, fin}}} \tag{7.H}$$

This is exactly the same as the result we found by taking the combined system. ■

This example shows the great advantage of the momentum balance; the details of the collision are very complicated when we wish to know the exact distance-time-shape history of the bullet in traversing the various feathers, bones, muscles, and internal organs of the duck, but from the momentum balance alone we can write down the final velocity of the bullet-duck system without knowing those details. It also shows that signs are important in the momentum balance. For the system chosen, the duck's initial velocity was −15 ft / s. If we had omitted that minus sign we would have calculated a final velocity of 31 ft / s, which would have been the final velocity if the bullet had been moving in the same direction as the duck (and overtaken it). That is the correct answer to a different problem. The signs in the momentum balance seem to be a permanent problem for students; pay attention to them!

This example also illustrates that some problems can be solved by the momentum balance but not by the energy balance. If we write the energy balance for Example 7.2, taking the combined duck-bullet as our system and neglecting the small change in volume of the system as the bullet enters the duck, we find it reduces to

$$\left[m\left(u+\frac{V^2}{2}\right)\right]_{\text{sys, fin}} = \left[m\left(u+\frac{V^2}{2}\right)\right]_{\text{duck, init}} + \left[m\left(u+\frac{V^2}{2}\right)\right]_{\text{bullet, init}} \tag{7.I}$$

Here we know the mass of the system and the initial kinetic and internal energies of its two parts, but these alone do not allow us to solve Eq. 7.I for either the final internal energy or the final kinetic energy. However, from the momentum balance we were able to find the final velocity, and we can then use it in Eq. 7.I to find the final internal energy. In this chapter we will see several other examples in which the momentum balance must be applied before the energy balance can be used (see Prob. 7.3).

7.3 SOME STEADY-FLOW APPLICATIONS OF THE MOMENTUM BALANCE

If we choose as our system some pipe, duct, channel, or jet with steady flow through it in one direction, e.g., the x direction, then Eq. 7.16 becomes

$$0 = \dot{m}(V_{x_{\text{in}}} - V_{x_{\text{out}}}) + \sum F_x \qquad \text{[steady flow]} \tag{7.17}$$

The application of this steady-flow one-dimensional momentum balance will be illustrated by several examples.

7.3.1 Jet-Surface Interactions

Many applications of Eq. 7.17 involve jets. A jet is a stream of fluid that is not confined within a pipe, duct, or channel; examples are the stream of water issuing from a garden hose and the exhaust gas stream from a jet engine. If any jet is flowing at a subsonic velocity, its pressure will be the same as the pressure of the surrounding fluid. If a jet enters or leaves a system or device at subsonic speed, it will enter and leave at the pressure of the surrounding fluid, although its pressure may be different inside the device. Sonic and supersonic jets are discussed in Chap. 8.

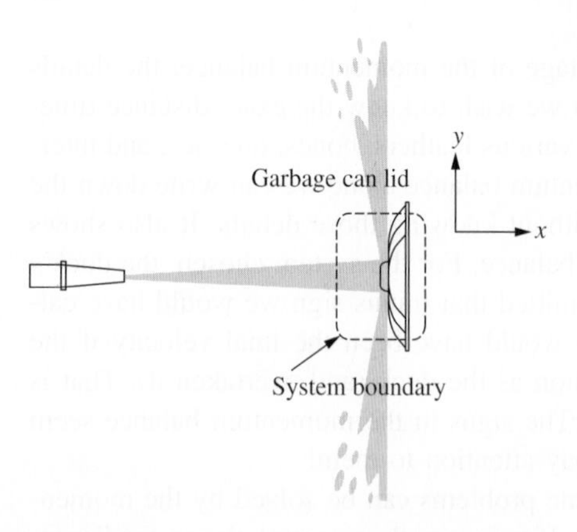

FIGURE 7.2
Interaction of a jet with a surface perpendicular to it.

Example 7.3. The police are using fire hoses to disperse an unruly crowd. The fire hoses deliver 0.01 m^3 / s of water at a velocity of 30 m / s. A member of the crowd has picked up a garbage can lid and is using it as a shield to deflect the flow. She is holding it vertically, so the jet splits into a series of jets going off in the y and z directions, with no x component of the velocity; see Fig. 7.2. What force must she exert to hold the garbage can lid?

By applying Eq. 7.17 in the x direction and taking the lid and the adjacent fluid as our system, as sketched in Fig. 7.2, we find

$$F_x = -0.01 \frac{\text{m}^3}{\text{s}} \cdot 998.2 \frac{\text{kg}}{\text{m}^3} \cdot \left[30 \frac{\text{m}}{\text{s}} - 0 \right] \cdot \frac{\text{N} \cdot \text{s}^2}{\text{kg} \cdot \text{m}} = -299.5 \text{ N}$$
$$= -67.3 \text{ lbf} \tag{7.J}$$

■

Here the pressure around the external boundary of the system is all atmospheric, so this force is simply the force exerted by the arms of the woman holding the lid. It is negative, because she is exerting this force in a direction opposite to the x axis. Here we could also have chosen as our system the fluid alone. Then to solve for the force we would have had to calculate the pressure exerted on it by the garbage can lid at every point of the system boundary. To do this we would have needed a detailed description of the flow. From such a detailed description, if it were available, we could calculate

$$F_x = \int_{\text{all area}} P\, dA \tag{7.K}$$

for the lid, finding the same answer. Thus, by a proper choice of system we can find the desired force without a detailed description of the flow; we can apply the momentum balance "from the outside."

Example 7.4. The member of the crowd in Ex. 7.3 now turns the lid around so that she can hold it by the handle. However, because of the shape of the lid the flow goes off as shown in Fig. 7.3, with an average x component of the velocity of -15 m / s. What force must she exert?

Applying Eq. 7.17 exactly as before, we find

$$F_x = -0.01 \frac{\text{m}^3}{\text{s}} \cdot 998.2 \frac{\text{kg}}{\text{m}^3} \cdot \left[30 \frac{\text{m}}{\text{s}} - \left(-15 \frac{\text{m}}{\text{s}} \right) \right] \cdot \frac{\text{N} \cdot \text{s}^2}{\text{kg} \cdot \text{m}}$$
$$= -449.3 \text{ N} = -101 \text{ lbf} \tag{7.L}$$

■

See Prob. 7.7!

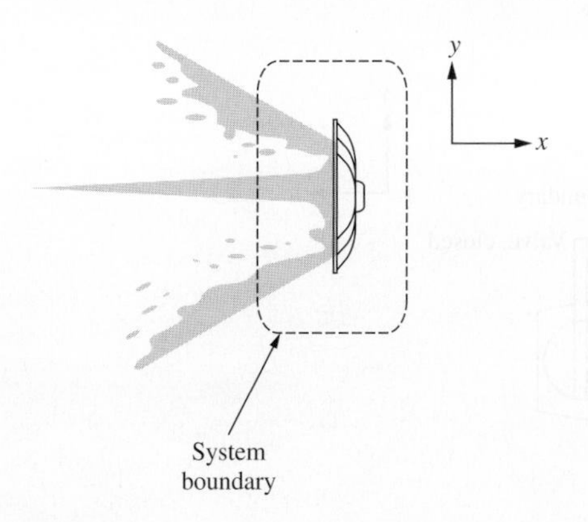

FIGURE 7.3
A curved surface turns the jet back toward its source.

7.3.2 Forces in Pipes

In the previous examples, the jets were open to the atmosphere, so their gauge pressure was always zero. Thus, we had no difficulty with deciding on the sign of pressure forces. The next two examples involves pressure forces inside pipes; they require us to consider the sign of the pressure forces. The easiest way to decide on the proper sign of a pressure force is to take the system boundary perpendicular to the axis in which we are

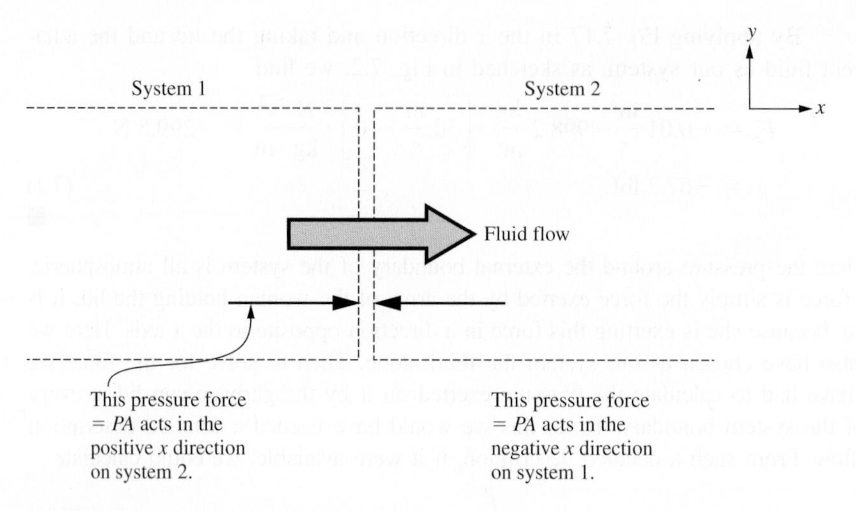

FIGURE 7.4
At the boundary between two systems, the pressure force acts inward on both systems.

applying the momentum balance, e.g., perpendicular to the x-direction for an x-directed momentum balance. If we do that, we will see that the pressure force acts *inward* on our system and simultaneously *outward* on the surroundings. The direction of the pressure force in a flowing fluid is independent of the direction of the flow. This is illustrated in Fig. 7.4.

If two systems adjoin each other, then an equal and opposite pressure force acts inward on each of them, as shown in Fig. 7.4. If you squeeze a coil spring between your thumb and forefinger, it exerts equal forces on thumb and forefinger, acting in opposite directions. It is the same with pressure forces on adjoining systems (or two systems we create by drawing a system boundary across a flow).

Example 7.5. A nozzle is attached to a fire hose by a bolted flange; see Fig. 7.5. What is the force tending to tear apart that flange when the valve in the nozzle is closed?

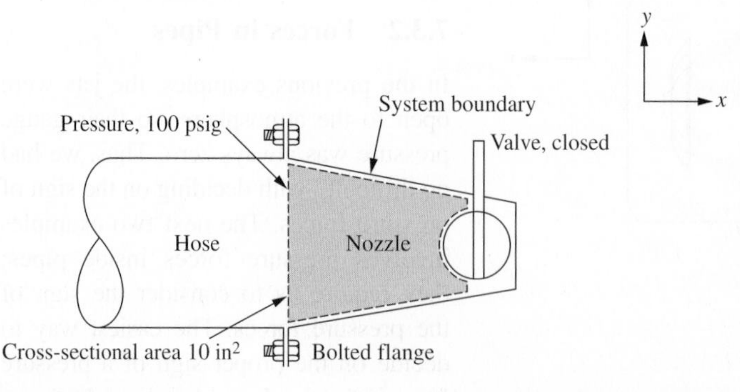

FIGURE 7.5
A hose nozzle, with the valve closed.

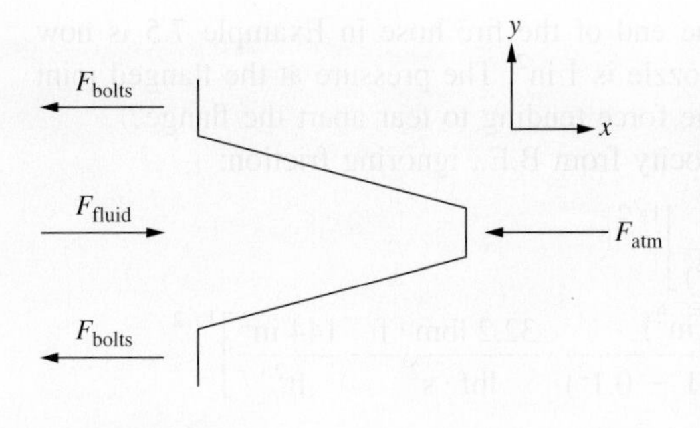

FIGURE 7.6
Forces acting on the nozzle in Fig. 7.5.

We take as our system the mass of fluid that is enclosed in the nozzle from the plane of the flange to the valve. Applying Eq. 7.17, we see that $\dot{m}$ is zero, so the summation of the x components of the forces on this body of fluid must be zero. In this case the summation of forces is the summation of the pressure forces in the x direction. In the plane of the flange the fluid outside the system exerts a pressure force on the system equal to PA. This is all in the x direction, because this surface is normal to the x axis. The x component of the pressure force exerted by the nozzle must be equal and opposite to this pressure force. The magnitude of these forces is

$$\begin{aligned} F_x = PA = (P_g + P_{atm})A &= (100 \text{ lbf / in}^2 \cdot 10 \text{ in}^2) + P_{atm} A \\ &= 1000 \text{ lbf} + P_{atm} A \end{aligned} \tag{7.M}$$

where P_g is the gauge pressure and P_{atm} is the atmospheric pressure.

Now we choose as a second system the nozzle itself. From Eq. 7.17 we see that for it, too, the sum of the x components of the forces must be zero. The forces acting on it are sketched in Fig. 7.6. We have previously calculated the force that it exerts on the fluid; by Newton's third law we know that the fluid exerts an equal and opposite force on it, which is given as F_{fluid} in the figure. The bolts also exert a force, as shown, and the atmosphere exerts a pressure on all those parts not exposed to the fluid. The atmospheric pressure force is not all in the x direction but, as we showed in Chap. 2, we could compute the x component of the atmospheric pressure force, which would be $-PA_x$, where A_x is the x projection of the net area exposed to the atmosphere. (The force is negative, because it acts opposite to the x direction.) Therefore, summing the forces shown in Fig. 7.6, setting the sum equal to zero, and solving for F_{bolts}, we find

$$\begin{aligned} F_{bolts} &= -F_{liq} - F_{atm} = -(1000 \text{ lbf} + P_{atm} A) - (-P_{atm} A) \\ &= -1000 \text{ lbf} = -4.448 \text{ kN} \end{aligned} \tag{7.N}$$

The bolt force shown is negative because it acts on the nozzle in the negative x direction. ■

We see that in this problem the atmospheric pressure terms canceled. Because this is a common occurrence, engineers ordinarily work such problems in gauge pressures and thereby only need to show pressure forces on those parts of the boundary of the system where the pressure is different from atmospheric. If we had done that here, we would have found exactly the same answer.

We could have solved this problem more easily by not using the momentum balance, but it illustrates the method, which will be useful in the next example.

Example 7.6. The valve on the end of the fire hose in Example 7.5 is now opened. The area of the outlet nozzle is 1 in^2. The pressure at the flanged joint is still 100 psig. Now what is the force tending to tear apart the flange?

We estimate the outlet velocity from B.E., ignoring friction:

$$V = \left[\frac{2(-\Delta P)}{\rho(1 - (A_2/A_1)^2)}\right]^{1/2}$$

$$= \left[\frac{2(100 \text{ lbf/in}^2)}{(62.3 \text{ lbm/ft}^3)(1 - 0.1^2)} \cdot \frac{32.2 \text{ lbm} \cdot \text{ft}}{\text{lbf} \cdot \text{s}^2} \cdot \frac{144 \text{ in}^2}{\text{ft}^2}\right]^{1/2}$$

$$= 122.6 \frac{\text{ft}}{\text{s}} = 37.4 \frac{\text{m}}{\text{s}} \tag{7.O}$$

And by material balance we know that the inlet velocity is $\frac{1}{10}$ of this value.

Again, we choose as our system the fluid enclosed in the nozzle. From Eq. 7.17 we have

$$0 = \dot{m}(V_{x_{\text{in}}} - V_{x_{\text{out}}}) + \sum F_x \tag{7.P}$$

From the mass balance for steady flow we have

$$\dot{m} = \rho A_{\text{out}} V_{\text{out}} = 62.3 \frac{\text{lbm}}{\text{ft}^3} \cdot 1 \text{ in}^2 \cdot 122.6 \frac{\text{ft}}{\text{s}} \cdot \frac{\text{ft}^2}{144 \text{ in}^2}$$

$$= 53.1 \frac{\text{lbm}}{\text{s}} = 24.1 \frac{\text{kg}}{\text{s}} \tag{7.Q}$$

So the net force on the fluid in the x direction is

$$\sum F_x = -\dot{m}(V_{x_{\text{in}}} - V_{x_{\text{out}}}) = -53.1 \frac{\text{lbm}}{\text{s}} \cdot \left(\frac{12.3 \text{ ft}}{\text{s}} - \frac{122.6 \text{ ft}}{\text{s}}\right) \cdot \frac{\text{lbf} \cdot \text{s}^2}{32.2 \text{ lbm} \cdot \text{ft}}$$

$$= 182 \text{ lbf} = 815 \text{ N} \tag{7.R}$$

This is the sum of the forces on the fluid, as sketched in Fig. 7.7.

As discussed in Chap. 5, the fluid leaving such a nozzle will be at the same pressure as the surrounding atmosphere if the flow is subsonic, which it is here. Therefore, if we use gauge pressure, then the pressure force on the system as the stream leaves the nozzle is zero, and the 182 lbf given is the algebraic sum of the pressure force exerted on the system at its left boundary and the x component of the force exerted by the nozzle. Thus,

$$182 \text{ lbf} = PA - F_{x_{\text{nozzle}}} \tag{7.S}$$

and the force exerted by the nozzle on the fluid is -818 lbf. By comparing

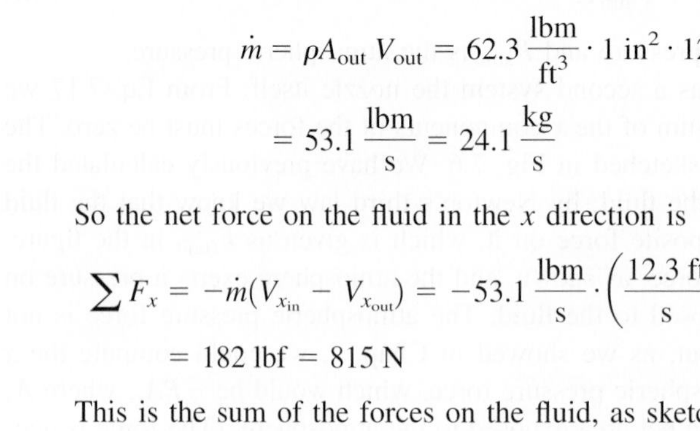

FIGURE 7.7
Forces acting on the fluid in the nozzle.

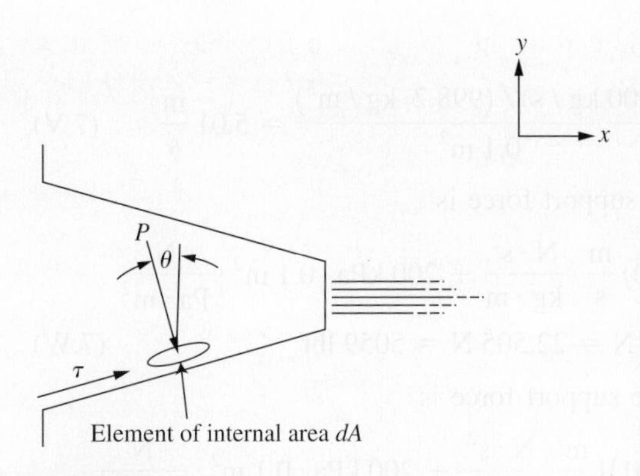

FIGURE 7.8
An alternative way of computing force on nozzle with fluid flowing.

this result with the one in Example 7.6, we can readily deduce that the force on the nozzle bolts in this case is −818 lbf. ■

Why is the force acting on the bolts less in this case? The pressure force acting on the system at its left boundary is the same as in the no-flow case (Example 7.5). However, when there is this flow, some of that force is being used to accelerate the fluid and hence is not being resisted by the nozzle bolts.

This example also illustrates how the momentum balance helps us solve problems from the outside without looking inside. We could also have found the force on the nozzle by determining the pressure and shear stress at every point on the internal surface of the nozzle; see Fig. 7.8. The x components of these pressure forces and shear forces are equal to the x component of the force on the nozzle:

$$-F_{x_{\text{nozzle}}} = \int (P \sin\theta + \tau \cos\theta)\, dA \tag{7.18}$$

Determining the local values of P and τ for all the internal surface of even a relatively simple device like this nozzle is a formidable task. For a complicated shape, it is beyond our current ability. Nevertheless, we computed the overall force (which we were seeking) by the momentum balance fairly easily.

Example 7.7. The pipe bend in Fig. 7.9 is attached to the rest of the piping system by two flexible hoses, which transmit no forces. Water enters in the $+x$ direction and leaves in the $-y$ direction. The flow rate is 500 kg / s, and the cross-sectional area of the pipe is constant = 0.1 m^2. The pressure throughout the pipe is 200 kPa gauge. Calculate F_x and F_y, the x and y components of the force in the pipe support.

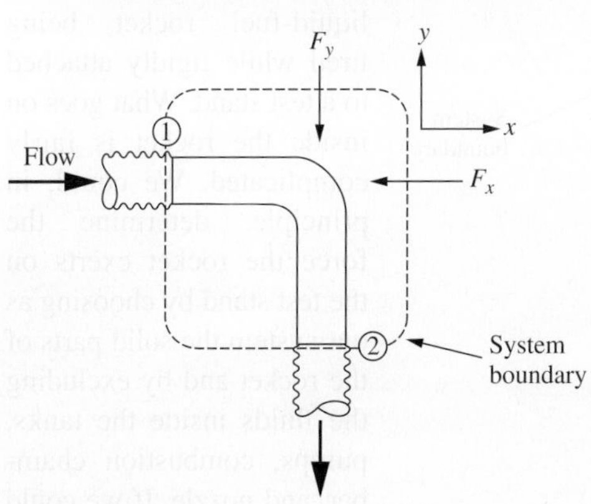

FIGURE 7.9
Forces on a pipe bend.

Applying Eq. 7.17 in the x direction to the system shown in Fig 7.9 and using gauge pressure, we find

$$-F_x = \dot{m}(V_{x_{\text{in}}} - V_{x_{\text{out}}}) + P_1 A_1 \tag{7.T}$$

and for the y direction

$$-F_y = \dot{m}(V_{y_{\text{in}}} - V_{y_{\text{out}}}) + P_2 A_2 \tag{7.U}$$

The velocity is constant,

$$V = \frac{Q}{A} = \frac{\dot{m} / \rho}{A} = \frac{(500 \text{ kg / s}) / (998.2 \text{ kg / m}^3)}{0.1 \text{ m}^2} = 5.01 \frac{\text{m}}{\text{s}} \tag{7.V}$$

so that the x component of the support force is

$$\begin{aligned} -F_x &= 500 \frac{\text{kg}}{\text{s}} (5.01 - 0) \frac{\text{m}}{\text{s}} \cdot \frac{\text{N} \cdot \text{s}^2}{\text{kg} \cdot \text{m}} + 200 \text{ kPa} \cdot 0.1 \text{ m}^2 \cdot \frac{\text{N}}{\text{Pa} \cdot \text{m}^2} \\ &= (2505 + 20{,}000) \text{ N} = 22{,}505 \text{ N} = 5059 \text{ lbf} \end{aligned} \tag{7.W}$$

whereas the y component of the support force is

$$\begin{aligned} -F_y &= 500 \frac{\text{kg}}{\text{s}} [0 - (-5.01)] \frac{\text{m}}{\text{s}} \cdot \frac{\text{N} \cdot \text{s}^2}{\text{kg} \cdot \text{m}} + 200 \text{ kPa} \cdot 0.1 \text{ m}^2 \cdot \frac{\text{N}}{\text{Pa} \cdot \text{m}^2} \\ &= (2505 + 20{,}000) \text{ N} = 22{,}505 \text{ N} = 5059 \text{ lbf} \end{aligned} \tag{7.X}$$

■

We see that the support force components are equal and point in the $-x$ and $-y$ directions, as sketched on the figure. In this example, and most piping-force problems, the pressure terms are larger than the fluid-acceleration terms. A few moments spent examining the signs of all the terms in this example is time well spent. Piping designers must support piping properly to deal with these forces, as well as those of thermal expansion. Failure to do so has led to serious process plant accidents, including the 1974 Flixboro disaster [1].

7.3.3 Rockets and Jets

Rockets are easy to analyze by means of the steady-flow one-dimensional momentum balance. Figure 7.10 shows a cutaway view of a liquid-fuel rocket being fired while rigidly attached to a test stand. What goes on inside the rocket is fairly complicated. We could, in principle, determine the force the rocket exerts on the test stand by choosing as our system the solid parts of the rocket and by excluding the fluids inside the tanks, pumps, combustion chamber, and nozzle. If we could determine the pressure and shear stress at every point in

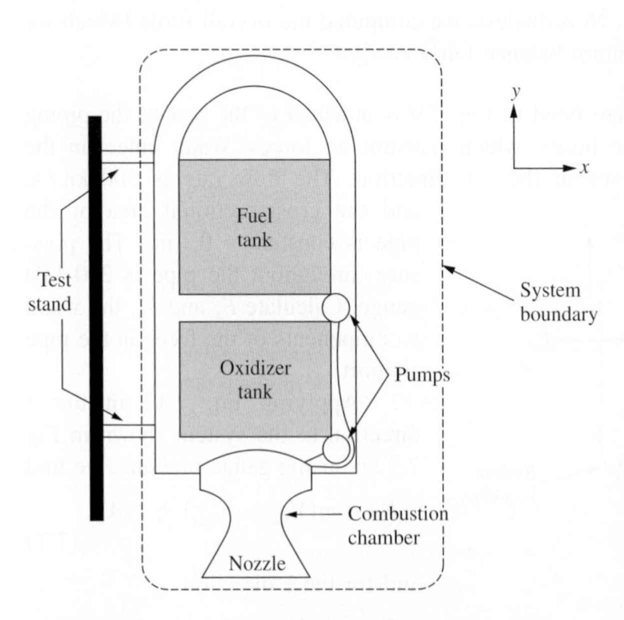

FIGURE 7.10
Simplified cross section of a liquid-fuel rocket.

the system, we could determine the total force by taking the integral

$$F = \int_{\text{int.ext.surf}} (P \sin \theta + \tau \cos \theta)\, dA \tag{7.19}$$

over the whole internal and external surfaces; that would be a giant task.

But if we want only to know what this force is, we can take the outside of the rocket as our system boundary and apply the momentum balance. With this boundary $[d(mV_y)/dt]_{\text{system}}$ is zero, because the system momentum is not changing with time.* There is no flow into the system, so Eq. 7.17 reduces to

$$F_y = V_{y_{\text{out}}}\, \dot{m}_{\text{out}} \tag{7.20}$$

The external forces acting on the system are the pressure forces around the entire boundary of the system shown and the force exerted by the test-stand support structure. The latter force is the one we are seeking, so we can split the F_y in Eq. 7.20 into two parts and rearrange to

$$F_{y_{\text{t.stand}}} = V_{y_{\text{out}}}\, \dot{m}_{\text{out}} - \begin{pmatrix} y \text{ component of pressure} \\ \text{force on system} \end{pmatrix} \tag{7.21}$$

In Sec. 5.5 we noted that for flows moving slower than the velocity of sound we can safely assume that a flow leaving an enclosed system and flowing into the atmosphere is at the same pressure as the atmosphere. We have made this assumption in the foregoing examples of this chapter. However, in the case of the rocket the flow leaving the system is generally supersonic, so we can no longer make this assumption. From Fig. 7.10 it is obvious that the pressure on the outside of the system is atmospheric everywhere except across the exit of the nozzle. Thus, the net y component of the pressure force on the system is

$$\begin{pmatrix} y \text{ component of pressure} \\ \text{force on system} \end{pmatrix} = A_{\text{exit}} \cdot (P_{\text{exit}} - P_{\text{atm}}) = A_{\text{exit}} \cdot P_{\text{exit, gauge}} \tag{7.22}$$

Since we have chosen the y direction as positive upward, both $F_{y_{\text{t.stand}}}$ and $V_{y_{\text{out}}}$ are negative. Multiplying Eq. 7.21 by -1, we find

$$-F_{y_{\text{t.stand}}} = +F_{y_{\text{rocket}}} = -V_{y_{\text{out}}}\, \dot{m}_{\text{out}} + A_{\text{exit}}\, P_{\text{exit, gauge}} \tag{7.23}$$

Here we show the force exerted by the rocket as the negative force exerted by the test stand, because they are equal and opposite. The force exerted by the rocket is referred to as the *thrust* of the rocket. It is commonly believed that rockets need something to push against in order to fly. Equation 7.23 shows that this is false. If it were true, rockets could not operate in the vacuum of outer space.

---

*Although the system as a whole is not moving in the y direction, some parts of it are, because of the internal fuel flows. Thus, the overall system has some y-directed momentum, but since this presumably is not changing with time, we have $[d(mV_y)/dt]_{\text{system}} = 0$. During the motor-starting period this simplification is not correct, but for a rocket standing still this term is always small compared with the other terms in the momentum balance.

Example 7.8. A rocket on a test stand is sending out 1000 kg / s of exhaust gas at a velocity of −3000 m / s (negative, because it is in the $-y$ direction). The exit area of the nozzle is 7 m^2, and the pressure at the exhaust-nozzle exit is 35 kPa gauge. What is the thrust of the rocket?

From Eq. 7.23 we find

$$F_{\text{rocket}} = \text{thrust} = -\left(-3000\,\frac{\text{m}}{\text{s}} \cdot 1000\,\frac{\text{kg}}{\text{s}}\right) + (7\ \text{m}^2 \cdot 35\ \text{kPa})$$
$$= 3\ \text{MN} + 0.245\ \text{MN} = 3.25\ \text{MN} = 0.73 \cdot 10^6\ \text{lbf} \tag{7.Y}$$

■

From this it is clear that, the higher the exhaust velocity, the greater the thrust per unit mass of fuel consumed. If the exit pressure were exactly atmospheric pressure, $P_{\text{exit, gauge}} = 0$, then the thrust would be directly proportional to the exhaust velocity. For horizontally firing rockets, such as artillery rockets and airplane-assist rockets, the atmospheric pressure remains constant during burning, and one could, in principle, design the rocket nozzle for $P_{\text{exh}} = P_{\text{atm}}$. However, this generally results in an impractically large nozzle (too much air resistance or too difficult to fabricate or launch), so the nozzle is usually designed for a P_{exh} significantly greater than the P_{atm} at the exit of the nozzle.

Vertically firing rockets, such as ballistic missiles and satellite launchers, must operate over a wide range of atmospheric pressures, from those at sea level to those in outer space. The nozzle is designed for some average pressure; therefore, P_{exh} could equal P_{atm} only at one particular altitude in the flight. Normally $P_{\text{exh}} > P_{\text{atm}}$ for the whole duration of the rocket flight.

Ignoring this complication for the moment, we can say that for $P_{\text{exh}} = P_{\text{atm}}$ the exhaust velocity is a simple, reliable, direct method of comparing the efficiency of various rocket engines. The early German workers in rocketry used it as a comparison basis. U.S. workers, on the other hand, have preferred to use the *specific impulse,* I_{sp}, as a comparison basis:

$$\begin{pmatrix}\text{Specific}\\ \text{impulse}\end{pmatrix} = I_{\text{sp}} = \frac{\text{lbf of thrust produced}}{(\text{fuel} + \text{oxidizer})\ \text{flow in lbm / s}} \tag{7.24}$$

If $P_{\text{exh}} = P_{\text{atm}}$, then

$$\text{Thrust} = I_{\text{sp}}\,\dot{m} = -V_{y_{\text{out}}}\,\dot{m} \tag{7.25}$$

indicating that I_{sp} must be exactly the same as $-V_{y_{\text{out}}}$ except for a conversion of units. By inspection this must be the conversion involving force, mass, length and time.

Example 7.9. For a rocket with an exhaust velocity of −3000 m / s and with $P_{\text{exh}} = P_{\text{atm}}$, what is the value of I_{sp}?

$$I_{\text{sp}} = -(V_{y_{\text{exh}}}) = -\left(-3000\,\frac{\text{m}}{\text{s}}\right) \cdot \frac{\text{N} \cdot \text{s}^2}{\text{kg} \cdot \text{m}} = 3\,\frac{\text{kN} \cdot \text{s}}{\text{kg}} = 305.9\,\frac{\text{lbf} \cdot \text{s}}{\text{lbm}} \tag{7.Z}$$

■

It is common practice in the U.S. rocket industry simply to write this as 305.9 s. This is incorrect, because 1 lbf does not equal 1 lbm, and they should not be canceled. Nonetheless, the cancellation is common in rocket publications.

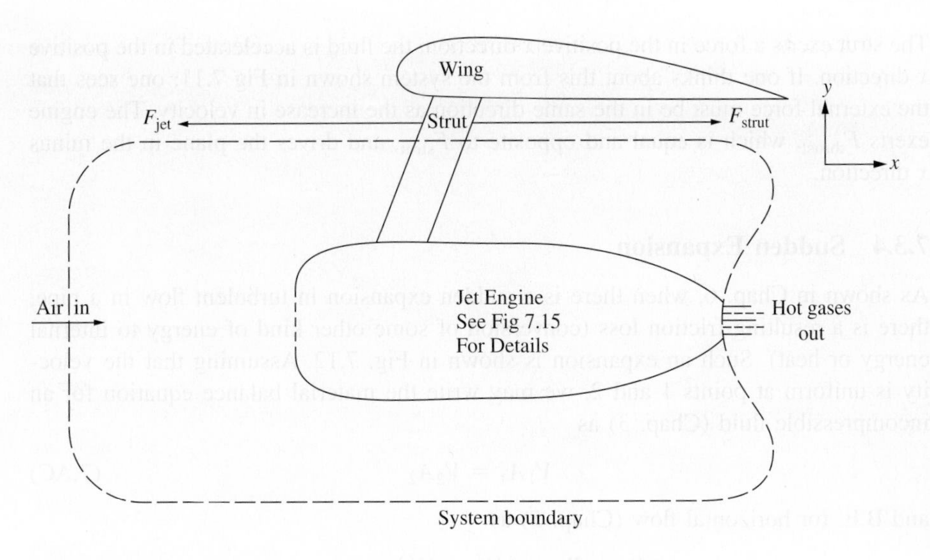

FIGURE 7.11
Simplified view of a jet engine. The fuel flows down the strut from wing tanks to engine.

Our discussion has concerned only rockets fixed to a test stand; we will consider moving rockets in Sec. 7.5.

Figure 7.11 shows a schematic of a jet engine, attached by a strut to the wing of a modern commercial airliner. Fuel flows in from tanks in the wing. Air flows in the front, is compressed, mixed with fuel and burned, then expanded to a high velocity and exhausted to the rear. The engine exerts a force, F_{jet}, called the *thrust* on the airplane, through the strut; the airplane exerts an equal and opposite force on the engine, F_{strut}, through the strut. Applying Eq. 7.17 to the system shown, for steady flow, we find

$$0 = \dot{m}_{\text{air}}(V_{x_{\text{in}}} - V_{x_{\text{out}}}) + \dot{m}_{\text{fuel}}(V_{x_{\text{in}}} - V_{x_{\text{out}}}) + F \tag{7.AA}$$

The fuel flow rate is much less than the air flow rate, so for simple analyses it is normally set equal to zero. The F in Eq. 7.AA is the force exerted by the strut on the engine, which is equal and opposite to the thrust of the engine, so

$$-F_{\text{strut}} = F_{\text{thrust}} \approx \dot{m}_{\text{air}}(V_{x_{\text{in}}} - V_{x_{\text{out}}}) \tag{7.26}$$

Example 7.10. A modern jet engine has an inlet velocity of almost zero and an exhaust velocity of about 1350 ft / s. Medium-sized ones produce a thrust of 20,000 lbf. What is the air flow rate required by such an engine?

Solving Eq. 7.26 for the mass flow rate, we find

$$\dot{m}_{\text{air}} \approx \frac{-F_{\text{strut}}}{(V_{x_{\text{in}}} - V_{x_{\text{out}}})} = \frac{-20{,}000 \text{ lbf}}{(0 - 1350) \text{ ft / s}} \cdot \frac{32.2 \text{ lbm} \cdot \text{ft}}{\text{lbf} \cdot \text{s}^2}$$

$$= 477 \frac{\text{lbm}}{\text{s}} = 216 \frac{\text{kg}}{\text{s}} \tag{7.AB}$$

■

The strut exerts a force in the positive x direction, the fluid is accelerated in the positive x direction. If one thinks about this from the system shown in Fig 7.11, one sees that the external force must be in the same direction as the increase in velocity. The engine exerts F_{thrust}, which is equal and opposite to F_{strut}, and drives the plane in the minus x direction.

7.3.4 Sudden Expansion

As shown in Chap. 6, when there is a sudden expansion in turbulent flow in a pipe, there is a resulting friction loss (conversion of some other kind of energy to internal energy or heat). Such an expansion is shown in Fig. 7.12. Assuming that the velocity is uniform at points 1 and 2, we may write the material balance equation for an incompressible fluid (Chap. 3) as

$$V_1 A_1 = V_2 A_2 \tag{7.AC}$$

and B.E. for horizontal flow (Chap. 5) as

$$\frac{P_2 - P_1}{\rho} + \left(\frac{V_2^2}{2} - \frac{V_1^2}{2} \right) = -\mathscr{F} \tag{7.AD}$$

Given V_1, A_1, and A_2, we can calculate V_2, but we cannot calculate $(P_2 - P_1)$ unless we know $-\mathscr{F}$. However, we can apply the steady-flow momentum balance, Eq. 7.17, to this flow. When we take the system as the fluid from point 1 to point 2, we find

$$\sum F_x = -\dot{m}(V_{x_1} - V_{x_2}) \tag{7.AE}$$

Here

$$\sum F_x = P_1 A_1 + P_{1a} A_{1a} - P_2 A_2 - \int \tau \, dA_w \tag{7.AF}$$

where P_{1a} is the average pressure over the annular area, and the integral $\int \tau \, dA_w$ is the total shear force at the walls of the pipe, due to viscous friction. For a sudden expansion the other terms in the momentum balance are large compared with $\int \tau \, dA_w$, so we will drop it.

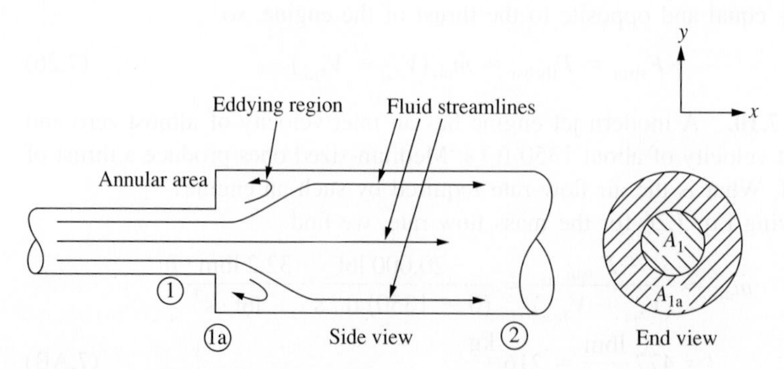

FIGURE 7.12
A sudden expansion in pipe flow.

Because of our previous discussions about the pressure of a fluid leaving a vessel, it is plausible that P_{1a} is approximately the same as P_1: i.e., the pressure is the same for the entire cross section at point 1. Making this assumption and substituting in Eq. 7.AF and then in Eq. 7.AE, we find

$$P_2A_2 - P_1A_2 = \dot{m}(V_{x_1} - V_{x_2}) \tag{7.AG}$$

But $\dot{m} = \rho V_{x_2} A_2$ and $V_{x_2} = V_{x_1}(A_1/A_2)$, so that

$$P_2 - P_1 = \rho V_{x_1}^2 \frac{A_1}{A_2}\left(1 - \frac{A_1}{A_2}\right) \tag{7.AH}$$

Substituting Eq. 7.AH in Eq. 7.AD and using Eq. 7.AC to eliminate V_{x_2}, we find

$$-\mathcal{F} = V_{x_1}^2 \frac{A_1}{A_2}\left(1 - \frac{A_1}{A_2}\right) - \frac{V_{x_1}^2}{2}\left(1 - \frac{A_1^2}{A_2^2}\right) \tag{7.AI}$$

which may be regrouped and factored to give

$$\mathcal{F} = \frac{V_{x_1}^2}{2} \cdot \left(1 - \frac{A_1}{A_2}\right)^2 \tag{7.27}$$

Comparing this equation with Eq. 6.25, which describes the same situation, we see that the two equations are the same if

$$K = \left(1 - \frac{A_1}{A_2}\right)^2 \tag{7.28}$$

This is the function plotted in Fig. 6.16. Experimental tests indicate that Eq. 7.28 is indeed a good predictor of experimental results, so the assumption of uniform pressure across the cross section at point 1 seems a good one.

It is interesting to compare what we did here with what we did in Sec. 6.3, where we applied a force balance to find Poiseuille's equation. That kind of force balance was usable in that case because there was no acceleration of any part of the fluid, so that the sum of the forces acting on any part the fluid was zero. Here the fluid is decelerated, so the sum of the forces acting on some part of it is not zero. The simple force balance used in Sec. 6.3 is a strongly restricted form of the momentum balance; with the complete momentum balance we can deal with much more complex flows, such as the one examined here, and in the next section.

7.3.5 Eductors, Ejectors, Aspirating Burners, Jet Mixers, and Jet Pumps

Figure 7.13 shows a cross section of a typical laboratory Bunsen burner. In it a jet of fuel gas flows upward. By momentum exchange with the air in the tube, it creates a slight vacuum at the level of the jet (labeled "2" on the figure). This sucks air in through the air inlet. The gas and air mix in the mixing tube, and flow out together into the flame. This is the most common type of gas burner, called an *aspirating burner*. The burners of all household gas furnaces, water heaters, and stoves are of this type, as are hand-held propane torches and small- to medium-sized industrial burners. (The largest industrial burners have combustion air driven by fans and thus do not "aspirate" it as this burner does.) This burner produces only enough vacuum to suck in the combustion air. The same basic device using high-pressure steam as motive fluid can produce industrially useful vacuums. Such devices, called *eductors* or *ejectors,* are widely used industrially as vacuum pumps. Modified forms are also used as *jet mixers* and *jet pumps.* All devices of this type work by exchanging momentum between a centrally located, high-velocity jet and a circumferential, slower-moving flow. The Bunsen burner leads to a simple analysis, because its body is a simple, cylindrical tube. The eductors, ejectors, and jet pumps and most simple burners are shaped like the venturis in Chapter 5. They are more efficient than the straight tube in a Bunsen burner but lead to a much more complex analysis. The simple analysis in this section is correct for any straight-tube version of this type of device and shows intuitively what goes on in the more complex geometries but is not directly applicable to them.

To analyze such a device we begin by making a momentum balance, choosing as our system the section of the mixing tube between 2 and 3 in Fig. 7.13. We assume

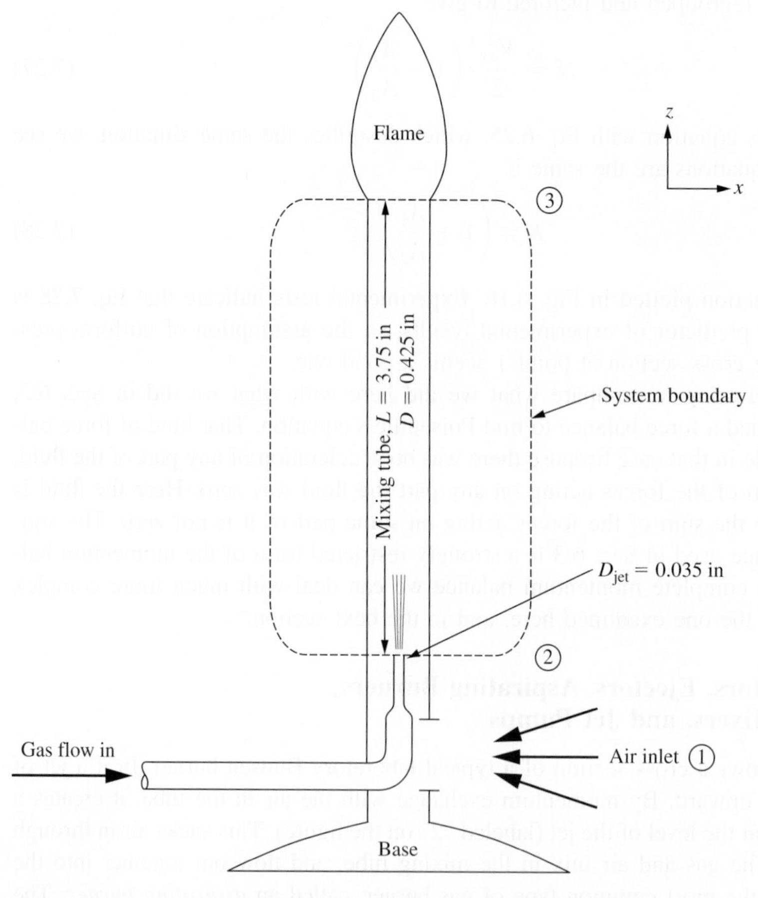

FIGURE 7.13
A simple Bunsen burner. Here the gas jet is shown above the air inlet, which makes the analysis simple. Often the gas jet is at the level of the air inlet, which works well but makes the analysis more complex.

steady flow in the positive z direction, dropping the direction subscripts, so that Eq. 7.17. becomes

$$0 = \dot{m}_{\text{air}}(V_3 - V_{\text{air, 2}}) + \dot{m}_{\text{gas}}(V_3 - V_{\text{gas, 2}}) + P_2A_2 - P_3A_3 - \tau_{\text{wall}}\pi D_3\,\Delta x \quad (7.29)$$

At point 3 the pressure must be atmospheric, whereas at point 2 it is not. We work the problem in gauge pressure, which makes the P_3A_3 term = zero. The term for the shear stress at the wall is negligible for low-velocity devices like the Bunsen burner in Fig. 7.13 (however, for high-velocity applications it is important). For the rest of this section we will assume $\tau \approx 0$ and drop the rightmost term in Eq. 7.29. Then

$$P_2 = -\frac{\dot{m}_{\text{air}}(V_3 - V_{\text{air, 2}}) + \dot{m}_{\text{gas}}(V_3 - V_{\text{gas, 2}})}{A_2}$$
$$= -\frac{\rho_{\text{air}}V_{\text{air, 2}}A_{\text{air, 2}}(V_3 - V_{\text{air, 2}}) + \rho_{\text{gas}}V_{\text{gas, 2}}A_{\text{gas, 2}}(V_3 - V_{\text{gas, 2}})}{(A_{\text{air, 2}} + A_{\text{gas, 2}})} \quad (7.30)$$

We also make a steady-flow material balance

$$\dot{m}_{\text{air in}} + \dot{m}_{\text{gas in}} = \dot{m}_{\text{air–gas mixture out}} \quad (7.31)$$

We then substitute ρAV for each of the $\dot{m}$ terms and rearrange to

$$V_3 = \frac{\dfrac{\rho_{\text{gas}}}{\rho_{\text{mixture out}}}V_{\text{gas}}A_{\text{gas}} + \dfrac{\rho_{\text{air}}}{\rho_{\text{mixture out}}}V_{\text{air}}A_{\text{air}}}{A_3} \quad (7.32)$$

If we know P_2, we can calculate the velocity of the fuel gas from B.E., and the known pressure in the gas supply line, and can also calculate the velocity of the air at point 2 from B.E. The equations to be solved simultaneously are one momentum balance, one mass balance and two B.E., as shown in Table 7.1. This set of

TABLE 7.1

Equations to be solved in a simple, cylindrical aspirating burner, eductor, jet mixer, or jet pump, for velocities well below sonic*

Equation type	Region	
Momentum balance	Point 2 to point 3	$P_2 = -\dfrac{\rho_{\text{air}}V_{\text{air, 2}}A_{\text{air, 2}}(V_3 - V_{\text{air, 2}}) + \rho_{\text{gas}}V_{\text{gas, 2}}A_{\text{gas, 2}}(V_3 - V_{\text{gas, 2}})}{(A_{\text{air, 2}} + A_{\text{gas, 2}})}$
Mass balance	Point 2 to point 3	$V_3 = \dfrac{\dfrac{\rho_{\text{gas}}}{\rho_{\text{mixture out}}}V_{\text{gas}}A_{\text{gas}} + \dfrac{\rho_{\text{air}}}{\rho_{\text{mixture out}}}V_{\text{air}}A_{\text{air}}}{A_3}$
B.E.	High pressure gas to point 2	$V_{\text{gas, 2}} = \left(2\dfrac{P_{\text{gas}} - P_2}{\rho_{\text{gas}}}\right)^{1/2}$
B.E.	Outside air to point 2	$V_{\text{air, 2}} = \left(2\dfrac{P_{\text{atm}} - P_2}{\rho_{\text{air}}}\right)^{1/2}$

*Pressures are gauge, not absolute.

equations is applicable to any cylindrical device of this type, whether the fluids are gases or liquids or slurries. If they are not gases, simply replace all the quantities with "gas" subscript with the quantity for the high-velocity (driver) stream and all those quantities with "air" subscript with the quantity for the low-velocity (driven) stream. If the gas velocities are close to the speed of sound, then the simple B.E. used here must be replaced by the high-velocity gas flow relations from Chap. 8. In the following example we assume that all the flowing fluids are ideal gases at low velocities.

Example 7.11. For the Bunsen burner shown in Fig. 7.13, with the dimensions shown, estimate P_2, V_3, and $\dot{m}_{\text{air}} / \dot{m}_{\text{gas}}$. The gas is assumed to be natural gas, which in the United States is distributed inside buildings at $P = 4$ in $H_2O = 0.145$ psig $= 1.00$ Pa gauge, and which has a density $\approx (16/29)$ times that of air. We also know that $A_{\text{tube}} = 0.142 \text{ in}^2$ and $A_{\text{jet}} = 0.00096 \text{ in}^2$.

We begin by guessing $P_{2,\text{ first guess}} = -0.01$ psig. Then by simple B.E. we compute that $V_{\text{air, 2, first guess}} = 35.2$ ft / s and by simple B.E. we know that the gas jet velocity ≈ 186 ft / s. The densities are computed from the ideal gas law by

$$\rho_{20^\circ\text{C}} = 0.075 \frac{\text{lbm}}{\text{ft}^3} \cdot \frac{P(\text{abs.})}{14.7 \text{ psia}} \cdot \frac{29(\text{g/mol})}{M} \tag{7.AJ}$$

The mass flow rates of the two inlet streams, calculated from $\dot{m} = \rho A V$, are $2.6 \cdot 10^{-3}$ and $5.15 \cdot 10^{-5}$ lbm / s for air and gas, respectively. From these we can compute that

$$M_{\text{air-gas mixture at outlet}} = 28.5 \frac{\text{lbm}}{\text{lbmol}} \tag{7.AK}$$

Then we substitute values in Eq. 7.32 (dropping the dimensions on the As, which are all in in^2, and on the velocities, which are all in ft / s), so that

$$V_{3,\text{ first guess}} = \frac{(16/28.5) \cdot (14.69/14.70) \cdot 186 \cdot 0.142 + 35.2 \cdot (29/28.5) \cdot (14.69/14.70) \cdot 0.00096}{(0.142 + 0.00096)}$$

$$= 36.0 \frac{\text{ft}}{\text{s}} \tag{7.AL}$$

Then from Eq. 7.29,

$$P_{2,\text{ calculated}} = -\frac{\begin{bmatrix} (0.075 \text{ lbm/ft}^3) \cdot (35.2 \text{ ft/s}) \cdot (0.142 \text{ in}^2) \cdot (36.0 - 35.2) \text{ ft/s} \\ + (0.041 \text{ lbm/ft}^3) \cdot (186 \text{ ft/s}) \cdot (0.00096 \text{ in}^2) \cdot (36.0 - 186) \text{ ft/s} \end{bmatrix}}{0.142 \text{ in}^2}$$

$$\cdot \frac{\text{lbf} \cdot \text{s}^2}{32.2 \text{ lbm} \cdot \text{ft}} \cdot \frac{\text{ft}^2}{144 \text{ in}^2} = -0.00044 \frac{\text{lbf}}{\text{in}^2} \tag{7.AM}$$

This is much less than the assumed -0.01 psig, so we use the numerical solution routine on the spreadsheet that we used to generate these values and find the solution, as shown in Table 7.2. We see that all the equations are

TABLE 7.2
Numerical solution to Example 7.11

Variable	Type	First guess	Solution
D_{tube}, in	Given	0.425	0.425
$D_{gas\ jet}$, in	Given	0.035	0.035
A_{tube}, in^2	Calculated	0.142	0.142
A_{jet}, in^2	Calculated	0.00096	0.00096
P_{gas}, in H_2O: gauge	Given	4	4
: psig	Given	0.14461	0.14461
$P_{2,\ guessed}$, psig	**Guessed**	**−0.01**	**−4.43E-05**
ρ_{air} at 2, lbm / ft^3	Calculated	0.0749	0.0750
ρ_{gas} at 2, lbm / ft^3	Calculated	0.04135	0.0414
V_{gas}, ft / s	Calculated	186.21	180.05
V_{air}, ft / s	Calculated	35.18	2.341
$\dot{m}_{air}$, lbm / s	Calculated	0.0026	0.000173
$\dot{m}_{gas}$, lbm / s	Calculated	5.14E-05	4.98E-05
M, air-gas mix at 3, lbm / lbmol	Calculated	28.5	24.5
$\rho_{gas\text{-}air\ mix}$ at 3, lbm / ft^3	Calculated	0.0738	0.0635
V_3, ft / s	Calculated	35.95	3.52
$P_{2,\ calculated}$, psig	Calculated	−0.00044	−4.43E-05
$P_{2,\ calc} / P_{2,\ guessed}$	**Check value**	0.044	**1.0000**
$\dot{m}_{air} / \dot{m}_{gas} = A / F$ ratio	Calculated	50.48	3.48

solved if

$$P_{2,\ solution} = -4.43 \cdot 10^{-5}\ \text{psig} = -0.0003\ \text{Pa} = -\ 0.0012\ \text{in}\ H_2O \quad (7.AN)$$

$$V_3 = 3.52 \frac{\text{ft}}{\text{s}} = 1.07 \frac{\text{m}}{\text{s}} \quad (7.AO)$$

The mass flow rates in Table 7.2 are computed from $\dot{m} = \rho AV$; we see that

$$\dot{m}_{air} / \dot{m}_{gas} = A / F \text{ ratio} = 3.48 \quad (7.AP)$$

■

From this example we see the following.

1. One could, in principle, use algebra to solve the four simultaneous equations explicitly, but the spreadsheet solution is quick, simple, and best of all, shows intermediate values that can be checked for plausibility.
2. The calculated air and air-gas mixture velocities are low. Most such burners have velocities of the magnitude shown here.
3. The A / F ratio is 3.48 (lbm / lbm). This is about 25 percent of the stoichiometric air-fuel ratio, which is a typical value for such burners. They all have adjustable shutters on the air inlet; one sets the shutters for the lowest air flow rate which gives a blue (non-smoky, non CO-producing) flame. Normally this requires about 25 percent of the air to be premixed with the natural gas.
4. The vacuum produced is minuscule. For vacuum pumps operating on the same principle, one substitutes high-pressure steam for the low-pressure natural gas,

finding velocities of several thousand ft / s for the central jet, and uses much smaller ratios of (driven fluid / driving fluid).

5. This is a very simple device; most of you have several of them in heaters in your house. To compute its behavior, we needed the momentum balance, a material balance, and two B.E.
6. See Probs. 7.31 to 7.34.

7.4 RELATIVE VELOCITIES

All the examples in the previous section concerned systems fixed in space. When a system is moving, the momentum balance still applies, but it is often convenient to introduce the idea of a relative velocity. Figure 7.14 shows a student on the ground throwing a ball to a student on a moving cart. The velocity of the ball, V, is 10 m / s. The cart is moving with a velocity V_{sys} of 5 m / s. As seen by the student who threw it, the ball is moving 10 m / s. As seen by the student who catches it, the ball is moving 5 m / s, because that is the velocity with which it is overtaking the cart. In general,

$$\mathbf{V} = \mathbf{V}_{sys} + \mathbf{V}_{rel} \tag{7.33}$$

where $\mathbf{V}$ is the velocity of a body or a stream of fluid relative to some set of fixed coordinates, $\mathbf{V}_{sys}$ is the velocity of the system (the cart in this case) relative to the same set of fixed coordinates, and $\mathbf{V}_{rel}$ is the velocity of the body or stream of fluid as seen by an observer riding on the moving system. When velocities are near the speed of light, this becomes more complicated, but such velocities seldom occur in fluid mechanics. We will see in this section that it is often practical to switch back and forth from the viewpoint of the fixed observer to that of the observer riding on some device or with some part of the flow, most often on a wave of some kind moving through a system.

Equation 7.33 is a vector equation; like all other vector equations it is simply a shorthand way of writing three scalar equations. In this text we will use only its scalar equivalents, such as

$$V_x = V_{x_{sys}} + V_{x_{rel}} \tag{7.34}$$

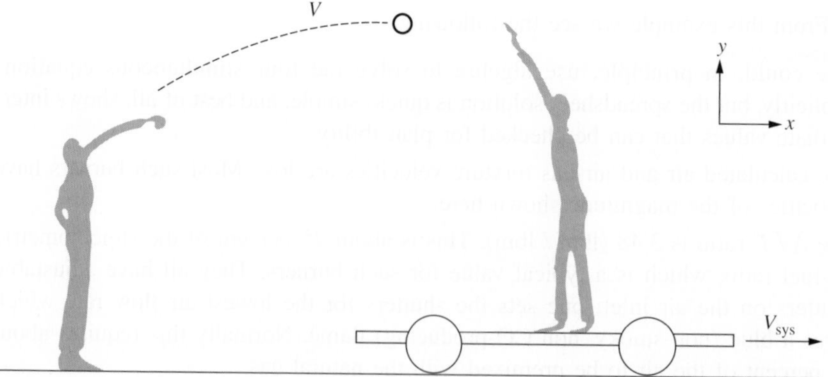

FIGURE 7.14
Relative velocities, illustrated by throwing a ball from a stationary pitcher to a catcher on a moving cart.

To illustrate the utility of this equation, let us consider a rocket in horizontal flight with no air resistance. We choose the rocket as our system and simplify by letting $P_{exh} = P_{atm}$. Then, since there is no flow into the system or any external force acting in the x direction, Eq. 7.13 becomes

$$d(mV_x)_{sys} = -V_{x_{out}}\, dm_{out} \tag{7.35}$$

Expanding the left side and substituting for $V_{x_{out}}$ from Eq. 7.34, we have

$$m_{sys}\, dV_{x_{sys}} + V_{x_{sys}}\, dm_{sys} = -(V_{x_{sys}} + V_{x_{rel,\,out}})\, dm_{out} \tag{7.36}$$

Because all the velocities are in the plus or minus x direction, we can drop the x subscripts. Now we note that $dm_{out} = -dm_{sys}$. Making this substitution and canceling like terms, we can divide by m_{sys} to find

$$dV_{sys} = V_{rel,\,out}\,(dm_{sys} / m_{sys}) \tag{7.37}$$

If the exhaust velocity relative to the rocket is constant (which is practically true of most rockets), then we can readily integrate this to

$$(V_{fin} - V_{init})_{sys} = V_{rel,\,out} \ln \frac{m_{fin}}{m_{init}} \tag{7.38}$$

This equation, often referred to as the *rocket equation* or the *burnout velocity equation,* indicates the limitation on possible speeds of various kinds of rockets.

Example 7.12. A single-stage rocket is to start from rest; $V_{init} = 0$. The mass of fuel is 0.9 of the total mass of the loaded rocket; $m_{fin} / m_{init} = 0.1$. The specific impulse of the fuel is 430 lbf · s / lbm, and the pressures are $P_{exh} = P_{atm}$. What speed will this rocket attain in horizontal flight if there is no air resistance?

From Eq. 7.25 we have

$$V_{rel,\,out} = -I_{sp} = -430\,\frac{\text{lbf}\cdot\text{s}}{\text{lbm}}\cdot 32.2\,\frac{\text{lbm ft}}{\text{lbf}\cdot\text{s}^2}$$

$$= -13{,}850\,\frac{\text{ft}}{\text{s}} = -4220\,\frac{\text{m}}{\text{s}} \tag{7.AQ}$$

Therefore,

$$V_{fin} = -13{,}850\,\frac{\text{ft}}{\text{s}}\cdot \ln 0.1 = 31{,}900\,\frac{\text{ft}}{\text{s}} = 9730\,\frac{\text{m}}{\text{s}} \tag{7.AR}$$

■

This example shows the probable maximum speed attainable with single-stage rockets using chemical fuels. It appears that the maximum I_{sp} for chemical fuels is about 430 lbf · s / lbm. Better structural design may reduce the value of m_{fin} / m_{init}, but it is unlikely to go much under 0.1 if there is any significant payload involved. For higher velocities, staged rockets are needed. Equation 7.38 does not include the effects of gravity or air resistance. These can be included, making the equation somewhat more complicated (Prob. 7.36). For more on rockets see Sutton [2] and Ley [3].

Another example of the utility of the relative-velocity concept concerns the interaction of a jet of fluid and a moving blade. Such interactions are the basis of

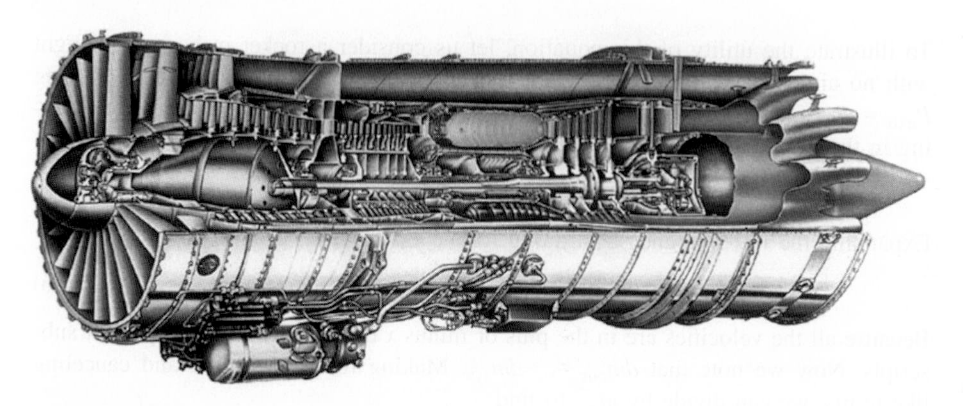

FIGURE 7.15
Cutaway of a modern jet engine. This is a Pratt and Whitney JT8D-219, whose basic parameters are shown in Example 7.10. Observe the large number of rotating and fixed blades that interact with the moving fluids. (Courtesy of Pratt and Whitney, a United Technologies Company.)

turbines and rotating compressors as used in turbojet and gas turbine engines and in the steam and water turbines that produce almost all of the world's electricity. Figure 7.15 shows a cutaway of a modern aircraft jet engine. In it the fluid interacts with multiple sets of blades, some of which do work on the fluid, increasing its pressure, and some of which extract work from the fluid, to be used in other parts of the engine. The rest of this section considers the interaction of a fluid with a single moving blade, one of the many blades in Figure 7.15.

In Examples 7.3 and 7.4 we saw how the interaction of a jet and solid surface produces a force on the surface. For this force to do work it must move through a distance. The work will be given by $dW = F\,dx$, and the power (or rate of doing work) is given by $\text{Po} = dW/dt = F\,dx/dt$. The latter is equal to the force times the velocity of the system, $\text{Po} = FV_{\text{sys}}$.

A curved blade is moving in the x direction and deflecting a stream of fluid; see Fig. 7.16. Consider this first from the viewpoint of an observer riding with the blade. As far as the observer can tell, the blade is standing still; no work is being done. Therefore, with no change in pressure and elevation, B.E. tells the observer that

$$\frac{V_{\text{out}}^2}{2} - \frac{V_{\text{in}}^2}{2} = -\mathcal{F} \qquad (7.\text{AS})$$

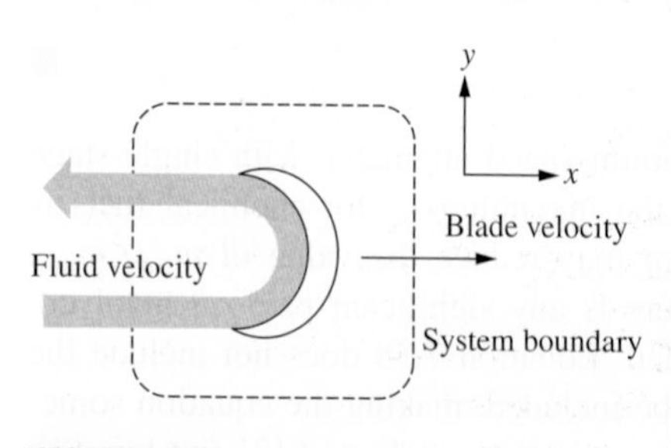

FIGURE 7.16
Simplest possible jet-blade interaction.

and that, if there is no friction, the outlet velocity is equal in magnitude to the inlet velocity but in a different direction. If there is friction, the outlet velocity will be less, but it cannot possibly be more, from the viewpoint of an observer riding on the blade. Here in the energy balance V^2 is a scalar, so we have no concern about the signs of V_{in} and V_{out}.

Now assume frictionless flow and that the blade in the figure is so shaped that the inlet and outlet streams have only x velocity and no y velocity. Then, by applying Eq. 7.17 and dropping the x subscript, because all the velocities are in the plus or minus x direction, we write

$$F = \dot{m}(V_{\text{out}} - V_{\text{in}}) \tag{7.AT}$$

This is the F exerted by the blade. The fluid exerts an equal and opposite force on the blade. Substituting Eq. 7.34 twice, we find

$$F = \dot{m}(V_{\text{rel, out}} + V_{\text{blade}} - V_{\text{rel, in}} - V_{\text{blade}}) = \dot{m}(V_{\text{rel, out}} - V_{\text{rel, in}}) \tag{7.39}$$

We see that the velocity of the blade cancels out of the force equation, so the force is the same whether viewed by an observer riding on the blade or by an observer standing still.

The work* done by the fluid per unit time (the power) is

$$\text{Po} = \frac{dW}{dt} = -F\frac{dx}{dt} = -FV_{\text{blade}} = -\dot{m}(V_{\text{rel, out}} - V_{\text{rel, in}})\,V_{\text{blade}} \tag{7.40}$$

and therefore the work done per unit mass of fluid is

$$\frac{dW}{dm} = (V_{\text{rel, in}} - V_{\text{rel, out}})\,V_{\text{blade}} \tag{7.41}$$

As shown above, from B.E. for frictionless flow we know that $V_{\text{rel, out}} = -V_{\text{rel, in}}$; therefore,

$$\frac{dW}{dm} = (2V_{\text{rel, in}})\,V_{\text{blade}} \tag{7.42}$$

Now suppose that the velocity of the jet is fixed. This would occur if it were a jet of water entering the power plant at the base of a dam with constant upstream water level (in which case we could calculate the jet velocity by B.E.) or if it were a jet of steam from a boiler with constant steam temperature and pressure (in which case we could calculate the jet velocity by the methods to be developed in Chap. 8). In Eq. 7.42 we replace $V_{\text{rel, in}}$ by $V_{\text{jet}} - V_{\text{blade}}$ and divide both sides by $V_{\text{jet}}^2/2$ to find

$$\frac{dW/dm}{V_{\text{jet}}^2/2} = 4\left(1 - \frac{V_{\text{blade}}}{V_{\text{jet}}}\right)\cdot\frac{V_{\text{blade}}}{V_{\text{jet}}} \tag{7.43}$$

The left side of Eq. 7.43 is the ratio of the work extracted from the fluid per pound of fluid to the kinetic energy per pound of the fluid in the jet. We may think of it as the fractional efficiency of the blade in converting jet kinetic energy into useful work (of the rotating turbine shaft). The right-hand side of Eq. 7.43 is plotted versus $V_{\text{blade}}/V_{\text{jet}}$ in Fig. 7.17.

*The work terms in this chapter are all exclusive of injection work and would have the symbol $W_{\text{n.f.}}$ in Chaps. 4, 5, and 6. Here we drop the subscript because it causes no confusion to do so.

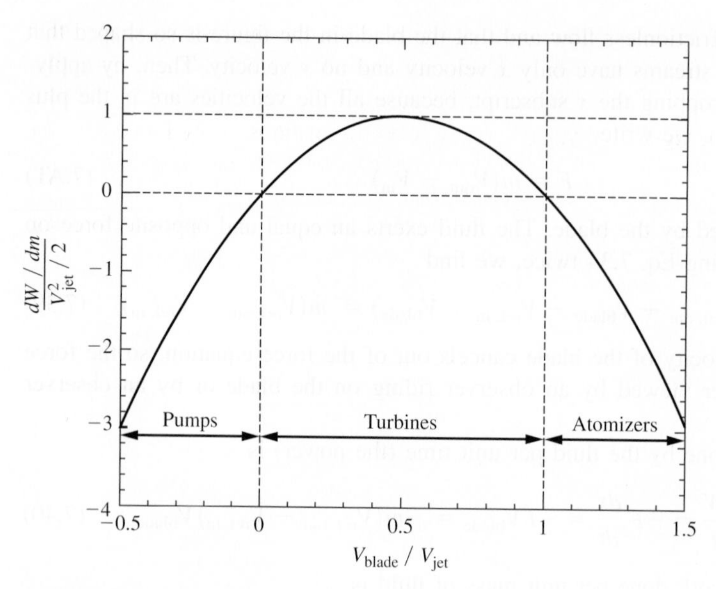

FIGURE 7.17
The ratio of work produced per pound to inlet kinetic energy per pound for a range of values of blade speed/jet speed.

From this figure we see:

1. For $V_{blade} / V_{jet} = 0$ the work extracted per pound = 0. The blade is standing still and resisting the flow but not extracting any work from it.
2. For $V_{blade} / V_{jet} = 1.00$, the work extracted per pound = 0. The blade is moving at the same speed as the jet (like a person walking through a revolving door that is turning at exactly the person's walking speed) and has no force interaction with the jet.
3. For V_{blade} / V_{jet} between 0 and 1.00 the work extracted is positive. For $V_{blade} / V_{jet} =$ 0.5 it is a maximum, and $(dW / dm) / (V_{jet}^2 / 2) = 1.00$. At this condition all of the kinetic energy in the jet is being extracted by the blade and converted to work. If we consider Fig. 7.16 from the viewpoint of the person riding on the blade, then the fluid is overtaking us at 0.5 times the jet speed and leaving at that speed, in the opposite direction. From the viewpoint of a fixed observer watching us ride by, the jet leaves the blade at zero velocity; all of its kinetic energy has been extracted. (In a practical turbine of this kind the jet leaves with a little y velocity to get out of the way of the next batch of fluid that follows it; if it left with only x velocity, it would run into the part of the jet behind it.)
4. At the left of the figure for $V_{blade} / V_{jet} < 0$, the blade is moving in the opposite direction from that shown in Fig. 7.16. In this case it is doing work on the jet. From the viewpoint of someone riding the jet, the exit velocity is still the same as the inlet velocity, but from the viewpoint of a stationary observer the exit velocity is greater than the initial jet velocity. The sign of the work has changed because, instead of the jet doing work on the blade, the blade is doing work on the jet. This

is the description of a pump or compressor; the fluid leaving the blade at a high velocity is passed through some kind of diffuser (Sec. 5.5) and slowed, thus increasing in pressure.

5. At the right of the figure for $V_{\text{blade}} / V_{\text{jet}} > 1.00$ the blade is going faster than the jet and picks up the fluid and expels it at a higher velocity. This occurs in rotating disk atomizers, widely used to produce small drops for spray dryers. The sign of the work is the same as for pumps, and the opposite of that for turbines, because the rotating blade does work on the fluid. For this application, the orientation of the blade in Fig. 7.16 is rotated 180°.

This is a simple discussion of the interactions of jets of fluid and moving blades. The interactions in real turbines, compressors, and atomizers are more complex, but almost all of these devices are based on the transfer of momentum between a moving jet of fluid and a moving blade, as sketched here.

7.5 STARTING AND STOPPING FLOWS

The previous examples have been for steady flows. The momentum balance is powerful enough to deal with unsteady flows as well. Several very simple examples will illustrate this power.

7.5.1 Starting Flow in a Pipe

Example 7.13. Figure 7.18 shows a large water reservoir that discharges through a long, horizontal pipe, at the end of which is a valve. What is the velocity-time behavior of this system when the valve is suddenly opened?

Here the steady-state velocity, V_∞, can be found by B.E. from point 1 to point 3, finding

$$V_\infty = \left[\frac{(P_2 - P_3)}{\rho} \cdot \frac{D}{2fL}\right]^{1/2} \tag{7.AU}$$

Using the values in Table A.3, we find that steady-state velocity in the pipe is 2.45 m / s (8.03 ft / s) and the steady-state friction factor is 0.0042.

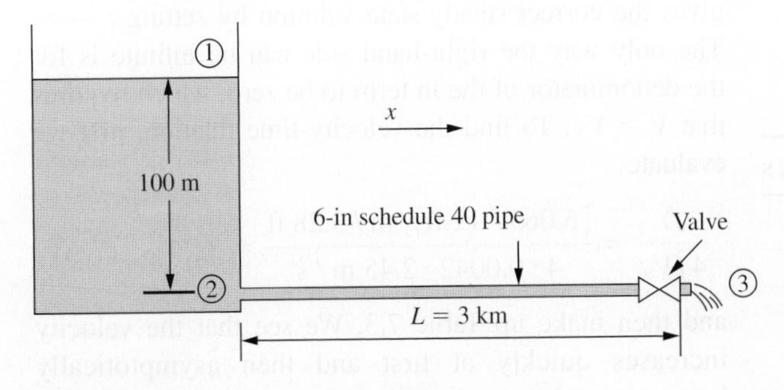

FIGURE 7.18
Long pipe with quick-opening and -closing valve.

To estimate the starting behavior, we take as our system the pipe from its entrance, point 2, to its exit, point 3. Here we can assume that the pressure at point 2 does not change during the starting of the flow and is given by $P_2 = \rho g(z_1 - z_2)$. Applying the x-directed momentum balance (Eq. 7.15), we assume that the density of the fluid does not change, so the mass of fluid in the system is constant and the mass flow rates and velocities in and out at any instant are equal. Then

$$m_{sys}\, dV_{sys} = \sum F\, dt = \left[(P_2 - P_3)\frac{\pi}{4} D^2 - \tau \pi D L \right] dt \tag{7.44}$$

Here the shear force acts in the direction opposite to the pressure force; at steady state they will be equal. Replacing τ by its expression in terms of the friction factor, and expressing the mass of the system in terms of its volume and density, we find

$$\rho \frac{\pi}{4} D^2 L\, dV = \sum F\, dt = \left[(P_2 - P_3)\frac{\pi}{4} D^2 - f\rho \frac{V^2}{2} \pi D L \right] dt \tag{7.45}$$

$$dV = \left[\frac{(P_2 - P_3)}{\rho L} - \frac{4f}{D} \cdot \frac{V^2}{2} \right] dt = \frac{D}{2f}(V_\infty^2 - V^2)\, dt \tag{7.46}$$

and

$$\frac{dV}{(V_\infty^2 - V^2)} = \frac{D}{2f} dt \tag{7.47}$$

Here f is not constant because the flow starts in the laminar region, so that f is initially large, then declines, then increases sharply during the transition, and then declines slowly in the turbulent region. But the term involving f is only significant near the end of the starting transient, so we can treat f as a constant and perform the indicated integration, finding

$$t = \frac{D}{4fV_\infty} \ln \frac{V_\infty + V}{V_\infty - V} + C \tag{7.48}$$

Here at $t = 0$, $V = 0$, so the ln term on the right is $\ln 1 = 0$, from which it follows that the constant of integration $C = 0$. We may also check to see that Eq. 7.48 gives the correct steady-state solution by setting $t = \infty$. The only way the right-hand side can be infinite is for the denominator of the ln term to be zero, which requires that $V = V_\infty$. To find the velocity-time relation, first we evaluate

$$\frac{D}{4fV_\infty} = \frac{[6.065\,/\,12\text{ ft}] \cdot \text{m}\,/\,3.28\text{ ft}}{4 \cdot 0.0042 \cdot 2.45\text{ m}\,/\,\text{s}} = 3.74\text{ s} \tag{7.AV}$$

and then make up Table 7.3. We see that the velocity increases quickly at first and then asymptotically increases to the steady-state value. ■

TABLE 7.3
Flow starting behavior in Example 7.13

Velocity, m / s	Time, s
0.1	0.31
1	3.24
2	8.57
2.4	17.11
2.44	23.1
2.449	31.8
2.45	Infinite

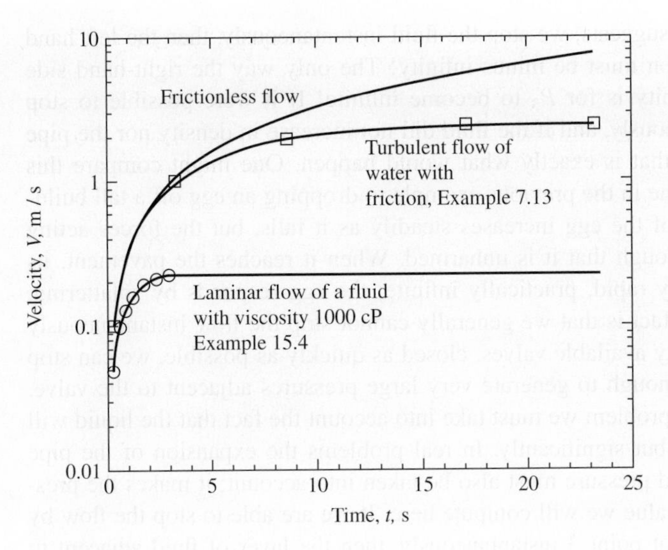

FIGURE 7.19
Flow-starting behavior for three situations. These all correspond to Fig. 7.18. The top curve corresponds to zero friction. The middle curve is the solution to Example 7.13; the square points shown are those from Table 7.3. The bottom curve is from Example 15.4, which is the same as Example 7.13 except that the water has been replaced by a fluid 1000 times as viscous as water.

Figure 7.19 shows the result of this example and compares it to two other results. The values from Table 7.3 are shown as squares, with a smooth curve through them. Above that curve is a frictionless curve, which results from setting $f = 0$ in Eq. 7.45. For the first few seconds it is identical to the curve from this example, indicating that the effects of friction on the starting behavior are negligible until the velocity begins to approach its final value. The lower curve shows the results of Example 15.4, in which the water in Example 7.13 has been replaced with a fluid 1000 times as viscous as water, which makes the flow laminar. That problem cannot be solved by the one-dimensional approach used here; it requires the two- and three-dimensional approach shown in Part IV of this book. We see that for this viscous a fluid the velocity is less than for Example 7.13 at all times, and that the final value is about a tenth of that for water. The starting transient is shorter for that case than for Example 7.13.

7.5.2 Stopping Flow in a Pipe; Water Hammer

Example 7.14. Repeat Example 7.13 for the case in which the fluid is flowing steadily and the valve at the end of the pipe is instantaneously closed. Here we begin by rearranging Eq. 7.44 to

$$\frac{dV_{\text{sys}}}{dt} = \frac{\left[(P_2 - P_3) \frac{\pi}{4} D^2 - \tau \pi D L \right]}{m_{\text{sys}}} \tag{7.49}$$

If, as the problem suggests, we stop the fluid instantaneously, then the left-hand side of this equation must be minus infinity! The only way the right-hand side can be minus infinity is for P_3 to become infinite! If it were possible to stop the fluid instantaneously, and if the fluid did not increase in density nor the pipe wall stretch, then that is exactly what would happen. One might compare this situation and the one in the previous example to dropping an egg off a tall building. The velocity of the egg increases steadily as it falls, but the forces acting on it are gentle enough that it is unharmed. When it reaches the pavement, its deceleration is very rapid, practically infinite; the egg responds by splattering. The observational fact is that we generally cannot stop the flow instantaneously but that with readily available valves, closed as quickly as possible, we can stop the fluid quickly enough to generate very large pressures adjacent to the valve.

To solve the problem we must take into account the fact that the liquid will compress, slightly but significantly. In real problems the expansion of the pipe due to the increased pressure must also be taken into account; it makes the pressure less than the value we will compute here. If we are able to stop the flow by shutting the valve at point 3 instantaneously, then the layer of fluid adjacent to the valve will be stopped. It will stop the next layer, and so the region of stopped fluid will propagate backward up the pipe to the reservoir. (This is analogous to the big freeway pileups that occur during heavy fogs. Someone slows down and is hit by a faster-moving car coming from behind. The first crash produces a pile of stopped, wrecked cars. This pile then enlarges in the upstream direction as more and more cars pile into the stopped wreckage.) The rate of propagation of the boundary between stopped and moving fluid (assuming rigid pipe walls) will be the local speed of sound. That is not proven here, but will seem clearer after we have discussed the speed of sound in Chap. 8. From Chap. 8 we can borrow the fact that for water the speed of sound is about $c = 5000 \text{ ft/s}$ (1520 m/s) so that the stopped layer of water will reach the reservoir in $(t = L/c = 3000 \text{ m}/1520 \text{ m/s})$ or about 2 s after the valve is closed.

To compute the pressure in the stopped fluid we take the viewpoint of the person riding on the interface between the moving fluid and the stopped fluid. Figure 7.20 shows this change of viewpoint and its consequences. We will apply this same logic several times again in this book. The upper part of the figure, from the viewpoint of a stationary observer, shows the wave passing from right to left, against the flow, with velocity $(c - V_1)$. The speed of sound, shown above, describes how fast a sound wave moves into a stationary fluid; here the fluid is moving, with the result shown. The lower part of the figure shows the viewpoint of someone riding the wave. From that viewpoint we are standing still, and the fluid is moving toward us with $V_{\text{upstream}} = c$. Changing the viewpoint does not change P or ρ at any point. Changing the viewpoint does change both the perceived upstream and downstream velocities. But it does not alter the velocity change, ΔV, across the wave. The merit of this change of viewpoint is that it changes an unsteady-state problem to a steady-state one. To see the advantage, try working out the rest of this problem and the problem in the next section from the viewpoint of the stationary observer.

Taking the viewpoint of the observer riding the wave, and using the system shown in the lower part of Fig. 7.20, we find that the x-directed, steady-state

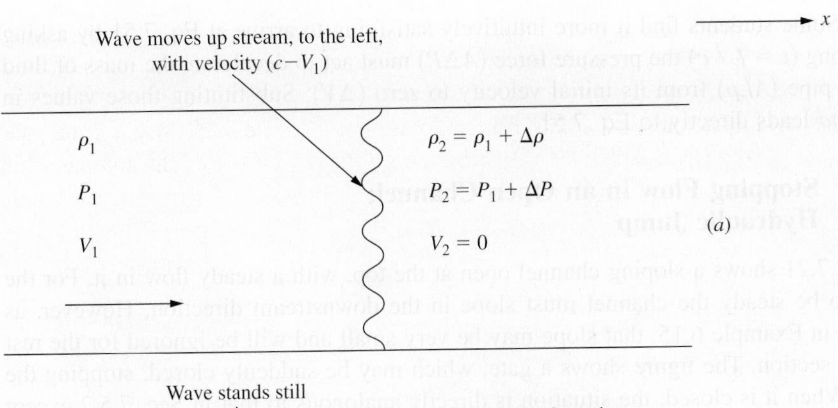

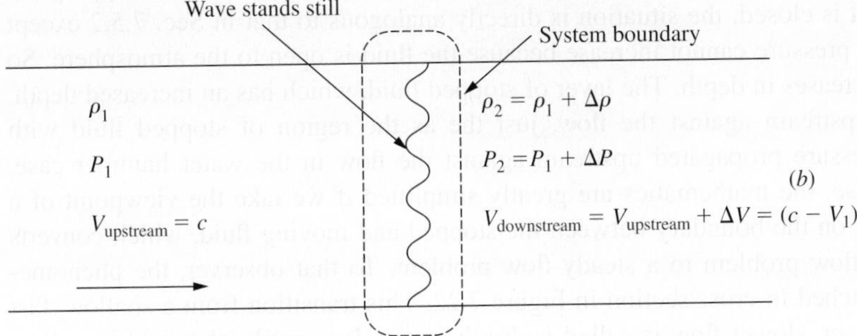

FIGURE 7.20
A wave as seen (*a*) passing by a stationary observer and (*b*) by an observer moving at the same speed as the wave.

momentum balance, Eq. 7.15, becomes

$$0 = \dot{m}_{\text{in}}(V_{\text{in}} - V_{\text{out}}) + \sum F = c\rho A\, \Delta V - A\, \Delta P \tag{7.50}$$

and

$$\Delta P = c\rho\, \Delta V \tag{7.51}$$

where ΔP is the pressure change across the moving boundary and ΔV is the velocity change across it, which in this case is the velocity of the fluid that has not been stopped yet, minus zero, or 2.45 m / s, from Example 7.13. Inserting numerical values, we find

$$\Delta P = c\rho\, \Delta V = 1520\,\frac{\text{m}}{\text{s}} \cdot 998.2\,\frac{\text{kg}}{\text{m}^3} \cdot 2.45\,\frac{\text{m}}{\text{s}} \cdot \frac{\text{N}\cdot\text{s}^2}{\text{kg}\cdot\text{m}} \cdot \frac{\text{Pa}}{\text{N}/\text{m}^2}$$
$$= 3.72\text{ kPa} = 539\text{ psi} \tag{7.AW}$$

■

This is a very large pressure, and it explains why this phenomenon, called *water hammer,* can be a serious problem, particularly in large hydroelectric structures. One often produces the same result at home by closing a faucet quickly; a pounding sound in the plumbing indicates that a high pressure has been generated. The treatment here is the simplest possible; many more interesting details and complications are shown in the books on this subject [4].

Some students find it more intuitively satisfying to arrive at Eq. 7.51 by asking how long ($t = L/c$) the pressure force ($A\Delta P$) must act to decelerate the mass of fluid in the pipe ($AL\rho$) from its initial velocity to zero (ΔV). Substituting those values in $F = ma$ leads directly to Eq. 7.51.

7.5.3 Stopping Flow in an Open Channel; Hydraulic Jump

Figure 7.21 shows a sloping channel open at the top, with a steady flow in it. For the flow to be steady the channel must slope in the downstream direction. However, as shown in Example 6.15, that slope may be very small and will be ignored for the rest of this section. The figure shows a gate, which may be suddenly closed, stopping the flow. When it is closed, the situation is directly analogous to that in Sec. 7.5.2 except that the fluid pressure cannot increase because the fluid is open to the atmosphere. So instead, it increases in depth. The layer of stopped fluid, which has an increased depth, propagates upstream against the flow, just the as the region of stopped fluid with increased pressure propagated upstream against the flow in the water hammer case. As in that case, the mathematics are greatly simplified if we take the viewpoint of a person riding on the boundary between the stopped and moving fluid, which converts an unsteady-flow problem to a steady-flow problem. To that observer, the phenomenon is as sketched in cross section in Figure 7.22. This transition from a shallow, fast flow to a deeper, slower flow is called *hydraulic jump*. It is easily observed in gutters during heavy rainstorms and at the bottom of chutes and spillways. Before we begin

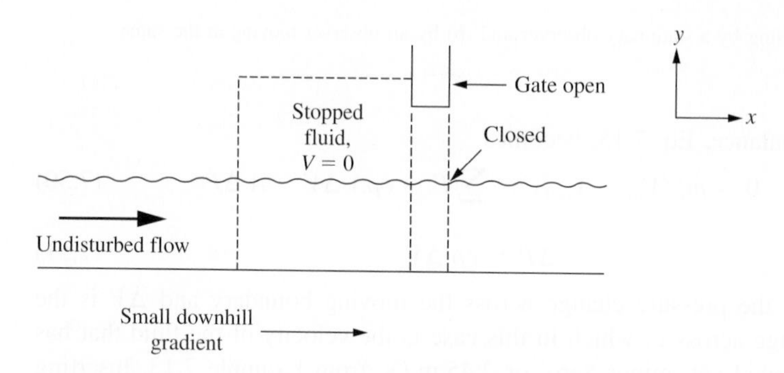

FIGURE 7.21
An open-channel flow stopped by closing a gate.

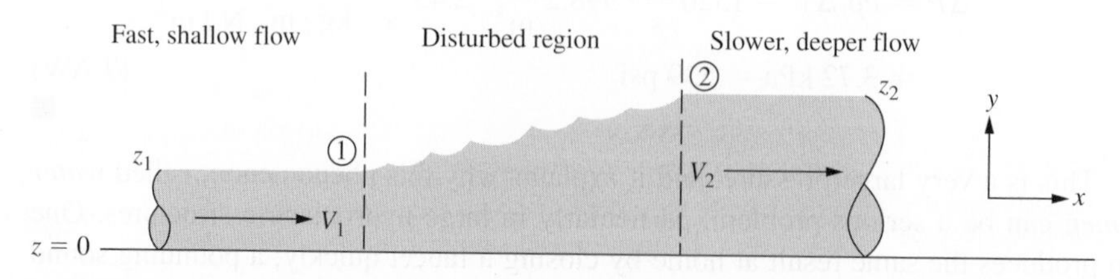

FIGURE 7.22
Hydraulic jump in a linear flow.

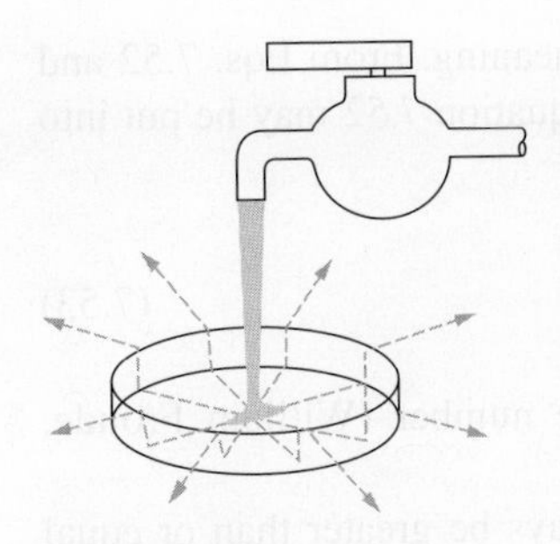

FIGURE 7.23
Hydraulic jump in a radial outward flow, as seen in any sink.

the analysis, I suggest that you look at this phenomenon in a kitchen sink. Figure 7.23 shows a jet of water from a faucet striking the bottom of a sink. It flows outward in a shallow, fast flow until it enters a hydraulic jump, which changes the flow to a deeper, slower flow than the flow near the faucet. This is easy see in any sink. The steady, one-dimensional version sketched in Fig. 7.22 is easiest to treat mathematically, so we return to it. The mathematics treat the wave as a step function, but as Fig. 7.22 (and observation of the jump in a sink) shows, the observed behavior is that the change in depth and velocity occurs over a short distance, not as a sharp step.

In Fig. 7.22 the cross section through the jump extends into and out of the paper. Consider a section l ft thick into the paper, and assume that the velocity across any section perpendicular to the flow is uniform. Then for steady flow of an incompressible fluid such as water, the mass balance gives

$$V_1 z_1 = V_2 z_2 \tag{7.AX}$$

In a typical problem we might know two of the four quantities here. This equation provides a relation for finding a third; one more is needed. B.E. written between states 1 and 2 shows

$$\frac{P_2 - P_1}{\rho} + g(z_2 - z_1) + \frac{V_2^2 - V_1^2}{2} = -\mathcal{F} \tag{7.AY}$$

Here, as in Sec. 5.11, the fluid does not all enter or leave at the same z or the same P, so we must use appropriate average values for the zs and Ps. If the friction term were negligible, this equation would supply the needed extra relation, but mathematical analysis and experimental tests indicate that friction is quite large; so, although B.E. supplies an additional relation between the unknowns in Eq. 7.AY, it does us no good because it introduces another unknown, $\mathcal{F}$.

Equation 7.17, however, can supply the needed relationship. Taking as our system the section of the fluid between points 1 and 2, we see that the only forces acting in the x direction are the shear force on the bottom, which is negligibly small and will be ignored, and the pressure forces on each side of the liquid in the system, which are each of the form

$$F = \int P\,dA = l\int_{z=0}^{z_{\text{surf}}} g\rho(z_{\text{surf}} - z)\,dz = lg\rho\frac{(z_{\text{surf}})^2}{2} \tag{7.AZ}$$

Since the flow at points 1 and 2 is all in the x direction, we may write Eq. 7.17 and drop the x subscripts to find

$$0 = l\rho z_1 V_1(V_1 - V_2) + \frac{l\rho g}{2}(z_1^2 - z_2^2) \tag{7.BA}$$

Equations 7.AX and 7.BA can be solved for z (see App. B.2), which gives us Eq. 7.52,

$$z_2 = \frac{-z_1}{2} \pm \sqrt{\left(\frac{z_1}{2}\right)^2 + \frac{V_1^2 z_1}{g}} \tag{7.52}$$

Here the minus sign before the radical has no physical meaning. From Eqs. 7.52 and 7.AY we can calculate the value of $\mathcal{F}$ (see Prob. 7.52). Equation 7.52 may be put into an interesting form by dividing through by z_1:

$$\frac{z_2}{z_1} = -\frac{1}{2} + \sqrt{\frac{1}{4} + 2\frac{V_1^2}{gz_1}} \tag{7.53}$$

The dimensionless group V_1^2 / gz_1 is called the Froude number (William Froude, 1810–1879); its significance is discussed in Chap. 9.

From the foregoing it is clear that z_2 / z_1 must always be greater than or equal to 1 (a value of 1 would correspond to a jump of negligible height, i.e., one that was vanishingly small). One may verify from Eq. 7.53 that $z_2 / z_1 = 1$ for a Froude number of 1, and that z_2 / z_1 is greater than 1 for any Froude number greater than 1. In studying normal shock waves in gases, we see that another dimensionless group, the Mach number, plays a similar role.

This topic is traditionally included in fluid mechanics books for the following reasons:

1. Hydraulic jump is readily observed in nature.
2. Hydraulic jump is an interesting example of a problem that cannot be solved without using the momentum balance.
3. Shock waves and hydraulic jumps are very similar, as we will see when we study shock waves in high-velocity gas flow. Hydraulic jumps are easily demonstrated in any kitchen sink and easily studied in any well-equipped hydraulics laboratory. Shock waves are much harder to demonstrate and study. Therefore, from visual observation and mathematical analysis of hydraulic jumps we can gain an intuitive understanding of shock waves. We will return to their similarity in Chap. 8.

Equations 7.52 and 7.AY are equally well satisfied whether a flow is from left to right or from right to left in Fig. 7.22. However, if we calculate $\mathcal{F}$ for both we see that right-to-left flow in Fig. 7.22 (deep, slow flow to shallow, fast flow) results in a negative value of $\mathcal{F}$. This is forbidden by the second law of thermodynamics, so the flow can be only in the sense indicated in the figure. We see here a strong parallel with what we will see concerning shock waves, in which the continuity, energy, and momentum equations are also satisfied by flow in either direction; but the second law of thermodynamics shows that only one direction is possible. We also see that a hydraulic jump is only possible if the upstream value of the Froude number is greater than 1; when we study normal shock waves we will see that a normal shock wave is only possible if the upstream Mach number is greater than 1.

Example 7.15. A steady water flow as shown in Fig. 7.22 has $V_1 = 4$ ft / s, and $z_1 = 0.0005$ ft (= 0.006 inches). What are the values of V_2 and z_2?

First we calculate

$$\text{Froude number} = \mathcal{F}r = \frac{V_1^2}{gz_1} = \frac{(4\ \text{ft / s})^2}{(32.2\ \text{ft / s}^2) \cdot 0.0005\ \text{ft}} = 993.8 \tag{7.BB}$$

and then

$$\frac{z_2}{z_1} = -0.5 + \sqrt{0.25 + 2 \cdot 993.8} = 45.08 \qquad \text{(7.BC)}$$

From this we easily compute that $z_2 = 0.0225$ ft $= 0.27$ in $= 6.86$ mm, and $V_2 = 0.09$ ft / s $= 0.017$ m / s. ■

These are representative values for the bathroom sink flow shown in Fig. 7.23. If you plug the sink so that the water cannot escape, you will see that as the water depth downstream of the jump increases, the radius at which the jump occurs will move inward. The shallow, fast flow decreases in velocity as it moves outward (which the material balance for steady flow says it must do if its depth remains constant, which it practically does), so the fluid is solving Eq. 7.53 for the place where V_1 corresponds to the value of (z_2 / z_1) set by the rising water level in the sink. Eventually the radius of the jump becomes small enough that the low, swift flow is "drowned," and the hydraulic jump disappears.

Returning to the straight, rectangular channel in Fig. 7.21 we can consider a more typical example, a fluid flow with $V_1 = 10$ ft / s as seen by an observer moving along with the jump, and $z_1 = 1$ ft. Straightforward calculations show that for this case the Froude number $= 3.1$, $V_2 = 3.24$ ft / s, and $z_2 = 3.25$ ft. Now if we switch back to the viewpoint of a stationary observer, we see that the value of z_2 is the same for a stationary or a moving observer, so our change in viewpoint did not affect that value. But the value of V_1, which we guessed, is the *sum* of the velocity of the upstream flow, as seen by a fixed observer, and the velocity at which the jump moves upstream. From the viewpoint of a stationary observer, the downstream flow is standing still, so we know that the jump must be moving upstream at $V_{\text{jump}} = 3.24$ ft / s and that, from the viewpoint of a stationary observer, the upstream velocity is $V_1 = 10 - 3.24 = 6.76$ ft / s. The corresponding problem in which we know the upstream velocity in Fig. 7.22 and want to know how fast the jump moves upstream requires a trial-and-error solution; see Prob. 7.53.

7.6 A VERY BRIEF INTRODUCTION TO AERONAUTICAL ENGINEERING

Chapters 5 and 6 were devoted to problems that could be most easily understood by applying the energy balance, and this chapter has been devoted to other problems, which could be most easily understood by applying the momentum balance. Some problems are most easily understood by applying *both* the energy and the momentum balances. A very interesting example is the elementary analysis of flight, which helps explain the behavior of airplanes and helicopters, and also birds and insects.

An airplane (or a bird or a flying insect) is a fluid-mechanical device; it flies by making a fluid, the air, move. Consider an airplane in constant-velocity, level flight; see Fig. 7.24. The airplane has no acceleration in either the x or the y directions; therefore, the sum of the forces acting on the airplane in each of these directions is zero. These forces are shown in the figure and given their common aeronautical-engineering names.

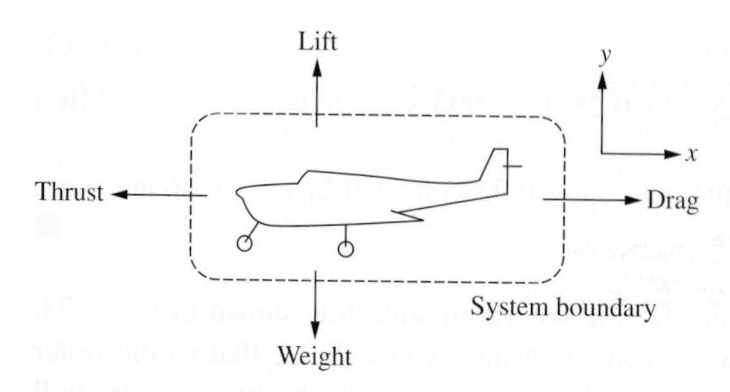

FIGURE 7.24
Airplane in constant-velocity, level flight.

The force that gravity exerts, the weight, acts downward. To counteract this, the air must exert an equal and opposite upward force, called the lift. It is the function of the airplane's wings to make the air exert this force.

To see how the wing does this, we make a momentum balance around the airplane and take as our system the airplane plus an envelope of air around it, large enough for the pressure on the outside of the envelope to be constant. This system boundary is also shown in Fig. 7.24. We base our coordinate system on the airplane, so the airplane appears to stand still and the air to flow toward it. Applying the y component of the momentum balance in constant-velocity flight, we see that there is no accumulation: we have $d(mV)_{\text{sys}} = 0$.* We have assumed that the pressure around the outside of the system is uniform; then the only external y-direction forces acting on the plane are the force of gravity and the force exerted by the air. Thus,

$$F = \text{weight of the plane} = \dot{m}(V_{y_{\text{out}}} - V_{y_{\text{in}}}) \tag{7.54}$$

Since not all the air comes in or goes out at the same velocity, the two V_y terms in this equation must be some appropriate average velocities, obtained by an integral of the flow per unit surface area over the entire surface of the system. However, we need not worry about this integration, if we merely think of these velocities as some appropriate average.

In the direction of the $+y$ axis, F_y is negative. The flow through $\dot{m}$ is positive, so $(V_{y_{\text{out}}} - V_{y_{\text{in}}})$ must be negative; the air must be accelerated in the $-y$ direction, downward. Thus, we see that, to stay in level flight, the airplane must accelerate the surrounding air downward. This is precisely what a swimmer does in treading water—by accelerating the water downward, the swimmer stays up.

Example 7.16. An airplane with a loaded mass of 1000 kg (and thus a weight of 9810 N) is flying in constant-velocity, horizontal flight at 50 m / s. Its wingspread is 15 m, and we assume that it influences a stream of air as wide as its wingspread and 3 m thick. How much average vertical downward velocity must it give this air? Assume that the air comes in at zero vertical velocity.

$$\dot{m} = \rho A V_{x_{\text{in}}} = 1.21 \frac{\text{kg}}{\text{m}^3} \cdot (15 \text{ m} \cdot 3 \text{ m}) \cdot 50 \frac{\text{m}}{\text{s}} = 2723 \frac{\text{kg}}{\text{s}} = 6000 \frac{\text{lbm}}{\text{s}} \tag{7.BD}$$

$$V_{y_{\text{out,avg}}} = \frac{F_y}{\dot{m}} = \frac{9810 \text{ N}}{(2723 \text{ kg / s})} \cdot \frac{\text{kg} \cdot \text{m}}{\text{N} \cdot \text{s}^2} = 3.60 \frac{\text{m}}{\text{s}} = 11.8 \frac{\text{ft}}{\text{s}} \tag{7.BE}$$

■

*In the most exact work, we would have to consider the decrease in mass due to the burning of fuel, but that is small enough to neglect here.

This is the approach "from the outside." Without knowing any details of the flow around the various parts of the airplane, we can find the average downward velocity it must give the air that it influences to stay in level flight. If we chose as our system the airplane itself, we would see that it has no flow in or out, excluding the negligible engine intake and exhaust, and therefore the sum of the forces on it must be zero. For the airplane as the system, these forces are the gravity force and the pressure force integrated over its entire surface. To find the latter it is necessary to analyze the flow in detail around the entire airplane. It can be done in the case of some simple structures, such as certain types of wings, by means of B.E.; this type of analysis is introduced in Chaps. 16 and 17. Briefly, the result of the detailed calculation is that the wing is shaped so that the pressure over its top surface is less than that over its bottom surface. From the balance of forces in the x direction we see that, in order to fly, the plane must overcome the air resistance, which is called drag. In constant-velocity, level flight the drag is equal and opposite to the forward force, or thrust, developed by the power plant.

In elementary and high-school science classes students are taught that the wing is curved so that the air flow over the top is faster than that over the bottom, and by B.E. the pressure is lower over the top of the wing, producing lift. It is true that the pressure is lower over the top of the wing, but if this were the correct explanation of how lift occurs, then flat-winged aircraft could not fly. We all have seen that flat-winged paper gliders fly very well. One could build full-sized airplanes with flat wings, and they would fly. However, a flat wing has much more drag than a properly curved one producing an equivalent lift. This was discovered by birds in their evolution and later by the pioneers in aviation. By careful analysis and much experimentation, wings have been built that have ratios of lift to drag as high as 20. The turbine and compressor blades shown in Fig. 7.15 are shaped like the wings of aircraft, because their function is the same, to turn an air flow and in so doing extract or impart work to it, with the lowest possible friction (drag). The design of airplane wings and of turbine or compressor blades uses the same mathematics as also does the design of sailboat sails.

Example 7.17. A light plane is being designed with an overall aircraft lift / drag ratio of 10. The available power plants have thrust / weight ratios of 2. What percentage of the total loaded weight of the aircraft will be power plant? Assume that the plane will be used only in constant-velocity, level flight.

Under these circumstances, lift equals gross weight and drag equals thrust; therefore, lift / drag is weight / thrust; then

$$\frac{\text{Weight}}{\text{Engine weight}} = \frac{\text{weight}}{\text{thrust}} \cdot \frac{\text{thrust}}{\text{engine weight}} = 10 \cdot 2 = 20 \qquad \text{(7.BF)}$$

■

Suppose that we wish to design a helicopter using a power plant with the same thrust / weight ratio. For a helicopter in hovering flight, thrust is vertically upward and equals the weight. Thus, with this engine 50 percent of the gross weight of the helicopter must be engine. This illustrates the fact that horizontal flight, with a wing, is much more efficient than hovering flight. But why is this so? From

Eq. 7.54, we see that the upward force is equal to the velocity change of the air times the mass flow rate of the air. Thus, we may lift a given load by making a small velocity change in a large flow rate of air or a large velocity change in a small flow rate of air.

Now consider the work that must be done to accelerate this quantity of air. We will apply B.E. to the system shown in Fig. 7.24. Here again there is negligible change in pressure in the air passing through the system and negligible change in elevation. Solving for the external work gives

$$\frac{dW}{dm} = -\mathcal{F} - \frac{\Delta V^2}{2} \tag{7.55}$$

This is the negative work which must be done on the air by the airplane's power plant. Here the ΔV^2 is the change in the square of the average value of the velocity, which is given by $V = (V_x^2 + V_y^2)^{1/2}$. The power is

$$\text{Po} = \frac{dW}{dm}\dot{m} = \dot{m}\left(-\mathcal{F} - \frac{\Delta V^2}{2}\right) \tag{7.56}$$

If we neglect the friction term, we see that the power required is proportional to the change in the square of the velocity.

Thus, from the momentum balance, Eq. 7.54, we see that an airplane with a given weight can be lifted by any flow that has the proper combination of $\dot{m}\,\Delta V_y$ but that the power to be supplied by the engine is proportional to $\dot{m}\,\Delta V^2$. So, to lift the maximum weight with the minimum power, one should make $\dot{m}$ as large as possible and thereby make ΔV_y and ΔV^2 as small as possible. This is the same as saying that the wings should be as long and thin as possible. However, long, thin wings are difficult to build and are not very satisfactory for high-speed flight. The fliers most interested in efficiency are soaring birds and human glider fliers; both have settled on the longest, thinnest wings that seem structurally feasible. Commercial aircraft designers have sacrificed some of this efficiency for better high-speed performance and sturdier wings. The first airplane to fly around the world without refueling had even longer and narrower wings than any glider or any bird; this feat became possible only when the development of fiber-reinforced plastics made it possible to build extremely long, thin, lightweight wings.

This same consideration of the energy and momentum effects of an aircraft shows why the helicopter is an inefficient weight-lifting device. In hovering flight a helicopter can move only as much air as it can suck into its blades, so it must give that air a very large velocity change. This leads to a high power requirement. It also explains why helicopters hover as little as possible; as soon as possible they move forward so that the amount of air influenced by their rotors is increased by their forward motion. The same arguments explain why bees and very small birds like hummingbirds can hover by wing beating, but larger birds cannot. Insects and hummingbirds have small values of mass / unit wing area, so they can stay up in inefficient hovering flight. Big birds have higher values of mass / unit wing area, and cannot hover by wing beating [5]. (Soaring birds like eagles seek out rising air currents: They do not hover by beating their wings.)

Finally this same consideration explains why the commercial aircraft industry is replacing simple jet engines with fan-jet or high-bypass engines, which move more air than a simple jet engine, making a smaller change in its velocity, and hence getting better fuel efficiency. Figure 7.15 shows this. About 40 percent of the air passing the first compressors passes through the combustion chambers; the remaining 60 percent passes through the channel around the outside of the engine. These flows mix at the tail, so that the overall air flow is greater and the change in velocity smaller than was seen in the earliest engines of this type, which did not have this "bypass".

7.7 THE ANGULAR-MOMENTUM BALANCE; ROTATING SYSTEMS

In the study of rotating systems it is convenient to define a quantity called the angular momentum of a body:

$$\begin{pmatrix}\text{Angular momentum}\\ \text{of a body, } L\end{pmatrix} = \begin{pmatrix}\text{mass of}\\ \text{a body, } m\end{pmatrix} \cdot \begin{pmatrix}\text{tangential}\\ \text{velocity, } \omega\end{pmatrix}; \qquad L = m\omega \qquad (7.57)$$

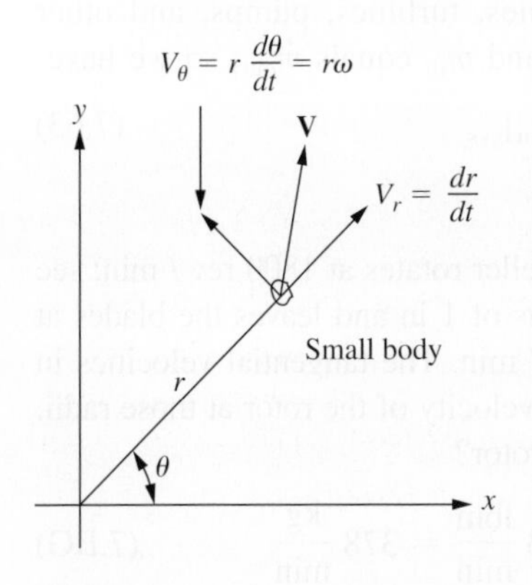

FIGURE 7.25
Velocity components in polar coordinates. Here **V** is the velocity vector, V_θ is the tangential component of the velocity, or simply the tangential velocity, and V_r is the radial component of the velocity, or simply the radial velocity.

The geometric significance of these terms is most easily seen by examining a body in motion and using polar coordinates; see Fig. 7.25.* We see from Eq. 7.57 that the angular momentum of a body depends not only on the body's velocity and mass but also on the point chosen for the origin of the coordinate system. This causes no confusion if we always make clear what choice of origin we make. Since the idea of angular momentum is used most often in rotating systems, generally it is easiest to choose the origin so that it coincides with the axis of rotation.

The r of a large body is not constant over the entire mass, so we must find the angular momentum by integrating over the entire mass:

$$L = \int_{\text{entire mass}} rV_\theta \, dm \qquad (7.58)$$

As shown in Fig. 7.25, the tangential velocity V_θ is equal to $r\omega$; therefore, for constant angular velocity over an entire body (i.e., a rotating rigid body), this

*This equation is really one component of a vector equation, which is written $\mathbf{L} = m\mathbf{r} \times \mathbf{V}$. The other components of this vector equation refer to motions of the axis of rotation, which are significant in systems such as gyroscopes but seldom important in fluid mechanics. Therefore, we refer the reader to texts on mechanics for the three-dimensional vector form of this equation.

simplifies to

$$L = \omega \int_{\text{entire mass}} r^2 dm = \omega I \tag{7.59}$$

where I is the angular moment of inertia, $\int_{\text{entire mass}} r^2\, dm$.

It can be readily shown that angular momentum, like linear momentum, obeys the balance equation, but with the difference that in place of force acting we have torque acting, where

$$\text{Torque} = \text{tangential force} \cdot \text{radius}; \qquad \Gamma = F_\theta r \tag{7.60}$$

So the angular momentum balance (for a fixed axis of rotation) becomes

$$dL = (rV_\theta)_{\text{in}}\, dm_{\text{in}} - (rV_\theta)_{\text{out}}\, dm_{\text{out}} + \Gamma\, dt \tag{7.61}$$

Again we can divide by dt to find the rate form:

$$\left(\frac{dL}{dt}\right)_{\text{sys}} = (rV_\theta)_{\text{in}}\, \dot{m}_{\text{in}} - (rV_\theta)_{\text{out}}\, \dot{m}_{\text{out}} + \Gamma \tag{7.62}$$

This equation, often referred to as the *moment-of-momentum equation,* is one of the basic tools in the analysis of rotating fluid machines, turbines, pumps, and other devices [6]. In steady-state flow $(dL/dt)_{\text{sys}}$ is zero and $\dot{m}_{\text{in}}$ equals $\dot{m}_{\text{out}}$; so we have

$$\Gamma = \dot{m}[(rV_\theta)_{\text{out}} - (rV_\theta)_{\text{in}}]_{\text{sys}} \tag{7.63}$$

which is *Euler's turbine equation.*

Example 7.18. A centrifugal water-pump impeller rotates at 1800 rev / min; see Fig. 7.26. The water enters the blades at a radius of 1 in and leaves the blades at a radius of 6 in. The total flow rate is 100 gal / min. The tangential velocities in and out may be assumed equal to the tangential velocity of the rotor at those radii. What is the steady-state torque exerted on the rotor?

$$\dot{m} = 100\,\frac{\text{gal}}{\text{min}} \cdot 8.33\,\frac{\text{lbm}}{\text{gal}} = 833\,\frac{\text{lbm}}{\text{min}} = 378\,\frac{\text{kg}}{\text{min}} \tag{7.BG}$$

$$(V_\theta)_{\text{in}} = r_{\text{in}}\omega, \qquad (V_\theta)_{\text{out}} = r_{\text{out}}\omega \tag{7.BH}$$

From Eq. 7.63, we write

$$\Gamma = \dot{m}\omega(r_{\text{out}}^2 - r_{\text{in}}^2) \tag{7.BI}$$

$$\Gamma = 833\,\frac{\text{lbm}}{\text{min}} \cdot \frac{2\pi \cdot 1800}{\text{min}} \cdot \left[\left(\frac{6}{12}\,\text{ft}\right)^2 - \left(\frac{1}{12}\,\text{ft}\right)^2\right] \cdot \frac{\text{lbf} \cdot \text{s}^2}{32.2\ \text{lbm} \cdot \text{ft}} \cdot \frac{\text{min}^2}{3600\ \text{s}^2}$$

$$= 19.8\ \text{ft} \cdot \text{lbf}$$

$$= 26.8\ \text{N} \cdot \text{m} \tag{7.BJ}$$

■

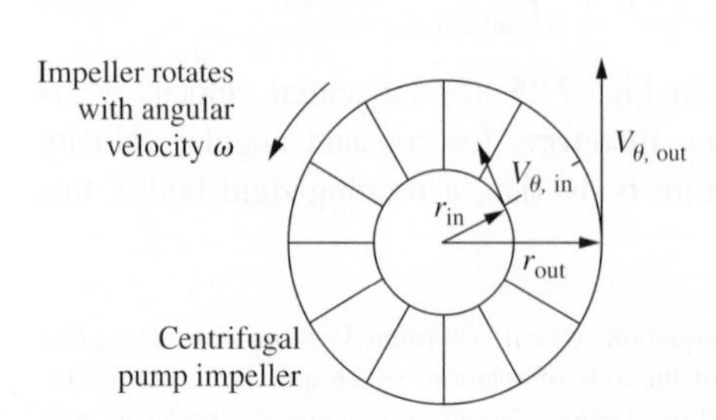

FIGURE 7.26
Centrifugal pump impeller.

This is the net torque acting on the rotor; the algebraic sum of the positive torque exerted by the shaft driving the rotor and the negative torque exerted by friction between the rotor and the surrounding fluid. If we wished to know the total torque applied to the shaft, we would need to know the frictional resistance; that is a much harder problem than this one.

7.8 SUMMARY

1. Momentum is the product of mass and velocity.
2. The momentum balance is simply the restatement of Newton's second law, $F = ma$, in a form that is convenient for fluid-flow problems.
3. The momentum balance is useful in allowing us to solve some fluid-flow problems from the outside, without having to know in detail what goes on inside.
4. The momentum balance, as we show, is applicable to unsteady flows like starting and stopping flows. B.E. is of no use for such flows!
5. The momentum balance is useful for flows in which two streams at different velocities mix and exchange momentum, e.g., the Bunsen burner. B.E. is of no use for such flows!
6. For rotating systems it is convenient to introduce an additional defined quantity called the angular momentum, which obeys a simple balance equation. It is used to analyze rotating systems like pumps and compressors.
7. In Part IV we will apply the momentum balance to three-dimensional flows and show some of its applications.

PROBLEMS

See the Common Units and Values for Problems and Examples, inside the back cover! An asterisk (*) on a problem number indicates that its answer is shown in App. D.

7.1.* The earth has a mass of roughly 10^{25} lbm. A person standing on the earth throws a 1 lbm rock vertically upward, in a direction perpendicular to the earth's motion about the sun, at a velocity of 20 ft / s.

(*a*) How much does the velocity of the earth increase in the direction opposite the throw?

(*b*) When the rock has fallen back to earth, what is the velocity of the earth compared with its velocity before the rock was thrown?

(*c*) When the rock has fallen back to earth, is the earth back on the same orbital path it had before the rock was thrown, or has its orbital path been shifted?

7.2. A 5 lbm gun fires a 0.05 lbm bullet. The bullet leaves the gun at a velocity of 1500 ft / s in the x direction. If the gun is not restrained, what is its velocity just after the bullet leaves? Work this problem two ways:

(*a*) Taking the gun as the system.

(*b*) Taking the combined gun and bullet as the system.

7.3. In Example 7.2, what fractions of the initial kinetic energy of the duck and of the bullet are converted to internal energy?

7.4. In movies and TV thrillers the hero shoots the villain, and the force of the bullet throws the villain into the air, causing his corpse to land several feet from his original position. If we assume that the bullet remains in the villain, what is the relation of the momentum transferred by the gun to the hand of the hero, and the momentum transferred by the bullet to the body of the villain? Are movies and TV a good place to learn one's physics?

7.5.*A fire hose directs a stream of water against a vertical wall. The flow rate of the water is 50 kg / s, and its incoming flow velocity is 80 m / s. The flow away from the impact point has zero velocity in the x direction. What is the force exerted by this stream on the wall?

7.6. Repeat Prob. 7.5, but instead of a fire hose the exhaust from a jet engine flows against the wall. Its velocity is 400 m / s, and its mass flow rate 200 kg / s.

7.7. In Examples 7.3 and 7.4 the analysis was simple because the jet was at right angles to the solid surface. You can observe with a garden hose that such jets, perpendicular to walls or sidewalks, go off radially in all directions, with more or less circular symmetry. You will also observe that there is a region near the jet that is much shallower than the rest, as shown in Fig. 7.23 and described in Sec. 7.5.3. A more interesting and complex problem is the flow of a jet against a flat surface that is not perpendicular to it; you can also observe this flow with a garden hose. The flow is more or less circularly symmetrical, but much more goes away in the direction away from the hose than in the direction toward the hose. But why does any of it flow back in the direction toward the hose?

We can understand this if we replace the three-dimensional problem (circular jet, moving in x, y, and z) directions with a two-dimensional jet (as might issue from a rectangular slot) that is constrained to move only in the x and y directions (by directing it into an open rectangular channel, which prevents flow in the z direction). This flow is sketched in Fig. 7.27. Friction is assumed to be negligible, so from B.E. (ignoring gravity) we see that both streams flowing along the wall must have the same velocity as that in the jet, V_1. In the figure the stream going off to the upper right, (2), is larger than that going to the lower left, (3), in accord with the observation described above. If the flow is frictionless, then there can be no shear stress on the wall, so the resisting force must act normal to the surface as shown. We could attempt to solve for these flows by writing the x and y components of the steady flow momentum balance, but that adds more terms and only makes the analysis harder (try it!).

Instead, we choose a new set of axes for our momentum balance, with one axis, the s direction, parallel to the plate and the other, the r direction, perpendicular to the plate, as sketched on Fig. 7.27. We now apply Eq. 7.17 in the s direction, finding

$$0 = \dot{m}_1 V_1 \cos\theta - \dot{m}_2 V_2 - \dot{m}_3 V_3 + F_s \qquad (7.64)$$

From the assumption of frictionless flow, we can see that $F_s = 0$, that the absolute magnitudes of V_1, V_2, and V_3 are the same, but that V_3 is in the minus s direction, so that it is equal to $-V_1$. Making these substitutions, and dividing by V_1, we find

$$0 = \dot{m}_1 \cos\theta - \dot{m}_2 + \dot{m}_3 \qquad (7.BK)$$

(*a*) Using the material balance to eliminate $\dot{m}_3$, show the equation for $\dot{m}_2 / \dot{m}_1$.
(*b*) Show the equation for the force exerted on the wall in the r direction.

7.8. In a steady-state methane-air flame at approximately atmospheric pressure the temperature is raised from 68°F to 3200°F. The incoming air-gas mixture and the products of

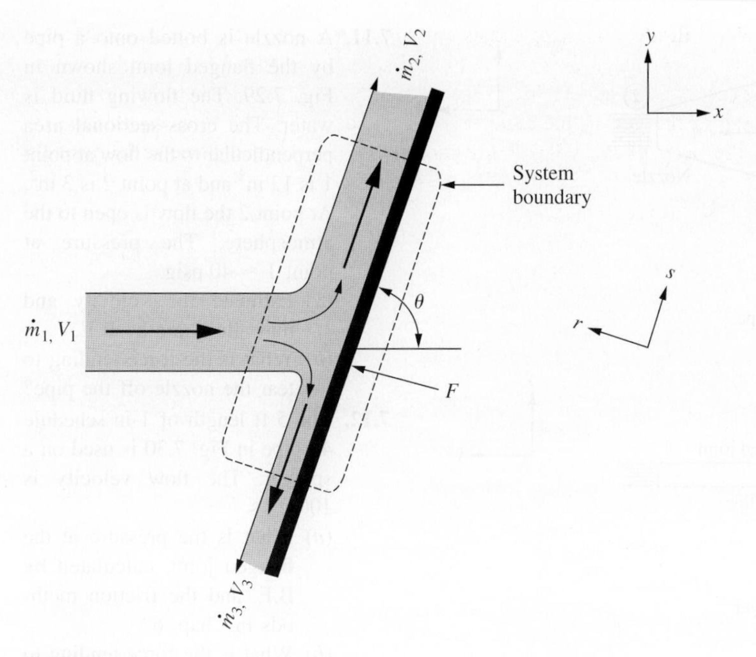

FIGURE 7.27
A jet impinging on a surface not perpendicular to it. We solve in the r-s coordinate system, instead of the x-y system.

combustion may both be considered ideal gases with a molecular weight of ≈28 g / mol. The flame is a thin, flat region perpendicular to the gas flow. If the flow comes into the flame at a velocity of 2 ft / s, what is the pressure difference from one side of the flame to the other? This problem and its consequences are discussed in Lewis and Von Elbe [7].

7.9. A new type of elevator is sketched in Fig. 7.28. The stream of water from a geyser will be regulated to hold the elevator at whatever height is required. If we assume that the maximum flow of the jet is 500 lbm / s at a velocity of 200 ft / s, what is the relation between the weight of the elevator and the maximum height to which the jet can lift it?

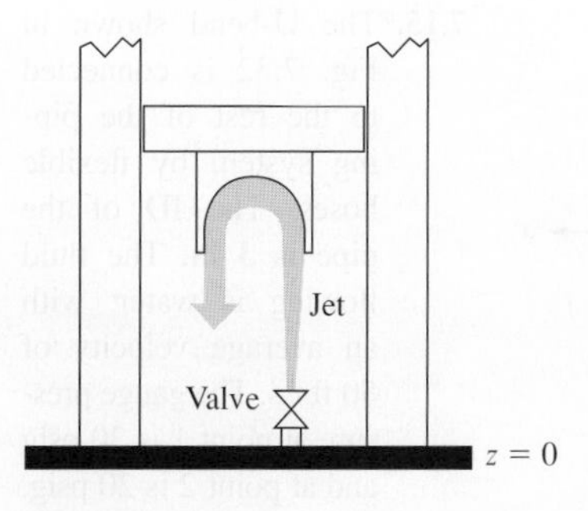

FIGURE 7.28
Hydraulic jet elevator.

7.10.*A sailboat is moving in the y direction. The wind approaches the boat at an angle of 45° to the y direction and is turned by the sails such that it leaves in exactly the minus y direction.

(*a*) If we assume that the average velocity of the incoming and outgoing wind is 10 m / s and that the mass flow rate of air being turned by the boat's sails is 200 kg / s, what are the x and y components of the force exerted by the boat's sails on the air?

(*b*) These are the opposite of the forces exerted by the air on the boat. The y component of the wind force drives the boat in its direction of travel. What does the x component do?

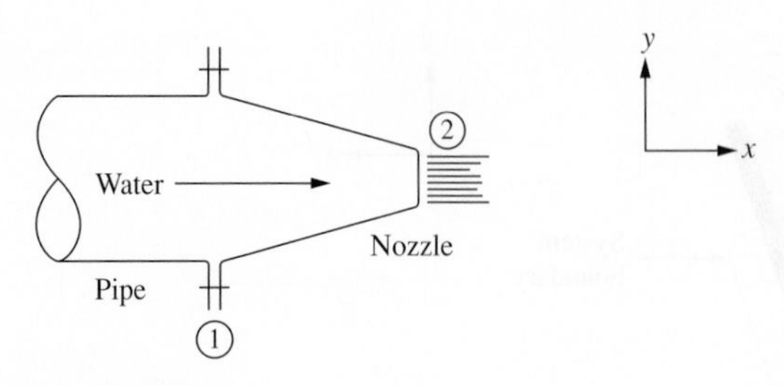

FIGURE 7.29
Nozzle, bolted to pipe.

7.11.*A nozzle is bolted onto a pipe by the flanged joint shown in Fig. 7.29. The flowing fluid is water. The cross-sectional area perpendicular to the flow at point 1 is 12 in^2 and at point 2 is 3 in^2. At point 2 the flow is open to the atmosphere. The pressure at point 1 $\approx$ 40 psig.

(a) Estimate the velocity and mass flow rate by B.E.

(b) What is the force tending to tear the nozzle off the pipe?

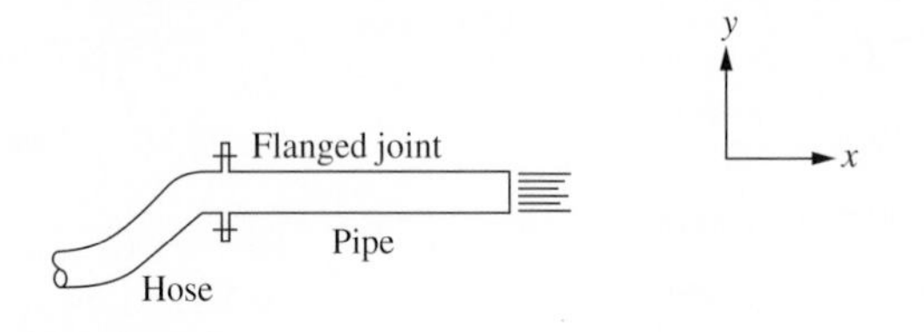

FIGURE 7.30
Pipe used as a sprayer.

7.12. The 5 ft length of 1-in schedule 40 pipe in Fig. 7.30 is used on a sprayer. The flow velocity is 100 ft / s.

(a) What is the pressure at the flanged joint, calculated by B.E. and the friction methods in Chap. 6?

(b) What is the force tending to tear the flange apart?

(c) How is this force transmitted by the fluid to the pipe?

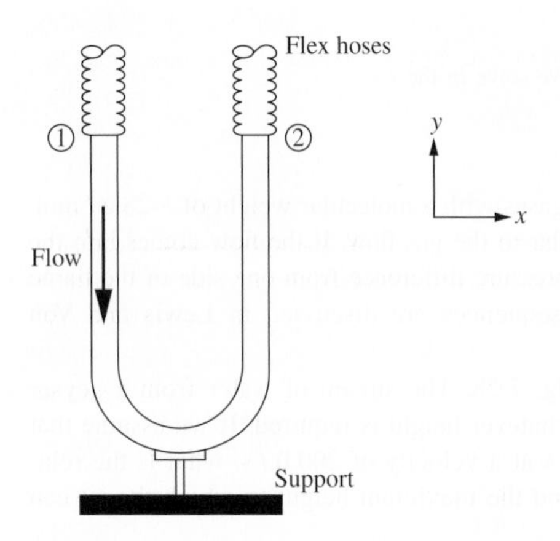

FIGURE 7.31
Vertical pipe U-bend.

7.13. Repeat Example 7.7 for 3-in schedule 40 pipe with water flowing at 8 ft / s and pressure 30 psig throughout.

7.14. The pipe U-bend in Fig. 7.31 is connected to a flow system by flexible hoses that transmit no force. The pipe has an ID of 3 in. Water is flowing through the pipe at a rate of 600 gal / min. The pressure at point 1 is 5 psig and at point 2 is 3 psig. What is the vertical component of the force in the support? Neglect the weight of the pipe and fluid.

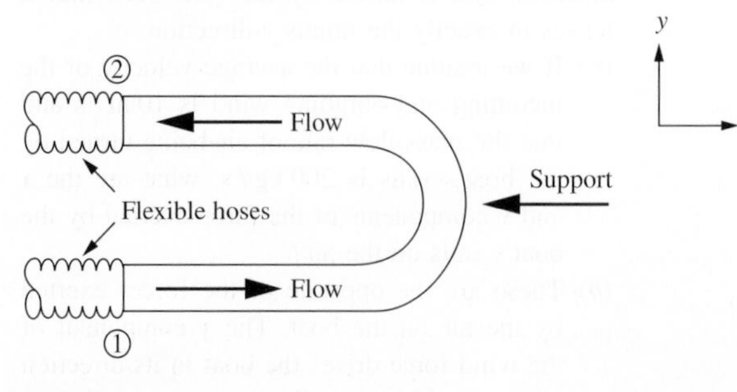

FIGURE 7.32
Horizontal pipe U-bend.

7.15.*The U-bend shown in Fig. 7.32 is connected to the rest of the piping system by flexible hoses. The ID of the pipe is 3 in. The fluid flowing is water, with an average velocity of 50 ft / s. The gauge pressure at point 1 is 30 psig and at point 2 is 20 psig. What is the horizontal component of the force in the support?

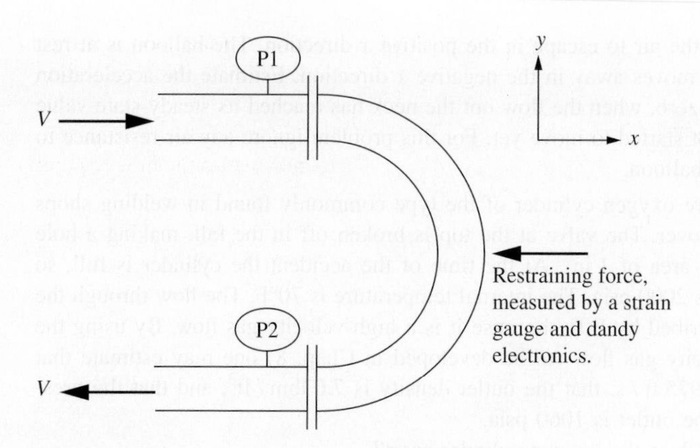

FIGURE 7.33
U-bend flow meter.

7.16. A new type of flow meter is sketched in Fig. 7.33. In it we read the two pressure gauges and the force on the pipe bend (using strain gauges and dandy electronics). From those three readings we compute the fluid velocity in the pipe. The cross-sectional area of the pipe and the couplings and of the bend is 1.000 in^2. The flowing fluid is water. P1 reads 20 psig, P2 reads 18 psig. The restraining force, measured by the strain gauge and the dandy electronics, is 45 lbf, acting in the minus x direction. The couplings between the straight sections of pipe and the bend are of a magical variety that transmit no forces and allow no leakage. For this problem, the acceleration of gravity is zero (we are in a space capsule). What is the fluid velocity in the pipe?

7.17.*A pump, together with the electric motor that drives it, is mounted on a wheeled cart that can be rolled about to various places in our plant for various pumping tasks. It is connected to the vessels to be pumped by flexible hoses that transmit no forces and is connected electrically by a flexible cord that transmits no forces. The pump inlet and outlet are parallel both to each other and to the x axis. The inlet pipe diameter is 4.00 in and the outlet pipe diameter is 3.00 in. The flow rate through the pump is 230 gal / min of water. The pressures at the inlet and outlet are 10 psig and 50 psig, respectively.

(*a*) How much force must we exert on the pump-motor-cart assembly to keep it from moving?

(*b*) In which direction must we exert this force?

7.18. A large rocket engine ejects 200 kg / s of exhaust gases at a velocity of 4000 m / s. The pressure of the exhaust gas is equal to the atmospheric pressure. What thrust does the engine produce?

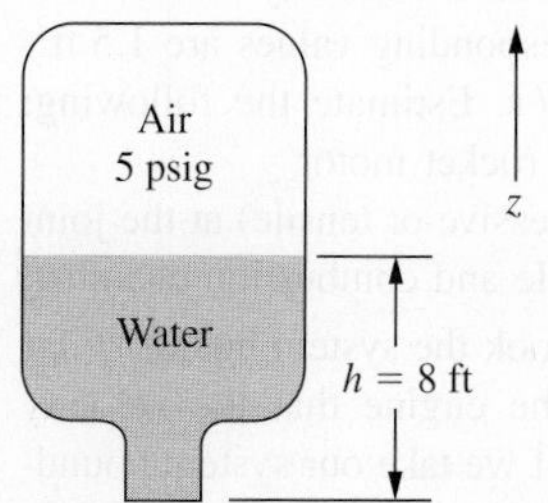

FIGURE 7.34
Compressed-air rocket.

7.19.*Calculate I_{sp} for the rocket in Example 7.8. Why is this different from the result in Example 7.9?

7.20. The compressed-air-driven water rocket shown in Fig. 7.34 is ejecting water vertically downward through a frictionless nozzle. The exit area of the nozzle is 1 in^2. When the pressure and elevation are as shown, how much thrust does the rocket produce?

7.21. A spherical toy balloon has diameter (when inflated) of 6 in and an internal pressure of 1 psig. The neck of the balloon has a diameter of 0.5 in. The skin of the balloon has a mass of 0.0075 lbm. At time zero we release the neck of the balloon, allowing the air to escape in the positive x direction. The balloon is at rest at time zero and then moves away in the negative x direction. Estimate the acceleration of the balloon at time zero, when the flow out the neck has reached its steady-state value but the balloon has not started to move yet. For this problem ignore any air resistance to the movement of the balloon.

7.22.*A typical high-pressure oxygen cylinder of the type commonly found in welding shops and laboratories falls over. The valve at the top is broken off in the fall, making a hole with a cross-sectional area of 1 in^2. At the time of the accident the cylinder is full, so the internal pressure is 2000 psia. The internal temperature is 70°F. The flow through the nozzle cannot be described by B.E., because it is a high-velocity gas flow. By using the methods of high-velocity gas flow (to be developed in Chap. 8) one may estimate that the outlet velocity is 975 ft / s, that the outlet density is 7.0 lbm / ft^3, and that the pressure in the plane of the outlet is 1060 psia.

(*a*) How much thrust does the oxygen cylinder exert?

(*b*) Is it a worthwhile safety practice to fasten these cylinders so they cannot fall over?

(*c*) The tops of high-pressure gas cylinders are protected either by collars (propane cylinders) or screwed-on thick-walled caps (gases like oxygen or hydrogen). Explain this practice.

7.23. The following is an incorrect solution to the preceding problem. Where is its error?

Incorrect solution. The pressure everywhere in the container is 2000 psig, except over the area of the outlet. So the pressure forces cancel, except for a section of 1 in^2, which has 2000 psig pointing away from the nozzle and 1060 psig pointing the other way at the nozzle. Thus, the net force is $(2000 - 1060)$ psig $\cdot$ 1 in^2 = 940 lbf.

7.24. A typical garden hose has an inside diameter of $\frac{3}{4}$ in. The water stream flowing from it has a velocity of 10 ft / s.

(*a*) If such a hose is left loose, will the end move about?

(*b*) Would it move about if it were perfectly straight, or must it be curved?

(*c*) What is the maximum plausible value of the force involved in any such motion?

7.25. The 3-ft-diameter, horizontal main water cooling line from a nuclear reactor breaks. The pressure inside the reactor is 1000 psia, and the water surface inside the reactor is 20 ft above the broken line. The exiting fluid (a steam-water mixture) has a density of 50 lbm / ft^3. Estimate the horizontal force on the pipe-reactor system due to the flow through the broken pipe. Assume frictionless flow and B.E.

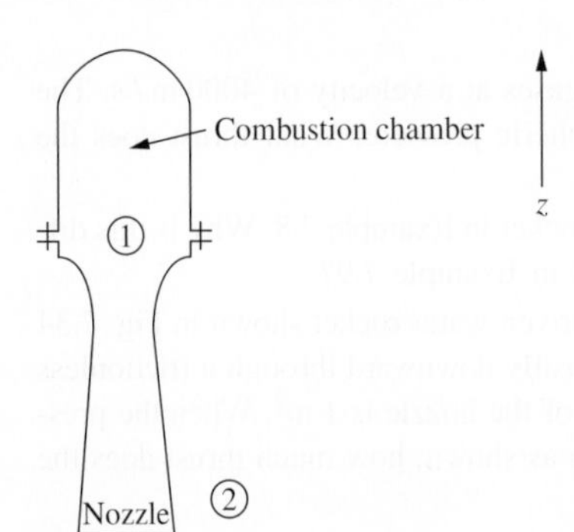

FIGURE 7.35
Solid-fueled rocket.

7.26.*The rocket motor sketched in Fig. 7.35 has the nozzle bolted to the combustion chamber, which contains the fuel. The flow rate is 300 lbm / s. At section 1 the cross-sectional area = 5 ft^2, the pressure is 300 psia, and the velocity is 300 ft / s. At section 2 the corresponding values are 1.5 ft^2, 40 psia, and 4600 ft / s. Estimate the following:

(*a*) The thrust of the rocket motor.

(*b*) The force (compressive or tensile) at the joint between the nozzle and combustion chamber.

7.27. In Example 7.10 we took the system boundary far enough away from the engine that the velocity there was negligible. If we take our system boundaries right at the inlet and outlet of the engine, then the outlet velocity will be unchanged, but the inlet velocity will become about 500 ft / s. The thrust of the engine is the same, independent of what system we choose (the engine does not care what we have in mind). Repeat Example 7.10, making this change in system. *Hint:* Consider the inlet pressure.

7.28. In Example 7.10 we ignored the mass flow of the fuel, which we asserted was negligible. Actual jet engines have $\dot{m}_{fuel} / \dot{m}_{air} \approx 0.02$. For the same air flow rate and inlet and outlet velocities, by what fraction does the computed thrust of the engine increase if we take this additional flow into account? Assume that the fuel crosses our system boundary in the y direction, so that its inlet x velocity $= 0$.

7.29. If the sudden expansion in Fig. 7.12 were replaced with a gradually outward-tapering transition, then the friction losses would be very small, often practically zero. Show how the momentum balance for that flow differs from the momentum balance for the sudden expansion.

7.30. Check several values of Fig. 6.16 to see if the sudden expansion curve was actually made up from Eq. 7.28.

7.31. (*a*) Set up the spreadsheet solution to Example 7.11, and show that the numerical solution does converge to the values shown in Table 7.2.

(*b*) Using the spreadsheet program, rerun the solution to Example 7.11 on the assumption that we should have used an orifice coefficient of 0.6 (see Sec. 5.8) in the B.E. for the gas flow. How does this change the calculated values of P_2, V_3, and $\dot{m}_{air} / \dot{m}_{gas}$?

7.32. (*a*) Using the spreadsheet prepared in the previous problem, repeat Example 7.11 for propane as a fuel. In the United States propane is distributed inside houses at 11 in of water, compared to the 4 in of water for natural gas. What are the calculated values of P_2, V_3, and $\dot{m}_{air} / \dot{m}_{gas}$?

(*b*) The values found in part (*a*) show that simple connection of a natural gas appliance to a normal propane supply line will produce a flame much larger and smokier than the appliance was designed to produce with natural gas. To solve the fuel-conversion problem, appliance manufacturers supply conversion kits. To convert a natural gas appliance to propane, the conversion kit normally replaces the fuel jets with ones with ≈45 percent as large a cross-sectional area as the jets for natural gas. Using the spreadsheet program used in the previous problem, rerun part (*a*) of this problem with a gas orifice area 45 percent as large as that in Example 7.12. What are the calculated values of P_2, V_3, and $\dot{m}_{air} / \dot{m}_{gas}$?

7.33. In our treatment of the Bunsen burner in Example 7.11,

(*a*) Calculate the Reynolds number of the gas jet, the incoming airflow, and the mixed flow (simplify by using the kinematic viscosity of air for all three flows).

(*b*) Which of these flows are laminar? Which are turbulent?

(*c*) Sketch the velocity distribution in such a flow. Here an intuitive sketch will do.

(*d*) In applying Eq. 7.29 we asserted that the rightmost term, which involves the shear stress at the wall was negligible. Check that assumption as follows; First, show that if we assume laminar flow we can substitute

$$\tau = f\rho\frac{V^2}{2} = \frac{16}{\mathscr{R}}\rho\frac{V^2}{2} = \frac{16\mu}{DV\rho}\rho\frac{V^2}{2} = \frac{8\mu V}{D} \tag{7.BL}$$

which makes the rightmost term $8\pi\mu V_3 \Delta x$. Second, evaluate the magnitude of this term for the values in Example 7.11. Your answer should have the dimension of lbf. Third, evaluate the magnitude of the $\dot{m}_{gas}(V_3 - V_{gas,2})$ term for the same example, also in lbf. Show the ratio of this "momentum of the gas jet" term to the wall shear term.

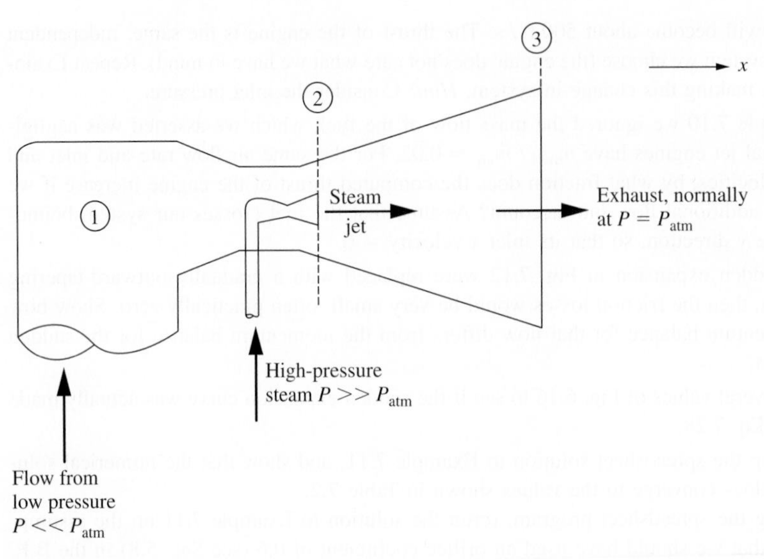

FIGURE 7.36
Steam-jet ejector or vacuum pump.

7.34. Figure 7.36 is a sketch of a *steam-jet ejector* of a type widely used to produce vacuums in process equipment. Conceptually, it is the same as the Bunsen burner in Fig. 7.14. A faster-moving central flow, produced in this case by high-pressure steam, exchanges momentum with a slower-moving surrounding flow, in this case the gas being removed at a vacuum; and the mixed stream discharges at a pressure intermediate between that of the incoming driver fluid and the lower-pressure driven fluid, in this case at atmospheric pressure. If we write the momentum balance (analogous to Eq. 7.29) for the system consisting of the inside of the device from 2 to 3, what term will appear in the momentum balance that does not appear in Eq. 7.29?

7.35. When a rocket is moving in the positive x direction with velocity V_1 and this velocity is equal and opposite to the velocity of the exhaust gas relative to the rocket, then the exhaust velocity relative to fixed surroundings is zero. Thus, according to Eq. 7.37, $d(mV) / dt = 0$. Does this mean that the rocket is not accelerating? Explain.

7.36. Show the equivalent of Eq. 7.38 for vertical flight with a constant value of the acceleration of gravity and zero air resistance.

7.37. The rocket in Example 7.12 is now fired vertically. The specific impulse and mass ratio are the same as in that example. The rocket consumes all its fuel in 1 min. Calculate its velocity at burnout, taking gravity into account.

7.38.*A rocket starts from rest on the ground and fires vertically upward. During the entire upward firing the velocity of the exhaust gas, measured relative to the rocket, is 4000 m / s. The pressure in the exit plane of the rocket nozzle is always exactly equal to the surrounding atmospheric pressure. The mass of the rocket and fuel before launching is 100,000 kg. The mass of the burned-out rocket is 20,000 kg. The entire burning process takes 50 s. What is the velocity of the rocket at burnout? Ignore air resistance.

7.39. A cylindrical tank, shown in Fig. 7.37, is sitting on a platform that in turn rests on absolutely frictionless wheels on a horizontal plane. There is no air resistance. At time zero the level in the tank is 10 ft above the outlet, and the whole system is not moving.

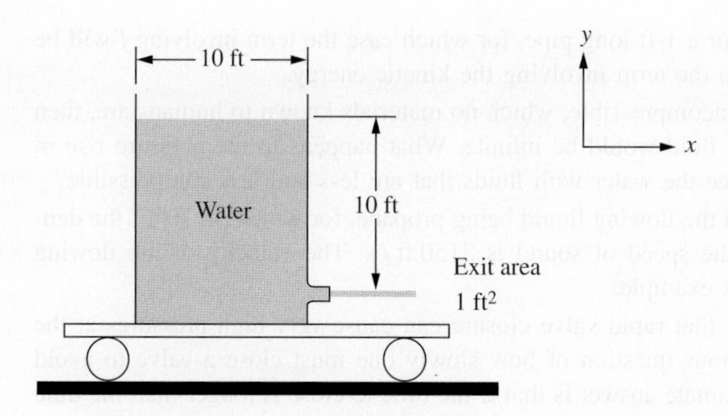

FIGURE 7.37
Draining tank on frictionless cart.

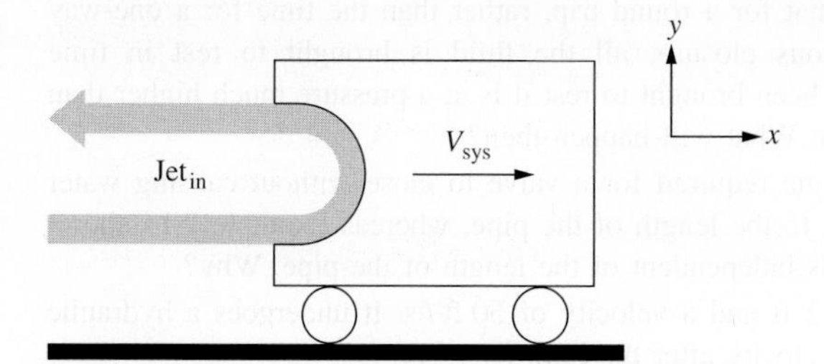

FIGURE 7.38
Starting a cart with a fire hose jet.

Then the outlet is opened and the system allowed to accelerate to the left. The flow through the outlet nozzle is frictionless. What is the final velocity, assuming that
(*a*) The mass of the tank and cart is zero?
(*b*) The mass of the tank and cart is 3000 lbm?

7.40. The cart in Fig. 7.38 has a mass of 2000 kg. It is resting on frictionless wheels on a solid, level surface and encounters no air resistance. At time zero it is standing still, and a jet from a fire hose is used to start it moving. The mass flow rate of the fluid from the fire hose is 100 kg / s, and its velocity relative to fixed coordinates is 50 m / s. The cup on the rear of the cart turns the jet around so that it leaves in the minus *x* direction with the same velocity relative to the cart with which it entered. Calculate the velocity-time behavior of the cart; assume the jet is unaffected by gravity. (This is not a very practical problem, but it is analogous to the more complex and interesting problem of starting a large turbine from rest. All such turbines must be occasionally shut down for maintenance; their starting and stopping behavior is more complex than their behavior running at a steady speed.)

7.41. Repeat Prob. 7.40 with the following change. Instead of the cart turning the jet around by 180° so that it flows out in the minus *x* direction, the cart only turns the jet by 90°, so that it flows out to the side at a right angle to the *x* axis. How long does it take the cart to reach a velocity of 40 m / s?

7.42. In Fig. 7.17 we show that the maximum efficiency for the simple blade-jet interaction in Fig. 7.16 occurs when the blade speed is exactly one-half of the jet speed. One can also show this by rewriting Eq. 7.42 as

$$\frac{dW}{dm} = 2(V_{\text{jet}} - V_{\text{blade}})V_{\text{blade}} \tag{7.BM}$$

Differentiate both sides of this equation with respect to V_{blade} and set the derivative equal to zero. Show that doing so leads to the same conclusion.

7.43. In Example 7.13 the pipe was long enough that we ignored the kinetic energy in the fluid leaving the downstream end of the pipe and the entrance loss.
(*a*) Rework the example, taking that kinetic energy and entrance loss into account, and show that the change is negligible.

(*b*) Rework the example for a 1-ft-long pipe, for which case the term involving *f* will be negligible compared to the term involving the kinetic energy.

7.44. If a fluid were absolutely incompressible, which no materials known to humans are, then the speed of sound in that fluid would be infinite. What happens to the pressure rise in Example 7.14 as we replace the water with fluids that are less and less compressible?

7.45.*Repeat Example 7.14, with the flowing liquid being propane, for which (at 70°F) the density is 31.1 lbm / ft^3 and the speed of sound is 2150 ft / s. The velocity of the flowing fluid is the same as in that example.

7.46. Example 7.14 makes clear that rapid valve closure can cause very high pressures at the valve. That raises the obvious question of how slowly one must close a valve to avoid water hammer. The approximate answer is that if the time to close is longer than the time for a sound wave to make a round trip from the valve to the reservoir, then no significant water hammer will occur. Again, see Parmakian [4] for more details.

(*a*) In Example 7.14, how long is this?

(*b*) Why is the time required that for a round trip, rather than the time for a one-way trip? *Hint:* For instantaneous closure, all the fluid is brought to rest in time ($t = L / c$). When it has all been brought to rest it is at a pressure much higher than the pressure in the reservoir. What will happen then?

7.47. Problem 7.46 shows that the time required for a valve to close without causing water hammer is linearly proportional to the length of the pipe, whereas Example 7.14 shows that the expected pressure rise is independent of the length of the pipe. Why?

7.48.*Water is flowing at a depth of 2 ft and a velocity of 50 ft / s. It undergoes a hydraulic jump. What are the depth and velocity after the jump?

7.49. Water is flowing at a depth of z_1. What is the minimum velocity at which this water could undergo a hydraulic jump? Why?

7.50.*Water is flowing steadily in a river 20 ft deep. If an obstruction is placed in the river, increasing the depth at the obstruction, what is the lowest river velocity at which the obstruction will cause a hydraulic jump to occur? If the velocity is less than this, what will happen?

7.51. In Eq. 7.AZ, we use gauge pressure, leaving out the atmospheric pressure force terms. Is this permissible? The areas on which the fluids exert pressure are not the same. Explain why that all works out satisfactorily.

7.52. Show that the friction heating in a hydraulic jump is given by

$$\mathcal{F} = \frac{g(z_2 - z_1)^3}{4z_1z_2} \tag{7.65}$$

7.53. Water is flowing in a horizontal gutter with velocity 10 ft / s and depth 0.1 ft. We now place a large brick in the gutter, which stops the flow. The flow is stopped by a hydraulic jump, which then moves upstream from the brick. The brick is large enough that there is no flow over it. The fluid between the brick and the jump has zero velocity. The jump moves upstream, with a velocity V_j. What is the numerical value of V_j?

This is a messy problem analytically. It is fairly easy on a spreadsheet if one takes the viewpoint of someone riding on the jump (the lagrangian viewpoint) and solves by trial and error for the jump velocity that satisfies the hydraulic jump equation in the moving frame of reference.

7.54.*Water flows through a hydraulic jump, entering at a velocity of 50 ft / s and a depth of 10 ft. How much does the temperature of the water increase in this jump? For water, $C_V = du / dt = 1.0$ Btu / lbm · °F.

7.55. Figure 7.23 shows a hydraulic jump, in radial geometry, which is easily demonstrated in a kitchen sink. If the sink drain is open, then the flow is steady and the depths and velocities

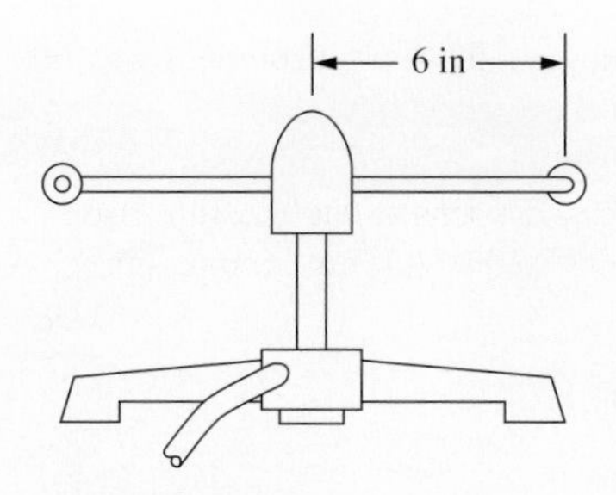

FIGURE 7.39
Rotating garden hose sprinkler.

in the sink do not change with time. If we now close the drain, without changing the flow from the faucet, what will happen? Describe the situation in terms of the mathematical description of hydraulic jumps.

7.56.*All other things being equal, is it easier for an airplane to take off on a short runway on a hot day or on a cold day? Why?

7.57. An ordinary garden hose sprinkler is sketched in Fig. 7.39. All the fluid enters at the axis ($r \approx 0$) and leaves through the nozzles ($r = 6$ in). If the total flow rate is 5 gal / min, and the rotor is held in place by someone's hand, how much torque will the rotor exert? The jets leaving the sprinkler each have a diameter of 0.25 in.

7.58. If the garden hose sprinkler in Prob. 7.57 is turned on and allowed to rotate freely, at what speed will it rotate? Assume that there is no air resistance or friction in the bearing of the sprinkler.

7.59. If we put the sprinkler in Prob. 7.57 under water and turn it on, will it rotate? Which way?

7.60. Why do helicopters have either one main propeller and a small tail propeller or two main propellers rotating in opposite directions? How is the corresponding problem solved in propeller-driven airplanes? In jet airplanes?

REFERENCES FOR CHAPTER 7

1. Kletz, T. *What Went Wrong? Case Histories of Process Plant Disasters.* 4th ed. Houston: Gulf Publishing, Houston, 1998, pp. 56–57.
2. Sutton, G. P. *Rocket Propulsion Elements.* New York: Wiley, 1959. This is, as the subtitle suggests, "an introduction to the engineering of rockets."
3. Ley, W. *Rockets, Missiles and Space Travel.* New York: Viking, 1957. This is an excellent history of the rocket business, which can be read and understood by any high-school graduate. The author was involved in much of the early German rocket work and has many amusing tales to tell about it.
4. Parmakian, J. *Waterhammer Analysis,* Englewood Cliffs, NJ: Prentice-Hall, 1955.
5. Adler, A. C. Vertical Takeoff. *Int. Sci. Tech., 48(12),* 1965, pp. 50–58.
6. Shepherd, D. B. *Principles of Turbomachinery.* New York: McMillan, 1956.
7. Lewis, B.; G. Von Elbe. *Combustion, Flames and Explosions of Gases.* New York: Academic Press, 1961, p. 206.

CHAPTER 8

ONE-DIMENSIONAL, HIGH-VELOCITY GAS FLOW

In this chapter we apply the ideas of the preceding chapters to the flow of gases at high velocities. No new principles are introduced in this chapter; those previously introduced will supply all our needs. Nevertheless, this is a separate chapter, because in high-velocity gas flow several phenomena occur that either are not present at all or are present in negligible amounts in the flow of liquids at their ordinary velocities and in the flow of gases at low velocities. For working purposes we define high velocity as a velocity in excess of about 200 ft / s (61 m / s). In Table 5.1 we showed that below this velocity the predictions of B.E. and those to be found in this chapter were approximately the same. Here we examine why that is no longer true for higher velocities.

The principal differences between high-velocity gas flows and the flows we have studied so far are the following:

1. In an expanding high-velocity gas flow the gas can convert significant amounts of internal energy into kinetic energy. This results in large decreases in gas temperature and in velocities higher than would be predicted by B.E., which has a constant-density assumption (see Table 5.1).
2. The changes in density that accompany high-velocity gas flows will complicate our mathematics. In typical situations we will have one more unknown and one more equation to deal with than in the corresponding constant-density flow. As the equations in this chapter become longer and more complex than those in the preceding chapters, remember that this is the reason for the added complexity.

3. The velocities in high-velocity gas flows are frequently equal to or greater than the local speed of sound. Information about small disturbances in fluid flow propagates through fluids at the speed of sound, so when the fluid is moving at the speed of sound or faster, certain kinds of information will be unable to travel upstream against the flow. This leads to some special phenomena in high-velocity gas flow, the most important of which are *choking* and shock waves. Choking is *very common* in chemical engineering practice.

8.1 THE SPEED OF SOUND

The speed of sound which will play an important part in what follows, is the speed at which a *small* pressure disturbance moves through a continuous medium. Sound, as our ears perceive it, is a series of small air-pressure disturbances oscillating in a sinusoidal fashion in the frequency range from 20 to 20,000 cycles per second. The magnitude of the pressure disturbances is generally less than 10^{-3} psia (7 Pa).

Suppose that we have a bar of steel 1 mi long. We tap the steel sharply on one end; our tap causes the near end of the bar to move 0.001 in. If the steel were *absolutely incompressible,* the far end of the bar would also move 0.001 in *instantly.* It does not; it moves about one-third of a second after we tap the near end. Nothing in this world is absolutely incompressible.

Consider a pipe full of some fluid, with pistons at each end. We tap one of the pistons. This causes the pressure adjacent to the piston to rise. This moves the next layer of fluid, whose pressure rises, and so on, causing a small pressure pulse to pass down the pipe. This is shown schematically in Fig. 8.1. In Chap. 7 (see Fig 7.20) we showed that it is easiest to analyze wave propagation of almost all kinds by taking the viewpoint of an observer moving at the same speed as the wave. If we take that view, we appear to be standing still and the walls of the pipe to be rushing past us. The fluid in the pipe also is rushing toward us and rushing away behind us. We measure the velocity, pressure, and density of the fluid ahead of us and behind us; the values of those ahead are slightly different from the values of those behind. This situation is shown in Fig. 8.2.

The pressure pulse is assumed to have the small volume shown in Fig. 8.2. The mass flowing into it is the same as the mass flowing out, so we can apply the steady-flow mass-balance equation,

$$\rho AV = (\rho + d\rho)A(V + dV) \tag{8.1}$$

Dividing by A, we expand the right-hand side and cancel the ρV term to get

$$0 = V\,d\rho + \rho\,dV + d\rho\,dV \tag{8.2}$$

The far right-hand term here is the product of two differentials and may be ignored (here, but not in Sec 8.5), so that

$$0 = V\,d\rho + \rho\,dV \tag{8.3}$$

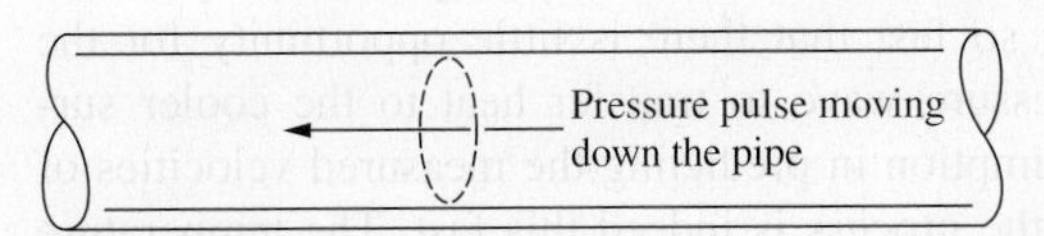

FIGURE 8.1
A pressure pulse passing down the fluid in a pipe.

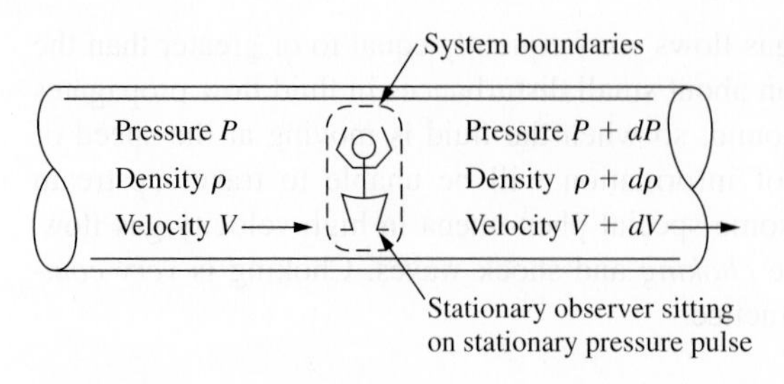

FIGURE 8.2
The pressure pulse in Fig. 8.1, from the viewpoint of an observer riding with it.

Applying the steady-flow momentum balance, Eq. 7.17, to the system shown in Fig. 8.2, we see that there is no accumulation, that there is a steady flow in and out, and that the only forces acting are the pressures on either side; so

$$0 = \dot{m}[V - (V + dV)] + A[P - (P + dP)] \quad (8.4)$$

Replacing $\dot{m}$ with ρAV and simplifying, we find

$$\rho \, dV = \frac{-dP}{V} \quad (8.5)$$

Substituting this value of $\rho \, dV$ in Eq. 8.3 and solving for V produces

$$V = \left(\frac{dP}{d\rho}\right)^{1/2} \quad (8.6)$$

This equation tell us how fast the fluid flows toward the observer riding on the small pressure pulse, which is the same as the velocity at which the pressure pulse moves past a stationary observer. In deriving this equation we never relied on our initial assumption that the material through which the pressure wave is passing is a fluid; the derivation is equally valid for a liquid, a solid, and a gas.

The equation was worked out for a single step change in pressure. Sound, as we experience it, is a sinusoidally varying pressure wave. However, it may be thought of as a combination of small step changes in pressure following each other. Thus, this equation, which we found for a single step change, is applicable also to any shape of pressure change, such as a sound wave.

Equation 8.6, as it stands, is ambiguous, because the derivative $dP/d\rho$ is ambiguous. The pressure P generally is a function not only of ρ but also of temperature. Newton [1] derived its equivalent and assumed that the derivative referred to a constant-temperature process, that is, that it was $(\partial P/\partial \rho)_T$. On this assumption he calculated the speed of sound in air and obtained an answer that was about 80 percent of the experimentally measured velocity. His assumption was plausible—we do not see the air being heated by sound passing through it—but it was wrong. Later workers decided that what really occurs is that a layer of gas is heated by being compressed and then cooled by expanding against the adjacent layer of gas. The net result is that the gas undergoes a practically reversible, adiabatic compression-expansion. For it to be adiabatic, it must occur so fast that there is little opportunity for the warmer gas in the middle of the pressure wave to transfer heat to the cooler surrounding gas. The success of this assumption in predicting the measured velocities of sound waves leads us to believe that the process is indeed this fast. The temperature rise in a sound wave is very small (see Prob. 8.2). In a reversible adiabatic process, the temperature is not constant, but the entropy is. Therefore, the final equation for the speed of sound is

$$V = \left(\frac{\partial P}{\partial \rho}\right)_S^{1/2} \tag{8.7-A}$$

The speed of sound is a special quantity, which we will want to keep separate from the local velocity of fluid flow. Therefore, we introduce another symbol, c, for it. This makes Eq. 8.7-A become

$$c = \left(\frac{\partial P}{\partial \rho}\right)_S^{1/2} \tag{8.7-B}$$

This equation is correct for solids, liquids, and gases. For solids and liquids, the easily measured $(\partial P / \partial \rho)_T$ is practically the same as the $(\partial P / \partial \rho)_S$ here; so we may write

$$c = \left(\frac{\partial P}{\partial \rho}\right)_S^{1/2} \approx \left(\frac{\partial P}{\partial \rho}\right)_T^{1/2} \quad \text{[solids and liquids]} \tag{8.8}$$

with satisfactory accuracy. Handbooks often tabulate, not $(\partial P / \partial \rho)_T$ of solids and liquids, but the bulk modulus K:

$$(\text{Bulk modulus}) = K = \rho\left(\frac{\partial P}{\partial \rho}\right)_T \tag{8.9}$$

or the isothermal compressibility, which is the reciprocal of the bulk modulus. In terms of the bulk modulus,

$$c \approx \left(\frac{\partial P}{\partial \rho}\right)_T^{1/2} = \left(\frac{K}{\rho}\right)^{1/2} \quad \text{[solids and liquids]} \tag{8.10}$$

Example 8.1. Calculate the speed of sound in steel and in water at 20°C.

For steel at that temperature, $K = 1.94 \cdot 10^{11}$ Pa and $\rho = 7800 \text{ kg / m}^3$. Thus we write

$$c = \left(\frac{1.94 \cdot 10^{11} \text{ Pa}}{7800 \text{ kg / m}^3} \cdot \frac{\text{N / m}^2}{\text{Pa}} \cdot \frac{\text{kg} \cdot \text{m}}{\text{N} \cdot \text{s}^2}\right)^{1/2} = 4.99 \frac{\text{km}}{\text{s}} = 16.4 \cdot 10^3 \frac{\text{ft}}{\text{s}} \tag{8.A}$$

For water at 20°C, $K = 3.14 \cdot 10^5 \text{ lbf / in}^2$ and $\rho = 62.3 \text{ lbm / ft}^3$, and

$$c = \left(\frac{3.14 \cdot 10^5 \text{ lbf / in}^2}{62.3 \text{ lbm / ft}^2} \cdot \frac{144 \text{ in}^2}{\text{ft}^2} \cdot 32.2 \frac{\text{lbm} \cdot \text{ft}}{\text{lbf} \cdot \text{s}^2}\right)^{1/2}$$

$$= 4.83 \cdot 10^3 \frac{\text{ft}}{\text{s}} = 1.6 \frac{\text{km}}{\text{s}} \tag{8.B}$$

■

For real gases, $(\partial P / \partial \rho)_S$ is a complicated function of pressure and temperature. However, the ideal gas law is a reasonable approximation of the behavior of most gases at low pressures and of low-boiling gases, such as air, up to reasonably high pressures. For a ideal gas it is shown in App. B.3 that

$$\left(\frac{\partial P}{\partial \rho}\right)_S = \frac{kP}{\rho} \quad \text{[ideal gases]} \tag{8.11}$$

TABLE 8.1
Values of the ratio of specific heats

Gas	*k*	Comment
Monatomic gases: He, Ar, Ne, Kr, Na, K	1.6667	Exactly
Diatomic gases: N_2, O_2, CO, NO, H_2, air	1.40	Not quite as exact, and decreases with increasing temperature
Triatomic gases: H_2O, CO_2, etc.	1.30 to 1.33	Less exact and more temperature dependent
More complex gases	Less than 1.3	Still more temperature dependent

Here, k is the ratio of specific heats, C_P / C_V, as defined in App. B.3 (this ratio is called γ in many texts). It is dimensionless; it values are given in Table 8.1.

In most engineering calculations we consider k a constant in a given problem even though for most gases it decreases slightly with increasing temperature. If we substitute for P / ρ in Eq. 8.11 from the ideal gas law and inset the result in Eq. 8.7-B, we find

$$c = \left(\frac{\partial P}{\partial \rho}\right)_S^{1/2} = \left(\frac{kP}{\rho}\right)^{1/2} = \left(\frac{kRT}{M}\right)^{1/2} \quad \text{[ideal gas]} \tag{8.12}$$

where M is the molecular weight and R is the universal gas constant.

This equation shows why Newton's calculated value of the speed of sound was only 80 percent of the observed value. If we substitute $(\partial P / \partial \rho)_T$ for $(\partial P / \partial \rho)_S$ in this equation, we find the same result, except that k is replaced with 1. Since k for air is 1.4, this incorrect substitution lowers the calculated value to 80 percent of the value shown in this equation.

Example 8.2. What is the speed of sound in air at 20°C = 68°F?

We will need $R^{1/2}$. In speed-of-sound calculations it is convenient to convert the most commonly used form of R as shown below:

$$R^{1/2} = \left(10.73 \frac{\text{lbf}}{\text{in}^2} \frac{\text{ft}^3}{\text{lbmol} \cdot \text{°R}} \cdot \frac{144 \text{ in}^2}{\text{ft}^2} \cdot \frac{32.2 \text{ lbm} \cdot \text{ft}}{\text{lbf} \cdot \text{s}^2}\right)^{1/2}$$
$$= 223 \frac{\text{ft}}{\text{s}} \cdot \left(\frac{\text{lbm}}{\text{lbmol} \cdot \text{°R}}\right)^{1/2} = 91.2 \frac{\text{m}}{\text{s}} \cdot \left(\frac{\text{g}}{\text{mol} \cdot \text{K}}\right)^{1/2} \tag{8.13}$$

For this example we have

$$c = \left(\frac{kRT}{M}\right)^{1/2} = R^{1/2}\left(\frac{kT}{M}\right)^{1/2}$$
$$= 223 \frac{\text{ft}}{\text{s}} \cdot \left(\frac{\text{lbm}}{\text{lbmol} \cdot \text{°R}}\right)^{1/2} \cdot \left(\frac{1.4 \cdot 528\text{°R}}{29 \text{ lbm / lbmol}}\right)^{1/2}$$
$$= 1126 \frac{\text{ft}}{\text{s}} = 344 \frac{\text{m}}{\text{s}} \tag{8.C}$$

■

The speed of sound of a ideal gas, as just shown, is a function of the temperature and not of the velocity. Keep in mind that the speed of sound is a property of the matter, not a property of the flow. If the temperature changes from point to point as the fluid flows, then the speed of sound will change from point to point, but at any point it is the same in a flowing gas as it would be in the same gas standing still at the same temperature. This argument applies equally well for solids and liquids. Observe also these magnitudes; $c \approx 3$ mi / s for steel, 1 mi / s for water, and $\frac{1}{5}$ mi / s for air. The "stiffer" the material, the faster sound moves in it; steel is stiffer than water, which is stiffer than air. An absolutely incompressible material would be infinitely stiff and thus have an infinite speed of sound.

Our discussion of the speed of sound has the built-in assumption that a sound wave is a *small* pressure pulse. The difference in velocity and pressure of the medium before and after the sound wave is very small. Another type of pressure pulse, in which these differences are large, called a shock wave, will be discussed in Sec. 8.5.

8.2 STEADY, FRICTIONLESS, ADIABATIC, ONE-DIMENSIONAL FLOW OF AN IDEAL GAS

Many of the most interesting features of high-velocity gas flow can be seen in the simplest of all cases, the steady, frictionless, adiabatic, one-dimensional flow of an ideal gas. We will study this type of flow in detail; other types will be treated more briefly, because they have so much in common with this one. The ideal gas assumption means that in the equations in this chapter P always stands for absolute pressure (psia or equivalent) and never gauge pressure (psig or equivalent). It also means that in those equations T always stands for absolute temperature (K or °R) and never for °F or °C. The problems and examples often start with °F or °C or psig, to conform with current practice in industry, but the values are always converted to psia or K or °R before the calculations begin.

The flow is assumed to be in some kind of a duct or pipe or closed channel of varying cross-sectional area A. As long as this cross-sectional area changes slowly with distance down the duct, the velocities in the direction perpendicular to the main flow will be small enough to neglect, and we can treat the flow as one-dimensional. The gas is assumed to be ideal and to have a constant heat capacity C_P.

The open-system energy balance (Chap. 4) between any two points R and 1 in such a duct, for steady flow without heat transfer or turbines or compressors, is

$$\left(h + gz + \frac{V^2}{2}\right)_R = \left(h + gz + \frac{V^2}{2}\right)_1 \tag{8.14}$$

It can readily be shown (Prob. 8.8) that the potential-energy changes Δgz are negligible for most high-velocity gas flows, so we will drop the gz terms from this equation. Next we assume that state R is some upstream reservoir, where the cross-sectional area perpendicular to the flow is very large; therefore, V_R is negligible. This condition is referred to in various texts as the *reservoir, stagnation,* or *total* condition. We will call it the reservoir condition and use the subscript R.

Substituting $V_R = 0$ in Eq. 8.14, we find

$$V_1^2 = 2(h_R - h_1) = 2C_P(T_R - T_1) = \frac{2Rk}{M(k-1)}(T_R - T_1) \tag{8.15}$$

Here we have substituted $C_P\,\Delta T$ for Δh and then substituted $Rk/M(k-1)$ for C_P. Both of these substitutions are justified in App. B.3.

We now divide both sides of this equation by RkT_1/M, finding

$$\frac{MV_1^2}{RkT_1} = \frac{2}{k-1}\left(\frac{T_R}{T_1} - 1\right) \tag{8.16}$$

But, as shown previously, RkT_1/M is the square of the speed of sound at state 1, or c_1^2, so the left side is $(V_1/c_1)^2$. The ratio V/c is called the Mach number, $\mathcal{M}$ (Ernst Mach, 1838–1916). We will see that this ratio plays a crucial role in the study of high-velocity gas flows (and is widely reported in the press describing the speed of supersonic aircraft). It is the ratio of the *local* flow velocity to the *local* speed of sound. For subsonic flows $\mathcal{M}$ is less than 1; for sonic flows it equals 1; for supersonic flows it is greater than 1. Making this definition, we can rearrange Eq. 8.16 to

$$\frac{T_R}{T_1} = \mathcal{M}_1^2\frac{k-1}{2} + 1 \tag{8.17}$$

Example 8.3. Air flows steadily and adiabatically from a reservoir in which its velocity is negligible and its temperature is 68°F = 20°C. What is the temperature of the gas at the point where the Mach number is 2.0?

Air is a diatomic gas, so as shown in Table 8.1, $k = 1.4$. From Eq. 8.17 we have

$$\frac{T_R}{T_1} = 2.0^2\left(\frac{1.4-1}{2}\right) + 1 = 1.80 \tag{8.D}$$

$$T_R = 68°\text{F} = 528°\text{R} = 293.15\ \text{K} \tag{8.E}$$

$$T_1 = \frac{T_R}{1.80} = \frac{528°\text{R}}{1.8} = 293°\text{R} = -167°\text{F} = 163\ \text{K} = -110°\text{C} \tag{8.F}$$

■

The startlingly low temperature indicated above shows clearly that the expanding gas is converting its internal energy into kinetic energy; the decrease in internal energy is indicated by the large decrease in temperature. One may also picture that the gas, as it expands, does work on the adjacent masses of gas, thus lowering its internal energy and temperature. The ratio T_R/T_1 depends only on k and $\mathcal{M}_1$, not on the identity of the gas or on the reservoir temperature.

Example 8.4. What is the velocity of the air in Example 8.3 at the point where the Mach number is 2.0?

We need to know the speed of sound at state 1. From Eq. 8.12 we have

$$c = 223\,\frac{\text{ft}}{\text{s}}\cdot\left(\frac{\text{lbm}}{\text{lbmol}\cdot°\text{R}}\right)^{1/2}\cdot\left(\frac{1.4\cdot 293°\text{R}}{29\ \text{lbm}/\text{lbmol}}\right)^{1/2} = 839\,\frac{\text{ft}}{\text{s}} = 256\,\frac{\text{m}}{\text{s}} \tag{8.G}$$

$$V_1 = c_1\mathcal{M}_1 = 839\,\frac{\text{ft}}{\text{s}}\cdot 2.0 = 1678\,\frac{\text{ft}}{\text{s}} = 511\,\frac{\text{m}}{\text{s}} \tag{8.H}$$

■

Equation 8.17 applies to any adiabatic, steady flow of an ideal gas, with or without friction (we will see its application to adiabatic flow with friction in Sec. 8.4).

Now we add the assumption that the flow is frictionless. Frictionless, adiabatic flow of any nonreacting gas is isentropic; so we may use the relations between temperature, pressure, and density for an isentropic change of an ideal gas, as developed in App. B. Using these, we find

$$\frac{P_R}{P_1} = \left(\frac{T_R}{T_1}\right)^{k/(k-1)} \quad \text{[isentropic, ideal gas]} \tag{8.18}$$

$$\frac{\rho_R}{\rho_1} = \left(\frac{T_R}{T_1}\right)^{1/(k-1)} \quad \text{[isentropic, ideal gas]} \tag{8.19}$$

Substituting the value of T_R / T_1 from Eq. 8.17, we find

$$\frac{P_R}{P_1} = \left(\mathcal{M}_1^2 \frac{k-1}{2} + 1\right)^{k/(k-1)} \quad \text{[isentropic, ideal gas]} \tag{8.20}$$

$$\frac{\rho_R}{\rho_1} = \left(\mathcal{M}_1^2 \frac{k-1}{2} + 1\right)^{1/(k-1)} \quad \text{[isentropic, ideal gas]} \tag{8.21}$$

Example 8.5. For the air in Example 8.3, the reservoir pressure is 2 bar, and the reservoir density is 2.39 kg / m³. What are the pressure and density at the point in the flow where $\mathcal{M} = 2.0$?

The term in parentheses in Eqs. 8.20 and 8.21 is precisely T_R / T_1, found in Example 8.3 to be 1.80; therefore,

$$\frac{P_R}{P_1} = 1.80^{1.4/(1.4-1)} = 1.80^{3.5} = 7.82 \tag{8.I}$$

$$P_1 = \frac{P_R}{7.82} = \frac{2 \text{ bar}}{7.82} = 0.256 \text{ bar} = 3.71 \text{ psia} \tag{8.J}$$

$$\frac{\rho_R}{\rho_1} = 1.80^{1/(1.4-1)} = 1.80^{2.5} = 4.35 \tag{8.K}$$

$$\rho_1 = \frac{\rho_R}{4.35} = \frac{2.39 \text{ kg / m}^3}{4.35} = 0.549 \frac{\text{kg}}{\text{m}^3} = 0.034 \frac{\text{lbm}}{\text{ft}^3} \tag{8.L}$$

■

At this point we can compare the calculated behavior of this flow with what one would predict by simple B.E. Continuing Example 5.1 for the initial density and upstream and downstream pressures in this example (see Prob. 8.29), we find a calculated velocity of 1253 ft / s = 382 m / s, or 75 percent of the value calculated here. The main reason for the difference is that this expanding flow is converting internal energy to kinetic energy, which is forbidden by the constant-density assumption in B.E. (Below about 200 ft / s, the calculated values are practically the same; see Example 5.1 and Prob. 8.29.)

Comparing these results with the temperature ratio T_R / T_1, we see that the pressure and the density change much more rapidly in frictionless, adiabatic flow than does the temperature.

Equations 8.11, 8.20, and 8.21 allow us to calculate the change of temperature, pressure, and density with a change in Mach number for isentropic, steady flow of an ideal gas. From the Mach number and the temperature we can calculate the velocity.

The other item of interest is the cross-sectional area perpendicular to flow. By applying the mass-balance equation for steady flow between states R and 1 and solving for A_R/A_1, we find

$$\frac{A_R}{A_1} = \frac{\rho_1 V_1}{\rho_R V_R} \tag{8.22}$$

We have defined the reservoir conditions such that A_R is infinite and V_R is zero. If we insert these values in Eq. 8.22, we see that both sides are large without bound (i.e., infinite). The reservoir condition, which is the most convenient reference condition for temperature, pressure, and density, is therefore a very poor reference condition for the cross-sectional area; we will choose a better one. In any such flow there is or could be a state at which the Mach number is exactly 1. Even if such a state does not exist for the flow in question, pretending that it exists will help us solve the problem. Let us refer to this state as the *critical state* and denote it by an asterisk. The mass-balance equation between some arbitrary state and the critical state is

$$\frac{A_1}{A^*} = \frac{\rho^* V^*}{\rho_1 V_1} \tag{8.23}$$

Substituting the values of the density ratio in terms of the temperature for isentropic flow and the velocity in terms of the Mach number and the speed of sound, and then eliminating the temperature ratio, we find (App. B.4)

$$\frac{A_1}{A^*} = \frac{1}{\mathcal{M}_1}\left[\frac{\mathcal{M}_1^2(k-1)/2+1}{(k-1)/2+1}\right]^{(k+1)/2(k-1)} \tag{8.24}$$

which is plotted in Fig. 8.3.

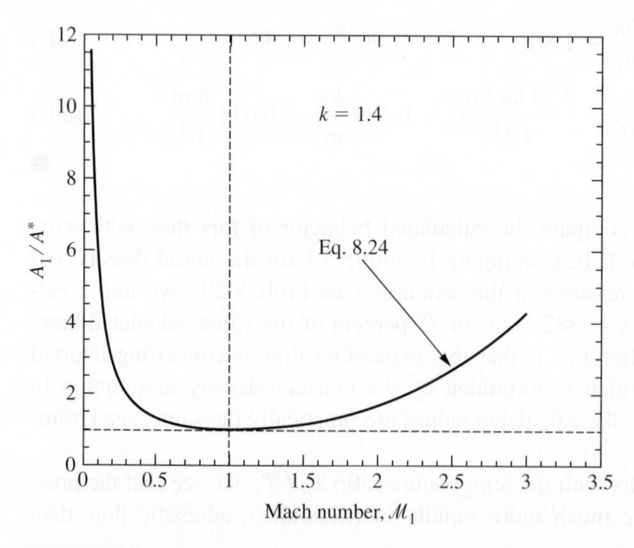

FIGURE 8.3
Equation 8.24 shows that to get a supersonic flow we must use a converging-diverging nozzle.

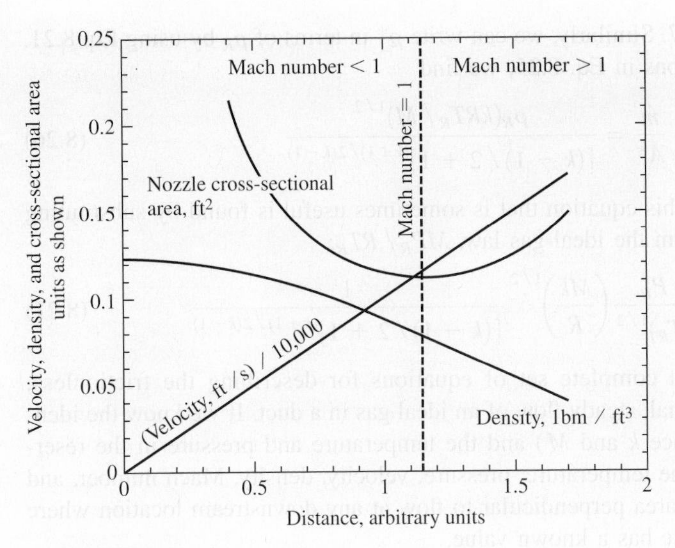

FIGURE 8.4
Density, velocity, and area variations through a nozzle. The numerical values correspond to Example 8.6. Here the cross-sectional area–distance relationship is chosen to make the velocity increase linearly. That is not necessarily how one designs a nozzle, but doing so makes a simple plot.

Figure 8.3 leads to the commonplace conclusion that, for Mach numbers less than 1, to get the fluid to go faster we must reduce the cross-sectional area perpendicular to flow. This is how garden hose nozzles work; the flow area decreases, so the velocity increases for a constant mass flow rate. However, when the Mach number is greater than 1, we are led to the startling conclusion that, to get the fluid to go faster, we must increase the area! If this is intuitively obvious to the reader, he has better intuition than the author does. Let us simultaneously plot ρ, A, and V against distance through such a nozzle; see Fig. 8.4.

Here we have let V be some small value at the inlet and increase linearly with distance. Because this is an expanding flow, the density decreases with distance. In the subsonic range V goes up faster than ρ goes down, so A must decrease to keep ρAV constant. However, as the fluid goes faster and faster, ρ drops more and more rapidly, until at $\mathscr{M} = 1$ it is decreasing just as rapidly as V is increasing. At supersonic flow, $\mathscr{M} > 1$, the density falls more rapidly than the velocity increases, and therefore A must increase. This is shown in the figure.

From the preceding information we can also calculate the mass flow rate through our duct. Since the flow is steady, the mass flow rate must be the same at all points:

$$\dot{m} = \rho AV \tag{8.25}$$

Here the product ρAV is the same for every point in the duct, including the point where the flow is sonic (the critical point). Writing it for that point, we divide both sides by A^* and note that V^* equals c^*, which we can write in terms of k, R, M, by using Eqs. 8.12 and 8.17. Similarly, we can write ρ^* in terms of ρ_R by using Eq. 8.21. Making these substitutions in Eq. 8.25, we find

$$\frac{\dot{m}}{A^*} = \frac{\rho_R (kRT_R / M)^{1/2}}{[(k-1)/2 + 1]^{(k+1)/2(k-1)}} \tag{8.26}$$

An alternative form of this equation that is sometimes useful is found by substituting for ρ_R its equivalent from the ideal-gas law, MP_R / RT_R:

$$\frac{\dot{m}}{A^*} = \frac{P_R}{(T_R)^{1/2}} \left(\frac{Mk}{R}\right)^{1/2} \frac{1}{[(k-1)/2 + 1]^{(k+1)/2(k-1)}} \tag{8.27}$$

We finally have a complete set of equations for describing the frictionless, adiabatic, one-dimensional, steady flow of an ideal gas in a duct. If we know the identity of the gas (and hence k and M) and the temperature and pressure in the reservoir, we can calculate the temperature, pressure, velocity, density, Mach number, and mass flow rate per unit area perpendicular to flow at any downstream location where any one of these variable has a known value.

Example 8.6. Air at 30 psia and 200°F flows from a reservoir into a duct. The flow is steady, adiabatic, and frictionless, with mass flow rate = 10 lbm / s. What are the cross-sectional area, temperature, pressure, and Mach number at the point in the duct where the velocity is 1400 ft / s?

We do not know the Mach number at the point in question; to use Eqs. 8.17, 8.20, and 8.21, we must first find it. From Eq. 8.15 rearranged we get

$$\begin{aligned} T_1 &= T_R - V_1^2 \left(\frac{k-1}{k}\right) \frac{M}{2R} \\ &= 660°\text{R} - \frac{(1400 \text{ ft/s})^2 \cdot (1.4 - 1) \cdot 29 \text{ lbm/lbmol}}{1.4 \cdot 2 \cdot 4.98 \cdot 10^4 (\text{ft}^2/\text{s}^2) \cdot (\text{lbm}/(\text{lbmol} \cdot °\text{R}))} \\ &= 660°\text{R} - 163°\text{R} = 497°\text{R} = 276 \text{ K} \end{aligned} \tag{8.M}$$

Thus,

$$c = 223 \frac{\text{ft}}{\text{s}} \cdot \left(\frac{\text{lbm}}{\text{lbmol} \cdot \text{R}}\right)^{1/2} \left(\frac{1.4 \cdot 497°\text{R}}{29 \text{ lbm/lbmol}}\right)^{1/2} = 1092 \frac{\text{ft}}{\text{s}} = 333 \frac{\text{m}}{\text{s}} \tag{8.N}$$

and

$$\mathcal{M}_1 = \frac{1400 \text{ ft/s}}{1092 \text{ ft/s}} = 1.282 \tag{8.O}$$

Now we can use Eqs. 8.20 and 8.21:

$$\frac{P_R}{P_1} = \left(\frac{T_R}{T_1}\right)^{k/(k-1)} = \left(\frac{660°\text{R}}{497°\text{R}}\right)^{3.5} = 2.70 \tag{8.P}$$

and

$$P_1 = \frac{P_R}{2.70} = \frac{30 \text{ psia}}{2.70} = 11.1 \text{ psia} = 76.5 \text{ kPa} \tag{8.Q}$$

For a flow rate of 10 lbm / s we can calculate the area at the critical condition from Eq. 8.27:

$$\frac{\dot{m}}{A^*} = \frac{30 \text{ lbf / in}^2 \cdot [(29 \text{ lbm / lbmol}) \cdot 1.4]^{1/2} \cdot 32.2 \text{ lbm} \cdot \text{ft / lbf} \cdot \text{s}^2}{223 \text{ ft / s} \cdot (\text{lbm / lbmol} \cdot {}^\circ\text{R})^{1/2} \cdot (660^\circ\text{R})^{1/2} \cdot [0.4/2 + 1]^{2.4/0.8}}$$

$$= 0.62 \frac{\text{lbm}}{\text{s} \cdot \text{in}^2} = 437 \frac{\text{kg}}{\text{s} \cdot \text{m}^2} \tag{8.R}$$

Therefore,

$$A^* = \frac{\dot{m}}{0.62 \text{ lbm / in}^2 \cdot \text{s}} = \frac{10 \text{ lbm / s}}{0.62 \text{ lbm / in}^2 \cdot \text{s}} = 16.1 \text{ in}^2 = 0.030 \text{ m}^2 \tag{8.S}$$

Now that we know A^*, we can calculate the area at which $V = 1400$ ft / s from Eq. 8.24:

$$\frac{A}{A^*} = \frac{1}{1.282} \cdot \left[\frac{(1.282^2 \cdot 0.4/2) + 1}{0.4/2 + 1}\right]^{2.4/2(0.4)}$$

$$= \frac{1}{1.282} \cdot \left(\frac{1.329}{1.20}\right)^3 = 1.059 \tag{8.T}$$

Therefore, $A = 1.059 \cdot A^* = 1.059 \cdot 16.1 \text{ in}^2 = 17.0 \text{ in}^2 = 0.011 \text{ m}^2$ ■

Certainly by now the reader has observed that this calculation involves a lot of algebra, much of it concerning computation of the quantity $[\mathcal{M}_1^2(k-1)/2 + 1]$ to various powers. For constant k this function obviously can be tabulated to the powers of interest for various values of $\mathcal{M}$, saving us much of the algebra. This is done for $k = 1.4$ in App. A.4. Before we all had computers, high-velocity gas flow problems were most often solved using those tables; currently, it is more common to solve standard problems in computer programs or on spreadsheets.

Appendix A.4 is based on Eqs. 8.17, 8.20, 8.21, and 8.24. In addition there is a column labeled V/c^*, calculated by Eq. B.6-12. This ratio is useful in problems in which we know the velocity and the reservoir conditions and want to find $\mathcal{M}$. We could solve for the local temperature and local speed of sound, as we did in Example 8.6, but that is tedious. We would like some velocity ratio like V/V_R to appear in the table, but that is an impossible choice, because V_R is zero. The logical choice is $V/V^* = V/c^*$; its use is illustrated below.

Example 8.7. Rework Example 8.6, using App. A.4.

First, we calculate c^*. To do this we need T^*, which we find by looking in the table for T/T_R at $\mathcal{M} = 1.0$, finding $T^*/T_R = 0.83333$; then

$$T^* = 660^\circ\text{R} \cdot 0.83333 = 550^\circ\text{R} \tag{8.U}$$

$$c^* = \left(\frac{kRT^*}{M}\right)^{1/2} = 223 \frac{\text{ft}}{\text{s}} \left(\frac{\text{lbm}}{\text{lbmol} \cdot {}^\circ\text{R}}\right)^{1/2} \cdot \left(\frac{1.4 \cdot 550^\circ\text{R}}{29 \text{ lbm / lbmol}}\right)^{1/2}$$

$$= 1149 \frac{\text{ft}}{\text{s}} = 350 \frac{\text{m}}{\text{s}} \tag{8.V}$$

Now we compute

$$\frac{V}{c^*} = \frac{1400 \text{ ft / s}}{1149 \text{ ft / s}} = 1.218 \tag{8.W}$$

We go to the V / c^* column in the table and read down until we find $V / c^* = 1.218$. (In this and all subsequent examples we could interpolate in App. A.4. Instead the values shown are computed from the same equations and program used to make up App. A.4 and are accurate to four decimal places). This corresponds to $\mathcal{M} \approx 1.2815$. In the same interpolated row on the table we read

$$\frac{T}{T_R} = 0.7528 \qquad \frac{P}{P_R} = 0.3701 \qquad \frac{A}{A^*} = 1.0587 \tag{8.X}$$

Thus,

$$T = 0.7532\, T_R = 0.7532 \cdot 660°\text{R} = 497°\text{R} \tag{8.Y}$$

$$P = 0.3708\, P_R = 0.3708 \cdot 30 \text{ psia} = 11.1 \text{ psia etc.} \tag{8.Z}$$

■

This is clearly a gigantic saving in effort over solving this kind of problem long-hand as we did in Example 8.6. The homework problems should convince the student of the utility of App. A.4. Appendix A.4 makes clear why the calculations have all been done in terms of the R and * conditions. How else could one make up such a table?

It is also instructive to consider the following variant of Example 8.6.

Example 8.8. Rework Example 8.6 for the point where the velocity is 4000 ft / s.

As in Example 8.7, $c^* = 1149$ ft / s; so $V / c^* = 4000 / 1149 = 3.48$. We now look in the V / c^* column of App. A.4 for 3.48. In that table the highest value of V / c^* is 1.6330. By looking in the original document from which App. A.4 was extracted, we find that for $\mathcal{M}$ up to 100 the value of V / c^* approaches 2.4489 as a limit; so 4000 ft / s must not be possible in this flow! If we return to our method of solution in Example 8.6, we see that substituting $V = 4000$ ft / s leads to

$$T_1 = 660°\text{R} - 1340°\text{R} = -680°\text{R} \qquad ??? \tag{8.AA}$$

■

From this we see that the energy balance sets a maximum possible velocity for an expanding flow. Once all the internal energy has been turned into kinetic energy, the gas can no longer accelerate. Obviously, there are other limits to the velocity. Air will turn into a liquid long before it reaches 0°R, and the assumption of constant C_P becomes false at very low temperatures. Furthermore, unless the air were very dry, it would be expected to form a fog at these extremely low temperatures. This explains why high-speed wind tunnels have either big air dryers or big air preheaters.

How is this maximum velocity to be reconciled with Eq. 8.17, which indicates that one can calculate a value of T_R / T_1 for any $\mathcal{M}$, no matter how large? The answer is that, as $\mathcal{M}$ goes higher and higher, it does so by driving the temperature lower and

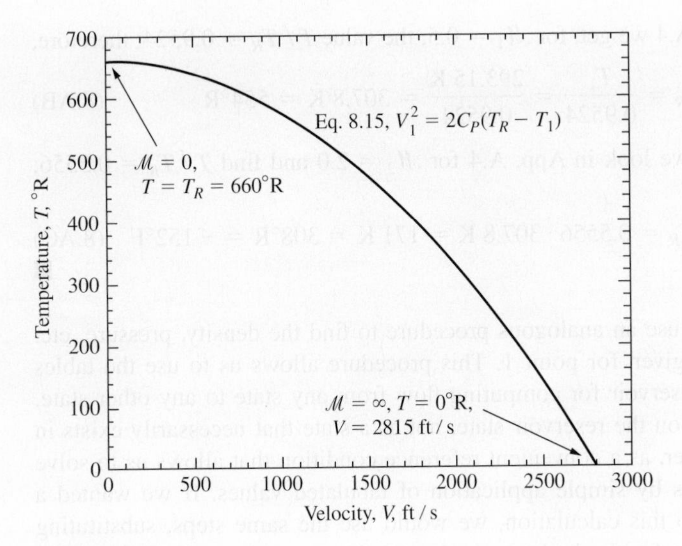

FIGURE 8.5
Variation of temperature with velocity for steady, adiabatic flow of a perfect gas with $k = 1.4$ and $T_R = 660°\text{R}$.

lower. Since we have $c = (kRT/M)^{1/2}$, we can raise the $\mathscr{M}$ by raising the velocity or by lowering the temperature or both. At high $\mathscr{M}$ the calculated temperature is very low.

This may be visualized by plotting Eq. 8.15 as T versus V; see Fig. 8.5. The figure is one-half of a parabola with a maximum value V_{max} equal to $(2C_P T_R)^{1/2}$. Again, remember that the assumptions made in deriving the equation for steady, adiabatic, frictionless flow of an ideal gas in a duct become inaccurate as T approaches 0 K.

Our discussion has entirely concerned flow from a reservoir. If the initial conditions are stated, not for a reservoir where $V = 0$, but for some other point, we can develop the equivalents of all the equations shown so far, with the initial velocity not taken equal to zero. That is even messier algebraically than what we have done so far! A much more convenient approach is to utilize App. A.4 to solve for the reservoir condition that corresponds to the given starting condition.

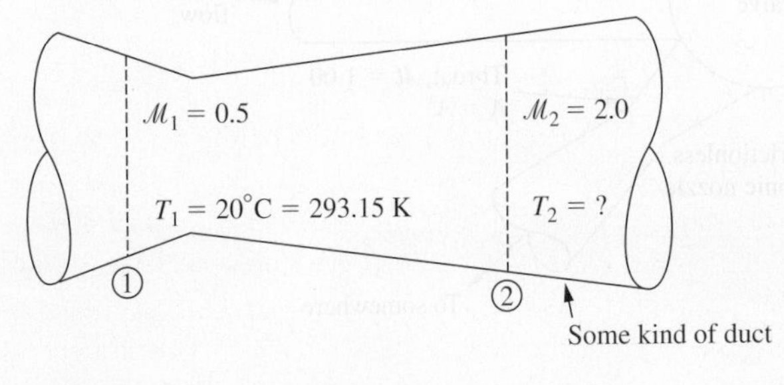

FIGURE 8.6
A supersonic nozzle, used in Example 8.9.

Example 8.9. Air is flowing in a duct in steady, frictionless, adiabatic flow. At the place in the duct where the Mach number is 0.5 the temperature is 20°C. What is the temperature at the point in the duct where the Mach number is 2.0? See Fig. 8.6.

From App. A.4 we get, for $\mathcal{M}_1 = 0.5$, the value $T/T_R = 0.9524$; therefore,

$$T_R = \frac{T_1}{0.9524} = \frac{293.15 \text{ K}}{0.9524} = 307.8 \text{ K} = 554°\text{R} \quad (8.\text{AB})$$

Now, to find T_2, we look in App. A.4 for $\mathcal{M}_2 = 2.0$ and find $T/T_R = 0.5556$; therefore,

$$T_2 = 0.5556 \cdot T_R = 0.5556 \cdot 307.8 \text{ K} = 171 \text{ K} = 308°\text{R} = -152°\text{F} \quad (8.\text{AC})$$

■

Clearly, we could use an analogous procedure to find the density, pressure, etc. at point 2 if they were given for point 1. This procedure allows us to use the tables based on flow from a reservoir for computing flow from any state to any other state. Thus, we ought to look on the reservoir state, not as a state that necessarily exists in a real reservoir but, rather, as a convenient reference condition that allows us to solve many common problems by simple application of tabulated values. If we wanted a computer program to do this calculation, we would use the same steps, substituting the appropriate equations for the look-ups in App. A.4.

The existence of two reference states, the reservoir and the critical, is the source of some confusion to students. The states are most easily visualized by means of Fig. 8.7. We assume that at any point in the flow we could introduce a frictionless, three-way valve. The flow is actually going through the valve in the direction of the outgoing flow, so the introduction of this imaginary, frictionless valve does not change the flow. Now, if we switched the valve so that the flow was diverted

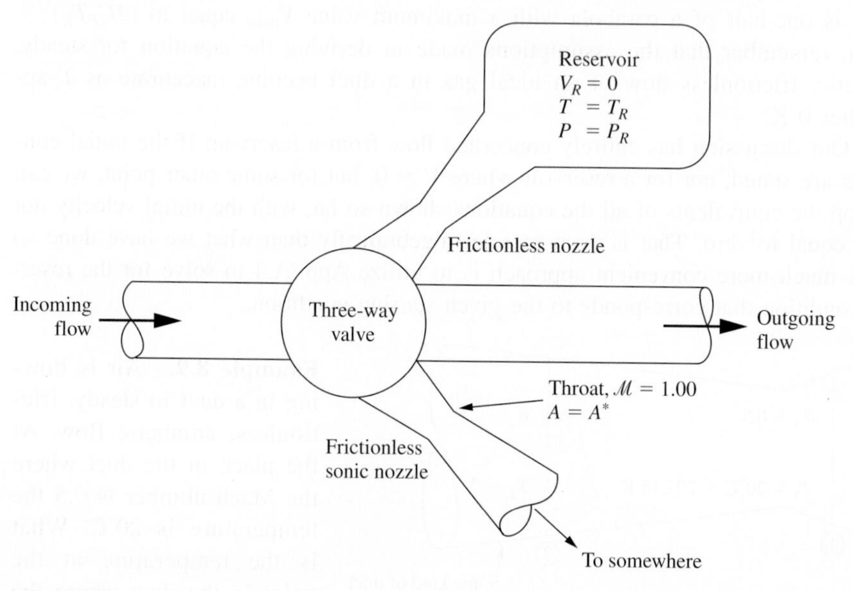

FIGURE 8.7
Visualization of the reservoir and critical conditions. These conditions need not exist in an actual flow, but they still can be conceptually useful in analyzing the flow.

frictionlessly to a reservoir, then the temperature, pressure, density, etc., that we measured in that reservoir would constitute the reservoir conditions corresponding to the incoming flow. Similarly, if we diverted the flow through the sonic nozzle, then the conditions we would measure at the throat of the nozzle would be the critical conditions corresponding to the incoming flow. Thus, the imaginary valve allows us to visualize what we would do experimentally to find the reservoir or critical conditions corresponding to the incoming flow. However, since the two nozzles that we should use in the figure are assumed frictionless, we need not actually perform the experiments; we can calculate what would happen in them by using the equations given in this section.

For a reversible, adiabatic flow, neither the reservoir conditions nor the critical conditions change from point to point in the system. Thus, a measurement or calculation of them for any point in the flow gives the values for the entire flow. However, in flow with friction (Sec. 8.4) or flow with heating or cooling (not treated in this text) or for normal shock waves (Sec. 8.5), the reservoir and critical conditions change from point to point in the flow. For such flows we must be conceptually prepared to use the apparatus shown in Fig. 8.7 at many points in the flow, and we must not assume that the values we find (or calculate) at one point would be the same as those at another.

8.3 NOZZLE CHOKING

Suppose that we connect a high-pressure reservoir full of air to a low-pressure reservoir full of air by means of a converging nozzle. We assume that by suitable pumps, etc., we can maintain the pressure in each reservoir at any value we select and that we have some method of measuring the mass flow rate of gas passing through the nozzle; see Fig. 8.8.

At first, the pressure in both reservoirs is the same high pressure, P_1. Since there is no pressure gradient across the nozzle, there is no flow. Then, holding the pressure in the high-pressure reservoir constant at P_1, we begin to lower the pressure P_2 in the low-pressure reservoir. For each value of P_2 we measure the mass flow rate $\dot{m}$, and we plot $\dot{m}$ versus P_2 / P_1. The results we would find are shown in Fig. 8.9 (here the results are shown as $\dot{m} / A_{\text{nozzle}}$ which is independent of the size of the apparatus in Fig. 8.7). That figure also shows the values one would calculate from B.E. assuming a constant fluid density.

We observe that the mass flow rate increases steadily as we lower P_2, until P_2 / P_1 equals 0.5283, and then that further lowering of P_2 does not increase the mass

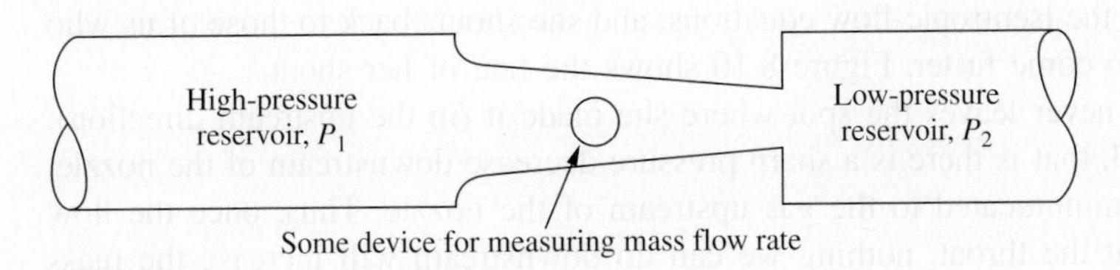

FIGURE 8.8
Device for measuring the mass flow rate in a converging nozzle.

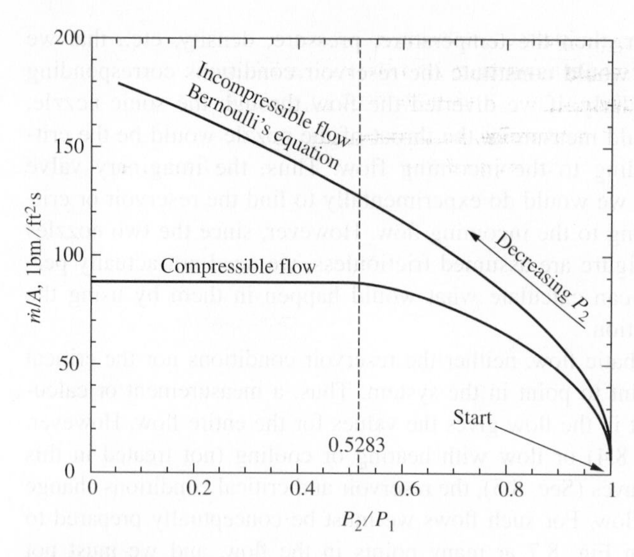

FIGURE 8.9
Mass flow rate per unit area (also sometimes called *mass velocity*) for an isentropic flow with $P_R = 30$ psia and $T_R = 660°\text{R}$. The Bernoulli's equation (B.E.) curve is computed from $V = \sqrt{2(-\Delta P)/\rho_R}$, and $\dot{m}/A = V\rho_R$. As expected, these curves are practically identical for low values of ΔP, but not for high values. The compressible flow solution shows nozzle choking, the B.E. solution does not.

flow rate. If we refer to App. A.4, we see that, if the assumptions of isentropic, one-dimensional, steady flow apply, then this pressure ratio corresponds exactly to sonic velocity ($\mathcal{M} = 1$) at the throat of the nozzle. Lowering the downstream pressure more does not increase the mass flow rate, because the flow at the narrowest point in the flow is sonic. We also observe that the value of $\dot{m}/A$ at the throat is exactly the value predicted by Eq. 8.27 if we take state 1 as the reservoir state. The flow rates computed by B.E. are practically the same at the start (P_2/P_1 close to 1.0, ΔP small). However, as ΔP becomes large the B.E. solution shows the mass flow rate steadily increasing (as it would for a liquid), quite different from what we observe for a gas.

Why does lowering the downstream pressure below 0.5283 times the upstream pressure not cause the gas to flow faster in the nozzle? Suppose we attach an observer to a balloon and let her ride along with the fluid through the nozzle. When she gets to the nozzle throat, she observes that the downstream pressure is lower than she had anticipated from the isentropic-flow equations, and she shouts back to those of us who are behind her to come faster. Figure 8.10 shows the fate of her shout.

The shout never leaves the spot where she made it (in the upstream direction). The sound signal, that is there is a sharp pressure decrease downstream of the nozzle, can never be communicated to the gas upstream of the nozzle. Thus, once the flow becomes sonic at the throat, nothing we can do downstream will increase the mass flow rate at that point. This situation, in which the flow at the throat is sonic, is called *choking*. One speaks of the nozzle as being choked because no more mass can get

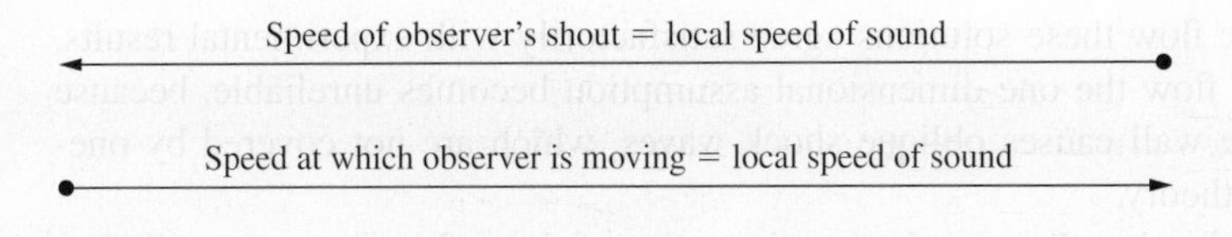

FIGURE 8.10
The shout from an observer riding on a flow at sonic velocity can never propagate upstream in the flow. (The flow is from left to right.)

through it without a change in upstream conditions. The adjustment to the lower pressure takes place downstream of the throat by a *rarefaction,* which is neither an isentropic nor a one-dimensional process and is not covered by the one-dimensional equations developed in this chapter.

This type of choking is very common in valves, orifices, and vacuum systems. For air, any time that the pressure ratio across a valve, orifice, or leak into a vacuum system is 0.5283 or less, choked flow is occurring. When one lets air out of an auto or bicycle tire the flow is normally choked. Most control valves for gas flow operate in the choked condition. Most safety relief valves on gaseous systems, when they open, operate in the choked condition. Autos at wide-open throttle on a level road have choked flow in their air-inlet system; that choking determines their maximum speed. Choked orifices are a common way of providing a small flow of a gas at a flow rate independent of the downstream pressure. Although these systems are not frictionless or one-dimensional, their behavior can be described reasonably well by the set of equations for one-dimensional, isentropic flow derived in Sec. 8.2. Of all of the ideas developed in this chapter, the idea of choking in gas flows is the one *most likely* to help an engineer understand the behavior of a system that others do not understand.

Here we can explain the fact, merely stated in Sec. 5.5, that when a fluid, liquid, or gas flows as a jet into another fluid, the pressure of the jet will be the same as that of the surrounding fluid if the flow is subsonic but not if the flow is sonic or supersonic. If the flow is subsonic and the pressure of the surrounding fluid is less than that of the jet, that information will propagate back along the flow, causing the flow to speed up until the pressures match. If the flow is sonic or supersonic, that information cannot propagate upstream, and the jet can have a pressure different from the surrounding pressure.

8.4 HIGH-VELOCITY GAS FLOW WITH FRICTION, HEATING, OR BOTH

In previous sections we omitted the possible influence of friction, heat transfer from the surroundings to the flowing gas, and chemical reaction (such as combustion) in the flowing gas. For the one-dimensional flow of ideal gases it is possible to deduce mathematical solutions for flow with friction in constant-area ducts, flow with heating in constant-area ducts, flow with combustion in constant-area ducts, and so forth. For subsonic flow these solutions agree satisfactorily with experimental results, but for supersonic flow the one-dimensional assumption becomes unreliable, because friction at the tube wall causes oblique shock waves, which are not covered by one-dimensional flow theory.

Rather than develop the complete mathematics of these flow here, we will simply indicate some of the salient results, referring the reader to other sources where the complete mathematics are shown. We will concentrate on the two types of flow of most practical interest to chemical engineers.

8.4.1 Adiabatic Flow with Friction

This type of flow occurs when a gas flows through a length of pipe at high velocity. If the pipe is insulated or the flow is rapid, the heat transfer to the fluid will be negligible, and the flow will be practically adiabatic.

For this type of flow the mass balance, energy, and ideal-gas equations take the same form as for steady, isentropic flow of an ideal gas. However, in Sec. 8.2 we used the isentropic relations from App. B; here we cannot, because the effect of friction is to increase the entropy of the flowing gas. In their place we will use the steady-flow one-dimensional momentum balance, Eq. 7.17, written for two points dx apart in the flow direction. For steady flow this becomes

$$0 = \rho AV(dV) - A\,dP - \tau_{\text{wall}}\pi D\,dx \tag{8.28}$$

We saw that for the flow of incompressible fluids the shear stress at the wall, τ_{wall}, could be represented in terms of the Fanning friction factor f by

$$\tau_{\text{wall}} = f\rho\frac{V^2}{2} \tag{8.29}$$

Experimental data indicate that the same relation holds reasonably well for flow of gases at high velocities and that the values of f determined from the friction-factor plot, Fig. 6.10, for various Reynolds numbers and pipe roughnesses apply equally well to compressible and incompressible fluids. Equation 8.29 may be substituted in Eq. 8.28 and, together with the energy, continuity, and ideal-gas equations, solved (see App. B.5) to show the change in pressure, temperature, etc. with distance down the pipe. The results are Eq. 8.17, which shows the relation of temperature to Mach number and is the same with or without friction, and

$$\frac{4f\,\Delta x}{D} - \frac{1}{k}\left(\frac{1}{\mathcal{M}_1^2} - \frac{1}{\mathcal{M}_2^2}\right) + \frac{k+1}{2k}\ln\left\{\frac{\mathcal{M}_2^2}{\mathcal{M}_1^2}\cdot\frac{1+[(\mathrm{k}-1)/2]\mathcal{M}_1^2}{1+[(\mathrm{k}-1)/2]\mathcal{M}_2^2}\right\} = 0 \tag{8.30}$$

The result is normally shortened by defining and substituting $N = 4f\,\Delta x / D$.

The most common and interesting problem of this type is that sketched in Fig. 8.11, in which gas flows through a converging nozzle, assumed isentropic, and then through a length of straight pipe where the friction is significant. This is the situation of a high-pressure relief valve or bursting disk discharging through a pipe to a flare or stack, and the situation of a high-pressure vessel discharging through a pipe that has broken some distance from it. One normally designs systems to avoid this flow, but it occurs in accidents and in many safety analyses.

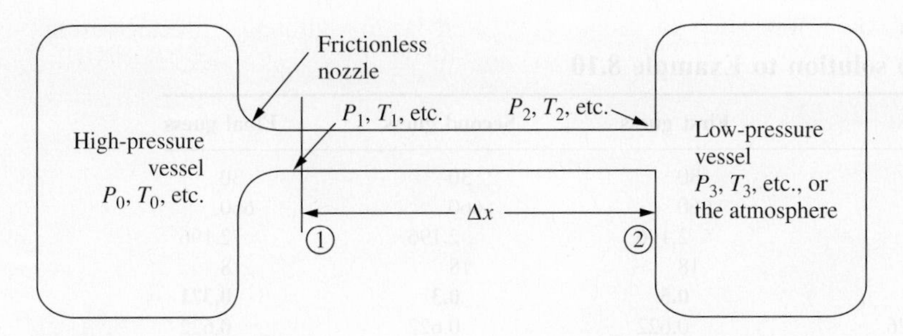

FIGURE 8.11
Flow from a high pressure to a lower pressure through a frictionless nozzle and an adiabatic pipe with friction.

In the type of apparatus shown in this figure, as the fluid flows down the pipe, friction causes its pressure to decrease. This pressure decrease makes the density decrease, and hence the velocity must increase. Since the friction effect is proportional to the velocity squared, the pressure gradient $(-dP/dx)$ is not the same for every foot of pipe, as it is for incompressible flows but, rather, it increases with distance down the pipe.

For the apparatus shown in Fig. 8.11, if the pressures P_0 and P_3 are the same, there will be no flow. If P_3 is then lowered while P_0 held constant, the flow rate will increase, and the Mach number at the outlet of the pipe will steadily increase until the outlet flow is at $\mathscr{M}_2 = 1$. Then further reduction of P_3 will not cause the flow rate to increase, because the flow at the end of the pipe will be choked, just as converging-diverging nozzles become choked (Sec. 8.3). When the flow at the outlet is subsonic, we have $P_2 = P_3$. Once the flow at the outlet becomes sonic, the choked situation exists, and P_3 may be lowered further without any change in P_2.

To solve for the flow in such a system we use the steps shown in Table 8.2.

Example 8.10. In Fig. 8.11, the gas is air, $P_0 = 30$ psia, $P_3 = 18$ psia, and $T_0 = 200°$F. The pipe is 1-in schedule 40, steel pipe 8 ft long. Find the mass flow rate.

To solve for the flow in such a system we use the steps shown in Table 8.2. For 1-in schedule 40 pipe the relative roughness is about 0.0017 and, from

TABLE 8.2
Procedure for solving Example 8.10

Preliminary step	**Estimate *f*, compute *N***
Nozzle, assumed frictionless	Guess $\mathscr{M}_1$, using that value, compute $\dot{m}_1/A^*$, A_1/A^*, $\dot{m}_1/A_1$
Straight pipe with friction	Compute $\mathscr{M}_2$, from Eq. 8.30; using that value, compute T_2, c_2, V_2, ρ_2, $\dot{m}/A_2$
Match	Does $\dot{m}/A_1 = \dot{m}/A_2$? If so, check value of *N*, accept solution. If not, make a new guess of $\mathscr{M}_1$, repeat solution.

TABLE 8.3
Step-by-step solution to Example 8.10

	First guess	Second guess	Final guess
P_0, psia	30	30	30
T_0, °R	660	660	660
N	2.196	2.196	2.196
P_2, psia	18	18	18
$\mathcal{M}_1$, guessed	**0.5**	**0.3**	**0.371**
$\dot{m}_1 / A^*$, Eq. 8.26	0.622	0.622	0.622
A_1 / A^*, Eq. 8.24	1.340	2.035	1.692
$\dot{m}_1 / A_1$ lbm / s in^2	0.464	0.306	0.367
$\mathcal{M}_2$, guessed	0.800	0.363	0.553
N, Eq. 8.30	0.997	2.196	2.196
T_2, °R, Eq. 8.17		643.052	621.983
c_2, ft / s, Eq. 8.12		1242.490	1221.966
V_2, ft / s		451.035	675.527
ρ_2 based on P_2 and T_2		0.076	0.078
$\dot{m}_2 / A_2$, lbm / s ft^2		34.122	52.837
$\dot{m}_2 / A_2$ lbm / s in^2		0.237	0.367
Ratio $\dfrac{\dot{m}_2 / A_2}{\dot{m}_1 / A_1}$		0.776	**0.999**

Fig. 6.10, for high Reynolds numbers the friction factor is about 0.006. Thus, we estimate that

$$N = \frac{4f\,\Delta x}{D} = \frac{4 \cdot 0.006 \cdot 8 \text{ ft}}{1.049 \text{ ft} / 12} = 2.196 \tag{8.AD}$$

We record the results of this calculation as we go in Table 8.3. The first four rows in that table show the three given values and the preceding value of N.

Then, for our first guess (column 2) we take $(\mathcal{M}_1)_{\text{first guess}} = 0.5$. Then we can compute the values in the next three rows for $(\mathcal{M}_1)_{\text{first guess}} = 0.5$. Next we must solve Eq. 8.30 for $\mathcal{M}_2$. This is inherently a trial and error, because the equation is transcendental. (It contains both $\mathcal{M}_2$ and its logarithm. Try solving it by hand; you will see!) The easy way to solve it numerically on a spreadsheet is to guess a starting value of $\mathcal{M}_2$ and then use the spreadsheet's numerical solution engine to find the value of $\mathcal{M}_2$ that makes N in Eq. 8.30 $= 2.196$. In the second column of Table 8.3 we guessed $(\mathcal{M}_2)_{\text{first guess}} = 0.8$. The next entry in the table shows that substituting that value in Eq. 8.30 leads to $N = 0.997$. The numerical solution engine was then asked to find the value of $\mathcal{M}_2$ that makes N in Eq. 8.30 $= 2.196$. It failed. Subsequent investigation shows that for $(\mathcal{M}_1)_{\text{first guess}} = 0.5$ there is no value of $\mathcal{M}_2$ between 0.5 and 1.0 that leads to $N = 2.196$. So $\mathcal{M}_1$ cannot equal 0.5 (for this value of N).

For our second guess (column 3 in Table 8.3) we take $(\mathcal{M}_1)_{\text{second guess}} = 0.3$. Continuing down that column we put in some guess for $\mathcal{M}_2$ and let the numerical solution engine replace it with the value ($\mathcal{M}_2 = 0.363$), which makes N in

Eq. 8.30 = 2.196. Then we find T_2 from Eq. 8.17 (which is limited to adiabatic flows, but not to frictionless ones). With that value of T, we find c_2 from Eq. 8.12, and V_2 from the local speed of sound and the Mach number. We find ρ_2 from P_2 and T_2, and

$$\dot{m}_2 / A_2 = \rho_2 V_2 \tag{8.AE}$$

Table 8.3 shows the values in lbm / (s · ft²), calculated by Eq. 8.AE, and the corresponding value in lbm / (s · in²). Finally, the ratio of $\dot{m} / A$ at 2 to that at 1 is computed. If we have chosen the right value of $\mathcal{M}_1$, this ratio should be 1.00. In column 3 it is 0.776, indicating that $(\mathcal{M}_1)_{\text{second guess}} = 0.3$ is not correct. Table 8.3, column 4 shows $(\mathcal{M}_1)_{\text{final guess}} = 0.371$. (There were some intermediate guesses, which are not shown in Table 8.3). For this final guess the calculated values of $\dot{m} / A$ at 1 and at 2 agree to within 0.1 percent. We accept this as a satisfactory solution, and read from the table that $\dot{m} / A$ in the pipe is 0.367 lbm / (s · in²). The cross-sectional area of 1-in schedule 40 pipe = 0.864 in², so that

$$\dot{m} = 0.864 \text{ in}^2 \cdot 0.367 \frac{\text{lbm}}{\text{s} \cdot \text{in}^2} = 0.317 \frac{\text{lbm}}{\text{s}} = 1.44 \frac{\text{kg}}{\text{s}} \tag{8.AF}$$

Finally we compute that $\mathcal{R} \approx 3.4 \cdot 10^5$ for which $f \approx 0.006$, which checks our starting assumption. ■

At the end of this *long and tedious* example we observe the following:

1. Because Eq. 8.30 is transcendental and cannot be solved algebraically, its solution is inherently trial and error.
2. That solution is inside the trial and error for $\mathcal{M}_1$. This leads to nested trial and errors.
3. The spreadsheet is a poor way to do this kind of nested trial and error solutions. The spreadsheet was chosen here because it shows the details of how such solutions proceed and shows the intermediate values.
4. Nested trial and error solutions are easy in programming languages like Fortran. If one is setting up to do a series of this type of calculations, such a program should be used.
5. For hand calculations several authors have prepared graphical solutions to this problem by solving for a variety of cases and plotted the results. The most widely used is Fig. 8.12. Its use is illustrated in the next example.

Example 8.11. Repeat Example 8.10, using Fig. 8.12.

Here the exit flow is subsonic (which we know from Example 8.10 and will see from Fig. 8.12) so that $P_2 = P_3$ and

$$\frac{P_2}{P_0} = \frac{P_3}{P_0} = \frac{18 \text{ psia}}{30 \text{ psia}} = 0.6 \tag{8.AG}$$

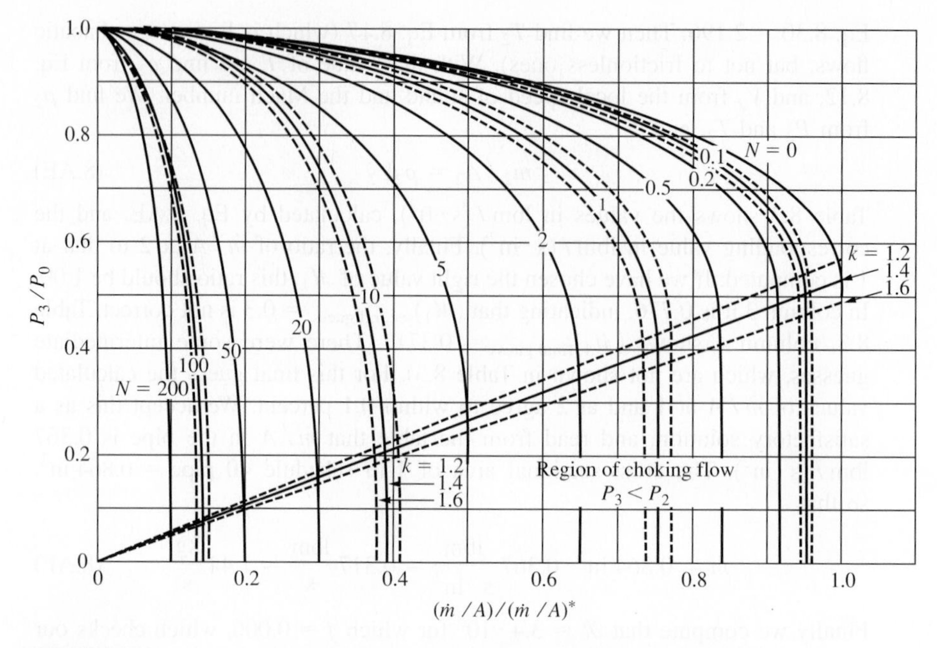

FIGURE 8.12
Pressure–mass flow rate relationship for the apparatus shown in Fig. 8.11. (Levenspiel, O., "The discharge of gases from a reservoir through a pipe," *AIChEJ 23*:402–403 (1977). Reproduced by permission of the publisher.) In the denominator, the $(\dot{m}/A^*)$ is that corresponding to frictionless flow from the reservoir shown with subscript 0 in Fig. 8.11. The solid lines are for $k = 1.4$. A few dotted lines are shown for other values of k. $N = 4f\,\Delta x/D$.

Entering Fig. 8.12 at this value and reading horizontally to the $N = 2$ curve ($N \approx 2.196$) we then read to the bottom, finding

$$\frac{(\dot{m}/A)}{(\dot{m}/A)^*} \approx 0.6 \tag{8.AH}$$

The value of $(\dot{m}/A)^*$ in this figure corresponds to adiabatic frictionless flow. From Example 8.6 we know that for these values of P_0 and T_0, $[(\dot{m}/A)^*]_{\text{frictionless flow}} = 0.62\ \text{lbm}/(\text{s}\cdot\text{in}^2)$, so that for $N = 2.0$

$$(\dot{m}/A)^* \approx 0.6 \cdot 0.62\,\frac{\text{lbm}}{\text{s}\cdot\text{in}^2} = 0.372\,\frac{\text{lbm}}{\text{s}\cdot\text{in}^2} \tag{8.AI}$$

which is within chart reading accuracy of the 0.367 found in the previous example for $N = 2.196$ ■

We can further illustrate the properties of Fig. 8.12 (or the equivalent spreadsheet calculations in Example 8.10 on which Fig. 8.12 is based) by asking what the flow rates are for a variety of values of P_3/P_0. The results are shown in Fig. 8.13, which is of the same type as Fig. 8.9, but for a nozzle and a pipe with friction instead of the simple nozzle in Fig. 8.9. For $P_3 = 30$ psia, the pressure ratio

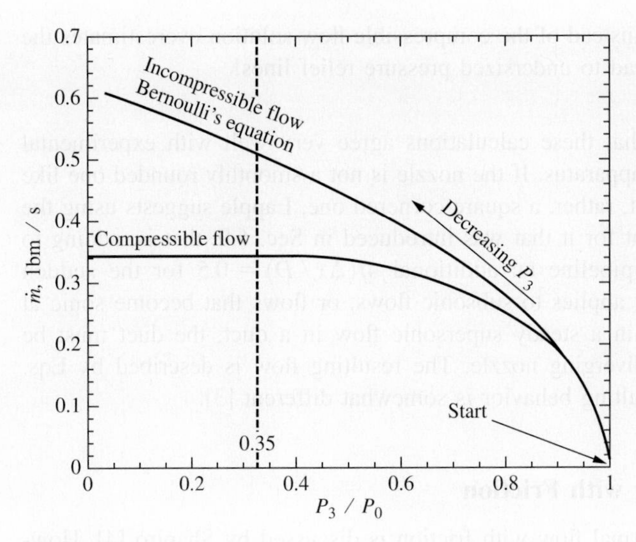

FIGURE 8.13
Flow as a function of (P_3 / P_0) for the device shown in Fig. 8.11 and the numerical values in Example 8.10. The B.E. curve is computed from $V = \sqrt{2(-\Delta P) / (1 + 4f\,\Delta x / D)\rho_R}$ with the same numerical values used in Example 8.10. As expected, these curves are practically identical for low values of ΔP, but not for high values. The compressible flow solution shows nozzle choking, the B.E. solution does not.

across the nozzle and pipe = 1.00, and the curves for all values of N converge at $(\dot{m} / A) / (\dot{m} / A)^* = 0$, i.e., zero flow. As we lower P_3, we move to the right on the $N = 2\,(\approx 2.196)$ curve on Fig. 8.12, showing larger and larger values of $(\dot{m} / A) / (\dot{m} / A)^*$, and thus higher values of $\dot{m}$. For $P_3 = 10.5$ psia the flow is choked; subsequent lowering of the downstream pressure will not increase the mass flow rate. Comparing Fig. 8.13 to Fig. 8.9 we see the following:

1. For both plots, the B.E. and compressible flow curves are $\approx$ identical for (P_3 / P_0) close to 1.00 (i.e., small values of ΔP). As (P_3 / P_0) becomes smaller (ΔP becomes larger) the two curves separate.
2. In both plots there is a minimum value of $(P_{\text{outlet}} / P_{\text{inlet}})$, below which the flow at the outlet is choked. Further reduction of $(P_{\text{outlet}} / P_{\text{inlet}})$ does not increase the mass flow rate.
3. In Fig. 8.9, that ratio $(P_{\text{outlet}} / P_{\text{inlet}})$ was 0.5283, here it is 0.35. One may check that the curve for $N = 0$ on Fig. 8.12, which corresponds to a frictionless nozzle and a pipe of zero length (i.e., no pipe), reaches the choked condition at $(P_{\text{outlet}} / P_{\text{inlet}}) = 0.5283$, as it must. Increasing the length of the pipe (increasing N) moves the choked condition to lower and lower values of $(P_{\text{outlet}} / P_{\text{inlet}})$.
4. The B.E. calculation substantially overestimates the mass flow rate for low values of $(P_{\text{outlet}} / P_{\text{inlet}})$. This type of calculation is routinely made in safety analysis;

using the B.E. solution instead of the compressible flow solution overestimates the flow rates, which can lead to undersized pressure relief lines!

Lapple [2] showed that these calculations agree very well with experimental tests run with this type of apparatus. If the nozzle is not a smoothly rounded one like that shown in Fig. 8.11, but, rather, a square-cornered one, Lapple suggests using the same contraction coefficient for it that was introduced in Sec. 6.9, that is, adding to the friction effect of the pipeline an additional $4f(\Delta x / D) = 0.5$ for the sudden entrance. All the foregoing applies to subsonic flows, or flows that become sonic at the choked outlet. To obtain a steady supersonic flow in a duct, the duct must be attached to a converging-diverging nozzle. The resulting flow is described by Eqs. 8.17, and 8.30, but the resulting behavior is somewhat different [3].

8.4.2 Isothermal Flow with Friction

The general case of isothermal flow with friction is discussed by Shapiro [4]. However, as he points out, at Mach numbers approaching the choking condition an infinite heat-transfer rate would be required to keep the flow isothermal. Thus, for common pipe sizes and lengths high-velocity gas flow is always much closer to adiabatic than to isothermal. The one exception of interest is the flow of natural gas and other gases through long-distance pipelines. These may be a hundred miles long between pumping stations and are normally buried in the ground, which supplies heat as needed to keep the flow isothermal.

The commonly used formulae for calculating the flow in these pipelines are based on the momentum equation, Eq. 8.28, the mass-balance equation, and the ideal gas law. Substituting Eq. 8.29 in Eq. 8.26 and dividing by A, we find

$$\rho V\,dV + dP = -4f\rho \frac{V^2}{2} \cdot \frac{dx}{D} \tag{8.31}$$

In the previous section we used all three terms in this equation. Here we can greatly simplify the calculations by noting that for long pipelines the first term, $\rho V\,dV$, is negligible compared with the others (Prob. 8.47) and can be dropped; then, from the mass-balance equation, V is replaced with $\dot{m} / \rho A$, and Eq. 8.31 is simplified to

$$dP = \frac{-4f}{2}\left(\frac{\dot{m}}{A}\right)^2 \frac{1}{\rho} \cdot \frac{dx}{D} \tag{8.32}$$

Replacing ρ in this equation with its ideal gas law value, we find

$$P\,dP = \frac{-4f}{2} \cdot \frac{RT}{DM}\left(\frac{\dot{m}}{A}\right)^2 dx \tag{8.33}$$

If $f =$ constant, this equation may be integrated and rearranged to

$$\dot{m} = \left[\frac{(P_1^2 - P_2^2)D^5 M(\pi / 4)^2}{4f\,\Delta x RT}\right]^{1/2} \tag{8.34}$$

If we substitute the empirical approximation $f = 0.0080\,/\,(\text{pipe diameter in inches})^{1/3}$ in this equation, we obtain the *Weymouth equation,* which was widely used in the design of early natural-gas pipelines. The historical trend of the gas pipeline industry is to use higher and higher gas pressures; as the pressure increases, the gas departs further and further from the ideal gas state. Later workers have corrected the Weymouth equation to take this departure into account [5].

As discussed in Sec. 6.13, the economic velocity in a pipeline is primarily dependent on the density of the fluid flowing. For long-distance natural-gas pipelines, the pressures are normally in the range of 500 to 1000 psia, so densities are of the order of 1 to 2 lbm/ft^3. From Table 6.4 we can estimate the economic velocity at about 20 ft / s, which is typical of these pipelines. Thus, this kind of flow does not really correspond to the subject of this section, high-velocity gas flow. However, it fits in naturally here, after we have developed the equations for high-velocity gas flow, which explains why it is placed here.

8.5 NORMAL SHOCK WAVES

The next three sections (8.5, 8.6, and 8.7) concern supersonic flows. They are of great practical interest to aeronautical engineers and to chemical engineers in the rocket propulsion industry. There are only a few practical applications of interest to chemical engineers outside that industry. These include the steam nozzles in jet ejectors (see Fig 7.35), which operate in supersonic flow, and the free discharge of high-pressure gases from ordinary valves whose throat area is less than their exit area and whose discharge is often supersonic. The one-stage steam turbines often used as drivers in process plants also operate with supersonic flows.

Suppose that a nozzle is steadily discharging a gas stream at $\mathcal{M} = 2.0$ into a low-pressure reservoir and that a valve in the end of this nozzle is suddenly shut. This will stop the flow. In Sec. 7.5.2 we discussed the analogous problem for a flow of a liquid in a pipe. There the flow was subsonic, and the boundary between the stopped and moving fluid propagated against the flow at the local speed of sound relative to the fluid. Here the fluid is moving faster than the local speed of sound, so how can the information that the valve is closed propagate upstream against the flow? This information cannot move upstream against a supersonic flow, if it moves only at the speed of sound; therefore, there must be some way of conveying this information as a pressure signal that moves faster than the local speed of sound. We said before that a sound wave was a *small* pressure disturbance that moved at the local speed of sound. Now we will consider shock waves, *large* pressure disturbances that can move faster than the local speed of sound.

Shock waves occur in nature in the air surrounding explosions (the shock wave causes much of the destruction of buildings, etc., in any bomb blast) and in the sudden closing of a valve in a duct with a high-velocity flow. Sonic booms are shock waves. Shock waves also can be produced in the laboratory in a duct or nozzle with supersonic flow. In such cases the shock wave will stand still in one place while the fluid flows through it. The latter is the easier to analyze mathematically, so we will use it as a basis for calculations. The nomenclature for a shock wave is shown in Fig. 8.14.

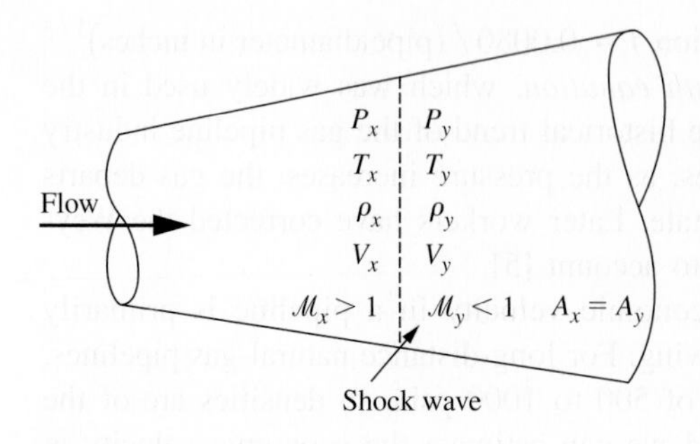

FIGURE 8.14
Nomenclature of a shock wave. The subscripts x and y refer to conditions upstream and downstream, respectively, of the shock wave.

The situation is quite similar to that shown in Fig. 8.2, as is much of the following mathematical development. Writing the equivalent of Eq. 8.1, the continuity or mass-balance equation, for this flow, we find

$$\rho_x V_x = \rho_y V_y \tag{8.35}$$

Equation 8.4, the momentum balance, becomes

$$V_x - V_y = \frac{P_y}{\rho_y V_y} - \frac{P_x}{\rho_x V_x} \tag{8.36}$$

These two equations alone were enough to solve for the speed of a sound wave, because we were able to neglect one term, $d\rho\, dV$, which was small in Eq. 8.2. It is not small here, so we need another relation, the energy balance, Eq. 8.17.

Equations 8.17, 8.35, and 8.36 can be solved to show the changes in temperature, pressure, and Mach number in a shock wave; see App. B.6. The results are

$$\mathcal{M}_y^2 = \frac{\mathcal{M}_x^2 + 2/(k-1)}{[2k/(k-1)]\mathcal{M}_x^2 - 1} \tag{8.37}$$

$$\frac{P_y}{P_x} = \frac{2k}{k+1}\mathcal{M}_x^2 - \frac{k-1}{k+1} = \frac{2k\mathcal{M}_x^2 - (k-1)}{k+1} \tag{8.38}$$

and

$$\frac{T_y}{T_x} = \frac{\mathcal{M}_x^2[(k-1/2)] + 1}{\left[\dfrac{k-1}{2}\right] \cdot \left[\dfrac{1 + \mathcal{M}_x^2[(k-1)/2]}{k\mathcal{M}_x^2 - [(k-1)/2]}\right] + 1} \tag{8.39}$$

These three functions are somewhat messy, but for a gas with a constant k they can be tabulated as a function of $\mathcal{M}_x$. Their values for $k = 1.4$ are shown in App. A.5 with some useful combinations of them.

If a shock wave is a large pressure disturbance and a sound wave is a small pressure disturbance, what is the dividing line between them? There is none; as the pressure disturbance of shock waves becomes smaller and smaller, these equations become closer and closer to those for a sound wave, until they become identical. In the case of a sound wave we had the flow in Fig. 8.2 coming toward the wave at $\mathcal{M}_x = 1.0$. The patient student may verify that substituting $\mathcal{M}_x = 1.0$ into Eqs. 8.37, 8.38, and 8.39 shows that there is no change in temperature, pressure, or velocity through the shock wave. Thus, an $\mathcal{M}_x = 1.0$ shock wave has the same properties as a sound wave, and we may consider a sound wave to be a minuscule shock wave which moves at the lowest possible velocity for any pressure disturbance ($\mathcal{M}_x = 1.0$).

Notice the form of the flow in a shock wave, shown in Fig. 8.14. The fluid enters at a supersonic velocity, a low pressure, and a low temperature and leaves at a subsonic velocity, a higher pressure, and a higher temperature. All of the foregoing and all of the derivations in App. B.6 place no restriction on the direction of the shock. In the derivations, we never made use of the fact that the flow was from supersonic to subsonic or vice versa. Considering these equations alone, we could conclude that the flow could be either from high velocity and low pressure to low velocity and high pressure or the reverse. The first might logically be called a *compression shock* and the second a *rarefaction shock.*

The second law of the thermodynamics, however, shows that only compression shocks are possible (see Prob. 8.51). A compression shock results in an increase in entropy whereas a rarefaction shock, if it existed, would result in a decrease in entropy, which is impossible in an adiabatic, steady-flow system. Thus, we conclude that this type of shock waves always requires a supersonic flow upstream and a subsonic flow downstream. This explains why we can stop the flow in a supersonic nozzle by closing a downstream valve; a compression shock passes upstream against a sonic or supersonic flow and stops the flow. But we cannot increase the speed in a choked flow by lowering the downstream pressure, because that would require a rarefaction shock to pass upstream, and rarefaction shocks are thermodynamically impossible.

In Sec. 7.5.3 we considered hydraulic jumps, which occur in open-channel flow. They are compared with shock waves in Table 8.4. From this comparison we see that there is a strong similarity between the two. The principal difference is that, because a shock wave occurs in a compressible fluid, we have an additional variable, the density, and we must add the energy balance to the mass and momentum balances needed to solve for hydraulic jump, with a resulting increase in mathematical complexity.

Our discussion has been restricted to shock waves in which the flow is perpendicular to the wave, the only kind that can occur in one-dimensional flow. Such waves are called *normal shock waves* or *normal shocks* because of this perpendicular relationship. In two-dimensional flow another kind occurs, called an *oblique shock,* in which the flow is not perpendicular to the shock wave. Oblique

TABLE 8.4
Comparison of hydraulic jumps with shock waves

	Hydraulic jump	**Shock wave**
Flowing material	Liquid	Gas
Type of flow	Open channel	Closed duct
Inlet flow	High velocity, low depth	High velocity, low pressure
Outlet flow	Lower velocity, higher depth	Lower velocity, higher pressure
Equations needed to solve	Balances of mass, momentum	Balances of mass, momentum, energy, and ideal gas law
Permissible direction of occurrence determined by	Second law of thermodynamics	Second law of thermodynamics
Upstream condition	Froude number > 1	Mach number > 1
Downstream condition	Froude number < 1	Mach number < 1

shock waves form at the leading edge of the wings of supersonic aircraft and cause "sonic booms"; for discussion of them see Ooosthuizen and Carscallen [3, Chap. 6].

8.6 RELATIVE VELOCITIES, CHANGING RESERVOIR CONDITIONS

Thus far we have discussed flows that move past a stationary observer. They are of considerable importance and are found in wind tunnels, turbine nozzles, high-pressure valves, etc. Equally important are systems in which the gas stands still and the system moves; examples are airplanes, missiles, and meteorites. In principle we could develop a separate set of equations for these systems, but it is much simpler to learn to apply the equations that we already have applied to stationary systems, and to use the tables in Apps. A.4 and A.5.

The application of the equations in the preceding sections to moving-coordinate systems is quite simple, once we grasp the idea that the reservoir conditions are a function of the frame of reference. This appears startling at first, but it must be so.

Example 8.12. An airplane is moving at $\mathcal{M} = 2$ in air at 0°C = 273.15 K and 50 kPa. What are the reservoir temperature and pressure of this air?

If we choose the ground as our frame of reference, then the velocity of the air relative to the ground is zero, so we conclude that the reservoir condition is 0°C = 273.15 K and 50 kPa. However, this would be a very impractical choice if we wished to analyze the performance of the airplane. We would rather choose a coordinate system based on the airplane. In such a case the air is moving at $\mathcal{M} = 2$ toward the plane, so it is certainly not standing still and hence is not in its reservoir state. To find the reservoir temperature we may slow the air down to zero velocity relative to the airplane by an adiabatic device and then measure its temperature. From Eq. 8.17 we see that the reservoir temperature is

$$T_R = 273.15\text{ K}\left(2.0^2\,\frac{1.4-1}{2}+1\right) = 491.7\text{ K} = 885°\text{R} = 425°\text{F} \tag{8.AJ}$$

If our device for slowing the air down were frictionless (e.g., a ideal diffuser, Sec. 8.6), then the pressure at its outlet would be the reservoir pressure. From Eq. 8.18 we can compute the result:

$$P_R = 50\text{ kPa}\cdot 1.80^{3.5} = 391\text{ kPa} = 56.8\text{ psia} \tag{8.AK}$$

■

Once we have made this shift of frame of reference and the corresponding shift in reservoir conditions, we can solve the problem by using the relations developed previously for steady flow in a duct.

We might also think about this problem in another way. Suppose we wanted to test the airplane standing still in a wind tunnel. We want the air to come in at $\mathcal{M} = 2$, at 0°C, and at 50 kPa. What must the conditions be in the reservoir that sends air to this wind tunnel? Obviously, they are the values computed above. From a fluid-mechanics standpoint it makes no difference whether the plane is moving relative to the air or the air relative to the plane.

From this example we see clearly that the reservoir condition is a function of the coordinate system chosen. Notice also the high reservoir temperature we compute. This is the temperature of the air in contact with the outside of the plane, where the velocity relative to the plane is zero. The high value indicates why supersonic aircraft must be refrigerated to keep the interior at a temperature suitable for human occupancy.

From the moving airplane the air appears to have the same temperature as it appears to have from a stationary position. Changing frames of reference changes only reservoir conditions, not local conditions. The temperature at any point, such as T_1, T_x, and T_y in the preceding equations is the same for an observer riding with the fluid or an observer standing still. The observer moving relative to the fluid may have some difficulty in measuring the temperature (see Prob. 8.56), but with suitable instruments one can get the same reading as the stationary observer.

The equations that we developed for normal shock waves standing still in a nozzle are also applicable to moving systems.

Example 8.13. An atom bomb blast in still air raises the pressure around the bomb to a high value. This high pressure causes a shock wave that flows outward. The geometry is spherical, the shock wave expands like a balloon. The pressure inside the system steadily falls as the shock wave moves out, but at any particular instant, the behavior of the shock wave is given to an excellent approximation by the steady-flow equations developed in Sec. 8.5. At the instant when the pressure inside the shock is 2.0 atm, how fast does the shock wave move, and what are the pressure, temperature, and velocity behind it? Outside the shock wave, the air conditions are 14.7 psia and 528°R.

We will use the notation of Fig. 8.14, with the subscript x standing for the still air into which the shock wave is advancing and the subscript y applying to the gas inside the expanding high-pressure region. Then we have $P_y / P_x = 2$, and from App. A.5 we find that $\mathcal{M}_x = 1.3630$. The speed of sound in the air (Example 8.2) is 1126 ft / s, so the shock wave moves into the still air at $V = \mathcal{M}_x c_x = 1.3630 \cdot 1126 \text{ ft / s} = 1535 \text{ ft / s} = 468 \text{ m / s}$. Behind the shock the pressure is $2 \cdot 14.7 = 29.4$ psia, as stated. From App. A.5 we find $T_y / T_x = 1.2309$; therefore,

$$T_y = 528°\text{R} \cdot 1.2309 = 649.9°\text{R} = 361 \text{ K} \tag{8.AL}$$

and so we can compute the speed of sound at y as 1249 ft / s and, from App. A.5, we find $\mathcal{M}_y = 0.7558$. Then

$$V_y = \mathcal{M}_y c_y = 0.7558 \cdot 1249 \text{ ft / s} = 944 \text{ ft / s} = 288 \text{ m / s} \tag{8.AM}$$

Thus, in the region just inside the wave, the air is moving 944 ft / s away, as seen by an observer riding on the shock wave. Switching to fixed coordinates, we see that the shock wave is moving 1535 ft / s, so the air adjacent to it on the high-pressure side is moving $(1535 - 944) = 591$ ft / s outward from the

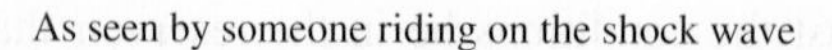

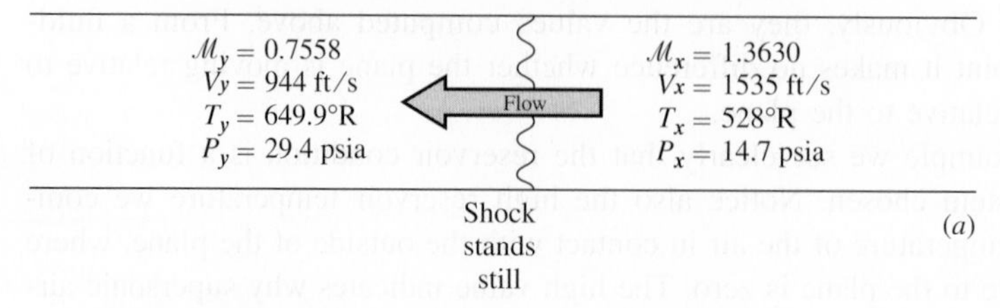

As seen by someone standing still

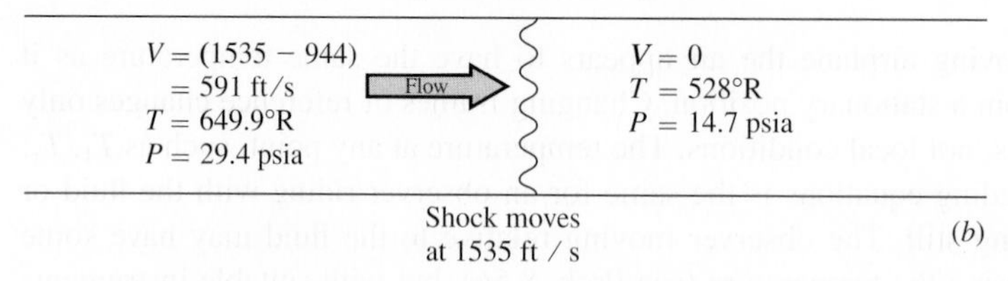

For both views, $(V_y - V_x) = 591$ ft / s

FIGURE 8.15
The moving shock wave in Example 8.13 as seen by an observer riding on the shock wave (*a*), and by an observer standing still and watching it go by (*b*). For both observers, the velocity change across the shock is 591 ft / s.

center. The relations between the velocities and other properties as seen by an observer riding with the wave and one watching from the ground are sketched in Fig. 8.15.

The simple picture here hides much of the complexity of the fluid mechanics of blast waves. It is implied here that the pressure inside the expanding spherical high-pressure region is constant. That is not correct. The pressure is lowest at the center and highest at the edge of the expanding wave [6]. A lightning strike heats the air it passes through enough to produce a similar shock wave. The shock wave is a sharp sound near the lightning stroke, but becomes a diffuse rumble as it passes through the nonuniform atmosphere and interacts with its ground reflections. ■

8.7 NOZZLES AND DIFFUSERS

Figure 8.3 is a plot of A / A^* versus $\mathcal{M}$ for steady, one-dimensional, frictionless, adiabatic flow of an ideal gas. It is, in effect, a design guide for a supersonic nozzle. If we want the Mach number to increase linearly with distance in steady, isentropic flow of an ideal gas, then the cross-sectional area–distance relation must be exactly the curve in Fig. 8.3. This is the diagram for a *converging-diverging nozzle,* commonly referred to as a *de Laval nozzle* after Carl de Laval, (1845–1913), who used it in the first practical steam turbine.

We have already discussed (see Fig. 8.4) the intuitive explanation of the necessity of the converging-diverging shape. In all our derivations for frictionless, adiabatic, steady flow of an ideal gas we never said whether the gas was speeding up or slowing down; we said only that the flow was steady, frictionless, and adiabatic. Therefore, all the equations work equally well for an accelerating flow and a decelerating flow. In Fig. 8.3 the flow could be from left to right, as we have tacitly assumed before, or from right to left. If the latter, the gas would enter the nozzle in supersonic flow and emerge in subsonic flow. When a nozzle is used this way, it is called a *diffuser.* There is no known way to obtain a steady, supersonic flow other than by means of a converging-diverging nozzle (there are several ways to produce unsteady supersonic flows, e.g., explosives). There is no known *isentropic* way to slow a steady, supersonic flow down to subsonic speeds other than by means of a *converging-diverging diffuser.* We can steadily convert a supersonic flow to a subsonic flow by means of a normal shock, which is not isentropic.

From Fig. 8.3 and the equations on which it is based, we would assume that we could change the Mach number by any amount over as a short a distance as we wished by changing the area rapidly, because there is no restriction on the equations as to dA/dl, where l is the length of the nozzle. However, if this rate of area change becomes too great, then our one-dimensional frictionless-flow assumptions become unreliable, and the observed flow no longer follows our isentropic equations. In engineering practice the converging section has a wide angle (that is, dA/dl has a large negative value), and the diverging section has a small angle. Such a nozzle is shown in Fig. 8.16.

We are now able to calculate the flow characteristics of such a nozzle for frictionless, adiabatic flow, using the equations of isentropic flow and the equations of

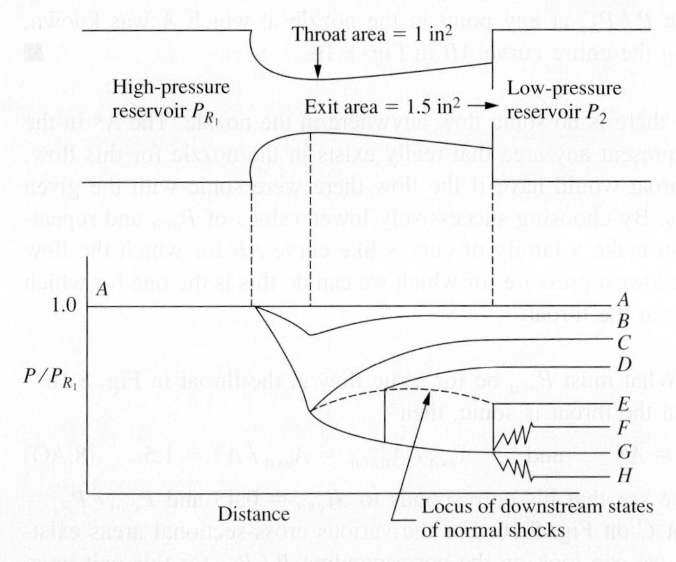

FIGURE 8.16
Pressue–distance plot for various downstream pressures in a converging-diverging nozzle. These curves are worked out in Examples 8.14 to 8.16.

normal shock waves. The results we calculate here are a *reasonable approximation* of the behavior of real nozzles, in which friction is always significant. We assume that the cross-sectional area perpendicular to flow at each point in the nozzle is known, and we prepare a plot of P / P_{R_1} versus distance for various flows. We begin by making the pressures in the high-pressure and low-pressure reservoirs the same. Then there is no flow, so the pressure is the same as the reservoir pressure throughout the nozzle; see line *AA* in Fig. 8.16.

Throughout our calculation we will hold the upstream reservoir pressure P_{R_1} constant. We begin to lower the downstream reservoir pressure P_2. As long as the exiting gas is in subsonic flow, the exit pressure will be the same as the downstream pressure; so, if we set the downstream pressure, we have set the pressure at the exit of the nozzle. From this pressure and the area of the nozzle we can calculate the pressure at the throat or at any other point whose cross-sectional area we know, as shown in the following example.

Example 8.14. We now set the downstream reservoir pressure at 0.9506 times the upstream reservoir pressure (point *B* in Fig. 8.16). What is the pressure at the throat of the nozzle?

We look in App. A.4 for $P / P_R = 0.9506$ and find that the exit Mach number is 0.27. We also observe that A_{exit} / A^* is 2.2385. Then we can find

$$\frac{A_{\text{throat}}}{A^*} = \frac{A_{\text{exit}}}{A^*} \cdot \frac{A_{\text{throat}}}{A_{\text{exit}}} = 2.2385 \cdot \frac{1 \text{ in}^2}{1.5 \text{ in}^2} = 1.4923 \qquad (8.AN)$$

We see in App. A.4 that this corresponds to a Mach number of about 0.43 at the throat, and therefore we have $P / P_R = 0.88$ at the throat. In the same way we could calculate P / P_{R_1} at any point in the nozzle at which *A* was known, thereby completing the entire curve *AB* in Fig. 8.16. ■

In this calculation there is no sonic flow anywhere in the nozzle. The A^* in the calculations does not represent any area that really exists in the nozzle for this flow; it is the area that the throat would have if the flow there were sonic with the given exit area and exit P / P_R. By choosing successively lower values of P_{exit} and repeating this example, we can make a family of curves like curve *AB* for which the flow is entirely subsonic. The lowest pressure for which we can do this is the one for which the flow is exactly sonic at the throat.

Example 8.15. What must P_{exit} be for sonic flow at the throat in Fig. 8.16?

If the flow at the throat is sonic, then

$$A_{\text{throat}} = A^* \quad \text{and} \quad A_{\text{exit}} / A_{\text{throat}} = A_{\text{exit}} / A^* = 1.5 \qquad (8.AO)$$

From App. A.4 we see that his corresponds to $\mathcal{M}_{\text{exit}} = 0.43$ and $P_{\text{exit}} / P_{R_1} = 0.88$. This is point *C* on Fig. 8.16. For the various cross-sectional areas existing in the nozzle, we can look up the corresponding P / P_{R_1} for this exit pressure and thus complete curve *AC*. ■

What happens if we lower P even more, say, to $P_2 / P_{R_1} = 0.8022$? If we use exactly the same procedure as in Ex. 8.14, we find that $A_{\text{throat}} / A^* = 0.8175$. This is physically impossible, because in such a nozzle the minimum area occurs at the point of $\mathcal{M} = 1$ (see Fig. 8.3), and this calculation indicates that the throat area is less than this minimum area. The impossible answer results from an incorrect assumption. We assumed isentropic flow through the whole nozzle in Examples 8.14 and 8.15; for subsonic flow throughout this is correct, but here it is not. If we start with the conditions shown by curve AC in Fig. 8.16, and if we lower P_2, the flow tends to go faster. However, the nozzle is choked, because the flow is sonic at the throat, so the mass flow rate $\dot{m}$ cannot increase. With the same mass flow rate at a lower pressure, the flow immediately downstream of the throat becomes supersonic. However, the downstream pressure along the whole nozzle is not low enough for the flow to be supersonic throughout, so the flow will convert back to subsonic somewhere in the nozzle via a normal shock.

Consider the situation in which we take the subsonic flow after a normal shock wave and slow it isentropically to zero velocity, that is, to the downstream reservoir condition. We previously showed by an energy balance that for a normal shock wave $T_{R_x} = T_{R_y}$. What about P? By rearranging Eq. 8.18 twice, we find

$$P_x = \frac{P_{R_x}}{\{\mathcal{M}_x^2[(k-1)/2] + 1\}^{k/(k-1)}}, \qquad P_y = \frac{P_{R_y}}{\{\mathcal{M}_y^2[(k-1)/2] + 1\}^{k/(k-1)}} \tag{8.40}$$

Substituting these in Eq. 8.38 and rearranging produces

$$\frac{P_{R_y}}{P_{R_x}} = \left\{\frac{\mathcal{M}_y^2[(k-1)/2] + 1}{\mathcal{M}_x^2[(k-1)/2] + 1}\right\}^{k/(k-1)} \cdot \left(\frac{2k\mathcal{M}_x^2 - (k-1)}{k+1}\right) \tag{8.41}$$

This is another messy equation, but fortunately it, too, is tabulated in App. A.5 for gases with $k = 1.4$.

The decrease in reservoir pressure has an interesting consequence. In Eq. 8.24 we showed that $\dot{m}/A^*$ was proportional to P_R. Thus, when P_R decreases, as it does in a normal shock wave, $\dot{m}/A^*$ also decreases. Equation 8.24 shows that

$$\frac{(\dot{m}/A^*)_x}{(\dot{m}/A^*)_y} = \frac{P_{R_x}}{P_{R_y}} \tag{8.42}$$

For steady flow $\dot{m}_x = \dot{m}_y$, and therefore $A_y^* / A_x^* = P_{R_x} / P_{R_y}$. We may look at the same phenomenon from the viewpoint of reversibility. Isentropic flow is reversible; the whole flow could run backward. A normal shock wave is irreversible; once it has occurred, the whole flow could not run backward: the same flow could not fit back through the same nozzle.

Thus, in this case, when the normal shock is present, although we have $A_{\text{throat}} / A^* = 1$, we cannot directly calculate the value of A_{exit} / A^*, because the value of A^* is not the same at these two points. At this point the reader might do well to reread what we said about A^* at the end of Sec. 8.2.

Example 8.16. Construct the curve analogous to curve AC on Fig. 8.16 for $P_2 / P_{R_1} = 0.8022$. In this case we do not know the downstream reservoir pressure P_{R_2}, so we cannot yet calculate the outlet Mach number. The key unknown here is the upstream Mach number at which the normal shock occurs, $\mathcal{M}_x$. Since the exit flow is subsonic, it must exit at $P_{\text{exit}} = 0.8022\ P_{R_1}$ and must have the A / A^* that corresponds to its Mach number. There is only one upstream Mach number at the shock that satisfies these conditions. We find it numerically, as shown on Table 8.5. As in previous numerical solutions the second column shows the calculations for our first guess, with a ratio at the bottom that must = 1.00 for the correct solution. In the third column we let the spreadsheet's numerical solution engine find the value of the guessed variable that makes that ratio = 1.00, thus solving the problem.

For our first guess we assume that $\mathcal{M}_x = 1.10$, from which we calculate from Eq. 8.37 $\mathcal{M}_y = 0.9188$. Then from Eq. 8.41 we calculate $P_{R_y} / P_{R_x} = 0.9989$. Here, because all the flow is isentropic except for the flow through the shock wave, we have $P_{R_1} = P_{R_x}$ and $P_{R_2} = P_{R_y}$. Then

$$\frac{P_2}{P_{R_2}} = \frac{P_2}{P_{R_1}} \cdot \frac{P_{R_x}}{P_{R_y}} = \frac{0.8022}{0.9989} = 0.8031 \tag{8.AP}$$

and

$$\frac{A_x^*}{A_y^*} = \frac{P_{R_y}}{P_{R_x}} = 0.9989 \tag{8.AQ}$$

so that

$$\left(\frac{A}{A^*}\right)_2 = \frac{A_2}{A_1^*} \cdot \frac{A_x^*}{A_y^*} = 1.5 \cdot 0.9989 = 1.4984 \tag{8.AR}$$

From the value of P_2 / P_{R_2} we compute $\mathcal{M}_{\text{exit}} = 0.5686$ from Eq. 8.18 and from Eq. 8.24 $(A / A^*)_2 = 1.2282$. Finally, the ratio of the two calculated values of $(A / A^*)_2$ is $(1.4984 / 1.2282) = 1.2200$. If we had guessed the correct

TABLE 8.5

Results of trial-and-error solution to Example 8.16

Trial number	1, based on guessed $\mathcal{M}_x$	2, solved numerically
P_2 / P_{R_1}, given	0.8022	0.8022
$\mathcal{M}_x$, guessed	**1.1000**	**1.4878**
$\mathcal{M}_y$, Eq. 8.38	0.9118	0.7055
P_{R_y} / P_{R_x} Eq. 8.42	0.9989	0.9336
P_2 / P_{R_2}	0.8031	0.8592
$(A / A^*)_2$, based on P_{R_y} / P_{R_x}	1.4984	1.4004
$\mathcal{M}_2$, based on P_2 / P_{R_2} and Eq. 8.17	0.5686	0.4706
$(A / A^*)_2$, based on $\mathcal{M}_2$ and Eq. 8.21	1.2282	1.4004
Ratio of two $(A / A^*)_2$ values	1.2200	**1.0000**

value of $\mathcal{M}_x$, this ratio would be 1.00. Then (column 3), we let the spreadsheet's numerical solution engine find the value of $\mathcal{M}_x$ that makes this ratio equal to 1.00.

For an assumed upstream Mach number of 1.4878, the two calculated values of $(A / A^*)_2$ are the same, so this is the correct solution. Now we can locate the shock in the duct: It occurs at the place where the upstream $\mathcal{M}_1 = 1.4878$. Using this value in Eq. 8.24, we find that the shock stands where $A / A_{\text{throat}} \approx 1.1680$. We calculate pressures to the left of this point by the isentropic-flow equations, using $A^* = A_{\text{throat}}$ with supersonic flow from here to the throat and subsonic on the upstream side of the throat. We calculate pressures to the right of this point by the subsonic isentropic-flow equations, using $A^* = A_{\text{throat}} / 0.9336$. The entire curve is sketched in Fig. 8.16 as curve *AD*. ■

In this example, we must keep track of three Mach numbers, $\mathcal{M}_x$, $\mathcal{M}_y$, and $\mathcal{M}_2$, time spent identifying those is time well spent. This same example can be solved by trial and error using the values in Table A.5; the solution that way is shown in the first two editions of this book. Now that we all have good computers the spreadsheet solution shown here seems more satisfactory.

If we continue to lower P_2, the shock wave will move to higher and higher values of $\mathcal{M}_x$, to the right on Fig. 8.16. The farthest to the right that is possible (with the shock inside the nozzle) is at the nozzle's very exit; we can calculate the upstream Mach number there from the area ratio (by interpolation in App. A.5), $\mathcal{M}_x = 1.854$, and $P_x / P_{R_1} = 0.1601$, so that $P_2 / P_{R_1} = 0.1601 (P_y / P_x) = 0.1601 \cdot 3.844 = 0.6154$. This condition is shown as curve *AE* on the figure. If we continue to lower the pressure in the downstream reservoir below that shown at *E*, then the shock wave cannot occur in the nozzle at all, and the flow must exit at $\mathcal{M}_2 = 1.854$, the pressure shown at *G*. If the pressure in the reservoir is more than that at *G* (for example, that at *F*), then there will be a shock wave outside the nozzle, which will bring the flow up to the pressure of the reservoir. This will not be a normal shock but, rather, a two- or three-dimensional shock. If the pressure is exactly equal to that at *G*, the flow will be isentropic throughout, and there will be no shock wave at all. If the pressure in the downstream reservoir is less than that at *G* (for example, that at *H*), then the pressure adjustment outside the nozzle will take place via a two- or three-dimensional rarefaction.

8.8 SUMMARY

1. The speed of sound is the speed of propagation of a *small* pressure disturbance. For ideal gases the speed of sound is given by $(kRT / M)^{1/2}$.
2. From an energy balance alone we can relate temperature to Mach number for steady, adiabatic flow (isentropic or nonisentropic). By using the pressure-temperature relation for an isentropic change in an ideal gas we can complete the mathematical description of steady, frictionless, adiabatic, ideal gas flow. The mass-balance equation is used to solve for the cross-sectional area perpendicular to the flow.

3. When flow in a nozzle or duct becomes sonic, further lowering of the downstream pressure will not cause the flow upstream of the sonic point to increase. This condition, called choking, is very common in gas flow in valves, orifices, etc.
4. In high-velocity subsonic flow with friction the effect of the friction is to lower the pressure, thus lowering the density, thus increasing the velocity, ultimately leading to sonic flow and choking.
5. A normal shock wave is a large pressure disturbance that travels faster than the local speed of sound. Normal shock waves are irreversible, causing an increase of the entropy of the fluid flowing through them.
6. The only known way of producing a steady supersonic flow is by means of a converging-diverging nozzle. Such a nozzle is also the only known isentropic way of converting a steady, supersonic flow to a subsonic flow. A supersonic flow may also be converted to a subsonic one by a shock wave, which is not isentropic. Unsteady supersonic flows can be produced in several ways, including explosions.
7. High-velocity gas flow is of great practical significance in aerodynamics, rocket and turbine design, high-speed combustion, ballistics, etc. The pressure differences needed to produce high-velocity gas flow are modest, so whenever the pressure of a gas is reduced in a pipe, valve, or other fitting, high-velocity flow is likely to occur.
8. This chapter treats only the simplest cases, showing why this kind of flow is different from the flow of liquids or the flow of gases at low velocities. The reader who wishes to find out more about this fascinating subject should consult Oosthuizen and Carscallen [3] or Shapiro [4].

PROBLEMS

See the Common Units and Values for Problems and Examples, inside the back cover. An asterisk (*) on a problem number indicates that its answer is shown in App. D. In all problems in this section, unless stated to the contrary, assume that gases are ideal gases with the properties shown in Table 8.6.

8.1.* Assuming a friction factor $f = 0.005$, calculate the pressure drop per foot due to friction for flow in a 2-in ID circular pipe at a velocity of 1000 ft / s for

(*a*) Water.

(*b*) Air. Use the constant-density formulae developed in Chap. 6.

TABLE 8.6

Gas properties to be used in all problems unless stated to the contrary

Gas	M, g / mol = lbm / lbmol	$k = C_P / C_V$
Air	29	1.4
Helium	4	1.667
Hydrogen	2	1.4
Steam	18	1.33

8.2. Assuming that a sound wave causes an isentropic pressure rise of 10^{-3} psia in air, calculate the temperature rise caused by such a wave. Isentropic relations for ideal gases (e.g., air) are shown in App. B.3.

8.3. The "loudness" of sounds is a nonlinear function of the pressure rise caused by a passing sound wave. The standard unit of reporting (and regulating) sound for industrial hygiene and public nuisance purposes is the *decibel, dB*. The definition is

$$\begin{pmatrix}\text{Sound loudness} \\ \text{expressed in } dB\end{pmatrix} = 10 \log \frac{I}{I_0} = 20 \log \frac{P}{P_0} \tag{8.43}$$

Here I is the *intensity* of the sound wave, its energy per unit area, normally expressed in W/m^2. The intensity seems to correspond best to human perceptions of the loudness of sounds.

The intensity is proportional to the square of the pressure rise, P, leading to the 20 in the definition of the dB for sound, which is different from the 10 in most electrical engineering applications. The most common datum definition is $P_0 = 2 \cdot 10^{-4}$ dyne / cm^2 $= 2 \cdot 10^{-5}$ Pa $= 2.9 \cdot 10^{-9}$ psi. Using this definition, estimate the pressure corresponding to the following:

(*a*) 1 *dB*, which is the lowest sound pressure that young adult ears can detect.

(*b*) 60 *dB*, which is the sound level in normal conversation at a distance of about 3 ft.

(*c*) 140 *dB*, which is the sound of a jackhammer, and a typical threshold of pain.

8.4.*Calculate the speed of sound in wood. For most woods, $K \approx 1.5 \cdot 10^6$ psi and $\rho \approx 60$ lbm / ft^2. Is the speed of sound likely to be the same "with the grain" as "across the grain" of the wood?

8.5. Calculate the speed of sound in acetic acid at 20°C, at which $K = 0.110 \cdot 10^{10}$ Pa. and $\rho \approx 1.049$ g / cm^3.

8.6.*Calculate the bulk modulus of water at 212°F from Keenan and Keyes' steam tables or equivalent.

8.7. Calculate the speed of sound in helium gas at 500°F.

8.8.*Uranium hexafluoride (the gas used in the gaseous-diffusion separation of uranium isotopes) has $M = 352$ g / mol. Assuming that for it $k = 1.2$, calculate its speed of sound at 200°F.

8.9. Air is flowing in a vertical nozzle 1 ft high. In the nozzle its velocity changes from 1 ft / s to 2000 ft / s. What is the ratio of the change of potential energy to the change in kinetic energy?

8.10.*Rework Example 8.3 for helium gas.

8.11. Rework Example 8.4 for hydrogen gas.

8.12.*Rework Example 8.5 for helium gas. Assume that $P_R = 30$ psia, and $T_R = 70$°F. Include the values of T_1 and P_1.

8.13. Air flows isentropically from a reservoir through a nozzle; $P_R = 60$ psia, and $T_R = 100$°F. At some point in the nozzle the Mach number = 0.60. At that point, what are the pressure, temperature, and velocity?

8.14.*Repeat the previous problem for helium.

8.15. Air is flowing from a reservoir into a nozzle in isentropic flow; $P_R = 60$ psia, and $T_R = 40$°F. At the point where the velocity is 1300 ft / s, what are the temperature and pressure?

8.16.*Helium is flowing from a reservoir through a nozzle in isentropic flow; $P_R = 14.7$ psia, and $T_R = 100$°F. At the point in the nozzle where the Mach number = 1.00, what are the pressure, temperature, velocity, and density?

8.17. The compressed air-line at a service station has in it air at $P = 125$ psia, and $T = 70$°F. We now open the valve and let some of this air flow into the atmosphere, through a valve that may be considered an isentropic nozzle. What is the temperature of the compressed air when it flows into the atmosphere? This is cold enough that it should cause condensation of the water in the atmosphere, forming a cloud. Do we normally see such a cloud? If not, why not?

8.18.*Air flows from a reservoir into a nozzle in isentropic flow; $T_R = 100°F$, and $P_R = 14.7$ psia. At the point in the nozzle where V is 1200 ft / s, what are P, T and $\mathcal{M}$?

8.19. Hydrogen is flowing steadily and isentropically through a nozzle; $P_R = 20$ psia, and $T_R = 60°F$. What is the Mach number at the point where the velocity is 6000 ft / s?

8.20.*Steam is flowing steadily and isentropically from a reservoir through a converging-diverging nozzle; $P_R = 50$ psia, and $T_R = 600°F$. What are the temperature, pressure, and velocity at the point where $\mathcal{M} = 2.0$?

8.21. Helium flows from a reservoir through a converging-diverging nozzle in steady, isentropic flow. At the point where $\mathcal{M} = 1.8$ the temperature is 500°R and the pressure is 20 psia. What are the reservoir pressure and temperature?

8.22.*Air at $P_R = 30$ psia, and $T_R = 70°F$ flows through a nozzle with a throat area of 1 in^2. What is the maximum mass flow rate that can be passed through this nozzle?

8.23. The required test flow rate for the pressure relief valves on uninsulated containers of liquefied gases [7] is

$$Q_a = 2 \cdot 0.00154 \cdot P \cdot Wc \tag{8.AS}$$

where Q_a is the air flow rate in standard cubic feet per minute (1 atm, 60°F) through a wide-open relief valve, P is the test pressure in psia, and Wc is the "water capacity" of the container, equal to the weight of water required to fill the container completely full.

The "20 lb" propane containers used in backyard barbecues (of which about 50 million are in use in the United States) have $Wc \approx 48$ lbm, and $P_{\text{test}} = 480$ psig. The propane industry normally equips such cylinders and smaller ones with a valve sized for a "40 lb" container, with $Wc = 96$ lbm, to avoid having to make a special valve for each size of cylinder smaller than "40 lb."

(*a*) What is the required Q_a for the pressure relief valve on these containers?

(*b*) If we model the wide-open relief valve as a simple orifice with $C_v = 0.6$, using B.E., how large would its diameter have to be to pass this flow of air at a pressure difference of (480 − 14.7) psig? Use the ideal gas approximation in Example 5.2.

(*c*) If we model this by the isentropic flow models of this chapter, how large would the diameter have to be?

(*d*) The answer in part (*c*) is for a frictionless nozzle. The actual design of these valves corresponds more closely to a square-edged orifice than a rounded nozzle. The orifice coefficient for choked flow through an orifice with downstream pressures $<<$ reservoir pressure is ≈ 0.84 [8]. Repeat part (*c*) with this orifice coefficient.

(*e*) How do these three simple estimates compare with the actual diameter used, which is approximately 0.156 in?

(*f*) Discuss why such a pressure relief valve is required on containers like these.

(*g*) If you have access to such a container, find the pressure relief valve. It will normally be marked 375 psi, which is the common setting for such valves.

8.24. A vacuum tank is connected to a vacuum pump through a valve. The tank has a volume of 100 ft^3; the pump has a volumetric flow rate of 10 ft^3 / min, independent of the density of the material flowing through it. The valve (when wide open) is the equivalent of a reversible nozzle with cross-sectional area 10^{-4} ft^2. All of the piping of the

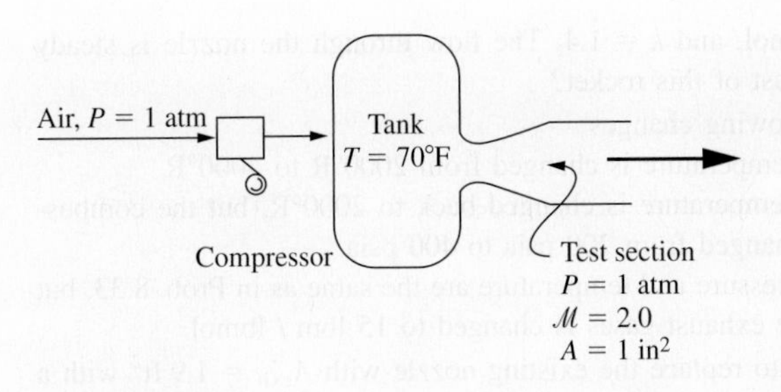

FIGURE 8.17
Flow diagram for a simple laboratory wind tunnel; see Prob. 8.26.

system is very large compared to the dimensions of the valve. The vacuum tank is full of air at a pressure of 1 atm. Heating coils hold the temperature of the air in the vacuum tank at 70°F. The valve is opened wide and the pump started. How long does it take the tank pressure to fall to 0.1 atm?

8.25.*Air P_R = 100 psia, and T_R = 100°F passes through an isentropic, converging nozzle into a second vessel whose pressure is 80 psia. The area of the nozzle at its minimum is 2 in^2. What is the mass flow rate?

8.26. A laboratory wants to design a steady-state, supersonic wind tunnel. This tunnel is to exhaust a steady flow of air at $\mathcal{M} = 2$ and $P = 1$ atm to the atmosphere. The exit area is to be 1 in^2. As sketched in Fig. 8.17, the whole system consists of a compressor, which takes in air from the atmosphere, a tank, which stores it at 70°F, and a converging-diverging nozzle. Calculate the required tank pressure, the mass flow rate, and the compressor horsepower needed to supply this tank and nozzle. For an ideal gas in an isothermal compressor the work per unit mass is $-W/m = (RT/M) \ln (P_2/P_1)$.

8.27. Check the results of Examples 8.3 and 8.5 by using App. A.4.

8.28. You have been commissioned to make up for helium a table analogous to App. A.4. Show that you know how to do this by calculating all the table entries corresponding to App. A.4 for $\mathcal{M} = 0.8$. (V/c^* is given by Eq. B.6-12.)

8.29. Section 5.6 shows a comparison of the velocities for flow through a converging nozzle, calculated by means of (*a*) Bernoulli's equation, assuming that air is a constant-density fluid, and (*b*) the equations for isentropic flow. Show the calculations in the latter case, and check your answers against those shown in Sec. 5.6.

8.30. Steam at P_R = 100 psia, and T_R = 600°F is expanded through a nozzle in steady, isentropic flow. Calculate the temperature, pressure, and Mach number at the point where the velocity is 2000 ft / s from the following:

(*a*) The Keenan and Keyes steam tables or their equivalent (which do not assume ideal gas behavior) and

(*b*) The equations for steady, isentropic flow of an ideal gas.

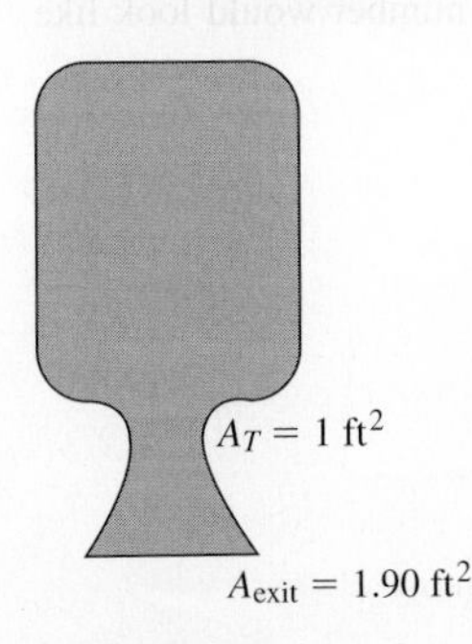

FIGURE 8.18
A simple rocket with a converging-diverging nozzle; see Prob. 8.33.

8.31.*In Example 8.9, assume that at the point where $\mathcal{M} = 0.5$ the pressure is 20 psia and the density 0.102 lbm / ft^3. What are the pressure and density at the point where $\mathcal{M} = 2.0$?

8.32. Rework Example 8.9, not by making the assumption of an upstream reservoir but, rather, by deriving the equivalent of Eq. 8.17 for two arbitrary states, state 1 and state 2, at each of which the velocity is not negligible.

8.33.*A rocket has the throat area and exit area shown in Fig. 8.18. It is to be fired while being held down by the test stand. In the combustion chamber the velocity is negligible, P = 200 psia,

$T = 2000°R$, $M = 20$ lbm / lbmol, and $k = 1.4$. The flow through the nozzle is steady and isentropic. What is the thrust of this rocket?

8.34. Repeat Prob. 8.33 with the following changes
(*a*) The combustion chamber temperature is changed from 2000°R to 4000°R.
(*b*) The combustion chamber temperature is changed back to 2000°R, but the combustion chamber pressure is changed from 200 psia to 400 psia.
(*c*) The combustion chamber pressure and temperature are the same as in Prob. 8.33, but the molecular weight of the exhaust gases is changed to 15 lbm / lbmol.

8.35. In Prob. 8.33 we have decided to replace the existing nozzle with $A_{\text{exit}} = 1.9$ ft^2 with a new nozzle, with a different exit area, in which the exit pressure will be exactly atmospheric pressure (14.7 psia). The chamber pressure and temperature and the throat area will not be changed.
(*a*) What is new exit velocity?
(*b*) What will the thrust of the rocket be with this new nozzle?

This problem gets you beyond the highest Mach numbers in App. A.4. You must use the equations on which that table is based.

8.36. In the vacuum system shown in Fig. 8.19 it has been decided that the pressure in the chamber, 0.01 mm Hg, is not low enough. Someone proposes that, if we use a bigger vacuum pump, we can get the pressure lower. Is he right? What do you recommend?

8.37. A process vessel has a pressure that fluctuates between 15 and 35 psia. The pressure never exceeds 35 psia. We want to design a system to admit air into this vessel at a steady rate of 1 lbm / h. Our compressed-air supply main is at $P = 100$ psia and $T = 70°F$. This temperature and pressure do not fluctuate. Your competitor for a forthcoming promotion has proposed that we install a conventional flow-rate control system consisting of an orifice meter, differential-pressure transducer, controller, and control valve in a line between the compressed-air main and the vessel. Have you a suggestion that is likely to win you the promotion?

8.38. For the data shown in Example 8.10, when the flow is choked at the pipe outlet, what are the pressure, temperature, and Mach number at the outlet of the nozzle (station 1 on Fig. 8.11)?

8.39. In Example 8.10, estimate the pressure at the point exactly halfway between the end of the frictionless nozzle and the end of the pipe.

8.40. Make a sketch to indicate what the lines of constant outlet Mach number would look like in Fig. 8.12.

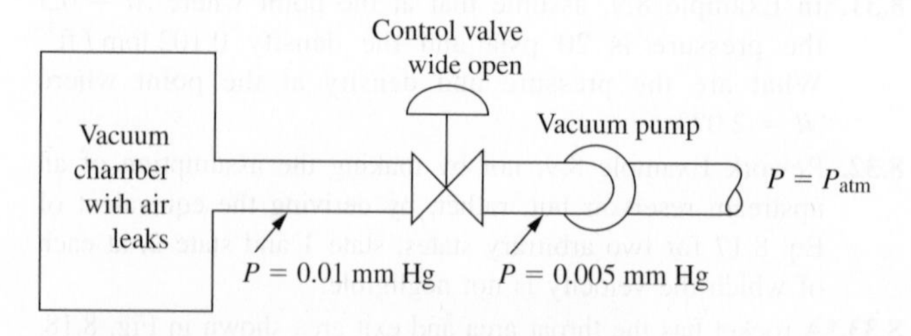

FIGURE 8.19
A vacuum chamber and vacuum pump; see Prob. 8.36.

8.41. On Figure 8.12, it appears that for any value of k the states corresponding to choked flow form practically straight lines on the coordinates of that figure. Is that

(*a*) A coincidence? i.e., the line does not have to be straight, but the numbers just fall out that way.

(*b*) Something that is required by the assumptions and the mathematics that went into making up that figure and that require the line to be straight?

8.42. For the case shown in Example 8.10, prepare a plot of pressure versus length from one reservoir to the other for

(*a*) $P_3 / P_0 = 0.8$.

(*b*) $P_3 / P_0 = 0.4$.

(*c*) $P_3 / P_0 = 0.2$.

8.43.*For the flow shown in Figs. 8.10 and 8.11, if $P_0 = 150$ psia and $P_3 = 14.7$ psia, what is the longest 1-in schedule 40 pipe for which the flow will not be choked at the outlet? Assume that $f \approx 0.0060$ independent of length.

8.44. The $N = 0$ line in Fig. 8.12 corresponds to an isentropic nozzle only. Check to see whether this line corresponds to the equations in Sec. 8.2 by calculating the mass flow rate per unit area for air corresponding to $\mathcal{M}_1 = 0.5$ and $\mathcal{M}_1 = 1.0$ from the equations in that section and comparing the results with the $N = 0$ line in Fig. 8.12.

8.45. A pressure vessel contains air at 30 psia at 100°F. The air is to be vented to the atmosphere through a frictionless nozzle and a length of 1-in schedule 40 steel pipe of undetermined length, which may take values between 4 and 800 ft.

(*a*) Prepare a sketch of the flow rate in lbm / s versus the length of this pipe.

(*b*) Repeat the calculation for a square-cornered pipe entrance instead of a frictionless nozzle.

8.46. Set up the spreadsheet program shown in Table 8.3. Verify the results shown there. Then use it to find the value of $\dot{m} / A$ for $P_3 = 12$ psia.

8.47. In developing Eq. 8.33 we dropped the $\rho V\, dV$ term in Eq. 8.31, asserting that it was much smaller than the others. For a typical long-distance constant-diameter pipeline the pressure is 750 psia at the outlet of a compressor station and 500 psia at the inlet of the next compressor station. The fluid may be considered an ideal gas with a constant temperature of 70°F and a molecular weight of 18 lbm / lbmol. If the velocity at the outlet of the first compressor station is 20 ft / s, what is the ratio of the first two terms in Eq. 8.31?

8.48. Sketch a plot of P_2 versus distance, as predicted by Eq. 8.33, for constant mass flow rate, friction factor, etc.

8.49. A natural-gas line has an inside diameter of 36 in. The compressor stations are located 60 mi apart; the pressure at the outlet of the first is 750 psia, and the pressure at the inlet of the second is 500 psia. The gas may be considered an ideal gas with molecular weight 18. The temperature is constant at 70°F. What is the mass flow rate, according to the Weymouth equation? How much different would the answer have been if we had used the constant-density B.E. (Sec. 6.5) with friction and based the friction calculation on the following:

(*a*) The upstream density and velocity.

(*b*) The average density and velocity from inlet to outlet.

8.50. For $\mathcal{M}_x = 1.50$ calculate $\mathcal{M}_y$, T_y / T_x, P_y / P_x, and ρ_y / ρ_x for a normal shock in a gas with $k = 1.4$. Compare your results with those in App. A.5.

8.51. Show that compression shocks are thermodynamically possible and rarefaction shocks are not. The procedure is as follows. Find the entropy change by substituting Eqs. B.6-18 and B.6-22 from App. B in Eq. B.3-29 to find

$$M\frac{s_y - s_x}{R} = \ln\left(\left\{\frac{\left[\left(\frac{2k}{k-1}\right)\mathcal{M}_x^2 - 1\right]\cdot\left[1+\left(\frac{k-1}{2}\right)\mathcal{M}_x^2\right]}{\mathcal{M}_x^2\,\dfrac{(k+1)^2}{2(k-1)}}\right\}^{k/(k-1)} \cdot\left[\frac{k+1}{2k\mathcal{M}_x^2 - (k-1)}\right]\right) \tag{8.44}$$

Then show that for $k < 1.67$ (that is, for all gases) $s_y - s_x$ is positive for $\mathcal{M}_x > 1$ and negative for $\mathcal{M}_x < 1$.

8.52. Why does App. A.5 not contain a column of figures for A_y^*/A_x^* for normal shock waves?

8.53.*An earth satellite is entering the upper atmosphere at 30,000 km / h. The air it is entering has $T = 225$ K. Estimate the temperature of the gas in contact with the surface of the satellite.

8.54.*A jet fighter plane is flying at $\mathcal{M} = 2$ in still air that has $T = 0°\text{F}$ and $P = 4.0$ psia. The air inlet to the jet engine is a converging-diverging diffuser. Inside the engine the flow is subsonic. $A_{\text{throat}} = 2.0\ \text{ft}^2$. What is the mass flow rate through this diffuser, assuming isentropic flow?

8.55. A wind tunnel has a flow with $\mathcal{M} = 2$ at $T = 300°\text{R}$. If we insert a thermometer at this point, what will it read? *Hint:* If we assume that the air has zero thermal conductivity, the solution is quite simple. If the air can conduct heat (as it actually does), then the problem is more complex, and we cannot give more than an approximate answer without experimental heat-transfer data.

8.56.*Air at 70°F and 10 psia is flowing at 500 ft / s in a pipe. A valve at the end of the pipe is suddenly closed. This causes a shock wave to form at the closed end of the duct and move up it. Calculate the speed at which this wave moves up the duct and the temperature and pressure in the closed end of the duct. *Hint:* The gas downstream of the shock is standing still. Regardless of which coordinate system is chosen, we have $V_x - V_y = 500$ ft / s. If we take the coordinate system to be based on the moving shock wave, then we can use the tabulated values in App. A.5 to solve (by trial and error) for the upstream Mach number that has this relation between the two velocities. This is conceptually the same as Prob. 7.53 for a hydraulic jump moving upstream. The solution method is the same, but because we have one more variable in high velocity gas flow than in liquid flow, the mathematics are somewhat messier. (An analytical solution to this type of problem is shown by Zucker [9].)

8.57. At some point in the nozzle shown in Fig. 8.16 the cross-sectional area is 1.5 times the throat area. If we assume that the flow is the isentropic, steady flow of an ideal gas with no shock waves and $k = 1.4$, what are the possible Mach numbers at that point when the Mach number at the throat is 1.0? What is the Mach number at that point when the Mach number at the throat is 0.1, 0.5, and 0.9?

8.58. A converging-diverging nozzle has an outlet area equal to 1.9 times the throat area. Using Apps. A.4 and A.5, prepare a plot of outlet Mach number versus P_{out}/P_R for all

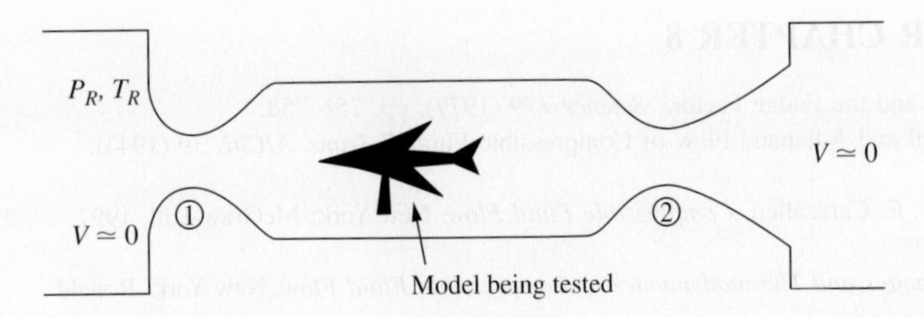

FIGURE 8.20
A wind tunnel; see Prob. 8.61.

possible outlet conditions, subject to the assumptions of isentropic flow, with or without normal shock waves.

8.59. In a converging-diverging nozzle the cross-sectional area at point x (downstream of the throat) is 1.5 times the cross-sectional area at the throat. Air with $T_R = 530°\text{R}$ is flowing steadily through the nozzle.
(*a*) For isentropic flow list all of the possible values of T at point x.
(*b*) For flow that is isentropic except for possible shock waves list all of the possible values for T at point x.

8.60. In a converging-diverging nozzle the exit area is 1.50 times the throat area. The flow of air is isentropic except for the possibility of shock waves. What is the Mach number at the throat when the Mach number at the exit is
(*a*) 0.30 and
(*b*) 0.50?

8.61. Air flows through a supersonic wind tunnel; see Fig. 8.20. The flow is steady and isentropic, except that somewhere in the system there is a normal shock wave. If $A_2 / A_1 = 1.01$ and $\mathcal{M} = 1.0$ at both A_1 and A_2, what is the upstream Mach number at the place where the normal shock occurs?

8.62.*Repeat Example 8.14 for an outlet pressure of 0.70 P_{R_1}. What is the lowest outlet pressure for which one can perform this calculation?

8.63. In the nozzle in Fig. 8.21 air is flowing, and there is a normal shock wave at A. The remainder of the flow is isentropic. For this shock wave we have $\mathcal{M}_x = 3.0$. What is P_{R_y} / P_{R_x}? Some of the following values from the National Advisory Committee for Aeronautics tables [10] may be useful. For $\mathcal{M}_x = 3$ we have $\mathcal{M}_y = 0.4752$, $P_y / P_x = 10.33$, $T_y / T_x = 2.679$, and $\rho_y / \rho_x = 3.857$.

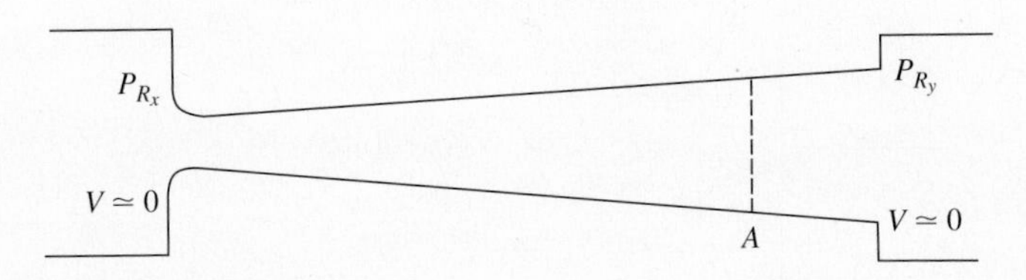

FIGURE 8.21
A nozzle with a shock wave; see Prob. 8.63.

REFERENCES FOR CHAPTER 8

1. Westfall, R. S. "Newton and the Fudge Factor." *Science 179* (1979), pp. 751–758.
2. Lapple, C. E. "Isothermal and Adiabatic Flow of Compressible Fluids." *Trans. AIChE 39* (1943), pp. 385–432.
3. Oosthuizen, P. H. and W. E. Carscallen. *Compressible Fluid Flow.* New York: McGraw-Hill, 1997, Chapter 9.
4. Shapiro, A. H. *The Dynamics and Thermodynamics of Compressible Fluid Flow.* New York: Ronald Press, 1958.
5. Katz, D. L., et al. *Handbook of Natural Gas Engineering.* New York: McGraw-Hill, 1959.
6. Taylor, G. I. "The Formation of a Blast Wave by a Very Intense Explosion," *Proc. Roy. Soc. Ser. A 201* (1950), pp. 159–187.
7. Compressed Gas Association, Inc. *GCA S-1.1-2001 Pressure Relief Device Standards Part 1, Cylinders for Compressed Gases.* 9th ed. Arlington, Virginia: CGA, p. 11.
8. Boyce, M. P. "Transport and Storage of Fluids." In *Perry's Chemical Engineers' Handbook.* 7th ed., D. W. Green and J. O. Maloney. New York: McGraw-Hill, 1997, pp. 10–16.
9. Zucker, R. D. *Fundamentals of Gas Dynamics.* Champaign, Illinois: Matrix Publishers, 1977, p. 175.
10. National Advisory Committee for Aeronautics. *Report 1135, Tables and Charts for Compressible Flow.* Washington, D.C.: U.S. Government Printing Office, 1953.

PART III

SOME OTHER TOPICS THAT CAN BE VIEWED BY THE METHODS OF ONE-DIMENSIONAL FLUID MECHANICS

The next six chapters cover a variety of topics, which can be considered in any order, and from which the student or instructor may choose or omit. Many of them are simple introductions to topics that fill large books. The goal in most of them is to show how these topics relate to the basic ideas of Parts I and II, and how the terms and symbols of those topics relate to the corresponding terms and symbols in Parts I and II.

CHAPTER 9

MODELS, DIMENSIONAL ANALYSIS, AND DIMENSIONLESS NUMBERS

9.1 MODELS

A model is an intellectual construct that represents reality and that can be manipulated to predict the consequences of future actions. Most of engineering is the application of mathematical models to practical problems. For example, $F = ma$ is a mathematical model of the relation between force, mass, and acceleration. Using it, engineers have been spectacularly successful in predicting the behavior of real physical systems. Much more complex mathematical models are regularly used; as the size and power of our computers have grown, the size and complexity of the mathematical models we can use has grown. Maintain your skepticism about your models; to quote the statistician George Box, "All models are wrong; some models are useful."

There are still many problems for which we have insufficient confidence in our mathematical models to risk large sums of money or human lives on their predictions without first testing those predictions with some kind of a physical model. For example, although we have made great strides in computational power, we still will not build a major new type of aircraft based on calculations alone, without wind tunnel tests of smaller-scale models of it to verify the computations. We know a great deal about chemical reactors, but not enough that we will build a full-scale plant to produce a chemical by a novel reaction scheme without some bench-scale or pilot-plant tests to verify that our mathematical models are reliable. We still cannot predict the behavior of complex structures under earthquake loads with total confidence, so we resort to testing of physical models there as well. Even automobile designers currently use wind-tunnel tests on models of new designs, to determine their air resistance! For problems like these, in which we still do not have completely reliable mathematical models, engineers have developed some valuable techniques for making useful predictions without a complete answer. These methods played a greater role in the precomputer era than they do now. For that reason, this chapter is more part of an engineer's historical-cultural background than of an engineer's current technical tool kit. However, the methods described here often provide a useful physical insight into problems that will complement the insight that mathematical models and computer solutions provide.

These methods are mostly based on studies using physical model tests. If we cannot calculate the behavior of a new shape of ship hull or airplane or a new type of chemical reactor, then we must build it and test it. If we can build and test a small model of the finished product instead of a full-sized airplane or ship or reactor, we will save time, money and, possibly, the lives of test pilots. (Have your failures on a small scale, in private; have your successes on a large scale, in public!) The enormous progress of the aircraft industry from the first powered flight in 1903 to the present is largely due to the fact that engineers have learned to test new designs by using small-scale models and to use the test results for designing full-scale airplanes. Similarly, the progress of the chemical industry is largely due to the chemical engineer's ability to scale up bench-size or pilot-size plants to commercial plants with confidence that the resulting plants will perform as predicted. With the advent of big computers we can do much more by calculation than we could just 10 years ago. For that reason we do more computing (mathematical model testing) and less physical model testing than we did in the past. Nonetheless, really important engineering decisions are most often made on the basis of a combination of computations and physical model tests, with the physical model tests serving to confirm the computations before the full-scale airplane, boat, or chemical plant is built.

However, just making and testing a scale model of the airplane, boat or chemical reactor is not the whole story, as indicated by the following quotation from J. B. S. Haldane [1]:

> The most obvious differences between different animals are differences of size, but for some reason the zoologists have paid singularly little attention to them. In a large textbook of zoology before me I find no indication that the eagle is larger than the sparrow, or the hippopotamus bigger than the hare, though some grudging admissions are made in the case of the mouse and the whale. But yet it is easy to show that a hare could not be as large as a hippopotamus, or a whale as small as a herring. For every type of animal there is a most convenient size, and a large change in size inevitably carries with it a change of form.
>
> Let us take the most obvious of possible cases, and consider a giant man sixty feet high—about the height of Giant Pope and Giant Pagan in the illustrated *Pilgrim's Progress* of my childhood. These monsters were not only ten times as high as Christian, but ten times as wide and ten times as thick, so that their total weight was a thousand times his, or about eighty to ninety tons. Unfortunately the cross sections of their bones were only a hundred times those of Christian, so that every square inch of giant bone had to support ten times the weight borne by a square inch of human bone. As the human thigh-bone breaks under about ten times the human weight, Pope and Pagan would have

broken their thighs every time they took a step. This was doubtless why they were sitting in the picture I remember. But it lessens one's respect for Christian and Jack the Giant Killer.

To turn to zoology, suppose that a gazelle, a graceful little creature with long thin legs, is to become large; it will break its bones unless it does one of two things. It may make its legs short and thick, like the rhinoceros, so that every pound of weight has still about the same area of bone to support. Or it can compress its body and stretch out its legs obliquely to gain stability, like the giraffe. I mention these two beasts because they happen to belong to the same order as the gazelle, and both are quite successful mechanically, being remarkably fast runners.*

9.2 DIMENSIONLESS NUMBERS

If, as shown in the preceding quotation, holding the shape constant while increasing the size does not guarantee equal performance by differently sized equipment or animals, what does? In the case of the bones of the giant, presumably one should keep constant the ratio stress/crushing strength. To do this, as the height of the giant is increased, the crushing strength of the giant's bones could be increased. Alternatively, the average density of the giant's body could be lowered, to keep the same bone stress while the height increased; or the giant could go to a planet with a lower acceleration of gravity. Combining all these possibilities, we see that, if the giant keeps the ratio of

$$\text{Bone ratio} = \frac{\text{height} \cdot \text{average body density} \cdot \text{acceleration of gravity}}{\text{crushing strength of bones}} \tag{9.1}$$

constant, then the giant can become any height with the same relative resistance to bone failure as a human has. This ratio is dimensionless, a pure number (given the fanciful name, Bone ratio).

Thus, we might suspect that dimensionless ratios like the Bone ratio would be important in predicting the behavior of a large piece of equipment from tests of a small model. Experience indicates that this is certainly the case. These dimensionless numbers have also proven invaluable in correlating, interpreting, and comparing experimental data. For example, if you were asked to compare a business venture in which you invest $4000 for a return of $400 per year with one in which you invest $6000 for a return of $650 per year, you would certainly make the comparison through the most common of all dimensionless numbers, the percentage.

Furthermore, there is often a real benefit in understanding if we can show our experimental or computational results in dimensionless form. The behavior of nature does not depend on what system of dimensions we humans choose to describe that behavior. Thus, the results of our observations, if they are correct, can be expressed in a form that is totally independent of the system of units we use. If they cannot, then we should wonder whether they are correct. Finding the best way to reduce observations made in any system of units to dimensionless form is always a good test of the experimental data, and often a good test of our understanding of that data.

---

*From J. B. S. Haldane, *Possible Worlds,* Harper & Row, New York, 1928; reprinted in J. R. Newman, *The World of Mathematics,* Simon and Schuster, New York, 1956, p. 952 *et seq.* Quoted by permission of the publisher.

So far in this book we have seen the following dimensionless numbers: $\mathcal{R}, f$, C_d, Fr, $\mathcal{M}$. In addition we have seen ratios of dimensions, e.g., $\Delta x / D$ and ε / D for pipe flow and z_2 / z_1 for hydraulic jumps. In Fig 7.17 we saw that the interactions of a jet of fluid with a moving blade could be best understood on a plot of a dimensionless energy ratio versus a dimensionless velocity ratio. In Chap. 8 all the parameters in App. A.4 and A.5 are presented as dimensionless ratios. This is a simple matter of convenience; presenting them any other way would be very cumbersome. Finally, you may not even think of e, π, θ, and $\sin\theta$, as dimensionless numbers, but they all are.

The application of dimensional analysis shown in Fig. 6.10, one of the most celebrated applications of dimensional analysis in fluid mechanics history, produces the following benefits:

1. It reduces a problem with six dimensioned variables to one with three dimensionless variables. That allows us to get all the available friction data for steady pipe flow onto one plot, and then finally into two dimensionless equations. This is a great economy of presentation and effort compared to the original tables and plots of individual experiments on which Fig. 6.10 is based.
2. It produces a solution that is universal, independent of what system of units the measurements were taken in or any of the other local idiosyncrasies of the experiments.
3. From an understanding of the physical meaning of the dimensionless variables (discussed in this chapter) one can develop more insight into the physics of the data represented on Fig. 6.10 than one could develop from plots or tables of those data in their original dimensional form.
4. If one is planning an experimental investigation of some new phenomenon, finding the right dimesionless variables first shows what experimental values will most satisfactorily clarify the phenomenon.

The important dimensionless numbers are almost all the product of individual variables to powers which are integers (positive or negative) or rational fractions. There are none that involve, for example, x^{π} or $x^{0.327\ldots}$. The product-of-integer or rational-fraction form is necessary to make the dimensions come out right. Sometimes one of the variables will be a difference, e.g., $\Delta P = P_2 - P_1$, which appears in the friction factor, or $\Delta T = T_{\text{wall}} - T_{\text{average}}$, which appears in the dimensional analysis of heat transfer problems. ΔP and ΔT are perfectly well-behaved variables. But some combination like $(\Delta P / \rho + g\,\Delta z)$, which is dimensionally homogeneous, does not appear in any of the useful dimensionless numbers.

9.3 FINDING THE DIMENSIONLESS NUMBERS

How does one go about finding which dimensionless numbers are important for a given type of model study or for changing our experimental findings from the dimensioned form in which we made the observations to a dimensionless form? Three general methods are in common use: the governing equation method, the method of force ratios, and Buckingham's π method.

9.3.1 The Method of Governing Equations

Suppose that the problem with which we are dealing concerns a complicated fluid flow system in which we suspect the B.E., along with other equations, would apply. Then we can write B.E. in differential form (without pump or compressor work) and integrate to find

$$\frac{P}{\rho} + gz + \frac{V^2}{2} + \mathcal{F} = \text{constant} \tag{9.2}$$

Let us assume that the $\mathcal{F}$ term is of the form given by the Poiseuille equation:

$$\mathcal{F} = \frac{(V_{\text{avg}} D_0^2 \pi / 4) \cdot [\Delta x \mu (128 / \pi)]}{D_0^4 \rho} \tag{6.14}$$

Each of the terms in Eq. 9.2 (or any correct equation) has the same dimensions; therefore, if we divide through by any one of them, the result will be a dimensionless equation.

If we divide by $V^2/2$, the result is

$$\frac{2P}{\rho V^2} + \frac{2gz}{V^2} + \frac{\mu}{\rho V D} 64 \frac{\Delta x}{D} = \frac{\text{const.}}{V^2} \tag{9.3}$$

The first term on the left in this equation is important enough to be given a name in fluid mechanics, the *pressure coefficient;* it is also sometimes called 1 / (Euler number)^2. It appears in problems in which there are significant changes in velocity and in pressure between different parts of the system. For example, Eq. 6.23 may be rewritten

$$f = \frac{2\Delta P}{\rho V^2} \cdot \frac{D}{\Delta x} \tag{9.4}$$

and Eq. 6.50 may be rewritten

$$C_d = \frac{F}{A} \cdot \frac{2}{\rho V^2} = \frac{2P_{\text{avg}}}{\rho V^2} \tag{9.5}$$

so the friction factor and drag coefficient introduced in Chap. 6 are special cases of this more general pressure coefficient.

The second term on the left in Eq. 9.3 also is important enough to have a name, 2 / (Froude number), or

$$\mathcal{F}r = \text{Froude number} = \frac{V^2}{gz} \tag{9.6}$$

(Some textbooks call the Froude number the square root of the value shown here.) In problems involving changes in velocity and free surfaces, the Froude number plays a very important role, e.g., in ship-model studies and open-channel flow. We saw in Chap. 7 that it is the key parameter in describing hydraulic jumps.

The third term on the left side of Eq. 9.3 is 1 / (Reynolds number), times 64 · (length / diameter). The Reynolds number is important whenever viscous forces are important and there are significant changes in velocity. We have seen its use in Chap. 6 and will see it again in later chapters.

We see from this example that simply by dividing B.E. by one of the terms, we can find three of the most frequently used dimensionless groups in fluid mechanics. Similarly, we saw in Chap. 8 that dividing both sides of Eq. 8.15 by part of one side leads naturally to the Mach number, which also is used very frequently.

These simple examples do not show all the possibilities of this method; for more on it, see Kline [2]. Interesting variants of this method are shown by Bird, Stewart, and Lightfoot [3, p. 97] and Hellums and Churchill [4].

9.3.2 The Method of Force Ratios

The method of force ratios, discussed below, is often referred to as the *method of similitude* or the *method of similarity.*

Most of the dimensionless groups in fluid mechanics may be thought of as ratios of lengths or ratios of forces. For example, in pipe flow we saw two length ratios, $\Delta x / D$ and ε / D, and two dimensionless groups—which, as we will see shortly, are expressible as force ratios—the Reynolds number and the friction factor. The length ratios are obviously important in model studies; a scale model has the same length / width or length / height as the original. The force ratios are important in model studies because, for two different-sized models to encounter the same kind of fluid behavior, the influences of gravity, viscosity, compressibility, etc., must be in the same proportion for both. The dimensionless groups generally may be thought of in some other way (e.g., the Mach number as a ratio of velocities), but they also can be seen as force ratios. This is easiest to see for the pressure coefficient. The pressure force exerted by a fluid on some planar body is

$$\text{Pressure force} = \int_{\text{all surface}} P\, dA = P_{\text{avg}} A = F_P \tag{9.7}$$

Similarly, the force required to stop a unit volume of the flow (which a body inserted into the flow does) is given by $F = ma$. Multiplying both sides by dx, we get

$$F\, dx = ma\, dx = m\frac{dV}{dt}dx = mV\, dV \tag{9.8}$$

but m is the mass of a unit volume equal to ρL^3, where L is the length of one side of the unit volume. Substituting this and integrating, we find

$$\int F\, dx = F_{\text{avg}}\,\Delta x = \int \rho L^3 V\, dV = L^3 \rho \Delta\left(\frac{V^2}{2}\right) \tag{9.9}$$

Here the final velocity is zero, so we may replace the $\Delta(V^2/2)$ in this equation with $-V^2/2$. Ignoring the sign and solving Eq. 9.9 for the average force, we have

$$F_{\text{avg}} = \frac{L^3}{\Delta x}\rho\frac{V^2}{2} \tag{9.10}$$

Now, $L^3 / \Delta x$ has the same dimensions as an area, so we may replace it with some area A. Thus, the force to stop the fluid, which is commonly called the *inertia force,* is

$$\text{Inertia force} = A\rho\frac{V^2}{2} = F_I \tag{9.11}$$

Dividing Eq. 9.6 by Eq. 9.11, we find

$$\frac{F_P}{F_I} = \frac{A\Delta P}{A\rho(V^2/2)} = \frac{\Delta P}{\rho(V^2/2)} = \text{pressure coefficient} \tag{9.12}$$

We can find the appropriate force ratios systematically by making up a list of forces and their dimensions.

To find the viscous force, we consider the shear force exerted by a Newtonian fluid flowing past a surface in laminar flow. From Chap. 1, we know that the shear stress τ is given by

$$\tau = \frac{F}{A} = \mu \frac{dV}{dy} \tag{1.4}$$

Solving for the force gives us

$$F = \mu A \frac{dV}{dy} \tag{9.13}$$

However, at the wall $V = 0$ and $y = 0$; so this becomes

$$\text{Viscous force} = \frac{\mu A V}{L} = F_V \tag{9.14}$$

Similarly, the force of gravity on a unit volume of fluid is

$$\text{Gravity force} = g\rho L^3 = F_G \tag{9.15}$$

and the surface tension force on a unit length of fluid surface is

$$\text{Surface tension force} = \sigma L = F_S \tag{9.16}$$

The elastic force for a spring is given by Hooke's law as $F = kx$, where k is the spring constant and x is the displacement. If we apply the same equation to the one-dimensional compression of a fluid, we find that the *spring constant* is $A(dP/dx)$; so,

$$\text{Elastic force} = A\frac{dP}{dx}\Delta x = \frac{A^2}{A}\frac{dP}{dx}\frac{\Delta x}{} = A^2\frac{dP}{dV}\Delta x \tag{9.17}$$

But $V = m/\rho$, so for a constant mass $dV = -(m/\rho^2)\,d\rho$. Ignoring the sign and substituting give

$$\text{Elastic force} = \frac{A^2\rho^2}{m}\Delta x\frac{dP}{d\rho} \tag{9.18}$$

Then we substitute $m = \rho V$ to find

$$\text{Elastic force} = \frac{A^2\rho^2}{\rho V}\Delta x\frac{dP}{d\rho} = A^2\frac{\Delta x}{V}\rho\frac{dP}{d\rho} \tag{9.19}$$

but $V/\Delta x$ has the dimensions of an area, so we may call it an area:

$$\text{Elastic force} = A\rho\frac{dP}{d\rho} = F_E \tag{9.20}$$

From this list of six kinds of forces we may make up a table of all the possible force ratios, Table 9.1. From this table we see that we can combine the six kinds of forces into 15 force ratios, of which 8 are important enough in fluid mechanics to have the common names shown. An entirely analogous table is also shown by

TABLE 9.1
Dimensionless numbers obtained as force ratios*,**

	F_I	F_V	F_P	F_E	F_S	F_G
	$A\rho \dfrac{V^2}{2}$	$\dfrac{\mu AV}{L}$	$A\,\Delta P$	$A\rho \dfrac{dP}{d\rho}$	σL	$g\rho\, L^3$
F_I	$\dfrac{F_I}{F_I} = 1$	$\dfrac{F_V}{F_I} = \dfrac{2\mu}{LV\rho}$ **1**	$\dfrac{F_P}{F_I} = \dfrac{2\,\Delta P}{\rho V^2}$ **2**	$\dfrac{F_E}{F_I} = \dfrac{2(dP/d\rho)}{V^2}$ **3**	$\dfrac{F_S}{F_I} = \dfrac{2\sigma}{L\rho V^2}$ **4**	$\dfrac{F_G}{F_I} = \dfrac{2gL}{V^2}$ **5**
F_V		$\dfrac{F_V}{F_V} = 1$	$\dfrac{F_P}{F_V} = \dfrac{\Delta PL}{\mu V}$ **6**	$\dfrac{F_E}{F_V} = \dfrac{\rho(dP/d\rho)\,L}{\mu V}$	$\dfrac{F_S}{F_V} = \dfrac{\sigma}{\mu V}$	$\dfrac{F_G}{F_V} = \dfrac{g\rho L^2}{\mu V}$
F_P	These numbers are merely reciprocals of numbers shown above and to the right.		$\dfrac{F_P}{F_P} = 1$	$\dfrac{F_E}{F_P} = \rho \dfrac{dP/d\rho}{\Delta P}$	$\dfrac{F_S}{F_P} = \dfrac{\sigma}{L\,\Delta P}$ **7**	$\dfrac{F_G}{F_P} = \dfrac{g\rho L}{\Delta P}$
F_E				$\dfrac{F_E}{F_E} = 1$	$\dfrac{F_S}{F_E} = \dfrac{\sigma}{L\rho(dP/d\rho)}$	$\dfrac{F_G}{F_E} = \dfrac{gL}{dP/d\rho}$
F_S					$\dfrac{F_S}{F_S} = 1$	$\dfrac{F_G}{F_S} = \dfrac{g\rho L^2}{\sigma}$ **8**
F_G						$\dfrac{F_G}{F_G} = 1$

***1.** 2 / Reynolds number. The Reynolds number is introduced in Chap. 6 and appears in many subsequent chapters.

2. Pressure coefficient = 1 / (Euler number)^2. The Fanning friction factor, *f*, introduced in Chap. 6 = (pressure coefficient). $D / 2\ \Delta x$ and the drag and lift coefficients introduced in Chap. 6 = 2 · (pressure coefficient).

3. 2 / (Mach number)^2 = Cauchy number. The Mach number plays a dominant role in Chap. 8.

4. 2 / Weber number. The Weber number plays a dominant role in all phenomena involving drops and sprays; see Chap. 14.

5. 2 / Froude number The Froude number is introduced in Chap. 7 in describing hydraulic jumps. It plays a major role in open channel flow phenomena and in the design of boat hulls.

6. Stokes number. Stokes law for the settling of spherical particles in laminar flow (Eq. 6.57) can be rewritten as Stokes number = 12. One seldom sees it that way, but this formulation shows that for gravity settling of spherical particles in laminar flow the Stokes number is a constant. For other shapes that constant has different values.

7. Capillary number. Figure 11.6 shows that the capillary number (in slightly modified form) governs the displacement of liquids from porous solids.

8. Eötvös number. The Eötvös number appears in studies of bubbles. For example, Tate's law, for the size of bubbles forming at an orifice, Eq. 14.14, can be written as Et = 6 $D_{\text{bubble}} / D_{\text{orifice}}$.

**Other dimensionless groups that appear in this book mostly are not named, and are the ratios of similar quantities; e.g., the relative roughness, ε / D, in Chap. 6 is the ratio of two lengths, and A / A^*, which appears in Chap. 8, is the ratio of two areas. In Chap. 8 most quantities are presented as ratios to some reference value of the same quantity; these are dimensionless groups, which are presented that way simply for convenience.

Kline [2] for the dimensionless ratios important in heat transfer, in which, instead of using force ratios, we use the ratios of energy quantities. A list of 27 dimensionless groups used in chemical engineering fluid mechanics, made up in a way similar to Table 9.1 is shown by Tilton [5, pp. 6–49].

The merit of Table 9.1 is that in using it we can quickly estimate which ratios are likely to be important for a given problem.

Example 9.1. Using Table 9.1, estimate which dimensionless ratios are probably important for (*a*) steady laminar flow in a horizontal pipe, (*b*) completely turbulent steady flow in a horizontal pipe, (*c*) resistance to an airplane in steady flight, (*d*) resistance to a ship in steady motion, (*e*) resistance to a submerged submarine in steady motion, and (*f*) the rise of a fluid in a capillary tube.

(*a*) For steady laminar flow in a pipe we would assume that the only important forces are the pressure force and the viscous force (in fact, we know from Chap. 6 that they are equal and opposite). Then from Table 9.1 we would conclude that the only important force ratio is probably the Stokes number. We would also assume that the length / diameter ratio would be important and hence that the Stokes number was some function of that ratio. If we further assume that the Stokes number is a constant times the ratio of length / diameter, we can then solve for ΔP / length:

$$\frac{\Delta P}{\text{Length}} = \text{constant} \cdot \frac{\mu V}{LD} \tag{9.A}$$

Here the L in the Stokes number is length perpendicular to the flow direction (in this case, the diameter), so that the right-hand side of the equation is $\rho V / D^2$. This result is the same as Poiseuille's equation (written for the average velocity).

Dimensional analysis will not tell us the value of the constant in Eq. 9.A (which is 4 if we take V as the average velocity), but it will suggest to us that for steady laminar flow in horizontal pipes a plot of pressure drop per unit length versus (viscosity · average velocity / diameter^2) might give the same straight line for all fluids, pipes, and velocities (which is experimentally verifiable).

(*b*) For completely turbulent flow, we would assume that viscous forces would be negligible compared with the pressure and inertia forces, so that the only important force ratio should be the pressure coefficient. The length / diameter ratio also should be important, as well as the pipe roughness. If we further assume that for a constant pipe roughness the pressure coefficient is a constant times the length / diameter ratio, we can solve for the pressure drop per unit length:

$$\frac{\Delta P}{\Delta x} = \frac{\text{constant}}{D} \rho \frac{V^2}{2} \quad \text{[for a given pipe roughness]} \tag{9.B}$$

Comparing this with the definition of the friction factor, we see that the constant here is $4f$. So far, we have not included the effect of variable pipe roughness. It would be plausible to assume that the constant in this equation is a function of pipe roughness, so that our final form would be

$$\frac{\Delta P}{\Delta x} = \text{some function of } \frac{\varepsilon}{D} \text{ and } \frac{\rho V^2}{2D} \tag{9.C}$$

It also would be plausible to assume that this function was a linear one, i.e.,

$$\frac{\Delta P}{\Delta x} = \text{some constant} \cdot \frac{\varepsilon}{D} \cdot \frac{\rho V^2}{2D} \qquad ??? \tag{9.D}$$

but experimental data (Fig. 6.10) indicate that nature disagrees with this plausible assumption. However, Eq. 9.C is a good description of the right-hand side of the friction factor plot; the friction factor for very high Reynolds numbers depends on the roughness alone, not on the Reynolds number.

(*c*) For an airplane, the forces that may be significant are pressure force, inertia force, viscous force, and elastic force. At low velocities the elastic force is probably negligible compared with the others, so we would conclude that the pressure coefficient (normally called a drag coefficient for airplanes) should be a function of Reynolds number and geometry. At high velocities the viscous forces are probably negligible compared with the elastic forces, so the pressure coefficient should depend on the Mach number and geometry. Both of these assumptions are experimentally verifiable. We might make either of two plausible assumptions about intermediate velocities: (i) The two ranges overlap, i.e., there is a range of velocities in which both the Reynolds and the Mach number affect the pressure coefficient, or (ii) these two ranges do not meet, i.e., there is a range of velocities in which neither the Reynolds number nor the Mach number affects the pressure coefficient, which depends on the geometry alone. From theoretical speculation alone we cannot decide between these two possibilities. Experiments indicate that the second is correct; over a large range of velocities the pressure coefficient is independent of the Reynolds and Mach numbers; it depends only on the shape of the airplane and its angle of attack relative to the oncoming airstream.

(*d*) For a ship the forces that may be significant are the pressure force, gravity force, viscous force, and inertia force. The gravity force enters this list because of the bow wave thrown up by ships; because of this wave the ship seems to be steadily traveling uphill. Then, from Table 9.1, we assume that the pressure coefficient should be a function of the Froude number and the Reynolds number as well as of the shape of the ship. This is experimentally verifiable.

(*e*) For a submerged submarine the gravity force is no longer important, because a submerged submarine causes no bow wave. Therefore, we should drop the Froude number from the list obtained in (*d*).

(*f*) For capillary rise the significant forces are gravity and surface tension of the fluid. Therefore, from Table 9.1 we see that the important force ratio is $g\rho L^2 / \sigma$, the Eötvös number. Here the L^3 in the gravity force indicates three perpendicular directions. The L in the surface-tension force cancels one of them, but the remaining L^2 does not refer to one dimension squared, but rather to two perpendicular dimensions. Setting this force ratio equal to a constant and solving for one of these dimensions, we find

$$L_1 = \text{constant} \frac{\sigma}{g L_2} \tag{9.E}$$

If the geometry we are discussing is a right cylinder with vertical axis, and if we choose L_1 as the height of capillary rise for perfect wetting of the surface by the fluid, then L_2 is a characteristic length at right angles, e.g., the radius of the tube. In that case it can be shown (see Chap. 14) that the constant in this equation is a dimensionless 2. ■

9.3.3 Buckingham's π Method

Another systematic approach to finding the dimensionless numbers is the method of Buckingham [6] often referred to as the *π theorem* or *Buckingham's π theorem*. This states that, if there is some relationship in which A (the dependent variable) is a function of $B_1, B_2, \ldots, B_n$ (the independent variables), then the relationships can be written

$$A = f(B_1, B_2, \ldots, B_n) \tag{9.21}$$

or, alternatively,

$$f(A, B_1, B_2, B_3, \ldots, B_n) = 0 \tag{9.22}$$

(For example, $F = ma$ or $[F - ma = 0]$.) Furthermore, if the quantities contain among them k independent dimensions, then it will be possible to rewrite Eq. 9.22 as

$$f(\pi_1, \pi_2, \ldots, \pi_m) = 0 \tag{9.23}$$

in which the π's are dimensionless quantities and the number of π's is $(n + 1 - k)$. Here the π's may be dimensionless numbers like the Reynolds number, or ratios of dimensions, like the L/D that appears in equations for the friction effect in pipe flow. This part of Buckingham's theorem shows how to decide how many such dimensionless groupings we should seek for a given problem.

The proof is given by Buckingham [6]; in brief it is that, if a relation such as Eq. 9.22 exists, and if the final relation contains more than one term (i.e., is not of the form $A = 0$), then each term must have the same dimensions. Thus, we know that some relation of the form of Eq. 9.22 exists and, in addition, we know that k independent equations exist among the dimensions of the $(n + 1)$ quantities in Eq. 9.22. This means that we can, in principle, eliminate k unknowns among these equations; hence, there are $(n + 1 - k)$ independent, dimensionless π's.

There are several restrictions in this logic, as follows:

1. The list of independent dimensions should not contain redundant dimensions. For example, if it contains length and force, then it should not contain energy (energy is dimensionally equivalent to the product of length and force). If we wish to include dimensions that are redundant in this sense, then the conversion factor for converting the redundant equations to each other should be included in the Bs. For example, if we wish to include length, force, and thermal energy in our list of dimensions, then we should also add to the list of *B*s the conversion factor between mechanical and thermal energy which is 778 ft · lbf / Btu = 4.184 J / cal = 1. Thus, we can add a redundant dimension and a conversion factor, and the number of π's, equal to $(n + 1 - k)$, will not change, because both n and k increase by 1. The other common example of this approach is the inclusion of force, mass, length, and time as independent dimensions. This is permissible *if* one adds $g_c = 32.2 \text{ lbm} \cdot \text{ft} / (\text{lbf} \cdot \text{s}^2) = 1.00 \text{ kg} \cdot \text{m} / (\text{N} \cdot \text{s}^2) = 1$ to the list of *B*s.

2. If two dimensions occur only in a specific ratio, then they are not independent and must be treated as one dimension. Suppose, for example, that our list of *A*s and *B*s consisted of two velocities, V_1 and V_2, and two forces, F_1 and F_2. By simple application of Buckingham's theorem we would conclude that $(n + 1)$ equals 4 and that k equals 3 (length, time, force); so there should be one π. But to conclude that there is only one π here is incorrect. Since length and time appear in our list of variables only in the combination length / time, there are really only two independent dimensions, force and length / time; so k is 2 and there are two π's, which are presumably V_1 / V_2 and F_1 / F_2. In this case it is obvious that the dimensions length and time are not independent. In less obvious cases, the recommended procedure is to find from the list the largest number of variables that cannot form a dimensionless group. In this example that number is two; we can select one velocity and one force, and they cannot be converted into a dimensionless group using the other members of the variable list. Any three from the list can form a dimensionless group, either V_1 / V_2 or F_1 / F_2. This largest number of uncombinable variables is equal to the number of independent dimensions. It can never be more than the total number of dimensions; as shown above, it may be less.

Example 9.2. We believe that some force is a function of a velocity, a density, a viscosity, and two lengths. How many dimensionless π's should be required to correlate the data for this problem?

Here we have three choices as to which set of dimensions to choose, all of which give the same result.

1. We may choose as our dimensions length, force, and time, in which case we must express the density in dimensions of [force · time^2 / length^4] and the viscosity in [force · time / length^2]. In this case $(n + 1)$ equals 6, and k equals 3. Can we find three parameters that cannot be combined into any dimensionless group? Yes, the force, one of the lengths, and the density cannot be so combined; the three dimensions are independent, and we can see that the number or π's is $6 - 3 = 3$.
2. We may choose as our dimensions length, mass, and time, in which case we must express the force in dimensions of [mass · length / time^2] and the viscosity in [mass / (length · time)]. Here again $(n + 1)$ equals 6 and k equals 3. The same group of a length, the force, and the density again cannot be internally combined into any dimensionless group; so the three dimensions are independent, and the number of π's is 3.

3. We may choose force, mass, length, and time as our dimensions, in which case we must add the force-mass conversion factor, $g_c = 32.2\ \text{lbm} \cdot \text{ft} / (\text{lbf} \cdot \text{s}^2) = 1.00\ \text{kg} \cdot \text{m} / (\text{N} \cdot \text{s}^2)$, to our list of parameters. Now we can select a length, the force, the density, and g_c as the four parameters that cannot be combined into any dimensionless group, so that all four dimensions are independent, and the number of π's is now $7 - 4 = 3$. ■

Having decided how many π's to look for, how do we look for them? Buckingham's theorem also provides an algorithm for selecting π's;

1. Select k variables that cannot be combined internally into a dimensionless group. These are the *repeating variables.*
2. Then the combination of the repeating variables with any of the remaining, non-repeating variables can form a dimensionless group; make up $(n + 1 - k)$ such groups from the variables. This often can be done by inspection, or it can be done systematically by the algorithm shown in Example 9.3.

Example 9.3. For the variables shown in Example 9.2 find a set of three π's.

This problem can be worked three ways, just as Example 9.2 could be worked three ways. We work here only with the first of the three in Example 9.2. The other two are Probs. 9.4 and 9.5.

We choose the independent dimensions as length, force, and time. We now construct a table of the variables, shown in Table 9.2, indicating their dimensions.

As our repeating variables we may select any three that cannot be combined internally to form a dimensionless group; for example A, B_2, and B_4 (a force, the density, and a length). Then the three dimensionless π's may be formed from the following groups: A, B_2, B_4, B_1; A, B_2, B_4, B_3; and A, B_2, B_4, B_5.

The first π must be some product of powers (plus or minus) of A, B_2, B_4, B_1. We form this product, using a, b, c, and d to indicate the unknown powers:

$$\pi_1 = (A)^a (B_2)^b (B_4)^c (B_1)^d \tag{9.F}$$

However, because the π's are dimensionless, the dimensions of this equation are

$$L^0 t^0 F^0 = (F)^a (Ft^2 / L^4)^b (L)^c (L / t)^d \tag{9.G}$$

TABLE 9.2

Dimensions of the variables in Example 9.3

Variable	Description	Dimensions
A	A force, F_1	F
B_1	Velocity, V	L / t
B_2	Density, ρ	Ft^2 / L^4
B_3	Viscosity, μ	Ft / L^2
B_4	Length, L_1	L
B_5	Length, L_2	L

This can be true only if the following three simultaneous equations are satisfied:

$$\text{Equation for length:} \qquad 0 = -4b + c + d \tag{9.H}$$

$$\text{Equation for time:} \qquad 0 = 2b - d \tag{9.I}$$

$$\text{Equation for force:} \qquad 0 = a + b \tag{9.J}$$

By straightforward algebra, we may show that this set of equations has the solution

$$a = -d/2, \qquad b = d/2, \qquad c = d \tag{9.K}$$

Now, we may arbitrarily select a value for d; we select 2 as the lowest value of d that avoids fractional exponents. Then our first π is

$$\pi_1 = (A)^{-1}(B_2)^1(B_4)^2(B_1)^2 = \frac{\rho L^2 V^2}{F} \tag{9.L}$$

If the force here is a pressure force, then the L^2/F is equivalent to (1 / a pressure) and π_1 is 2 / (pressure coefficient). If we had chosen $d = 1$ above, our resulting π_1 would be the square root of the π_1 we found here.

By equally straightforward mathematics (Prob. 9.3), we can set up the systems of equations for π_2 and π_3 and solve for them, finding

$$\pi_2 = \frac{\mu^2}{F\rho} \qquad \text{and} \qquad \pi_3 = \frac{L_1}{L_2} \tag{9.M}$$

Here π_3 is the ratio of the two lengths, and π_2 can be shown to be equal to 2 / (Reynolds number · pressure cofficient)^2. Examining this set of π's, most experienced engineers would select as their dimensionless groups π_3, $1/\pi_1$, and $(\pi_2/\pi_1)^{1/2}$, i.e., the length ratio, the pressure coefficient, and the Reynolds number. The justification for this procedure is that a different selection of the repeating variables will give a different set of π's. However, the different sets are all products or ratios of the first set to some power. For example, if the repeating variables in Example 9.3 were chosen as V, ρ, and L_1, then the same procedure leads to $\pi_1 = 2$ / (pressure coefficient), $\pi_2 = 1$ / (Reynolds number), and $\pi_3 = L_1/L_2$ (see Prob. 9.6). ■

9.4 DIMENSIONLESS NUMBERS AND PHYSICAL INSIGHT

It is extremely unlikely that any reader of this book will use the above procedures to discover a significant new dimensionless number. White [7, p. 280], in addition to presenting a brief history of this subject, directs the reader to 24 books on the subject, including one with over 300 dimensionless numbers. Rather, most readers of this book will find the dimensionless numbers and the thought behind them most useful in correlating and understanding experimental results, in planning future experimental programs, and in thinking about what is going on physically in some unfamiliar process.

Thinking of problems in terms of their dimensionless numbers is often very helpful. We have seen that flow in a circular pipe is very different at low Reynolds numbers than at high Reynolds numbers, that the behavior of an open channel flow that is blocked is very different for a Froude number < 1.00 than for Froude numbers > 1.00, and that gas flow is very different for Mach numbers < 1.00 than for Mach numbers > 1.00. A Mach numbers < 0.2 means that the effects of changes of fluid density are unimportant, and we may safely use B.E. A Mach number $>$ about 0.5 means the opposite and means that our intuitive picture of low-Mach-number flows is not a reliable guide to this flow. Experienced engineers think about many physical phenomena this way; students are encouraged to learn to do so as well.

Dimensional analysis is powerful, but not all-powerful. It shows us what three dimensionless groups are best for presenting all pipe-flow friction data, Fig. 6.10. But it does not tell us what that plot will look like. By dimensional analysis alone we would never have discovered the difference between laminar and turbulent flows. Only careful experimental observation told us that. Currently, we can solve almost all laminar flow problems by computational fluid dynamics (Chap. 20); even though dimensional analysis gives us insight into laminar flow problems, we rarely need it to solve them. The same is not true for turbulent flow problems, where we can often learn important things about the flows by dimensional analysis.

One can make up all sorts of dimensionless numbers, only a few of which are useful. For example, the mass of this book, divided by the mass of the universe forms a perfectly well-defined dimensionless number. What possible use has it? What problem would we use it in?

9.5 JUDGMENT, GUESSWORK, AND CAUTION

Dimensional analysis and physical model studies are needed only in those problems that cannot currently be completely or rigorously solved with mathematical models. Therefore, we cannot hope to obtain a complete and certain solution by these methods. Furthermore, as the examples show, applying these methods requires judgment and good guesswork. Because the results are only tentative, they should be applied with caution.

However, the gain in simplifying a problem may be very great. As stated by Kline,*

> The use of dimensionless parameters reduces the number of independent coordinates required. A convenient way to realize the importance of such a reduction is to recall that a function of one independent coordinate can be recorded on a single line; two independent coordinates a page; three require a book; and four, a library.

The gain in increased understanding of the problem, to be had by seeing it in its dimensionless form, may be even more valuable than the gain in simplifying the problem.

*From S. J. Kline, *Similitude and Approximation Theory,* McGraw-Hill, New York, 1965, p. 17. Quoted by permission of the publisher.

9.6 SUMMARY

1. Although many flow situations can be represented by a simple, closed, analytical equation (e.g., laminar flow in a pipe), many others currently cannot (e.g., turbulent flow in a pipe).
2. In the latter case the experimental data often can be correlated and simplified by the use of dimensionless ratios (e.g., the friction factor, Reynolds number, relative roughness plot).
3. In designing full-size equipment from tests on small-scale physical models, it is necessary to "scale up," keeping the values of the pertinent dimensionless groups the same for model and full-size equipment.
4. The common methods of finding the pertinent dimensionless groups are the method of governing equations, the method of force ratios, and Buckingham's π method.
5. The greatest benefit of dimensional analysis is often the insight it gives us into the physical nature of the flows.
6. All these methods require judgment and good guesswork. The results must be applied with caution.

PROBLEMS

See the Common Units and Values for Problems and Examples inside the back cover! An asterisk (*) on a problem number indicates that the answer is in App. D.

9.1. Estimate which of the dimensionless force ratios in Table 9.1 should be important for each of the following kinds of flow:

(*a*) Breakup of a jet of liquid into droplets.
(*b*) Formation of a vortex in the free surface of the drain of a bathtub.
(*c*) Same as (*b*), but for a bathtub full of molasses.
(*d*) The shape of a bubble rising in a viscous liquid.
(*e*) The breakaway of a bubble from a submerged horizontal orifice.
(*f*) The simultaneous horizontal flow of two fluids through a porous medium at low flow rates, e.g., the flow of oil and gas into an oil well.
(*g*) same as (*f*), but flow in the vertical direction.

9.2. In the example in the text in which the list of variables consists of two velocities and two forces, show what the possible forms of the final equation are if we assume that there is only one π and if we assume that there are two π's.

9.3.*Find π_2 and π_3 in Example 9.3.

9.4. Rework Example 9.3, using L, m, and t as the independent dimensions.

9.5. Rework Example 9.3, using L, m, t, and F as independent dimensions (and adding g_c to the list of variables).

9.6. Rework Example 9.3, using V, ρ, and L_1 as the repeating variables.

9.7. Rework Example 9.3 by the method of force ratios.

9.8. Make a list of the dimensionless groups that appear in the common introductory courses in thermodynamics, heat transfer and material balances. Indicate whether these are ratios, and if so, of what.

9.9.*An airplane scale model is to be tested. The problem of interest is one in which the landing behavior is to be tested. Since this is a low-speed problem, the Reynolds number is to be held constant between model and full-scale airplane. The model is to be one-tenth of the length of the airplane with the same shape. The landing speed of the full size airplane is 60 mi/h. How fast should the air in the wind tunnel move past the model to have the same Reynolds number? Suppose we could test the model in water. What should the water velocity be for the same Reynolds numbers?

9.10. We are designing a new, large centrifugal pump (Chap. 10). We wish to test its behavior on a small model. If the pump has an impeller 1 m in diameter that turns at 1800 rpm, and we want to test it with a model with a 0.1 m diameter impeller and the same fluid the pump will use. We want the same Reynolds number, taking the velocity as the velocity of the tip of the impeller. What rotational speed will be required? Is it practical to hold the Reynolds number constant in such pump model tests?

9.11. We are designing a new type of racing boat hull. In the model tests in a towing basin we will use a model one-tenth the size of the actual boat. We know that the drag of the boat (its resistance to moving through the water) is a function of the Reynolds and Froude numbers. Can we simultaneously hold both of these constant between model and real boat in our tests? If not, how will we test the model?

9.12. In Examples 9.2 and 9.3, we now believe that in addition to the variables included, we should also add a new variable, a frequency of vibration or oscillation, ω, with dimension (1 / time). Repeat those two examples with this added variable.

9.13. What dimensionless groups should influence the speed of a large surface wave in the deep ocean, i.e., a tsunami? Of a small surface wave in a shallow pond?

9.14. A small, spherical particle is settling by gravity in a liquid. Using the method of force ratios, estimate what dimensionless groups should describe this process.

9.15. Repeat Prob. 9.14, for a large particle.

9.16. Equation 6.56 shows the terminal velocity of a spherical particle settling in a liquid. Show that if one multiplies both sides by $(D\rho_{\text{fluid}} / \mu_{\text{fluid}})^2$, the left side becomes $\mathcal{R}_p^2$. The right side is described as $(4Ar / 3C_d)$ where Ar is the Archimedes number [8]. Show the equation for the Archimedes number. Can it be expressed as the ratio or product of other numbers in Table 9.1?

REFERENCES FOR CHAPTER 9

1. Haldane, J. B. S. "On Being the Right Size." in *Possible Worlds and Other Essays,* New York: Harper & Row, 1927. Reprinted in J. R. Newman, *The World of Mathematics*. New York: Simon and Schuster, 1956. This short essay is an absolute masterpiece, which everyone should read for pleasure if not for information
2. Kline, S. J. *Similitude and Approximation Theory*. New York: McGraw-Hill, 1965.
3. Bird, R. B., E. N. Stewart, W. E. Lightfoot. *Transport Phenomena*. 2nd ed. New York: Wiley, 2002.
4. Hellums, J. D., S. W. Churchill. "Simplification of the Mathematical Description of Boundary and Initial Value Problems." *AIChE J. 10:* 110–114 (1964).
5. Tilton, J. N. "Fluid and Particle Dynamics." In *Perry's Chemical Engineers' Handbook.* 7th ed. R. H. Perry, D. W. Green, J. O. Maloney. New York: McGraw-Hill, 1997, pp. 6–52.
6. Buckingham, E. "Model Experiments and the Form of Empirical Equations." *Trans. ASME 37,* (1915), pp. 263–296. See also E. Buckingham, *Dimensional Analysis*. Cambridge, Massachusetts: Harvard University Press, 1921.
7. White, F. M. *Fluid Mechanics*. 4th ed. New York: McGraw-Hill, 1999.
8. Karamanev, D. G. "Equations for Calculation of the Terminal Velocity and Drag Coefficient of Solid Spheres and Gas Bubbles." *Chem. Eng. Comm. 147:* 75–84 (1996).

CHAPTER 10

PUMPS, COMPRESSORS, AND TURBINES

In Chaps. 4, 5, and 6 we have written energy-balance equations that involve a $dW_{n.f.}$ term (see Sec. 4.8 for a definition of $dW_{n.f.}$). For steady-flow problems this term generally represents the action of a pump, fan, blower, compressor, turbine, etc. Here we will discuss the fluid mechanics of the devices that actually perform that $dW_{n.f.}$. Figure 10.1 shows the flow diagram for all the devices discussed in this chapter. The names of the various devices that fit this diagram are shown in Table 10.1. The boundaries between the names are approximate; for liquids, there is less variation than for gases. The naming shown is not absolute; the hand-operated air compressor used to fill bicycle tires is called a bicycle pump.

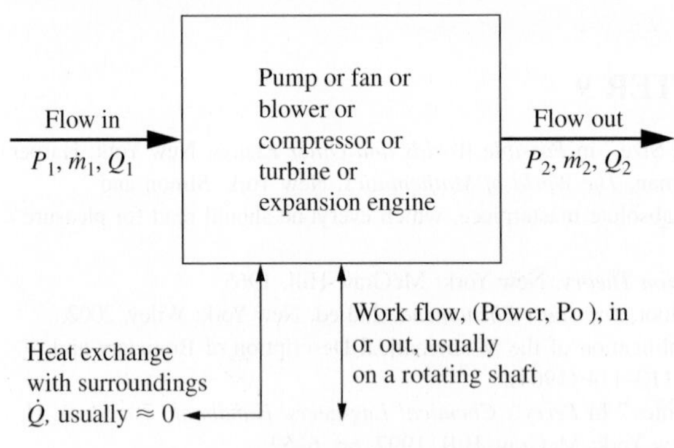

FIGURE 10.1
Flow diagram for any pump, compressor, turbine, or expansion engine. A fluid flows in at the left and out at the right. Work flows in or out on a rotating or reciprocating shaft. If the inlet pressure is less than the outlet pressure, this is a pump or compressor, and work flows in. If the inlet pressure is greater than the outlet pressure then this is a turbine or expansion engine, and work flows out. For steady flow $\dot{m}_{in} = \dot{m}_{out}$. If the density does not change through the device (practically true for liquids, almost never true for gases), then $Q_{in} = Q_{out}$.

From Fig. 10.1 one sees that the same diagram describes both devices that consume work and increase the pressure

TABLE 10.1
Names of various kinds of fluid machinery, based on usage

Fluid passing through device	Work-consuming device, $P_2 > P_1$	Work-producing device, $P_2 < P_1$
Liquid	Pump	Turbine or expansion engine
Gas		
$\Delta P < 0.1$ psi	Fan	Turbine or expansion engine
$0.1 < \Delta P < 1$ psi	Blower	Turbine or expansion engine
$\Delta P > 1$ psi	Compressor	Turbine or expansion engine
Inlet P less than 1 atm	Vacuum pump	
Other	Bicycle pump	

of the fluid passing through and devices that lower that pressure to produce work. Many such devices can work both ways. Most are designed to be efficient one way, for example, an efficient pump works poorly as an expansion engine or turbine. But for some applications (pumped storage [1] and tidal power plants [2]) the same device operates as an efficient pump running one way part of the time and as an efficient turbine running the opposite way part of the time.

10.1 GENERAL RELATIONS FOR ALL PUMPS, COMPRESSORS, AND TURBINES

To determine the performance of a device in Fig. 10.1 we measure the inlet and outlet pressures and the steady-flow mass flow rate $\dot{m}$ or the volumetric flow rate $Q = \dot{m} / \rho$ for various downstream pressures. The result, normally presented as a plot of ΔP versus Q, is called a *pump or compressor curve* or *map*. As shown below, it is common to divide the pressure increase by ρg and define

$$\begin{pmatrix}\text{Pump or} \\ \text{compressor head}\end{pmatrix} = h = \frac{\Delta P}{\rho g} \tag{10.1}$$

This matches the head form of B.E. (Sec. 5.4). In such a pump test, in addition to h and Q, we normally simultaneously measure the power input to the device or power output from it. Most pump or compressor maps present the data as a plot of h versus Q and show the power input or output plotted versus Q on the same figure (Figs. 10.3, 10.8, and 10.9, discussed later).

If we write B.E. (Eq. 5.5), from the inlet of the pump, compressor, or turbine in Fig. 10.1 to its outlet and solve for the work input to the pump, we find

$$\frac{dW_{\text{n.f.}}}{dm} = \Delta\left(\frac{P}{\rho} + gz + \frac{V^2}{2}\right) + \mathcal{F} \tag{10.2}$$

The term on the left is the work flow per unit mass into the pump; it is positive for a pump or compressor and negative for any power-producing device such as a turbine or expansion engine. The first term on the right of the equal sign represents the "useful" work done by the pump: increasing the pressure, elevation, or velocity of the fluid. The second term represents the "useless" work done, either in heating the fluid or in heating the surroundings. For most such devices we take points 1 and 2 as shown in Fig. 10.1, for which the elevation change is negligible. We observe that the change in kinetic energy is generally much less than $\Delta P / \rho$, so that (except in the most careful work) we drop the gz and $V^2 / 2$ terms from Eq. 10.2.

The normal definition of pump efficiency is

$$\text{Pump efficiency} = \eta = \frac{\text{useful work}}{\text{total work}} = \frac{\Delta P / \rho}{dW_{\text{n.f.}} / dm} \tag{10.3}$$

This gives the pump efficiency in terms of a unit mass of fluid passing through the pump. It is common to multiply the top and bottom of this equation by the mass flow rate $\dot{m} = dm / dt = Q\rho$, which makes the denominator exactly equal to the power, Po, supplied to the pump:

$$\eta = \frac{\dot{m} \cdot \Delta P / \rho}{(dm / dt) \cdot (dW_{\text{n.f.}} / dm)} = \frac{Q \cdot \Delta P}{\text{Po}_{\text{supplied}}} \tag{10.4}$$

This form is applicable only to fluids whose density is nearly constant, i.e., liquids. For gases we must substitute $\int (dP / \rho)$ for $\Delta P / \rho$ in B.E. and do the integration, finding

$$\eta = \frac{\dot{m} \int_{\text{in}}^{\text{out}} (dP / \rho)}{\text{Po}_{\text{supplied}}} \tag{10.5}$$

As in Chap. 5, for small changes in P (e.g., most fans and blowers) we can use Eq. 10.4 for gases with negligible errors. But for large changes in P (most compressors and vacuum pumps), we must use Eq. 10.5.

Example 10.1. A pump is pumping 50 gal / min of water from a pressure of 30 psia to a pressure of 100 psia. The changes in elevation and velocity are negligible. The motor that drives the pump supplies 2.80 hp. What is the efficiency of the pump?

The right-most form of Eq. 10.4 is almost always the more convenient, so

$$\eta = \frac{(50 \text{ gal / min}) \cdot (70 \text{ lbf / in}^2)}{2.8 \text{ hp}} \cdot \frac{\text{hp} \cdot \text{min}}{33{,}000 \text{ ft} \cdot \text{lbf}} \cdot \frac{231 \text{ in}^3}{\text{gal}} \cdot \frac{\text{ft}}{12 \text{ in}} = 0.73 \tag{10.A}$$

■

From this calculation we see that both the numerator and denominator in Eq. 10.4 have the dimension of horsepower (or kW). We may think of the numerator (referred to as the *hydraulic horsepower* of the pump) as the useful work expressed in horsepower (or kW). The pump efficiency = (hydraulic horsepower) / (total horsepower supplied to the pump), or the same in kW. In Example 10.1 we may calculate that the hydraulic horsepower is 2.04 hp.

10.2 POSITIVE-DISPLACEMENT PUMPS AND COMPRESSORS

A pump or compressor is a device that does work *on* a liquid; $dW_{\text{n.f.}} / dm$ is positive in the new sign convention used in this book, but negative in the traditional sign convention used in older books. Most pumps and compressors are one of the following:

1. Positive-displacement (P.D.).
2. Centrifugal.
3. Special designs intermediate in characteristics between these two.

In addition there are nonmechanical pumps (i.e., electromagnetic, ion, diffusion, jet, etc.), which will not be considered here.

10.2.1 P.D. Pumps

P.D. pumps work by allowing a fluid to flow into some enclosed cavity from a low-pressure source, trapping the fluid, and then forcing it out into a high-pressure receiver by decreasing the volume of the cavity. These are extremely common; examples are the oil and fuel injector pumps on most automobiles, the pumps on most hydraulic systems, the hand-operated dispensers for liquid soap, cosmetics, and catsup, and the hearts of most animals. The most important pump in my life is a P.D. pump, my heart. P.D. pumps and compressors have been used for thousands of years. Before about 150 years ago all pumps and compressors were P.D. Since then our improved ability to make rotating machinery has made possible other types of pumps (centrifugal, axial flow, regenerative), which have replaced P.D. pumps in many applications. The P.D. devices still move our blood, and they can produce small, high-pressure flows like the oil pumps in our autos most economically. They have many other applications. There are more P.D. pumps and compressors in the world than all other types of pumps and compressors combined.

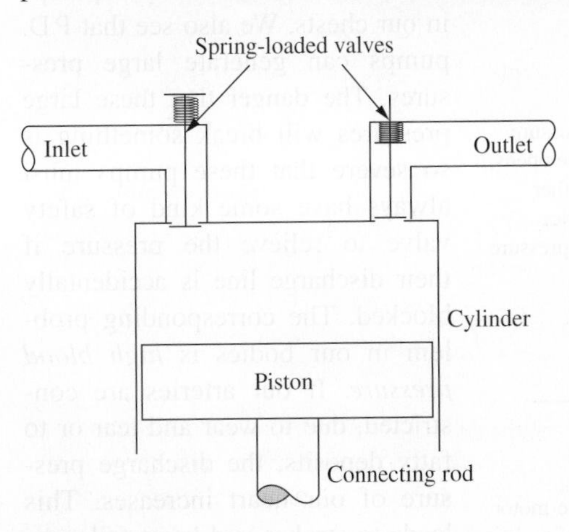

FIGURE 10.2
Positive-displacement pump of the piston-and-cylinder type.

Figure 10.2 shows the cross-sectional view of a simple piston-and-cylinder P.D. pump. The connecting rod moves the piston up and down in a cyclical fashion. The operating cycle of such a pump is as follows, starting with the piston at the top.

1. The piston starts downward, creating a slight vacuum in the cylinder.
2. The pressure of the fluid in the inlet line is high enough relative to this vacuum to force open the

left-hand one-way valve, whose spring has been designed to let the valve open under this slight pressure difference.

3. Fluid flows in during the entire downward movement of the piston.
4. The piston reaches the bottom of its stroke and starts upward. This raises the pressure in the cylinder higher than the pressure in the inlet line, so the inlet valve is pulled shut by its spring.
5. The pressure continues to rise until it is higher than the pressure in the outlet line. If the fluid were totally incompressible, this pressure rise would be instantaneous. For most liquids it is practically instantaneous. For gases (discussed below), it is not instantaneous.
6. When the pressure in the cylinder is higher than the pressure in the outlet line, the one-way outlet valve is forced open.
7. The piston pushes the fluid out into the outlet line.
8. The piston starts downward again; the spring closes the outlet valve, because the pressure in the cylinder has fallen, and the cycle begins again.

Suppose that we test such a pump, for various downstream pressures. We begin with all downstream valves open, so that the downstream gauge pressure is zero. We slowly close a downstream valve, raising the downstream pressure. For a given speed of the pump's motor, the results for various discharge pressures are shown in Fig. 10.3. (The input and output pressures and velocities of a P.D. pump vary cyclically. All of the values shown in this section are average pressures or average velocities.)

From Fig. 10.3 we see that P.D. pumps are practically constant-volumetric-flow-rate devices (at a fixed drive motor speed). To increase the flow rate for a fixed-geometry P.D. pump, one must increase the motor speed. When we exercise and need more blood supply we increase the number of beats per minute of the P.D. pumps in our chests. We also see that P.D. pumps can generate large pressures. The danger that these large pressures will break something is so severe that these pumps must always have some kind of safety valve to relieve the pressure if their discharge line is accidentally blocked. The corresponding problem in our bodies is *high blood pressure*. If our arteries are constricted, due to wear and tear or to fatty deposits, the discharge pressure of our heart increases. This leads to strokes and heart failures.

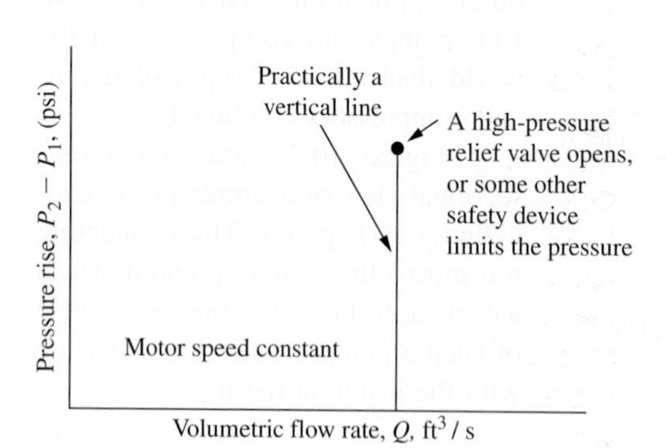

FIGURE 10.3
Pump performance curve for a P.D. pump, with the motor running at constant speed. Most pump curves plot the pressure rise or the pump head, $h = (P_{out} - P_{in}) / \rho g$, vertically and the volumetric flow rate, Q, horizontally.

For a perfect P.D. pump and an absolutely incompressible

fluid the volumetric flow rate equals the volume swept out per unit time by the piston, or

$$\text{Volumetric flow rate} = \text{piston area} \cdot \text{piston travel} \cdot \text{cycles / time} \quad (10.6)$$

For an actual pump the flow rate will be slightly less because of various fluid leakages (around the piston, the wrong way through the valves). The curve on Fig. 10.3 would be vertical for zero leakage, but bends slightly to the left of vertical for real pumps.

For large P.D. pumps the efficiency can be as high as 0.90; for small pumps it is less. One may show (Prob. 10.3) that for the pump in this example the energy that was converted into friction heating and thereby heated the fluid would cause a negligible temperature rise. The same it not true of gas compressors, as discussed in Sec. 10.2.2.

If we connect our P.D. pump to a sump, as shown in Fig. 10.4, and start the motor, what will happen? A P.D. pump is generally operable as a vacuum pump. Therefore, the pump will create a vacuum in the inlet line. This will make the fluid rise in the inlet line.

If we write the head form of B.E. (Eq. 5.6) between the free surface of the fluid (point 1) and the inside of the pump cylinder (point 2), there is no pump work over this section; so

$$z_2 - z_1 = h = \frac{P_1 - P_2}{\rho g} - \frac{V_2^2}{2g} - \frac{\mathcal{F}}{g} \quad (10.7)$$

If, as shown in Fig. 10.4, the fluid tank is open to the atmosphere, then $P_1 = P_{\text{atm}}$. The maximum possible value of h corresponds to $P_2 = 0$ psia. If there is no friction and the velocity at 2 is negligible, then

$$h_{\max} = \frac{P_{\text{atm}}}{\rho g} \quad (10.8)$$

For water under normal atmospheric pressure and room temperature this height, called the *suction lift* is about 34 ft $\approx$ 10 m.

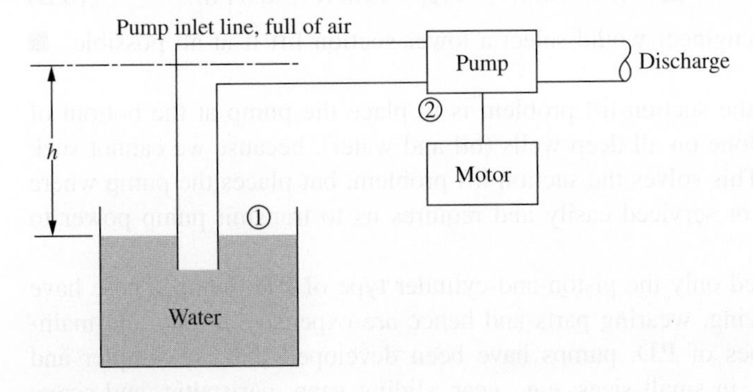

FIGURE 10.4
Suction lift. P.D. pumps work as vacuum pumps and can easily lift liquid modest distances above a reservoir. Centrifugal pumps have a harder time doing this, as discussed in the text.

The actual suction lift obtainable with a P.D. pump is less than that shown by Eq. 10.8, because

1. There is always some line friction, some friction effect through the pump inlet valve, and some inlet velocity.
2. The pressure on the liquid cannot be reduced to zero without causing the liquid to boil. All liquids have some finite vapor pressure. For water at room temperature it is about 0.3 psia or 0.02 atm. If the pressure is lowered below this value, the liquid will boil.
3. This assumes zero valve leakage in the pump. All pumps of the type shown in Fig. 10.2 have some valve leakage so that they cannot produce a very high vacuum.

Example 10.2. We wish to pump 200 gal / min of water at 150°F from a sump. We have available a P.D. pump that can reduce the absolute pressure in its cylinder to 1 psia. We have an $\mathcal{F} / g$ (for the pipe only) of 4 ft. The friction effect in the inlet valve may be considered the same as that of a sudden expansion (see Sec. 5.5) with inlet velocity equal to the fluid flow velocity through the valve, which here is 10 ft / s. The atmospheric pressure at this location is never less than 14.5 psia. What is the maximum elevation above the lowest water level in the sump at which we can place the pump inlet?

The lowest pressure we can allow in the cylinder (P_2) is 3.72 psia, the vapor pressure of water at 150°F. If the pressure were lower than this, the water would boil, interrupting the flow. The density of water at 150°F is 61.3 lbm / ft^3. Thus,

$$h_{\max} = \frac{(14.5 - 3.7)\text{lbf / in}^2}{61.3 \text{ lbm / ft}^3 \cdot 32.2 \text{ ft / s}^2} \cdot \frac{144 \text{ in}^2}{\text{ft}^2} \cdot 32.2 \frac{\text{lbm} \cdot \text{ft}}{\text{lbf} \cdot \text{s}^2} - \frac{(10 \text{ ft / s})^2}{2 \cdot 32.2 \text{ ft / s}^2} - 4 \text{ ft}$$

$$= 25.4 \text{ ft} - 1.6 \text{ ft} - 4 \text{ ft} = 19.8 \text{ ft} = 6.04 \text{ m} \tag{10.B}$$

An experienced engineer would select a lower suction lift if at all possible. ■

One solution to the suction-lift problem is to place the pump at the bottom of the inlet line. That is done on all deep wells (oil and water), because we cannot suck the fluid out of them. This solves the suction lift problem, but places the pump where it cannot be observed or serviced easily and requires us to transmit pump power to the bottom of the well.

We have discussed only the piston-and-cylinder type of P.D. pump. These have a large number of moving, wearing parts and hence are expensive to buy and maintain. Several other types of P.D. pumps have been developed that are simpler and cheaper than this type, in small sizes, e.g., gear, sliding vane, peristaltic, and screw pumps. Although the mechanical arrangements are different, the operating principle and the performance of these are similar to those of piston-and-cylinder types. Of these designs, the sliding-vane pumps are the easiest to understand. Figure 10.5 shows a cutaway of one such pump, and illustrates how it works.

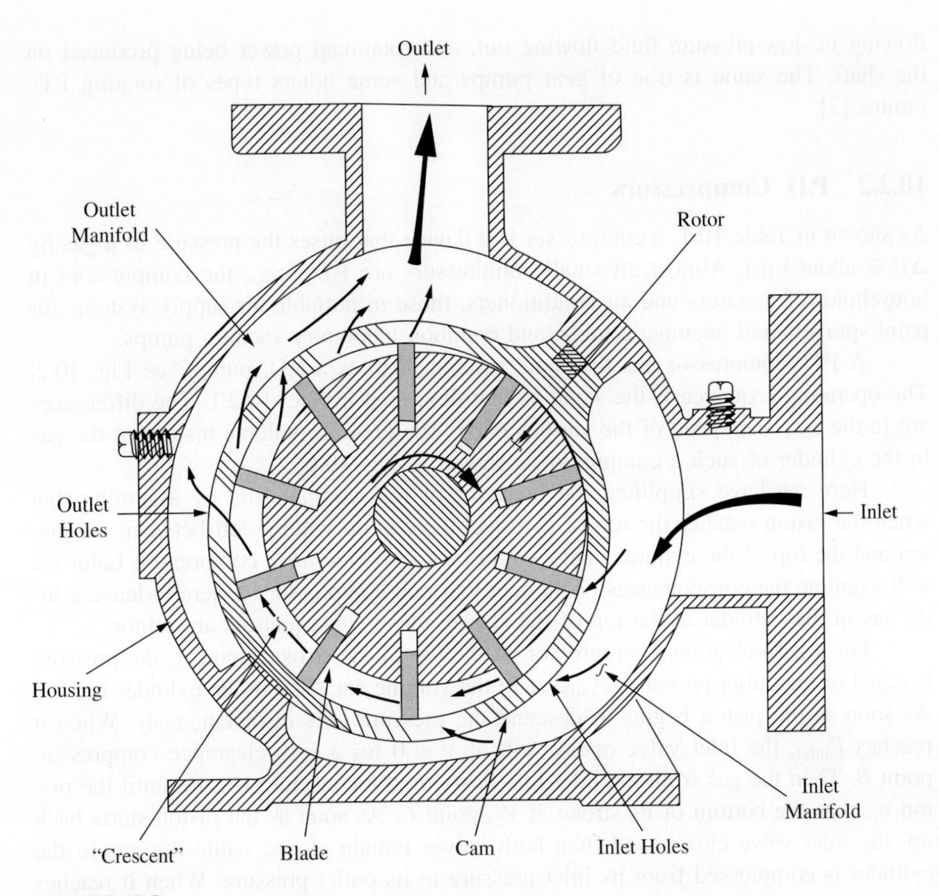

FIGURE 10.5
Cross-sectional view of a sliding-vane P.D. pump. The circular rotor is placed off-center in the circular casing (called a cam on this drawing). The rotor rotates clockwise. The blades move in and out in slots in the rotor, driven against the casing by centrifugal force and/or hydraulic pressure below them, and/or springs. As the rotor rotates, it traps a mass of liquid in the space in the crescent between two vanes. Then, further rotation forces that fluid into the outlet line. The movement of the vanes (in and out) is exactly analogous to the movement of the valves in Fig. 10.2. This device is functionally equivalent to Fig. 10.2, but because of its rotary motion it is smaller, simpler, and more robust. (Courtesy of the Corken Pump Division of the Ibex Corporation.)

This type of P.D. pump is very common. Its mode of operation is the same as that of the piston-and-cylinder P.D. pump in Fig. 10.2. It sucks the fluid into an expanding cavity, traps that fluid, then reduces the size of the cavity, thus forcing the fluid out into the discharge line at a pressure higher than the inlet line. Its pump curve (like Fig. 10.3) is not quite as vertical as that sketched in Fig. 10.3, because there is some leakage around the tips and sides of the sliding vanes. But the curve is close enough to vertical that such pumps almost always have a pressure relief (bypass) valve to prevent them from developing dangerously high pressures. This is one of the most common pumps in the hydraulic systems used on construction equipment. With very modest modifications it can be run backward, with high-pressure hydraulic fluid flowing in, low-pressure fluid flowing out, and rotational power being produced on the shaft. The same is true of gear pumps and some others types of rotating P.D. pumps [3].

10.2.2 P.D. Compressors

As shown in Table 10.1, a compressor is a device that raises the pressure of a gas by $\Delta P \geq$ about 1 psi. Almost all small compressors are P.D., e.g., the compressors in household refrigerators and air conditioners, those in portable air supply systems for paint sprayers and pneumatic tools, and common laboratory vacuum pumps.

A P.D. compressor has the same general form as a P.D. pump; see Fig. 10.2. The operating sequence is the same as that described in Sec. 10.2.1. The differences are in the size and speed of the various parts. The pressure-volume history of the gas in the cylinder of such a compressor is shown in Fig. 10.6.

Here we have simplified the behavior of real compressors by assuming that when the piston reaches the top of its travel there is no volume left between the piston and the top of the cylinder. This would be a *zero-clearance* compressor. Later we will examine the consequences of the fact that real compressors generally leave a little gas in the cylinder at the top of the stroke; those consequences are minor.

For a zero-clearance compressor, at the top of the stroke, point A, the pressure is equal to the outlet pressure, P_{outlet}, and the volume enclosed in the cylinder is zero. As soon as the piston begins to descend the pressure falls instantaneously. When it reaches P_{inlet}, the inlet valve opens, still at $V = 0$ for a zero-clearance compressor, point B. Then the gas from the inlet line flows in at a constant pressure until the piston reaches the bottom of its stroke at V_2, point C. As soon as the piston starts back up, the inlet valve closes, and then both valves remain closed while the gas in the cylinder is compressed from its inlet pressure to its outlet pressure. When it reaches that pressure the outlet valve opens ($P = P_{\text{outlet}}$, $V = V_1$), point D. Then as the piston

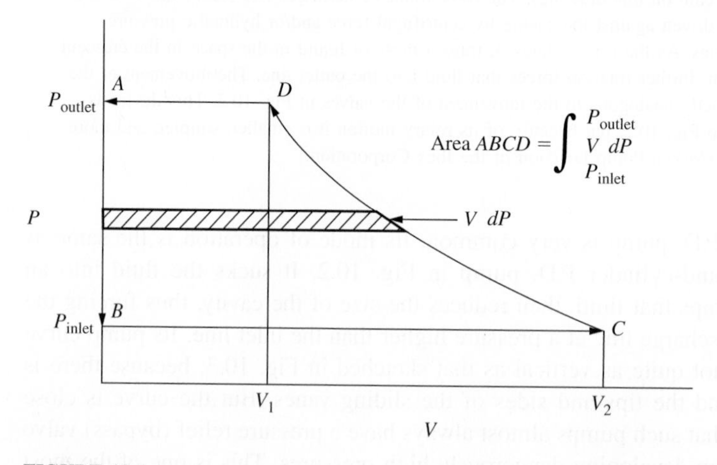

FIGURE 10.6
Pressure–volume history of one cycle of a zero-clearance piston-and-cylinder P.D. compressor.

continues to rise the gas is forced out into the outlet line at constant P, completing the cycle at point A.

The work of any single piston process is given by

$$W = \int F\,dx = \int PA\,dx = \int P\,dV \tag{10.9}$$

The work done by the compressor on the gas is the gross work done on the gas (area under curve CDA) minus the work done by the gas on the piston as the gas flowed in (area under the curve BC); thus, the net work is the area enclosed by the curve $ABCD$. It is equal to the gross work required to drive the compressor only if there is no friction, no gas leakage, etc. (i.e., 100 percent efficient operation).

This area on a PV diagram is the algebraic sum of three areas: the area $P_{\text{inlet}} V_1$, which is the work done by the gas in the inlet line driving back the piston; the area $\int_C^D P\,dV$, which is the work input of the compression step; and the area $P_{\text{outlet}} V_2$, which is the work to drive the gas out of the cylinder into the outlet line. One can see that, although for any one of these three steps the work is given by a $P\,dV$ integral, the algebraic sum of three such $P\,dV$ integrals adds up to a $V\,dP$ integral (Eq. 10.10):

$$W_{\substack{\text{done on the}\\ \text{gas, per cycle}}} = \int_{\text{inlet}}^{\text{outlet}} V\,dP \quad \text{[zero-clearance compressor]} \tag{10.10}$$

This integral is the work done *by* the compressor *on* the gas. If we take the compressor as our system, then it is negative, because it is work flowing out of the system; an equal or greater amount of positive-sign work must flow in from whatever is driving the compressor. (One can come to the same result directly from B.E., in which we included the injection work in our steady-flow energy balance.)

Compressors are most often used to compress gases that can be reasonably well represented by the ideal gas law, $PV = nRT$. If a compressor works slowly enough and has good cooling facilities, then the gas in the cylinder will be at practically a constant temperature throughout the entire compression process. Then we may substitute nRT/P for V in Eq. 10.10 and integrate:

$$\Delta W_{\substack{\text{done on the}\\ \text{gas, per cycle}}} = \int_{P_1}^{P_2} V\,dP = nRT \int_{P_1}^{P_2} \frac{dP}{P} = nRT \ln \frac{P_2}{P_1} \quad \text{[isothermal]} \tag{10.11}$$

However, in most compressors the piston moves too rapidly for the gas to be cooled much by the cylinder walls. If the piston moves very rapidly, the gas will undergo what is practically a reversible, adiabatic process, i.e., an isentropic process. In that case, we may rearrange Eq. B.3-23 of App. B to

$$PV^k = \text{constant} = P_1 V_1^k \quad \text{[adiabatic]} \tag{10.12}$$

Inserting this in Eq. 10.10 and making the algebraic manipulations shown in App. B.7, we find

$$\Delta W_{\substack{\text{done on the}\\ \text{gas, per cycle}}} = \int_{P_1}^{P_2} V\,dP = \frac{nRT_1 k}{k-1}\left[\left(\frac{P_2}{P_1}\right)^{(k-1)/k} - 1\right] \quad \text{[adiabatic]} \tag{10.13}$$

Most often we divide both sides of Eq. 10.11 and 10.13 by the number of mols compressed, n, to find the work per mol.

Example 10.3. A 100 percent efficient compressor compresses air from 1 atm to 10 atm. The inlet temperature is 68°F = 20°C. Calculate the required work per pound mol for (*a*) an isothermal compressor and (*b*) an adiabatic compressor,

(*a*) For an isothermal compressor:

$$\frac{\Delta W}{n} = 1.987 \frac{\text{Btu}}{\text{lbmol} \cdot \text{R}} \cdot 528°\text{R} \cdot \ln 10 = 2416 \frac{\text{Btu}}{\text{lbmol}} = 5.62 \frac{\text{kJ}}{\text{mol}} \quad (10.\text{C})$$

(*b*) For an adiabatic compressor:

$$\frac{\Delta W}{n} = 1.987 \frac{\text{Btu}}{\text{lbmol} \cdot °\text{R}} \cdot 528°\text{R} \cdot \frac{1.4}{0.4} \cdot (10^{0.4/1.4} - 1)$$

$$= 3418 \frac{\text{Btu}}{\text{lbmol}} = 7.96 \frac{\text{kJ}}{\text{mol}} \quad (10.\text{D})$$

The difference between the answers to parts (*a*) and (*b*) is due to the rise in temperature of the gas in the adiabatic compressor. From Eq. B.3-16 we can calculate that the adiabatic compressor's outlet temperature is 1.93 times its inlet temperature, or 1019°R = 559°F. Thus, by the ideal gas law we know that its exit volume is 1.93 times the exit volume for the isothermal compressor, and the integral of $V\,dP$ must be larger. ■

Equations 10.11 and 10.13 indicate that the result in Example 10.3 is a general result; i.e., the required work per mol for an adiabatic compressor is always greater than that for an isothermal compressor with the same inlet and outlet pressures. Therefore, it is advantageous to try to make real compressors as nearly isothermal as possible. One way to do this is to cool the cylinders of the compressor. Almost all P.D. compressors have cooling jackets or cooling fins on their cylinders. Students may observe that the cylinders of the compressors in service stations and on paint-sprayer compressors always have cooling fins, to reduce the outlet temperature and thus the required power input. Another way to reduce the power demand is by staging and intercooling; see Example 10.4.

Example 10.4. Rework Example 10.3 by using a two-stage adiabatic compressor in which the gas is compressed adiabatically to 3 atm, then cooled to 68°F, and then compressed from 3 atm to 10 atm.

Here

$$\frac{\Delta W}{n} = 1.987 \frac{\text{Btu}}{\text{lbmol} \cdot °\text{R}} \cdot 528°\text{R} \cdot \frac{1.4}{0.4} \cdot \left[3^{0.4/1.4} - 1 + \left(\frac{10}{3}\right)^{0.4/1.4} - 1 \right]$$

$$= 2862 \frac{\text{Btu}}{\text{lbmol}} = 6.66 \frac{\text{kJ}}{\text{mol}} \quad (10.\text{E})$$

The power requirement is more than that of the isothermal compressor in Example 10.3, but less than that of the adiabatic compressor in that example. ■

This example illustrates the advantage of staging and intercooling. There is a limit to the amount of work saving possible; with an infinite number of stages with intercooling an adiabatic compressor would have the same performance as an isothermal compressor (Prob. 10.13). Thus, the behavior of an isothermal compressor represents the best performance obtainable by staging. The optimal number of stages is found by an economic balance between the extra cost of each additional stage and the improved performance as the number of stages is increased.

In Example 10.4 the interstage pressure was arbitrarily selected as 3 atm. It can be shown (Prob. 10.9) that for a two-stage compressor (with the same inlet temperature to both stages) the optimal interstage pressure (that which requires the least amount of total work) is given by

$$P_{\text{interstage}} = (P_{\text{discharge}} \cdot P_{\text{inlet}})^{1/2} \tag{10.14}$$

This is the interstage pressure that makes the pressure ratio $P_{\text{out}} / P_{\text{in}}$ the same for each stage. By similar calculations it can be shown (Prob. 10.10) that for more than two stages the optimal interstage pressures are those that have the same pressure ratio for each stage. The power-saving that results from staging and intercooling is great enough that even compressors as small as those on modest-sized portable compressed air systems have more than one stage. In Eq. 10.12, the value of k is that of the gas for an adiabatic compressor. One may substitute $k = 1$ in this equation and see that it reduces to Eq. 10.11, the equation for an isothermal compressor. Real compressors (with cooling) operate somewhere between these extremes, normally closer to adiabatic than to isothermal. The common practice is to use a value of k somewhat smaller than that for the gas; this is called a *polytropic* value of k, chosen to make Eq. 10.13 match the experimental behavior of real compressors.

All the foregoing concerned zero-clearance compressors, ones in which no gas is left in the cylinder at the end of the discharge stroke. For mechanical reasons it is impractical to build a compressor with zero clearance. So in real compressors there is always a small amount of gas in the top of the cylinder, which is repeatedly compressed and expanded. If the compression and expansion are reversible, either adiabatic or isothermal, then that gas contributes as much work on the expansion step as it requires on the compression step, and thus it contributes nothing to the net work requirement of the compressor. For real compressors the compression and expansion of the gas in the clearance volume does contribute slightly to the inefficiency of the compressor; compressor designers make the clearance volume as small as practical.

Equations 10.11 and 10.13 were derived with the assumption of a zero-clearance compressor. They apply equally well to real compressors if we understand the n in them to represent the net number of mols passing through, not the total number of mols present in the cylinder at the start of the compression stroke. Because we normally analyze the work of compressors on a (work / mol or unit mass processed) basis, this causes no difficulty. In these derivations and in Fig. 10.6 we assumed zero pressure drop through the inlet and outlet valves. All real compressors of this type have some pressure drop through these valves; it is a significant part of the friction heating that causes the efficiency to be less than 100 percent. The typical value of $P_{\text{out}} / P_{\text{in}}$ for a one-stage P.D. compressor is 3 to 5. If a higher ratio is needed, the common practice is to used staging, as shown in Example 10.4, with as many stages as needed to meet the desired value of P_{out}/P_{in}.

In general we use a P.D. compressor for any small gas flow with P_{out}/P_{in} greater than about 1.1, and also for large gas flows when the final pressure is very high, as in very high pressure chemical reactors.

10.3 CENTRIFUGAL PUMPS AND COMPRESSORS

Starting about 1850, industrial countries learned to build high-speed rotating machinery. Before that time no such devices existed. Rotating high-speed pumps and compressors can be centrifugal, axial flow, or some other kind. They have almost completely displaced P.D. devices for high flow rates, and for most medium-pressure, medium-flow-rate pumping operations. For those, they are simpler, smaller, cheaper, and more robust than the P.D. devices they replace [4]. They are the most common type of pump in chemical engineering processes.

10.3.1 Centrifugal Pumps

A centrifugal pump raises the pressure of a liquid by moving it outward in a centrifugal force field, and by giving it a high kinetic energy and then converting that kinetic energy into injection work. The water pump on most automobiles is a typical centrifugal pump. As shown in Fig. 10.7, it consists of an impeller (i.e., a wheel with blades) attached to a rotating shaft, some form of housing with a central inlet and a peripheral outlet, and a back cover with some kind of seal to let the shaft in without letting the liquid leak out.

In such a pump the fluid flows in the central inlet into the "eye" of the impeller, is spun outward by the rotating impeller, and flows out through the peripheral outlet.

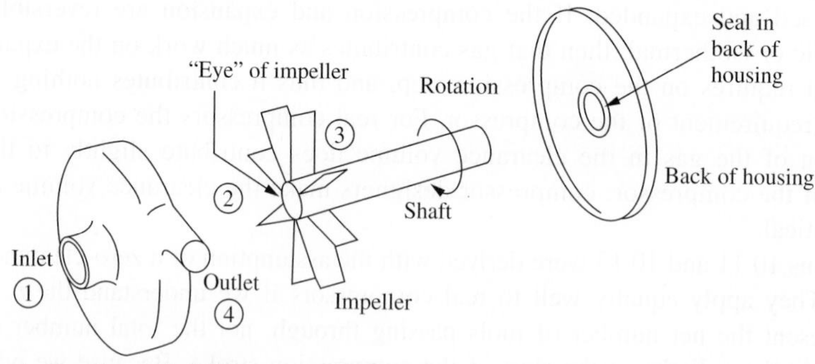

FIGURE 10.7
Exploded view of a very simple centrifugal pump, such as an automobile water pump. Real pumps are more complex, but almost all have a rotor with some kind of blades, a shaft to drive the rotor, a housing with flow in at its center and discharge at its periphery. All simple ones also have a back plate, with a seal where the shaft passes out of the pump to the motor that drives it. Numbers 1 to 4 refer to discussion in the text.

To analyze such a pump on a *very simplified basis,* we consider it as three parts: an inlet, a centrifuge, and a diffuser. In the inlet, the fluid flows into the pump and is picked up by the impeller. Applying B.E. between the inlet pipe (point 1) and the eye of the impeller (point 2) we find

$$P_2 - P_1 = \frac{\rho}{2}(V_1^2 - V_2^2) - \rho \mathcal{F}_{1\text{-}2} \tag{10.15}$$

The impeller acts as a centrifuge, described in Chap. 2. Integrating Eq. 2.34 from the eye of the impeller (point 2) to the tip of the impeller (point 3) we find that

$$P_3 - P_2 = \int_{r_2}^{r_3} \rho\omega^2 r\, dr = \frac{\rho\omega^2}{2}(r_3^2 - r_2^2) \tag{10.16}$$

The third part of the centrifugal pump is the diffuser (see Sec. 5.5), in which the fluid flows from the tip of the impeller to the outlet pipe. Applying B.E. between the tip of the impeller (point 3) and the outlet line (point 4), we find

$$P_4 - P_3 = \frac{\rho}{2}(V_3^2 - V_4^2) - \rho \mathcal{F}_{3\text{-}4} \tag{10.17}$$

Adding these three equations, dividing by ρg, and canceling like terms, we find

$$h = \frac{P_4 - P_1}{\rho g} = \frac{\omega^2}{2g}(r_3^2 - r_2^2) + \frac{(V_3^2 - V_4^2) + (V_1^2 - V_2^2)}{2g} - \frac{(\mathcal{F}_{3\text{-}4} + \mathcal{F}_{1\text{-}2})}{g} \tag{10.18}$$

The division by ρg converts this expression to the equivalent of the head form of B.E. (Sec. 5.4), which is commonly used with pumps. We then simplify this equation by noting that

$$V_2 = \omega r_2 \quad \text{and} \quad V_3 = \omega r_3 \tag{10.19}$$

which allows us to group two terms and find

$$h = \frac{P_4 - P_1}{\rho g} = \frac{\omega^2}{g}(r_3^2 - r_2^2) + \frac{(V_1^2 - V_4^2)}{2g} - \frac{(\mathcal{F}_{3\text{-}4} + \mathcal{F}_{1\text{-}2})}{g} \tag{10.20}$$

Each of the terms in Eq. 10.20 has the dimensions of a length; h is called the *pump head,* which is the height to which the pump will lift a fluid at a given flow rate. In some centrifugal pumps the inlet and outlet lines have the same diameter so that the velocity terms in Eq. 10.20 cancel. However, for reasons discussed below, the inlet line is often larger than the outlet line, so this term remains.

Example 10.5. A centrifugal pump has the following dimensions: inlet pipe 2-in pipe size, (diameter = 2.067 in), outlet pipe 1.5-in pipe size, (diameter = 1.61 in), impeller inner diameter = 2.067 in, outer diameter = 6.75 in, rotational velocity = 1750 rpm. What is the pump head, according to Eq. 10.20, assuming zero friction, for a volumetric flow rate of 100 gpm?

Here we can write

$$\omega = \frac{1750}{\text{min}} \cdot 2\pi \cdot \frac{\text{min}}{60\ \text{s}} = 183.26\,\frac{1}{\text{s}} \tag{10.F}$$

and

$$V_1 = \frac{100 \text{ gal / min}}{(\pi / 4) \cdot (2.067 \text{ in})^2} \cdot \frac{231 \text{ in}^3}{\text{gal}} \cdot \frac{\text{min}}{60 \text{ s}} \cdot \frac{\text{ft}}{12 \text{ in}} = 9.56 \frac{\text{ft}}{\text{s}} = 2.91 \frac{\text{m}}{\text{s}} \quad \text{(10.G)}$$

and similarly, $V_2 = 15.76$ ft / s = 4.81 m / s. The impeller inner and outer radii are 0.086 ft and 0.336 ft. Substituting these values in Eq. 10.20, with zero friction, we find

$$h = \frac{(183.26 / \text{s})^2}{32.2 \text{ ft / s}^2}[(0.336 \text{ ft})^2 - (0.086 \text{ ft})^2] + \frac{[(15.75 \text{ ft / s})^2 - (9.56 \text{ ft / s})^2]}{2 \cdot 32 \text{ ft / s}^2}$$
$$= 74.76 \text{ ft} - 2.44 \text{ ft} = 72.3 \text{ ft} = 30.4 \text{ m} \quad \text{(10.H)}$$

■

This value and the calculated values for other flow rates are shown in Fig. 10.8, along with the experimental values for a pump with the same dimensions as that in this example. We see that the experimental values are about half those shown by Eq. 10.20 and that the difference increases with increasing flow rate. This simply shows that the effects of friction are substantial in pumps of this type and that they increase with increasing flow rate. The experimental values are taken from Fig. 10.9, which shows a wealth of information about actual centrifugal pumps.

From Fig. 10.9 we see the following:

1. This is a pump-head flow rate curve, just like Fig. 10.8. However, it shows curves for four different diameter impellers, A, B, C, and D. The reader may verify that the Experimental pump curve on Fig. 10.8 was copied from the "C 6 $\frac{3}{4}$" curve on Fig. 10.9. Pump manufacturers make only a modest number of pump external housings, and fit each one with a series of different-sized impellers, thus providing a suitable-sized pump for a wide variety of demands, with far fewer different models than they would need if they produced a completely different pump for each service. Figure 10.9 shows the pump curves for a family of four different pumps, each with its own impeller size but sharing all other parts in common.
2. The plot shows lines of constant power requirement. Thus, for Example 10.5, we can interpolate on Fig. 10.9 that the required power input is about 1.15 hp. This power

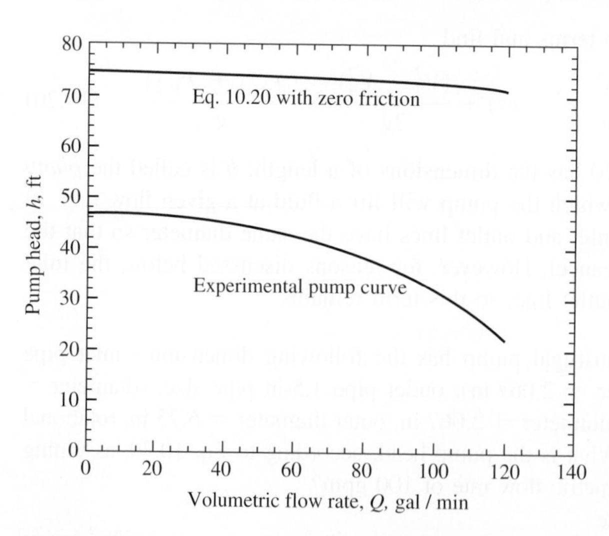

FIGURE 10.8
Comparison of the *pump curves* (head versus volumetric flow rate plots) for a frictionless centrifugal pump, as computed by Eq. 10.20, and a real pump of the same dimensions, copied from Fig. 10.9.

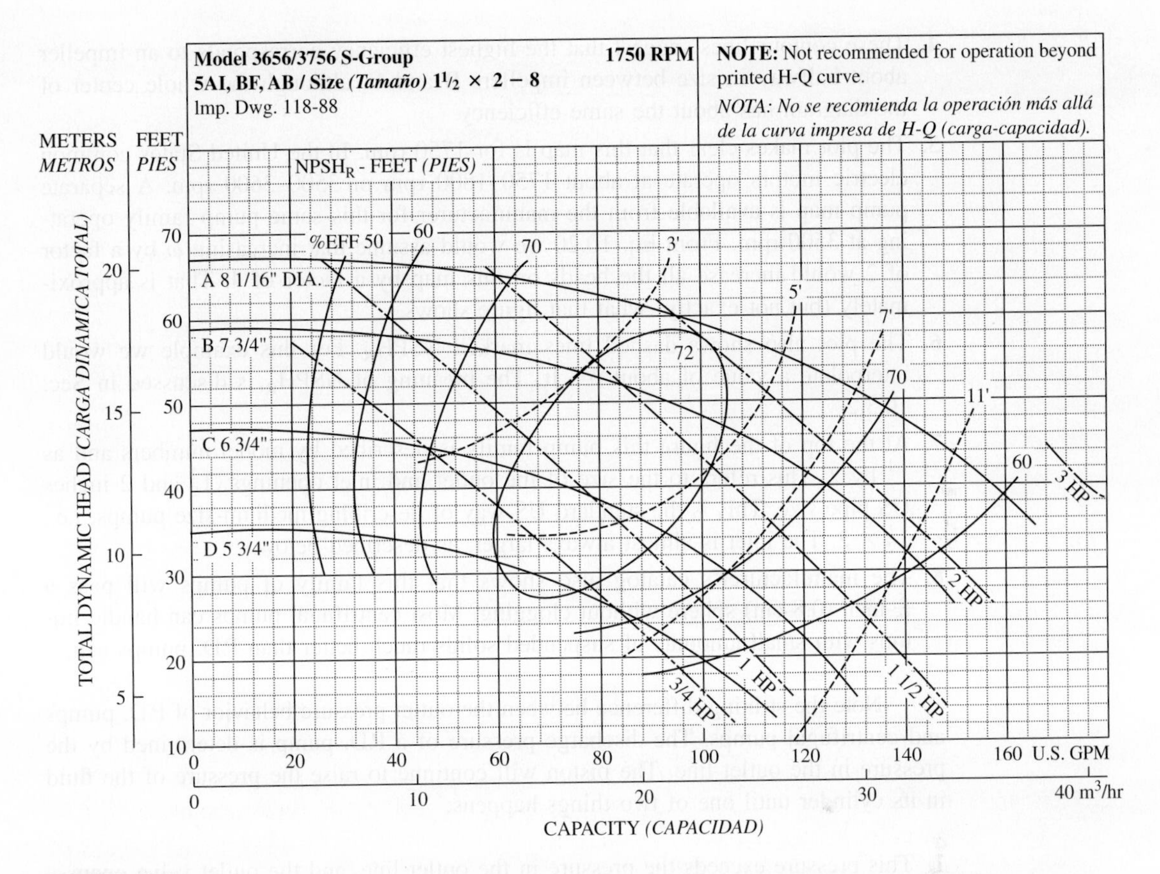

FIGURE 10.9
Pump map for a family of four centrifugal pumps (four different size impellers in the same size casing). This map shows results in English and some metric units and in English and Spanish. The designation $1\frac{1}{2} \times 2$ refers to the sizes of the outlet and inlet pipes, $1\frac{1}{2}$ and 2-in U.S. pipe size. This is the common way of describing centrifugal pumps in the United States. The horsepower lines are only applicable for a fluid with SG 1.00, e.g., water. The plot is discussed in the text. (Courtesy of Goulds Pumps.)

input is based on a fluid with SG = 1.00. At $Q = 100$ gpm we read that $h \approx 32$ ft, so the pressure rise through the pump (for water) at this flow rate is

$$\Delta P = \rho g h = 62.3 \frac{\text{lbm}}{\text{ft}^3} \cdot 32.2 \frac{\text{ft}}{\text{s}^2} \cdot 32 \text{ ft} \cdot \frac{\text{lbf} \cdot \text{s}^2}{32.2 \text{ lbm} \cdot \text{ft}} = 1994 \frac{\text{lbf}}{\text{ft}^2}$$

$$= 13.8 \text{ psi} = 95.5 \text{ kPa} \tag{10.I}$$

From Eq. 10.4 we may compute

$$\eta = \frac{(100 \text{ gal / min}) \cdot (13.8 \text{ lbf / in}^2)}{1.15 \text{ hp}} \cdot \frac{\text{hp} \cdot \text{min}}{33{,}000 \text{ ft} \cdot \text{lbf}} \cdot \frac{231 \text{ in}^3}{\text{gal}} \cdot \frac{\text{ft}}{12 \text{ in}}$$

$$= 0.70 = 70\% \tag{10.J}$$

3. The plot shows contour lines of equal efficiency. Based on these, we would estimate an efficiency in this example of about 71 percent. The power requirement and efficiency curves are only approximate, so that 70 percent approximately equals 71 percent.

4. These contour lines suggest that the highest efficiency corresponds to an impeller about halfway in size between impellers B and C, although the whole center of the diagram has about the same efficiency.
5. The plot makes clear that this map is for 1750 rpm. In the United States common electric motors operate at about 1750–1800 rpm or 3500–3600 rpm. A separate pump map is available from the manufacturer for this same pump family operating at 3500 rpm. From Eq. 10.20 we would assume that increasing ω by a factor of 2 would increase all the heads on this map by a factor of 4. That is approximately (but not exactly) what that figure shows.
6. The plot also shows dashed lines marked $NSPH_R$. For this example we would interpolate a value of about 6.5 ft. The meaning of $NSPH_R$ is discussed in Sec. 10.3.2.
7. At the top of the figure this pump family is described by model numbers and as $1\frac{1}{2}$ by 2. This refers to the size of the outlet and inlet openings, $1\frac{1}{2}$ and 2 inches US pipe size. This is the standard US way of describing medium-size pumps, i.e., $1\frac{1}{2} \times 2$. The inlet is almost always larger, as described below.
8. The manufacturer's catalog also shows that this family of pumps will pass a $\frac{5}{16}$-inch (0.8-cm) sphere without clogging. Most centrifugal pumps can handle liquids with small amounts of suspended solids much better than P.D. pumps can.

Note the striking difference between the outlet pressure behavior of P.D. pumps and centrifugal pumps. The discharge pressure of a P.D. pump is determined by the pressure in the outlet line. The piston will continue to raise the pressure of the fluid in its cylinder until one of two things happens:

1. This pressure exceeds the pressure in the outlet line, and the outlet valve opens.
2. Something breaks, or a high-pressure (bypass) valve opens.

For a centrifugal pump, on the other hand, there is a maximum pressure rise across the pump that is set by r_1, r_2, and ω of the impeller and the fluid density. If the difference between the inlet and outlet pressures becomes greater than this, fluid will flow backward through the pump—in the outlet and out the inlet, even while the pump is running.

Suppose that we repeat the suction-lift experiment shown in Fig. 10.4 for a centrifugal pump. Reading values from Fig. 10.9, at zero flow ($Q = 0$) and 1750 rpm, we see that this pump has a head of 47 ft, so if the pump is full of water it will pump from a sump below it up to 34 ft minus the friction head loss with no trouble. But suppose instead of being full of water, the pump is full of air. This pump will work (poorly) as an air compressor. But with a head of 47 ft, full of air, the pressure rise across the pump is

$$\Delta P = \rho g h = 0.075 \frac{\text{lbm}}{\text{ft}^3} \cdot 32.2 \frac{\text{ft}}{\text{s}^2} \cdot 47 \text{ ft} \cdot \frac{\text{lbf} \cdot \text{s}^2}{32.2 \text{ lbm} \cdot \text{ft}}$$

$$= 3.5 \frac{\text{lbf}}{\text{ft}^2} = 0.024 \text{ psi} = 0.17 \text{ kPa} \quad (10.K)$$

This will raise the water in the inlet line about 0.7 inch above the reservoir level. Thus, to get a centrifugal pump going, it is not enough to start the motor. One must also replace the air in the pump and inlet lines with liquid. This is called *priming.* This pump, and all common centrifugal pumps are not *self-priming.* To start these pumps one must normally get a liquid into the pump. Once it is running it keeps its prime and has no such problem. Various stratagems have been devised to retain liquid in the pump when it shuts down (e.g., a one-way *foot valve* on the pipe inlet from the reservoir in Fig. 10.4 that prevents the liquid from draining out of the pump when it is shut off), and several other designs called "self-priming pumps."

10.3.2 NPSH

In Example 10.2 we saw that in calculating the permissible suction lift for a P.D. pump it was necessary to account for the pressure drop across the inlet valve. The difficulty to be avoided there was boiling of the liquid as its pressure was reduced through the inlet valve. The analogous problem with centrifugal pumps is boiling in the eye of the impeller. As the fluid enters the eye, it has a small velocity in the axial direction and negligible rotational velocity. To be picked up by the blades of the impeller, the fluid must be brought up to the rotational speed of the impeller blades. We again apply B.E. between the inlet pipe (point 1) and the point on the blades of the impeller where the pump work starts to increase the pressure of the fluid (point 2). If we assume that the friction effects are negligible, ignore changes in elevation, and ignore the velocity of the inlet fluid, we conclude that

$$P_2 = P_1 - \rho \frac{V_2^2}{2} \tag{10.21}$$

Thus, the pressure falls and boiling may occur, as discussed in Sec. 5.9. Centrifugal pumps very often are used to pump boiling liquids: e.g., at the bottom of distillation columns, flash drums, evaporators, reflux drums, reboilers, condensers, etc. (see Fig. 10.10). The elevation, h, in Fig. 10.9 must be large enough to prevent this boiling.

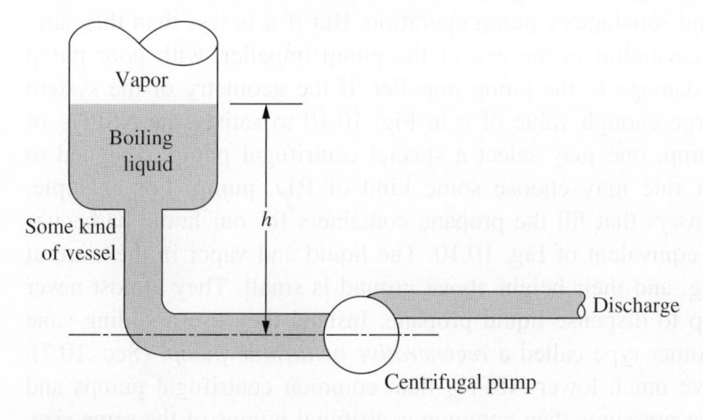

FIGURE 10.10
A centrifugal pump used to pump a boiling liquid, introducing the idea of Net Positive Suction Head (NPSH).

Example 10.6. For the flow in Example 10.5, how much must h be in Fig. 10.10 for the pressure in the eye of the pump impeller to be the same as the pressure at the vapor-liquid surface in the vessel? Assume frictionless flow.

Writing B.E. from the free surface in the vessel (point 1) to the eye of the impeller (point 2) and setting the two pressures equal, we find

$$h = \frac{1}{\rho g} \cdot \left(\frac{V^2_{\text{eye of impeller}}}{2} + \mathcal{F}_{\text{whole system}} \right) \tag{10.22}$$

From the frictionless assumption, we may drop one term. The velocity at the eye of the impeller is the vector sum of the rotational velocity at that point and the radial velocity due to the flow outward through the impeller. By simple algebra we find that the rotational velocity is 15.8 ft / s. To find the radial velocity we assume that it is practically equal to the inlet velocity, 9.56 ft / s. Their vector sum (square root of sum of squares) = 18.45 ft / s, so

$$h = \frac{1}{(62.3 \text{ lbm / ft}^3) \cdot (32.2 \text{ ft / s}^2)} \cdot \frac{(18.46 \text{ ft / s})^2}{2} \cdot \frac{32.2 \text{ lbm} \cdot \text{ft}}{\text{lbf} \cdot \text{s}^2} = 5.28 \text{ ft} \tag{10.L}$$

■

This distance is called the *Net Positive Suction Head, NPSH.* On Fig. 10.9 we see dashed curves labeled NPSH_R. The R subscript indicates "required." From those curves we interpolate, finding that for this flow rate this pump has $\text{NPSH}_\text{R} \approx 6.5$ ft, which is close to the value shown in Eq. 10.L but is presumably somewhat larger because of the $\mathcal{F}$ term, which we ignored.

This is only the value for the pump. To find the value for the pump plus the piping system between the free surface in Fig. 10.10 and the pump, we must add to this height a friction term of the form

$$h_{\substack{\text{additional to overcome} \\ \text{friction in piping}}} = \frac{\mathcal{F}_{\text{in piping to pump inlet}}}{\rho g} \tag{10.M}$$

If the actual distance h in Fig. 10.10 is greater than the sum of these values, then we would expect smooth and satisfactory pump operation. But if h is less than this sum, then we would expect cavitation in the eye of the pump impeller, with poor pump performance and rapid damage to the pump impeller. If the geometry of the system does not allow for a large enough value of h in Fig. 10.10 to satisfy the NSPH_R of a typical centrifugal pump, one may select a special centrifugal pump, designed to have a low NSPH_R, or one may choose some kind of P.D. pump. For example, the local propane dispensers that fill the propane containers for our home barbecues are the thermodynamic equivalent of Fig. 10.10. The liquid and vapor in them are at the equivalent of boiling, and their height above ground is small. They almost never have a centrifugal pump to dispense liquid propane. Instead they use a sliding-vane pump (Fig. 10.5) or another type called a *regenerative* or *turbine pump,* (Sec. 10.7). Both of these types have much lower NSPH_R than common centrifugal pumps and can deliver higher output pressures than common centrifugal pumps of the same size. The NPSH requirement also explains why most centrifugal pumps have larger inlets than their outlets (2 in versus 1.5 in on Fig. 10.9); the larger inlet slows the inlet flow, reducing the $NPSH_R$ compared to that of a pump with a smaller inlet.

In summary on centrifugal pumps, we may say:

1. They are very widely used because they are modest-sized (for their flow rates), relatively simple, and robust.
2. The pressure rise they develop (for equal inlet and outlet diameters) would be

$$\Delta P = \rho_{\text{fluid}} \left[\begin{pmatrix} \text{rotor tip} \\ \text{speed} \end{pmatrix}^2 - \begin{pmatrix} \text{rotor inlet} \\ \text{speed} \end{pmatrix}^2 \right]$$

$$\approx \rho_{\text{fluid}} \left(\omega \frac{D_{\text{impeller}}}{2} \right)^2 \qquad \text{[frictionless]} \tag{10.23}$$

for frictionless flow. For common pumps at their design flow rates,

$$\Delta P_{\text{real, at design flow rate}} \approx 0.5 \Delta P_{\text{frictionless}} \tag{10.24}$$

which we will use subsequently in this chapter and which should be used for estimating purposes in the problems. Equation 10.24 is correct for the small- to medium-sized centrifugal pumps most often seen in industry. It is not true for very large centrifugal pumps in which the 0.5 is replaced by values up to 0.8; see Prob. 10.20.
3. To get high values of ΔP for a given fluid we must go to high values of ω or of D. Often there is a practical limit on these. Very high driver and pump speeds need very good bearings, balancing, and servicing. For deep well pumps, on which much of irrigated agriculture depends, the pump must be small enough to fit down the deep well, whose diameter is set by cost and the size of available deep well drills (typically less than a foot). To meet this requirement, deep well pumps are normally a series of up to 20 centrifugal pumps like the ones described here, attached head-to-tail. The discharge of the first is the inlet to the second, etc. Thus, the modest pressure rise of one stage is multiplied by the number of stages to produce an overall pressure rise suitable for that service (often pumping up from several thousand feet). These are cleverly packaged together, to fit on a common shaft, driven from above, and able to be hung on the shaft down the well.
4. The simple paddle impeller in Fig. 10.7 is used in some applications, but most commercial pumps have much more sophisticated impellers, with a variety of designs. For high-head, low-volumetric flow rates the impeller has a large diameter and is very thin in the axial direction. For low-head, high-volumetric flow rates the impeller becomes similar to an ordinary fan blade. The intermediate cases make the transition from one of these extremes to the other.
5. Large pumps often are the equivalent of two pumps like Fig. 10.7, back-to-back, so that the flow comes in from both sides and exits from a common peripheral exit pipe. This is called a *double-suction* pump. This design eliminates the axial thrust in the shaft, which is modest for small pumps but becomes troublesome for large ones.
6. A large centrifugal pump may have an efficiency greater than 0.90. However, such high-efficiency pumps are expensive and are justified only for very high capacity

TABLE 10.2
Comparison of positive displacement and centrifugal pumps

Characteristic	Positive displacement	Centrifugal
Normal flow rate	Low, up to perhaps 100 gpm	From small (automobile coolant pumps) to huge
Normal pressure rise per stage	Large, can be dangerous, requires high-pressure relief or bypass	Proportional to square of (rpm · impeller diameter), small for small pumps, larger for large ones
Self-priming (i.e., able to work as a weak vacuum pump and suck in liquid)	Yes	No
Number of moving and wearing parts	Many	Few
Number of basically different designs	Many	Only one, with modest variations
Outlet flow rate	Pulsing	Steady
Works well with high-viscosity fluids	Yes	No
$NPSH_R$	Low	Significant, increases with increasing flow rate
Ability to handle liquids with suspended solids	Poor	Fair

applications (e.g., the aqueduct from the Colorado River to Los Angeles). For most applications, pumps are designed for simpler, cheaper construction and for efficiencies of 0.50 to 0.80. The efficiency of centrifugal pumps decreases rapidly as the viscosity of the pumped fluid increases, for which reason they are seldom used for fluids more viscous than ≈300 cP. Sliding-vane and similar P.D. pumps work better for viscous fluids.

7. Figure 10.7 shows that there must be a seal where the shaft enters the pump from outside. Most often this is either a packed seal [Fig. 6.18(*b*)] or a mechanical seal [an advanced version of that in Fig. 6.18(*c*)]. If the fluid pumped is water, these seals are normally set loose, allowing a little water to leak out all the time. This lubricates the seal, prolonging its life; the modest water leak causes no problems. For liquids that are flammable (gasoline) or toxic (benzene), modest leaks are not tolerable, and much more attention must be paid to the pump seal (Prob. 10.25).

Table 10.2 compares the properties of P.D. and centrifugal pumps.

10.3.3 Centrifugal Compressors

The P.D. compressor has been a common industrial tool for a century. However, it is a complicated, heavy, expensive, low-flow-rate device. The need to supercharge aircraft reciprocating engines and the development of turbojet and gas turbine engines demanded the development of lightweight, efficient, low-cost, high-flow-rate compressors. By the 1950s high-flow-rate, high-efficiency centrifugal industrial compressors became available. They have largely replaced P.D. compressors in natural gas pipelines, in large air conditioners, and in large high-pressure chemical processes. Their introduction revolutionized the ammonia production industry, which shut down its P.D. compressors and the plants based on them.

Example 10.7. A modern multistage centrifugal compressor has an impeller diameter of about 2 ft, and operates at about 10,000 rpm ($\omega = 1047 / \text{s}$). If the fluid in the first stage of the compressor is air at 1 atm, what is the estimated pressure rise in the first stage?

Treating this as a centrifugal pump, we combine Eqs. 10.23 and 10.24 so that

$$\Delta P = 0.5 \cdot 0.075 \frac{\text{lbm}}{\text{ft}^3} \cdot \left(\frac{1047}{\text{s}} \cdot \frac{2\ \text{ft}}{2} \right)^2 \cdot \frac{\text{lbf} \cdot \text{s}^2}{32.2\ \text{lbm} \cdot \text{ft}} \cdot \frac{\text{ft}^2}{144\ \text{in}^2}$$
$$= 8.9\ \text{psi} = 61\ \text{kPa} \tag{10.N}$$

This assumes an incompressible fluid. It is only approximately correct, because as the pressure and temperature rise, the density increases. But it illustrates that with centrifugal compressors this large and at rotational speeds this high, one does get substantial pressure increases. ■

If one continued the analysis of such a compressor, one would see that in each subsequent stage the inlet density was larger than in the preceding one, and thus the pressure increase per stage would steadily increase. In applications like natural gas pipelines and ammonia plants the inlet pressure is often much higher than 1 atm, so the pressure increase per stage is comparably greater.

10.4 AXIAL FLOW PUMPS AND COMPRESSORS

The centrifugal compressors just described are better for raising the pressure of a gas already at a modest pressure to a high pressure than for compressing from a low pressure. Axial flow compressors, described next, are better than centrifugals at low pressures and have the advantage that for equal flow rates their diameters are smaller, which makes them easier to build into jet engines. The jet engine cutaway, Fig. 7.15, shows, at the left, a simple axial flow compressor, which consists of a wheel with blades that rotates inside a concentric tube. The blades are curved and at an angle to the axis, so that as they turn they push the fluid down the tube. This is the functional equivalent of the portable fans that move air around in offices and houses, the fixed fans that draw air out of our kitchens and bathrooms, and the smaller ones that cool our computers. The student should examine such a fan while it is both standing still and moving.

The first row of blades at the left is fixed; it turns the flow at practically constant speed. The second row does work on the fluid, increasing its speed. Then the third, non-moving row slows the fluid, converting kinetic energy to increased pressure. This series of moving blades, which increase the velocity, and fixed blades, which slow the fluid and increase the pressure, is common in axial flow compressors. After the first (large diameter) compressor in Fig. 7.15 are two smaller compressors, with multiple rows of moving and fixed blades (six and seven stages), that raise the pressure to $\approx$15 atm at the entrance to the combustion chambers. In these compressors the blades become shorter and the flow passage narrower in the flow direction. As the density of the fluid increases the area perpendicular to the flow is reduced, to keep an approximately constant velocity.

The advantages of the axial flow compressor over centrifugal compressors are the small cross-sectional area perpendicular to the gas flow, which makes it easy to build into a streamlined airplane, and the lower velocities, which lead to lower friction losses and slightly higher efficiency.

Centrifugal and axial flow compressors generally handle very large volumes of gases in small pieces of equipment, so the heat transfer from the gases is negligible. Thus, their performance is well described by the equations for adiabatic compressors (see Eq. 10.20). Efficiencies (Sec. 10.1) are normally from 80 percent to 90 percent. For high-flow applications like natural gas pipelines or ammonia plants, these compressors have almost completely replaced P.D. compressors.

In principle, axial flow pumps are the same as axial flow compressors. In practice, they generally have only one stage (because of the higher density of liquids) and are used for very high flow rate, low-head applications, like storm water removal. They have few applications in chemical engineering.

10.5 COMPRESSOR EFFICIENCIES

The discussion of compressors (centrifugal and axial flow) has been largely from the mechanical viewpoint; this viewpoint is helpful in understanding the fluid mechanics of these devices. In elementary thermodynamics books, one considers turbines and compressors from a first- and second-law viewpoint. That viewpoint is sketched here.

For any steady-flow compressor or turbine in which changes in potential and kinetic energy are negligible,

$$-\frac{dW_{\text{n.f.}}}{dm} = h_{\text{in}} - h_{\text{out}} + \frac{dQ}{dm} \quad \begin{bmatrix}\text{steady-flow} \\ \text{compressor or turbine}\end{bmatrix} \tag{10.25}$$

We may readily arrive at Eq. 10.13 by substituting the relation for an isentropic process for an ideal gas into Eq. 10.25. We may find Eq. 10.11 by substituting the isothermal relation for an ideal gas in Eq. 10.25 and using the entropy balance to solve for dQ / dm. Thus, we may find exactly the same results by a fluid mechanical view of what happens inside the compressor or by a thermodynamic view of the compressor as the system.

In defining the efficiency of a pump (Sec. 10.1), we compared the useful work to the total work. For an incompressible fluid this is most easily done by means of B.E., which is restricted to constant-density fluids, leading to Eq. 10.4. The definition for a pump could be restated as

$$\text{Efficiency} = \frac{\text{work required for the best possible device doing this job}}{\text{work required by this real device}} \tag{10.26}$$

This statement can be used for compressors, as well. In the case of an adiabatic compressor, the best possible device is a reversible, adiabatic compressor for which the

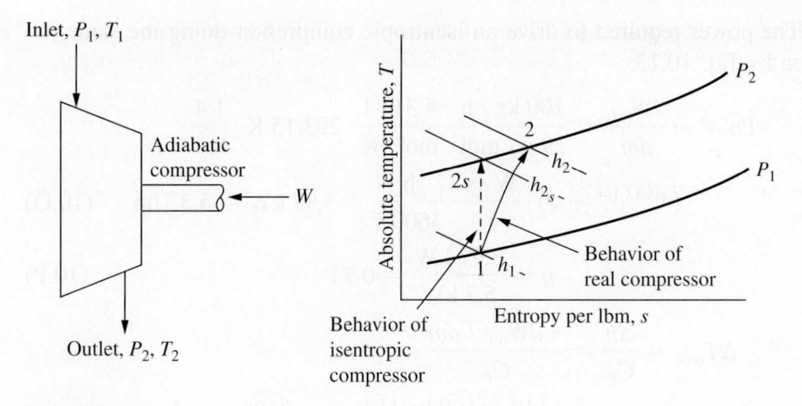

FIGURE 10.11
A rotating compressor (centrifugal or axial) shown as a schematic, and on a *T-s* diagram. The 2 state is the outlet of a real compressor; the 2*s* state is the outlet of a corresponding reversible, isentropic compressor.

inlet and outlet entropies are the same, called an *isentropic compressor.* Thus, we normally define

$$\text{Compressor efficiency} = \eta = \frac{\text{work of isentropic compressor}}{\text{work of real compressor}} \tag{10.27}$$

Consider the steady flow, adiabatic compressor shown in Fig. 10.11. The energy balance for this process (taking the compressor as the system and assuming that changes in kinetic and potential energies are negligible) is

$$-\frac{dW}{dm} = h_{\text{in}} - h_{\text{out}} = h_1 - h_2 \tag{10.28}$$

If we wish to compare the work done by this compressor with that done by a reversible compressor, we immediately see that we cannot compare it with a reversible compressor doing exactly the same thing because the real compressor has a higher outlet entropy, temperature, and enthalpy than would the outlet stream from a reversible compressor ($h_2 > h_{2_s}$ in Fig. 10.11). Thus, we could compare the real compressor with a reversible one having the same outlet enthalpy, having the same outlet temperature, or having the same outlet pressure. The latter seems to be the most logical choice, since real compressors are generally regulated by controlling the outlet pressure; this is the choice that has been universally made in defining the efficiency of compressors. So Eq. 10.27 becomes

$$\eta_{\text{pump or compressor}} = \frac{W_{\text{isentropic}}}{W_{\text{real}}} = \frac{h_{2_s} - h_1}{h_2 - h_1} \tag{10.29}$$

Example 10.8. An adiabatic compressor is compressing air from 20°C and 1 atm to 4 atm. The air flow rate is 100 kg / h, and the power required to drive the compressor is 5.3 kW. What are the efficiency of the compressor and the temperature of the outlet air? What would the outlet air temperature be if the compressor were 100 percent efficient?

The power required to drive an isentropic compressor doing the same job is given by Eq. 10.13:

$$\text{Po} = \dot{m}\frac{dW_{\text{n.f.}}}{dm} = \frac{100\ \text{kg}/\text{h}}{29\ \text{g}/\text{mol}} \cdot \frac{8.314\ \text{J}}{\text{mol}\cdot\text{K}} \cdot 293.15\ \text{K} \cdot \frac{1.4}{0.4}$$

$$\cdot\,(4^{0.4/1.4} - 1) \cdot \frac{\text{W}\cdot\text{s}}{\text{J}} \cdot \frac{\text{h}}{3600\ \text{s}} = 3.97\ \text{kW} = 5.32\ \text{hp} \qquad (10.\text{O})$$

$$\eta = \frac{3.97\ \text{kW}}{5.3\ \text{kW}} = 0.75 \qquad (10.\text{P})$$

$$\Delta T_{\text{real}} = \frac{\Delta h}{C_P} = \frac{-dW_{\text{n.f.}}/dm}{C_P}$$

$$= \frac{5.3\ \text{kW}/(100\ \text{kg}/\text{h})}{(29.1\ \text{J}/\text{mol}\cdot\text{K})\cdot(\text{mol}/29\ \text{g})} \cdot \frac{3600\ \text{s}}{\text{h}} \cdot \frac{\text{J}}{\text{W}\cdot\text{s}}$$

$$= 190\ \text{K} = 342°\text{F} \qquad (10.\text{Q})$$

$$T_{\text{out}} = 20°\text{C} + 190\ \text{K} = 210°\text{C} = 410°\text{F} \qquad (10.\text{R})$$

For an isentropic compressor,

$$\Delta T_{\text{isentropic}} = \frac{\Delta h}{C_P} = \frac{-dW_{\text{n.f.}}/dm}{C_P}$$

$$= \frac{3.97\ \text{kW}/(100\ \text{kg}/\text{h})}{(29.1\ \text{J}/\text{mol}\cdot\text{K})\cdot(\text{mol}/29\ \text{g})} \cdot \frac{3600\ \text{s}}{\text{h}} \cdot \frac{\text{J}}{\text{W}\cdot\text{s}}$$

$$= 142\ \text{K} = 142°\text{C} = 256°\text{F} \qquad (10.\text{S})$$

$$T_{\text{out}} = 162°\text{C} = 324°\text{F} \qquad (10.\text{T})$$

■

This example shows that compressor inefficiency raises the outlet temperature of the compressor. One may look upon this compressor inefficiency as being a type of friction heating. The extra work above that which would have been required for a 100 percent efficient compressor goes either to heat the gas passing through the compressor or to heat the surroundings.

10.6 PUMP AND COMPRESSOR STABILITY

In most chemical engineering situations, a pump or compressor moves a fluid through some kind of pipe or duct, with elbows, expansions or contractions, and valves. Most often it pumps against a higher static pressure or elevation than that at its inlet. Normally, there will be some kind of control valve (manual, like your bathroom faucet, or pneumatic or electric, driven by the plant's process control system), which allows us to set and to change the flow through the system. Figure 10.12 shows how these interact, on P-Q coordinates. From upper left to lower right we show a typical centrifugal pump curve, e.g., Fig. 10.9. The system resistance curve consists of a static component (the pressure difference from pump inlet to pump outlet with the pump turned off, due to the higher pressure in the place to which the fluid is going than the pressure in the supply line) and a fluid friction component, calculated by the

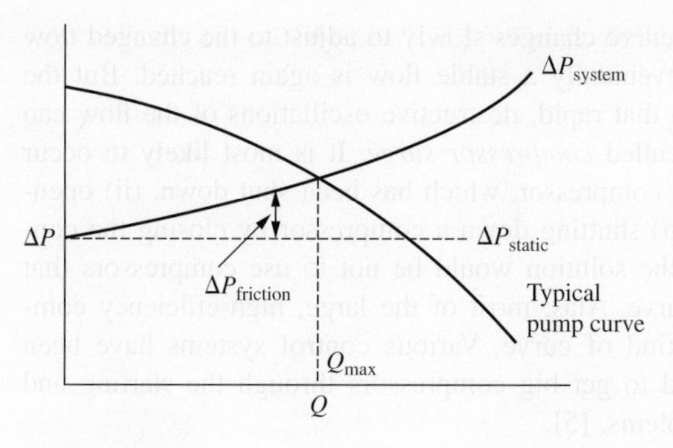

FIGURE 10.12
Pump curve and system resistance curve. In steady operation the flow will correspond to the intersection of the two curves. This pump curve is stable at all flow rates.

methods in Chap. 6. The latter increases from left to right, practically as the square of Q for turbulent flow. The sum of these two is the ΔP_{system} curve shown on the figure. It intersects the pump curve at the wide-open-flow-valve rate. Closing the control valve makes its value of K (see Sec. 6.9 and Table 6.7) increase, which rotates the $\Delta P_{friction}$ part of the ΔP_{system} curves counterclockwise about $Q = 0$, thus moving the intersection of the curves to the left and reducing the flow rate. (This also increases the pressure at the pump outlet; the pressure reduction through the control valve increases as we close it, to offset this increase.)

The pump system sketched in Fig. 10.12 is completely stable for all settings of the control valve. But the one sketched in Fig. 10.13 is not. It shows a pump curve with a maximum. Many large compressors, centrifugal and axial, have curves of this type. If we start with the control valve wide open, the behavior is the same as in Fig. 10.12, with the system operating at point A. At this point, if some disturbance increases Q slightly, moving to the right on the figure, then the pressure delivered by the compressor decreases, reducing Q, thus forcing the system back to point A. Similarly, a small decrease in Q raises the pressure delivered by the compressor, moving the system back to point A.

At any point on the compressor curve to the left of its maximum, the system is unstable. Consider point B. At this point, if some disturbance increases Q slightly, moving to the right on the figure, then the pressure delivered by the compressor increases, thus further increasing the flow through the system and moving the system away from point B. Similarly, a small decrease in Q lowers the pressure delivered by the compressor, moving the system away from point B. Thus, point B is unstable; if a compressor and its connected system are at that point, then any disturbance will cause it to move away from B, in the direction

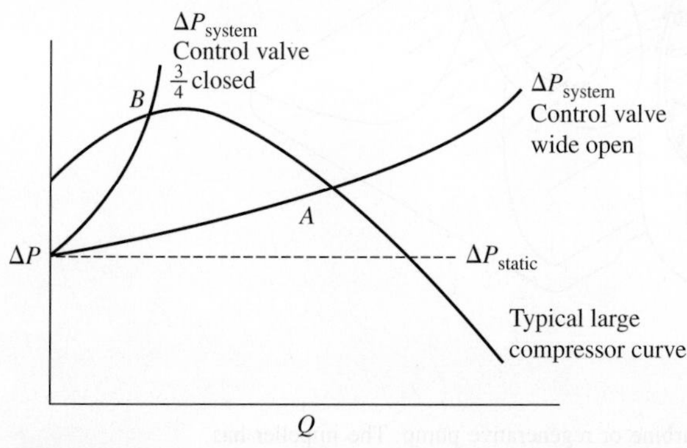

FIGURE 10.13
Same as Fig. 10.12, but for a pump curve that has a maximum. To the right of the maximum, the system is stable, but to the left it is unstable. This shape of pump curve is common for large compressors, leading to the problem of compressor surge.

of the disturbance. The resistance curve changes slowly to adjust to the changed flow (increased or decreased) so that eventually a stable flow is again reached. But the compressor's response is rapid, so that rapid, destructive oscillations of the flow can occur in the compressor. This is called *compressor surge.* It is most likely to occur when one is (i) starting up a large compressor, which has been shut down, (ii) opening the control valve slowly, or (iii) shutting down a compressor by closing the control valve. One would think that the solution would be not to use compressors that have this type of pressure-flow curve. Alas, most of the large, high-efficiency compressors in the world have this kind of curve. Various control systems have been devised to avoid this problem and to get big compressors through the starting and stopping phase without surge problems, [5].

10.7 REGENERATIVE PUMPS

Human ingenuity has produced all sorts of pumps, in addition to P.D., centrifugal, and axial flow. Of these, the type that the student is most likely to encounter is the *regenerative pump,* also called a *turbine pump.* Most auto fuel pumps are of this type, as are many low flow rate, high-head pumps in the process industries. Figure 10.14 shows a

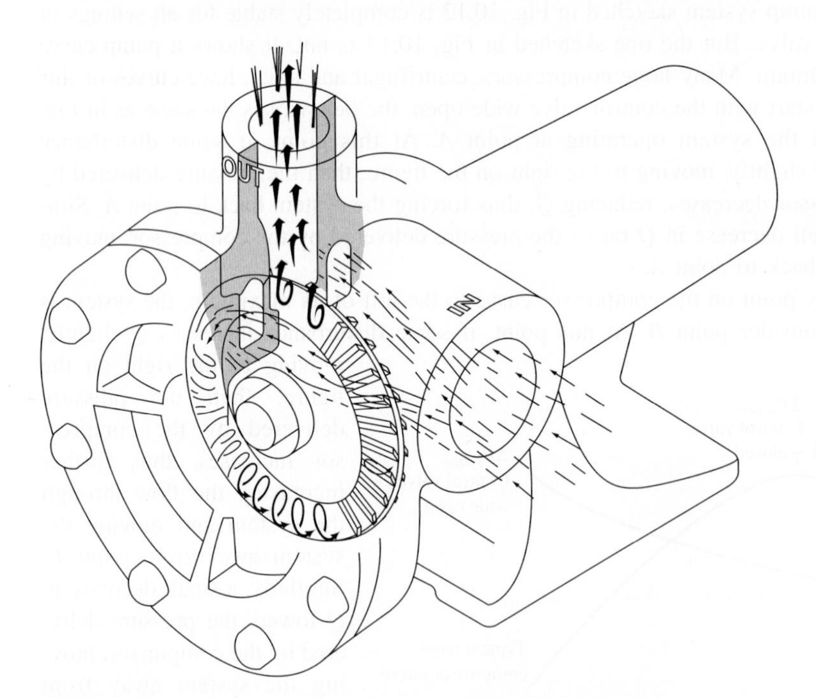

FIGURE 10.14
Partially cutaway schematic view of a turbine or regenerative pump. The impeller has grooves cut near its periphery, making it appear to have teeth. It rotates in a groove between the front and back pieces of the pump. The fluid flows in the entrance at the right and flows once around, counterclockwise, leaving by the exit at the top. The fluid circulates between the impeller and the outer part of the groove as shown by the arrows. (Courtesy of the Corken Pump Division of the Ibex Corporation.)

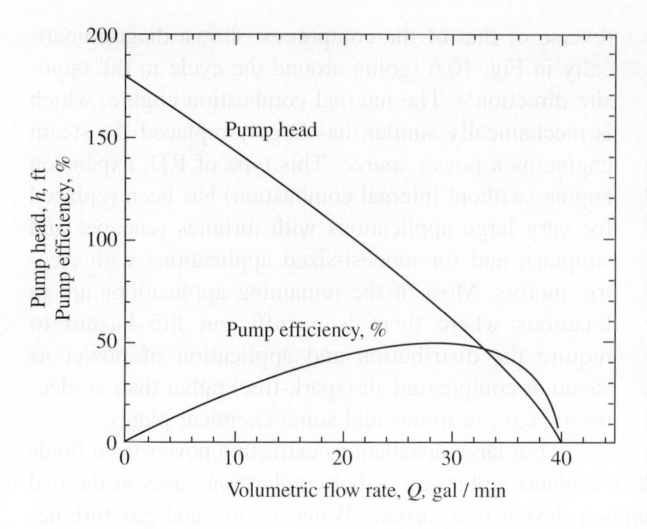

FIGURE 10.15
Pump curve for a turbine or regenerative pump with a 4.5-in diameter rotor at a speed of 1750 rpm [6].

schematic cutaway of a turbine pump. The impeller is a flat wheel with sections cut out on both sides of its periphery to make tooth-like blades. The impeller rotates in a groove between the back and front covers of the pump. The fluid enters the right opening and flows around once in the groove, exiting at the top. The moving impeller drives the fluid around the groove from inlet to outlet. In its passage around the groove the fluid circulates between the blades on the periphery of the impeller and the walls of the groove and increases substantially in pressure. This is not a centrifugal pump, because there is no net outward flow from inlet to outlet, and it is not a P.D. pump because the fluid is not trapped or squeezed at any point. Instead, it is one in which the fluid is literally pushed, against a pressure gradient by the moving blades, and carried along by turbulence and internal vortices produced by centrifugal force.

Figure 10.15 shows a pump map of a small pump of this type with rotor diameter 4.4 in and speed 1750 rpm [6]. On it we see that the head-flow curve is neither practically horizontal like a centrifugal nor practically vertical like a P.D. pump. Instead, it slopes at about a 45-degree angle from left to right. The head at zero flow ("shutoff head") is 180 ft. Extrapolating downward to a 4.4-in impeller on Figure 10.9, we estimate that a centrifugal pump with that size impeller at 1750 rpm would have a shutoff head of about 20 ft. Thus the pump curve in Fig. 10.15 has a shutoff head about 9 times as large as that of a common centrifugal pump of the same size and speed. This explains the uses of these pumps; when a high pressure rise is needed for a small volumetric flow rate, a pump of this type will produce the flow with a smaller and cheaper device than a centrifugal pump, with fewer moving parts and lower cost and maintenance problems than a P.D. pump. (There are some applications in which vane pumps, Fig. 10.5 and turbine pumps are about equal in performance, and in which both types compete, e.g., propane dispensing.) The theory of regenerative pumps [7] is not easily reduced to the simple form used for centrifugal pumps.

10.8 FLUID ENGINES AND TURBINES

The P.D. pump shown in Fig. 10.2 can be used as a fluid engine with simple changes in valve timing. This is the form of the steam engine, which supplied most of the world's mechanical power in the 19th and early 20th centuries; its operation is the

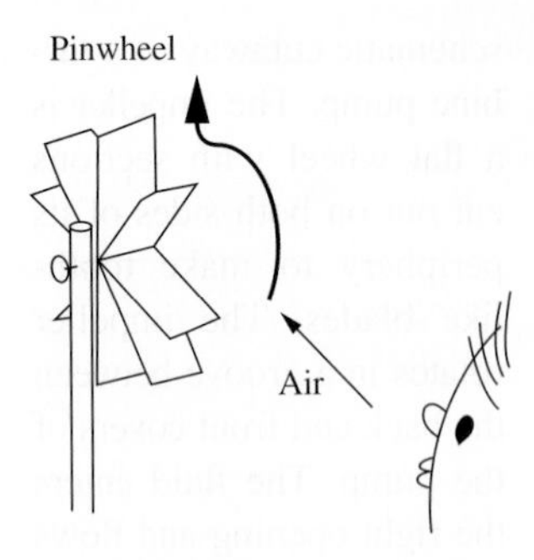

FIGURE 10.16
Child's pinwheel, the simplest impulse turbine.

reverse of that of the compressor shown diagrammatically in Fig. 10.6 (going around the cycle in the opposite direction!). The internal combustion engine, which is mechanically similar, has largely replaced the steam engine as a power source. This type of P.D. expansion engine (without internal combustion) has been replaced for very large applications with turbines (cheaper and simpler), and for modest-sized applications with electric motors. Most of the remaining applications are in locations where there is a sufficient fire hazard to require the distribution and application of power as steam or compressed air (spark-free) rather than as electricity, e.g., in mines and some chemical plants.

For large installations extracting power from fluids (generally, water in hydroelectric plants and steam or hot combustion gases in thermal power plants), the most common device is a turbine. Water, steam, and gas turbines have the same principles of operation but very different sizes, shapes, and speeds.

A turbine consists of a wheel with blades attached to its periphery and the associated casing, etc.; for many kinds of turbines these blades are called *buckets.* The blades change the direction of the flow, which results in a force on the blades. This force turning the wheel produces power. The simplest turbine is the child's pinwheel; see Fig. 10.16. In the pinwheel a high-velocity jet strikes the blades and is slowed down. This type of device, in which the fluid undergoes its pressure reduction in the fixed nozzle (the child's lips in the figure) and flows through the turbine at practically constant pressure, is called an *impulse turbine.* Its behavior is discussed in Sec. 7.5, where it is shown that for the most efficient operation the blade speed should be one-half of the jet speed, resulting in the exit fluid's having negligible velocity (relative to fixed coordinates).

In a *reaction turbine,* the fluid enters the blades with a negligible velocity and leaves at a high velocity relative to the blades. The simplest reaction turbine is the rotating garden sprinkler; see Fig. 10.17. This is called a reaction turbine for the same reason that a rocket motor is often called a reaction motor: The force exerted is described by Newton's third law, "Action equals reaction." A turbine of this type could be constructed by attaching two rockets to the ends of a shaft in place of the water jets shown in Fig. 10.17. In a reaction turbine the pressure reduction takes place in a *moving* nozzle.

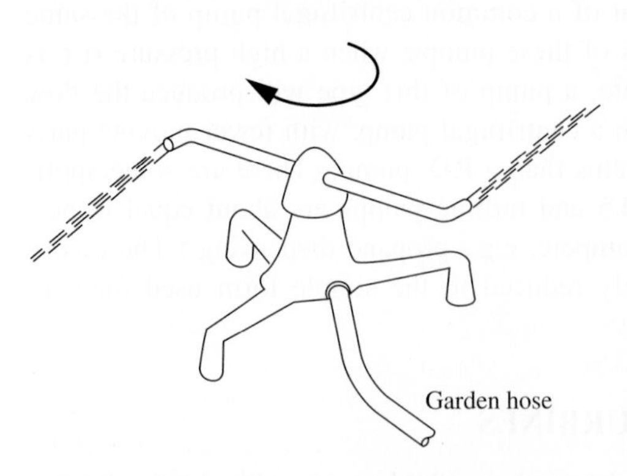

FIGURE 10.17
Rotating garden sprinkler, the simplest reaction turbine.

The simple reaction turbine is most easily analyzed by the angular momentum balance, Eq. 7.62, which, as shown in Sec. 7.7 for a steady flow turbine, reduces to Euler's turbine equation, Eq. 7.63. To find the power produced per unit time, we multiply both sides of Eq. 7.63 by ω,

$$\text{Po} = \frac{dW_{\text{n.f.}}}{dt} = \Gamma\omega = \dot{m}\omega[(rV_\theta)_{\text{out}} - (rV_\theta)_{\text{in}}] \tag{10.30}$$

Here the fluid enters at the center ($r_{\text{in}} = 0$), so the far right term is zero. Then, solving for the work per unit mass, we find

$$\frac{dW_{\text{n.f.}}}{dm} = \omega(rV_\theta)_{\text{out}} \tag{10.31}$$

But ωr_{out} is the tangential velocity of the nozzle, so

$$\frac{dW_{\text{n.f.}}}{dm} = V_{\theta_{\text{nozzle}}} V_{\theta_{\text{out}}} \tag{10.32}$$

The velocity $V_{\theta_{\text{out}}}$ is equal to $V_{\theta_{\text{nozzle}}} + V_{\text{rel}}$, where V_{rel} is the velocity of the jet as measured by an observer riding on the nozzle; so

$$\frac{dW_{\text{n.f.}}}{dm} = V_{\theta_{\text{nozzle}}}(V_{\text{rel}} + V_{\theta_{\text{nozzle}}}) \tag{10.33}$$

($V_{\theta_{\text{nozzle}}}$ and V_{rel} have opposite signs!) In the simple garden sprinkler shown in Fig. 10.17, V_{rel} is independent of $V_{\theta_{\text{nozzle}}}$; it depends on the pressure in the garden hose and the friction in the system. For constant V_{rel}, we may find the most efficient value of $V_{\theta_{\text{nozzle}}}$, i.e., the one that gives the largest value of $dW_{\text{n.f.}} / dm$, by differentiating Eq. 10.33 with respect to $V_{\theta_{\text{nozzle}}}$ and setting the derivative equal to zero. The result is $V_{\theta_{\text{nozzle}}} = \frac{1}{2} V_{\text{rel}}$; i.e., the nozzle moves at one-half of the jet speed, as measured by an observer riding on the nozzle, and in the opposite direction.

Comparing the simple impulse turbine in Fig. 10.16 with the simple reaction turbine in Fig. 10.17, we see that the former is the more efficient. Either may be considered to consist of a nozzle that converts internal energy and injection work ($\Delta P / \rho$) into kinetic energy and a device that uses this kinetic energy to produce work. In either case the kinetic energy of the fluid leaving the system is wasted. As shown in Sec. 7.5, we can, in principle, build an impulse turbine for which the outlet velocity (based on fixed coordinates) is zero. However, for a reaction turbine like that in Fig. 10.17, the maximum efficiency corresponds to $V_{\text{out}} = \frac{1}{2} \cdot V_{\text{rel}}$; so the outlet kinetic energy will be one-fourth of the total available kinetic energy. For this reason the pure reaction turbine of Fig. 10.17 is inefficient and is never used in industrial practice. Some large water turbines solve this problem by placing a diffuser on the outlet, thus recovering most of the kinetic energy in the waste stream and improving the efficiency.

Most modern water and steam turbines take part of the pressure reduction in a set of fixed nozzles and part of the pressure reduction in the moving wheel. Thus, they are part impulse and part reaction. However, since pure impulse turbines are in current use, common usage is to reserve the term impulse turbine only for a turbine that is 100 percent impulse, with no reaction, and to call any turbine that is not 100 percent impulse a reaction turbine. Most reaction turbines are less than 50 percent reaction, the remainder being impulse.

Liquid turbines are large, slow-moving devices, and gas and steam turbines are small, fast-moving devices. This is easiest to see for an impulse turbine, for which we previously showed that the optimum velocity for the blade is one-half that of the jet. So the speed of a single-stage impulse turbine rotor is set by the available jet speed. At Hoover Dam the fluid drops about 700 ft. Applying B.E. from the water surface to the turbine nozzle, we can solve for the maximum possible (frictionless) jet velocity:

$$V = (2gh)^{1/2} = \left(2 \cdot 32.2 \frac{\text{ft}}{\text{s}^2} \cdot 700 \text{ ft}\right)^{1/2} = 212 \frac{\text{ft}}{\text{s}} = 65 \frac{\text{m}}{\text{s}} \tag{10.U}$$

Therefore, the blade of the turbine should travel about 106 ft / s. It is convenient to build the rotor about 10 ft in diameter, so the rate of rotation is about

$$\begin{pmatrix}\text{Angular velocity} \\ \text{expressed as rpm}\end{pmatrix} = \frac{106 \text{ ft / s}}{10\pi \text{ ft}} \cdot \frac{60 \text{ s}}{\text{min}} = \text{approx. } 200 \text{ rpm} \tag{10.V}$$

Now consider a steam turbine; as shown by the methods in Chap. 8, the reversible, adiabatic expansion of steam through a nozzle from about 100 psia to atmospheric pressure produces a flow with a velocity of about 3000 ft / s. Thus, the blade should move at about 1500 ft / s. Here also it would be desirable to use a large-diameter wheel, but at these high rotational speeds the centrifugal force tending to pull the turbine wheel apart becomes so great that only a small-diameter wheel can survive. The largest single-stage steam turbine wheels are about 2 or 3 ft in diameter. For a 3 ft wheel and a blade speed of 1500 ft / s the rotational speed is

$$\begin{pmatrix}\text{Angular velocity} \\ \text{expressed in rpm}\end{pmatrix} = \frac{1500 \text{ ft / s}}{3\pi \text{ ft}} \cdot \frac{60 \text{ s}}{\text{min}} = \text{approx. } 15{,}000 \text{ rpm} \tag{10.W}$$

The first successful steam turbine, developed by de Laval [8], was a simple one-stage impulse turbine, as described above, and turned at about 20,000 rpm.

The turbines on most jet and gas turbine engines are of the simple impulse variety described above and turn at about 20,000 rpm, as do the turbine parts of airplane and auto turbochargers. However, this is an inconvenient speed for connecting to a generator that is producing 60-cycle current; it must run at 3600 / n rpm, where n is any integer. The most common U.S. generator speeds are 1800, 1200, and 600 rpm. A high-speed turbine could be connected to such a generator through a gear reducer, but the more economical solution seems to be to build a turbine of lower speed. In it many wheels are attached to a common shaft; the steam flows through a nozzle, then a bladed wheel, then another nozzle, etc. Each combination of nozzle and wheel is, in effect, a separate turbine; common terminology calls such a nozzle-and-wheel combination a *stage* and thus refers to the multiwheel turbines as *multistage* turbines. Each stage has a small pressure reduction and thus a small jet velocity and a small tip speed for efficient operation. In current steam turbine practice the first few stages usually are pure impulse, followed by stages that are 50 percent impulse and 50 percent reaction. In Fig. 7.15 at the right are two turbines, the first a three-stage turbine, the second a one-stage turbine. These drive concentric shafts that drive the two compressors at the left at different speeds.

For high-head, low-flow-rate, water power application, the most economical turbine is the pure-impulse Pelton wheel. For high flow rates and lower heads the friction effects in the Pelton wheel cut down its efficiency, and a radially inward-flowing, part-reaction Francis turbine seems to be the most economical. For very high flow rates and very low heads the most economical is the Kaplan turbine, which looks quite like a ship's propeller, with adjustable pitch blades. The Francis and Kaplan turbines can be designed to function efficiently at many more revolutions per minute than an impulse turbine with the same input head; this is an advantage, because generators of very slow speed (e.g., 50 rpm) are expensive to build. Details of the various kinds of steam and gas turbines may be found in [8, 9]; water turbines, in [10]; pumps, in [11]; and compressors, in [12].

Before about 1900, steam and water power devices drove most machinery, including that in chemical engineering. The convenience and flexibility of electric drive motors has replaced these devices for most chemical engineering applications. However, for driving large fans and compressors, chemical engineers will often encounter steam turbines, and the cogeneration plants that are part of major chemical engineering complexes often contain such turbines. Chemical engineers in the electric power industry will certainly encounter them.

10.9 FLUID ENGINE AND TURBINE EFFICIENCY

Fluid engine and turbine efficiency is defined as the inverse of pump or compressor efficiency:

$$\text{Efficiency} = \frac{\text{work actually delivered}}{\text{maximum possible work}} \tag{10.34}$$

For an incompressible fluid (e.g., water) the common definition of the maximum possible work is the work that would be delivered if the fluid left the system with zero velocity and if the $\mathscr{F}$ term in B.E. were zero. For gases (e.g., steam) the common definition of the maximum work is that work that would have been obtained for zero outlet velocity and isentropic operation. Although the forms of these maximum-work definition appear different, they can be shown to be the same, because the $\mathscr{F}$ term in B.E. is related to the irreversible entropy increase.

10.10 SUMMARY

1. P.D. pumps and compressors work by trapping a fluid in a cavity and then squeezing it out at a higher pressure; they are generally high-pressure-rise, low-flow-rate devices.
2. Centrifugal pumps and compressors work by moving the fluid out in a centrifugal force field and giving the fluid kinetic energy, and then converting this to injection work. They are generally high-flow-rate, low-pressure-rise devices.
3. The problem of boiling in the inlet line (or "cavitation") limits how high any kind of pump may be placed above the reservoir on which it is drawing. For a boiling liquid the pump must be placed below the boiling surface by an amount equal to the NPSH_R.

4. Compressors are normally practically adiabatic and result in a significant temperature rise for the gas. The work of compression is less for an isothermal than for an adiabatic compressor; for this reason almost all compressors are cooled, and many compressors are staged, with intercooling to decrease the work requirement.
5. Turbines work by impulse, in which a fluid is accelerated by a fixed nozzle and slowed by moving blades, or by reaction, in which a fluid is accelerated in a moving nozzle. Most turbines are either pure impulse or part impulse and part reaction.

PROBLEMS

See the Common Units and Values for Problems and Examples, inside the back cover! An asterisk (*) on a problem number indicates that its answer is in App. D.

10.1.*How many gallons per minute should be delivered by a pump with a piston area of 10 in^2 and a piston stroke of 5 in and a speed of 1 Hz?

10.2. Calculate the hydraulic horsepower for pumping 500 gal / min from an inlet pressure of 5 psig to an outlet pressure of 30 psig
(*a*) For water.
(*b*) For gasoline.

10.3. (*a*) For the pump discussed in Example 10.1 calculate the temperature rise for the fluid passing through the pump. Assume that there is no heat transfer to the surroundings.
(*b*) We normally assume that liquids are practically incompressible, so that there is no density change in passing through a pump like the one in Example 10.1. Using the values from part (*a*) and App. A.6 and A.9, estimate the ratio of ρ_{out} / ρ_{in} for this pump. Is the effect of pressure more or less important than that of temperature? Is the incompressible assumption reasonable in this example?

10.4. We wish to pump mercury from a sump with a P.D. pump. Assuming that there is no friction and that the vapor pressure of mercury is negligible, what is the maximum height above the sump at which we can locate our pump?

10.5. Sketch the equivalent of Fig. 10.6 for an incompressible fluid.

10.6. Why is it impractical to try to build a zero-clearance compressor?

10.7.*How many horsepower are required to compress 20 lbmol / h of helium ($k = 1.666$) from 1 atm at 68°F to 10 atm using;
(*a*) An isothermal compressor?
(*b*) An adiabatic compressor?
(*c*) A two-stage, adiabatic compressor with optimum interstage pressure and intercooling to 68°F?

10.8. Prepare a plot of work per pound mol versus pressure ratio, P_2 / P_1 for an ideal gas, with $k = 1.4$, being compressed from an inlet condition of 68°F. Cover the pressure ratio range of 1 to 20. Show the curves both for an adiabatic and for an isothermal compressor.

10.9. Prove that the interstage pressure given by Eq. 10.13 gives the minimum work per pound-mol for a given P_{inlet} and P_{outlet}. *Hint:* Write the equation for the total work of a two-stage, intercooled compressor, and differentiate it with respect to the interstage pressure.

10.10. Show by induction how it follows from Eq. 10.14 that for a multistage compressor the optimum interstage pressures are those that result in equal pressure ratios for each stage.

10.11.*Rework Example 10.4 using the optimum interstage pressure instead of the interstage pressure selected for the example.

10.12. We wish to compress air from 1 atm at 68°F to 10 atm. We will use a two-stage, adiabatic compressor with an intercooler between stages. For the intercooler we have cooling water cold enough to cool the gas to 100°F. What is the optimum interstage pressure in this case? Write out the general formula for the optimum interstage pressure in terms of the inlet temperatures to the two stages.

10.13. Show that, as the number of stages of a multistage, intercooled compressor becomes very large, the work requirement approaches as a limit the work requirement of an isothermal compressor with the same inlet temperature and overall pressure ratio.

10.14. Rework Example 10.4 for a three-stage compressor, using the optimum interstage pressures.

10.15. Show that, as k approaches 1, Eq. 10.13 (adiabatic compressor) approaches as a limit Eq. 10.11 (isothermal compressor). *Hint:* Represent $(P_2 / P_1)^{(k-1)/k}$ by its series expansion $y^x = 1 + x \ln y + (x \ln y)^2 / 2! + (x \ln y)^3 / 3! + \dots$ and use $\ln(1 + x) = x - x^2 / 2 + x^3 / 3 + \dots$

10.16. Show the steps between Eqs. 10.18 and 10.20.

10.17. For the flow in Example 10.5, the experimental values are $h \approx 32$ ft, Po = 1.15 hp, and $\eta = 71\%$. Those are for pumping water. If we use this pump to pump gasoline at $Q = 100$ gpm, what are the predicted values of h, Po and η?

10.18. A centrifugal pump is pumping mercury. The inlet pressure is 200 psia. The pump impeller is 2 in in diameter, and the pump is rotating at 20,000 rpm. Estimate the outlet pressure based on Eq. 10.24.

10.19.*A centrifugal pump is tested with water and found at 1800 rpm to deliver 200 gal / min at a pressure rise of 50 psi. The mechanical efficiency is 75 percent. We wish to pump mercury in this pump at the same rpm and the flow rate. Estimate the pressure rise and pump horsepower required for this operation, assuming that the pump remains 75 percent efficient.

10.20. Figure 10.8 and Eq. 10.24 indicate that the actual head developed by centrifugal pumps is about 50 percent of the value we would compute by simple theory. That is true for the small- to medium-sized pumps most common in industry. However, for really large pumps, it is not true. Rich [1] presents a pump map like Fig. 10.9 for a huge centrifugal pump in a pumped-storage project. From it we can read the values in Table 10.3. This pump turns at 257 rpm; the inner and outer impeller diameters are 4.656 and 9.375 ft. Using those values, estimate the ratio of the observed head to that estimated from Eq. 10.23, and estimate the efficiency for the four values of Q shown in Table 10.3.

TABLE 10.3

Data (read from a pump curve) for Prob. 10.20

Volumetric flow rate, Q, ft^3 / s	Pump head, h, ft	Power input, hp
1600	1220	259,000
2000	1160	295,000
2400	1090	327,000
2800	1015	354,000

10.21. We wish to design a centrifugal blower for air. It will take in air at 1 atm, 68°F and deliver it at a gauge pressure of 2 psig. The impeller will rotate at 3600 rpm. What is the minimum impeller diameter, assuming Eq. 10.24?

10.22.*A manufacturer supplies a line of centrifugal pumps that use a gear-speed increaser to drive the impeller at 20,000 rpm with an 1800 rpm motor. One of the pumps delivers a head of 4500 ft. Estimate its impeller diameter, based on Eq. 10.24. List the most important mechanical design problems for such a pump. What are the advantages of such a pump?

10.23. Old hands in the chemical process industries pour cold water onto the suction side of any malfunctioning centrifugal pump. Why?

10.24.*Suppose that, instead of using a P.D. pump in Example 10.2, we used a centrifugal pump, which for 200 gal / min had a reported $NPSH_R$ of 10 ft. What would be the maximum elevation above the sump at which we could place the pump, assuming that we have a way to prime it? The $NSPH_R$ is for the pump only, not including the friction in the lines. There is no inlet valve like that in Example 10.2.

10.25. Seal leakage is a problem for any pump in which a drive shaft enters from outside; see Sec. 6.10.2. For some applications no leakage can be tolerated (chemical engineers deal with some nasty materials!). For these applications we use sealless pumps [13] in which the driving force enters a totally closed container as a rotating magnetic field or a rotating electromagnetic field. Sketch what such pumps might look like. Discuss their probable advantages and disadvantages.

10.26. Equation 10.23 suggests that the pump head of a centrifugal pump is proportional to the square of the diameter. Figure 10.9 shows the head-flow rate performance of four different-sized impellers in the same casing.

(*a*) If we consider only the values at zero flow (shutoff), do the reported heads agree with this suggestion?

(*b*) For a flow rate of 80 gpm, do the reported heads agree with Eq. 10.23 as well as, better than, or worse than at zero flow? Suggest an explanation.

10.27. Figure 10.9 relates the following variables: h, Q, D_{impeller}, Po, the efficiency, and $NSPH_R$. However, it is for only one value of ω, and the Po values are for one value of ρ (that of water). Thus, the variable list for a general equivalent of Fig. 10.9 is h, g, ρ, Q, Po, D_{impeller}, η, $NPSH_R$, and ω. Two other variables we might consider are the viscosity, μ, and the surface roughness of the metal parts, ε. However, for this problem we assume that their effect is negligible. (If they are included then the Reynolds number and relative roughness will appear in our analysis.)

(*a*) To use traditional dimensional analysis (Chap. 9) on this problem, we must treat gh as one variable and not as two separate variables. Explain why.

(*b*) By the methods of dimensional analysis, determine how many dimensionless variables are needed to relate gh to ρ, Q, D_{impeller}, and ω. First show by dimensional analysis that ρ does not belong on this variable list and should be dropped. Then show how many dimesionless variables there should be, and what values they should have if only one contains gh and only one contains Q.

(*c*) By the methods of dimensional analysis, determine how many dimensionless variables are needed to relate Po to ρ, Q, D_{impeller}, and ω. Then show what values they should have if only one contains Po and only one contains Q.

(*d*) The efficiency is itself dimensionless. Show how it can be expressed in terms of the dimensionless variables in the previous sections of this problem.

(*e*) By dimensional reasoning, show what the dimesionless variable containing $NPSH_R$ should be.
White [14, p. 724] shows an example of two geometrically similar pumps of different sizes, indicating that both of their pump maps, redrawn based on these dimensionless groups, are practically identical.

10.28. The two variables that describe a pump's performance are those plotted on pump maps (Figs. 10.3, 10.9, 10.10, 10.13, 10.14, and 10.16) namely, h and Q. If one wants to know what type of pump to select for some task, one wants to know these variables. To simplify matters, one would like to have one variable, combining h and Q, that told what kind of pump was needed. These two variables alone cannot form a dimensionless group, but if we add g and ω, they form a dimensionless variable called the *specific speed*

$$\begin{pmatrix}\text{Specific}\\ \text{speed}\end{pmatrix} = N_s = \frac{\omega\sqrt{Q}}{(gh)^{3/4}} \tag{10.35}$$

This quantity is different at different points on a pump map, but if we select the point of highest efficiency on the map, then the map has one value of N_s which characterizes the pump. A high-flow-rate, low-pressure-rise pump will have a large value of N_s and a low-flow-rate, high-pressure-rise pump will have a small value. Many texts show that centrifugal pump impeller shape is a unique function of N_s. The same idea is applied to hydraulic turbines; the shape of the most satisfactory turbine is a unique function of N_s.

(*a*) Show that N_s is dimensionless.
(*b*) Show the value of N_s at the maximum efficiency point for the $6\frac{3}{4}$-inch diameter impeller in Fig. 10.9.
(*c*) It is common practice in the United States not to use the dimensionless value of N_s but rather to show its value in $(\text{rpm}) \cdot (\text{gal / min})^{1/2} / (\text{ft of head})^{3/4}$. Show the numerical value of the answer to part (*b*) in these units.

10.29. It has been suggested that for short-term service we could make a simple reaction turbine by attaching two solid-fuel rockets to the ends of a rotor. What would the optimum speed for maximum power production be for this type of device? Why does the answer differ from that for the garden-sprinkler type shown in Fig. 10.17?

10.30. One of the highest-head water power plants in the world is that at Dixence, Switzerland, with a net head of 5330 ft. The water from this plant drives an impulse turbine (Pelton wheel) with a diameter to the middle of the blades of 10.89 ft. The wheel turns at 500 rpm [10]. What is the ratio of blade speed to jet speed for this turbine? How does this compare with the optimum discussed in Sec. 7.4?

10.31. Most P.D. compressors are driven by constant-speed motors, because variable-speed motors are much more expensive. This poses a problem in controlling the compressor flow rate. One way to control a compressor with a constant-speed motor is to vary the clearance volume by means of "clearance pockets," which are connected to or disconnected from the head of the compressor by remotely operated valves. For a 100 percent efficient, adiabatic compressor, what is the effect of such a pocket on the pressure ratio, flow rate, and power requirement?

10.32. See Prob. 10.31. An alternative procedure is to use an inlet or outlet valve that can be stopped in the open position by some remote controller. For a 100 percent efficient, adiabatic compressor what is the effect of a stuck-open inlet valve on pressure ratio, flow rate, and power requirement? Such devices are normally called "valve unloaders."

REFERENCES FOR CHAPTER 10

1. Rich, G. A., H. A. Mayo Jr. "Pumped Storage." In *Pump Handbook,* 3rd ed., I. J. Karrasic; J. P. Messina; P. Cooper; C. C. Heald. New York: McGraw-Hill, 2001, Section 9.13.
2. Charlier, R. H. *Tidal Energy.* New York: Van Nostrand Reinhold, 1982.
3. Mobley, R. K. *Fluid Power Dynamics.* Woburn, Massachusetts: Newnes-Butterworth-Heineman, 2000.
4. To see why the P.D. pumps have disappeared in high flow systems, visit the Museum of Science and Technology in Melbourne. Part of it is an old sewage pumping plant. The original pumps were huge P.D., driven by huge steam engines. The staff run the steam engines (on compressed air) for the visitors. They are marvels of engineering complexity and beauty. Next to them is a small electric motor and small centrifugal pump, which do the same pumping task as the huge complex steam engine-P.D. pump combination.
5. Gravdahl, J. T., O. Egeland. *Compressor Surge and Rotating Stall.* London: Springer, 1999.
6. Abramson, E. I. "The Modern Turbine Pump—Choice of Plant Operators for Specialized Jobs." *Power 99(4),* (1955), pp. 120–123.
7. Wilson, W. A., M. A. Santalo; J. A. Oelrich. "A Theory of the Fluid-Dynamic Mechanism of Regenerative Pumps." *ASME Trans. 77,* (1955), pp. 1303–1316.
8. Stodola, A. *Steam and Gas Turbines.* Transl. L. C. Loewenstein. New York: McGraw-Hill, 1927, p. 558.
9. Skrotski, B. G. A. "Steam Turbines." *Power 106(6),* (1962), pp. s-1, s-42.
10. Franzini, J. B., E. J. Finnemore. *Fluid Mechanics with Engineering Applications,* 9th ed., New York: McGraw-Hill, 1997.
11. Karassik, I. J., J. P. Messina; P. Cooper; C. C. Heald. *Pump Handbook,* 3rd ed., New York: McGraw-Hill, (2001).
12. Bloch, H. P. *A Practical Guide to Compressor Technology.* New York: McGraw-Hill, 1996.
13. Buse, F. W., S. A. Jaskiewicz. "Sealless Pumps." In *Pump Handbook,* 3rd ed., I. J. Karassik; J. P. Messina; P. Cooper; C. C. Heald. New York: McGraw-Hill, 2001, Section 2.2.7.
14. White, F. M. *Fluid Mechanics.* 4th ed., New York: McGraw-Hill, 1999.

CHAPTER 11

FLOW THROUGH POROUS MEDIA

A porous medium is a continuous solid phase that has many void spaces, or pores, in it. Examples are sponges, cloths, wicks, paper, sand and gravel, filters, bricks, plaster walls, many naturally occurring rocks (e.g., sandstones and some limestones), and the packed beds used for distillation, absorption, etc. In many such porous solids the void spaces are not connected, so there is no possibility that fluid will flow through them. For example, expanded polystyrene hot-drink cups, life preservers, and iceboxes have many pores, but because of the "closed cell" structure of the plastic these pores are not interconnected. Thus, these porous media form excellent barriers to fluid flow. On the other hand, a pile of sand has fewer pores than an expanded polystyrene drinking cup, but its pores are all connected, so fluids can easily flow through it. Porous media with no interconnected pores are described as impermeable to fluid flow, and those with interconnected pores as permeable (we will give a mathematical definition of permeability in Sec. 11.1). The flow of fluids in *permeable* porous media is of great practical significance in groundwater hydrology, oil and gas production, filters, distillation and packed absorption columns, and fluidized beds.

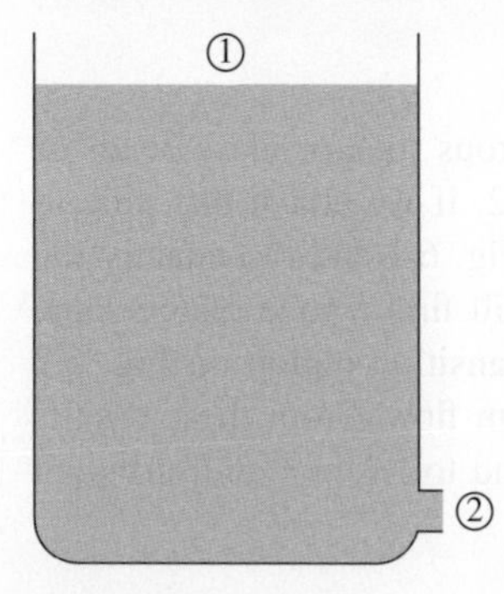

FIGURE 11.1
A tank full of a porous medium.

To view the similarities and differences between this kind of flow and the flows we have previously discussed, consider the gravity flow of water through some vessel in Fig. 11.1. In a vessel of this type, as we saw in

Chaps. 5 and 6, the flow could be described by B.E. There are no pumps or turbines, and the pressure change from point 1 to point 2 is negligible, so B.E. simplifies to

$$g\,\Delta z + \frac{\Delta V^2}{2} = -\mathcal{F} \tag{11.1}$$

In Chap. 5 we considered numerous cases in which the friction term in Eq. 11.1 was negligible. In Chap. 6 we considered how we could calculate the friction term (normally based on generalizations of experimental data) in those cases in which the friction term was not negligible.

Now, suppose the entire vessel is filled with some porous solid, such as sand (this vessel now resembles the sand filters that are frequently used to clarify muddy water). Equation 11.1 still describes the situation exactly as it did before, because it is based on the steady-flow energy balance for a constant-density fluid, and it has no built-in assumption that the flow is occurring in an open vessel rather than a porous medium. The significant differences that we will see, if we compare the two situations, are these:

1. In most porous-medium flows the friction term is much, much larger than it would be in the analogous flow in an empty vessel, and it is not directly calculable from the results in Chap. 6.
2. For most porous-medium flows, even though V_2 does not equal V_1, both velocities are so small that ΔV^2 is negligible compared with $\mathcal{F}$.
3. If the tank in Fig. 11.1 does not contain sand and is originally full of one fluid, e.g., air, when we admit a second fluid, e.g., water, it will quickly flush out all of the first fluid. However, if the tank contains sand, and its voids are originally full of air, then admitting water will not flush out all the air. Some significant part of the air (probably 10 percent to 30 percent) will be trapped permanently in the pores. This kind of behavior is of great significance in filtration, groundwater hydrology, and oil recovery.

In this chapter we will examine the friction term in B.E. for flow in a porous medium and examine the phenomena of incomplete displacement of one fluid from a porous medium by another. We will also examine competitive countercurrent flow in porous media and look briefly at filtration and fluidization.

11.1 FLUID FRICTION IN POROUS MEDIA

Consider a porous medium consisting of sand or some porous rock or glass beads or macaroni or cotton cloth contained in a pipe; see Fig. 11.2. If we attach this pipe to the apparatus for the pressure-drop experiment shown in Fig. 6.1 and run exactly the same tests on it that we described there for a pipe, we will find results of the same form as those shown in Fig. 6.2, except that the abrupt transition region on Fig. 6.2 will be replaced with a smooth curve for a porous-medium flow. From these results we can guess that the two end parts of the curve correspond to laminar and turbulent flows; this is experimentally verifiable.*

*The term "laminar" literally means in "shells" or "laminae." This is not an accurate description of the flow in a porous medium, in which the width of the individual flow channels changes from point to point. A better term is "streamline flow," indicating that the individual fluid particles follow streamlines, which do not cross or mix, as they would in turbulent flow. However, the name laminar is the more widely used and will be used here.

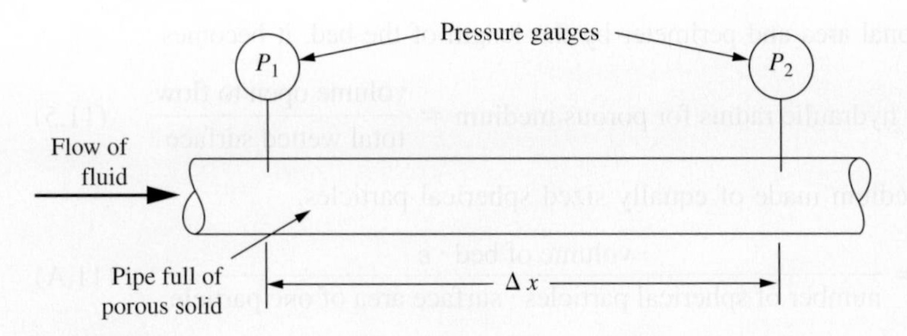

FIGURE 11.2
Measurement of pressure gradient for flow in a pipe full of porous medium.

For flow in a pipe we were able to calculate the laminar-flow portion of the curve from a simple force balance and to make a simple correlation for the turbulent-flow portion of the curve. For porous media this has been done successfully only for media consisting of uniformly sized, spherical particles. Here we examine that solution, because it provides useful insights into flow in more complex media and it allows us to define and discuss many of the terms in common use in the porous-media literature. We begin with some definitions.

At any one cross section perpendicular to the flow, the average velocity may be based on the entire cross-sectional area of the pipe, in which case it is called the *superficial velocity* V_S,

$$\begin{pmatrix}\text{Superficial}\\ \text{velocity}\end{pmatrix} = V_S = \frac{Q}{A_{\text{pipe}}} = \frac{\dot{m}}{\rho A_{\text{pipe}}} \tag{11.2}$$

or it may be based on the area actually open to the flowing fluid, in which case it is called the *interstitial velocity* V_I,

$$\begin{pmatrix}\text{Interstitial}\\ \text{velocity}\end{pmatrix} = V_I = \frac{Q}{\varepsilon A_{\text{pipe}}} = \frac{\dot{m}}{\varepsilon \rho A_{\text{pipe}}} \tag{11.3}$$

where ε is the *porosity,* or *void fraction:*

$$\begin{aligned}\begin{pmatrix}\text{Porosity or}\\ \text{void fraction}\end{pmatrix} = \varepsilon &= \frac{\text{total volume of system} - \text{volume of solids in system}}{\text{total volume of system}} \\ &= \frac{A\,\Delta x(1 - \text{average fraction of cross section occupied by solids})}{A\,\Delta x} \\ &= \text{average fraction of cross section not occupied by solids}\end{aligned} \tag{11.4}$$

From a theoretical standpoint the interstitial velocity is the more important; it determines the kinetic energy and the fluid forces and whether the flow is turbulent or laminar. From a practical standpoint the superficial velocity is generally more useful; it shows the flow rate in terms of readily measured external variables. Both are in common use.

Previously it was indicated that for noncircular conduits Fig. 6.10 (the friction-factor plot) could be used if we replaced the diameter in both the friction factor and the Reynolds number with four times the hydraulic radius. The hydraulic radius is the cross-sectional area perpendicular to flow, divided by the wetted perimeter. For a uniform duct, this is a constant. For a packed bed it varies from point to point; but if we multiply both cross-sectional area and perimeter by the length of the bed, it becomes

$$\text{HR} = \text{hydraulic radius for porous medium} = \frac{\text{volume open to flow}}{\text{total wetted surface}} \tag{11.5}$$

For a porous medium made of equally sized spherical particles,

$$\text{HR} = \frac{\text{volume of bed} \cdot \varepsilon}{\text{number of spherical particles} \cdot \text{surface area of one particle}} \tag{11.A}$$

and

$$\text{Number of particles} = \frac{\text{volume of bed} \cdot (1 - \varepsilon)}{\text{volume of one particle}} \tag{11.B}$$

so that, for spherical particles only,

$$\text{HR} = \frac{\text{volume of bed} \cdot \varepsilon}{\text{volume of bed} \cdot (1 - \varepsilon) \cdot (\text{surface} / \text{volume})}$$

$$= \frac{\varepsilon}{(1 - \varepsilon)\left[(\pi D_p^2) / (\pi D_p^3 / 6)\right]} = \frac{D_p}{6} \cdot \frac{\varepsilon}{1 - \varepsilon} \quad \text{[spherical particles]} \tag{11.6}$$

Here D_p is the particle diameter. If we now insert 4 times this definition of the hydraulic radius into the definitions of the pipe flow friction factor and the pipe flow Reynolds number, we find

$$f = \mathscr{F} \frac{4\left[D_p \varepsilon / 6(1 - \varepsilon)\right]}{4 \Delta x} \cdot \frac{2}{V_I^2} = \frac{\mathscr{F}}{3} \cdot \frac{D_p}{\Delta x} \cdot \frac{\varepsilon}{1 - \varepsilon} \cdot \frac{1}{V_I^2} \tag{11.7}$$

$$\mathscr{R} = \frac{V_I 4\left[D_p \varepsilon / 6(1 - \varepsilon)\right] \rho}{\mu} = \frac{2 D_p \varepsilon V_I \rho}{3 \mu (1 - \varepsilon)} \tag{11.8}$$

It is customary to replace V_I in these equations with V_S / ε, so

$$f = \frac{\mathscr{F}}{3} \cdot \frac{D_p}{\Delta x} \cdot \frac{\varepsilon^3}{(1 - \varepsilon)} \cdot \frac{1}{V_S^2} \tag{11.9}$$

and

$$\mathscr{R} = \frac{2 D_p V_S \rho}{3 \mu (1 - \varepsilon)} \tag{11.10}$$

As in the case of flow in pipes, there are several different friction factors in common use for flowing porous media, all differing by a constant. The choice between these is completely arbitrary; in this text we will drop the $\frac{1}{3}$ in Eq. 11.9 and the $\frac{2}{3}$ in Eq. 11.10 to find our working forms of the friction factor and Reynolds number for porous media:

$$f_{\text{porous medium}} = \mathscr{F} \frac{D_p}{\Delta x} \frac{\varepsilon^3}{(1 - \varepsilon)} \frac{1}{V_S^2} = f_{\text{P.M.}} \tag{11.11}$$

$$\mathscr{R}_{\text{porous medium}} = \frac{D_p V_S \rho}{\mu (1 - \varepsilon)} = \mathscr{R}_{\text{P.M.}} \tag{11.12}$$

Having made definitions, we now inquire whether the pipe flow friction factor plot will predict the pressure drop for flow in porous media. For the laminar-flow region in pipes, Poiseuille's equation may be rewritten $f = 16 / \mathcal{R}$. Here the f and $\mathcal{R}$ are consistent with the definitions given in Eqs. 11.9 and 11.10. When we convert to the definitions given in Eqs. 11.11 and 11.12, this becomes $f_{\text{P.M.}} = 72 / \mathcal{R}_{\text{P.M.}}$ (see Prob. 11.1). There is one obvious error in this derivation, namely, the tacit assumption that the flow is in the x direction. Actually, the flow is a zigzag; it must detour around one particle and then around another. If we assume that this zigzag proceeds with an average angle of 45° to the x axis, then the actual flow path is $\sqrt{2}$ times as long as the flow path shown in Eq. 11.11, and the actual interstitial velocity is $\sqrt{2}$ times the interstitial velocity used in Eq. 11.11. If we make these changes, we conclude that to agree with the pipe friction factor plot, laminar flow in a porous medium made of uniform-sized spheres should be described by $f_{\text{P.M.}} = 144\, \mathcal{R}_{\text{P.M.}}$ (see Prob. 11.2). Experimental data indicate that the constant is about 150; that is, the 45° assumption made above is slightly incorrect (but it is still pretty impressive to come within 4 percent of the best experimental value by applying pipe flow results from Chap. 6!). For laminar flow we find experimentally

$$f_{\text{P.M.}} = \frac{150}{\mathcal{R}_{\text{P.M.}}} \tag{11.13}$$

or, rearranged,

$$\mathcal{F} = 150 \frac{V_S \mu (1-\varepsilon)^2}{D_p^2 \varepsilon^3} \cdot \frac{\Delta x}{\rho} \qquad \text{[laminar flow]} \tag{11.14}$$

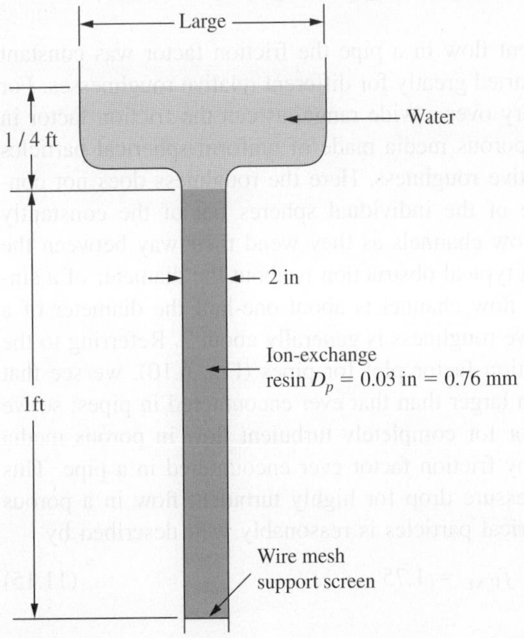

Equation 11.14 is known as the *Blake-Kozeny equation* or the *Kozeny-Carman equation;* it describes the experimental data for steady flow of Newtonian fluids through beds of uniform-sized spheres satisfactorily for $\mathcal{R}_{\text{P.M.}}$ less than about 10.

Example 11.1. Figure 11.3 shows a water softener in which water trickles by gravity through a bed of spherical ion-exchange resin particles, each 0.03 in (0.76 mm) in diameter. The bed has a porosity of 0.33. Calculate the volumetric flow rate of water.

Applying B.E. from the top surface of the fluid to the outlet of the packed bed and ignoring the kinetic-energy term and the pressure drop through the support

Large
Water
1 / 4 ft
2 in
Ion-exchange resin $D_p = 0.03$ in $= 0.76$ mm
1ft
Wire mesh support screen

FIGURE 11.3
Gravity drainage of fluid through a porous medium.

screen, which are both small, we find

$$g(\Delta z) = -\mathcal{F} \tag{11.C}$$

Substituting from Eq. 11.14 and solving for V_S, we find

$$\begin{aligned} V_S &= \frac{g(-\Delta z)D_p^2\varepsilon^3\rho}{150\mu(1-\varepsilon)^2\Delta x} \\ &= \frac{32.2 \text{ ft / s} \cdot 1.25 \text{ ft} \cdot (0.03 \text{ ft / } 12)^2 \cdot 0.33^3 \cdot 62.3 \text{ lbm / ft}^3}{150 \cdot 1.002 \text{ cP} \cdot (1-0.33)^2 \cdot 1 \text{ ft} \cdot 6.72 \cdot 10^{-4} \text{ lbm / (ft} \cdot \text{s} \cdot \text{cP)}} \\ &= 0.0124 \text{ ft / s} = 0.00379 \text{ m / s} \end{aligned} \tag{11.D}$$

Therefore,

$$Q = AV_S = \left(\frac{2}{12}\text{ ft}\right)^2 \cdot \frac{\pi}{4} \cdot 0.0124 \frac{\text{ft}}{\text{s}} = 0.00027 \frac{\text{ft}^3}{\text{s}} = 7.6 \frac{\text{cm}^3}{\text{s}} \tag{11.E}$$

Before accepting this as the correct solution, we check the Reynolds number, finding

$$\mathcal{R}_{\text{P.M.}} = \frac{(0.03 \text{ ft / } 12) \cdot 0.0124 \text{ ft / s} \cdot 62.3 \text{ lbm / ft}^3}{1.002 \text{ cP} \cdot 0.67 \cdot 6.72 \cdot 10^{-4} \text{ lbm / (ft} \cdot \text{s} \cdot \text{cP)}} = 4.29 \tag{11.F}$$

This is less 10, for which we can safely use Eq. 11.13. ■

If there had been no porous medium in the lower part of the apparatus in Fig. 11.3, then the exit velocity would have been given by Torricelli's equation, equal to about 9 ft / s. Here the calculated velocity is $\frac{1}{250}$ as large. Fluid friction effects in porous media are large!

We saw that for fully turbulent flow in a pipe the friction factor was constant for a given relative roughness but varied greatly for different relative roughnesses. For pipes the relative roughness can vary over a wide range, as can the friction factor in fully turbulent flow. However, for porous media made of uniform spherical particles there can be little variation in relative roughness. Here the roughness does not consist of rough spots on the surface of the individual spheres but of the constantly changing shape of the individual flow channels as they wend their way between the individual particles. The height of a typical obstruction is about the diameter of a single particle; the width of a typical flow channel is about one-half the diameter of a single particle. Therefore, the relative roughness is generally about 2. Referring to the upper right-hand corner of the friction factor plot for pipes (Fig. 6.10), we see that this relative roughness is very much larger than that ever encountered in pipes; so we would expect that the friction factor for completely turbulent flow in porous media should be very much larger than any friction factor ever encountered in a pipe. This is experimentally verifiable; the pressure drop for highly turbulent flow in a porous medium made up of uniform, spherical particles is reasonably well described by

$$f_{\text{P.M.}} = 1.75 \tag{11.15}$$

which can be rearranged to

$$\mathcal{F} = 1.75 \frac{V_S^2 \, \Delta x}{D_p} \cdot \frac{1-\varepsilon}{\varepsilon^3} \qquad \text{[turbulent flow]} \tag{11.16}$$

Equation 11.16, known as the *Burke-Plummer equation,* is satisfactory for $\mathscr{R}_{\text{P.M.}}$ greater than about 1000.

Example 11.2. We now wish to apply a sufficient pressure difference to the water flowing through the packed bed in Fig. 11.3 for the water superficial velocity to be 2 ft / s. What pressure gradient is required?

Applying B.E. as before, we find

$$\frac{\Delta P}{\rho} + g\,\Delta z = -\mathscr{F} \tag{11.G}$$

Here, however, the gravity term is negligible compared with the others, so, substituting from Eq. 11.16, we find

$$\begin{aligned}\frac{-\Delta P}{\Delta x} &= \frac{1.75\rho V_S^2}{D_p}\cdot\frac{1-\varepsilon}{\varepsilon^3}\\ &= \frac{1.75\cdot 62.3\ \text{lbm}/\text{ft}^3\cdot(2\ \text{ft}/\text{s})^2\cdot 0.67}{(0.03\text{ft}/12)\cdot 0.33^3\cdot 32.2\ \text{lbm}\cdot\text{ft}/(\text{lbf}\cdot\text{s}^2)\cdot 144\ \text{in}^2/\text{ft}^2}\\ &= 701\ \text{psi}/\text{ft} = 15.9\ \text{MPa}/\text{m}\end{aligned} \tag{11.H}$$

Here we check, finding $\mathscr{R}_{\text{P.M.}} = 690$, so this is at the low end of the turbulent flow region (see Fig. 11.4, discussed below). ■

This startlingly high calculated pressure drop (701 psi / ft) illustrates the fact that turbulent flows very seldom occur in porous media composed of particles this small. Since this particle size is typical of those encountered in most soils or underground aquifers or petroleum reservoirs and in most industrial filters, we see why almost all flows of fluids in the earth or in industrial filters are laminar.

For the transition region from laminar to turbulent flow in pipes there is no simple friction-loss correlation, because the flow may be laminar or turbulent or oscillate between these two. This is not the case in a porous medium, because the flow does not switch all at once from laminar to turbulent. The reason is that the flow is not through one channel, but through a large number of parallel channels, of varying sizes. As the flow rate is increased from a low value, it is originally entirely laminar, and then the largest channels switch to turbulent flow. As the flow rate is further increased, more and more channels become turbulent, until at very high flow rates there is turbulent flow in all the channels. This leads to a smooth transition from all-laminar to all-turbulent flow.

Thus, in a plot of $f_{\text{P.M.}}$ versus $\mathscr{R}_{\text{P.M.}}$, there is one smooth curve for transition from laminar to turbulent flow; Fig. 11.4 is such a plot. Ergun [1] showed that, if we add the right-hand sides of the Kozeny-Carman and Burke-Plummer equations, the result fits the data in the transition region reasonably well, i.e.,

$$f_{\text{P.M.}} = 1.75 + \frac{150}{\mathscr{R}_{\text{P.M.}}} \tag{11.17}$$

which can be rearranged to

$$\mathscr{F} = 1.75\frac{V_S^2}{D_p}\frac{(1-\varepsilon)\Delta x}{\varepsilon^3} + 150\frac{V_S\mu(1-\varepsilon)^2\,\Delta x}{D_p^2\varepsilon^3\rho} \tag{11.18}$$

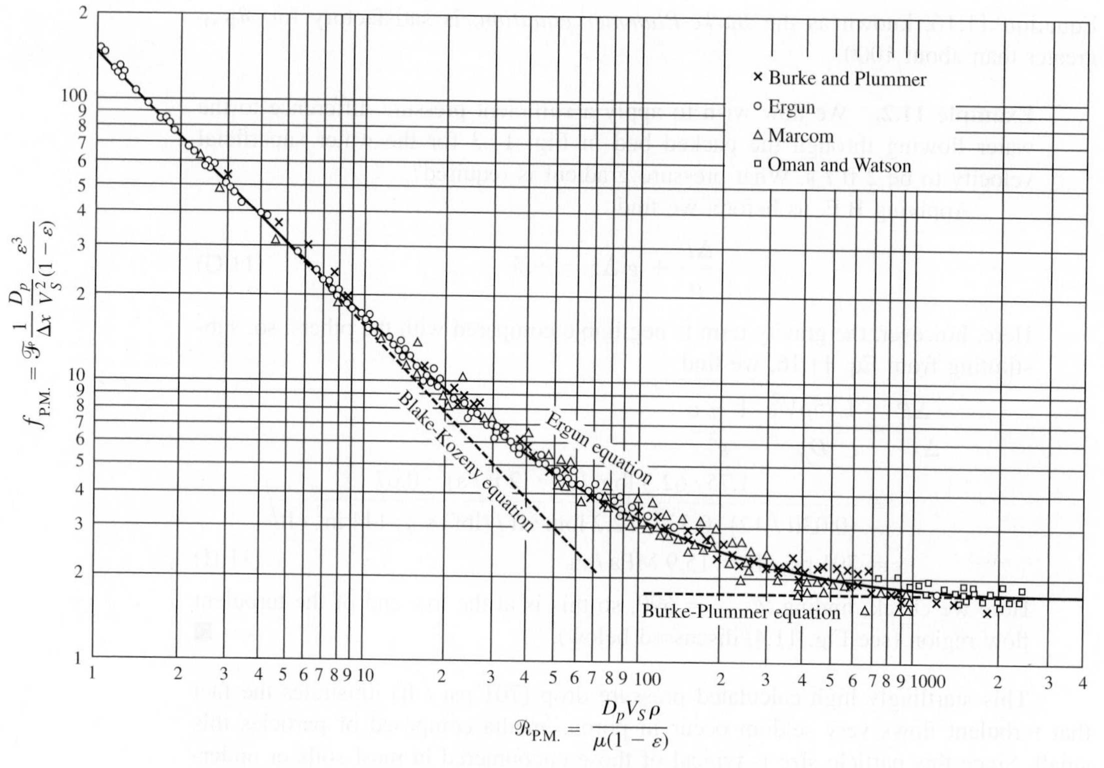

FIGURE 11.4
Pressure drop data for flow through porous media. (From R. B. Bird, E. N. Stewart, and W. E. Lightfoot, *Transport Phenomena,* 2nd ed. New York: John Wiley (2002), p. 192, who redrew it from S. Ergun, "Fluid flow through packed columns," *CEP 48*:89–94 (1952). Reproduced by permission of the publisher.)

which is known as the *Ergun equation.* It fits the data satisfactorily for all Reynolds numbers because the second term becomes negligible at high Reynolds numbers, giving Eq. 11.14. At small Reynolds numbers the second term becomes so large that the first term is negligible in comparison, giving Eq. 11.16 (see Prob. 11.3).

The foregoing is all for beds of uniformly-sized spherical particles. For other shapes of uniform particles some efforts have been made to relate the results to those shown here through defining an empirical "sphericity" factor [2, 3]. However, results have not been successful enough to allow one to calculate the behavior of such porous media accurately without experimental test, because even for a completely uniform set of nonspherical particles the porosity is a strong function of how the bed is assembled, and if the particles are loose, the porosity can be significantly altered by simply shaking the bed, etc.

There has been very little progress in calculating the flow of water, oil, or gas in naturally occurring rocks without experimental test because naturally occurring rocks are much, much less uniform than the beds of uniformly-sized spheres described by Fig. 11.4. Thus, in the study of groundwater movement and in petroleum reservoir engineering it is customary to simplify Eq. 11.14 to

$$\mathcal{F} = \frac{V_S}{k} \cdot \frac{\mu}{\rho} \Delta x \tag{11.19}$$

where k is called the *permeability.* Equation 11.19 is *Darcy's equation,* and the unit of permeability is called the *darcy:*

$$1 \text{ darcy} = \frac{1\,(\text{cm / s}) \cdot \text{cP}}{\text{atm / cm}} = 0.99 \cdot 10^{-12} \text{ m}^2 = 1.06 \cdot 10^{-11} \text{ ft}^2 \tag{11.20}$$

Example 11.3. In a test of a horizontal-flow filter with compressed air at 1 atm, the following data were obtained: filter area: 1 ft^2; pressure difference across filter: 2 psi; length of filter medium in flow direction: $\frac{1}{2}$ in; flow rate: 1 ft^3 / min. Calculate the permeability of the filter medium and estimate the pressure drop necessary to force 1 ft^3 / min of water through it.

The permeability is given by rearrangement of Eq. 11.19:

$$k = \frac{V_S \mu \, \Delta x}{\rho \mathcal{F}} = \frac{V_S \mu \, \Delta x}{\Delta P} = \frac{Q \mu \, \Delta x}{A \, \Delta P}$$

$$= \frac{1 \text{ ft}^3 / \text{min} \cdot 0.018 \text{ cP} \cdot \text{ft} / 24}{1 \text{ ft}^2 \cdot 2 \text{ lbf / in}^2} \cdot \frac{\text{ft}^2}{144 \text{ in}^2}$$

$$\cdot\, 2.09 \cdot 10^{-5} \frac{\text{lbf} \cdot \text{s}}{\text{ft}^2 \cdot \text{cP}} \cdot \frac{\text{min}}{60 \text{ s}} \cdot \frac{\text{darcy}}{1.06 \cdot 10^{-11} \text{ ft}^2}$$

$$= 0.080 \text{ darcy} = 80 \text{ millidarcy} \tag{11.I}$$

For water we could use this value directly in Eq. 11.19. However, it is easier to solve Eq. 11.19 for the pressure difference and then apply it twice, once to the air flow and once to the water flow, and take the ratio. All the terms cancel except

$$\frac{\Delta P_{\text{water}}}{\Delta P_{\text{air}}} = \frac{\mu_{\text{water}}}{\mu_{\text{air}}} = \frac{1.0 \text{ cP}}{0.018 \text{ cP}} = 55.5 \tag{11.J}$$

so that the required pressure drop for water is 111 psi = 765 kPa. ■

The velocities encountered in groundwater and petroleum-reservoir flow are generally small enough for the kinetic-energy terms in B.E. to be neglected. Furthermore, the flow is almost always laminar, so that $\mathcal{F}$ is described by Darcy's equation (Eq. 11.19). Thus, B.E. for this situation becomes

$$\Delta\left(\frac{P}{\rho} + gz\right) = -\frac{\mu}{k} \cdot \frac{V \, \Delta x}{\rho} \tag{11.21}$$

which for constant ρ, k, μ, and g may be rewritten as

$$\frac{d(P + \rho g z)}{dx} = -\frac{\mu}{k} V_x \tag{11.22}$$

If we write a similar equation for the y direction, differentiate each with respect to x and y, respectively, and then add them, we find

$$\frac{\partial^2(P+\rho g z)}{\partial x^2}+\frac{\partial^2(P+\rho g z)}{\partial y^2}=-\frac{\mu}{k}\left(\frac{\partial V_x}{\partial x}+\frac{\partial V_y}{\partial y}\right) \tag{11.23}$$

but from the constant-density mass-balance equation (see Chaps. 15 and 16) the right-hand side of Eq. 11.23 is zero; so

$$\frac{\partial^2\phi}{\partial x^2}+\frac{\partial^2\phi}{\partial y^2}=0, \quad \phi=P+\rho g z \tag{11.24}$$

This is Laplace's equation, which describes "potential flow." It is widely used in heat flow and electrostatic field problems; an enormous number of solutions to Laplace's equation are known for various geometries. These can be used to predict the two-dimensional flow in oil fields, underground water flow, etc. The same method can be used in three dimensions, but solutions are more difficult. The solutions to the two-dimensional Laplace equation for the common problems in petroleum reservoir engineering are summarized by Ahmed [4]. The analogous solutions for groundwater flow are shown in the numerous texts on hydrology, e.g., [5]. See Chap. 16 for more on potential flow.

11.2 TWO-FLUID COCURRENT FLOW IN POROUS MEDIA

So far we have assumed that all the pore space in the porous medium was occupied by the same fluid, such as air or water or oil. However, there are very important problems in which two immiscible fluids are present in the same pore space, e.g., the simultaneous flow of oil and gas or of oil and water, which occurs in petroleum reservoirs or the air-blowing of a filter cake to drive out a valuable filtrate or to lower the moisture content of a valuable filter cake.

In the experimental apparatus shown in Fig. 11.2 if we fill all the pores with water and then force air through the system, we find that the fraction of water in the outlet stream behaves as shown in Fig. 11.5. Initially only water will flow out of the downstream end of the apparatus. Its volumetric flow rate will equal the volumetric flow rate of air entering the apparatus. Then air will "break through" and appear at the outlet end. For a brief period both air and water will emerge from the outlet, the volume fraction of water steadily decreasing. Finally, no more water will emerge (except that which is being removed as water vapor in the air, due to evaporation inside the apparatus). If, after we have reached the point of

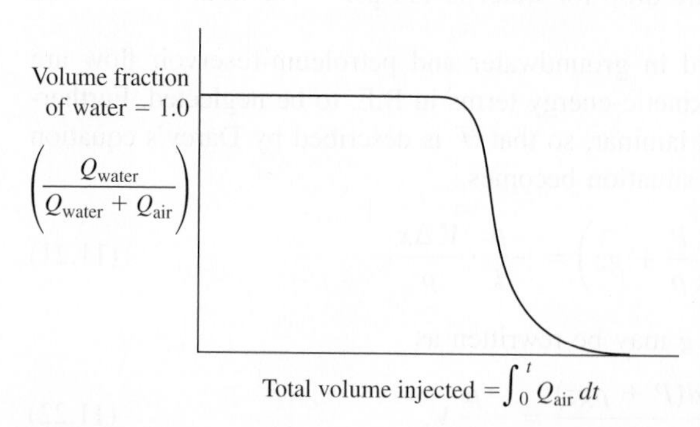

FIGURE 11.5
Breakthrough of air displacing water from a porous medium.

no more water flow in Fig. 11.5, we open the apparatus and determine the amount of water present inside, we find that 10 percent to 30 percent of the pores in the sample are full of water.

Why does this water not flow out? It is held in place by surface forces. This can be made plausible by comparing the surface of a gallon of fluid in a cylindrical container with a gallon of fluid in a typical sandstone. An ordinary 1-gal paint can has a surface area of 1.5 ft^2. A gallon of fluid contained in the pores of a sandstone made of spherical grains of diameter 0.01 in and porosity 0.3 has 216 ft^2 of surface area. One may think of the fluid in the pores of such a sandstone as being spread out as a film of average thickness 0.007 in = 0.18 mm.

If only one fluid is present, then this is all solid-fluid surface, and the interaction is simply one of adhesion at the surface. If two fluids are present, then in addition to two kinds of solid-fluid interface there will be a fluid-fluid interface. The pressure difference that can exist across such a fluid-fluid interface (called the *capillary pressure*) is of the form*

$$\begin{pmatrix}\text{Capillary}\\ \text{pressure}\end{pmatrix} = \Delta P = \frac{\sigma}{r} \tag{11.25}$$

Here σ is the surface tension, and r is the radius of the pore space. For a water-air interface σ is about $4 \cdot 10^{-4}$ lbf / in. For 1 gal of water in an ordinary 1-gal can r is about 3 in, so the pressure difference between the air and the water is about $2 \cdot 10^{-4}$ psi, which is negligible in most problems. If, however, the surface is inside a pore of radius 0.001 in, then the pressure difference is about 0.8 psi, which frequently is not negligible (see Chap. 14 for more on surface tension effects).

These surface forces prevent the complete displacement of one phase from a porous medium by another. The displacing fluid (air in the above example) tends to move first into the largest of the available flow channels in the porous medium and thus bypass some of the displaced fluid. When this bypassed displaced fluid, which is still flowing, is reduced in quantity so much that it breaks up into droplets surrounded by the displacing fluid rather than moving as a continuous filament, then it stops moving (in petroleum terminology it becomes *immobile*). If we examine microscopically a sand that has had water displaced from it by air, as described above, we see that the retained water exists, not as continuous filaments, but as small layers or drops, normally held in the junctions between various grains of the sand.

From this physical description we concluded that a particle of fluid stops moving when the displacing force (which equals the pressure gradient times the length of the droplet times its cross-sectional area perpendicular to the flow) is balanced by the surface force (which equals the surface tension divided by the radius of the drop times its cross-sectional area). Equating these, we find that the fluid particle should stop moving when

$$\frac{\Delta P}{\Delta x} LA = \frac{\sigma}{r} A \tag{11.26}$$

---

*This assumes a "contact angle" of zero degrees. See Scheidegger [6] for a discussion of this equation without this simplifying assumption.

or, rearranging,

$$\frac{\Delta P}{\Delta x}\frac{Lr}{\sigma} = 1 \tag{11.27}$$

For a naturally occurring stone it is impossible to measure L (the length of the drop) or r in Eq. 11.27. Furthermore, this analysis only applies to spherical droplets, whereas microscopic observation shows that the drops are seldom spherical. However, the basic idea is sound. To show this, we note that Lr has the same dimension as the permeability; low permeabilities go with low values of L and r and high permeabilities with high values of L and r. Thus, we could reasonably expect that the fraction of the void space full of displaced fluid, when the displaced fluid stops moving, should be some function of the dimensionless group

$$\frac{k}{\sigma}\frac{\Delta P}{\Delta x} = \text{capillary number} \tag{11.28}$$

Figure 11.6 shows a correlation of measured *residual saturation* (residual saturation is the fraction of pore space occupied by displaced fluid when the displaced fluid stops flowing) as a function of this capillary number. For high permeabilities (e.g.,

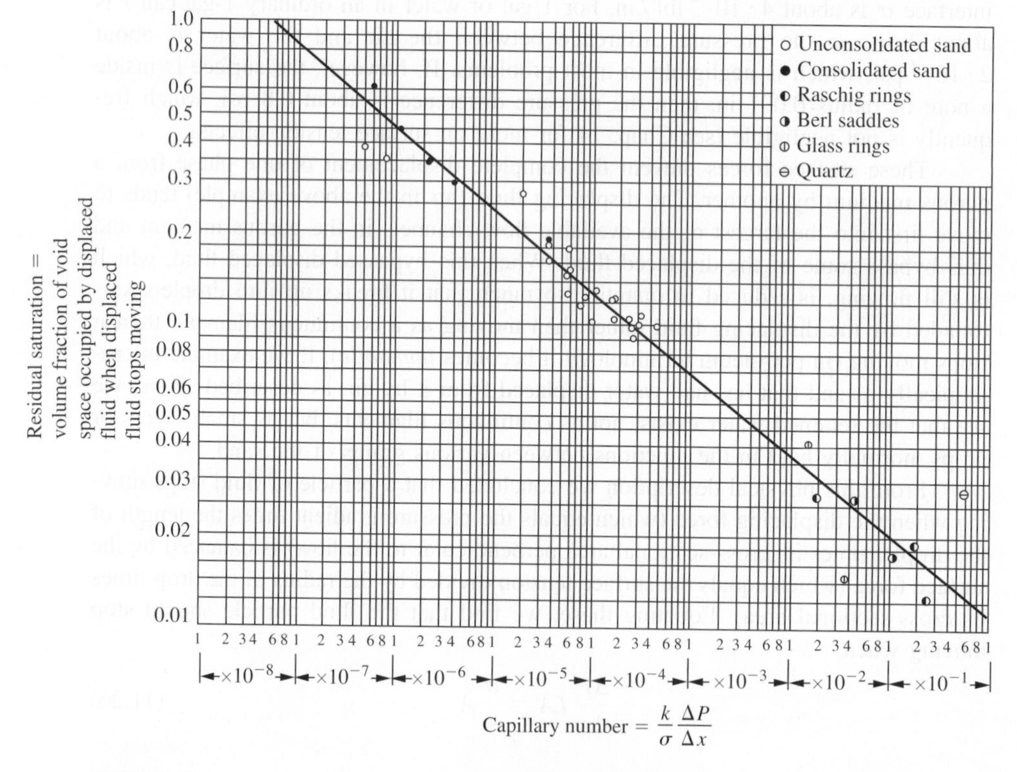

FIGURE 11.6
Residual saturation as a function of capillary number. (From L. E. Brownell and D. L. Katz, "Flow of fluids through porous media," *CEP 43*:601–612 (1950). Reproduced by permission of the publisher.)

large-sized pores, as might exist in a pile of bricks or coarse gravel) the residual saturation is very low, perhaps 2 percent or 3 percent, whereas for low permeabilities (e.g., a very fine-grained sandstone or shale) the residual saturation is very high, perhaps 30 to 60 percent. The scatter of the data shows that this correlation should be used only for order-of-magnitude estimates of the fraction of immobile fluid in a porous medium.

This kind of two-phase *cocurrent* flow is very important in oil fields, where the two phases may be oil and water or oil and natural gas; in hydrology, where the two phases flowing (near the earth's surface) are air and water; or in filters, where air is blowing residual liquid out of the filter cake. The treatment here is much simplified; for more detailed treatments see Rushton, Ward and Holdich [3] or Ahmed [4]. These flows are complicated by the effect of viscosity. If the displacing fluid is more viscous than the displaced fluid, then viscosity damps out any unevenness in the boundary between displacing and displaced fluids. But if the displacing fluid is less viscous, then viscosity increases such unevenness, and the less viscous displacing fluid "fingers" through the displaced fluid, leading to early breakthrough and much less complete displacement than if the displacing fluid is more viscous than the displaced fluid. Unfortunately, in cases of industrial significance (recovery of petroleum, air-blowing of filter cakes) the displacing fluid is often less viscous so that this "fingering" is a very serious limitation on the effectiveness of the processes [7].

11.3 COUNTERCURRENT FLOW IN POROUS MEDIA

When the particles in a porous medium are small, then if a fluid is moving in one direction through the medium, any other fluid that is simultaneously moving through the medium will almost certainly move in the same direction. However, when the particles are large (say $\frac{1}{4}$ in or larger), it is entirely possible for two immiscible fluids to move in opposite directions through the same medium at the same time. This is the basis for the gas-liquid and liquid-liquid contacting devices that use a bed of particles to increase the surface of contact between the fluids. Such devices, usually called *packed towers* or *packed beds,* are commonly used for absorption, distillation, humidification, etc.

The important feature of the two-fluid flow in such systems is the competition of the fluids for the area available to flow. Normally the denser fluid (e.g., water) runs down the surface of the particles by gravity, while the less dense fluid (e.g., air) flows upward because it is introduced at the bottom of the column at a pressure higher than that at which it is withdrawn from the top. Typical pressure-difference results for such a system are shown in Fig. 11.7.

In Fig. 11.7 minus the pressure gradient from the bottom to the top of the tower is plotted against gas superficial mass flow rate up the column for various liquid superficial flow rates down the column. Consider first curve *A*. This is for flow of gas only; its slope (on a log-log plot) is 1.8, indicating that the flow is in the transition region on Fig. 11.4, in which $f_{\text{P.M.}}$ is proportional to $\mathscr{R}_{\text{P.M.}}$ to the -0.2 power. Curve *B* is for no liquid flow, but for the packing having been wetted and drained. The slope of the curve is the same as for curve *A*, but at each flow rate the pressure drop is about

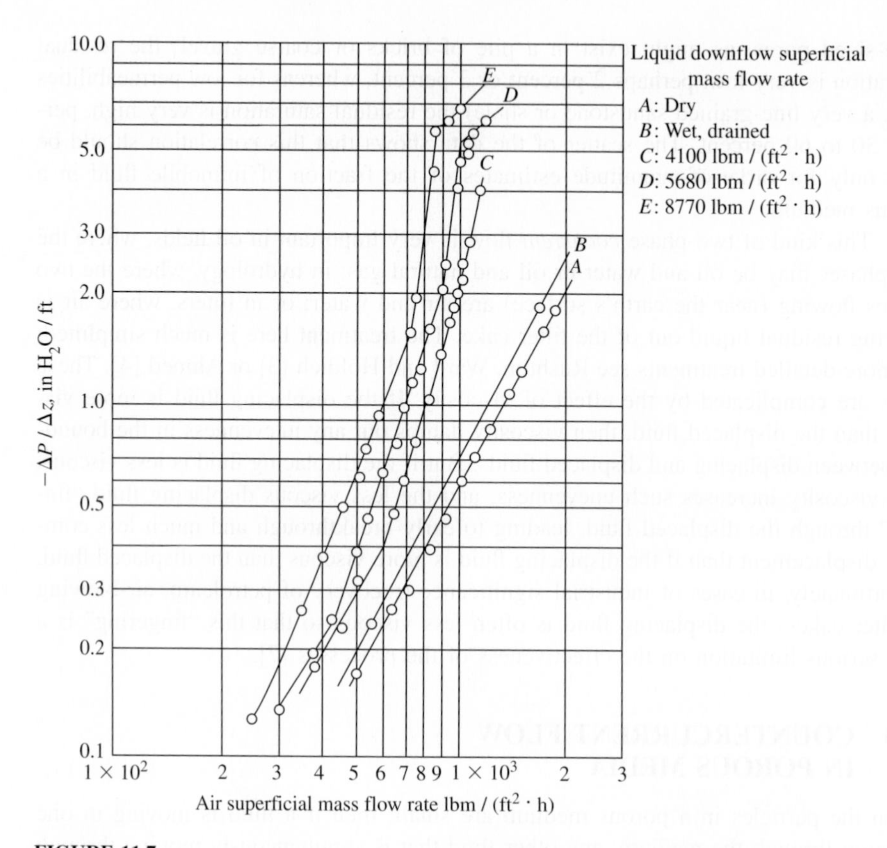

FIGURE 11.7

Pressure drop in competitive countercurrent air–water flow in a porous medium, that consisted of raschig rings, which are thin-walled cylinders 1 in long and 1 in in diameter, with both ends open. (From B. J. Lerner and E. S. Grove Jr., "Critical conditions of two-phase flow in packed columns," *Ind. Eng. Chem. 53:*216 (1951). Copyright 1951 by the American Chemical Society. Reproduced by permission of the publisher.)

25 percent higher, because some of the flow passages are now blocked by retained liquid. With some of the passages blocked, the interstitial velocity of the gas must increase because less area is available for its flow. That raises the pressure drop due to friction. Curve *C* shows a typical curve for the competitive flow of gas and liquid. At low gas flow rates the form is similar to curves *A* and *B,* but the pressure drop is higher because more of the passages are blocked by liquid. However, at an air flow rate of about 600 lbm / (h · ft²) the curve turns sharply upward. Visual observation indicates that at this point fluid begins to be held up in passages in which it previously flowed downward. This blocks these passages to the flow of the gas. Continued flow-rate increase causes more and more of these passages to be thus blocked, and the pressure rises steeply; this behavior is sometimes called *loading.*

Curves *D* and *E* show similar behavior, but they also show that for higher liquid flow rates a new region can be entered, in which the steep rise in the pressure-drop curve moderates. In this region, called *flooding,* the liquid fills the column, and it becomes the continuous phase instead of the dispersed one. The gas rises through it as bubbles rather than as a continuous gas stream.

Because devices with this kind of flow are widely used in distillation and absorption, the empirical methods for estimating their performance are summarized in books on those topics [8, 9]. In a packed bed inside some container the porosity is always greater near the walls than in the rest of the bed because the particles cannot pack together as well there. This complicates the design and operation of such devices.

11.4 SIMPLE FILTER THEORY

Filters are widely used to purify gases and liquids or to separate valuable products from gases or liquids. We can learn something about their behavior by applying the results previously found in this chapter. There are two kinds of filters: surface filters, in which the collected particles form a coherent cake on the filter surface (e.g., coffee filters, colanders, most industrial baghouses, most industrial thickeners, plate-and-frame or drum filters); and depth filters, in which the particles are collected throughout the entire depth of the filter (e.g., cigarette filters, auto oil filters, most home furnace filters).

11.4.1 Surface Filters

The flow through a surface filter is shown schematically in Fig. 11.8. A slurry (a fluid containing suspended solids) flows through a filter medium (most often a cloth, but sometimes a paper, porous metal, or bed of sand). The solid particles in the slurry deposit on the face of the filter medium, forming the "filter cake." The liquid, free from solids, flows through both cake and filter medium. Applying B.E. from point 1 to point 3, we see that there is no change in elevation. There is a slight change in velocity (due to the solid particles left behind in the filter cake in all cases and due to the pressure drop if the fluid is a gas), but this is generally negligible, and there is no pump or compressor work; so $\Delta P/\rho = -\mathcal{F}$. The flow is laminar in almost all filters, so that the pressure drop due to friction is given by Eq. 11.19. Solving Eq. 11.19 for the superficial velocity, we find

$$V_S = \frac{Q}{A} = \frac{-\Delta P}{\mu}\frac{k}{\Delta x} \tag{11.29}$$

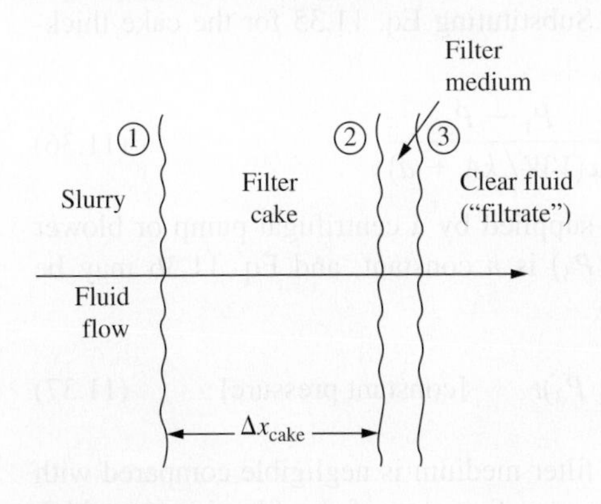

FIGURE 11.8
Flow in a surface filter.

Here there are two resistances in series with the same flow rate through them. Letting the subscript "f.m." indicate "filter medium," we write Eq. 11.29 twice and equate the identical flow rates (see Fig. 11.8):

$$V_S = \frac{P_1 - P_2}{\mu}\left(\frac{k}{\Delta x}\right)_{\text{cake}}$$

$$= \frac{P_2 - P_3}{\mu}\left(\frac{k}{\Delta x}\right)_{\text{f.m.}} \tag{11.30}$$

Solving for P_2, we get

$$P_2 = P_1 - \mu V_S\left(\frac{\Delta x}{k}\right)_{\text{cake}} = P_3 + \mu V_S\left(\frac{\Delta x}{k}\right)_{\text{f.m.}} \tag{11.31}$$

and then solving this equation for V_S, we find

$$V_S = \frac{P_1 - P_3}{\mu[(\Delta x / k)_{\text{cake}} + (\Delta x / k)_{\text{f.m.}}]} = \frac{Q}{A_{\text{filter}}} \tag{11.32}$$

This equation describes the instantaneous flow rate through a filter; it is analogous to Ohm's law for two resistors in series, so the $\mu\ \Delta x / k$ terms are called the *cake resistance* and the *cloth resistance.*

The resistance of the filter medium is normally assumed to be a constant independent of time; so $(\Delta x / k)_{\text{f.m.}}$ is replaced with a constant, a. If the filter cake is uniform, then its instantaneous resistance is proportional to its instantaneous thickness. However, this thickness is related to the volume of filtrate that has passed through the cake by the material balance:

$$\begin{aligned} \Delta x_{\text{cake}} &= \left(\frac{\text{mass of cake}}{\text{area}}\right) \cdot \left(\frac{1}{\rho_{\text{cake}}}\right) \\ &= \left(\frac{1}{\rho_{\text{cake}}}\right) \cdot \left(\frac{\text{volume of filtrate}}{\text{area}}\right) \cdot \left(\frac{\text{mass of solids}}{\text{volume of filtrate}}\right) \end{aligned} \tag{11.33}$$

Customarily we define

$$W\eta = \left(\frac{\text{mass of solids collected}}{\text{volume of filtrate}}\right) \cdot \left(\frac{1}{\rho_{\text{cake}}}\right) = \left(\frac{\text{volume of cake}}{\text{volume of filtrate}}\right) \tag{11.34}$$

Here W is the volume of cake per unit volume of clear liquid processed and η is the collection efficiency for solids, which is normally assumed to ≈ 1.00 and dropped from Eq. 11.34, so that

$$\Delta x_{\text{cake}} = \frac{V}{A} W \quad \text{and} \quad \frac{d(\Delta x_{\text{cake}})}{dt} = V_S W \tag{11.35}$$

Here V is the volume of filtrate $= \int Q\ dt$.Substituting Eq. 11.35 for the cake thickness in Eq. 11.32, we find

$$\frac{Q}{A} = \frac{1}{A}\frac{dV}{dt} = \frac{P_1 - P_3}{\mu(VW / kA + a)} \tag{11.36}$$

For many industrial filtrations, the filter is supplied by a centrifugal pump or blower at practically constant pressure, so $(P_1 - P_3)$ is a constant, and Eq. 11.36 may be rearranged and integrated to

$$\left(\frac{V}{A}\right)^2 \cdot \frac{\mu W}{2k} + \frac{V}{A}\mu a = (P_1 - P_3)t \qquad \text{[constant pressure]} \tag{11.37}$$

For many filtrations the resistance a of the filter medium is negligible compared with the cake resistance (except perhaps for the very first part of the filtration, in which the cake is forming), so that the second term on the left of Eq. 11.37 may be dropped;

in such cases the volume of filtrate is proportional to the square root of the time of filtration.

Because of this buildup of cake, the filter must be cleaned at regular intervals; the optimum time between cleanings is discussed by Chen [10].

For a few industrial filtrations, the filter is supplied by a P.D. pump, which is practically a constant-flow-rate-device. Such a pump feeds the filter at a pressure that is steadily increasing during the filtration. Equation 11.36 shows that, for constant k and negligible a, the pressure increases linearly with time, because the cake thickness increases linearly with time.

In many real filtration problems the k in Eq. 11.36 is not constant but is a function of pressure. This occurs because many filtrates, such as the iron hydroxides and aluminum hydroxides use in water clarification, are weak-structured gels or flocs. In the loose state they have a relatively low flow resistance, but under pressure they collapse and form denser structures that have a higher flow resistance. The common practice in describing such cakes is to write

$$\text{Cake specific resistance} = \frac{1}{k} = \alpha P^s \tag{11.38}$$

Substituting Eq. 11.38 in Eq. 11.36 and letting P_3 be zero (i.e., atmospheric pressure), we find

$$\frac{Q}{A} = \frac{1}{A} \cdot \frac{dV}{dt} = \frac{P}{\mu[(\alpha P^s V W / A) + a]} \tag{11.39}$$

If a is negligible (the usual case), then Eq. 11.39 indicates that at a given cake thickness increasing the pressure will (i) linearly increase the flow rate if s is 0 (as for sand), (ii) have no effect at all on the flow rate if s is 1 (as for some gelatinous hydroxides), and (iii) have some intermediate effect if s is between 0 and 1. There are cases in which s is less than 1 at low pressures and greater than 1 at higher pressures, so that there is a pressure that gives maximum flow rate with lower flow rates for lower or higher pressures. Much of the art of filtration consists of selecting additives, "filter aids," that lower α or s [3, Sec. 5.3].

11.4.2 Depth Filters

Surface filters are used when the concentration of solids to be removed from the fluid is high; the solids form a cake, which actually does the filtering. Once the cake forms, the filter medium serves only to support the cake but does practically no filtering. Depth filters are used when the concentration of solids is low and the goal is to produce a very clean fluid stream at the outlet. Often the depth filter is the final cleanup step, following a surface filter that takes out most of the particles to be removed. This is the case with the depth filters that make the final cleanup of air going to operating rooms or to microchip fabrication facilities. Whereas surface filters can be cleaned by removing the cake from the surface, most depth filters are thrown away when they become loaded with solids, e.g., cigarette filters and auto oil filters.

The fluid mechanics of depth filters do not lend themselves to as simple a mathematical treatment as the one shown above for surface filters [11, Sec. 9.2.2].

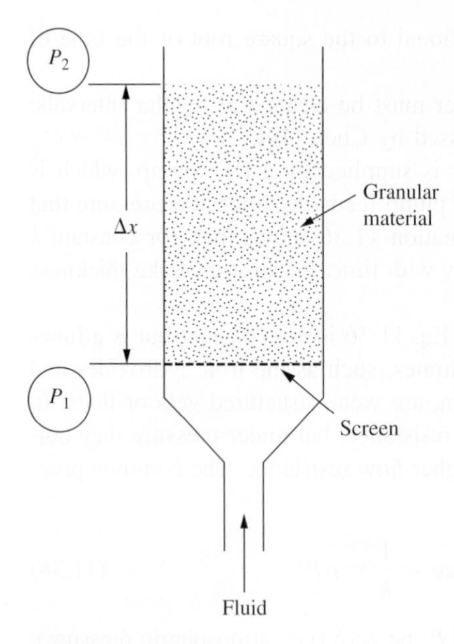

FIGURE 11.9
A fluidized bed formed by upflow of fluid through a porous medium.

11.5 FLUIDIZATION

The previous treatment of flow in porous media will help us understand fluidization, which plays a major role in many chemical processes. Figure 11.9 shows an apparatus for creating a fluidized bed. A granular material such as sand is resting on some sort of support screen in a tube. What would happen if we were to flow a fluid, such as air or water, upward through this material?

The behavior will be described by B.E., with the laminar-flow friction-heating term given by Eq. 11.14:

$$\frac{\Delta P}{\rho} + g\,\Delta z = -\mathscr{F} \qquad (11.40)$$

Here we have dropped the $\Delta V^2/2$ term, which is negligible. Now we can multiply through by ρ and assume that we are dealing with air, in which the $\rho g\,\Delta z$ is also small and can also be dropped. Thus, we see that the pressure difference across the bed of granular material, $(P_2 - P_1)$, is linearly proportional to the volumetric flow rate.

Now, if we steadily increase the fluid flow rate, what will happen when the pressure force on the entire bed, [$(P_2 - P_1)$ times the cross-sectional area], is slightly greater than the weight of the bed? If the bed is made up of some solid, porous material such as a block of sandstone, then the entire block will be expelled from the tube, exactly as a bullet is expelled from a gun by the high-pressure gases behind it. However, if the solid is a mass of individual particles, like sand grains, they can decrease their resistance to flow by increasing their porosity. Substituting Eq. 11.14 in Eq. 11.40, we see that for gas-solid systems

$$\begin{aligned} -\Delta P &= g\,\Delta z(1-\varepsilon)\rho_{\text{particles}} \\ &= \frac{150\,V_S\mu(1-\varepsilon)^2\,\Delta x}{(D_p)^2\varepsilon^3} \qquad \text{[gas-solid systems]} \end{aligned} \qquad (11.41)$$

As V_s increases, ε may increase and hold ΔP constant (Δx will also increase, but its effect is much less than the effect of a change in ε). Thus, the experimental result for such a test is shown in Fig. 11.10. For velocities less than V_{mf}, the *minimum fluidizing velocity,* the bed behaves as a packed bed. However, as the velocity is increased past V_{mf}, not only does the bed expand (Δx increases) but also the particles are seen to move apart and become able to slide past each other, and the entire particle-fluid mass becomes a fluid that can be poured from one vessel to another and be pumped, etc. As the velocity is further increased, the bed becomes more and more expanded, and the solid content becomes more and more dilute.

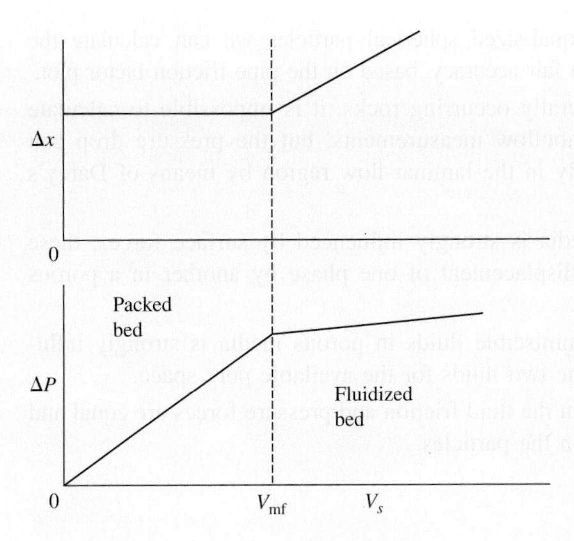

FIGURE 11.10
The transition from a packed to a fluidized bed.

Finally, the velocity becomes as large as the terminal settling velocity of the individual particles, so the particles are swept out of the system. Thus, the velocity range for which a fluidized bed can exist is from V_{mf} (whose value can be calculated from Eq. 11.41 for spherical particles and analogous equations for non-spherical ones) to V_t (whose value can be calculated as shown in Sec. 6.15).

The above description of the particles in the bed increasing their value of ε to accommodate an increasing flow rate, as sketched in the upper part of Fig. 11.10, is actually observed if the fluidizing fluid is a liquid. However, if the fluidizing fluid is a gas (by far the more common case industrially), the behavior is more complex. In that case as the fluid flow rate is increased past V_{mf} the increased gas flow forms bubbles, which contain virtually no particles and which rise through the bed of fluidized particles very much as bubbles rise through liquids. These bubbles greatly complicate all the aspects of the behavior of fluidized beds and of the chemical reactions carried out in them.

These fluidized beds have proved very useful in chemical technology when one wishes to move a granular solid through a series of processing steps in a continuous fashion. Their most dramatic application is in *fluidized-bed catalytic cracking,* which is a standard petroleum refining operation, but there are dozens of other applications. This brief treatment only shows how they are formed; for more on their properties and uses see Kunii and Levenspiel [2] and Pell [12].

If one continues to increase the velocity of the fluidizing stream, eventually it will exceed the settling velocity of the largest particle in the bed, and then the entire bed will be conveyed upward. This ceases to be a fluidized bed, but becomes a *pneumatic transport* pipe, which is widely used to move granular solids like grain, portland cement or plastic pellets [2, Chap. 15]. As a rule of thumb, fluidized beds operate with gas superficial velocities of 1 to 3 ft / s (0.3 to 1 m / s); pneumatic transport operates with superficial velocities of 30 to 60 ft / s (10 to 20 m / s).

11.6 SUMMARY

1. Flow through a porous medium can be laminar, turbulent, or intermediate. The transition from laminar to turbulent flow is smoother and more reproducible than for flow in a pipe. Generally, the flow in filters, the flow of groundwater, and the flow in petroleum and natural-gas fields are laminar.

2. For porous media made of equal-sized spherical particles we can calculate the laminar-flow pressure drop with fair accuracy, based on the pipe friction factor plot.
3. For most filter cakes and naturally occurring rocks, it is impossible to calculate the pressure drop based on nonflow measurements, but the pressure drop can be correlated very satisfactorily in the laminar-flow region by means of Darcy's equation.
4. Two-phase flow in porous media is strongly influenced by surface forces; these generally lead to incomplete displacement of one phase by another in a porous medium.
5. Countercurrent flow of two immiscible fluids in porous media is strongly influenced by the competition of the two fluids for the available pore space.
6. Fluidized beds are formed when the fluid friction and pressure forces are equal and opposite to the gravity force on the particles.

PROBLEMS

See the Common Units and Values for Problems and Examples inside the back cover! An asterisk (*) on a problem number indicates that the answer is shown in App. D.

11.1. Show that $f = 16/\mathscr{R}$ is equivalent to $f_{\text{P.M.}} = 72/\mathscr{R}_{\text{P.M.}}$.

11.2. Show that, if one assumes that $V_l = \sqrt{2}(V_l)_{x\text{-dir}}$ and that the length of the flow path is $\sqrt{2}\,\Delta x$, then the friction factor should be twice as high as shown in $f_{\text{P.M.}} = 72/\mathscr{R}_{\text{P.M.}}$.

11.3. Show the relative sizes of the two terms in the Ergun equation, Eq. 11.17, at $\mathscr{R}_{\text{P.M.}}$ values of 0.1, 1, 10, 100, 1000, and 10,000.

11.4. Calculate $\mathscr{R}_{\text{P.M.}}$ in Example 11.2.

11.5.*Calculate the permeability of the bed of ion-exchange particles in Example 11.1.

11.6. For the flows in Examples 11.1 and 11.2, calculate the magnitudes of the $\Delta V^2/2$ terms omitted in B.E., and compare these with the magnitude of the $\mathscr{F}$ terms.

11.7.*For the apparatus in Example 11.1 estimate the residual saturation of air if the apparatus is originally full of air and then filled with water so that water is allowed to percolate through as shown in Fig. 11.3.

11.8. In Fig. 11.7 what value would the abscissa have if the packing were filled completely with water that was not moving? From this value, what can you say about the fraction of the pore space occupied by liquid in the "flooded" regime at the top of curves *D* and *E*?

11.9.*(*a*) Calculate the $\mathscr{R}_{\text{P.M.}}$ at the low end of curve *A* on Fig. 11.7. Here use $D_p = 1$ in and $\varepsilon = 0.5$. Does this Reynolds number correspond to the laminar, transition, or turbulent range of flow rates?

(*b*) Using the same values of D_p and ε, estimate the pressure gradient for 1000 $\text{lbm}/(\text{ft}^2 \cdot \text{h})$ according to Eq. 11.16 and compare the result with the value shown in Fig. 11.7. Is the agreement satisfactory?

11.10. The results of a constant-pressure filtration test are a table of volume of filtrate per unit area of filter versus time. The data are presumed to agree with Eq. 11.37. Show how one may plot these data so that they will form a straightline plot and how one finds the values of *k* and *a* from the slope and intercept of this line.

11.11. Equation 11.41 correctly predicts V_{mf}, the minimum fluidizing velocity, for beds of spherical particles, but it does not correctly predict the relationship of V_s to ε after the bed has started to expand. Why not?

11.12. Write the equation analogous to Eq. 11.41 for water flowing upward through a bed of sand. In this case the $g\,\Delta z$ term in Eq. 11.40 cannot be neglected.

11.13.*For spherical sand particles with $D_p = 0.03$ in and $\rho_{\text{particles}} = 150\ \text{lbm}/\text{ft}^3$ estimate the minimum fluidizing velocity for air and for water. Assume $\varepsilon = 0.3$. In the case of the water we must rederive Eq. 11.41, taking into account the buoyant force on the particles.

11.14. Equations 11.14 and 11.16 show that the frictional resistance for flow in porous media depends strongly on the porosity, ε. The porosity depends on the shape of the particles and on how they are arranged. Show the following:

(*a*) If we pack cubes of the same size, one above the other and side by side (rectangular array), the porosity = 0.

(*b*) If we stack spheres the same way (egg-crate packing), the porosity is 0.476.

(*c*) If we maintain the rectangular array in any one plane, but shift the planes so that the individual spheres in one plane fall into the holes between the spheres in the plane below (orthorhombic array), the porosity decreases to 0.3954.

(*d*) If we use the closest packing in each plane, and then place the planes so that they have the closest packing to the planes above and below (rhombohedral array), the porosity further decreases to 0.2595.

These arrays, and data on the porosities of other types of particles, are shown in Brown [13, p. 215].

REFERENCES FOR CHAPTER 11

1. Ergun, S. "Fluid Flow Through Packed Columns." *Chem. Eng. Prog. 48* (1952), pp. 89–94.
2. Kunii, D., O. Levenspiel. *Fluidization Engineering,* 2nd ed. Boston: Butterworth-Heinemann, 1991.
3. Rushton, A., A. S. Ward, R. G. Holdich. *Solid-Liquid Filtration and Separation Technology,* 2nd ed. Weinheim: Wiley-VCH, 1996.
4. Ahmed, T. *Reservoir Engineering Handbook,* 2nd ed. Boston: Gulf Professional Publishing, 2001.
5. Kashef, A. I. *Groundwater Engineering.* New York: McGraw-Hill, 1986.
6. Scheidegger, A. E. *The Physics of Flow Through Porous Media.* 3rd ed. Toronto: University of Toronto Press, 1974.
7. Wooding, R. A., H. J. Morel-Seytoux. "Multiphase Fluid Flow Through Porous Media." *Ann. Rev. Fluid Mech. 8* (1976), pp. 233–274.
8. Fair, J. R., D. E. Steinmeyer, W. R. Penny, B. B. Crocker. "Gas Absorption and Gas-Liquid System Design." In *Perry's Chemical Engineers' Handbook,* 7th ed. D. W. Green, J. O. Maloney. New York: McGraw-Hill, 1997, Sec. 14.
9. Kister, H. Z. *Distillation Design.* New York: McGraw-Hill, 1992, Chap. 8.
10. Chen, N. H. "Liquid-Solid Filtration: Generalized Design and Optimization Calculations." *Chem Eng. 85(17)* (1978), pp. 97–101.
11. de Nevers, N. *Air Pollution Control Engineering.* 2nd ed. New York: McGraw-Hill, 2000, p. 120.
12. Pell, M. "Gas-Solid Operations and Equipment." In *Perry's Chemical Engineers' Handbook,* 7th ed. D. W. Green, J. O. Maloney. New York: McGraw-Hill, 1997, Sec. 17.
13. Brown, G. G., et al. *Unit Operations.* New York: John Wiley, 1950.

CHAPTER 12

GAS-LIQUID FLOW

In the preceding chapters, as in most fluid mechanics problems, the flowing fluid was composed of one homogeneous phase, such as water, oil, air, or steam. There are, however, many interesting and important flows that involve the simultaneous flow of two quite different materials through the same conduit, such as the flow in a coffee percolator (Fig. 2.17), the flow in most vaporizers or boilers or condensers, and the flow in the carburetor of an automobile (Fig. 5.35). These flows can be of gas-liquid mixtures, liquid-liquid mixtures, gas-solid mixtures, or liquid-solid mixtures. Although there are industrially important examples of each of these combinations, the most important and interesting seems to be the gas-liquid case, which we will discuss in this chapter. Brief summaries of what is known about gas-solid and liquid-solid flows are given by Tilton [1, pp. 6–26].

In all of these flows the influence of gravity is much greater than in the one-phase flows we have considered previously. For laminar or turbulent flow of water in a tube, the velocity distribution and friction effect ($\mathcal{F}$) would be the same on earth as in the zero-gravity environment of an earth satellite. Neither the velocity distribution nor $\mathcal{F}$ is influenced at all by changing the pipe's position from the horizontal to the vertical. Gravity does not significantly influence the flow pattern or $\mathcal{F}$, because it works equally on each particle of the fluid. This is not the case for two-phase flows, because normally the phases have different densities and thus are affected to different extents by gravity. All of the flows described in this chapter would *not* be the same in a zero-gravity environment as on earth and, as shown in Secs. 12.1 and 12.2, horizontal and vertical two-phase flows have very different velocity distributions and $\mathcal{F}$ values.

12.1 VERTICAL, UPWARD GAS-LIQUID FLOW

Many of the salient features of multiphase flow, and the terms used to describe it can be illustrated by considering the simultaneous gas-liquid flow up a vertical pipe with the apparatus sketched in Fig. 12.1.

Assume that at first only water flows; then, from B.E.,

$$P_1 - P_2 = \rho g(z_2 - z_1) + \rho \mathscr{F} \tag{12.1}$$

For zero flow rate we can calculate $(P_1 - P_2) = \rho g \, \Delta z = 8.7$ psi $= 60$ kPa. As we start the flow of water only, $(P_1 - P_2)$ will increase, because of the increase in $\mathscr{F}$ with increasing velocity.

Now let us hold the water flow rate constant at some modest average velocity, such as 2 ft / s, and slowly increase the air velocity from zero to some large value. This will cause $\mathscr{F}$ to increase, since the overall linear velocity is increased. However, now there will be bubbles of gas in the pipe; the density in Eq. 12.1 is no longer the density of water but the average density of the gas-liquid mixture in the vertical pipe. At low flow rates the density goes down much faster than $\mathscr{F}$ goes up; so $(P_1 - P_2)$ decreases steadily as we increase the gas flow rate. Finally, a point is reached where further increase in the gas velocity causes $\mathscr{F}$ to increase faster than ρ decreases; $(P_1 - P_2)$ will increase with increasing gas flow rate. A typical plot of experimental data for such a system is shown in Fig. 12.2.

The pressure gradient curve on Fig. 12.2 has the shape described above, with a distinct kink in it near a volumetric flow rate ratio of 10. This is typical of such systems and indicates that the change described above does not take place smoothly. If the system is made of glass, so that the flow pattern can be observed, it will be seen that several distinctly different flow patterns are formed as the air flow rate is increased. This is illustrated in Fig. 12.3. At the lowest air flow rates small bubbles rise through the liquid. As the air flow rate is increased, large single bubbles are formed, which practically fill the tube, driving slugs of liquid between them. At higher rates these slugs become frothy, and finally at high gas flow rates the liquid is present either as an annular film on the walls or as a mist in the gas.

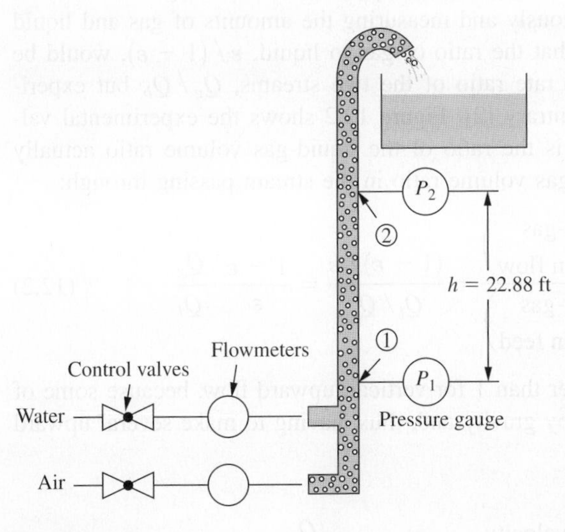

FIGURE 12.1
Apparatus for vertical, upward gas-liquid flow.

Two additional ideas are widely used to explain and correlate the results of experiments like that shown in Fig. 12.1: *holdup* and *slip*. One can measure what fraction of the tube is occupied by gas, ε, and what fraction by liquid, 1 − ε, in such a flow (experimentally this

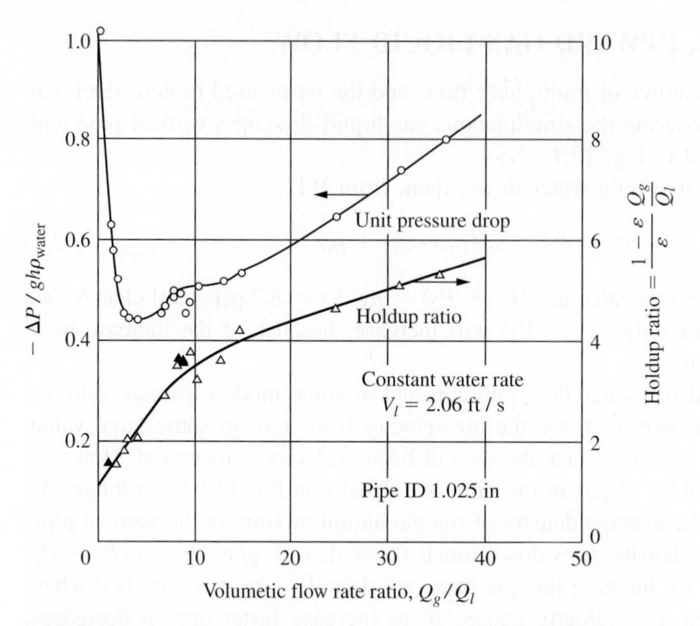

FIGURE 12.2
Typical experimental results for vertical, upward air–water flow in the apparatus shown in Fig. 12.1, with constant liquid flow rate. Here V_L is Q_l / A. The test section is 22.88 ft long, 1.025 in ID, made of transparent plastic tubing. The pressure at the midpoint of the test section is 36.0 psia and $T \approx 70°F$. (From G. W. Govier, B. A. Radford and J. S. C. Dunn, "The upward vertical flow of air–water mixtures," *Can. J. Chem. Eng. 35:*58–70 (1957). Reproduced by permission of the publisher.)

is done by placing two quick-closing valves in the tube, establishing the flow, and then closing the valves simultaneously and measuring the amounts of gas and liquid so trapped). One might assume that the ratio of gas to liquid, $\varepsilon / (1 - \varepsilon)$, would be the same as the volumetric flow rate ratio of the two streams, Q_g / Q_l, but experimental evidence indicates the contrary [2]. Figure 12.2 shows the experimental values for the *holdup ratio,* which is the ratio of the liquid-gas volume ratio actually present in the pipe to the liquid-gas volume ratio in the stream passing through:

$$\text{Holdup ratio} = \frac{\begin{pmatrix}\text{liquid-gas} \\ \text{ratio in flow}\end{pmatrix}}{\begin{pmatrix}\text{liquid-gas} \\ \text{ratio in feed}\end{pmatrix}} = \frac{(1-\varepsilon)/\varepsilon}{Q_l / Q_g} = \frac{1-\varepsilon}{\varepsilon} \cdot \frac{Q_g}{Q_l} \tag{12.2}$$

This holdup ratio is always greater than 1 for vertical, upward flow, because some of the liquid is always falling back by gravity and, thus, having to make several upward trips to get out.

Then we define

$$\begin{matrix}\text{Average velocity} \\ \text{of gas in tube}\end{matrix} = V_{\text{avg,gas}} = \frac{Q_g}{A_{\text{tube}}\varepsilon} \tag{12.3}$$

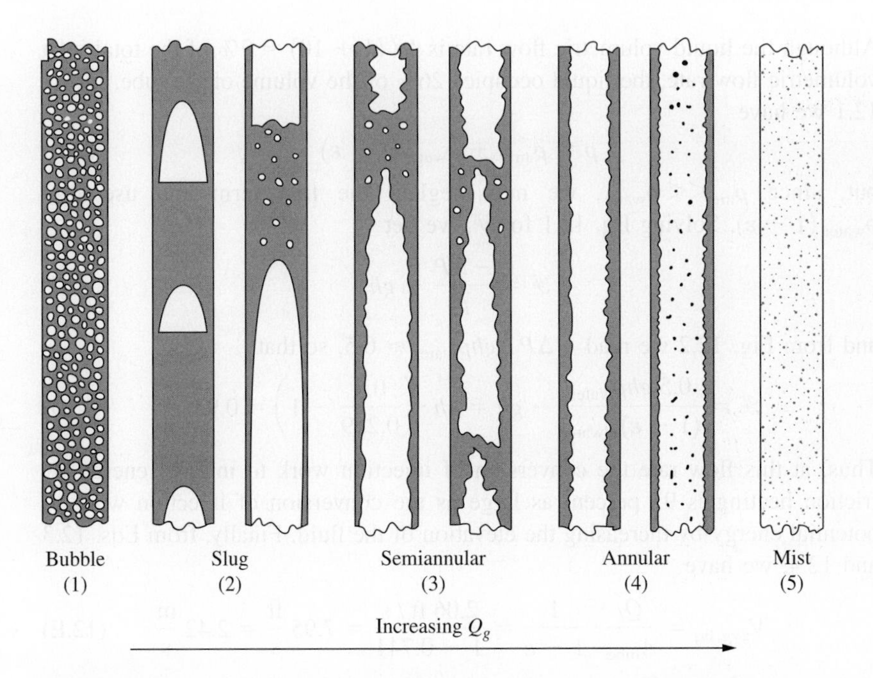

FIGURE 12.3
Two-phase flow patterns in vertical tubes. The liquid flow rate is upward at a small, constant velocity. The gas flow rate upward increases steadily from left to right. The "annular" pattern shown is often referred to as *climbing film flow.* (From D. J. Nicklin and J. F. Davidson, "The onset of instability in two-phase slug flow," in *Symposium in Two-Phase Flow,* Institution of Mechanical Engineers, London, 1962. Reproduced by permission of the publisher.)

and, analogously

$$\begin{matrix}\text{Average velocity} \\ \text{of liquid in tube}\end{matrix} = V_{\text{avg,liq}} = \frac{Q_l}{A_{\text{tube}}(1-\varepsilon)} \tag{12.4}$$

The difference between these two is defined as the *slip velocity:*

$$V_{\text{slip}} = V_{\text{avg,gas}} - V_{\text{avg,liq}} = \frac{1}{A_{\text{tube}}}\left(\frac{Q_g}{\varepsilon} - \frac{Q_l}{1-\varepsilon}\right) \tag{12.5}$$

If the holdup ratio were 1.00, then the term in parentheses on the right in Eq. 12.5 would be zero. For actual upward flows the slip velocity is always positive, because the gas mostly moves upward, and the liquid partly moves upward and partly flows backward downhill because of gravity, lowering its net average upward velocity.

Example 12.1. From the data shown in Fig. 12.2 for $Q_g / Q_l = 10$, calculate the values of $\mathscr{F}$, ε, and V_{slip}.

We start by reading that at this value of Q_g / Q_l the holdup ratio is ≈ 3.5, so that

$$\frac{1-\varepsilon}{\varepsilon} = \frac{3.5}{Q_g / Q_l} = \frac{3.5}{10} = 0.35, \qquad \varepsilon = 0.741 \tag{12.A}$$

Although the liquid volumetric flow rate is $1/(1+10) \approx 9\%$ of the total inlet volumetric flow rate, the liquid occupies 26% of the volume of the tube. In Eq. 12.1 we have

$$\rho = \rho_{\text{air}}\varepsilon + \rho_{\text{water}}(1-\varepsilon) \tag{12.B}$$

but, since $\rho_{\text{air}} << \rho_{\text{water}}$, we may neglect the first term and use $\rho \approx \rho_{\text{water}}(1-\varepsilon)$. Solving Eq. 12.1 for $\mathscr{F}$, we get

$$\mathscr{F} = \frac{-\Delta P}{\rho} - gh \tag{12.C}$$

and from Fig. 12.2 we read $-\Delta P / gh\rho_{\text{water}} \approx 0.5$, so that

$$\mathscr{F} = \frac{0.5\; gh\rho_{\text{water}}}{(1-\varepsilon)\rho_{\text{water}}} - gh = gh \cdot \left(\frac{0.5}{0.259} - 1\right) = 0.93\; gh \tag{12.D}$$

Thus, at this flow rate the conversion of injection work to internal energy by friction heating is 93 percent as large as the conversion of injection work to potential energy by increasing the elevation of the fluid. Finally, from Eqs. 12.3 and 12.4, we have

$$V_{\text{avg, liq}} = \frac{Q_l}{A_{\text{tube}}} \cdot \frac{1}{1-\varepsilon} = \frac{2.06\text{ ft/s}}{1-0.741} = 7.95\,\frac{\text{ft}}{\text{s}} = 2.42\,\frac{\text{m}}{\text{s}} \tag{12.E}$$

$$V_{\text{avg, gas}} = \frac{Q_g}{A_{\text{tube}}\varepsilon} = \frac{Q_g}{Q_l} \cdot \frac{Q_l}{A_{\text{tube}}} \cdot \frac{1}{\varepsilon} = \frac{10 \cdot 2.06\text{ ft/s}}{0.741} = 27.8\,\frac{\text{ft}}{\text{s}} = 8.48\,\frac{\text{m}}{\text{s}} \tag{12.F}$$

$$V_{\text{slip}} = 27.8\,\frac{\text{ft}}{\text{s}} - 7.9\,\frac{\text{ft}}{\text{s}} = 19.9\,\frac{\text{ft}}{\text{s}} = 6.07\,\frac{\text{m}}{\text{s}} \tag{12.G}$$

Both liquid and gas average velocities are larger than the values they would have if they occupied the tube alone, because they have less area available to them individually. ■

Vertical cocurrent flow of this type is extremely common in engineering. It occurs in almost all flowing and gas-lifted oil wells and in gas-lift pumps. Much of the research on this type of flow has been connected with boiling of liquids in vertical tubes. In a vertical tube in which a liquid (e.g., water) is being boiled, it is entirely possible to have all the flow patterns shown in Fig. 12.3 present in the same tube at the same time. In that case the fluid enters the bottom of the tube as all liquid. As it passes up the tube, more and more of it is converted to a vapor by boiling, so that at the top of the tube it may be all vapor (this is generally avoided in boilers, because it results in a dangerously high metal temperature at the top of the tube). Thus, the various patterns shown in Fig. 12.3 exist at various elevations in the same tube, the bottom of the tube corresponding to the left of Fig. 12.3 and the top corresponding to the right.

Because the various flow patterns shown in Fig. 12.3 are so different from each other, it is unlikely that there will ever be completely successful single description of the friction effect and holdup of vertical, gas-liquid flows that is comparable, for example, to the friction factor plot for pipe flow. The various flow forms are so different physically that they cannot be expected to obey the same mathematical relationships. Several such overall correlations are available, e.g., that of Hughmark and Pressburg [3] in which, for a variety of fluids, they correlated liquid holdup $(1-\varepsilon)$

with the group,

$$\left(\frac{\dot{m}_l}{\dot{m}_g}\right)^{0.9} \cdot \frac{\mu_l^{0.19}\, \sigma^{0.205}\, \rho_g^{0.70}}{(Q_l / A + Q_g / A)^{0.435}\, \rho_l^{0.72}} \tag{12.H}$$

They found that they could represent all the available data for the holdup in such flow on a plot of $(1 - \varepsilon)$ versus this group with an average error of about 12 percent. This type of relation has no theoretical basis, but is widely used.

The other approach is to try to find separate equations for the various different flow patterns shown in Fig. 12.3. This approach requires some kind of "flow map" to indicate which flow pattern will be observed for the conditions of interest, and then suitable relations for each of the individual flow patterns. There has been considerable progress in this approach, [4, 5]; see Prob. 12.5. One major complication in trying to correlate the behavior of such flows is the fact that for a given choice of fluids, pipe size, and flow rates the quantities ε and $\mathcal{F}$ are strongly influenced by the design of the gas-liquid mixer. If in Fig. 12.1 we switch from the simple pipe tee mixer to one in which the liquid is introduced through porous walls, this makes no change in our observations in the slug flow region, but it makes a significant change in the observations in the annular and mist flow regions [6, p. 248]. This effect becomes unimportant for very long pipes, but is important for short ones.

12.2 HORIZONTAL GAS-LIQUID FLOW

If the apparatus shown in Fig. 12.1 is turned so that the pipe is horizontal, then the behavior will be quite different from that for the vertical flow case. The observed flow patterns are shown on Fig. 12.4. For a constant liquid flow rate, as we increase the gas flow rate from zero, we will see that at first the gas flows in bubbles along the top of the pipe. Then the bubbles grow in size and length. Finally, they become so large that they coalesce into a continuous stream of gas flowing over a continuous layer of liquid. Further increase in gas flow rate causes the gas to raise waves on the liquid surface. These waves grow until they eventually reach across the tube, in which case they are propelled as slugs of liquid with interspersed slugs of gas. Then this pattern switches to the annular pattern observed in vertical gas-liquid flow and finally to a mist flow.

Knowledge of the flow pattern can be quite important. If, for example, we are vaporizing a liquid in a horizontal tube under high heat fluxes, then in annular flow the tube wall will always be covered with liquid and presumably be safe from excessive metal temperatures. On the other hand, if the flow is stratified, then the top part of the tube will be covered by gas, which is much less effective in conducting heat away from the surface; so the inside metal temperatures may become very high, even to the point of melting the tube. Although several correlations have been proposed for determining what kind of flow patterns will exist [1, p. 6-26]; [6, pp. 199–277] there is none now known that is universally applicable. Furthermore, the transitions from one kind of flow to another do not occur at sharply defined conditions but may take place over a range of conditions and, as in vertical gas-liquid flow for a given set of fluids and flow rates, the flow pattern can be completely changed and the value of $\mathcal{F}$ doubled by simply changing the type of gas-liquid mixer.

All available experimental data indicate that $\mathcal{F}$ is always higher for two-phase horizontal flow than for single-phase flow under similar conditions. This is principally

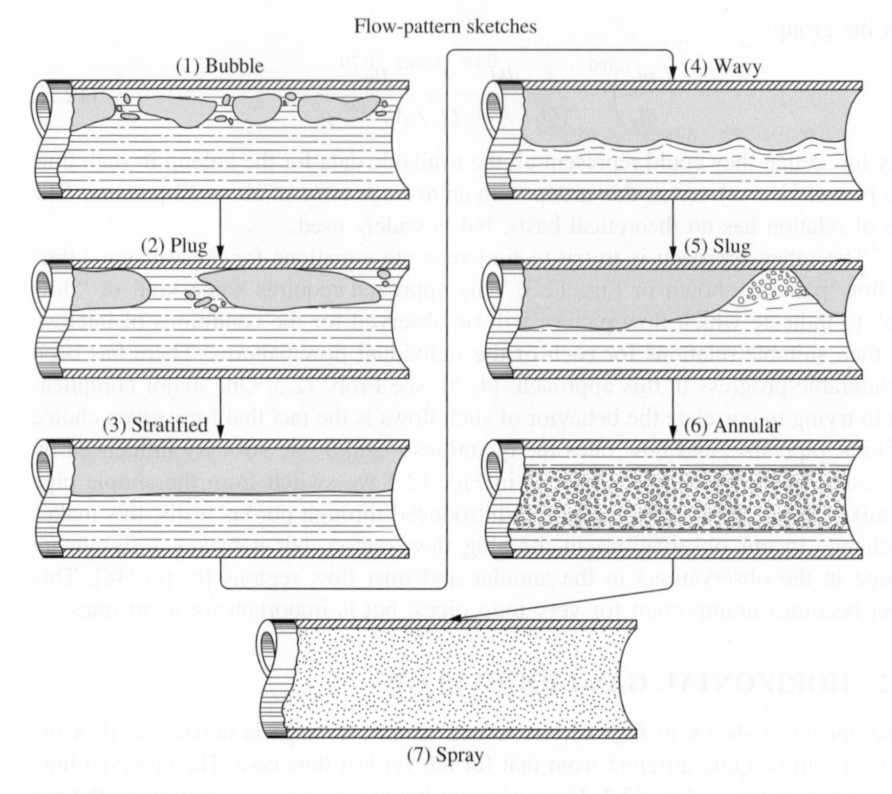

FIGURE 12.4
Flow patterns in horizontal gas–liquid flows. These are for a constant liquid flow rates, with the gas flow rate increasing as shown by the arrows. (From R. S. Brodkey, *The Phenomena of Fluid Motions,* Addison-Wesley, Reading, MA, 1967, as redrawn from G. Alves, "Concurrent liquid-gas flow in a pipeline contactor," *CEP 50:*449–456 (1954). Reproduced by permission of the publisher.)

due to the movement of the two phases relative to each other in the tube, which does not contribute to flow along the tube but, rather, to the conversion of other forms of energy into internal energy. For horizontal gas-liquid flow, numerous empirical correlations have been proposed, in which the most widely used is that of Lockhart and Martinelli [7]. The comparison of those correlations with experiment has been extensively reviewed by Scott [6] and indicates that it is not reliable to more than about ± 50 percent in most cases but that it is as good as any other single correlation.

12.3 TWO-PHASE FLOW WITH BOILING

When a liquid at its boiling point flows in a pipe, the pressure decrease due to friction will cause the pressure of the liquid to fall below its saturation pressure, and the liquid will boil. This type of flow, called *flashing flow,* is important in the design of boilers, steam condensate lines, etc.

The principal difference between this type of flow and those discussed above is that, as the pressure falls, more and more vapor is formed, so the volumetric flow rates, average velocities, and pressure drops per foot are not constant for an entire pipe but vary with length. The velocity and pressure gradient increase rapidly as the amount of vapor increases. Furthermore, in this type of flow it is very common to have the kind of choked condition found in high-velocity gas flows (Sec. 8.3). However, the observed sonic velocities for gas-liquid mixtures are much lower than for gas alone, so that this choking occurs at velocities much lower than the sonic velocity of the vapor alone [8]. This kind of flow is of great interest in the design of steam boilers and vaporizing furnaces of all kinds [1].

12.4 SUMMARY

1. In gas-liquid flows numerous different flow patterns are possible, depending on the gas and liquid flow rates and properties and on the direction of the flow relative to the direction of gravity.
2. For such flows the pressure drop due to friction heating of the fluids, $\mathcal{F}$, is always greater than that for single-phase flows under comparable circumstances.
3. There are numerous empirical correlations available for predicting pressure drop, holdup, and slip velocity for such systems. These are not nearly as reliable as the pressure-drop correlations for single-phase flow. For some of the flow patterns we have useful theoretical equations.
4. Countercurrent, vertical gas-liquid flow in packed beds is discussed in Sec. 11.3.

PROBLEMS

See the Common Units and Values for Problems and Examples inside the back cover! An asterisk (*) on a problem number indicates that its answer is shown in App. D.

12.1. From the data given in Fig. 12.2 prepare a plot showing ρ_{avg}, $10 \cdot \rho_{\text{avg}} / \rho_{\text{water}}$ and $\mathcal{F} / g\rho_{\text{water}}$ as functions of Q_g / Q_l.

12.2. From the data in Fig. 12.2 prepare a plot of slip velocity versus Q_g / Q_l.

12.3.*For the flow discussed in Example 12.1 calculate the values of $\mathcal{F}$ that would exist if only the liquid were flowing and if only the gas were flowing. Compare these with the observed value of $\mathcal{F}$ for the simultaneous two-phase flow. Assume $\varepsilon / D = 0$, i.e., a smooth tube.

12.4. Figure 12.5 shows a flow map for upward air-water flow. All such maps are approximate, and those of various authors often disagree. There is only fair agreement between authors about naming the various regions. With all these caveats, use Fig. 12.5 to estimate which of the various flow regimes shown on it appear on Fig. 12.2, and at which values of Q_g / Q_l the boundaries between regimes occur.

12.5. Wilkes [9] summarizes the available relations for estimating ε and dP / dz for the various flow regimes shown in Figs. 12.3 and 12.5, showing logical bases and references.
(*a*) For the bubble flow region he suggests

$$\varepsilon = \frac{Q_g}{Q_g + Q_l + V_b A} \tag{12.I}$$

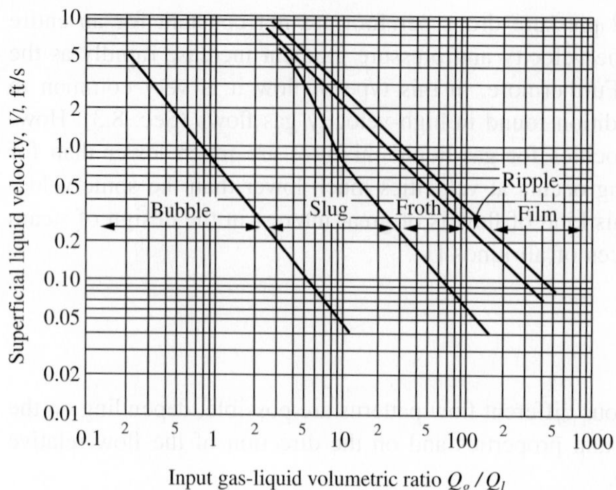

FIGURE 12.5
Flow map for co-current air-water flow in upflow in vertical pipes. All such flow maps are approximate, and some contradict each other. (From G. W. Govier, B. A. Radford, and J. S. C. Dunn, *Can. J. Chem. Eng.* *35*:58–70 (1957). Reproduced by permission of the publisher.)

where V_b is the terminal rising velocity for individual bubbles, and

$$\frac{-dP}{gh\rho_{\text{water}}} = (1 - \varepsilon) \tag{12.J}$$

Test these relationships for the data in Fig 12.2 at $(Q_g / Q_l) = 1.0$, which is in the bubble flow region. From that figure, at this value of Q_g / Q_l, I read the two curves as 0.58 and 1.5. V_b varies with bubble size, but for a wide range of sizes it is $\approx$25 cm / s $\approx$ 0.8 ft / s.

(*b*) For the slug flow region, Wilkes suggests

$$\varepsilon = \frac{Q_g}{1.2(Q_g + Q_l) + 0.35A\sqrt{gD}} \tag{12.K}$$

Where D is the diameter of the tube. The pressure drop should be somewhat larger than that estimated by Eq. 12.K, but no simple relation exists. Test these relations for the data in Fig. 12.2 at $(Q_g / Q_l) = 5.0$ which is in the slug flow region. From that figure, at this value of Q_g / Q_l, I read the two curves as 0.45 and 2.5.

Wilkes also suggests an estimation method for annular flow, which is too long to include here.

12.6. If we wished to perform the same liquid lifting task as shown in Example 12.1, we could use the apparatus sketched in Fig. 12.6. For such an apparatus, with the flow rates given in Example 12.1, calculate the necessary pressures at the inlets to the turbine and to the pump (assume that these pressure are the same). Compare these with the necessary pressure at the base of the gas-liquid column in Example 12.1. Assume that the overall vertical elevation change is 20 ft for both gas and liquid in both cases. Q_g is based on the inlet pressure.

12.7. For the coffee percolator sketched in Fig. 2.17, estimate on the basis of B.E. the maximum possible mass flow rate per unit area of the steam-water mixture in the riser tube.

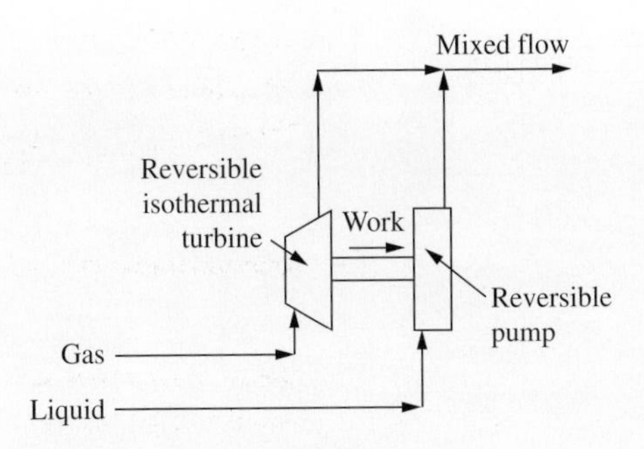

FIGURE 12.6
A turbine-pump equivalent of the device in Fig. 12.1.

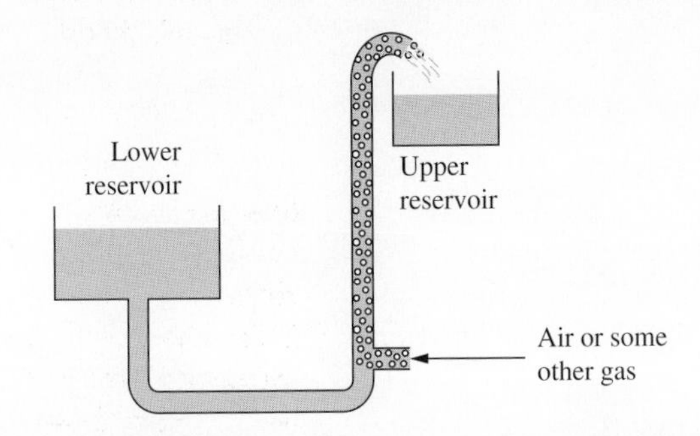

FIGURE 12.7
Air lift or gas lift.

12.8. Figure 12.7 shows the flow diagram of an air-lift or gas-lift pump. In it a fluid is pumped from the lower to the upper reservoir by way of a vertical pipe into the bottom of which air or gas is introduced. Explain why such pumps are widely used in the processing of the highly radioactive solutions that occur in nuclear fuel processing. Suggest other applications.

12.9. For the pump in Fig. 12.7 sketch a plot of liquid flow rate versus air flow rate for constant geometry and fluid properties. *Hint:* See Fig. 12.2.

12.10.*In any gas-liquid flow, pressure drop will cause the gas to expand, causing Q_g to increase in the flow direction. This effect is very significant in a boiling mixture, but can be significant in any such flow. For the flow in Example 12.1, how much does Q_g change from the inlet to the outlet of the test section? Show the value of $(Q_g)_{top} / (Q_g)_{bottom}$. Assume zero evaporation, so that $\dot{m}_l$ and $\dot{m}_g$ are constant through the test section.

REFERENCES FOR CHAPTER 12

1. Tilton, J. N. "Fluid and Particle Dynamics." In *Perry's Chemical Engineers' Handbook.* 7th ed. Ed. R. H. Perry, D. W. Green, J. O. Maloney. New York: McGraw-Hill, 1997, pp. 6–52.
2. Govier, G. W, B. A. Radford, J. S. C. Dunn. "The Upward Vertical Flow of Air-Water Mixtures." *Can. J. Chem. Eng. 35,* (1957) pp. 58–70. See also later articles by Govier and his co-workers in the same journal: *36,* 195–202, and *38,* 62–66.
3. Hughmark, G. A., B. S. Pressburg. "Holdup and Pressure Drop with Gas-Liquid Flow in a Vertical Pipe," *AIChE J. 7,* (1961), pp. 677–682.
4. Delhaye, J. M. "Two Phase Flow Patterns." In *Two-Phase Flow and Heat Transfer in the Power and Process Industries,* New York: Hemisphere-McGraw-Hill, 1981.
5. Barnea, D., Y. Taitel. "Flow Pattern Transition in Two-Phase Gas-Liquid Flows." In *Encyclopedia of Fluid Mechanics, Volume 3, Gas-Liquid Flows.* ed. N. P. Cheremisinoff, Houston: Gulf Publishing, 1986.
6. Scott, D. S. "Properties of Cocurrent Gas-Liquid Flow." *Advan. Chem Eng.* 4th ed. Ed. T. B. Drew, J. W. Hoopes, Jr. and T. Vermeulen. New York: Academic Press, 1963.
7. Lockhart, R. W., R. C. Martinelli. "Proposed Correlation of Data for Isothermal Two-Phase, Two-Component Flow in Pipes." *CEP 45,* (1949), pp. 39–48.
8. Wallis, G. B. *One-Dimensional Two-Phase Flow,* New York: McGraw-Hill, 1969, p. 143.
9. Wilkes, J. O. *Fluid Mechanics for Chemical Engineers.* Upper Saddle River, NJ: Prentice-Hall, 1999, Sec. 10.4.

CHAPTER 13

NON-NEWTONIAN FLUID FLOW IN CIRCULAR PIPES

The introduction to this topic is given in Sec. 1.5.3. In this brief chapter some comparisons will be made between the behavior of Newtonian and non-Newtonian fluids in pipe flow, and references will be given for the student who wishes to pursue the subject further.

13.1 THE ROLE OF STRUCTURE IN NON-NEWTONIAN BEHAVIOR

Almost all non-Newtonian fluids contain suspended particles or dissolved molecules that are large compared with the size of typical fluid molecules (a typical polymer molecule may be many thousands of times as large as a water molecule). Most non-Newtonian behavior is believed to be associated with the "long-range structure" due to such larger constituents, where "long-range" implies long compared with the diameter of a small molecule like water. For example, a Bingham fluid is assumed to have a three-dimensional structure that will resist small shearing stresses but that comes apart when subjected to a stress higher than its yield stress. Pseudoplastic fluids (by far the most common type of non-Newtonian fluid) mostly have dissolved or dispersed particles (e.g., dissolved long-chain molecules or suspended small particles) that have a random orientation in the fluid at rest but that line up when the fluid is sheared. They offer more resistance to deformation in the random position, so the viscosity drops as they become aligned. Dilatant fluids are almost all slurries of solid particles in which there is barely enough liquid to keep the solid particles from touching each other. Their behavior is explained by assuming that at low shear rates the fluid between the particles is able to lubricate the sliding of one particle past another but that at high shear rates this lubrication breaks down.

Thixotropic fluids are assumed to have alignable particles, like pseudoplastic fluids (most thixotropic fluids are pseudoplastic), but with a finite time required for the particles to become aligned with the flow. An additional factor in thixotropic behavior is probably the existence of weak bonds between molecules (e.g., hydrogen bonds or entanglements of polymer chains). The bonds are gradually destroyed by shearing (some authors suggest that ordinary pseudoplastic fluids are really thixotropic fluids whose particles align or whose bonds break much faster than can be observed on currently available viscometers). Rheopectic fluids are rare and generally only show rheopectic behavior under very mild shearing. It has been suggested that mild shearing may help particles in the fluid to fit together better, thus forming a tighter structure and increasing the viscosity. Viscoelastic fluids normally contain long-chain molecules, that can exist in coiled or extended forms and that can connect one to another. When stretched, these molecules straighten out but, when the flow stops, they tend to revert to their coiled position, causing the elastic behavior.

These descriptions are in accord with most observed behavior of these fluids and thus provide a mental picture of what is going on within the fluid. However, they are by no means rigorous descriptions of the microscopic internal behavior of such fluids.

13.2 MEASUREMENT AND DESCRIPTION OF NON-NEWTONIAN FLUIDS

Much of the past and present research in non-Newtonian fluids has consisted of measuring their stress-rate-of-strain curves (such as Fig. 1.6) and seeking mathematical descriptions of these curves. The study of the flow behavior of materials is called *rheology* (from Greek words meaning "the study of flow"), and diagrams like Fig. 1.6 are often called *rheograms*. Non-Newtonian fluids also have two- and three-dimensional behavior quite different from the behavior of Newtonian fluids in similar circumstances. Viscoelastic and time-dependent fluids can have truly bizarre behavior. This chapter confines itself to steady, one-dimensional pipe flows of two very common types of non-Newtonian fluids. Much more is said about the other behaviors in [1–4].

As shown in Sec. 1.5.3, the basic definition of viscosity is in terms of the sliding-plate experiment shown in Fig. 1.4. For Newtonian fluids it was shown in Sec. 6.3 (Example 6.2) that the viscosity could be determined easily by a capillary-tube viscometer. It can be shown both theoretically and experimentally that the viscosity determined by such a viscometer for a Newtonian liquid is exactly the same as the viscosity one would determine on a sliding-plate viscometer. Since capillary-tube viscometers are cheap and simple to operate, they are widely used in industry for Newtonian fluids.

For non-Newtonian fluids that are not time-dependent or viscoelastic, it is possible to convert capillary-viscometer measurements to the equivalent sliding-plate measurements, but this involves some mathematical manipulations. For time-dependent (e.g., thixotropic) fluids, this does not seem to be possible. Thus, most studies of the behavior of non-Newtonian fluids use some variant of the sliding-plate viscometer. The most common is the concentric-cylinder viscometer; see Fig. 1.5. Cone-and-plate viscometers are also widely used, but not discussed here [2, p. 517]. In reading the non-Newtonian literature, observe that most authors use μ as the symbol for viscosity only for Newtonian fluids; η is used as the symbol for viscosity of non-Newtonian fluids.

The data from a plot such as Fig. 1.6 can be used more easily if they can be represented by an equation. The Bingham fluid can be easily represented by

$$\tau \leq \tau_{\text{yield}}; \frac{dV}{dy} = 0; \qquad \tau \geq \tau_{\text{yield}}; \tau = \tau_{\text{yield}} + \mu_0 \frac{dV}{dy} \tag{13.1}$$

where μ_0 is the slope of the straight line on Fig. 1.6. Table 13.1 shows some experimental values for fluids that can be represented reasonably well by Eq. 13.1.

From this table we see that the examples are common substances with which we are familiar. The yield stresses are small; compare them to that of steel at the bottom of the table.

In many cases the experimental curves for both dilatant and pseudoplastic fluids can be reasonably well represented by the *power law,* also called the *Ostwald–de Waele equation:*

$$\tau = K\left(\frac{dV}{dy}\right)^n \tag{13.2}$$

Here K and n are constants whose values are determined by fitting experimental data. For Newtonian fluids $n = 1$ and $K = \mu$. For pseudoplastic fluids n is less than 1, and for dilatant fluids it is greater than 1. The power law has little theoretical basis; its virtues are that it represents a considerable amount of experimental data with reasonable accuracy and that it leads to relatively simple mathematics. Table 13.2 shows some experimental values for fluids that can be represented reasonably well by Eq. 13.2.

TABLE 13.1
Parameters for Bingham plastics*

Material	Yield stress, τ_{yield} (Pa)	Plastic viscosity μ_0(kg / m · s)
Catsup (30°C)	14	0.08
Mustard (30°C)	38	0.25
Margarine (30°C)	51	0.72
Mayonnaise (30°C)	85	0.63
Toothpaste	200	10
Paint	8.7	0.095
Steel (for comparison)	$20 \text{ to } 50 \cdot 10^7$	—

*Assume 20°C unless another temperature is specified. These values are approximate; repeated testing would find similar but not identical values.

TABLE 13.2
Parameters for power law fluids*

Material	n (dimensionless)	K (kg / m · s^{2-n})
Applesauce (24°C)	0.41	0.66
Banana puree (24°C)	0.46	6.5
Human blood (body T)	0.89	0.00384
Soups and sauces	0.51	3.6 to 5.6
Tomato juice (5.8% solids, 32°C)	0.59	0.22
Concentrated tomato juice (30% solids, 32°C)	0.40	18.7
4% Paper pulp in water	0.575	20.7
33% Powdered lime in water	0.171	7.16
23.3% Clay in water	0.229	5.56
0.67% Carboxymethylcellulose in water	0.716	0.30
15% Carboxymethylcellulose in water	0.554	3.13
10% Napalm in kerosine	0.52	4.28
54.3% Portland cement in water	0.153	2.51
50% Powdered coal in water	0.2	0.58
Water (for comparison)	1.00	0.001002

*Assume 20°C unless another temperature is specified. These values are approximate; repeated testing would find similar but not identical values.

Again we see some familiar materials, but also slurries of paper pulp, lime, clay, portland cement, and coal, and also one solution (napalm in kerosine). Comparing these with water (bottom of the table), we see that for all of these materials K is substantially larger than that of water; at almost any shear rate these materials are more viscous than water.

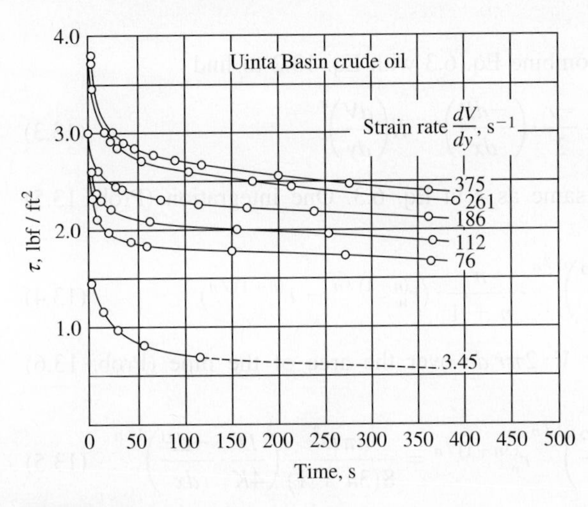

FIGURE 13.1
Stress-time curve for various strain rates for a typical thixotropic fluid, obtained in an apparatus like that shown in Fig. 1.5. This type of crude oil is waxy and forms a slush of solid wax and liquid oil slightly above its melting point, as shown here. (Courtesy of the late E. B. Christiansen.)

Many other equations have been used to represent these stress strain-rate curves (see Probs. 13.2 and 13.3). These two simple ones lead to fairly simple mathematics and allow us to describe the flow of many interesting fluids in pipes with reasonable accuracy.

For time-dependent fluids (thixotropic or rheopectic) there are no simple relations now available for showing the stress-strain-rate-time dependence. Figure 13.1 is a typical stress-time curve for a thixotropic fluid, showing lines of constant strain rate (dV/dy). The change with time occurs mostly in the first 60 s, after which the change with time is minor. For most engineering applications it would be safe to treat this fluid as a simple pseudoplastic fluid with properties corresponding to those of the right-hand side of Fig. 13.1.

For viscoelastic fluids no simple relations are known at all, and current thought is that it may never be possible to describe these fluids adequately by simple scalar equations but only by tensorial equations [1].

13.3 STEADY LAMINAR FLOW OF NON-NEWTONIAN FLUIDS IN HORIZONTAL CIRCULAR TUBES

Most fluids with pronounced non-Newtonian behavior have such high viscosities that in most industrially interesting situations their flow is laminar. This section considers only laminar, steady, pipe flow, which is the non-Newtonian flow problem that chemical engineers are most likely to encounter in their work. We saw in Sec. 6.3 that for any fluid, in laminar or turbulent flow, the shear stress at any point in a horizontal circular pipe for steady, horizontal flow is given by Eq. 6.3:

$$\tau = \frac{-r(P_1 - P_2)}{2\,\Delta x} = \frac{-r}{2} \cdot \left(\frac{-dP}{dx}\right) \tag{6.3}$$

For laminar flow of Newtonian fluids we substituted Newton's law of viscosity for the shear stress and integrated twice to find Poiseuille's equation.

13.3.1 Power Law

For a power-law fluid we can combine Eq. 6.3 with Eq. 13.2 to find

$$\tau = \frac{-r}{2} \cdot \left(\frac{-dP}{dx}\right) = K\left(\frac{dV}{dy}\right)^n \tag{13.3}$$

Here the y in Eq. 13.3 is the same as r in Eq. 6.3. One integration (Prob. 13.5) produces

$$V = \left(\frac{1}{2K} \cdot \frac{-dP}{dx}\right)^{1/n} \cdot \frac{n}{n+1} \cdot \left(r_w^{(n+1)/n} - r^{(n+1)/n}\right) \tag{13.4}$$

Then an integration of $V\,dA = V \cdot 2\pi r\,dr$ over the area of the pipe (Prob. 13.6) produces

$$Q = \frac{n\pi}{3n+1}\left(\frac{1}{2K} \cdot \frac{-dP}{dx}\right)^{1/n} r_w^{(3n+1)/n} = \frac{n\pi D^3}{8(3n+1)}\left(\frac{D}{4K} \cdot \frac{-dP}{dx}\right)^{1/n} \tag{13.5}$$

where r_w is the radius of the tube or pipe and $D = 2r_w$.

Example 13.1. We wish to pump applesauce at an average velocity of 1 ft / s in a 0.5 ft I.D. pipe. What pressure gradient ($-dP/dx$) will we need? For comparison, what pressure gradient would we need for water?

Here we assume that the flow is laminar (to be checked later). To use Eq. 13.5 we need

$$Q = VA = 1\frac{\text{ft}}{\text{s}}\cdot\frac{\pi}{4}(0.5\,\text{ft})^2 = 0.196\frac{\text{ft}^3}{\text{s}} = 0.00556\frac{\text{m}^3}{\text{s}} \tag{13.A}$$

Then we solve Eq. 13.5 for $(-dP/dx)$,

$$\frac{-dP}{dx} = \left(Q\cdot\frac{8(3n+1)}{n\pi D^3}\right)^n\frac{4K}{D} \tag{13.6}$$

Table 13.2 gives $n = 0.41$, $K = 0.66\,(\text{kg}/\text{m}\cdot\text{s}^{2-n})$ for applesauce. It will be easier to work in SI, so we note that $D = 0.5\text{ ft} = 0.152\text{ m}$. Then

$$\frac{-dP}{dx} = \left(0.00556\frac{\text{m}^3}{\text{s}}\cdot\frac{8(3\cdot 0.41+1)}{0.41\cdot\pi\cdot(0.152\,\text{m})^3}\right)^{0.41}\cdot\frac{4\cdot[0.66\,\text{kg}/\text{m}\cdot\text{s}^{(2-0.41)}]}{0.152\,\text{m}}$$

$$= 61.3\frac{\text{Pa}}{\text{m}} = 0.27\frac{\text{psi}}{100\text{ ft}} \tag{13.B}$$

As described later, this flow is laminar. If we wish to compare the pressure gradient with that of water at the same flow rate in the same pipe, we may look in App. A.4 and see that the required pressure gradient is $\approx$0.03 psi / 100 ft. But at this velocity in this pipe, water would be in turbulent flow. The meaningful comparison is to water at this velocity in laminar flow. From Eq. 6.9 we can compute that the required pressure gradient would be 0.0019 psi / 100 ft. So the laminar flow of applesauce requires 145 times the pressure gradient for water in the same flow if turbulence were suppressed. ■

Applesauce is a highly non-Newtonian fluid; its power law coefficient is 0.41, compared to 1.00 for Newtonian fluids. One consequence of this is that if we repeated this calculation for an average velocity of 2 ft / s, the flow would still be laminar, and the calculated pressure drop would increase by only 33 percent. For laminar flow of a Newtonian fluid, doubling the average velocity doubles the pressure drop, and for turbulent flow doubling the average velocity increases the pressure drop by a factor of 3.5 to 4.

13.3.2 Bingham Plastic

For a Bingham plastic, if the shear stress at some point in the pipe is less than the yield stress, τ_{yield}, then at that point $dV/dy = 0$. From Eq. 6.3 we know that the shear stress τ in steady flow in a circular pipe goes from a maximum at the wall to zero at the center. So for any flow of a Bingham plastic in a circular tube, there must be a central core in which $dV/dy = 0$. If the shear stress at the wall is greater than τ_{yield}, then there will be a region with a velocity gradient between the tube wall and a moving, rod-like center. If the shear stress at the wall is less than τ_{yield}, there will be no flow. You can observe this kind of flow with most products that are dispensed in a tube that you squeeze, e.g., toothpaste, hair goo, all sorts of creams, gels and lotions, and oil paint pigments. When you squeeze such a tube with a very gentle squeeze, there is no flow, i.e., $\tau_{\text{wall}} < \tau_{\text{yield}}$. As you make the squeeze more vigorous, $\tau_{\text{wall}} > \tau_{\text{yield}}$ and the fluid begins to flow. The extruded flowing fluid retains its shape (normally cylindrical) as the fluid is extruded; most of the flow is rod-like, only a small layer near the wall has a velocity gradient. Some fluids in squeeze tubes, e.g., most glues, are not Bingham plastics; when you squeeze those tubes the velocity is highest at the center of the flow, and the ejected fluid forms a semispherical drop.

If we set τ_{yield} equal to the negative of the local shear stress in Eq. 6.3 (opposite directions, opposite signs) we find

$$-\tau_{\text{yield}} = \frac{-r}{2} \cdot \left(\frac{-dP}{dx}\right) = \frac{-D_b}{4} \cdot \left(\frac{-dP}{dx}\right) \tag{13.7}$$

which gives the location of the boundary, D_b, between the rod-like core and the surrounding fluid with a velocity gradient. For the rest of this section we can simplify by observing that in any pipe flow the shear stress at the wall is given by

$$\tau_w = \frac{-r_w}{2} \cdot \left(\frac{-dP}{dx}\right) = \frac{-D}{4} \cdot \left(\frac{-dP}{dx}\right) \tag{13.8}$$

Combining these, we have

$$D_b = \frac{4\tau_{\text{yield}}}{(-dP\,/\,dx)} = \frac{\tau_{\text{yield}}}{\tau_w} D \tag{13.9}$$

We can write the velocity distribution in this flow in two parts. From $r = 0$ (the tube center) to $r = r_b$ the velocity is constant, independent of r. From r_b to the wall the velocity is given by substituting Eq. 13.1 into Eq. 6.3,

$$\tau = \tau_{\text{yield}} + \mu_0 \frac{dV}{dr} = \frac{-r}{2} \cdot \left(\frac{-dP}{dx}\right) = \frac{r}{r_w} \tau_w \tag{13.10}$$

$$\frac{dV}{dr} = \frac{-1}{\mu_0}\left(\frac{r}{r_w}\tau_w - \tau_{\text{yield}}\right) \tag{13.11}$$

Integrating from r_w (where $V = 0$) to r and replacing τ_{yield} by $\tau_w r_b / r_w$ give

$$\begin{aligned} V &= \frac{1}{\mu_0}\left(\frac{\tau_w}{r_w}\frac{r^2}{2} - \tau_{\text{yield}} r\right)\Bigg|_{r_w}^{r} \\ &= \frac{\tau_w r_w}{2\mu_0}\left[1 - \left(\frac{r}{r_w}\right)^2\right] - \frac{\tau_{\text{yield}} r_w}{\mu_0}\left(1 - \frac{r}{r_w}\right) \\ &= \frac{\tau_w r_w}{2\mu_0}\left[1 - \left(\frac{r}{r_w}\right)^2\right] - \frac{\tau_w r_b}{\mu_0} \cdot \left(1 - \frac{r}{r_w}\right) \\ &= \frac{\tau_w r_w}{2\mu_0}\left[1 - \left(\frac{r}{r_w}\right)^2\right] - \frac{2r_b}{r_w} \cdot \left(1 - \frac{r}{r_w}\right) \end{aligned} \tag{13.12}$$

For r less than r_b the velocity is constant, the same as that at r_b found by substituting r_b for r in Eq. 13.12. Simplifying, we write

$$V_{\text{boundary}} = \frac{\tau_w r_w}{2\mu_0}\left(1 - \frac{r_b}{r_w}\right)^2 \tag{13.13}$$

Then we find the volumetric flow rate by

$$Q = \int_0^{r_b} V_{\text{Eq. 13.13}} \cdot 2\pi r\,dr + \int_{r_b}^{r_w} V_{\text{Eq. 13.12}} \cdot 2\pi r\,dr \tag{13.14}$$

After some algebra (Prob. 13.11), we find

$$Q = \frac{\pi r_w^4}{8\mu_0} \cdot \left(\frac{-dP}{dz}\right) \cdot \left[1 - \frac{4}{3}\left(\frac{\tau_{\text{yield}}}{\tau_w}\right) + \frac{1}{3}\left(\frac{\tau_{\text{yield}}}{\tau_w}\right)^4\right] \tag{13.15}$$

The first term on the right of this equation is the same as Poiseuille's equation (Eq. 6.8) for Newtonian fluids. If we set $\tau_{\text{yield}} = \tau_w$ in this equation and solve, we find $Q = 0$, i.e., no flow. For $\tau_{\text{yield}} > \tau_w$ the equation gives meaningless answers because we have assumed in deriving it that $\tau_{\text{yield}} < \tau_w$.

Example 13.2. Repeat Example 13.1 for mustard. From Table 13.1 we find that for mustard $\tau_{\text{yield}} = 38$ Pa and $\mu_0 = 0.25$ kg / m · s. In Example 13.1 we could solve directly for $(-dP / dx)$. Here we can solve Eq. 13.15 for $(-dP / dx)$, but τ_w appears on the right to the first and fourth powers, and from Eq. 13.10 we know that τ_w depends on $(-dP / dx)$. So this problem requires a trial-and-error solution. We begin by computing the value of Q for a specified value of $(-dP / dx)$. Then we repeat (on a spreadsheet or computer program), finding the value of $(-dP / dx)$ that produces the specified value of Q.

As a first step let us assume that $(-dP / dx)$ is the same value as we found in Example 13.1, 61.3 Pa / m = 0.27 psi / 100 ft. Then

$$\tau_w = \frac{D_w}{4} \cdot \left(\frac{-dP}{dx}\right) = \frac{0.153\text{ m}}{4} \cdot 61.3\,\frac{\text{Pa}}{\text{m}} = 2.33\text{ Pa} \tag{13.C}$$

This is less than the 38 Pa yield stress, so for this pressure gradient the mustard will not flow at all. For our next guess we increase the assumed pressure gradient by a factor of 100, making $\tau_w = 233$ Pa and $\tau_{\text{yield}} / \tau_w = 0.163$,

$$Q = \frac{\pi(0.0762\text{ m})^4}{8(0.25\text{ kg / m}\cdot\text{s})} \cdot \left(6130\,\frac{\text{Pa}}{\text{m}}\right) \cdot \left(1 - \frac{4}{3}(0.163) + \frac{1}{3}(0.163)^4\right)$$

$$= 0.254\,\frac{\text{m}^3}{\text{s}} \tag{13.D}$$

This is much larger than the specified $Q = 0.00556$ m^3 / s. Now we use our spreadsheet to find the value of $(-dP / dx)$ that meets the specified flow, finding $(-dP / dx) = 1276$ Pa / m = 5.57 psi / 100 ft. ■

This example shows what we all know, that mustard is harder to flow than applesauce, and that for small pressure gradients mustard will not flow at all. After these two examples we can compare the velocity profiles for laminar flow of these two kinds of fluids with that for a Newtonian fluid. Figure 13.2 shows the velocity profiles

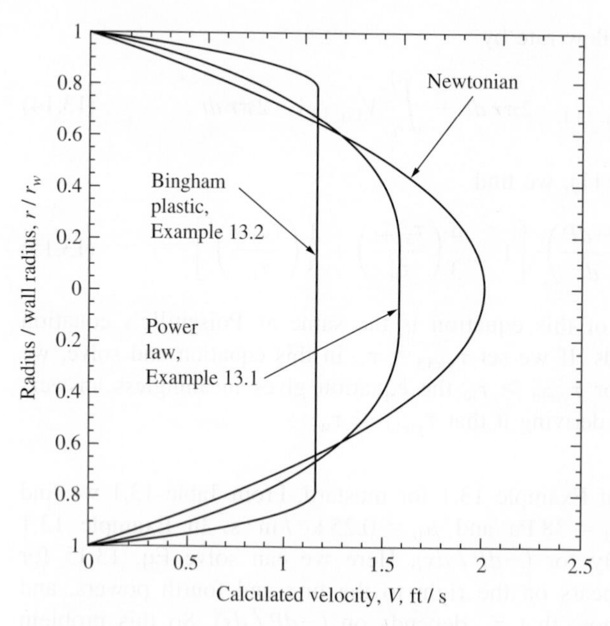

FIGURE 13.2
Calculated velocity distributions for laminar flow in a circular pipe, for flow of three kinds of fluid. In each case $V_{avg} = 1$ ft / s. The power law and Bingham plastic curves correspond to Examples 13.1 and 13.2. The Newtonian curve is based on Poiseuille's equation.

for Examples 13.1 and 13.2, together with the profile for a Newtonian fluid, all with an average velocity of 1 ft / s. We see the central core for the Bingham plastic, over which the velocity is constant, and see that the power law fluid has a velocity profile intermediate between that of the Bingham plastic and the Newtonian fluid. Looking back to Fig. 1.6, we see that the pseudoplastic curve lies between the Newtonian and Bingham plastic curves. As we reduce the value of n in a power law fluid, the velocity profile shown becomes closer and closer to the curve for a Bingham plastic.

13.4 TURBULENT STEADY PIPE FLOW OF NON-NEWTONIAN FLUIDS, AND THE TRANSITION FROM LAMINAR TO TURBULENT FLOW

The previous examples show that there is not one simple description of pipe flow for non-Newtonian fluids, but rather there is a separate description for the two kinds of fluid we consider (and if you study the subject more you will find other descriptions for other types). From this we might surmise that the same is true for turbulent flow, and for deciding when a flow is laminar or turbulent; that surmise is correct.

13.4.1 Power Law

For Newtonian fluids we can expect laminar flow for Reynolds numbers of about 2000 or less. For non-Newtonian fluid we must redefine the Reynolds number. For power-law fluids, the most common procedure is to write Eq. 6.20 (the friction factor form of Poiseuille's equation) as

$$f_{\text{power law}} = 16 / \mathcal{R}_{\text{Power law}} \tag{13.16}$$

Calculating the friction factor from Eq. 13.6, substituting, and solving, we find

$$\mathcal{R}_{\text{Power law}} = \frac{8\rho V_{\text{avg}}^{(2-n)} D^n}{K[2(3n+1)/n]^n} \tag{13.17}$$

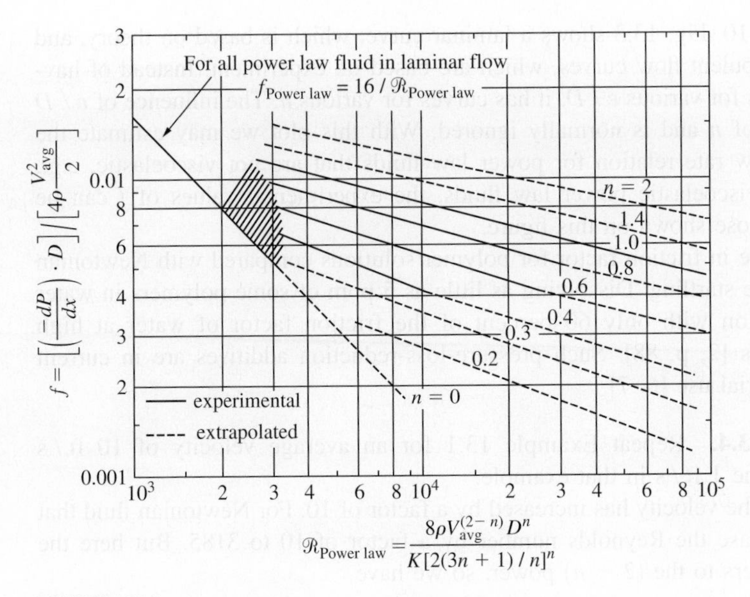

FIGURE 13.3
Friction factor–Reynolds number plot for power-law fluids (see Fig. 6.10 for Newtonian fluids). The friction factor is the same as in Chap. 6, but the normal $\mathcal{R}$ is replaced by $\mathcal{R}_{\text{Power law}}$ as shown in the text. Based on Dodge and Metzner [5]. The solid lines represent measured values, the dashed lines extrapolations.

Many authors further substitute $K = K'[4n/(3n+1]^n$ so that the final expression becomes

$$\mathcal{R}_{\text{Power law}} = D^n \rho V_{\text{avg}}^{(2-n)} / 8^{(n-1)} K' \tag{13.E}$$

Figure 13.3 [5] is the equivalent of Fig. 6.10 (the friction factor Reynolds number plot) for power-law fluids. It shows that the transition region between laminar and turbulent flow occurs at values of $\mathcal{R}_{\text{Power law}}$ between 2000 and 4000. We will discuss the turbulent region later.

Example 13.3. Calculate f and $\mathcal{R}_{\text{Power law}}$ for the flow in Example 13.1. Assume the density of applesauce is 1000 kg/m^3.

The Fanning friction factor for horizontal pipe flow of applesauce in Example 13.1 is

$$f = \frac{(-dP/dx) \cdot D}{4\rho\, V_{\text{avg}}^2 / 2}$$

$$= \frac{(61.3 \text{ Pa/m}) \cdot 0.152 \text{ m}}{4 \cdot (1000 \text{ kg/m}^3) \cdot (0.305 \text{ m/s})^2 / 2} \frac{\text{kg}}{\text{Pa} \cdot \text{m} \cdot \text{s}^2} = 0.0502 \tag{13.F}$$

$$\mathcal{R}_{\text{Power law}} = \frac{8 \cdot (1000 \text{ kg/m}^3)(0.304 \text{ m/s})^{(2-0.41)} (0.152 \text{ m})^{0.41}}{[0.66 \text{ kg/m} \cdot \text{s}^{(2-0.41)}] \cdot [2 \cdot (3 \cdot 0.41 + 1)/0.41]^{0.41}}$$

$$= 318.5 \tag{13.G}$$

The reader may verify that these values satisfy Eq. 13.15. The calculated $\mathcal{R}_{\text{Power law}} < 2000$, so the flow in Example 13.1 is indeed laminar. ■

Like Fig. 6.10, Fig. 13.3 shows a laminar curve, which is based on theory, and transition and turbulent flow curves, which are based on experiment. Instead of having various curves for various ε / D, it has curves for various n. The influence of ε / D is less than that of n and is normally ignored. With this plot we may estimate the pressure drop–flow rate relation for power law fluids that are not viscoelastic, e.g., applesauce. For viscoelastic power law fluids, the experimental values of f can be much less than those shown on this figure.

The decrease in friction factor for polymer solutions compared with Newtonian fluids can be quite startling. Dissolving as little as 5 ppm of some polymers in water produces a solution with only 60 percent of the friction factor of water at high Reynolds numbers [2, p. 88]. Such pressure-loss-reduction additives are in current large-scale industrial use [6, 7].

Example 13.4. Repeat Example 13.1 for an average velocity of 10 ft / s instead of the 1 ft / s in that example.

Here the velocity has increased by a factor of 10. For Newtonian fluid that would increase the Reynolds number by a factor of 10 to 3185. But here the velocity enters to the $(2 - n)$ power, so we have

$$\mathcal{R}_{\text{Power law}} = \frac{8 \cdot (1000\ \text{kg} / \text{m}^3)(3.04\ \text{m} / \text{s})^{(2-0.41)} (0.152\ \text{m})^{0.41}}{[0.66\ \text{kg} / \text{m} \cdot \text{s}^{(2-0.41)}] \cdot [2 \cdot (3 \cdot 0.41 + 1) / 0.41]^{0.41}} = 12{,}390 \tag{13.H}$$

This simply reflects the fact that the definition of $\mathcal{R}_{\text{Power law}}$ is somewhat artificial, meant to force the laminar flow curve into the familiar form. One may also think of this by considering how we defined the apparent viscosity in Fig. 1.7. There, as the velocity increased the viscosity decreased. So in this example (and all power law fluids with $n < 1$) increasing the flow rate increases $\mathcal{R}_{\text{Power law}}$ both by increasing V and by decreasing μ.

Reading Fig. 13.3 for this $\mathcal{R}_{\text{Power law}}$ and $n = 0.41$, we find $f \approx 0.004$, from which

$$\frac{-dP}{dx} = 4f \frac{\rho}{D} \cdot \frac{V_{\text{avg}}^2}{2} = 4 \cdot 0.004 \frac{(1000\ \text{kg} / \text{m}^3)}{0.152\ \text{m}} \cdot \frac{(3.04\ \text{m} / \text{s})^2}{2} \cdot \frac{\text{Pa} \cdot \text{m} \cdot \text{s}^2}{\text{kg}}$$

$$= 0.488 \frac{\text{kPa}}{\text{m}} = 2.16 \frac{\text{psi}}{100\ \text{ft}} \tag{13.I}$$

■

13.4.2 Bingham Plastics

The laminar-turbulent transition and the turbulent flow of Bingham plastics is described by Fig. 13.4, [8], (which requires some explanation!).

We needed to develop a new Reynolds number for power law fluids because there was no obvious way to define a value of μ to substitute in the standard Reynolds number. For Bingham plastics there is, namely, μ_0. Figure 13.4 simply substitutes μ_0 for μ in the Reynolds number. The friction factor is the same as that shown in Fig. 6.10. But there are lines of another parameter on the figure, related to the τ_{yield}.

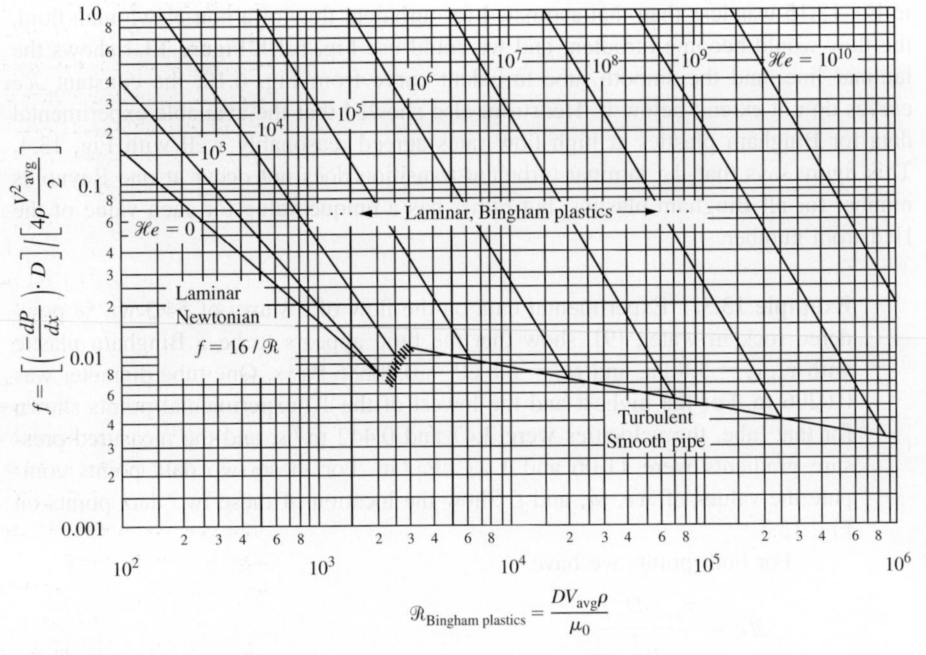

FIGURE 13.4
Friction factor–Reynolds number plot for Bingham plastics (see Fig. 6.10 for Newtonian fluids). The friction factor is the same as in Chap. 6, but the normal $\mathcal{R}$ is replaced by $\mathcal{R}_{\text{Bingham}}$ as shown in the text. Based on Hedstrom [8]. See discussion in the text.

The parameter is

$$\begin{pmatrix}\text{Hedstrom}\\ \text{number}\end{pmatrix} = \mathcal{H}e = \begin{pmatrix}\text{Bingham}\\ \text{number}\end{pmatrix} \cdot \begin{pmatrix}\text{Reynolds}\\ \text{number}\end{pmatrix}$$

$$= \frac{\tau_{\text{yield}} D}{\mu_0 V_{\text{avg}}} \cdot \frac{D V_{\text{avg}} \rho}{\mu_0} = \frac{\tau_{\text{yield}} D^2 \rho}{\mu_0^2} \tag{13.18}$$

The Bingham number is found by the standard methods of dimensional analysis (Chap. 9), as one of the proper correlating parameters for this flow. It is the ratio of the yield stress of the fluid to the viscous stress in a laminar flow (see Table 9.1). It could be used as the additional parameter in Fig. 13.4, but the plot works better and is easier to use if one multiplies the Bingham number by the Reynolds number, producing $\mathcal{H}e$, because $\mathcal{H}e$ depends only on pipe diameter and fluid properties, not on the fluid flow rate, which the Bingham number does.

If we calculate the Reynolds and Hedstrom numbers and the friction factor for Example 13.2, we find 92.9, 14,100, and 4.2. We see that these are above the upper left corner on Fig. 13.4, but if we extrapolate the curves there we find that these values match them. We could repeat Example 13.2 for a variety of flow rates and pipe diameters, and thus make up all the $\mathcal{H}e$ curves on Fig. 13.4. In making up this figure, Hedstrom [8] assumed that when the friction factor computed from the flow rate in Eq. 13.15 was less than that estimated for turbulent flow of a non-Newtonian fluid, the flow would become turbulent and we could use Fig. 6.10. Figure 13.4 shows the laminar line, and the smooth tube turbulent curve from Fig. 6.10; the constant $\mathcal{He}$ curves do not extend below it. Hedstrom also showed that the available experimental data for Bingham plastics at high flow rates agreed reasonably well with Fig. 13.4. This figure says that the laminar-turbulent transition does not occur at one Reynolds number for all Bingham plastics, but rather has a unique value for each value of the Hedstrom number.

Example 13.5. Experimental data on the flow of a slurry of 54.3 wt. % powdered rock in water, [9], show that the fluid appears to be a Bingham plastic with $\tau_{\text{yield}} = 3.8$ Pa, and $\mu_0 = 6.8$ cP $= 0.00686$ Pa · s. One tube diameter was 0.0206 m. For the highest and the lowest of the 10 experimental points shown for that tube, the velocities were 3.47 and 0.442 m / s, and the measured pressure gradients were 11.06 and 1.27 kPa / m. For these two data points compute the values of $\mathcal{He}$, $\mathcal{R}$, and f. Show the location of these two data points on Fig. 13.4.

For both points we have

$$\mathcal{He} = \frac{\tau_{\text{yield}} D^2 \rho}{\mu_0^2}$$

$$= \frac{3.8\ \text{Pa} \cdot (0.0206\ \text{m})^2 \cdot 1530\ \text{kg / m}^3}{(0.00686\ \text{Pa} \cdot \text{s})^2} \cdot \frac{\text{Pa} \cdot \text{m} \cdot \text{s}^2}{\text{kg}} = 52{,}400 \qquad \text{(13.J)}$$

For the first point we have

$$\mathcal{R} = \frac{0.0206\ \text{m} \cdot (3.47\ \text{m / s}) \cdot (1530\ \text{kg / m}^3)}{0.0686\ \text{Pa} \cdot \text{s}} \cdot \frac{\text{Pa} \cdot \text{m} \cdot \text{s}^2}{\text{kg}} = 15{,}900 \qquad \text{(13.K)}$$

and

$$f = \frac{(-dP / dx) \cdot D}{4\rho\, V_{\text{avg}}^2 / 2}$$

$$= \frac{(11{,}069\ \text{Pa / m}) \cdot 0.0206\ \text{m}}{4 \cdot (1530\ \text{kg / m}^3) \cdot (3.47\ \text{m / s})^2 / 2} \cdot \frac{\text{kg}}{\text{Pa} \cdot \text{m} \cdot \text{s}^2} = 0.0062 \qquad \text{(13.L)}$$

For the last point we have $\mathcal{R} = 2030$ and $f = 0.044$.

We can sketch the $\mathcal{He} = 52{,}400$ curve on Fig. 13.4, finding that it crosses the turbulent smooth pipe curve at $\mathcal{R} \approx 5000$, so that for this fluid in this diameter pipe, the laminar-turbulent transition would be expected to occur at $\mathcal{R} \approx 5000$. The first point would be expected to lie on the smooth curve turbulent line. For $\mathcal{R} \approx 15{,}900$ we read that line as $f \approx 0.007$, which is close (but not identical) to the experimental value of 0.0062. For the last point we would expect the friction factor could be calculated exactly the same as in Example 13.2. Repeating that calculation for this flow, we find a calculated friction factor of 0.039, which again is close to but not identical to the experimental value of 0.044. In making up Fig. 13.4, if we drew the $\mathcal{He} = 52{,}400$ curve we would use the calculated value, showing the experimental value slightly below it. ■

13.5 SUMMARY

1. Although intuitive explanations and numerous equations are available to describe the behavior of non-Newtonian fluids, no general, universally applicable theory or equation is currently known. For time-dependent and viscoelastic fluids our knowledge consists mostly of descriptions of observed behavior.
2. For laminar flow of non-Newtonian fluids in circular pipes we can readily calculate the behavior for the power-law and Bingham plastic representations of the non-Newtonian viscosity.
3. For turbulent flow, we have data and correlations for power-law fluids and Bingham plastics, both based on the turbulent flow behavior of Newtonian fluids. The transition Reynolds number for power-law fluids is approximately the same as for Newtonian fluids, with a redefined Reynolds number. For Bingham plastics the transition Reynolds number depends on the τ_{yield} of the fluid, in addition to the Reynolds number. Some dilute polymer solutions have surprisingly low friction factors.
4. The behavior of non-Newtonian fluids is currently a very active research topic. More detailed summaries of results to date can be found in [1, 10].

PROBLEMS

See the Common Units and Values for Problems and Examples inside the back cover. An asterisk (*) on a problem number indicates that the answer is shown in App. D.

13.1. Show that for $n = 1$, the power law fluid becomes a Newtonian fluid. Show that for $\tau_{\text{yield}} = 0$, a Bingham plastic becomes a Newtonian fluid.

13.2. For pseudoplastic fluids (the most common types of non-Newtonian fluid), the fluid frequently appears to be a Newtonian fluid with very high viscosity (μ_0) at low shear rates and then again to be a Newtonian fluid of lower viscosity (μ_∞) at higher shear rates, with a transition between, as sketched in Fig. 13.5.

(*a*) Show that the Reiner-Phillipoff equation,

$$\tau = \left(A + \frac{B - A}{1 + (\tau / C)^2}\right)\frac{dV}{dy} \tag{13.19}$$

where *A*, *B*, and *C* are data-fitting constants, corresponds to this behavior, and show what constants in that equation correspond to μ_0 and μ_∞.

(*b*) Repeat part (*a*) for the Carreau equation,

$$\mu = \mu_\infty + \frac{\mu_0 - \mu_\infty}{[1 + (\lambda\, dV / dy)^2]^p} \tag{13.20}$$

where λ and p are data-fitting constants, based on experimental data.

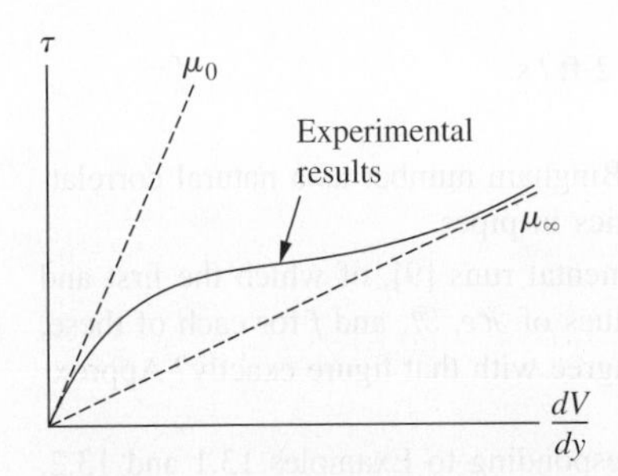

FIGURE 13.5
Behavior of some pseudoplastic fluids, showing the two limiting viscosities.

13.3. Show that the Ellis equation,

$$\tau = \frac{dV / dy}{A + B\tau^C} \tag{13.21}$$

where A and B are data-fitting constants, corresponds, practically, to a Newtonian fluid at low shear rates and a power-law fluid at high shear rates. Show what constant or combination of constants in the Ellis equation corresponds to μ_0 in Fig. 13.5.

13.4.*The data for 200 s in Fig. 13.1 can be reasonably represented by a power-law expression. Find the constants in that expression. *Hint:* The power law can be represented as a straight line with slope n on logarithmic paper.

13.5. Show the integration of Eq. 13.3 to find Eq. 13.4. Use a definite integration from r_w to r.

13.6. Show the integration of Eq. 13.4 over the cross-sectional area of the pipe to find Eq. 13.5.

13.7. Show that for a Newtonian fluid ($n = 1$ and $K = \mu$), Eq. 13.4 becomes the same as Eq. 6.8 and that Eq. 13.5 becomes the same as Poiseuille's equation (Eq. 6.9).

13.8. Set up the spreadsheet to solve Example 13.1. Verify the pressure gradient found in that example. Use the spreadsheet to solve the problem for $V_{avg} = 2$ ft / s. Compare the result to the statement in the paragraph following the example.

13.9. Verify the statements in Example 13.1 about the pressure gradient for water in the same pipe at the same velocity for laminar and for turbulent flow.

13.10. Table 13.2 for power law fluids shows no values of $n > 1.00$, i.e., no dilatant fluids. The reason is that these are scarce; no values for such fluids appear in the published tables from which Table 13.2 was prepared. However, if such a fluid exists, it must follow the treatment in Sec. 13.3.1. Suppose that a new form of applesauce is discovered, which has the same K as in Table 13.2 but has $n = 1.5$.

(*a*) Using the spreadsheet you developed in Prob. 13.8, repeat Example 13.1 for this new form of applesauce.

(*b*) Compare the required pressure gradient with that shown in Example 13.1.

(*c*) Sketch the curve on Fig. 13.2 for this new form of applesauce.

13.11. Show the following:

(*a*) The integration of Eq. 13.11 to find Eq. 13.12.

(*b*) The algebra leading from Eq. 13.12 to Eq. 13.13.

(*c*) The integration leading to Eq. 13.15.

13.12. Sketch the value of $\left[1 - \frac{4}{3}(\tau_{yield} / \tau_w) + \frac{1}{3}(\tau_{yield} / \tau_w)^4\right]$ as a function of τ_{yield} / τ_w for values from 0 to 2. Comment on the meaning of the values for $\tau_{yield} / \tau_w < 1$, $= 1$, and > 1.

13.13.*Set up the spreadsheet to solve Example 13.2. Verify the calculated pressure gradient in that solution. Use the spreadsheet to solve the problem for $V_{avg} = 2$ ft / s.

13.14. Show the algebra between Eq. 13.16 and 13.E.

13.15. Repeat Example 13.3 for an average velocity of 2 ft / s.

13.16.*Repeat Example 13.4 for $V_{avg} = 5$ ft / s.

13.17. Show the dimensional analysis that leads to the Bingham number as a natural correlating parameter for laminar flow of Bingham plastics in pipes.

13.18. Table 13.3 shows the complete set of 10 experimental runs [9], of which the first and last are shown in Example 13.5. Compute the values of $\mathcal{He}$, $\mathcal{R}$, and f for each of these, and plot them in the form of Fig. 13.4. Do they agree with that figure exactly? Approximately? Not at all?

13.19.*Figure 13.2 shows the velocity distributions corresponding to Examples 13.1 and 13.2. To verify that it is made correctly, calculate the following:

(*a*) The maximum (centerline) velocity and the velocity at $r / r_w = 0.5$ in Example 13.1.

(*b*) The value of r_b / r_w and the velocity at that point in Example 13.2.

TABLE 13.3
Experimental data for flow of a 54.3 wt. % rock slurry in water, in an 0.812 in = 2.06 cm diameter pipe*

Run number	V_{avg}, m / s	$(-dP / dx)$, kPa / m
23-1	3.47	11.06
23-2	3.23	10.01
23-3	2.97	8.43
23-4	2.26	5.40
25-5	1.81	3.50
23-6	1.37	2.04
23-7	0.893	1.41
23-8	1.20	1.59
23-9	0.360	1.10
23-10	0.442	1.27

*From [9].

13.20. One of the most economical ways to transport large quantities of small solid particles is via a slurry pipeline [10]. The Black Mesa Pipeline [11] transports 660 tons / h of coal ground to 8 mesh (2.4 mm) as a 50 wt. % slurry in water with an estimated SG of 1.26 for 273 mi across northern Arizona. The pipeline ID is 18 in, and the slurry flow rate 4200 gal / min. There are four pumping stations, which divide the 273 mi of pipeline into four sections. The slurry behaves as a power law fluid, as shown in Table 13.2. Estimate:

(*a*) The pressure drop in a section (273 / 4) mi long.

(*b*) The pumping power required for all four stations, assuming 70 percent efficient pumps.

(*c*) The total electric cost per ton · mile of coal, with electricity at \$0.05 / kwh. How does this compare with typical railroad rates of \$0.01 / ton · mile for unit trains on long hauls? Is comparing electric cost only to railroad cost a fair comparison? If not, why not?

REFERENCES FOR CHAPTER 13

1. Schowalter, W. R. *Mechanics of Non-Newtonian Fluids.* New York: Pergamon, 1978.
2. Bird, R. B., R. C. Armstrong, O. Hassager. *Dynamics of Polymeric Liquids, 1, 2,* New York: John Wiley, 1986.
3. Bird, R. B., E. N. Stewart, W. E. Lightfoot. *Transport Phenomena.* 2nd ed. New York: Wiley, 2002.
4. Morrison, F. *Understanding Rheology.* Oxford: Oxford University Press, 2001.
5. Dodge, D. W., A. B. Metzner. "Turbulent Flow of Non-Newtonian Systems," *AIChEJ. 5,* (1959), pp. 189–205.
6. Berman, N. S. "Drag Reduction by Polymers." *Ann. Rev. Fluid Mech. 10,* (1978), pp. 47–64.
7. Savins, J. G. "Drag Reduction Characteristics of Solutions of Macromolecules in Turbulent Pipe Flow." *Soc. Petr. Eng. J. 4* 203–214 (1964), pp. 203–214.
8. Hedstrom, B. O. A. "Flow of Plastic Materials in Pipes." *Ind. Eng. Chem. 44* 651–656 (1952), pp. 651–656.
9. Wilhelm, R. W., D. M. Wroughton, W. F. Loeffel. "Flow of Suspensions Through Pipes." *Ind. Eng. Chem. 31* (1939), pp. 622–629.
10. Shook, C. A., M. C. Roco. *Slurry Flow; Principles and Practice,* Boston: Butterworth-Heinemann, 1991.
11. Anon. "Potential Energy Sources Pose Mining Problems." *Chem. Eng. News 52* (15) (1974), pp. 16–17.

CHAPTER 14

SURFACE FORCES

The introduction to this chapter begins with Sec. 1.5.5. There we discussed how surface forces arise. Such forces are present in any system in which there is a two-phase interface, i.e., solid-liquid, solid-gas, liquid-gas, or liquid-liquid. Thus, they are present in all of the examples treated so far in this text; in those examples they are generally small and can be neglected without measurable error. But they must be taken into account in very small systems or in systems in which other forces are small or zero. To see why surface forces are important in small systems, let us consider the pressure difference due to surface tension in a spherical drop or bubble. (Surface tension values are in Table 1.1).

Figure 14.1 shows such a drop, cut in half along its equator. If the system is at rest, then the surface force, which tends to shrink the bubble, is equal and opposite to the pressure-difference force, which tends to expand it. Here $\Delta P = P_{\text{inside}} - P_{\text{outside}}$. Equating these two forces and solving for ΔP, we find

$$\Delta P = \frac{4\sigma}{D} \tag{14.1}$$

As long as D is large, this pressure difference is negligible. For example, if

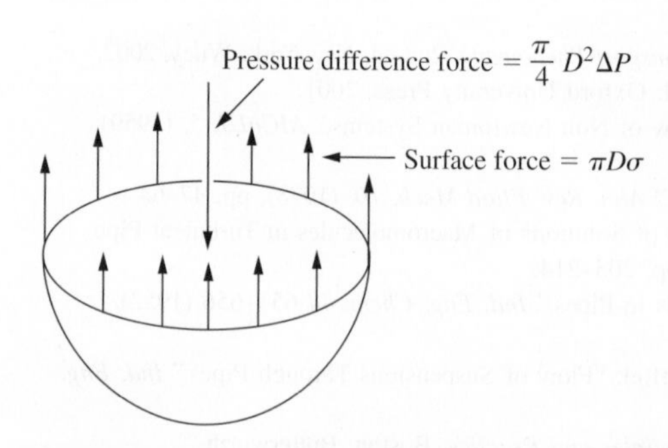

FIGURE 14.1
Pressure difference due to surface forces.

this is a drop of water at 20°C = 68°F with a 1 in diameter, then

$$\Delta P = \frac{4 \cdot 0.000415 \text{ lbf / in}}{1 \text{ in}} = 0.00166 \text{ psi} = 11.4 \text{ Pa} \tag{14.A}$$

which is negligible for most engineering applications. But if the drop has a diameter of 10^{-5} in, then the pressure difference is 166 psi = 1.14 MPa, with is seldom negligible.

One may appreciate why surface forces become important in small systems by observing that the surface forces are proportional to the diameter of the body, whereas the pressure forces are proportional to the projected area (i.e., proportional to the diameter squared), and the gravity and inertia forces are proportional to the volume (i.e., the diameter cubed). For other shapes some other dimension replaces the diameter, but the idea is the same. Hence if we hold the shape of a system constant and increase all its dimensions, the inertia and gravity forces grow fastest, the pressure force at an intermediate rate, and the surface forces slowest.

Surface tension forces are also important in problems in which the other forces are negligible. For example, in an ordinary liquid-storage tank on earth the shape and position of the liquid is determined by gravity and pressure forces, as discussed in Chap. 2. However, in rockets and earth satellites, which have zero gravity, the position and shape of a liquid in a tank is largely determined by surface forces [1, 2].

14.1 SURFACE TENSION AND SURFACE ENERGY

Surface tension, as discussed in Sec. 1.5.5, has the dimensions of force / length, lbf / in, or N / m. (Historically, surface tensions have been reported in handbooks in dyne / cm; 1 dyne / cm = 0.001 N / m = $5.7 \cdot 10^{-6}$ lbf / in.) However, if we multiply the top and bottom of this ratio by length, we obtain (force · length) / length^2, e.g., ft · lbf / ft^2 or J / m^2, which is equivalent to energy / area. Thus, one may conclude that the surface tension is connected with the surface energy per unit area. However, it is an experimental fact that as we prepare a film, as is shown in Fig. 1.11, and stretch it by increasing the weight of the frame, the film cools. Some of the internal energy of the fluid is used to create new surface. We may visualize this as the fluid below the surface making new surface by pushing molecules from the bulk into the surface. Thus, the surface energy is not merely the energy put into the system by the external force moving through the distance but also the energy derived from the bulk fluid, which suffers a decrease in internal energy. If the process is performed slowly, so that external heat may flow in to hold the temperature constant, then the increase in surface energy is equal to the work done ($F\,dx = \sigma\,dA$) plus the heat added. It may be shown [3] from free-energy arguments that the correct expression for the surface energy is

$$\frac{E_s}{A} = \sigma - T\frac{d\sigma}{dT} \tag{14.2}$$

This equation is equivalent to the statement that the surface tension is exactly equal to the Helmholz free energy per unit area. Thus, we may make use of the thermodynamic equilibrium statement that isolated systems tend toward the condition of lowest free energy to show that the stable state of a system is the one with minimum free energy, including the contribution of the surface free energy.

In most fluid mechanics problems it is simplest to think of surface tension as a force per unit length, but in problems involving several liquids and a solid it is sometimes more convenient to work in terms of surface energy. It is hard to visualize a solid having surface tension but easier to visualize it having surface energy.

For almost all pure liquids, the surface tension decreases with increasing temperature, becoming zero at the critical point.

14.2 WETTING AND CONTACT ANGLE

Before discussing how we measure surface tension we must discuss wetting. Fluids wet some solids and do not wet others. Figure 14.2 shows some of the possible wetting behaviors of a drop of liquid placed on a horizontal, solid surface (the remainder of the surface is covered with air, so there are two fluids present).

Figure 14.2(*a*) represents the case of a liquid that spontaneously wets a solid surface well, e.g., water on a very clean glass or mercury on very clean copper. The angle θ shown is the angle between the edge of the liquid surface and the solid surface, measured inside the liquid. This angle is called the *contact angle* and is a measure of the quality of the wetting. For perfect wetting, in which the liquid spreads as a thin film over the surface of the solid, θ is zero.

Figure 14.2(*b*) shows the case of moderate wetting, in which θ is less than 90° but not zero. This might be observed for water on dirty glass or mercury on a slightly oxidized copper.

Figure 14.2(*c*) represents the case of no wetting. If there were exactly zero wetting, θ would be 180°. However, the gravity force on the drop flattens the drop, so that a 180° angle is never observed. This might represent water on Teflon or mercury on clean glass. Many surface treatments, e.g., modern breathable rainwear and stain-resistant fabrics, make the fabric surface repellent to water, so that a water drop on them looks like Fig. 14.2(*c*).

We normally say that a liquid wets a surface if θ is less than 90° and does not wet it if θ is more than 90°. Values of θ less than 20° are considered strong wetting, and values of θ more than 140° are examples of strong nonwetting.

Most of the methods of measuring surface and interfacial tensions described in the next section assume that the liquid wets some part of the apparatus perfectly; that is, $\theta = 0$.

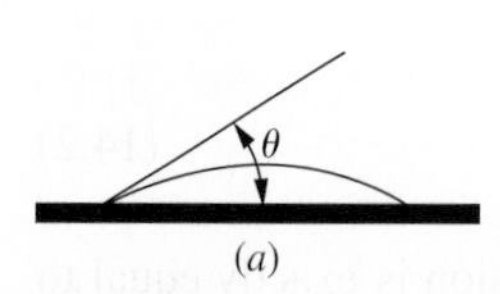

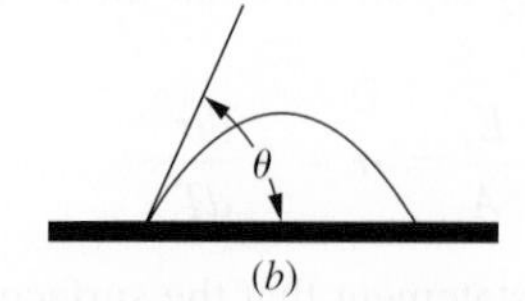

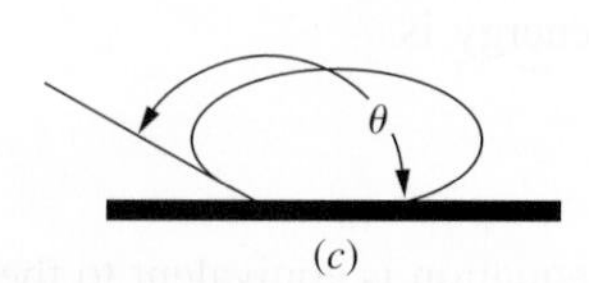

FIGURE 14.2
Wetting and contact angle.

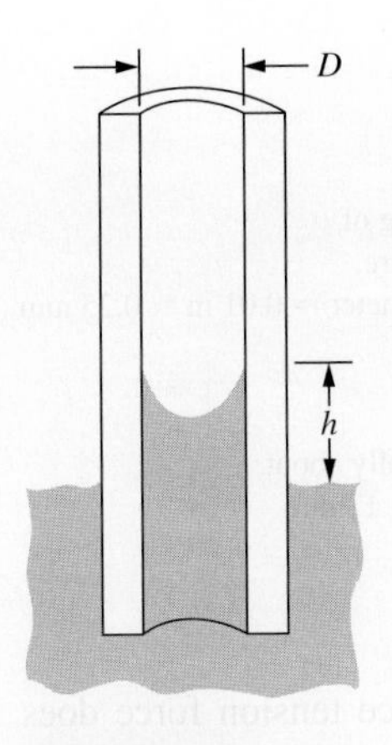

FIGURE 14.3
Capillary rise in a circular tube.

14.3 THE MEASUREMENT OF SURFACE TENSION

One of the simplest measurements of surface tension is by means of capillary rise. A small-diameter glass tube is inserted in a bath of liquid; see Fig. 14.3. The fluid is assumed to wet the surface of the tube perfectly, so that the contact angle θ is 0. For small-diameter tubes the free surface of the liquid in the tube is practically a hemisphere, so the film pulls up uniformly around the perimeter, and the net surface force upward is

$$F_s = \pi D \sigma \tag{14.3}$$

This is opposed by the gravity force on the column of fluid, which is equal to the weight of the fluid that is above the free surface and which equals

$$F_g = \frac{\pi}{4} D^2 h g \rho_l \tag{14.4}$$

Here ρ_l is the density of the liquid. Equating these forces and solving for the surface tension, we find

$$\sigma = \frac{h g \rho_l D}{4} \tag{14.5}$$

This equation leaves out the buoyant force due to the air, which is generally small compared with the fluid's weight and also the fact that the top surface is not flat. However, these omissions are not as serious as the difficulties of knowing the small diameter of the tube accurately and of getting the inside of the tube very clean so that there will be perfect wetting and θ will be zero.

To solve these problems, we sometimes use the drop-weight method; see Fig. 14.4. In this method we allow drops to fall slowly (one every 2–5 min) from the tip of a buret or hypodermic needle. The drops are caught and weighed. If the liquid wets the buret perfectly, then at the instant that the drop breaks away its weight must be exactly equal to the surface force holding it up, or

$$V_{\text{drop}} \rho_l g = \pi D \sigma \tag{14.6}$$

$$\sigma = \frac{V_{\text{drop}} \rho_l g}{\pi D} \tag{14.7}$$

Again, we neglect the buoyant force due to the air, but this is negligible in most cases. The drop-weight method solves the problem of measuring D accurately, because measuring the outside diameter of a small cylinder is much easier than measuring its inside diameter. Furthermore, the surface, which must be very clean, is an exterior surface, which is easier to clean than an interior surface The difficulty with this method is that the liquid surface is not always vertical but may take one of the

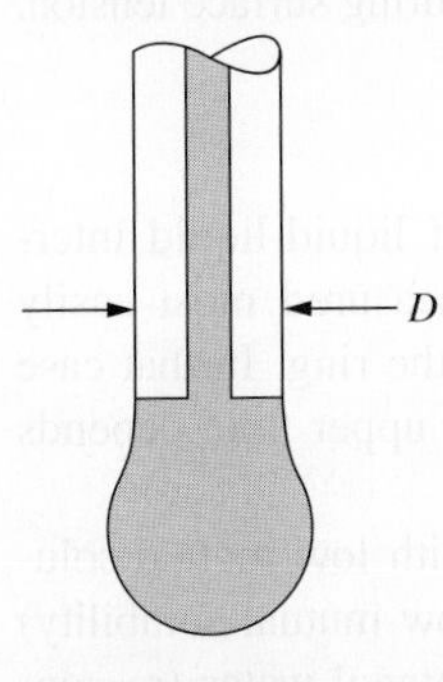

FIGURE 14.4
Drop-weight method of measuring surface tension.

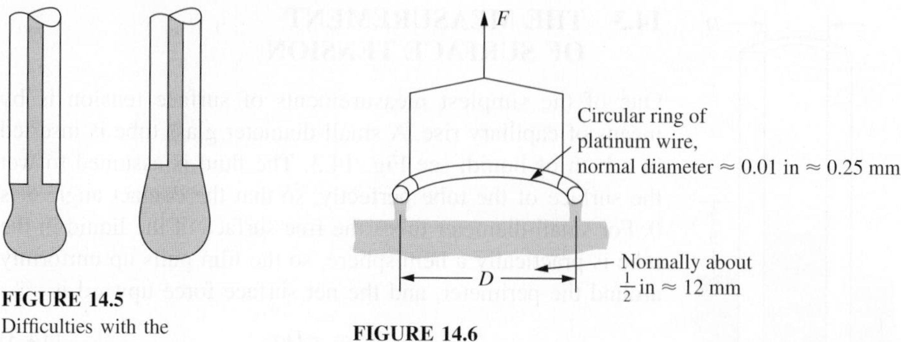

FIGURE 14.5
Difficulties with the drop-weight method.

FIGURE 14.6
Du Nouy tensiometer ring.

shapes shown in Fig. 14.5. In either of these cases all the surface tension force does not act in the vertical direction, so the volume of the drop at breakaway will be less than that shown by Eq. 14.6. Experimentally, we must multiply the calculated surface tension by a correction factor, whose value is generally about 1.5, to take this shape change into account [3, p. 45].

The most common routine laboratory method of measuring surface tension involves the Du Nouy tensiometer. As sketched in Fig. 14.6, a small ring is fabricated from thin platinum wire and then immersed in the liquid. The ring is drawn out of the liquid by applying a force F on a hanger or stirrup. This force is applied through a balance (normally a torsion balance), and the force required to remove the ring is measured. If the fluid wets the ring perfectly ($\theta = 0$), and if the films at the inside and outside of the ring point vertically downward at the instant of breakaway, then

$$F = 2(\pi D \sigma) \quad \text{or} \quad \sigma = \frac{F}{2\pi D} \tag{14.8}$$

Experimentally, it is found that the assumptions of perfect wetting, etc. are not exactly correct and that the surface tension calculated by Eq. 14.8 is too large by a factor of about 1.1. However, with this kind of apparatus the cleaning problem is much easier than the case of the other two, and the dimensions are easily measured. For these and other reasons of convenience this is the most common laboratory device for measuring surface tension.

14.4 INTERFACIAL TENSION

Surface forces exist not only at gas-liquid interfaces but also at liquid-liquid interfaces. The latter are called *interfacial tensions.* They can be measured most easily with a Du Nouy tensiometer (Fig. 14.6), if the denser fluid wets the ring. In that case the force required to pull the ring from the lower fluid up into the upper fluid depends on the interfacial tension.

In general, interfacial tensions are greater for liquid pairs with low mutual solubilities than for those with high ones. Thus, hexane-water (very low mutual solubility) has an interfacial tension two-thirds that of air-water, whereas butanol-water (reasonably large mutual solubility) has an interfacial tension only a few percent of that of air-water. For miscible liquid pairs, such as ethanol-water, there can be no interfacial tension because there can be no interface.

In all surface and interfacial tension measurements we must take extreme care to keep the surfaces clean. Many impurities tend to collect at interfaces, and very small quantities of them can make large changes in interfacial properties. Soaps, detergents, and wetting agents are prime examples of *surface-active agents.* These normally consist of tadpole-shaped molecules with a polar, water-soluble head and a nonpolar, oil-soluble tail. Here "oil" means any organic liquid not miscible with water. Small quantities of detergent are soluble in either water or oil, but their preferred position is at the water-oil interface, where they can put their water-soluble heads in the water and their oil-soluble tails in the oil. Thus, the concentration of such agents at water-oil interfaces is much higher than in the bulk fluids surrounding the interfaces. Their high concentration there changes the chemical and physical properties of the interfaces. The principal function of such soaps and detergents is to disperse oils and fats into microscopic droplets, called *micelles,* and to keep them from coalescing. Thus, they disperse oils in water and allow them to be washed off surfaces.

14.5 FORCES DUE TO CURVED SURFACES

A plane surface exerts forces in the plane of the surface, as shown in Fig. 1.10. A curved surface can also exert a net force a right angles to itself, as shown in Fig. 14.7.

Here we see a small piece of curved liquid surface perpendicular to the z axis, whose projection on the x-y plane is the small rectangle $\Delta x \cdot \Delta y$. The forces exerted by the surface along its four edges are $F = \sigma\ \Delta y$ for the two edges perpendicular to the x axis and $F = \sigma\ \Delta x$ for the two edges perpendicular to the y axis. Assuming that this piece of surface is symmetrical about the x and y axes, we can see that the x and y components of the force on the surface are zero, because the plus and minus parts cancel.

To find the z component of this force, we sum the z components of the four surface forces on the edges to find

$$F_{z_{\text{total}}} = \sum_{\text{4 sides}} F_z = 2\sigma\ \Delta y \sin\alpha_1 + 2\sigma\ \Delta x \sin\alpha_2 \tag{14.9}$$

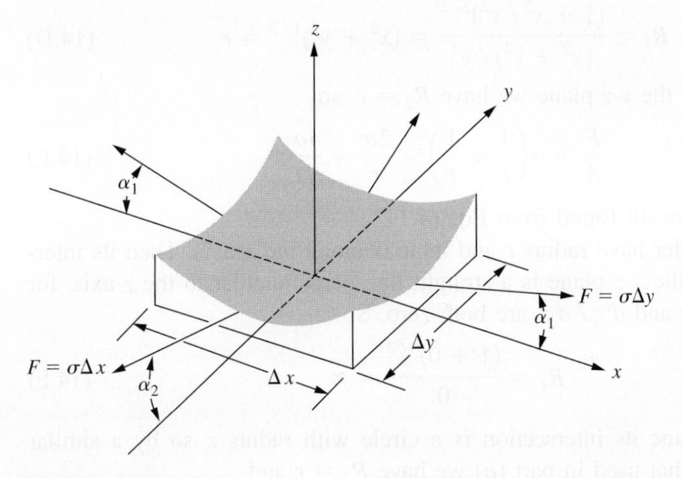

FIGURE 14.7
Forces due to curved surfaces.

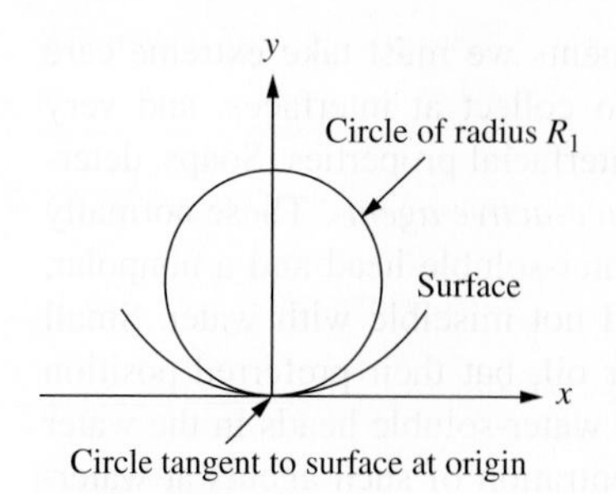

FIGURE 14.8
Radius of curvature at a point.

For small angles α_1 and α_2, we have

$$\sin \alpha_1 \approx \tfrac{1}{2}\,\Delta x / R_1 \qquad \text{and} \qquad \sin \alpha_2 \approx \tfrac{1}{2}\,\Delta y / R_2 \tag{14.10}$$

where R_1 and R_2 are the radii of curvature of the surface in the x-z and y-z planes. One may visualize these radii by imagining a circle drawn tangent to the surface at the origin; its radius is the radius of curvature in that plane; see Fig. 14.8.

Substituting Eq. 14.10 in Eq. 14.9, we find

$$F_z = 2\sigma\left(\Delta y \frac{1}{2}\frac{\Delta x}{R_1} + \Delta x \frac{1}{2}\frac{\Delta y}{R_2}\right) = \sigma A\left(\frac{1}{R_1} + \frac{1}{R_2}\right) \tag{14.11}$$

Here A is the area of the surface $\Delta x \cdot \Delta y$.

Example 14.1. Evaluate the force per unit area due to surfaces in the shape of (*a*) a sphere, (*b*) a cylinder, and (*c*) a plane.

(*a*) Let the sphere have radius r and have its center at the origin. Then in the x-y plane it forms a circle, whose equation is

$$x^2 + y^2 = r^2, \qquad \frac{dy}{dx} = -\frac{x}{y}, \qquad \frac{d^2y}{dx^2} = \frac{x^2 + y^2}{y^3} \tag{14.B}$$

In the calculus it is proven that the radius of curvature of any curve on the x-y plane is given by

$$R = \frac{[1 + (dy/dx)^2]^{3/2}}{d^2y/dx^2} \tag{14.C}$$

So here

$$R_1 = \frac{(1 + x^2/y^2)^{3/2}}{(x^2 + y^2)/y^3} = (x^2 + y^2)^{1/2} = r \tag{14.D}$$

Similarly, for the y-z plane we have $R_2 = r$, so

$$\frac{F}{A} = \sigma\left(\frac{1}{r} + \frac{1}{r}\right) = \frac{2\sigma}{r} = \frac{4\sigma}{D} \tag{14.E}$$

which is the result found from Eq. 14.1.

(*b*) Let the cylinder have radius r and its axis along the x axis. Then its intersection with the x-z plane is a straight line perpendicular to the z axis, for which dz/dx and d^2z/dx^2 are both zero. So

$$R_1 = \frac{(1 + 0)^{3/2}}{0} = \infty \tag{14.F}$$

In the y-z plane its intersection is a circle with radius r, so by a similar argument to that used in part (*a*) we have $R_2 = r$ and

$$\frac{F}{A} = \sigma\left(\frac{1}{r} + \frac{1}{\infty}\right) = \frac{\sigma}{r} = \frac{2\sigma}{D} \tag{14.G}$$

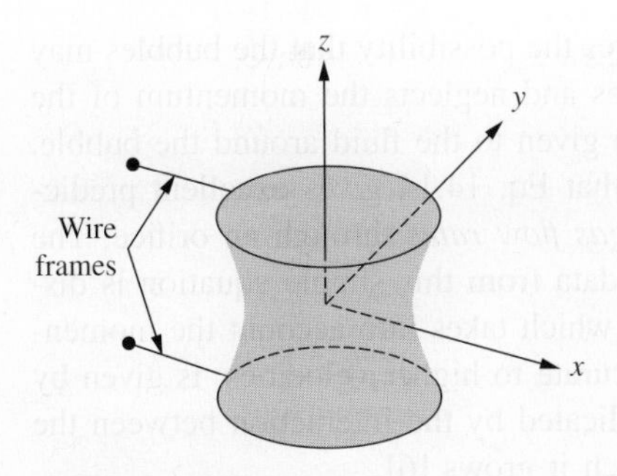

FIGURE 14.9
Soap bubble suspended between two wire frames.

(*c*) Let the plane be parallel to the *x*-*y* plane, so that its intersections with both the *x*-*z* and *y*-*z* planes are straight lines perpendicular to the *z* axis, for which, by the argument in part (*b*), we have $R_1 = R_2 = \infty$. Thus,

$$\frac{F}{A} = \sigma\left(\frac{1}{\infty} + \frac{1}{\infty}\right) = 0 \quad \text{(14.H)}$$

■

For static surfaces the force described by Eq. 14.11 is normally balanced by a pressure-difference force. If the fluid is moving, this force can serve to accelerate the fluid, causing oscillations, breakup of jets, etc.

For a fluid at rest with both sides exposed to the same pressure, Eq. 14.11 indicates that $[1/R_1 + 1/R_2]$ must be zero. This is obviously true of a plane surface, but it is also true of many complicated surfaces, such as the one shown in Fig. 14.9. This is the shape taken by an open-ended soap film attached to two circular wire loops. Here the center of curvature in the *x*-*y* plane is on the *z* axis, and the center of curvature in the *x*-*z* plane is outside the film. Since these are on opposite sides of the film, R_1 and R_2 have opposite signs. To make $[1/R_1 + 1/R_2]$ equal zero, we must make them have the same absolute value; we can show (Prob. 14.8) that this is the description of a catenoid curve.

14.6 SOME EXAMPLES OF SURFACE-FORCE EFFECTS

In many industrial devices bubbles are formed by forcing a gas through an orifice into a stagnant pool of liquid. For gas being forced through a circular, horizontal orifice this situation is shown in Fig. 14.10.

If the bubble is spherical, then the buoyant force on the bubble is

$$F_{\text{buoyant}} = \pi D^3 \frac{(\rho_l - \rho_g)g}{6} \tag{14.12}$$

If the liquid wets the orifice, including the vertical part of the hole, as shown in Fig. 14.10, then the surface force at the bottom of the hole acts vertically downward with magnitude

$$F_{\text{surface}} = \pi D_0 \sigma \tag{14.13}$$

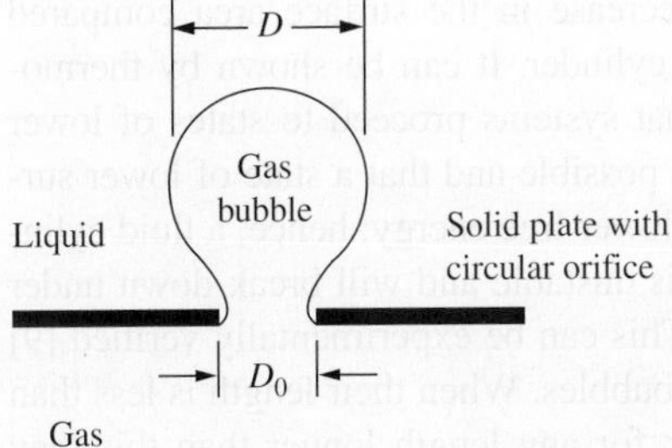

FIGURE 14.10
Gas bubble growing at a circular orifice.

The bubble breaks away from the orifice when the buoyant force exceeds the surface force. Assuming that breakaway begins when these forces are just equal, we can equate these two forces and solve for the bubble diameter:

$$D = \left[\frac{6D_0\sigma}{(\rho_l - \rho_g)g}\right]^{1/3} \tag{14.14}$$

which is called *Tate's Law.* This treatment ignores the possibility that the bubbles may interact with the preceding or following bubbles and neglects the momentum of the gas flowing into the bubble and the momentum given to the fluid around the bubble. Nevertheless, it can be shown experimentally that Eq. 14.14 gives excellent prediction of the observed bubble diameter for *low gas flow rates* through an orifice. The criterion for the departure of the experimental data from this simple equation is discussed by Soo [4]. A more complex treatment, which takes into account the momentum of the gas and the liquid, and which is accurate to higher velocities, is given by Hayes et al. [5]. This behavior is greatly complicated by the interaction between the growing bubble and the gas chamber from which it grows [6].

As shown in Fig. 14.10, this treatment assumes that the gas does not wet the orifice and that instead, a thin film of liquid wets the inside of the orifice. This is commonly observed for gas-liquid systems and also for some liquid-liquid systems. However, if the fluid flowing through the orifice wets the surface of the orifice, then much larger bubbles result [7].

An unconfined jet of liquid will break up into drops; this is observable in the jet leaving a faucet or a garden hose; see Fig. 5.22. A cylindrical jet of liquid leaves a nozzle. As the fluid falls, it speeds up, because it is being accelerated by gravity. This causes the jet to decrease in cross-sectional area to satisfy the material balance. Finally the jet breaks up into liquid droplets. This breakup is caused by surface forces; the cylindrical column of fluid can rearrange into a system with less surface area by changing into spherical droplets.

This breakup is possible for a cylinder whose length is 4 / 9 that of the diameter (Prob. 14.11), but such a cylinder is metastable: to pass from the cylindrical shape to the spherical one, it would have to pass through intermediates states with more surface. Rayleigh [8] analyzed the problem as follows. He assumed that there existed such a cylinder of length L (Fig. 14.11), which had superimposed on its cylindrical surface a cosine wave disturbance of wavelength L. Thus, the radius at any point is given by

$$r = r_0 + a \cos \frac{2\pi}{L} z \tag{14.15}$$

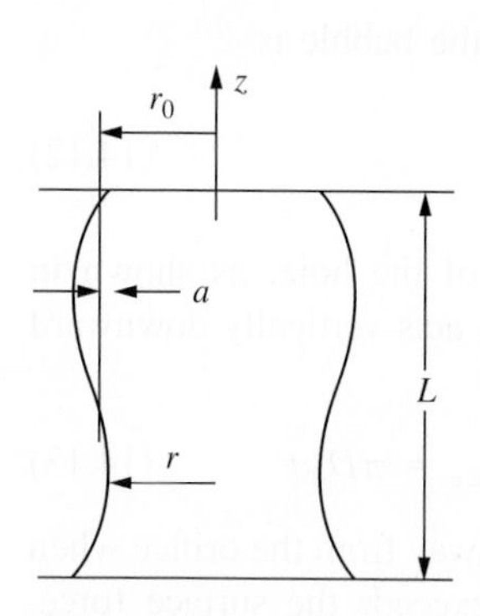

FIGURE 14.11
Small displacements on a cylinder. Here a is larger than it would appear in nature, to make the displacements visible.

In his analysis a is always assumed very small, so that the curvature shown in Fig. 14.11 is quite exaggerated. By assuming that a was very small he could show (Prob. 14.12) that, if L was greater than $2\pi r_0$, a small disturbance of this form resulted in a decrease in the surface area compared with the undisturbed cylinder. It can be shown by thermodynamic reasoning that systems proceed to states of lower free energy whenever possible and that a state of lower surface area is a state of lower free energy; hence, a fluid cylinder longer that $2\pi r_0$ is unstable and will break down under a small disturbance. This can be experimentally verified [9] with cylindrical soap bubbles. When their length is less than $2\pi r_0$, they are stable; for any length longer than this they break down as a result of any minor vibration.

Rayleigh also showed in the same paper that, although any disturbance with wavelength greater than $2\pi r_0$ would grow and break up a cylindrical jet of liquid, the fastest-growing disturbances should be those with wavelength 9 r_0. One may demonstrate this breakup in a kitchen sink by adjusting the flow from the faucet so that it has the smallest flow for which a continuous column of fluid will reach the bottom of the sink. Then tapping the faucet with a hand will cause the faucet to vibrate, introducing such disturbances and causing the jet to break up into droplets.

This simple analysis only indicates what size of jet is unstable and how the surface forces cause the breakup of the jets. It does not indicate how fast such breakup proceeds. Rayleigh's analysis of the most favored wavelength for breakup was based on the assumption that the viscosity of the fluid was negligible. For thick jets of liquids such as water, the experimental data [10] indicate that the breakup occurs most frequently at lengths corresponding 10 r_0 to 12 r_0 rather than the 9 r_0 predicted by Rayleigh.

For a very viscous fluid like maple syrup the breakup is very slow (which is why we can pour a long, thin stream of maple syrup onto our pancakes without its breaking up into droplets). Jet breakup is very important in such processes as spray drying, vaporization of liquid fuels in combustion, spray painting, and insecticide spraying. For more on this subject see [10, 11, 12].

In all of the preceding it has been assumed that the surface tension is uniform on the entire surface. If there is a gradient in the surface tension, then the surface will tend to flow in the direction of the higher surface tension and to drag adjacent liquid with it, causing bulk fluid motion. Such gradients in the surface tension can arise from differences in composition, differences in temperature, or differences in electric charge. An easily observed example of surface flow caused by concentration gradients is the formation of "wine tears," Fig. 14.12. To see these, place any liquor or wine with at least 12 percent alcohol (beer, with an alcohol content <4 percent, will not work!) in a clean glass and swirl the glass to coat the sides with fluid. Because the liquid wets the glass, a film of liquid remains on the side of the glass. Alcohol is constantly being evaporated from all the exposed surface of the liquid. In the bulk liquid in the glass the concentration is kept practically constant, because alcohol diffuses in from below to replace that being evaporated. In the films along the walls there is no comparable source of alcohol resupply, so the concentration of alcohol there falls. In alcohol-water solutions the surface tension increases with increasing water concentration, so the surface tension is higher on the film upon the walls of the glass than in the bulk liquid. This causes the film to flow up the walls and drag fluid with it from below. At the top of this film the fluid accumulates, forming the "tears," which run back down the glass.

FIGURE 14.12
Wine tears, which can be demonstrated with wine or any stronger alcohol-water solution in a clean glass. Red wine is easier to see than white. The wine should be sloshed in the glass, to wet the walls of the glass above the liquid surface.

The same kind of motion can occur at the interface between two immiscible phases when there is a chemical reaction taking place between components present in both phases or there is diffusion of one substance from one phase to another.

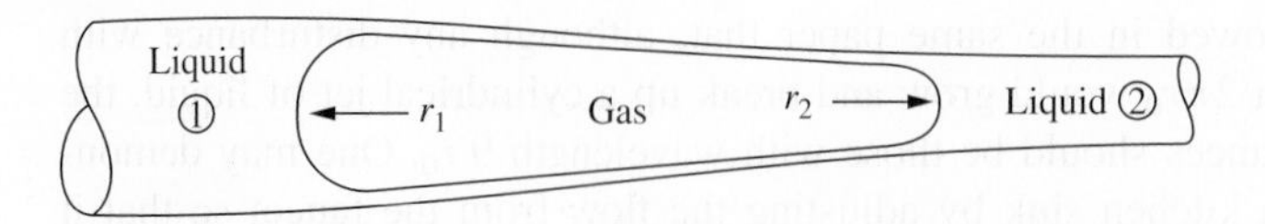

FIGURE 14.13
The Jamin effect. The bubble can stay in this location only if $P_1 > P_2$.

This type of behavior occurs in several chemical engineering systems and is called *interfacial turbulence* or the *Marangoni effect* [13].

Consider a gas or air bubble that is at rest in a converging tube, as shown in Fig. 14.13. The tube is assumed small enough for the ends of the bubble to be hemispherical with radii r_1 and r_2. The bubble is assumed at rest, so we may assume a uniform pressure P_i inside the bubble. From Eq. 14.1 we can then calculate the pressures at points 1 and 2, finding

$$P_1 = P_i - \frac{2\sigma}{r_1}, \qquad P_2 = P_i - \frac{2\sigma}{r_2}, \qquad P_1 - P_2 = 2\sigma\left(\frac{1}{r_2} - \frac{1}{r_1}\right) \tag{14.16}$$

Thus, in the absence of viscous forces this bubble can stay in the converging tube only when there is a pressure difference, as indicated by Eq. 14.16, from its large end to its small end. It can only be driven into the converging direction by a pressure difference greater than that computed by Eq. 14.16. If it is driven into the converging direction the values of the radii will decrease, until it finds a location where the surface tension forces will exactly balance the pressure difference force, and then the bubble will remain in place.

This ability of a bubble to resist a pressure gradient, which would tend to move it, can be very annoying in laboratory glassware and in the small-diameter lines used for measuring instruments. Such bubbles regularly form, often from the evolution of dissolved air from water, and block the tubing. The same effect is of great technical and economic significance in two-phase flow in porous media, such as groundwater flow in the presence of air and oil flow in the presence of water or gas. Such flows may be conceived of as occurring in a series of interconnecting, irregularly shaped tubes of very small diameter. The flow normally occurs as a result of a pressure difference. When one of the fluids breaks up into globules or bubbles, it can become trapped, just like the bubble in Fig. 14.13. It is experimentally observed in such systems that once one of the phases becomes discontinuous (i.e., breaks up into bubbles or drops), it then stops moving, and no amount of flushing with the other fluid will make it move. Although other factors are involved, this surface-tension factor is one of the major causes of this result. This effect is known in the petroleum engineering literature as the *Jamin effect* [14].

Chemical engineers are likely to encounter two other devices that depend strongly on surface tension, ink jet printers (Prob. 14.16) and mercury porosimeters (Prob. 14.17). Surface tension governs the whole field of nucleation, bubble formation and bubble collapse [15, Chap. 14]. Wetting and surface tension play a crucial role in biology, particularly in small systems. A duck, thoroughly washed to remove the water-repellent oils on its feathers, will sink and drown in water.

14.7 SUMMARY

1. Surface forces are likely to be important in small systems and in systems in which other forces are small or negligible.
2. We may think of surface tension as a force per unit length or as an energy per unit area. The surface tension is exactly equal to the surface Helmholz free energy per unit area.
3. When solid surfaces are involved in surface phenomena, it is necessary to take into account the wetting or nonwetting properties of the solid-fluid boundaries. These are normally expressed by means of the contact angle.
4. Surface forces are also present at the interfaces between immiscible liquids; these are called interfacial tensions. Such tensions are strongly influenced by impurities in the fluids, called surface-active agents.
5. The study of bubbles, drops, sprays, coatings, and interfaces between fluids generally requires the study of the surface forces involved.

PROBLEMS

See the Common Units and Values for Problems and Examples inside the back cover! An asterisk (*) on a problem number indicates that the answer is shown in App. D.

14.1. Some automatic dishwashers add a "wetting agent" to the final rinse water to prevent the formation of droplet marks on glassware. Explain how these agents do this.

14.2.*The surface tension of a liquid was measured with a capillary-rise tube and found to be A. Later tests show that this liquid does not wet the glass perfectly but makes a contact angle of $\theta = 30°$. Estimate the true value of the surface tension of this liquid.

14.3. As discussed in Sec. 14.4, we may use a Du Nouy tensiometer to measure the interfacial tension of immiscible liquid pairs in the normal way, if the more dense fluid wets the ring. Sketch how to set up to use this instrument to measure such tensions if the less dense fluid wets the ring.

14.4. Calculate the correction factor to be applied to Eq. 14.7 for the case in which the film at the surface is slanted 10° from the vertical. Does it make any difference whether the film slants inward or outward?

14.5.*A common statement in chemistry laboratory manuals is "20 drops from a buret equals approximately 1 cm^3." Assuming this applies only to dilute aqueous solutions whose surface tension is approximately equal to that of water, calculate the diameter of the tip of a standard buret. Does your calculated D match your observation of such devices? Would the number of drops per cm^3 be the same for benzene ($\sigma_{20°C} = 28.9$ dyne / cm, $\rho_{20°C} = 0.8765$ g / cm^3)?

14.6. Experiments [3, p. 45] indicate that the surface tension calculated from drop-weights on burets by means of Eq. 14.7 must be multiplied by a factor that ranges up to 1.5 to agree with those obtained by the most reliable methods. Assuming that the need for this correction factor is entirely due to the shape of the droplet just before breakaway, calculate the angle that the surface of the drop makes with the vertical.

14.7.*Two perfectly flat glass plates are assembled as shown in Fig. 14.14. The space between the plates has the form of a wedge with zero thickness on one side and thickness B on the other side; $A >> B$. The lower edges of the plates are now immersed in a pan of water. Calculate the shape of the water layer drawn up between the plates by surface

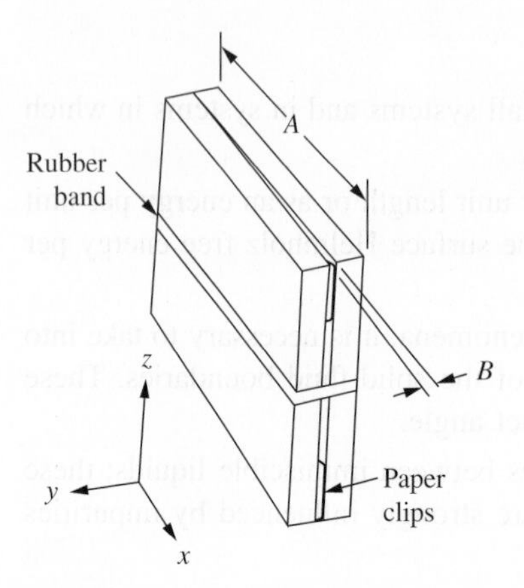

FIGURE 14.14
Capillary rise in a wedge-shaped space.

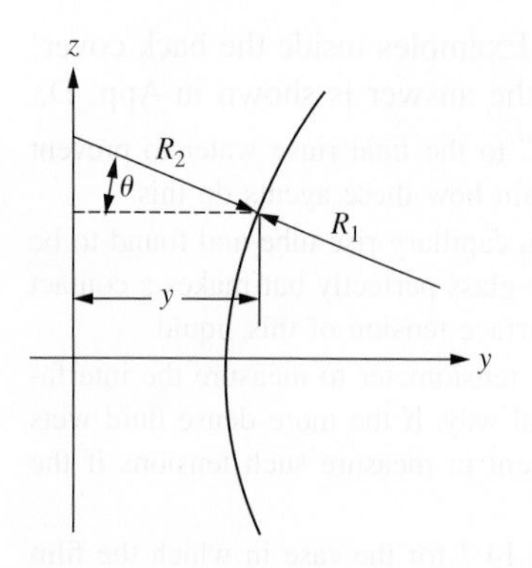

FIGURE 14.15
Figure for Prob. 14.8.

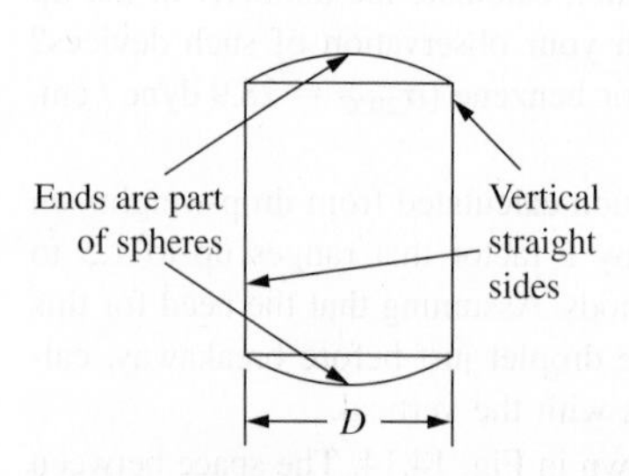

FIGURE 14.16
Soap bubble attached to two circular rings.

tension in terms of *A, B,* and σ. This result is quite easy to verify experimentally; the glass plates used for microscope slides are easy to use.

14.8. Show that the surface shown in Fig. 14.9 is a catenoid surface (one described by rotating a catenary curve about the *z* axis). *Hint:* See Fig. 14.15. From the condition that the film exerts no force in the direction normal to it, we know that the two radii of curvature at right angles to each other and normal to the film are equal and of opposite sign. These are R_1 and R_2. Equate these, and then show that $R_2 = y[1 + (dy/dz)^2]^{1/2}$. Then show that the resulting differential equation is satisfied by $y = a \cosh(z/a)$, which is the equation of a catenary curve. Here *a* is an arbitrary constant.

14.9. One can prepare a closed-end cylindrical soap bubble by blowing the bubble between two solid rings, as shown in Fig. 14.16. Because of the pressure inside the bubble, which is necessary to hold the long surface in the cylindrical shape, the top and bottom surfaces will bow out into spherical segments. If the rings have diameter *D*, what is the radius of these spherical segments?

14.10. Two flat, rectangular pieces of glass are held parallel and vertical, about $\frac{1}{32}$ in apart. Their lower edges are immersed in water. The water rises in the space between them because of surface tension; the plates pull together. After they have pulled together, they can be easily separated by being slid parallel to their surface, but they are very difficult to separate by being pulled perpendicular to the surface. Why? What is the magnitude of the force pulling them together? Does this force change if they are now removed from the water and placed on a horizontal table?

14.11. (*a*) Consider a cylindrical column of length L and diameter D_c. Show that, if this column were rearranged into a sphere of equal volume, the ratio of the surface of the new sphere to the cylindrical surface of the cylinder (total surface less surface of ends) would equal $S_s/S_{\text{cyl}} = (3/2)^{2/3} (D_c/L)^{1/3}$ and, hence, that such a rearrangement causes a decrease in surface if L is greater than $4\,D_c/9$, no change if L equals $4\,D_c/9$, and an increase if L is less than $4\,D_c/9$.

(*b*) If we take the area of the ends into account in calculating the surface of the cylinder, what minimum length is required so that the cylinder can rearrange into a sphere with a smaller surface area?

14.12. Below is an excerpt from Rayleigh's classic paper [8] on the instability of jets, in which he describes the maximum stable length; see Fig. 14.11. (Here we have changed from the symbols in the original paper to those used in this book.)

"Let us, then, taking the axis of z along the axis of the cylinder, suppose that at time t, the surface of the cylinder is of the form

$$r = r_0 + a \cos \frac{2\pi}{L} z, \tag{1}$$

where a is a small quantity variable with the time . . .

"If we denote the surface corresponding (on the average) to the unit length along the axis by A, we readily find

$$A = 2\pi r_0 + \frac{1}{2} \pi r_0 \left(\frac{2\pi}{L} \right)^2 a^2. \tag{2}$$

"In this, however, we have to substitute for r_0 (which is not strictly constant) its value obtained from the condition that V, the volume enclosed per unit of length, is given. We have

$$V = \pi r_0^2 + \frac{1}{2} \pi a^2, \tag{3}$$

whence

$$r_0 = \sqrt{\left(\frac{V}{\pi} \right)} \cdot \left(1 - \frac{1}{4} \frac{\pi a^2}{V} \right). \tag{4}$$

"Using this in (2), we get with sufficient approximation

$$A = 2\sqrt{(\pi V)} + \frac{a\pi^2}{2r_0} \left[\left(\frac{2\pi r_0}{L} \right)^2 - 1 \right]; \tag{5}$$

or, if A_0 be the value of A for the undisturbed condition,

$$A - A_0 = \frac{\pi a^2}{2r_0} \left[\left(\frac{2\pi r_0}{L} \right)^2 - 1 \right]. \tag{6}$$

From this we infer that, if $(2\pi r_0 / L) > 1$, the surface is greater after displacement than before."*

And, conversely, if $2\pi r_0 / L < 1$, the surface is less after displacement.

Fill in the missing steps in the derivation. *Hint:* In finding Eq. 1, he started with the equation for the length of an element on the surface of the cylinder, parallel to the axis, and then used the binomial theorem to perform the necessary integration. Similarly, to obtain Eq. 4 from Eq. 3, he used the binomial theorem. In both cases this is the form of $(1 + x)^{1/2} = 1 + \frac{1}{2} x +$ other terms. When x is small, the other terms may be neglected, which he has done here.

---

*From Lord Rayleigh (John William Strutt), "On the Instability in Jets," *Proc. London Math Soc. 10,* 4–13 (1879). Reprinted in *Scientific Papers of Lord Rayleigh,* Dover, New York, 1964, p. 362. Quoted by permission of the publisher.

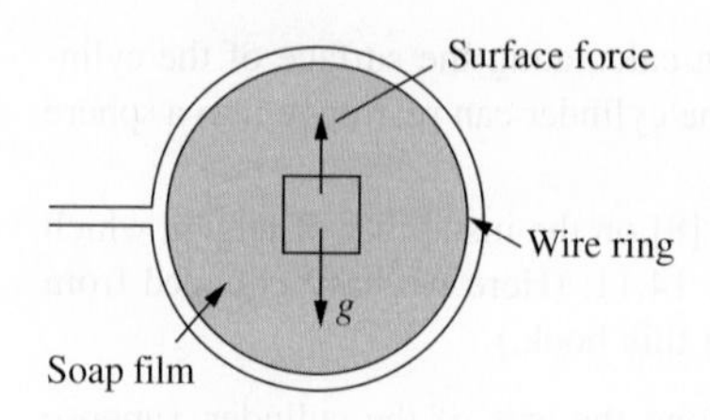

FIGURE 14.17
Vertical liquid film, held in a ring.

14.13. It is frequently stated that the breakup of a jet of fluid is a way of preparing a high-surface area of the fluid for vaporization, chemical reaction, etc. However, the analysis in Sec. 14.6 indicates that the breakup occurs because the fluid is decreasing its surface area. How are these ideas to be reconciled?

14.14.*Calculate the pressure difference required to make an air bubble move through a water-wet orifice whose diameter is 0.001 in.

14.15. If we dip a wire ring into a soap solution and hold it as shown in Fig. 14.17, we have a vertical soap film. From a force balance on a small section of the film we see that it is acted on by gravity force downward and, if it is to stay in place, it must be acted on by an upward surface force. How can this upward surface force be generated? From this consideration can we conclude that it is impossible to form such a film from an absolutely pure liquid? This topic is discussed in detail by Ross [16].

14.16. Ink jet printers work by spitting small drops of ink out of nozzles onto paper, in a very well controlled way. The drops are expelled by pressure generated by either a piezoelectric crystal actuator or by a vapor bubble formed by the sudden application of heat. A typical nozzle has a diameter of $10\ \mu = 10^{-5}$ m $= 0.00039$ in. The inks are complex and expensive fluids, which for this problem only may be assumed to have the physical properties of water, and to wet the nozzle with $\theta = 0$.
(*a*) Estimate the pressure required to eject a drop.
(*b*) If the drops are spherical with the same diameter as the nozzle, what is their volume?
(*c*) How does this compare to the observed volume, approximately 10 pL?
(*d*) What is the probable cause of this disagreement?
(*e*) If a drop has volume $\approx$ 10 pL, how large a dot will it make on the paper? Assume that the drop is spherical and expands in diameter by a factor of 3 as it flattens onto the paper. How does your answer compare with the reported printing quality of 300 dots per inch?

14.17. Adsorbents and catalyst supports are normally solids, typically the size of a pea, with many internal pores, some quite small. One should examine a charcoal briquette with a magnifying glass to get an idea of the structure. It is often worthwhile to measure sizes of the internal pores. One of the most widely used devices for this measurement is a mercury porosimeter. In such a device a sample of the solid is placed in a chamber, the chamber and sample are evacuated, and then mercury is introduced. When the mercury has filled the chamber and surrounded the sample, it is then forced into the pores by increasing its pressure. A record is made of the volume forced into the sample as a function of the pressure, from which the distribution of pore sizes can be computed. The process is the inverse of that shown in Fig. 14.3. The mercury does not wet the solid, so the interface points toward the liquid surface. Other than that, the mathematics are the same. Estimate the mercury pressure needed to force mercury into pores with diameters of 10 μ, 1 μ, 0.1 μ and 0.01 μ.

REFERENCES FOR CHAPTER 14

1. Serebryakov, V. N. "Control of the Dynamics of a Two-Phase Liquid-Gas Weightless Medium with the Aid of Surface Effects." Available in *NASA Tech. Transl., NASA TT F-469,* 1967.
2. Masica, W. J., J. D. Derdul, D. A. Petrash. "Hydrostatic Stability of the Liquid-Vapor Interface in a Low-Acceleration Field." *NASA Tech. Note, NASA TN-2444,* 1964.

3. Davies, J. T., E. K. Rideal. *Interfacial Phenomena.* New York: Academic Press, 1961.
4. Soo, S. L. *Fluid Dynamics of Multiphase Systems.* Waltham, Massachusetts: Blaisdell, 1967, p. 100.
5. Hayes, W. B. I., B. W. Hardy, C. D. Holland. "Formation of Gas Bubbles at Submerged Orifices." *AIChEJ 5,* (1959), pp. 319–324.
6. Park, Y., A. L. Tyler, N. de Nevers. "The Chamber-Orifice Interaction in the Formation of Bubbles." *Chem. Eng. Sci. 32,* (1977), pp. 907–916.
7. Halligan, J. E., L. E. Burkhart. "Determination of the Profile of a Growing Droplet." *AIChEJ 14,* (1968), pp. 411–414.
8. Lord Rayleigh (John William Strutt). "On the Instability of Jets." *Proc. London Math. Soc. 10,* (1879), pp. 4–13. Reprinted in *Scientific Papers of Lord Rayleigh.* New York: Dover, 1964.
9. Boys, C. V. *Soap Bubbles and the Forces Which Mould Them.* First published 1902; New York: Doubleday, (1959), p. 75.
10. Meyer, W. E., W. E. Ranz. "Sprays." In *Encyclopedia of Chemical Technology,* Vol. 12. 1st ed., ed. R. E. Kirk, D. F. Othmer. New York: Interscience, 1954, pp. 703–721.
11. Giffen, E., A. Muraszew. *The Atomization of Liquid Fuels.* New York: Wiley, 1953.
12. Castleman, R. A. J. "The Mechanism of the Atomization of Liquids." *Natl. Bur. Stds. J. Res. 6,* (1931), pp. 369–376.
13. Sternling, C. V., L. E. Scriven. "Interfacial Turbulence; Hydrodynamic Instability and the Marangoni Effect." *AIChEJ 5,* (1959), pp. 514–523.
14. Muskat, M. *Physical Principles of Oil Production.* New York: McGraw-Hill, 1949.
15. de Nevers, N. *Physical and Chemical Equilibrium for Chemical Engineers.* New York: John Wiley, 2002.
16. Ross, S. "Mechanisms of Foam Stabilization and Antifoam Action." *Chem. Eng. Prog. 63,* (1967), pp. 41–47.

PART IV

TWO- AND THREE-DIMENSIONAL FLUID MECHANICS

Before powerful digital computers became available, two- and three-dimensional fluid mechanics were of great use to aeronautical and civil engineers, who developed most of the hand-calculation methods for estimating their behavior. But there were few practical chemical engineering problems for which these calculation methods were of much use. Now that our computers can solve numerically problems for which no analytical (hand) solutions are known, chemical engineers are finding more and more use of two- and three-dimensional fluid mechanics, normally through Computational Fluid Dynamics (CFD). Many chemical engineers have access to very user-friendly CFD programs that can solve complex fluid mechanical problems for which no hand solutions are possible.

This section shows the basic equations for two- and three-dimensional fluid mechanics and some simple (hand) applications (Chap. 15). Then it shows the classic concepts of potential flow and the boundary layer (Chaps. 16 and 17). These two are rarely used in chemical engineering calculations but provide valuable insight for reexamining and reinterpreting the one-dimensional flows in Part III. Chapter 18 discusses turbulence, and how we apply turbulence theory in CFD. Chapter 19 shows a little about mixing, and Chap. 20 is a very brief introduction to CFD.

CHAPTER 15

TWO- AND THREE-DIMENSIONAL FLUID MECHANICS

The previous sections of this book showed that we can solve a wide array of problems of great practical interest using one-dimensional fluid mechanics. Most of the flows of interest are really two- or three-dimensional, but we can approximate them, e.g., by substituting V_{average} for the true velocity distribution, and produce calculations that match and predict experiments very well. If we wish to understand those flows in more detail, then we need to examine their three-dimensional forms. Furthermore, most of the flows we have considered so far have been geometrically simple, e.g., flow in a pipe. Many industrially interesting flows are not geometrically simple, e.g., those in the furnace in Fig. 1.15. Before we had big computers, we designed those furnaces with one-dimensional approximations; now we can use CFD to take their complex geometry into account. Finally, there are flows, which are not basic chemical engineering but are of interest to all broadly educated engineers, that are inherently two- or three-dimensional, e.g., the atmosphere, the oceans, flow around airplanes and ships. For these we need the methods in this section of the book.

In one-dimensional fluid mechanics we mostly take the view from the outside, solving for the volumetric flow rate(s) entering and leaving some system and the pressures and elevations at various entries and exits. In two- or three-dimensional fluid mechanics we look inside, asking what are the local values of the velocity as a function of x, y, z, and t at every point in the system.

15.1 NOTATION FOR MULTIDIMENSIONAL FLOWS

In Chap. 7 we introduced the velocity vector as

$$\mathbf{V} = V_x\mathbf{i} + V_y\mathbf{j} + V_z\mathbf{k} \tag{7.A}$$

We will continue that notation throughout this book. However, in the fluid mechanics literature one often sees

$$\mathbf{V} = u\mathbf{i} + v\mathbf{j} + w\mathbf{k} \tag{15.1}$$

in which we have replaced V_x, V_y, and V_z with u, v, and w. The student should be prepared to translate between these two notation systems as needed.

As discussed in Chap. 7, a vector equation is a shorthand way of writing three scalar component equations. In two- and three-dimensional fluid mechanics it is common to express the equations only in vector form. In this book we will show all important equations both in scalar component form and in vector form. Appendix C summarizes the vector notation used in this book.

15.2 MASS BALANCES FOR MULTIDIMENSIONAL FLOWS

To find the mass balance equation for an arbitrary point in space, we define the coordinates and the components of the local fluid velocity, as shown in Fig. 15.1. Our system is a small open-faced cube. We may think of it as being a wire frame with flow in or out through all six of its faces. The frame itself is fixed in space and does not move.

The rate form of the mass balance is

$$\begin{pmatrix}\text{Accumulation rate of}\\ \text{mass in the system}\end{pmatrix} = \begin{pmatrix}\text{all mass flow}\\ \text{rates in}\end{pmatrix} - \begin{pmatrix}\text{all mass flow}\\ \text{rates out}\end{pmatrix} \tag{15.2}$$

The mass in the system at any instant is $\rho\ \Delta x\ \Delta y\ \Delta z$. The flow into the system through face 1 is

$$\dot{m}_1 = \rho_1 V_{x_1}\ \Delta y\ \Delta z \tag{15.3}$$

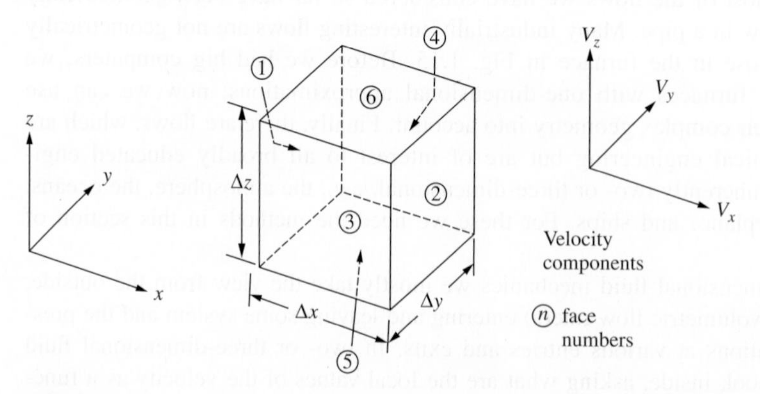

FIGURE 15.1
Notation for the three-dimensional mass and momentum balances.

and the flow out of the system through face 2 is

$$\dot{m}_2 = \rho_2 V_{x_2} \, \Delta y \, \Delta z \tag{15.4}$$

Writing the analogous terms for faces 3, 4, 5, and 6, and inserting all in Eq. 15.2 we find

$$\Delta x \, \Delta y \, \Delta z \frac{\partial \rho}{\partial t} = \Delta y \, \Delta z (\rho_1 V_{x_1} - \rho_2 V_{x_2}) + \Delta x \, \Delta z (\rho_3 V_{y_3} - \rho_4 V_{y_4}) + \Delta x \, \Delta y (\rho_5 V_{z_5} - \rho_6 V_{z_6}) \tag{15.5}$$

We now divide through by $-\Delta x \, \Delta y \, \Delta z$:

$$-\frac{\partial \rho}{\partial t} = \frac{(\rho_2 V_{x_2} - \rho_1 V_{x_1})}{\Delta x} + \frac{(\rho_4 V_{y_4} - \rho_3 V_{y_3})}{\Delta y} + \frac{(\rho_6 V_{z_6} - \rho_5 V_{z_5})}{\Delta z} \tag{15.6}$$

Now we let Δx, Δy, and Δz each approach zero simultaneously, so that our cube shrinks to a point. Taking the limit of the three ratios on the right-hand side of this equation, we find the partial derivatives,

$$-\frac{\partial \rho}{\partial t} = \frac{\partial (\rho V_x)}{\partial x} + \frac{\partial (\rho V_y)}{\partial y} + \frac{\partial (\rho V_z)}{\partial z} = \nabla \cdot (\rho \mathbf{V}) \tag{15.7}$$

Here we show the result both in the algebraic symbols and in the identical vector shorthand. Equation 15.7 shows clearly that **∇ · any vector** is a scalar, called the *divergence* of that vector. The time rate of decrease of density at any point equals the net rate of mass flow away from the point, or the rate at which the vector (density · **velocity**) field is spreading out (or diverging) at that point. (One often sees **∇ · any vector** written as **div · vector.**) See Fig. 15.2. Equation 15.7 is not a vector equation, even though its rightmost term contains vector notation. The rightmost term is the dot product of **∇** with a vector, which produces a scalar. Intuitively, we know this must be true because this is mass balance and mass is a scalar; it has magnitude, but not direction.

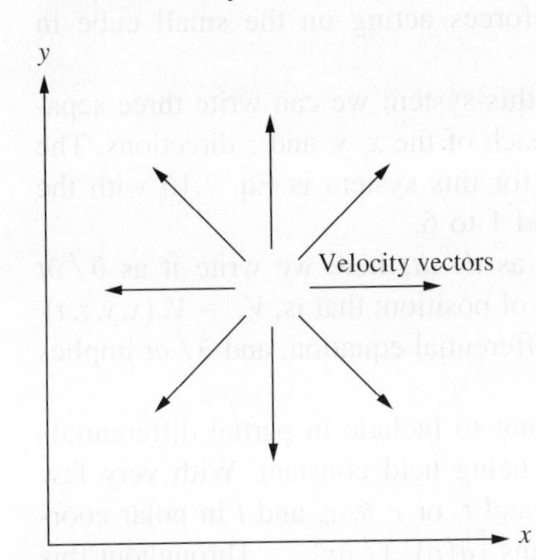

FIGURE 15.2
If the velocity vectors all point away from (*diverge* from) some point (as, for example, in an explosion), then the density at that point must be decreasing with time, as required by Eqs. 15.7 and 15.8. For a constant-density fluid, these equations show that **div · V** must be zero, so the diverging pattern shown here cannot occur for a constant-density fluid or a constant-density solid.

If the density is constant, or the density changes are small enough to be neglected, then Eq 15.7 simplifies to

$$0 = \frac{\partial V_x}{\partial x} + \frac{\partial V_y}{\partial y} + \frac{\partial V_z}{\partial z} = \nabla \cdot \mathbf{V} \quad \text{[constant density]} \tag{15.8}$$

In going from Eq. 15.6 to Eq. 15.7, by letting Δx, Δy, and Δz approach 0, we have shrunk our system to a single point. Thus, Eq. 15.7 is the mass balance for *any point* in space; it is often called the *general continuity equation.* Equation 15.8 is the mass balance for any point in space that contains a constant-density fluid. Equation 15.7 is one of the basic or *governing equations* for two- or three-dimensional flows.

One of the advantages of showing this and the other basic equations in vector form is that the vector form is independent of the coordinate system used, and the rules for expressing the scalar components in alternative coordinate systems are well known. Appendix C shows Eq. 15.7 in rectangular, cylindrical, and spherical coordinates.

15.3 MOMENTUM BALANCES FOR MULTIDIMENSIONAL FLOWS

In Chap. 7 we showed the rate form of the general momentum balance as

$$\frac{d(m\mathbf{V})_{\text{sys}}}{dt} = \mathbf{V}_{\text{in}}\dot{m}_{\text{in}} - \mathbf{V}_{\text{out}}\,\dot{m}_{\text{out}} + \sum \mathbf{F} \tag{7.14}$$

and its x-component as

$$\frac{d(mV_x)_{\text{sys}}}{dt} = V_{x_{\text{in}}}\dot{m}_{\text{in}} - V_{x_{\text{out}}}\dot{m}_{\text{out}} + \sum F_x \tag{7.16}$$

These equations are perfectly applicable to multidimensional flows, but in such flows $V_{x_{\text{in}}}$, $V_{x_{\text{out}}}$, and $\sum F_x$ are more complex than in one-dimensional flows. To make the working form of the momentum balance for multidimensional flows, we apply the general momentum balance to the flows and forces acting on the small cube in Fig. 15.1.

Our system is the small cube shown. For this system we can write three separate, independent momentum balances, one for each of the x, y, and z directions. The rate form of the x-directed momentum balance for this system is Eq. 7.16 with the flows and forces applied to all 6 faces, numbered 1 to 6.

Previously we have written the first term as d/dt; here we write it as $\partial/\partial t$ because V_x is a function not only of time but also of position; that is, $V_x = V_x(x, y, z, t)$. Thus, our momentum balance will be a partial differential equation, and $\partial/\partial t$ implies d/dt at some fixed location.

It is common practice in fluid mechanics not to include in partial differentials the subscripts that indicate which variables are being held constant. With very few exceptions the independent variables are x, y, z, and t, or r, θ, z, and t in polar coordinates. Thus, the symbol $\partial(\rho V_x)/\partial x$ really means $\left[\partial(\rho V_x)/\partial x\right]_{y,z,t}$. Throughout this text any partial derivative that does not have a subscript indicating what is being held constant will be assumed to have as independent variables x, y, z, and t or, in polar coordinates, r, θ, z, and t.

The mass of fluid in the system is equal to the system volume times the density of the fluid in the system; so we may write

$$m_{\text{sys}} = \rho\,\Delta x\,\Delta y\,\Delta z \tag{15.9}$$

The mass flow rate in through face 1 is

$$\dot{m}_1 = \rho_1 A V_{x_1} = \rho_1\,\Delta y\,\Delta z V_{x_1} \tag{15.10}$$

and the mass flow rate out through face 2 is

$$\dot{m}_2 = \rho_2 A V_{x_2} = \rho_2 \, \Delta y \, \Delta z V_{x_2} \tag{15.11}$$

Thus, the contribution of faces 1 and 2 to the first term on the right of the equal sign is

$$(\dot{m}V_x)_1 - (\dot{m}V_x)_2 = \Delta z \, \Delta y(\rho_1 V_{x_1}^2 - \rho_2 V_{x_2}^2) \tag{15.12}$$

The mass flow rate in through face 3 is

$$\dot{m}_3 = \rho_3 A V_{y_3} = \rho_3 \, \Delta x \, \Delta z V_{y_3} \tag{15.13}$$

and the flow in through face 4 is similar. The contributions from faces 3 and 4 to the first term on the right of the equal sign are

$$(\dot{m}V_y)_3 - (\dot{m}V_y)_4 = \Delta z \, \Delta y(\rho_3 V_{y_3}^2 - \rho_4 V_{y_4}^2) \tag{15.14}$$

Faces 5 and 6 contribute the following

$$(\dot{m}V_z)_5 - (\dot{m}V_z)_6 = \Delta z \, \Delta y(\rho_5 V_{z_5}^2 - \rho_6 V_{z_6}^2) \tag{15.15}$$

The forces acting on the cube are the force of gravity on the entire body of the fluid and the normal and shear forces on the faces of the cube. The gravity force in the x direction is

$$(\text{Gravity force})_x = mg \cos\theta = \rho g \, \Delta x \, \Delta y \, \Delta z \cos\theta \tag{15.16}$$

where θ is the angle between the gravity vector and the x axis.

The only normal forces on the cube that have components in the x direction are the normal forces on faces 1 and 2. We will denote a normal force in the x direction by the symbol σ_{xx}; then the normal force contribution may be written

$$\text{Normal force} = \Delta y \, \Delta z(\sigma_{xx_1} - \sigma_{xx_2}) \tag{15.17}$$

(In most circumstances the normal force is practically the same as the pressure. But in some circumstances it is not, so we will keep the general term.) The shear forces on faces 1 and 2 have no component in the x direction and do not enter the x-directed momentum balance. The shear forces on faces 3 and 4 will be given the symbol τ_{xy} and those on faces 5 and 6 the symbol τ_{xz}, so the shear force contribution is

$$\text{Shear force} = \Delta x \, \Delta z(\tau_{xy_4} - \tau_{xy_3}) + \Delta x \, \Delta y(\tau_{xz_6} - \tau_{xz_5}) \tag{15.18}$$

Making all these substitutions in Eq. 7.16, we divide by $\Delta x \, \Delta y \, \Delta z$ and find

$$\begin{aligned} \frac{\partial}{\partial t}(\rho V_x) = {} & \left(\frac{\rho_1 V_{x_1}^2 - \rho_2 V_{x_2}^2}{\Delta x} + \frac{\rho_3 V_{x_3} V_{y_3} - \rho_4 V_{x_4} V_{y_4}}{\Delta y} + \frac{\rho_5 V_{x_5} V_{z_5} - \rho_6 V_{x_6} V_{z_6}}{\Delta z} \right) \\ & + \rho g \cos\theta + \left(\frac{\sigma_{xx_1} - \sigma_{xx_2}}{\Delta x} + \frac{\tau_{xy_4} - \tau_{xy_3}}{\Delta y} + \frac{\tau_{xz_5} - \tau_{xz_5}}{\Delta z} \right) \end{aligned} \tag{15.19}$$

Now we let Δx, Δy, and Δz simultaneously approach zero and take the limit, which makes the difference terms on the right become minus partial derivatives, finding

$$\begin{aligned} \frac{\partial}{\partial t}(\rho V_x) = {} & -\left[\frac{\partial(\rho V_x^2)}{\partial x} + \frac{\partial(\rho V_x V_y)}{\partial y} + \frac{\partial(\rho V_x V_z)}{\partial z} \right] + \rho g \cos\theta \\ & + \left(\frac{-\partial \sigma_{xx}}{\partial x} + \frac{\partial \tau_{xy}}{\partial y} + \frac{\partial \tau_{xz}}{\partial z} \right) \end{aligned} \tag{15.20}$$

This is one form of the x component of the three-dimensional momentum balance. We can find a different, equally useful form by using the mass balance for a three-dimensional flow, Eq. 15.7, to eliminate several of the terms. First we expand the left side of Eq. 15.20 and also expand the first term in brackets on the right, choosing our terms on the right in a special way:

$$\rho \frac{\partial V_x}{\partial t} + \underline{V_x \frac{\partial \rho}{\partial t}}$$
$$= -\left[\underline{V_x \frac{\partial (\rho V_x)}{\partial x}} + \rho V_x \frac{\partial V_x}{\partial x} + \underline{V_x \frac{\partial (\rho V_y)}{\partial y}} + \rho V_y \frac{\partial V_x}{\partial y} + \underline{V_x \frac{\partial (\rho V_z)}{\partial z}} + \rho V_z \frac{\partial V_x}{\partial z} \right]$$
$$+ \text{ remaining terms} \tag{15.21}$$

The underlined terms in this equation are exactly minus V_x times the terms in Eq. 15.7 and, therefore, the underlined term on the left is exactly equal to the underlined terms on the right. Dropping the underlined terms on both sides of Eq. 15.21 and rearranging, we find

$$\rho \left(\frac{\partial V_x}{\partial t} + V_x \frac{\partial V_x}{\partial x} + V_y \frac{\partial V_x}{\partial y} + V_z \frac{\partial V_x}{\partial z} \right)$$
$$= \rho g \cos \theta + \left(\frac{-\partial \sigma_{xx}}{\partial x} + \frac{\partial \tau_{xy}}{\partial y} + \frac{\partial \tau_{yz}}{\partial z} \right) \tag{15.22}$$

which is the other widely used form of the x component of the three-dimensional momentum balance.

In going from Eq. 15.19 to Eq. 15.20 and letting Δx, Δy, and Δz simultaneously go to zero, we have shrunk our system to a point, so that the Eq. 15.20 is the momentum balance for some point in space. It applies to any point in any kind of flow that does not include magnetic or electrostatic forces. If the latter are significant, they will enter in forms similar to that of the gravity term.

Equations 15.20 and 15.22, as they stand, are of little practical use, because their right-hand terms are written in a form that is not readily evaluated. In Sec. 15.4 we will examine one way of evaluating those terms.

Equations 15.20 and 15.22, which are alternative forms of the x-directed momentum balance, represent two ways of regarding fluid mechanics problems. In Eq. 15.20 the left-hand side is the time rate of change of momentum of the fluid contained in an infinitesimal volume at some fixed point in space. The right-hand terms, in order, are the increase of momentum due to flow of matter into and out of this volume and the net forces on the system due to gravity, normal, and shear forces. This viewpoint, that of an observer fixed in space, is called the *eulerian* viewpoint.

Equation 15.22, which is Eq. 15.20 rearranged and simplified, has as its left side the time rate of change of the x momentum of an infinitesimal element of fluid, as seen by an observer who is not fixed in space but who rides with the fluid. The entire left side of Eq. 15.20 is often abbreviated DV_x / Dt and is called the *Stokes derivative,* the *substantive derivative, convective derivative,* or the *derivative following the motion.* In this case attention is focused on a specific piece of fluid rather than on a region of space. In this approach to the subject there can be no flow of matter into or out of the piece of matter we are watching. Thus, the left side is the increase in momentum of this piece of matter as it moves along, and the terms on the right are the gravity, normal, and shear forces acting on this fluid particle as it moves. This is called the *lagrangian viewpoint.* Both the eulerian and the lagrangian viewpoints are used in fluid mechanics, the choice between them depending on the problem at hand.

Equations 15.20 and 15.22 show only the x components of the momentum balance. One may produce the corresponding y and z components by repeating the foregoing derivation for those directions. The three resulting equations can be summarized in vector shorthand as

$$\frac{\partial}{\partial t}(\rho\mathbf{V}) = -[\mathbf{V}\cdot\nabla\rho\mathbf{V} + \rho\mathbf{V}\cdot\nabla\mathbf{V}] + \rho\mathbf{g} + \begin{pmatrix}\textbf{terms involving}\\ \textbf{normal and shear}\\ \textbf{stresses}\end{pmatrix} \tag{15.23}$$

which is the eulerian viewpoint. Equation 15.20 is the x component of Eq. 15.23. The vector form of the lagrangian viewpoint of the momentum balance is

$$\rho\left(\frac{\partial\mathbf{V}}{dt} + \mathbf{V}\cdot\nabla\mathbf{V}\right) = \rho\mathbf{g} + \begin{pmatrix}\textbf{terms involving}\\ \textbf{normal and shear}\\ \textbf{stresses}\end{pmatrix} \tag{15.24}$$

Equation 15.22 is the x component of Eq. 15.24. We will say more about these forms in the next section.

We could develop the analogous three-dimensional energy balance equation. It is used in flows with heating or combustion, and some high-velocity gas flows, mostly in computational fluid dynamics (CFD, Chap. 20). But for the flows considered in this book it is not very useful, and it is not included. Most of the historical and current work on two- and three-dimensional fluid mechanics has begun with the conservation of mass and momentum equations shown here (called the *governing equations*), and applied simplifications and mathematical techniques to produce useful solutions.

15.4 THE NAVIER-STOKES EQUATIONS

To use Eqs. 15.22 and 15.24 it is normally necessary to replace the normal-stress and shear-stress terms with terms involving measurable properties, such as viscosities, pressures, and velocities. No one has yet found a way to do this without introducing *very severe* restrictive assumptions. The commonly used set of assumptions is the following;

1. The fluid has constant density.
2. The flow is laminar throughout.
3. The fluid is Newtonian (Eq. 1.4).
4. The three-dimensional stresses in a flowing, constant-density Newtonian fluid have the same form as the three-dimensional stress in a solid body that obeys Hooke's law (that is, a perfectly elastic, isotropic solid).

If we make these assumptions, then it can be shown that

$$\frac{-\partial \sigma_{xx}}{\partial x} + \frac{\partial \tau_{xy}}{\partial y} + \frac{\partial \tau_{xz}}{\partial z} = \frac{-\partial P}{\partial x} + \mu\left(\frac{\partial^2 V_x}{\partial x^2} + \frac{\partial^2 V_x}{\partial y^2} + \frac{\partial^2 V_x}{\partial z^2}\right) \tag{15.25}$$

The derivation of this equation is shown in numerous texts [1, 2]. The intuitive meaning of three of the terms on the right is obvious, the fourth is a bit harder to see. The $\partial P / \partial x$ term is the result of pressure force on the infinitesimal cube. The $\mu(\partial^2 V_x / \partial y^2)$ and $\mu(\partial^2 V_x / \partial z^2)$ terms represent the net shear force on the cube due to changes of the x velocity in the y and z directions. One can see how these arise by assuming that the shear forces are independent of each other and by substituting Newton's law of viscosity, Eq. 1.4, in the $\partial \tau_{xy} / \partial y$ and $\partial \tau_{xz} / \partial z$ terms in Eq. 15.22.

The $\mu(\partial^2 V_x / \partial x^2)$ term is more subtle; it arises because, according to the fourth assumption listed above, the normal force is not the same as the pressure. For a fluid at rest the normal force is the same in all directions and is equal to the pressure. For a moving viscous fluid it is not the same in all directions because of the interactions of the various perpendicular shear forces. The pressure is defined as the average of the normal forces in three perpendicular directions; this is the pressure that appears in Eq. 15.25. The $\mu(\partial^2 V_x / \partial x^2)$ term results from this difference between the pressure and the normal force. It may be visualized by considering a piece of taffy being pulled while one end is held fixed. As the taffy stretches in one direction, it contracts in the two perpendicular directions. Thus, although the applied force is in one direction, there are resulting normal forces in the directions perpendicular to the direction of pulling. The normal force is tensile in the direction of pulling and compressive in the perpendicular directions; in any one direction it is not equal to the pressure. When taffy is pulled, with one end fixed, there is an increase in velocity with distance from the fixed end; therefore $(\partial^2 V_x / \partial x^2)$ is positive, leading to a force of the type shown here. In most applications of Eq. 15.25 the latter term drops out, either because of the geometry of the system or because it is assumed negligible compared with the other terms.

Substituting Eq. 15.25 in Eq. 15.22 yields

$$\rho\left(\frac{\partial V_x}{\partial t} + V_x\frac{\partial V_x}{\partial x} + V_y\frac{\partial V_x}{\partial y} + V_z\frac{\partial V_x}{\partial z}\right) = \rho g \cos\theta$$

$$- \frac{\partial P}{\partial x} + \mu\left(\frac{\partial^2 V_x}{\partial x^2} + \frac{\partial^2 V_x}{\partial y^2} + \frac{\partial^2 V_x}{\partial z^2}\right) \tag{15.26}$$

which is the differential momentum balance for the x direction, subject to the list of assumptions given above. Analogous balances can be made for the y and z directions; the three together make up the *Navier-Stokes equations.* Their vector shorthand form (with all terms divided by ρ) is

$$\frac{D\mathbf{V}}{Dt} = \mathbf{g} - \frac{\nabla \cdot P}{\rho} + \frac{\mu}{\rho}\nabla^2\mathbf{V} \tag{15.27}$$

Equation 15.26 is the x component of Eq. 15.27.

As with the mass balance, the Navier-Stokes equations can be expanded from their vector form in any coordinate system. The expansions in rectangular and cylindrical coordinates are shown in App. C. The great advantage of the vector shorthand

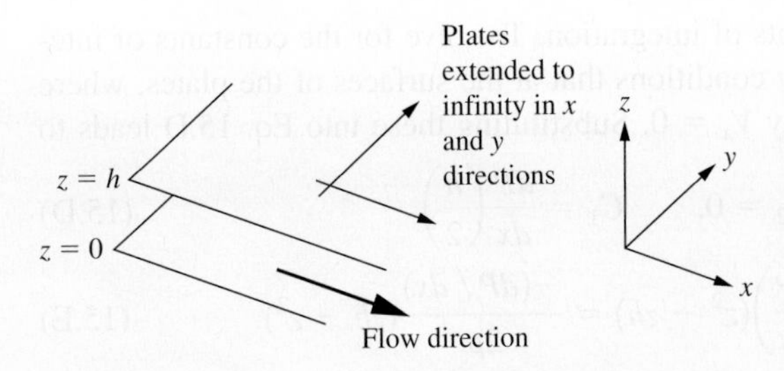

FIGURE 15.3
Notation for flow between horizontal, parallel plates.

and its expansion in various coordinate systems is that, whereas most of us can work out the rectangular coordinate momentum balance on an intuitive basis, as was shown here, doing the same in cylindrical or spherical coordinates is much less intuitive. (Try it, and compare your results to those shown in App. C; you will be impressed!) The corresponding equations for fluids with variable density also are shown in numerous texts [2, 3, 4]. If we set $\mu = 0$ in the Navier-Stokes equations, thus dropping the rightmost term, we find the *Euler equation,* which is often used for three-dimensional flow situations where viscous effects are negligible.

Example 15.1. To illustrate the application of the Navier-Stokes equations we investigate the flow between two stationary, infinite, parallel plates a distance h apart; see Fig. 15.3.

Before we can begin with the Navier-Stokes equations, we must make the assumptions listed previously, namely, that we are only considering laminar flow of a constant-density, Newtonian fluid. Then, from the geometry of the problem we can make the following assumptions:

1. There is no flow in the z or y directions: $V_y = V_z = 0$.
2. The velocity at any z is not a function of x or y; that is, the fluid appears to flow in sheets that are parallel to the plates. This is equivalent to $\partial V_x / \partial x = \partial V_x / \partial y = 0$.
3. The direction of gravity is perpendicular to the plates, so that $\cos\theta = 0$. Making these simplifications in Eq. 15.26, we find

$$\rho\left(\frac{\partial V_x}{\partial t}\right) = -\frac{\partial P}{\partial x} + \mu\frac{\partial^2 V_x}{\partial z^2} \tag{15.28}$$

This equation can be solved for various kinds of time-dependent flows. In this example we limit ourselves to finding the steady-flow solution, for which the term to the left of the equal sign is zero. For steady-state flow, subject to the assumptions given above, the pressure depends on x alone, and V_x depends on z alone, so we may replace all the partial derivatives with ordinary derivatives. We can further assume that the pressure gradient dP / dx is a constant and then separate the variables and integrate twice:

$$\frac{dP}{dx} = \mu\frac{d}{dz}\left(\frac{dV_x}{dz}\right); \qquad \frac{dP}{dx}\int dz = \mu\int d\frac{dV_x}{dz} \tag{15.A}$$

$$\frac{dP}{dx}z = \mu\frac{dV_x}{dz} + C_1; \qquad \frac{dP}{dx}\int z\,dz = \mu\int dV_x + C_1\int dz \tag{15.B}$$

$$\frac{dP}{dx}\frac{z^2}{2} = \mu V_x + C_1 z + C_2 \tag{15.C}$$

where C_1 and C_2 are constants of integration. To solve for the constants of integration, we use the boundary conditions that at the surfaces of the plates, where $z = 0$ and $z = h$, the velocity $V_x = 0$. Substituting these into Eq. 15.D leads to

$$C_2 = 0, \qquad C_1 = \frac{dP}{dx}\left(\frac{h}{2}\right) \tag{15.D}$$

$$V_x = \frac{1}{2\mu}\left(\frac{dP}{dx}\right)(z^2 - zh) = \frac{-(dP/dx)}{2\mu}(zh - z^2) \tag{15.E}$$

This gives the velocity distribution. To find the volumetric flow rate for a section of distance l in the y direction we must integrate:

$$Q = \int V_x\,dA = \frac{-(dP/dx)l}{2\mu}\int_{z=0}^{z=h}(zh - z^2)\,dz$$

$$= \frac{-(dP/dx)l}{2\mu}\left[\frac{z^2 h}{2} - \frac{z^3}{3}\right]_{z=0}^{z=h} = \frac{-(dP/dx)lh^3}{12\mu} \tag{15.F}$$

which is Eq. 6.28. ■

This example illustrates the advantages and disadvantages of the Navier-Stokes equations. We could have found Eq. 15.28 just as easily from the force balance shown in Sec. 6.3 or from a one-dimensional momentum balance, as we did from the Navier-Stokes equations. However, in more complicated problems in spherical or cylindrical geometry it is often difficult to set up the proper force balance or momentum balance, so it is convenient to start with the Navier-Stokes equations as a list of all the terms and then to drop out terms when necessary to find a solution, as was done here. Three more related examples illustrate this.

15.4.1 Three Examples of Laminar Flow in a Circular Tube

In Sec. 6.3 we used a simple force balance around a rod-like element, symmetrical about the center, to find Poiseuille's equation for steady, laminar flow of a Newtonian fluid, well downstream from the entrance in a circular pipe. Here we examine that same problem, but consider not only the steady flow well downstream from the entrance but also the entrance region flow and the starting flow.

We begin with the decision to use cylindrical coordinates (which is the only logical choice—if you do not believe that, try to work the following problems in any other coordinate system!), and with the assumption that the flow is all in the z direction, so that the r and θ components of the velocity are everywhere zero. With this set of assumptions we can see from App. C that only the z-component equation (Eq. C.20) has nonzero terms. It is

$$\rho\left(\frac{\partial V_z}{\partial t} + V_r\frac{\partial V_z}{\partial r} + \frac{V_\theta}{r}\frac{\partial V_z}{\partial \theta} + V_z\frac{\partial V_z}{\partial z}\right) =$$

$$-\frac{\partial P}{\partial z} + \mu\left[\frac{1}{r}\frac{\partial}{\partial r}\left(r\frac{\partial V_z}{\partial r}\right) + \frac{1}{r^2}\frac{\partial^2 V_z}{\partial \theta^2} + \frac{\partial^2 V_z}{\partial z^2}\right] + \rho g_z \tag{C.20}$$

In the following examples we will assume that the z direction is perpendicular to the direction of gravity, so that the last term in Eq. C.20 disappears.

Example 15.2. Show which terms remain in Eq. C.20 for steady, laminar flow of a constant-density Newtonian fluid in the axial direction in a cylindrical pipe, far downstream of the entrance.

It is intuitively obvious (or is intuitively obvious after someone tells you) that in a steady, one-dimensional flow in a cylindrical pipe the flow has only a z component, and that this varies with r but not with z or θ. Because V_z depends only on r, we can see that the following terms are zero: $\partial V_z / \partial t$, $V_r \partial V_z / \partial r$, $(V_\theta / r) \partial V_z / \partial \theta$, $V_z \partial V_z / \partial z$, $(1/r^2) \partial^2 V_z / \partial \theta^2$, and $\partial^2 V_z / \partial z^2$. The remaining terms are

$$0 = -\frac{\partial P}{\partial z} + \mu \frac{1}{r}\frac{\partial}{\partial r}\left(r \frac{\partial V_z}{\partial r}\right) \tag{15.G}$$

If we further assume that the pressure gradient $(-\partial P / \partial z)$ is a constant, we can separate variables and integrate, finding

$$\int \left(\frac{\partial P}{\partial z}\right) r\, dr = \left(\frac{\partial P}{\partial z}\right)\frac{r^2}{2} + C_1 = \mu \left(r \frac{\partial V_z}{\partial r}\right) \tag{15.H}$$

or

$$\mu \frac{\partial V_z}{\partial r} = \frac{C_1}{r} + \left(\frac{\partial P}{\partial z}\right)\frac{r}{2} \tag{15.I}$$

We know that the velocity cannot be infinite at the center of the tube ($r = 0$), so the constant of the first integration, C_1, must be zero. Furthermore, V_z depends only on r, and P depends only on z, so that all the partial derivatives in Eq. 15.I become total derivatives, and

$$\mu \frac{dV_z}{dr} = \left(\frac{dP}{dz}\right)\frac{r}{2} = \left(-\frac{P_1 - P_2}{\Delta z}\right)\frac{r}{2} \tag{6.5}$$

■

This must surely seem like a lot of work to find Eq. 6.5, which we obtained easily by a simple force balance on a properly chosen free body. One would scarcely go to this much trouble for this simple case. But see the next two examples.

Example 15.3. Show which terms remain in Eq. C.20 for laminar flow of a constant-density Newtonian fluid in the axial direction in a cylindrical pipe, far downstream of the entrance, for the assumption that the velocity is everywhere zero at times less than t_0 and that at t_0 the pressure gradient is applied instantly. After a long time the solution must be the same as in Poiseuille's equation (Eq. 6.8). Here we are interested in the starting transient.

One of the terms on the left (part of the substantive derivative) does not disappear, and we have

$$\rho \frac{\partial V_z}{\partial t} = -\frac{\partial P}{\partial z} + \mu \frac{1}{r}\frac{\partial}{\partial r}\left(r \frac{\partial V_z}{\partial r}\right) \tag{15.29}$$

This is a much more complex equation to solve than Eq. 15.G. Bird, Stewart and Lightfoot [5, p. 126] show the solution, which takes three pages and involves an infinite series of Bessel functions. The same authors show a much less detailed solution on p. 150 of their second edition. The results are summarized in Fig. 15.4, in which Eq. 15.29 has been divided by ρ. ■

Example 15.4. In Example 7.13 we estimated how long it took a long pipe attached to a water tank to come to its steady-state velocity. Repeat that example for a fluid with $\mu = 1000$ cP and $\rho = 1000\ \text{kg}/\text{m}^3$. This high value of the viscosity is chosen to make the flow laminar, so that we can use Fig. 15.4.

Here the pressure at the pipe inlet is 0.981 MPa, so the pressure gradient is 327 Pa / m. Using this value and the pipe dimensions in Eq. 6.8, we find that at steady flow, centerline velocity is 0.48 m / s and the steady flow average velocity is half of that, or 0.24 m / s. The Reynolds number corresponding to this velocity is 1470, so the flow is laminar.

From Fig. 15.4 we read that the centerline velocity will be 0.2 times the maximum velocity when $\nu t / r_0^2 = 0.05$ where r_0 is the tube radius. This velocity thus occurs at

$$t = \frac{\nu t}{r_0^2} \cdot \frac{r_0^2}{\nu} = 0.05 \frac{(0.0770\ \text{m})^2}{(1000 \cdot 0.001\ \text{Pa} \cdot \text{s}) / (1000\ \text{kg}/\text{m}^3)} = 0.296\ \text{s} \quad (15.J)$$

We can read the other curves from Fig. 15.4 and prepare a plot of $V_{\text{centerline}}$ versus t. The more interesting plot is one of V_{avg} versus t. We can see from Fig. 15.4 that the upper curves in the figure are close to parabolas, whereas the lowest few are flattened in the center, compared to parabolas. Ignoring that detail we can say that for parabolic velocity profile in a circular pipe the average

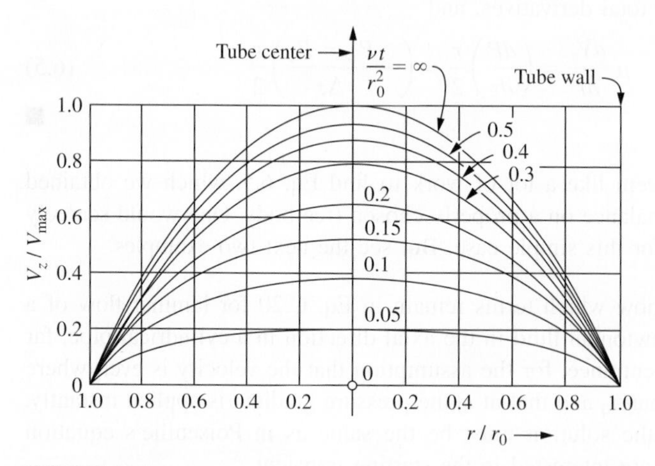

FIGURE 15.4
Flow-starting behavior for laminar flow in a circular tube, with the pressure gradient suddenly applied at $t = 0$. The uppermost curve corresponds to the steady-state solution. (Based on P. Szymanski, *J. Math. Pures Appl.*, Series 9, 11, 67–107; (1931).) In this figure r_0 is the tube radius.

velocity is 0.5 times the centerline velocity, so we would say that at 0.296 s, the average velocity was

$$V_{\text{avg}} \approx 0.5 \cdot 0.2 \cdot 0.48 \text{ m/s} = 0.048 \text{ m/s} = 0.157 \text{ ft/s} \qquad (15.K)$$

This value plus the ones corresponding to all the other curves in Fig. 15.4 are shown in Fig. 7.19, where they are compared to the result from Example 7.13. ■

We see that because of the 1000-fold increase in viscosity, the average velocity is about a tenth of that for turbulent flow of water, and that it comes to practically steady flow much faster. From Fig. 15.4 we see that for the lowest curves the velocity profile is flattened relative to the steady-flow curve, but as the flow approaches its steady velocity the velocity profile approaches the steady-flow velocity profile.

Example 15.5. Show the derivation of Eq. 15.29 from the force balance approach used to find Poiseuille's equation in Sec. 6.3.

In Sec. 6.3, which considered only steady flow behavior, we could choose a free body that was a cylinder symmetrical about the center. Here if we were to try that choice, we would find

$$\sum F = ma \qquad -\Delta P \cdot \pi r^2 + \mu \frac{\partial V}{\partial r} \cdot \pi r \, \Delta x = \pi r^2 \rho \, \Delta x \frac{\partial V}{\partial t}$$

$$-\frac{\Delta P}{\Delta x} + \frac{\mu}{r}\frac{\partial V}{\partial r} = \rho \frac{\partial V}{\partial t} \qquad \textbf{(INCORRECT!)} \qquad (15.L???)$$

We may compare that result with Eq. 15.29 and see that it is similar but not the same. The reason it is wrong is that it assumes that the central rod-like region about which the force balance is made all has the same acceleration, while in fact the acceleration is a strong function of r.

However, if we chose our free body element as a cylindrical shell of infinitesimal thickness (see Fig. 15.5), we will get the right result; see Prob. 15.7. ■

Example 15.6. To make the unsteady-state force balance for the start of a laminar flow in a pipe, one must choose as the force-balance element a cylindrical shell, with inner diameter r, and outer diameter $r + \Delta r$, and then let Δr approach zero, taking into account that the shear rate $\partial V_z / \partial r$ changes from one side of the infinitesimal shell to the other. See Prob. 15.7. ■

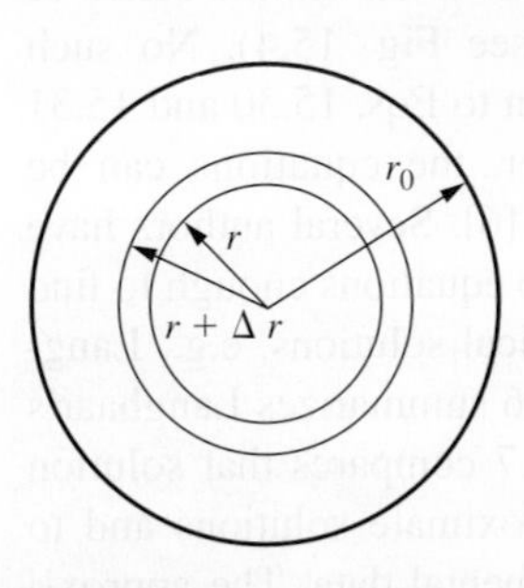

FIGURE 15.5
The proper element on which to make a force balance for starting flow in a circular tube.

We also see that if we apply the simple force balance correctly, we will get the correct result. However, for more complex geometries, it is harder and harder to do that properly, and the possibility that we will overlook some term and find an incorrect equation like Eq. 15.L??? becomes greater. Starting with the full equations in App. C and canceling terms is more reliable.

Example 15.7. Show which terms remain in Eq. C.20 for steady laminar flow of a constant-density Newtonian fluid in the axial direction in a cylindrical pipe, in the *entrance region,* (i.e., the region at the start of the pipe) with the assumption that at $z = 0$, the velocity profile is flat, i.e., V_z = constant independent of r.

Here we know that far downstream from the entrance the flow must be as described by Poiseuille's equation. At the tube entrance the velocity profile is flat, and the part of the flow that is slowed down by friction at the wall grows from the sides of the tube, eventually meeting in the center. To conserve mass, the fluid in the center must speed up as the fluid near the wall is retarded by friction, so that far downstream the velocity at the center is exactly twice the inlet velocity.

We also know from the continuity equation that our assumption that the r component of the velocity is zero cannot be correct. From the cylindrical version of the mass balance equation (Eq. C.13), we know for constant density, steady flow, and zero value of the θ component of the velocity, we have

$$\frac{1}{r}\frac{\partial}{\partial r}(rV_r) + \frac{\partial V_z}{\partial z} = 0 \tag{15.30}$$

which says that as the centerline velocity increases, there must be a radial inflow to conserve mass.

In the cylindrical momentum balance, Eq. C.20, the time derivative disappears, but two components of the substantive derivative remain, so that we have

$$\rho\left(V_r\frac{\partial V_z}{\partial r} + V_z\frac{\partial V_z}{\partial z}\right) = -\frac{\partial P}{\partial z} + \mu\left[\frac{1}{r}\frac{\partial}{\partial r}\left(r\frac{\partial V_z}{\partial r}\right) + \frac{\partial^2 V_z}{\partial z^2}\right] \tag{15.31}$$

which we must solve simultaneously with Eq. 15.30. ■

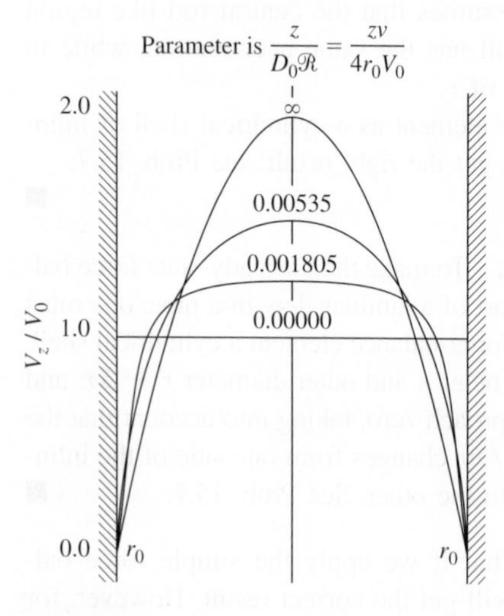

FIGURE 15.6
Summary of Langhaar's approximate solution for the flow in the entrance region in steady, laminar flow of a constant-density Newtonian fluid. (The original data are from H. L. Langhaar, *Trans ASME, 64,* 1942, p A.55.)

In Example 15.3 we could find a simple, closed-form solution (Poiseuille's equation). In Example 15.4, a complete solution is known in the form of an infinite series of Bessel functions (see Fig. 15.4). No such closed-form solution to Eqs. 15.30 and 15.31 is known. However, the equations can be solved numerically [6]. Several authors have simplified these two equations enough to find approximate analytical solutions, e.g., Langhaar [7]. Figure 15.6 summarizes Langhaar's solution. Figure 15.7 compares that solution to some other approximate solutions and to Nikuradse's experimental data. The approximate solutions are a fair, but not perfect, match of the experimental data. Figure 15.8 compares the friction factors for Poiseuille's equation, Langhaar's solution, and some

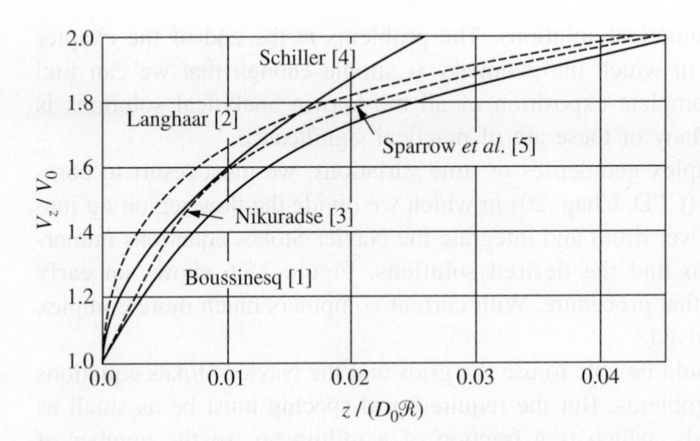

FIGURE 15.7
Comparison of the experimental data of Nikuradse with the approximate solutions of various authors for the centerline velocity in the entrance region in steady laminar flow of a constant-density Newtonian fluid. (The reference numbers are [1] J. Boussinesq, *Compt. Rend, 113,* 1891, pp. 9 and 49, [2] H. L. Langhaar, *Trans ASME, 64,* 1942, p. A.55, [3] J. Nikuradse, *Monographie, Zen. f. wiss. Berich.*, Berlin, 1942, [4] L. Schiller, *Physik, Z. 23,* 1922, p. 14 and also *ZAMM, 2,* 1922, p. 96, and [5], E. M. Sparrow, S. H. Lin and T. S. Lundgren, *Phys. Fluids, 7,* 1964, pp. 338–347.)

experimental data. The entrance region average f values are substantially greater than the "standard" values calculated from Poiseuille's equation.

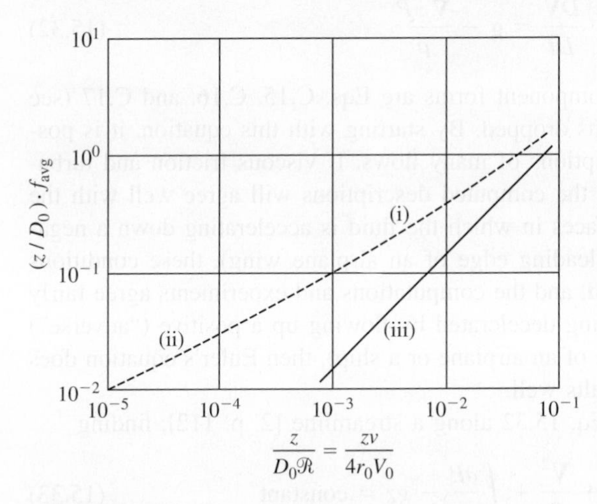

FIGURE 15.8
Comparison of the average friction factor from the entrance ($z = 0$) to length z to that calculated from Poiseuille equation. Here (i) is Langhaar's solution, (ii) is a curve fit of experimental data, and (iii) is Poiseuille's equation. (Based on J. G. Knudsen and D. L. Katz, "Fluid Mechanics and Heat Transfer" New York, McGraw-Hill, 1958.)

15.5 WHAT GOOD IS ALL OF THIS?

Examples 15.1 through 15.7 show that for simple geometries (e.g., steady laminar flow in a tube or a slit) we can throw out enough terms in the Navier-Stokes equations to find a form we can integrate and solve for local velocities, which we can then integrate to find the volumetric flow rate. If we add change with time, then the remaining terms become complex enough that we cannot do that analytically, but sometimes we can find series solutions or other closed-form solutions. Finally, as the geometry becomes even more complex, we must use numerical solutions. The problems at the end of the chapter show many of the cases in which the geometry is simple enough that we can find analytical solutions. A complete exposition of all the known analytical solutions is given by Churchill [8]. Many of these are of practical significance.

For even more complex geometries or time variations, we must resort to computational fluid dynamics (CFD, Chap. 20) in which we divide the flow region up into grids (over space and/or over time) and integrate the Navier-Stokes equations numerically over those grids to find the desired solutions. Figure 15.6 shows an early example of the results of that procedure. With current computers much more complex problems are regularly solved.

In principle, we should be able to use the grids and the Navier-Stokes equations to solve turbulent flow problems. But the required grid spacing must be as small as the smallest turbulent eddy, which is a fraction of a millimeter, so the number of required grid points becomes unmanageably large, even for our biggest supercomputers. Instead, we have developed approximate methods (Chaps. 19 and 20) that allow us to produce approximate solutions of the Navier-Stokes equations for turbulent flows. These approximate solutions are good enough that industrial users will pay the cost of finding them.

15.6 EULER'S EQUATION, BERNOULLI'S EQUATION AGAIN

If we consider a fluid with negligible viscosity, then the last term in Eq. 15.27 becomes zero, and we have

$$\frac{D\mathbf{V}}{Dt} = \mathbf{g} - \frac{\boldsymbol{\nabla} \cdot P}{\rho} \tag{15.32}$$

which is *Euler's equation.* Its component forms are Eqs. C.15, C.16, and C.17 (see App. C), with the rightmost terms dropped. By starting with this equation, it is possible to find the complete descriptions of many flows. If viscous friction and turbulence play negligible roles, then the computed descriptions will agree well with the experiments. For flows near surfaces in which the fluid is accelerating down a negative pressure gradient (e.g., the leading edge of an airplane wing), these conditions are reasonably well approximated, and the computations and experiments agree fairly well. However, if the fluid is being decelerated by flowing up a positive ("adverse") pressure gradient (as in the wake of an airplane or a ship), then Euler's equation does not predict the experimental results well.

It is possible to integrate Eq. 15.32 along a streamline [2, p. 112], finding

$$\int \frac{\partial \mathbf{V}}{\partial t} \cdot d\mathbf{r} + \frac{\mathbf{V}^2}{2} + \int \frac{\partial P}{\rho} - gz = \text{constant} \tag{15.33}$$

where $\mathbf{r}$ is the position vector, measured from the origin, which for steady flow $(\partial \mathbf{V} / \partial t = 0)$ and constant density becomes the familiar B.E.,

$$\frac{\mathbf{V}^2}{2} + \frac{P}{\rho} - gz = \text{constant} \tag{15.34}$$

This derivation applies only along a streamline. By obtaining it from a momentum balance, we have no easy way to incorporate a friction term or a pump or compressor term, which appeared naturally in Chap. 5 when we found the same result from an energy balance. The fact that the velocity is a vector here causes little problem because it appears squared (which is the equivalent of the dot product of the vector with itself, or a scalar equal to the square of the absolute value of the vector).

15.7 TRANSPORT EQUATIONS

Shear stress is normally defined as having dimensions of (force / length^2). If we consider a very simple flow like steady laminar shearing between two plates (see Fig. 1.4), we can see that for a Newtonian fluid

$$\tau = \mu \frac{dV}{dy} \tag{1.5}$$

However, we can also make the following dimensional manipulation:

$$\begin{pmatrix}\text{Shear}\\ \text{stress}\end{pmatrix} = \frac{\text{force}}{\text{length}^2} \cdot \frac{\text{mass} \cdot \text{length}}{\text{force} \cdot \text{time}^2} = \frac{\text{mass} \cdot \text{velocity}}{\text{area} \cdot \text{time}} = \frac{\text{momentum}}{\text{area} \cdot \text{time}} \tag{15.35}$$

or

$$\text{Momentum flux} = \mu \frac{dV}{dy} \tag{15.M}$$

Most often, we see Eq. 15.M written with a minus sign. Since which direction we assign to the shear stress is arbitrary, we normally write

$$\text{Momentum flux} = -\mu \frac{dV}{dy} \tag{15.36}$$

This is a simple renaming and sign change, which seems of little use. However, if we look into heat and mass transfer books, we will find

$$\text{Conductive heat flux} = -k \frac{dT}{dy} \tag{15.37}$$

which is *Fourier's law of heat conduction.* Here k is the *thermal conductivity* and T is the temperature. Similarly,

$$\text{Molecular diffusive mass flux} = -\mathscr{D} \frac{dc}{dy} \tag{15.38}$$

which is *Fick's law of diffusion.* Here $\mathscr{D}$ is the molecular diffusivity and c is the concentration of the diffusing species. Comparing these three equations, we can see that if we know the solution to any one of them for some geometry, then we know the solution to the other two for the same geometry, because we need only rename the variables to switch from one to another. This idea was popularized in chemical engineering by Bird, Stewart and Lightfoot [5] and given the name *Transport Phenomena.* The solutions to these equations for a wide variety of geometries are known, as

TABLE 15.1
Parallel forms of transport equations for laminar and turbulent flows*

Flux of	Parallel forms for laminar flow or in solids	Parallel forms for turbulent flow
Momentum	$\frac{\rho V}{A} = -\mu \frac{dV}{dy}$ μ = viscosity (or molecular viscosity)	$\frac{V}{A} = -\frac{\mu_{eddy}}{\rho}\frac{dV}{dy} = -\nu_{eddy}\frac{dV}{dy}$ μ_{eddy} = turbulent eddy viscosity
Heat	$\frac{\dot{Q}}{A} = -k\frac{dT}{dy}$ k = thermal conductivity (or molecular thermal conductivity)	$\frac{1}{C_P\rho}\frac{\dot{Q}}{A} = -\frac{k_{eddy}}{C_P\rho}\frac{dT}{dy} = -\alpha_{eddy}\frac{dT}{dy}$ k_{eddy} = turbulent eddy thermal conductivity; α_{eddy} = thermal eddy thermal diffusivity
Mass	$\frac{\dot{m}}{A} = -\mathscr{D}\frac{dc}{dy}$ $\mathscr{D}$ = diffusivity (or molecular diffusivity)	$\frac{\dot{m}}{A} = -\mathscr{D}_{eddy}\frac{dc}{dy}$ $\mathscr{D}_{eddy}$ = turbulent eddy diffusivity

*The laminar forms are shown to emphasize the simple parallels. The turbulent forms are rearranged to show the coefficients ν_{eddy}, α_{eddy}, and $\mathscr{D}_{eddy}$ each of which has the same dimensions (length^2 / time, e.g., m^2 / s).

summarized by Carslaw and Jaeger [9]. Normally the solution is shown in vector form; e.g., Eq. 15.36 is shown as a simplification of Eq. 15.27.

The equations presented above are applicable only for laminar flows and for heat conduction or diffusion in solids. However, they can be extended to turbulent flows if we introduce empirical coefficients shown in Table 15.1.

If there exist simple relations between the eddy viscosity, eddy thermal conductivity, and eddy molecular diffusivity, then we can make the same kind of comparison of fluid flow, heat transfer, and mass transfer problems in turbulent flow that we do for laminar flow.

Reynolds' analogy (Sec. 6.6) suggests that for simple turbulent flows there should exist simple ratios between the three eddy properties listed above. Experiments show that Reynolds' analogy is *only approximately correct,* but it has proven very useful in studying the relations between turbulent fluid flow, heat transfer, and mass transfer.

To feel comfortable with the Transport Phenomena approach, one must think of shear stress as being momentum flux. Most of us are more comfortable thinking of shear stress as force per unit area. But returning to Fig. 1.4, we can see that for the upper plate to slide there must be a force applied to the moving plate in the positive x direction, and to prevent the lower plate from moving there must be an equal and opposite force applied to it. The product of force and velocity is equal to (momentum / time), so that if we think of these forces as forces per unit area, we see that they are the equivalent of a (flow of momentum / unit area · time), as Eq. 15.36 requires.

15.8 SUMMARY

1. We can solve most of the simple, practical chemical engineering problems in fluid mechanics by using the one-dimensional flow simplifications in Parts I, II, and III of this book.
2. We can have a deeper understanding of those problems, at the cost of more mathematics, if we study them as two- or three-dimensional flow problems.

3. Many problems of practical interest can only be understood as two- or three-dimensional flow problems.
4. Before we had big computers, we could only solve a few of these. Some examples are in this chapter, some in the next two. Now that we have big computers we can solve many more, using CFD.
5. The basic tools of two- and three-dimensional fluid mechanics are the three-dimensional mass and momentum balances in partial differential equation form.
6. The Navier-Stokes equations are the simplification of the differential three-dimensional momentum balance for laminar flow of constant-density Newtonian fluids.

PROBLEMS

See the Common Units and Values for Problems and Examples, inside the back cover! An asterisk (*) on a problem number indicates that its answer is shown in App. D.

15.1. Show which terms of the mass balance equation (Eq. 15.7) remain for:
(*a*) Steady compressible flow in the x-y plane.
(*b*) Unsteady incompressible flow in the x-y plane.
(*c*) Unsteady compressible flow in the y direction only.
(*d*) Steady compressible flow in plane polar coordinates [i.e., polar coordinates in which there is no motion in the axial (z) direction].
(*e*) Describe the equipment you could use in a laboratory to demonstrate these four kinds of flow.

15.2. Show which terms of mass balance in cylindrical coordinates, Eq C.13, remain for:
(*a*) Steady flow of a constant-density fluid in a straight circular pipe, (e.g., Chaps. 5 and 6).
(*b*) Steady flow of a fluid with variable density in a straight circular pipe (e.g., Chap. 8).

15.3. The x component of the flow in a converging nozzle with its axis on the x axis is described by

$$V_x = V_0 \cdot (1 + x / L) \tag{15.N}$$

Here x is the distance from the nozzle inlet and L is the total length of the nozzle.
(*a*) Compute the value of the x acceleration, dV_x / dt, as a function of x.
(*b*) Compute the time required for a fluid particle to travel from $x = 0$ to $x = L$.

15.4. (*a*) Show the equation for V_r in Prob. 15.3. Assume that the flow is axisymmetric and that there is zero velocity in the θ direction.
(*b*) Sketch the flow in Prob. 15.3.

15.5. In Fig. 15.3 instead of parallel plates, the bottom plate is horizontal, but the top plate slopes down in the x direction, with slope $dz / dx = -\alpha$. The space between the plates is a wedge, with height h at $x = 0$, and decreasing height in the flow direction. The flowing fluid has a constant density.
(*a*) Show the mass balance equation for this flow.
(*b*) If the inlet velocity is V_0, and V_x is independent of z at any x show the equations for V_x and V_z as functions of x.

15.6. The left-hand term of Eq. 15.22 is often called the Stokes derivative or the derivative following the motion. Show that if $V_x = V_x(x, y, z, t)$, then

$$\frac{dV_x}{dt} = \frac{\partial V_x}{\partial t} + V_x \frac{\partial V_x}{\partial x} + V_y \frac{\partial V_x}{\partial y} + V_z \frac{\partial V_x}{\partial z} \tag{15.O}$$

That is, the Stokes derivative is indeed the total (not partial) derivative of V_x with respect to time for a small element of fluid.

15.7. Repeat Example 15.1 for the case in which the plate at $z = h$ is moving steadily in the x direction with velocity V_{wall} while the other plate is stationary. This type of flow with the plate moving is called *Couette flow;* it is somewhat similar to the flow of oil in a bearing, such as main bearings of your auto. Sketch values of the velocity profile for

$$\frac{h^2}{2\mu V_{\text{wall}}}\left(-\frac{dP}{dx}\right) = 0, \quad +3, \quad \text{and} \quad -3 \tag{15.P}$$

15.8.*In Chap. 6 we showed that for laminar flow in a circular tube, the average velocity was one-half the centerline velocity. What is the corresponding relation for the flow in Example 15.1?

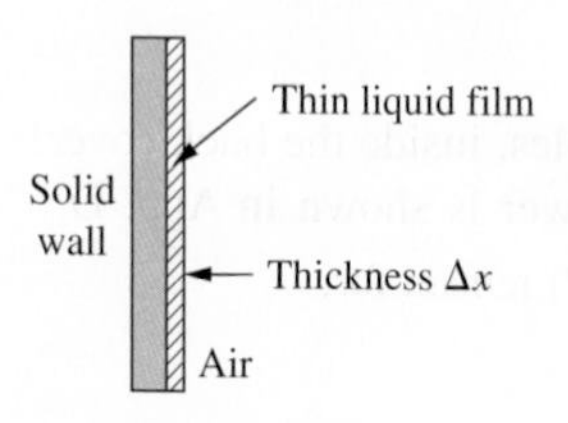

FIGURE 15.9
A thin liquid film flowing down a wall.

15.9. A constant-density Newtonian fluid is flowing as a thin film down a vertical wall in laminar flow; see Fig. 15.9. Find the velocity distribution and the volumetric flow rate per unit width of wall by using the Navier-Stokes equations (z component) on the assumptions that there is no flow in the x or y directions, that the z component of the velocity is zero at the solid wall, and that there is no shear stress at the liquid-air surface, and the flow is steady-state. (Waves may appear on the fluid surface in this situation; ignore that possibility for this problem).

15.10. In Prob. 15.9, we are now blowing air upward next to the fluid film at a high-enough velocity that the assumption of no shear stress between the liquid and the air is no longer a good assumption. Instead, we now assume that there is a shear stress at the gas-liquid interface, in the upward direction as seen by the liquid, with magnitude τ_{air} (where τ_{air} has dimensions of force / area). Repeat Prob. 15.9 for this modified circumstance.

15.11. Repeat the solution to Prob. 15.9, showing both the velocity distribution and the volumetric flow rate, for the case in which the flow under the influence of gravity is between two parallel vertical plates, separated by the distance Δx. In Prob. 15.9 the pressure gradient in the flow direction was zero, because the film was open to the air, which is at practically a constant pressure. Here also, assume that the pressure gradient in the flow direction is zero, because the vertical plates must end somewhere, and the pressure at the top and the bottom of the space between them may be assumed to be the same.

15.12. Repeat Prob. 15.9 with this change: instead of the wall being vertical, it forms an angle θ with the horizontal. *Hint:* Take the flow direction as the x direction, and the normal to the surface as the z direction. In this case the $\mathbf{g}$ vector no longer points in the $-z$ direction as it normally does, so you must use the proper value of this vector.

15.13.*We are designing a new paint. It will be sprayed on a wall, with a uniform thickness of 0.003 in. Like all paints it will harden by evaporation of its solvent, so that its viscosity will increase rapidly with time as the solvent evaporates. We are only concerned with the problem of the paint running down the wall (called "sagging" in the paint industry) for the first few minutes. The density of the paint, as applied will be 80 lbm / ft^3. *For this problem only* you may assume that paint is a Newtonian fluid.

Our paint experts tell us that there will be no visible sagging if, when the paint is first applied, the vertical velocity due to gravity of downward flow of the paint surface exposed to the air is less than or equal to 0.1 in / min. To meet this specification, what viscosity must the paint have, as applied?

15.14. Problem 15.9 is a very simplified version of the real problem of technical interest, which is the flow of a fluid like a paint or a coating down a vertical wall, as for example the behavior of the paint on an auto body that is dipped in a paint bath and then lifted out and suspended so that the excess paint can drip off. A closer approximation of that problem is as follows. At time zero, the vertical surface has a paint layer with uniform thickness Δx. At time zero we turn on gravity and allow the paint to flow down the vertical wall, without bringing in any new paint. As the flow progresses, the paint layer at the top of the wall becomes thinner, while that at the bottom of the wall becomes thicker. (At the bottom of the wall paint drips off, but we need not consider that.) The mathematical problem is to compute the thickness of the film at any position on the wall (at any value of z) at any time after time zero. The vertical direction is z and the direction perpendicular to the surface is x. The surface is large enough in the y direction that the problem is two-dimensional in z and x. The density and viscosity of the fluid are constant. (In the real problem the solvent evaporates, so that both viscosity and density change rapidly with time, but the problem is difficult enough without taking that into account here!) The fluid is assumed to be Newtonian (although real paints are highly non-Newtonian!) with viscosity μ and density ρ. The vertical wall has height h.

(*a*) Show the proper form of the material balance (continuity equation) for this problem.

(*b*) Show the proper form of the momentum balance (z-component Navier-Stokes equation) for this problem.

15.15. A tank of fluid has a long, narrow rectangular slot in its bottom. A fluid flows steadily out of the slot, in the form of a thin sheet. From the slot the sheet of fluid falls through the air, eventually falling on a solid surface. This is the arrangement used to put coatings on various products, which move on a conveyer belt under the falling sheet of fluid.

For this situation, write out the z component (i.e., vertical component) of the Navier-Stokes equations. Indicate which terms are zero or negligibly small. Indicate what additional information would be needed to solve for the velocity as a function of z and x.

15.16. A constant-density Newtonian fluid of infinite extent is adjacent to a solid wall. At time zero the wall is suddenly set in motion, with velocity V_0. (This is roughly what would happen in the water near the side of a speedboat that started from rest at full throttle.) Write the differential equation for V_x as a function of y and t for this problem by starting with Eq. 15.26 and canceling the terms that will be zero. List the boundary conditions. The resulting equation reappears several times in the following chapters.

15.17. Show that if one attempts to solve the starting problem for laminar flow in a circular tube, using the cylindrical shell element shown in Fig. 15.5, and follows the analogy of the derivation in Sec. 6.3, one finds Eq. 15.12.

15.18. Repeat Example 15.3 for flow between parallel plates instead of in a circular pipe. Make the same assumptions as in that example. Take z as the flow direction and x as the direction perpendicular to the plates.

15.19. Figure 15.4 (the starting behavior of a pipe in laminar Newtonian flow) does not include the pipe length explicitly. Does it contain it implicitly? If so, how?

15.20.*Repeat Example 15.4 for $V / V_{max} = 0.8$. Compare your result to the value shown on Fig. 7.19.

15.21. A fluid with kinematic velocity 10^{-4} m^2 / s (about 100 times as viscous as water) flows through a well-rounded entrance into a circular tube with diameter 0.01 m, in steady flow. $V_{avg} = 1$ m / s.

(*a*) Estimate how far downstream the centerline velocity will be 95 percent of the centerline velocity at infinite distance.

(*b*) Estimate the pressure drop from the entrance to that point. The fluid density is the same as that of water.

(*c*) Estimate the pressure drop for an equal length of pipe, far enough downstream to be out of the entrance region.

15.22. (*a*) Show the derivation of Eq. 1.AH. Starting with Eq. C.19, drop all the terms that are zero, finding

$$0 = \frac{\partial}{\partial r}\left[\frac{1}{r}\frac{\partial}{\partial r}(rV_\theta)\right] \tag{15.Q}$$

Integrate twice, and evaluate the constants of integration from the two boundary conditions, $V_\theta = 0 \text{ @ } r = R$ and $V_\theta = \omega kR \text{ @ } r = kR$.

(*b*) Show the corresponding equation for the other version of the couette viscometer, in which the outer cylinder rotates, and the inner one does not. For this case the integrated form above is the same, but the constants of integration are evaluated from $V_\theta = \omega R \text{ @ } r = R$ and $V_\theta = 0 \text{ @ } r = kR$.

(*c*) Show the derivation of Eq. 1.AI and 1.AJ, from the equation you find in part (*a*). Here $\sigma = r\dfrac{\partial \omega}{\partial r}$.

(*d*) In Ex. 1.2 and Prob. 1.10 at the wall of the outer, non-moving cylinder, what are the shear stress, shear rate, and the viscosity you would calculate as their ratio?

REFERENCES FOR CHAPTER 15

1. Schlichting, H., K. Gersten. *Boundary-Layer Theory,* 8th ed. Berlin: Springer, 2000.
2. Prandtl, L., O. G. Tietjens. *Fundamentals of Hydro- and Aeromechanics,* trans. L. Rosenhead. New York: Dover, 1957.
3. Schlichting, H. *Boundary Layer Theory,* 7th ed., trans. J. Kestin. New York: McGraw-Hill, 1979, p. 615.
4. Bird, R. B., E. N. Stewart, W. E. Lightfoot. *Transport Phenomena,* 2nd ed. New York: Wiley, 2002.
5. Bird, R. B., E. N. Stewart, W. E. Lightfoot. *Transport Phenomena,* New York: Wiley, 1960.
6. Christiansen, E. B., H. E. Lemmon. "Entrance Region Flow," *AIChEJ 11,* 999 (1965), 995–999.
7. Langhaar, H. L. "Steady Flow in the Transition Length in a Straight Tube." *ASME Trans. 64,* (1942), A55–58.
8. Churchill, S. W. *Viscous Flows: The Practical Use of Theory,* Boston: Butterworths (1988).
9. Carslaw, H. S., J. C. Jaeger. *Conduction of Heat in Solids,* 2nd ed., Oxford, UK: Oxford University Press, 1959.

CHAPTER 16

POTENTIAL FLOW

All the basic material on two- and three-dimensional flow (Chap. 15) was known by 1900. But without big computers, it could only be applied to laminar flows in very simple geometries. To deal with problems of technical interest, which involved two- and three-dimensional flows, two useful simplifications of the ideas in Chap. 15 were developed, potential flow and the boundary layer. Currently, chemical engineers make little use of potential flow, but much more use of the boundary layer. Thus, this chapter is mostly part of a chemical engineer's cultural background, not of her/his active technical tool kit.

16.1 THE HISTORY OF POTENTIAL FLOW AND BOUNDARY LAYER

In the late 19th century two schools of thought existed on fluid mechanics. One group, called the *hydraulicians,* looked at experimental data and attempted to generalize them into useful design equations. Their equations were generally empirical, without much theoretical content. The other group, called the *hydrodynamicists,* started with the equations in Chap. 15 and App. C and tried to apply them to practical problems. It was quickly apparent to the hydrodynamicists that, if they retained the viscous-friction terms or the change of density terms, then the resulting differential equations would be so cumbersome that solutions would seldom if ever be possible. So they ignored the viscous-friction and density-change terms by hypothesizing a *perfect fluid* with zero viscosity and constant density. For this perfect fluid they were then able to calculate the complete behavior of many kinds of flows. For flows that did not involve solid surfaces, such as deep-water waves or tides, these mathematical solutions agreed very well with observed behavior. But the hydraulicians found that the perfect-fluid solutions did not agree with observed behavior in the problems that concerned them: flow in channels, flow in pipes, forces on solid bodies caused by flow past them, etc. By 1900 the two schools had gone their separate ways, the hydrodynamicists publishing learned mathematical papers with little bearing on engineering problems and the hydraulicians solving engineering problems by trial and error, intuition, and experimental tests. A wit of the period said, "Hydrodynamicists calculate that which cannot be observed; hydraulicians observe that which cannot be calculated."

In 1904 Ludwig Prandtl [1] suggested a way to bring the two schools together by introducing a new concept, called the boundary layer. If a fluid flows past the leading edge of a flat surface, there will develop a velocity profile, as shown in Fig. 16.1. According to the laws of perfect-fluid flow, the surface should not influence the flow in any way; the velocity should be V_∞ everywhere in the flowing fluid. According to the equations of Chap. 15 and App. C, there should exist a velocity gradient in the y direction extending out to infinity. Prandtl's suggestion to reconcile these views was that the flow be conceptually divided into two parts along the line shown. In the region close to the solid surfaces the effects of viscosity are too large to be ignored. However, this is a fairly small region; outside it the effects of viscosity are small and can be neglected. Thus, outside this region the laws of perfect-fluid flow should be satisfactory.

Prandtl called the region where the viscous forces cannot be ignored the *boundary layer.* He arbitrarily suggested that it be considered that region in which the x component of the velocity, V_x, is less than 0.99 times the free-stream velocity, V_∞. Then, to obtain a complete solution to a flow problem in two or three dimensions, one should use the Chap. 15 and App. C equations inside the boundary layer and the equations of perfect-fluid flow outside the boundary layer. At the edge of the boundary layer the pressures and velocities of the two solutions must be matched.

This division does not correspond to any physically obvious boundary. The edge of the boundary layer does not correspond to any sudden change in the flow but rather corresponds to an arbitrary mathematical definition. Even with this simplification the calculations are very difficult, and in general only approximate mathematical solutions are possible. Nonetheless, this idea clarified numerous unexplained phenomena and provided a much better intellectual basis for discussing complicated flows than had

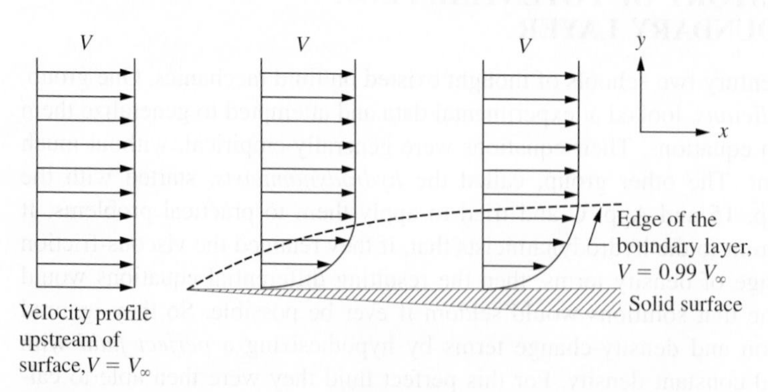

FIGURE 16.1
The idea of the boundary layer.

previously existed. As a result, the boundary layer has become a standard idea in the minds of fluid mechanicians. Once it became accepted in fluid mechanics, an analogous idea was tried in heat transfer and in mass transfer, generally with useful results.

From the preceding, it is clear that the ideas of perfect-fluid flow and of the boundary layer are intimately tied together. Both are generally needed for completely describing physically interesting flows, although sometimes one alone is sufficient. We will consider perfect-fluid flows in this chapter and the boundary layer in the next. First we must introduce the idea of streamlines.

16.2 STREAMLINES

In one-dimensional flow the direction of flow at every point in the flow is the same, although the velocity at every point may not be the same (e.g., laminar flow in a tube). In two- and three-dimensional flows both the velocity and direction change from place to place. For unsteady (i.e., time-varying) flows they also change from one instant to the next. For steady flow, we can map out the velocity and direction at any point; see Fig. 16.2, in which the velocity at any point is represented by an arrow showing the relative velocity and direction of the flow at various points.

If we follow the history of a fluid particle starting at A, we see that it moves, not in a straight line, but rather along a curve whose direction at any point is tangent to the flow direction. Such a curve, showing the path of any fluid particle in steady flow, is called a *streamline.* Obviously, there is a streamline passing through every point in the flow; so, if all the streamlines were drawn in Fig. 16.2, the entire flow area would be printed black. Therefore, it is common practice to draw only a few streamlines, from which the intermediate ones can be readily interpolated. In steady flow there is no flow across (i.e., perpendicular to) a streamline.

For unsteady-flow problems the direction and magnitude of the velocities in Fig. 16.2 are changing with time; so the direction and velocity of the streamline through A are changing with time. A fluid particle that was on the streamline through A at time t_1 may not be on the new streamline through A at time $t_1 + \Delta t$. To handle such problems, two other concepts are introduced: the *streakline,* which is the line made by a dye injected into a fluid at one point and which thus marks the position of all the particles of fluid that have passed that point, and the *pathline,* which gives the instantaneous velocity and direction of a single particle of fluid at various times. In steady flow, streaklines, pathlines, and streamlines are the same. Since we will deal only with steady flow, we will refer only to streamlines in the rest of this chapter.

If we use the alternative view of a streamline—a line across which there is no flow—then it is clear that the boundaries of solid objects immersed in the flow must be streamlines. For real fluid flows the fluid adjacent to the boundary of a solid body does not move relative to the body; it clings to

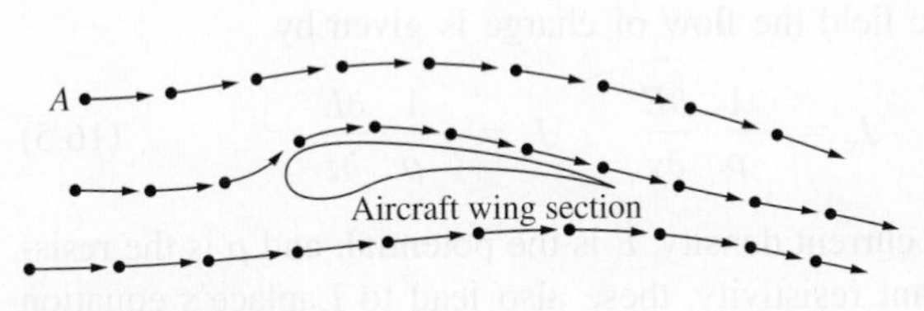

FIGURE 16.2
Point values of the flow velocity and direction for steady, two-dimensional flow, and the idea of a streamline.

the wall. Thus, in real fluids the wall is a streamline of zero velocity. In the theory of perfect-fluid flow, the imaginary perfect fluid has no tendency to cling to walls because it has no viscosity. Thus, the streamline adjacent to a solid body in perfect-fluid flow is one with finite velocity. This leads to the idea that we may divide a perfect-fluid flow along a streamline and substitute a solid body for the flow on one side of the streamline without changing the mathematical character of the flow on the other side of that streamline. Thus, to compute the flow around some solid body in perfect-fluid theory, we need only find the flow that has a streamline with the same shape as the solid body and then conceptually substitute the solid body for that part of the flow; this does not affect the rest of the flow. Several examples of this procedure will be shown.

16.3 POTENTIAL FLOW

In the region outside the boundary layer, where the fluid may be assumed to have no viscosity, the mathematical solution takes on the form known as potential flow. This form is analogous to the flow of heat in a temperature field or to the flow of charge in an electrostatic field. The basic equations of heat conduction (Fourier's law) are these:

$$q_x = -k\frac{\partial T}{\partial x} \qquad q_y = -k\frac{\partial T}{\partial y} \qquad q_z = -k\frac{\partial T}{\partial z} \tag{16.1}$$

Here q_x is the flow of heat per unit time per unit area (heat flux) in the x direction, etc., k is the thermal conductivity, and T is the temperature. An energy balance for some arbitrary region in space (analogous to the procedure shown in finding Eq. 15.7) yields

$$\rho C_V \frac{\partial T}{\partial t} = \frac{\partial[-k(\partial T/\partial x)]}{\partial x} + \frac{\partial[-k(\partial T/\partial y)]}{\partial y} + \frac{\partial[-k(\partial T/\partial z)]}{\partial z} \tag{16.2}$$

Here ρ is the density and C_V, the heat capacity. For constant k this simplifies to

$$-\frac{\rho C_V}{k}\cdot\frac{\partial T}{\partial t} = \frac{\partial^2 T}{\partial x^2} + \frac{\partial^2 T}{\partial y^2} + \frac{\partial^2 T}{\partial z^2} \tag{16.3}$$

For steady state the left term is zero, so that the steady-state heat conduction equation becomes

$$0 = \frac{\partial^2 T}{\partial x^2} + \frac{\partial^2 T}{\partial y^2} + \frac{\partial^2 T}{\partial z^2} = \nabla^2 T \tag{16.4}$$

This is *Laplace's equation,* shown here both in algebraic form and in the vector shorthand. Solutions to Laplace's equation are known for many geometries [2].

Similarly, in an electrostatic field the flow of charge is given by

$$J_x = -\frac{1}{\rho}\cdot\frac{\partial E}{\partial x} \qquad J_y = -\frac{1}{\rho}\cdot\frac{\partial E}{\partial y} \qquad J_z = -\frac{1}{\rho}\cdot\frac{\partial E}{\partial z} \tag{16.5}$$

Here J_x is the x component of the current density, E is the potential, and ρ is the resistivity. For steady-state and constant resistivity, these also lead to Laplace's equation in the form

$$0 = \frac{\partial^2 E}{\partial x^2} + \frac{\partial^2 E}{\partial y^2} + \frac{\partial^2 E}{\partial z^2} = \nabla^2 E \tag{16.6}$$

Because the flow of heat and electric charge obey Laplace's equation (under certain restrictions), the hydrodynamicists introduced a similar formulation for the flow of a liquid; they defined a *velocity potential* ϕ by the equations

$$V_x = x \text{ component of velocity} = \frac{-\partial\phi}{\partial x} \tag{16.7}$$

$$V_y = y \text{ component of velocity} = \frac{-\partial\phi}{\partial y} \tag{16.8}$$

and

$$V_z = z \text{ component of velocity} = \frac{-\partial\phi}{\partial z} \tag{16.9}$$

By applying the steady-state mass balance for a constant-density fluid, Eq. 15.8, we find that this definition also leads to Laplace's equation:

$$0 = \frac{\partial^2\phi}{\partial x^2} + \frac{\partial^2\phi}{\partial y^2} + \frac{\partial^2\phi}{\partial z^2} = \nabla^2\phi \tag{16.10}$$

From Eqs. 16.7, 16.8, and 16.9 it is clear that ϕ must have the dimension of ft^2/s or m^2/s.

The advantage of the formulation of flow problems in terms of the velocity potential is the great simplification that this formulation allows. If we are trying to determine the general solution to some steady-flow three-dimensional problem, then we will have $V_x = f_1(x, y, z)$, $V_y = f_2(x, y, z)$, and $V_z = f_3(x, y, z)$, three unknown functions of three independent variables. If the problem can be formulated in terms of velocity potential, then we can find all three of these functions from $\phi = \phi(x, y, z)$; so the problem is reduced from that of finding three functions to that of finding one.

What physical meaning should one attach to the velocity potential? For steady flow of an ideal, frictionless fluid, the velocity potential has no physical meaning whatsoever. To illustrate this, consider the steady flow of a frictionless, constant-density fluid in a horizontal pipe; see Fig. 16.3. (Such a frictionless fluid, once started in motion by some external force, would continue moving forever because there is no force to stop it.) For such a frictionless fluid the velocity is uniform over the cross section perpendicular to the flow. From B.E. we can see that there is no change with distance of pressure, velocity, or elevation, and by straightforward arguments we can show that there is no change of temperature, refractive index, dielectric constant, or any other measurable property. But from Eq. 16.7 we know that, because V_x is constant, there is a steady decrease of ϕ in the x direction. Intuitively we would like to

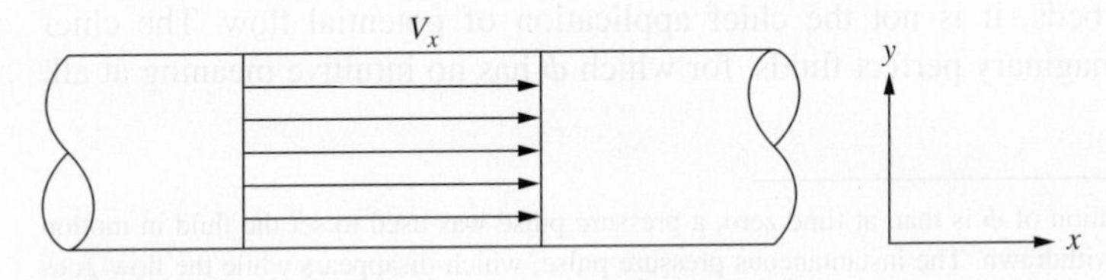

FIGURE 16.3
Steady flow of a frictionless fluid in a horizontal pipe.

identify ϕ with the pressure, but for the flow shown in Fig. 16.3, if the pressure were steadily decreasing in the flow direction, then a frictionless flow would have to accelerate steadily in that direction, which a constant-density fluid cannot do in a constant diameter pipe. Thus ϕ, the velocity potential for the steady flow of a perfect fluid, is not a function of *any measurable physical property* of the fluid: It has *no physical meaning*.*

We need not be disturbed by this lack of physical significance of ϕ. There are other such quantities in engineering, such as $i = (-1)^{1/2}$. Clearly, there can be no physical interpretation of imaginary voltages, currents, etc.; nonetheless, the treatment of alternating currents is easier if one uses i. We should take a similar view of ϕ; it has no real physical meaning but is a useful mathematical device for solving some problems.

An alternative meaning of ϕ appears in the study of the flow of real, viscous fluids through porous media. In Chap. 11 we saw that the $V^2/2$ term in B.E. is often negligible and that the friction-loss term is of the laminar form

$$\mathscr{F} = \frac{\text{viscosity} \cdot \text{velocity} \cdot \text{length}}{\text{permeability} \cdot \text{density}} \tag{16.11}$$

Here the permeability k is a property of the porous medium (discussed at greater length in Chap. 11). If we make this substitution, B.E. becomes

$$\Delta\left(\frac{P}{\rho} + gz\right) = -\frac{\mu}{k\rho} V_x \, \Delta x \tag{16.12}$$

If we now multiply through by $(k\rho/\mu\,\Delta x)$, take the limit of both sides as Δx becomes infinitesimal, and rearrange, we find

$$\frac{k}{\mu} \cdot \frac{d(P + \rho g z)}{dx} = -V_x \tag{16.13}$$

This is the same as Eq. 16.7, if

$$\phi = \frac{k}{\mu}(P + \rho g z) \tag{16.14}$$

This interpretation is intuitively quite satisfying. For the flow of a real, viscous fluid through a uniform, porous medium the density, viscosity, and permeability are normally constant; so, for constant elevation, the velocity in the x direction, $-\partial\phi/\partial x$, is proportional to the negative pressure gradient, $-\partial P/\partial x$. Although this porous-medium meaning of potential flow is intuitively satisfying and of considerable practical use in petroleum reservoir engineering, groundwater hydrology, and the study of filters and packed beds, it is not the chief application of potential flow. The chief application is for imaginary perfect fluids, for which ϕ has no intuitive meaning at all.

---

*One physical interpretation of ϕ is that, at time zero, a pressure pulse was used to set the fluid in motion and was then instantly withdrawn. The instantaneous pressure pulse, which disappears while the flow goes on forever, is the gradient of ϕ. See [3, p. 155].

TABLE 16.1
Comparison of systems obeying the Laplace equation

$$\frac{\partial^2 \phi}{\partial x^2} + \frac{\partial^2 \phi}{\partial y^2} + \frac{\partial^2 \phi}{\partial z^2} = \nabla^2 \phi = 0$$

System	What is flowing?	ϕ	Lines of constant ϕ
Steady-state temperatures	Heat (i.e., thermal energy)	Temperature	Isotherms
Steady-state electric field	Charge (i.e., electrons in opposite direction)	Potential (voltage)	Equipotentials
Steady-state perfect-fluid flow	Perfect fluid (zero viscosity, constant density)	No physical meaning whatsoever	Equipotentials
Steady-state viscous flow in porous media	Real, viscous, constant-density fluid	$\frac{k}{\rho}(P + \rho g z)$	Equipotentials (or, for constant elevations, isobars: or for constant pressure, elevation contour lines)

These potential-flow systems are compared in Table 16.1. All four potential flows can be expressed by Eq. 16.10. One consequence of Eq. 16.10 is that the flow is always perpendicular to the equipotential lines. Thus, in fluid-mechanics terminology the steady-flow streamlines are always perpendicular to the lines of constant ϕ. This property is contrary to our experience in watching balls roll down hills. They start from rest, rolling perpendicular to the contour lines but, unless the contour lines are straight and parallel, the balls eventually cross them at some other angle. Balls do this because they have inertia and try to keep going straight when the hill curves. Flows that obey Laplace's equation generally involve no inertia. Electricity and heat have no inertia. In viscous flow in a porous medium the inertia term $V^2 / 2$ is so small that it can be ignored. In perfect-fluid flow the inertia can be significant, but the nonphysical character of ϕ and the irrotational character (described in Sec. 16.4) allow this flow with inertia to fit an inertia-free formula.

To gain some feeling for the idea of a potential flow, we will show what kinds of flows are described by various choices of ϕ, restricting our attention to two-dimensional flows, because they are mathematically much easier than three-dimensional flows. In general, ϕ will be $\phi = \phi(x, y)$, but not every such function satisfies Laplace's equation, so not every such function represents a potential flow. The student may verify that $\phi = x^2$, $\phi = x^2 + y^2$, $\phi = e^x$, and $\phi = \sin x$ do not satisfy Laplace's equation, so they cannot represent potential flows because they violate the mass balance for a constant-density fluid (see Prob. 16.4).

Example 16.1. To illustrate some functions that do satisfy Laplace's equation, map out the flows described by the equations

$$\phi_1 = -Ax \tag{16.A}$$

$$\phi_2 = -(Ax + By) \tag{16.B}$$

and

$$\phi_3 = C \ln(x^2 + y^2)^{1/2} \tag{16.C}$$

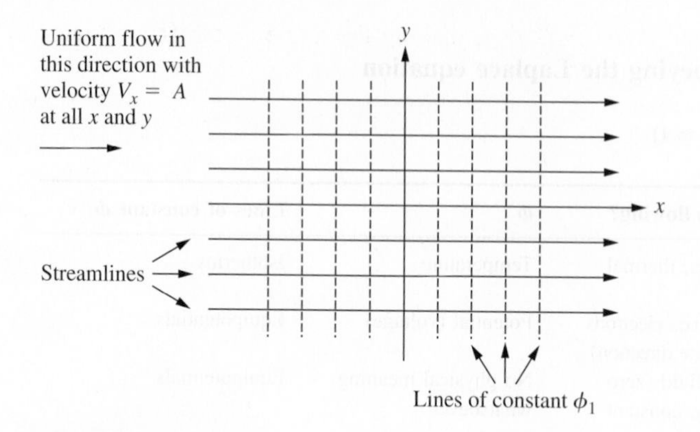

FIGURE 16.4
Flow described by the velocity potential $\phi_1 = -Ax$.

Here A and B have dimensions of velocity (ft / s or m / s), and C has dimensions of velocity times distance (ft^2 / s or m^2 / s). For ϕ_1, ϕ_2, and ϕ_3 the student may verify that Laplace's equation is satisfied.

For ϕ_1:

$$V_x = \frac{-\partial \phi_1}{\partial x} = A \qquad V_y = \frac{-\partial \phi_1}{\partial y} = 0 \tag{16.D}$$

and ϕ_1 describes a uniform, steady flow of velocity A in the positive x direction. This velocity is the same over the entire region described; this might be the description of a wind blowing over the ocean at a steady, uniform velocity of A; see Fig. 16.4.

For ϕ_2:

$$V_x = \frac{-\partial \phi_2}{\partial x} = A \qquad V_y = \frac{-\partial \phi_2}{\partial y} = B \tag{16.E}$$

This flow is shown in Fig. 16.5. From these two examples it is clear that any equation of the form $\phi = Ax + By + C$ represents a uniform, constant-

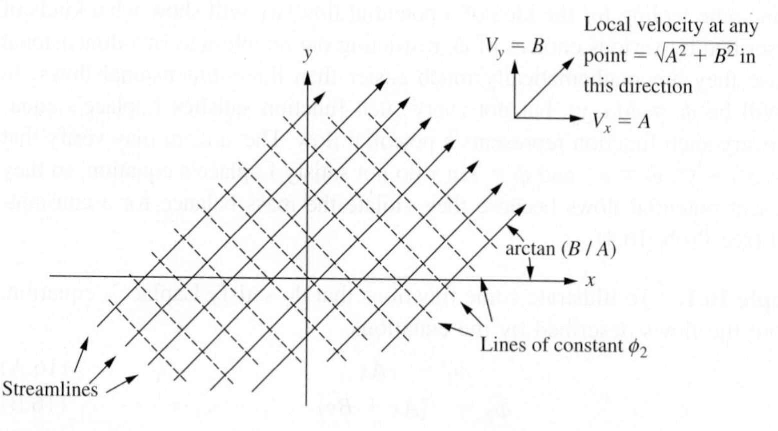

FIGURE 16.5
Flow described by the velocity potential $\phi_2 = -(Ax + By)$.

velocity flow with velocity $(A^2 + B^2)^{1/2}$, making the angle arctan (B/A) with the x axis. Such uniform, constant-velocity flows are not of much practical interest alone, but we will see later how they are combined with other flows to solve more interesting problems.

For many potential functions it is easier to work in plane polar coordinates than in rectangular coordinates. In polar coordinates Eqs. 16.7 and 16.8 take the form

$$\text{Radial velocity} = V_r = \frac{-\partial\phi}{\partial r} \tag{16.15}$$

$$\text{Tangential velocity} = V_\theta = r\frac{\partial\theta}{\partial t} = r\omega = \frac{-1}{r}\cdot\frac{\partial\phi}{\partial\theta} \tag{16.16}$$

and Laplace's equation takes the form

$$\frac{\partial\phi}{r\,\partial r} + \frac{\partial^2\phi}{\partial r^2} + \frac{\partial^2\phi}{r^2\,\partial\theta^2} = \nabla^2\phi = 0 \tag{16.17}$$

Although the algebraic expression looks different from that in Cartesian coordinates, the vector shorthand form is the same for any coordinate system (one of the merits of the vector shorthand!).

In polar coordinates ϕ_3 is expressed

$$\phi_3 = C\ln r \tag{16.F}$$

and the velocity components are

$$V_r = \frac{-C}{r} \qquad V_\theta = 0 \tag{16.G}$$

Thus, the streamlines are radially inward lines, and the lines of constant potential are circles, as shown in Fig. 16.6. If C is positive, the flow is radially inward; if it is negative, the flow is radially outward. This flow is of practical significance in the petroleum industry; it describes the flow into an oil well in a thin, horizontal stratum; see Fig. 16.7. ■

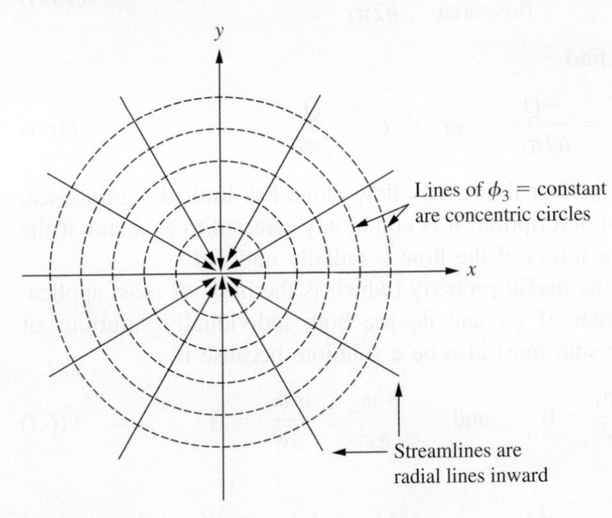

FIGURE 16.6
Flow described by the velocity potential $\phi_3 = C\ln r$.

Equation 16.G shows that the radial flow velocity becomes infinite at $r = 0$; thus, this equation cannot describe any real flow at $r = 0$. In Fig. 16.7, Eq. 16.G describes the flow from the oil-bearing stratum up to the well. Inside

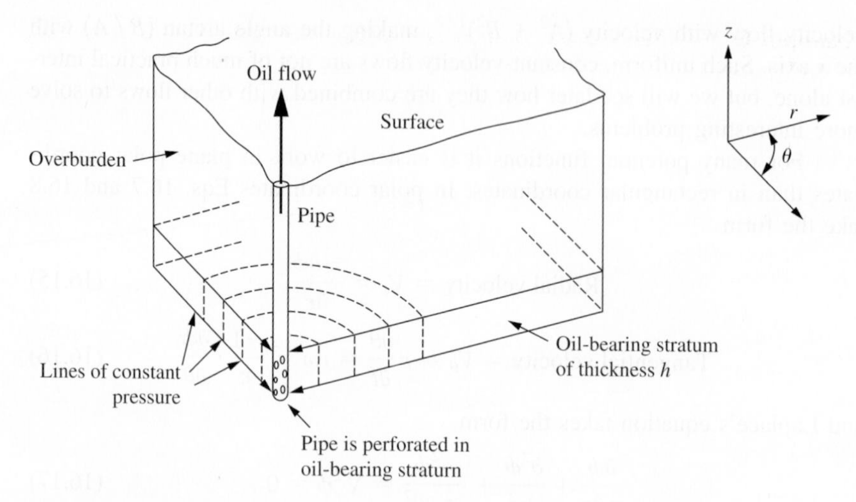

FIGURE 16.7
Flow from a thin, horizontal stratum into an oil well. The flow is two-dimensional and obeys $\phi_3 = C \ln r$ except very close to the origin, where it flows in the z (vertical direction) up the pipe.

the well the flow turns and moves in the direction perpendicular to the plane of Fig. 16.7 and is clearly not described by Eq. 16.G.

From Eq. 16.G we can calculate the value of C for any known flow up the well. If the oil-bearing stratum is h ft thick and is producing Q ft^3 / h of oil, then the steady-state flow inward across any cylindrical surface surrounding the well is Q. The radial velocity is

$$V_r = \frac{-Q}{\text{flow area}} = \frac{-Q}{h2\pi r} \tag{16.H}$$

Substituting in Eq. 16.G, we find

$$\frac{-C}{r} = \frac{-Q}{h2\pi r} \quad \text{or} \quad C = \frac{Q}{2\pi h} \tag{16.I}$$

In terms of frictionless perfect fluids, this flow alone has limited significance; but it has the same conceptual description. It is commonly referred to as a *sink* if the flow is radially inward or as a *source* if the flow is radially outward.

Laplace's equation has the useful property (which is the basis of most applications to perfect-fluid flow) that, if ϕ_1 and ϕ_2 are both individually solutions of Laplace's equation, then their sum must also be a solution, because if

$$\frac{\partial^2 \phi_1}{\partial x^2} + \frac{\partial^2 \phi_1}{\partial y^2} = 0 \quad \text{and} \quad \frac{\partial^2 \phi_2}{\partial x^2} + \frac{\partial^2 \phi_2}{\partial y^2} = 0 \tag{16.J}$$

then

$$\frac{\partial^2 \phi_1}{\partial x^2} + \frac{\partial^2 \phi_2}{\partial x^2} + \frac{\partial^2 \phi_1}{\partial y^2} + \frac{\partial^2 \phi_2}{\partial y^2} = 0 = \frac{\partial^2 (\phi_1 + \phi_2)}{\partial x^2} + \frac{\partial^2 (\phi_1 + \phi_2)}{\partial y^2} \tag{16.K}$$

Example 16.2. Illustrate this property by adding the two potential functions, ϕ_1 and ϕ_3, discussed previously.

$$\phi_4 = \phi_1 + \phi_3 = -Ax + C\ln(x^2 + y^2)^{1/2} \tag{16.L}$$

which has velocity components

$$V_x = -\left(-A + \frac{Cx}{x^2 + y^2}\right) \qquad V_y = \frac{-Cy}{x^2 + y^2} \tag{16.M}$$

and which is sketched in Fig. 16.8. ■

There are several physical systems that display the flow pattern in Fig. 16.8. The flow can be in the direction of the arrows shown or in the reverse direction (to reverse the direction of the flow, we need only reverse the signs of A and C in Eq. 16.L):

1. If there is a steady flow from left to right in a thin, porous stratum (as would exist in a horizontal oil stratum with a linear pressure gradient) and some of the fluid is being withdrawn from a well in a direction perpendicular to the page, then the flow pattern is as shown, and the equipotential lines are isobars.
2. If the flow direction is reversed, then this is the pattern of a fluid being injected into a stratum in which there is steady flow from right to left.
3. The lines of flow direction marked A and B in Fig. 16.8 close on the point at which there is zero flow (i.e., $y = 0$ and $x = C/A$). If we divide the flow along

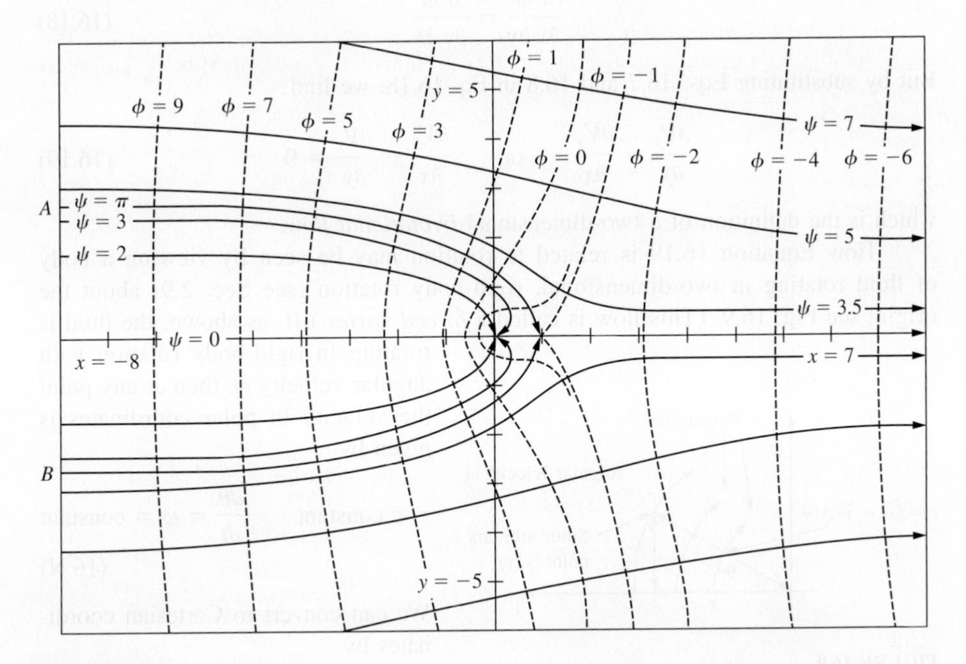

FIGURE 16.8
Flow described by the velocity potential $\phi_4 = -Ax + C\ln(x^2 + y^2)^{1/2}$. Drawn to scale for $A = 1$ and $C = 1$. The markings "$\psi = 7$" etc., on the streamlines are discussed in Section 16.5.

the curve AB and consider only the flow to the right of this curve, then this represents the potential flow outside some two-dimensional body shaped like AB. If the flow is from right to left, then this is quite similar to the flow over the leading edge of an airplane's wing or the upstream side of a rounded bridge abutment.

The latter interpretation is the one normally sought in the study of perfect-fluid flows; we wish to find the flow pattern around some arbitrary body. This is normally done by judicious combinations of steady flows, sources, sinks, etc. When a combination is found that produces a streamline with the shape of the body in question, the flow outside that streamline is a representation of the flow around the body. The flow inside that line (i.e., inside line AB in Fig. 16.8) normally has no meaning and is ignored. Figure 16.8 also shows curves of another function, ψ, whose properties are discussed in Sec. 16.5.

Similar flow maps for a wide variety of functions and their corresponding geometries are given in Porzikidis [4] and Kirchoff [5].

16.4 IRROTATIONAL FLOW

Equations 16.7, 16.8, and 16.9, which define the velocity potential, have an interesting consequence: Any flow that obeys them must be *irrotational.* If $\phi = \phi(x, y)$ has continuous derivatives, then the order of differentiation is immaterial, and

$$\frac{\partial^2 \phi}{\partial x\, \partial y} = \frac{\partial^2 \phi}{\partial y\, \partial x} \tag{16.18}$$

But by substituting Eqs. 16.7 and 16.8 in Eq. 16.18, we find

$$\frac{\partial V_x}{\partial y} = \frac{\partial V_y}{\partial x} \quad \text{or} \quad \frac{\partial V_y}{\partial x} - \frac{\partial V_x}{\partial y} = 0 \tag{16.19}$$

which is the definition of a two-dimensional *irrotational flow.*

How Equation 16.19 is related to rotation may be seen by viewing a body of fluid rotating in two-dimensional, rigid-body rotation (see Sec. 2.9) about the origin; see Fig. 16.9. (This flow is called a *forced vortex.*) If, as shown, the fluid is rotating in rigid-body rotation with angular velocity ω, then at any point the velocity in polar coordinates is given by

$$r = \text{constant} \qquad \frac{d\theta}{dt} = \omega = \text{constant} \tag{16.N}$$

We can convert to Cartesian coordinates by

$$x = r\cos\theta \qquad y = r\sin\theta \tag{16.O}$$

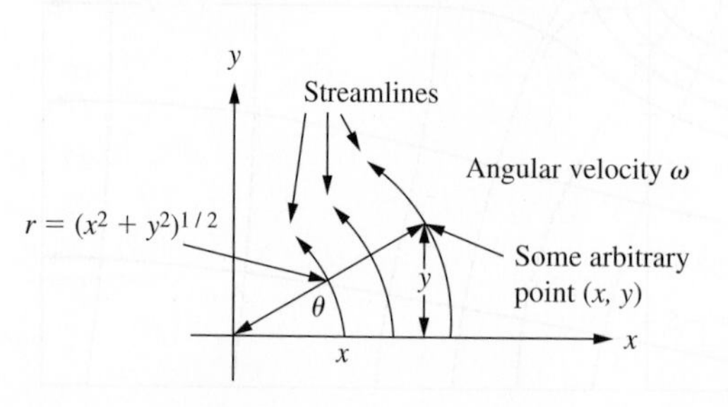

FIGURE 16.9
Solid-body rotation (called a *forced vortex*).

so that

$$V_x = \frac{dx}{dt} = -r\sin\theta\frac{d\theta}{dt} = -r\frac{y}{r}\omega = -y\omega \tag{16.P}$$

and

$$V_y = \frac{dy}{dt} = r\cos\theta\frac{d\theta}{dt} = r\frac{x}{r}\omega = x\omega \tag{16.Q}$$

so that

$$\frac{\partial V_y}{\partial x} - \frac{\partial V_x}{\partial y} = \omega + \omega = 2\omega \tag{16.R}$$

These calculations were carried out for any rigid-body rotation, so we see that $[(\partial V_y / \partial x) - (\partial V_x / \partial y)]$ is exactly twice the angular velocity. This quantity is given the name *vorticity* in theoretical fluid mechanics:

$$\text{Vorticity} = \zeta = 2\omega = \frac{\partial V_y}{\partial x} - \frac{\partial V_x}{\partial y} \qquad \text{[two-dimensional flow]} \tag{16.20}$$

This set of derivatives also has a name in the vector shorthand (see App. C),

$$\zeta = \frac{\partial V_y}{\partial x} - \frac{\partial V_x}{\partial y} = \mathbf{\nabla \times V} = \mathbf{curl\ V} = \mathbf{rot\ V} \tag{16.21}$$

The names *curl* (English usage) or *rot* (short for rotation, German usage) indicate that this function has to do with the rotation of the fluid. For three-dimensional flows there are more partial derivatives than the two shown here. Since this function describes rotation, if a flow is irrotational (described in more detail below) then at every point in the flow

$$\text{Vorticity} = \zeta = 2\omega = \frac{\partial V_y}{\partial x} - \frac{\partial V_x}{\partial y} = \mathbf{\nabla \times V} = \mathbf{curl\ V} = \mathbf{rot\ V} = 0 \tag{16.22}$$

It is a property of potential flows that this combination of derivatives is zero at every point in a potential flow, so that potential flows are everywhere irrotational. In vector shorthand, vorticity is a vector whose direction is that of the axis of rotation and whose magnitude is twice the angular velocity. For a flow in the x-y plane, the axis of rotation must be perpendicular to that plane, i.e., in the z direction. For three-dimensional flows the magnitude and direction (2 times angular velocity and direction of the axis of rotation) are found by Eq. C.10.

We see that for simple, rigid-body rotation $[(\partial V_y / \partial x) - (\partial V_x / \partial y)]$ is not zero. Thus, it is impossible to find any potential function ϕ that, when substituted in Eqs. 16.7 and 16.8, will describe such a flow. This does not mean that there can be no potential flows that have circular motion. Only those circular motions that have zero vorticity are irrotational and hence can be potential flows. For a flow to be irrotational requires that the two derivatives, $\partial V_y / \partial x$ and $\partial V_x / \partial y$, be equal. This is illustrated by the potential flow described by

$$\phi_5 = -xy \qquad V_x = \frac{-\partial\phi_5}{\partial x} = y \qquad V_y = \frac{-\partial\phi_5}{\partial y} = x \tag{16.S}$$

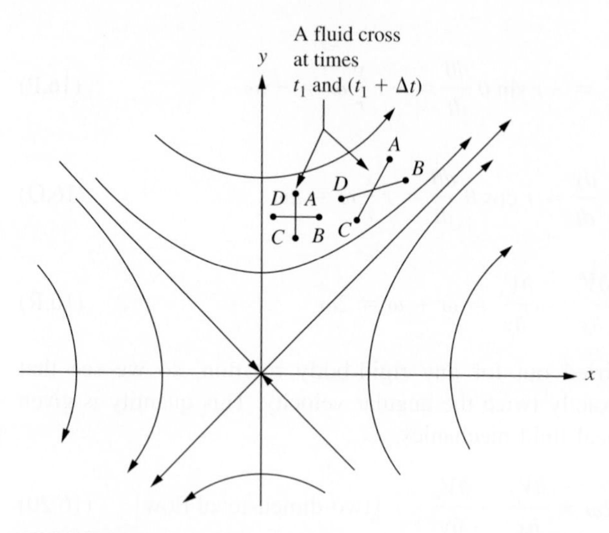

FIGURE 16.10
Flow described by the velocity potential $\phi_5 = -xy$. The lines $x = y$ and $x = -y$ are streamlines, which can be replaced in perfect fluid flow by solid surfaces, so that this is the flow into a corner, reproduced in all four quadrants.

This flow satisfies Laplace's equation and does have the irrotational property (Eq. 16.22); it is sketched in Fig. 16.10.

On this flow we have marked four particles of fluid at *A, B, C,* and *D*. At time t_1 these are at the corners of a cross. Now, if we follow them to time $t_1 + \Delta t$, we see that they have deformed into a flattened figure. The line *AC* is seen to be rotating in a clockwise direction, because the x component of the velocity increases in the y direction, and point *A* moves to the right faster than point *C* does. However, line *BD* is rotating counterclockwise, because the y component of the velocity increases in the x direction. We may show that these lines are rotating in the opposite directions at the same speed, so that, although the fluid is being deformed by the flow, it has no net rotation.

The flow shown in Fig. 16.10 is representative of the flow into a square corner. This may be seen by noting that the lines $y = x$ and $y = -x$ are both streamlines; thus, there is no flow across them.

To see that this irrotational property can exist in a flow in which the streamlines are circles, consider the potential flow described by

$$\phi_6 = -A \arctan \frac{y}{x} = -A\theta \tag{16.T}$$

Here both the rectangular and polar forms of ϕ_6 are shown. In polar coordinates the velocity components are

$$V_r = \frac{-\partial \phi_6}{\partial r} = 0 \qquad V_\theta = \frac{-1}{r} \cdot \frac{\partial \phi_6}{\partial \theta} = \frac{A}{r} \tag{16.U}$$

So at any point the velocity component toward the origin (V_r) is zero; thus, the streamlines are circles, and the equipotential lines are rays passing through the origin. This flow is sketched in Fig. 16.11.

In this figure we see that, if we mark a fluid cross *ABCD* at time t_1 and then look at it again at time $t_1 + \Delta t$, line *BD* has rotated in the counterclockwise direction, but line *AC* has rotated in the clockwise direction, because the fluid at *C* is moving much faster than the fluid at *A*. Thus, although the streamlines in this flow are all circles, the individual particles of fluid are not rotating. We can demonstrate this by

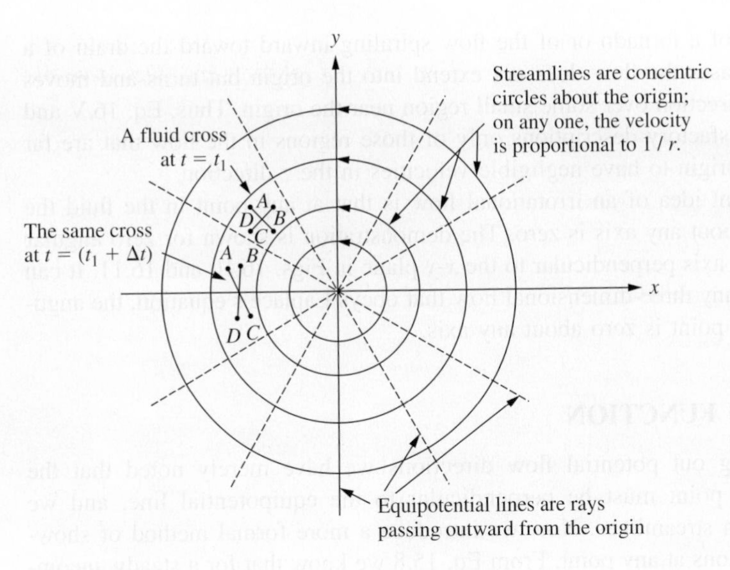

FIGURE 16.11
Flow described by the velocity potential $\phi_6 = -A \arctan(y/x) = -A\theta$ (called a *free vortex*). The circumferential velocity increases toward the center, proportional to $(1/r)$.

placing a float on such a flow and observing that it moves in a circle but maintains its x-y orientation, just as a compass needle would if the compass were moved in a circle. For the flow in Fig. 16.9, such a float would keep the same side radially outward at all times, making one rotation for every trip around the origin.

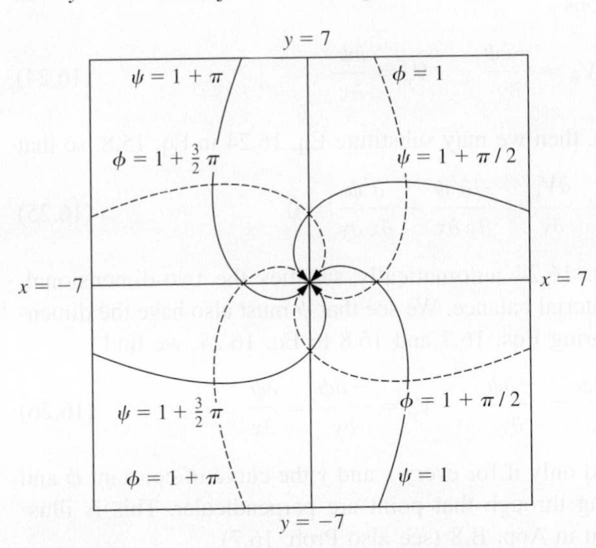

FIGURE 16.12
Flow described by the velocity potential $\phi_7 = -A\theta + C \ln r$. Drawn to scale for $A = 1$ and $C = 1$. The markings $\psi = 1 + \pi$, etc. are discussed in Section 16.5. Each streamline and potential line make an infinite number of circles about the origin as r approaches zero; this drawing only shows them entering the region in which they begin their infinite number of circles.

The flow shown in Fig. 16.11 is called a *free vortex;* it is not common alone in nature. However, the combination of a free vortex with the sink shown in Fig. 16.6, obtained by adding the potential functions, produces

$$\phi_7 = \phi_3 + \phi_6 = -A\theta + C \ln r \quad (16.V)$$

which describes a flow in which fluid spirals into a central sink; see Fig. 16.12.

Figure 16.12 is a fair description of most of the

flow into the eye of a tornado or of the flow spiraling inward toward the drain of a bathtub. In both cases the flow does not extend into the origin but turns and moves in the $+z$ or $-z$ direction over some small region near the origin. Thus, Eq. 16.V and Fig. 16.12 are satisfactory descriptions only of those regions in the flow that are far enough from the origin to have negligible velocities in the z direction.

The important idea of an irrotational flow is that at any point in the fluid the angular velocity about any axis is zero. The demonstration is shown for zero angular velocity about any axis perpendicular to the x-y plane in Figs. 16.10 and 16.11. It can be shown that for any three-dimensional flow that obeys Laplace's equation, the angular velocity at any point is zero about any axis.

16.5 STREAM FUNCTION

So far in mapping out potential flow directions we have merely noted that the streamline at any point must be perpendicular to the equipotential line, and we have sketched such streamlines. Now we introduce a more formal method of showing the flow directions at any point. From Eq. 15.8 we know that for a steady, incompressible, two-dimensional flow

$$\frac{\partial V_x}{\partial x} + \frac{\partial V_y}{\partial y} = 0 \qquad [\text{two-dimensional flow}] \tag{16.23}$$

We now arbitrarily define a new function ψ, called the *stream function,* or *Lagrange stream function,* by the equations

$$V_x = \frac{-\partial \psi}{\partial y} \qquad V_y = \frac{\partial \psi}{\partial x} \tag{16.24}$$

If ψ has continuous derivatives, then we may substitute Eq. 16.24 in Eq. 15.8, so that

$$\frac{\partial V_x}{\partial x} + \frac{\partial V_y}{\partial y} = \frac{-\partial^2 \psi}{\partial y\, \partial x} + \frac{\partial^2 \psi}{\partial x\, \partial y} = 0 \tag{16.25}$$

So any flow that satisfies Eq. 16.24 automatically satisfies the two-dimensional, steady-flow, incompressible-material balance. We see that ψ must also have the dimension of ft^2 / s or m^2 / s. Comparing Eqs. 16.7 and 16.8 to Eq. 16.24, we find

$$V_x = \frac{-\partial \phi}{\partial x} = \frac{-\partial \psi}{\partial y} \qquad V_y = \frac{-\partial \phi}{\partial y} = \frac{\partial \psi}{\partial x} \tag{16.26}$$

Equations 16.26 can be satisfied only if for every x and y the curve of constant ϕ and the curve of constant ψ passing through that point are perpendicular. This is illustrated by Fig. 16.13 and proven in App. B.8 (see also Prob. 16.7).

Since streamlines also have the property of being everywhere perpendicular to equipotentials, all the streamlines that appear in Figs. 16.3, 16.4, 16.5, 16.7, 16.8, and 16.10 are lines of constant ψ.

The stream function has an intuitive explanation, shown in Fig. 16.14. If, as discussed above, curves of constant ψ are streamlines, then there can be no flow across a curve of constant ψ. In Fig. 16.14 the entire flow that is passing between the curves

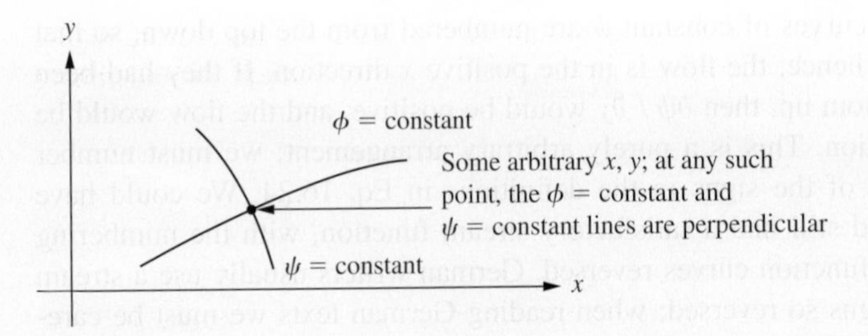

FIGURE 16.13
At any point in an irrotational flow, the lines of constant stream function and constant velocity potential are perpendicular; see App. B.8.

$\psi = 1$ and $\psi = 2$ at A must also be passing between them at B. However, the space between them at B is less than that at A, so there is less area available to the flow at B; hence, by mass balance the velocity must be greater. From Eq. 16.24 we see that the velocity in the x direction can be greater at B than at A only if the curves of constant ψ are closer together at B; that is $-\partial\psi / \partial y$ is greater at B than at A. We may thus think of the curves of constant ψ as being the boundaries of flow channels; as they squeeze together, the flow between them must go faster.

If we imagine the flow shown in Fig. 16.14 as being h ft deep in the z direction, then the flow passing between lines $\psi = 2$ and $\psi = 1$ at A is

$$Q = \int_{\psi=2\text{ ft}^2/\text{s}}^{\psi=1\text{ ft}^2/\text{s}} V_x \, dA = \int_{\psi=2\text{ ft}^2/\text{s}}^{\psi=1\text{ ft}^2/\text{s}} V_x h \, dy = h \int_{\psi=2\text{ ft}^2/\text{s}}^{\psi=1\text{ ft}^2/\text{s}} -\left(\frac{\partial\psi}{\partial y}\right)_x dy$$

$$= h \int_{\psi=1\text{ ft}^2/\text{s}}^{\psi=2\text{ ft}^2/\text{s}} d\psi = h[\psi]_{1\text{ ft}^2/\text{s}}^{2\text{ ft}^2/\text{s}} = h(2-1)\frac{\text{ft}^2}{\text{s}} = h\frac{\text{ft}^2}{\text{s}} \tag{16.W}$$

(Here h has the dimension of a length, so that the dimension of the volumetric flow rate is indeed correct.) Thus, if on a flow field curves of constant ψ are drawn at equal intervals of ψ, then the volumetric flow between each two such successive curves must be the same for each two (e.g., in Fig. 16.14 the volumetric flow between $\psi = 1$ and $\psi = 2$ is the same as the volumetric flow between $\psi = 2$ and $\psi = 3$, etc.).

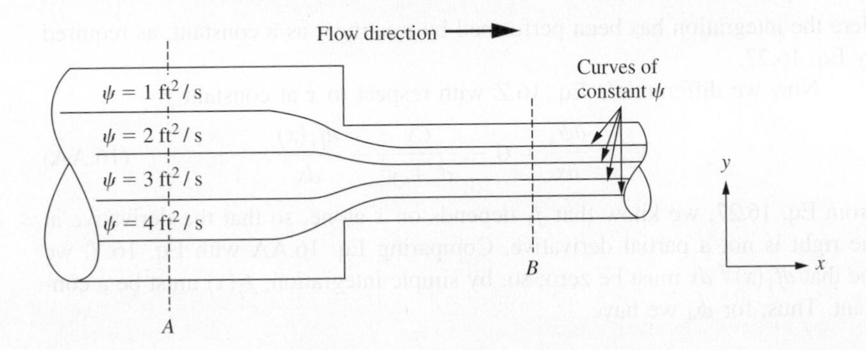

FIGURE 16.14
Stream function for steady flow in a channel transition of constant depth. The velocity must increase as the stream function lines crowd together at the right of the figure.

In Fig. 16.14 the curves of constant ψ are numbered from the top down, so that $\partial\psi / \partial y$ is negative and hence, the flow is in the positive x direction. If they had been numbered from the bottom up, then $\partial\psi / \partial y$ would be positive, and the flow would be in the negative x direction. This is a purely arbitrary arrangement; we must number them this way because of the signs in the definitions in Eq. 16.24. We could have reversed those signs and still had a satisfactory stream function, with the numbering direction of the stream function curves reversed. German writers usually use a stream function that has the signs so reversed; when reading German texts we must be careful to observe this sign difference.

If we know the potential function that describes some flow, we can compute the stream function, and conversely. The calculation is based on the following general integration property of partial derivatives. If $A = A(B, C)$, where A, B, and C are any mathematical functions, then

$$A = \left[\int \left(\frac{\partial A}{\partial B} \right)_C dB \right]_{C=\text{const.}} + f_1(C) \tag{16.27}$$

Here $f_1(C)$ is some function of C alone. If we now let A be ψ, and if we know ϕ, we can immediately write down the necessary partial derivatives and integrate.

Example 16.3. To illustrate the use of Eq. 16.27, find the ψ that corresponds to ϕ_4 (Eq. 16.L).

Using Eqs. 16.26, we find for ϕ_4,

$$V_x = -\frac{\partial \phi_4}{\partial x} = -\frac{\partial \psi_4}{\partial y} = -\left(-A + \frac{Cx}{x^2 + y^2} \right) \tag{16.X}$$

and

$$V_y = -\frac{\partial \phi_4}{\partial y} = +\frac{\partial \psi_4}{\partial x} = -\frac{Cy}{x^2 + y^2} \tag{16.Y}$$

Letting B in Eq. 16.27 be y in Eqs. 16.X and 16.Y, we find

$$\psi_4 = \int \left(-A + \frac{Cx}{x^2 + y^2} \right) dy + f_1(x) = -Ay + C \arctan \frac{y}{x} + f_1(x) \tag{16.Z}$$

Here the integration has been performed by treating x as a constant, as required by Eq. 16.27.

Now we differentiate Eq. 16.Z with respect to x at constant y:

$$V_y = \frac{\partial \psi_4}{\partial x} = 0 - \frac{Cy}{x^2 + y^2} + \frac{df_1(x)}{dx} \tag{16.AA}$$

From Eq. 16.27, we know that f_1 depends on x alone, so that the derivative at the right is not a partial derivative. Comparing Eq. 16.AA with Eq. 16.Y, we see that $df_1(x) / dx$ must be zero; so, by simple integration, $f_1(x)$ must be a constant. Thus, for ϕ_4 we have

$$\psi_4 = -Ay + C \arctan \frac{y}{x} + \text{some constant} \tag{16.AB}$$

■

One may show (Prob. 16.8) that, if we choose to make the B in Eq. 16.27 the x in Eqs. 16.X and 16.Y, we obtain the same result. By plotting various lines of constant ψ given by Eq. 16.AB, we may see that the lines of constant ψ given by Eq. 16.AB are the streamlines in Fig. 16.8.

Example 16.4. Repeat Example 16.3 for ϕ_3, which is shown in rectangular coordinates in Eq. 16.C and in polar coordinates in Eq. 16.F. It is easier to work in polar coordinates. The stream function in polar coordinates takes the form

$$V_r = \frac{-1}{r} \cdot \frac{\partial \psi_3}{\partial \theta} \tag{16.28}$$

$$V_\theta = \frac{\partial \psi_3}{\partial r} \tag{16.29}$$

Thus, from Eq. 16.C we have

$$V_r = \frac{-\partial \phi_3}{\partial r} = \frac{-1}{r} \cdot \frac{\partial \psi_3}{\partial \theta} = \frac{C}{r} \tag{16.AC}$$

$$V_\theta = \frac{-1}{r} \cdot \frac{\partial \phi_3}{\partial \theta} = \frac{\partial \psi_3}{\partial r} = 0 \tag{16.AD}$$

So, letting B in Eq. 16.27 be θ and C be r, we find

$$\psi_3 = \int -C\, d\theta + f_1(r) = -C\theta + f_1(r) \tag{16.AE}$$

$$\left(\frac{\partial \psi_3}{\partial r}\right)_\theta = \frac{df_1(r)}{dr} \tag{16.AF}$$

but from Eq. 16.AD we know that this is zero, so $f_1(r)$ must be a constant and

$$\psi_3 = -C\theta + \text{constant} \tag{16.AG}$$

So the streamlines are rays passing to the origin at various angles, as shown in Fig. 16.6. ■

An interesting property of the stream function and the velocity potential is that, if for a given flow $\phi_8 = f_1(x, y)$ and $\psi_8 = f_2(x, y)$, then there is another flow given by $\phi_9 = f_2(x, y)$ and $\psi_9 = f_1(x, y)$. This second flow has exactly the same map as the first, except that the labels are reversed; the streamlines on one are equipotentials on the other, and conversely. For example, Figs. 16.6 and 16.11 bear this relation one to the other. This property is also a property of the real and imaginary parts of any analytic complex function. Because the stream function and velocity potential have this property in common with the real and imaginary parts of an analytic complex function, they can be manipulated by the rules for complex functions. In particular, they obey the rules of *conformal mapping,* a complex-function procedure that is widely used both in heat flow and in potential flow of fluids [6, p. 66].

The discussion has all been for the application of the stream function to potential flows. However, it can also sometimes be used to advantage for flows that are not irrotational and for which a potential function cannot exist (see Probs. 16.10 and 17.4). This is possible because the stream function is used in these cases simply as a way of combining the mass balance with the other pertinent equations.

The entire discussion of the stream function has been for two-dimensional flow. The definition of a satisfactory stream function for three-dimensional flows is more difficult. However, if the flow is symmetrical about some axis, e.g., the uniform flow around some body of revolution, then it is possible to define a different stream function that is convenient for that problem. This three-dimensional stream function is called *Stokes' stream function* [6, p. 125], to distinguish it from the *Lagrange stream function,* the one discussed in this chapter.

16.6 BERNOULLI'S EQUATION FOR TWO-DIMENSIONAL, PERFECT-FLUID, IRROTATIONAL FLOWS

The velocity-potential stream-function methods shown in the preceding sections allow us to calculate the flow velocity and direction at any point in a two-dimensional, perfect-fluid, irrotational flow. Sometimes this is all the information sought. More often the desired information is the force on some body immersed in the flow, for example, the lift and drag of a wing section or the drag on a particle settling through a fluid.

In Chap. 5 we derived B.E. from the energy balance equation. Since the energy balance has no one-dimensional restriction on it, the same approach must apply to two- and three-dimensional flows. However, in our derivation of B.E. we restricted our attention to systems with only one flow in and out. How can we apply this idea to a two-dimensional flow field in which there is a continuously varying velocity over some region of space? In Fig. 16.15 such a region is shown with no sources, sinks, or solid bodies, but with streamlines.

At A we draw a closed curve (e.g., a circle) around a streamline, perpendicular to the flow. Then we draw streamlines from every point in that closed curve, in the direction of flow. Since streamlines cannot cross or separate except at a source, a sink, or the edge of a solid body, we can draw a closed curve farther downstream (e.g., at B) through which all the streamlines that pass through the curve at A must also pass. This curve at B probably will not have the same shape as the curve at A, but it will exist. Such a set of streamlines, which form a closed curve in any plane perpendicular to the flow, is called a *stream tube*. Let us choose the tube from A to B as our system. Then, for steady flow there is only flow into or out of the stream tube at the ends A and B.

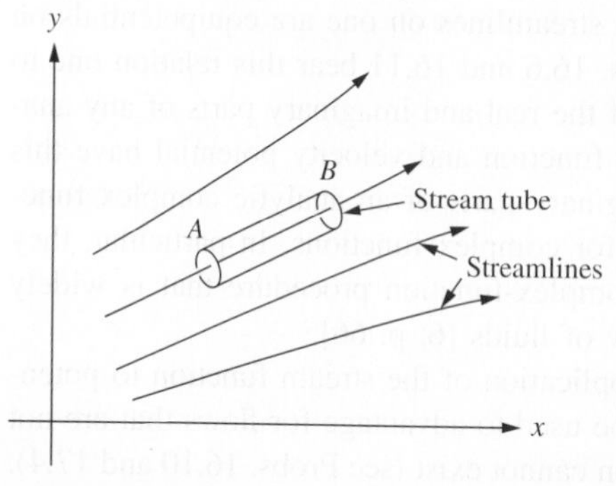

FIGURE 16.15
A stream tube is a small region of space surrounding a streamline.

In deriving B.E. we assumed that the flow into and out of the system was of uniform velocity, etc. This will not be true in general of any stream tube, because the velocity may be different from one streamline to the next. However, if we make our stream tube smaller and smaller, then the nonuniformity of the flow across its entrance (and exit) becomes negligible. In the limit the stream tube can be thought of as being so thin that it shrinks down to just a streamline. For such a stream tube B.E., as derived in Chap. 5, is obviously applicable. This result is true for any kind of incompressible flow; if the flow is frictionless (i.e., an ideal fluid), then the friction term may be dropped and B.E. integrated from some point in the streamline to any adjacent point, to find

$$d\left(\frac{P}{\rho} + gz + \frac{V^2}{2}\right) = 0$$

or

$$\frac{P}{\rho} + gz + \frac{V^2}{2} = \text{constant} \qquad \begin{bmatrix} \text{along a streamline,} \\ \text{frictionless, constant-} \\ \text{density flow} \end{bmatrix} \tag{16.30}$$

Here V is the velocity relative to our fixed coordinate system, given by

$$V = (V_x^2 + V_y^2)^{1/2} = (V_r^2 + V_\theta^2)^{1/2} \tag{16.31}$$

This result allows us to evaluate the pressure on the surface of any body immersed in the flow and hence the force, if only we know the pressure at some upstream position and the total velocity field. (It is proven in various hydrodynamics texts that for *irrotational flow from a single reservoir* the constant in Eq. 16.30 is the same for all streamlines, whereas for rotational flows it is not the same but is only constant along one particular streamline.)

16.7 FLOW AROUND A CYLINDER

To illustrate the idea of potential flow and how one can use it to calculate forces, we calculate the pressure distribution on the surface of a cylinder that is immersed in a perfect-fluid flow perpendicular to it. If this is a very long cylinder, then there will be negligible change in the flow in the direction of the cylinder's axis, and hence the flow will be practically two-dimensional. To find the flow field, we must make a judicious combination of a steady flow, a source, and a sink. Consider first a source and a sink with equal flow rates located some distance A apart on the x axis; see Fig. 16.16. The flow between them is given by

$$\phi_{10} = -C\ln(x^2 + y^2)^{1/2} + C\ln[(x - A)^2 + y^2]^{1/2} \tag{16.AH}$$

Figure 16.16 is a fair representation of the flow between an injection and a production well in a porous medium and hence has some application in petroleum reservoir engineering and hydrology. If the C in Eq. 16.AH is a constant, then, as A gets smaller and smaller, the flow at any point (except a point directly between the source and the sink) must become smaller and smaller, approaching zero, because more and more of the flow will take the direct path from the source to the sink. If we replace the C in Eq. 16.AH with C/A, then the total flow increases just as the distance between sink and source (A) decreases. In this case the flow at any point does not go to zero as A goes to zero; we find the value of ϕ_{10} as A goes to zero by *L'Hôpital's rule:*

$$\phi_{10 \text{ as } A\to 0} = \frac{\displaystyle\lim_{A\to 0} \frac{d}{dA}\left(\frac{-C}{2}\ln\frac{x^2 + y^2}{(x - A)^2 + y^2}\right)}{\displaystyle\lim_{A\to 0}\frac{d}{dA}A} = \frac{-Cx}{x^2 + y^2} \tag{16.AI}$$

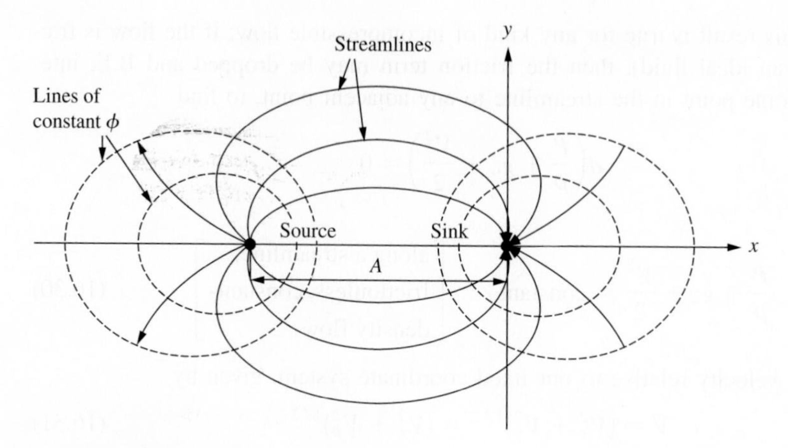

FIGURE 16.16
Flow represented by $\phi_{10} = -C \ln(x^2 + y^2)^{1/2} + C \ln[(x - A)^2 + y^2]^{1/2}$, consisting of a source and a sink, separated by distance A.

This limiting case of a source and a sink at zero distance apart is called a *doublet.* If we now combine this doublet flow with a uniform flow given by $\phi_{11} = -Dx$, we find

$$\phi_{12} = -Dx - \frac{Cx}{x^2 + y^2} \tag{16.AJ}$$

At this point it is convenient to switch to polar coordinates, in which case Eq. 16.AJ becomes

$$\phi_{12} = -Dr\cos\theta - C\frac{r\cos\theta}{r^2} = -\left(\frac{C}{r} + Dr\right)\cos\theta \tag{16.AK}$$

so that

$$V_r = \frac{-\partial\phi_{12}}{\partial r} = -\left(\frac{C}{r^2} - D\right)\cos\theta \tag{16.AL}$$

and

$$V_\theta = \frac{-1}{r}\cdot\frac{\partial\phi_{12}}{\partial\theta} = -\left(\frac{C}{r^2} + D\right)\sin\theta \tag{16.AM}$$

This flow is sketched in Fig. 16.17, in which one of the streamlines is a circle. From Eq. 16.AL, this is the circle for which $V_r = 0$; that is, $r = (C/D)^{1/2}$. Thus, this is a perfect-fluid flow that has a circular streamline. In perfect-fluid theory we can substitute a solid body for any streamline without affecting the flow outside that streamline; so the flow for $r > (C/D)^{1/2}$ is the same as the perfect-fluid flow that would exist outside a circular cylinder oriented perpendicular to the flow.

Example 16.5. Based on the equation for potential flow around a cylinder, use B.E. to estimate the pressure at any point on the surface of the cylinder. We assume that far to the left in Fig. 16.17 the flow is undisturbed by the cylinder,

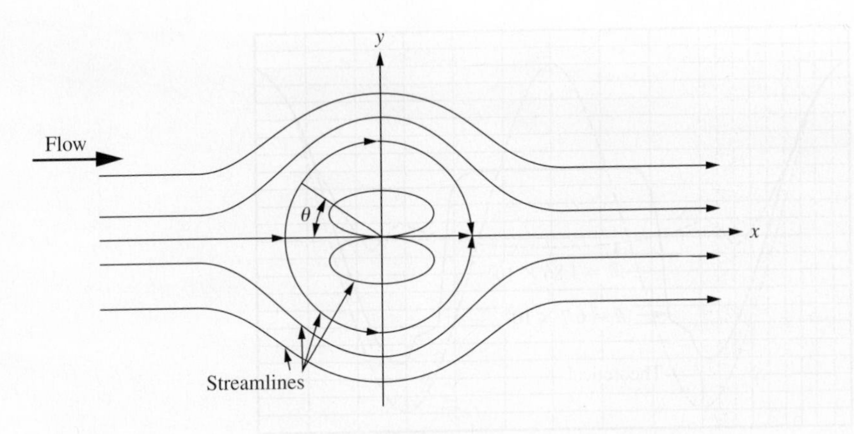

FIGURE 16.17
Streamlines for $\phi_{12} = -[(C/r) + Dr]\cos\theta$. The circular streamline has radius $\sqrt{C/D}$. The flow inside that streamline is normally ignored; the flow outside that streamline is the perfect-fluid flow perpendicular to a long circular cylinder.

so that it is moving from left to right at a uniform velocity V_0 with a uniform pressure P_0 and that the changes in elevation in the entire flow are negligible. Then from Eqs. 16.30 we know that the pressure at any point in the flow is given by

$$P = P_0 + \rho\left(\frac{V_0^2}{2} - \frac{V^2}{2}\right) \tag{16.AN}$$

Here the velocity is given by Eq. 16.31. Along the surface of the cylinder $V_r = 0$; so $V = V_\theta$. Substituting from Eq. 16.AM for $r = (C/D)^{1/2}$, we find

$$(V_\theta)_{\substack{\text{at the surface}\\\text{of the cylinder}}} = -2D\sin\theta \tag{16.AO}$$

We can compare ϕ_{12} to ϕ_1 and see that as x becomes very large (positive or negative) the second term in ϕ_{12} approaches zero, so that the velocity far away from the cylinder, which we have defined as V_0, must equal D. We also see that at the surface of the cylinder $V_r = 0$, so that

$$P_{\substack{\text{surface of the}\\\text{cylinder}}} = P_0 + \rho\frac{V_0^2}{2}(1 - 4\sin^2\theta) \tag{16.AP}$$

■

Now that we have found the perfect-fluid solution for the pressure at various points on the surface of a cylinder, we should ask whether nature really behaves this way. Figure 16.18 shows the pressure at various values of θ calculated from Eq. 16.AP and also the measured pressures at the same angles at two different flow rates. For both of these flow rates there is fair agreement between the observed pressures and Eq. 16.AP along the front of the cylinder (0° to 90° and 270° to 360°), but there is very poor agreement for the back of the cylinder. The explanation is in terms of separation, described in the following section.

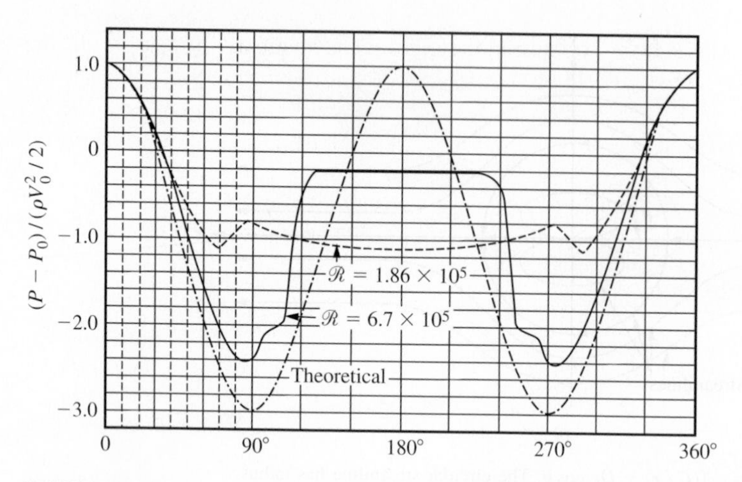

FIGURE 16.18
Pressure on the surface of a cylinder; comparison of the perfect-fluid calculation and experimental observations. The theoretical curve is Eq. 16.AP; the other curves correspond to Reynolds numbers of 1.86×10^5 and 6.7×10^5. The reason for the great difference between these two curves is given in Sec. 17.6. (From H. Muttray, "Die experimentalen Tatsachen des Widerstandes ohne Auftrieb"—"The Experimental Data of Drag Without Lift"—in *Handbuch der Experimentalphysik Hydro- und Aero-Dynamik,* Leipzig: Akademische verlagsgesellschaft m.b.h. p. 316 (1932); based on data by Flaschbart.) $\theta = 0$ corresponds to the front stagnation point, which would be 180° in Fig. 16.17.

16.8 SEPARATION

Figure 16.17 shows the perfect-fluid solution for the flow around a cylinder. The flow splits at the upstream face, flows smoothly around the cylinder, and rejoins at the downstream face. Equation 16.AP shows that at $\theta = 0$ and $\theta = 180°$ the flow velocity is zero; these are the points where the flow divides; hence, the flow has no velocity here. Such zero-velocity points are commonly called *stagnation points.*

The actual flow pattern of a real fluid flowing around a cylinder can be observed by putting dye markers or bits of lint in a fluid. The observed pattern is sketched in Fig. 16.19. To the left (upstream of the cylinder) the flow is very similar to that

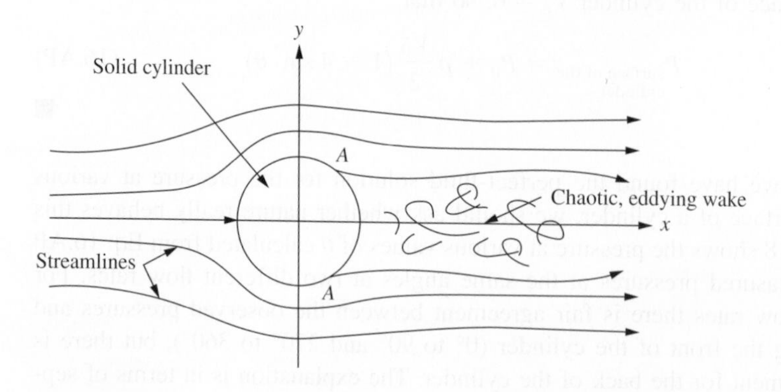

FIGURE 16.19
Real fluid flow around a cylinder at high Reynolds number.

described by Eq. 16.AP for a perfect fluid. However, downstream, instead of closing together behind the cylinder as a perfect fluid would, the flow pulls away from the cylinder at the points marked A, leaving the back of the cylinder covered by an eddying wake. This departure of the streamline from the body around which it is flowing is called *separation.*

Separation is caused by friction. In a real fluid the viscous friction in the boundary layer near the wall of the cylinder causes the fluid to go slower than the corresponding perfect fluid would. On the front of the cylinder this has relatively little effect on the flow pattern, because there the fluid is speeding up and decreasing in pressure. However, on the rear of the cylinder the perfect-fluid flow is one in which the fluid is slowing down and increasing in pressure. For the fluid to reach the rear stagnation point it cannot have lost any momentum due to the effects of viscous friction. Since the real fluid has lost some momentum due to viscous friction, it does not have enough momentum to overcome the *adverse pressure gradient* on the rear of the cylinder, so it flows away, as illustrated in Fig. 16.20.

For the designer of an aircraft, a racing car, or a structure, separation is very troublesome. Comparing the pressures before and behind the cylinder in Fig. 16.18, we see that the average pressure behind the cylinder, although high enough to cause the flow to separate, is not nearly as high as the pressure that would have existed if the flow had not separated. Thus, the cylinder has a large net pressure force acting on it. For an aircraft this would be a drag force, which would require the expenditure of power to overcome. This power is ultimately used up in friction heating in the eddying wake behind the cylinder. Much of the ingenuity of modern aircraft designers has gone into designing special wing structures (flaps, slots, spoilers, jets) to prevent separation [7].

For a *bluff body* like a cylinder or a flat plate perpendicular to the flow, separation is almost inevitable. On the other hand, for a *streamlined body* like an airplane wing or a fish, separation normally does not occur, and the flow pattern is very much like that shown by perfect-fluid theory. Figure 16.21 shows a comparison of the predicted and observed pressures on a "streamlined" body; the agreement between the predictions of perfect-fluid theory and experiment are quite good except at the very rear. Thus, the aeronautical engineers can predict the lift of a "streamlined" wing with fair accuracy by using perfect-fluid theory. However, perfect-fluid theory predicts zero

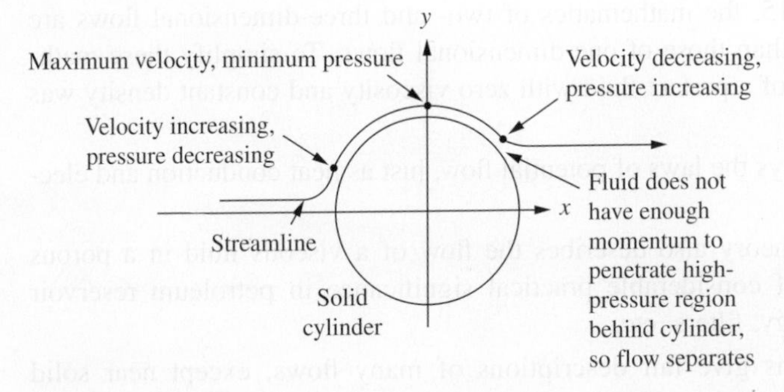

FIGURE 16.20
Separation in flow around a cylinder at high Reynolds number.

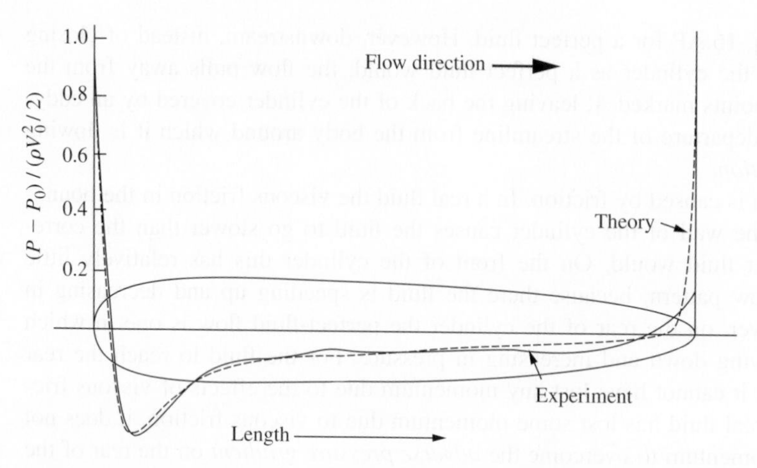

FIGURE 16.21
Comparison of experimental pressures with those calculated from perfect-fluid theory for a streamlined body. (From H. Muttray, "Die experimentalen Tatsachen des Widerstandes ohne Auftrieb"—"The Experimental Data of Drag Without Lift"—in *Handbuch der Experimentalphysik Hydro- und Aero-Dynamik,* Leipzig, Akademische verlagsgesellschaft m.b.h. p. 316 (1932); based on data by Fuhrmann.)

drag, whereas all real-fluid flows show drag; so perfect-fluid theory alone is of little use in drag predictions.

One may define a streamlined body as one for which the real-fluid streamlines cling to the body all along its length without separation; normally such shapes are blunt in front and gradually taper in back.

The development of supercomputers has made it possible to apply the idea of potential flow outside the boundary layer and viscous flow inside the boundary layer to very complex structures and situations. Currently, commercial aircraft designs are "tested" with supercomputers using the combination of potential flow and the boundary layer, much more quickly and cheaply than they could be tested with scale models in wind tunnels or in flight tests [8].

16.9 SUMMARY

1. As shown in Chap. 15, the mathematics of two- and three-dimensional flows are much more difficult than those of one-dimensional flows. To simplify these mathematics, the concept of a perfect fluid with zero viscosity and constant density was invented.
2. This perfect fluid obeys the laws of potential flow, just as heat conduction and electrostatic fields do.
3. This potential-flow theory also describes the flow of a viscous fluid in a porous medium, which is of considerable practical significance in petroleum reservoir engineering, hydrology, filters, etc.
4. Perfect-fluid solutions give fair descriptions of many flows, except near solid boundaries.

5. To handle complex flows involving solid boundaries, Prandtl introduced the idea of using perfect-fluid theory far from the solid surface and taking viscosity into account only in a thin boundary layer near the surface of the solid.
6. Potential flows are irrotational.
7. For many real flows the streamlines separate from the body around which they are flowing. This results in the formation of eddying wakes, low pressure behind the body, and large drag forces. This is not predictable by perfect-fluid theory, but it can be approached through boundary-layer theory.

PROBLEMS

See the Common Units and Values for Problems and Examples inside the back cover. An asterisk (*) on a problem number indicates that the answer is given in App. D.

16.1. Describe the heat conduction and electrostatic fields that correspond to Figs. 16.4, 16.6, 16.8, and 16.16.

16.2. The flow described by Fig. 16.6 may be thought of as an oil well drawing fluid from a horizontal stratum. The oil stratum is 10 ft thick, and the flow is 100 ft^3 / h of a fluid with $\rho = 50$ lbm / ft^3 and $\mu = 10$ cP. The permeability of the stratum is 10^{-11} ft^2. The pressure at the inside of the well (radius 0.25 ft) is 1000 psia. Prepare a sketch of the pressure versus distance from the center of the well.

16.3. Map out the following potential flows. In each case verify that Laplace's equation is satisfied, calculate the equation of the stream function, and indicate to what physical situation these flows might correspond:

(*a*)
$$\phi_{13} = A(x^2 - y^2) \tag{16.AQ}$$

(*b*)
$$\phi_{14} = Ar^n \cos n\theta / n \text{ for } n = \tfrac{1}{2}, \tfrac{2}{3}, 1, \tfrac{3}{2}, \text{ and } 2 \tag{16.AR}$$

(*c*)
$$\phi_{15} = Bx + C \ln(x^2 + y^2) - C \ln[(x - A)^2 + y^2] \tag{16.AS}$$

16.4. Map out the flows predicted by $\phi_{16} = x^2$, $\phi_{17} = x^2 + y^2$, $\phi_{18} = -e^x$, and $\phi_{19} = \sin x$, and show that each of these does indeed result in a flow in which mass is not conserved (for a constant-density fluid), as discussed in Sec. 16.3.

16.5. Show the derivation of Eqs. 16.15, 16.16, and 16.17. *Hint:* These transformations are made in terms of the "chain rules," as shown in any text on advanced calculus.

16.6. (*a*) Show that the vorticity (Eq. 16.20) in plane polar coordinates is given by

$$\zeta = V_\theta / r + \partial V_\theta / \partial r - (1 / r)(\partial V_r / \partial \theta) \tag{16.32}$$

(*b*) Using Eq. 16.32, show that the flow described by Eq. 16.N is not irrotational, but that the flow described by Eq. 16.T is irrotational.

16.7. Show that Eqs. 16.26 can be simultaneously satisfied only if both ϕ and ψ each satisfy Laplace's equation, Eq. 16.4. *Hint:* Differentiate V_x with respect to y at constant x and V_y with respect to x at constant y, and then subtract one equation from the other.

16.8. Show that if we choose B in Eq. 16.27 to be x in Eqs. 16.X and 16.Y, then we find the same ψ shown in Eq. 16.AB. *Hint:* Use the identity that $\arctan x = \pi / 2 - \arctan(1 / x)$.

16.9. Find the stream functions that correspond to ϕ_2, ϕ_5, ϕ_6, and ϕ_7.

16.10. Although the stream function is most often used with the potential function, it can be used for viscous flows for which no potential function exists. For example, the laminar flow of a viscous fluid along a sloping flat plate is given by the equation

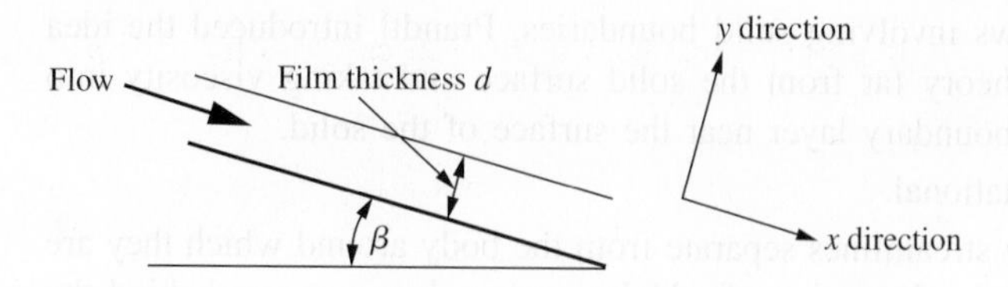

FIGURE 16.22
Laminar flow of a viscous liquid down an inclined plane; see Prob. 16.10.

$V_x = [(\rho g d^2 \cos\beta)/2\mu][1 - (y/d)^2]$, as described in Bird, Stewart, and Lightfoot [9, p. 45]; see Fig. 16.22. Calculate the stream function for this flow. Show that this flow cannot be represented by any potential function.

16.11. Show that one cannot calculate a ϕ from the ψ calculated in Prob. 16.10 by the method shown in Sec. 16.5 for finding ϕ from ψ. To do this, calculate ϕ from ψ, and show that the resulting function does not give back the correct values of $\partial\phi/\partial x$ and $\partial\psi/\partial y$ on differentiation. The reason this cannot be done is that Eqs. 16.26 hold only (as may be shown) if both ϕ and ψ satisfy Laplace's equation, and the ψ found in Prob. 16.10 does not satisfy Laplace's equation.

16.12.*The potential flow of a frictionless fluid with $\rho = 60\ \text{lbm}/\text{ft}^3$ is described by

$$\phi_{19} = \left(-10\frac{1}{\text{s}}\right)(x^2 - y^2) \tag{16.AT}$$

The pressure at $x = 0$, $y = 0$ is 20 psia. Calculate the pressure at $x = 1$ ft and $y = 1$ ft; at $x = 5$ ft and $y = 0$.

16.13. In Prob. 16.3(*a*) you showed that equation ϕ_{13} described the frictionless potential flow into a rectangular corner. At x_0 and $y = 0$ (i.e., location x_0 on the positive x axis) $P = 1.00$ atm and $V = 1$ ft / s in the minus x direction. Show the equation for the pressure as a function of position along the positive y axis (i.e., for all points for which $x = 0$ and $y \geq 0$.

16.14. A tornado is described by ϕ_7. At a distance 1 mi from the eye of the tornado (i.e., the origin of the coordinate system) the radial velocity is 1 mi / h inward toward the origin and the tangential velocity is 1 mi / h.

(*a*) What are the radial and tangential velocity components 50 ft from the center of the tornado?

(*b*) If the pressure is 14.7 psia 1 mi from the center of the tornado, what is the pressure 50 ft from the center?

16.15.*In Prob. 16.14, if we put a marker (e.g., a neutrally buoyant balloon) in the air at a point 1 mi from the center of the tornado, how long will it take it to move to a point 50 ft from the center?

16.16.*Figure 16.16 represents the flow from an injection well to a production well in a porous medium. How much of the flow between the two wells passes outside the streamlines that pass through the points $(A/2, A/2)$ and $(A/2, -A/2)$?

16.17. Figure 16.8 is drawn for $A = 1$ and $C = 1$. Sketch the equivalent of this figure

(*a*) For $A = 1$ and $C = 10$.

(*b*) For $A = 10$ and $C = 1$.

16.18. Figure 16.12 is drawn for $A = 1$ and $C = 1$. Sketch the equivalent of this figure

(*a*) For $A = 1$ and $C = 10$.

(*b*) For $A = 10$ and $C = 1$.

16.19. In Chap. 2 we showed that if we put a real fluid in an open can on a turntable and rotated it, after a long time the fluid would be in solid-body rotation, which corresponds to the forced vortex in Fig. 16.9. Suppose that we replaced the real fluid with a perfect fluid and repeated the experiment. After a long time, what would the flow in the can be? What does this tell you about how one would produce vorticity in a perfect fluid?

REFERENCES FOR CHAPTER 16

1. Prandtl, L. "Ueber Fluessigkeitsbewegung mit kleiner Reibung" ("Concerning Fluid Movement with Small Friction"), *Verhlandl III Intern. Math.-Kongr.* Heidelberg, 1904. Also reprinted in *Vier Abhandlungen zur Hydrodynamik.* Goetingen, 1927.
2. Carslaw, H. S., J. C. Jaeger. *Conduction of Heat in Solids.* 2nd ed. Oxford: Oxford University Press, 1959.
3. Prandtl, L., O. G. Tietjens. *Fundamentals of Hydro- and Aeromechanics,* trans. L. Rosenhead. New York: Dover, 1957.
4. Porzikidis, C. *Little Book of Streamlines.* New York: Academic Press, 1999.
5. Kirchoff, R. H. *Potential Flows; Computer Graphic Solutions.* New York: Marcel Dekker, 1985.
6. Lamb, H. *Hydrodynamics.* New York: Dover, 1945.
7. Hazen, D. C., C. J. Lynch. "Better Airplane Wings." *Intern. Sci. Technol. 50(2),* 1966, pp. 38–46.
8. Tinoco, E. N., A. W. Chen. "CFD Applications to Engine/Airframe Integration." In "Numerical Methods for Engine-Airframe Integration." *Progress in Astronautics and Aeronautics 102,* 1986, pp. 219–255.
9. Bird, R. B., E. N. Stewart, W. E. Lightfoot. *Transport Phenomena,* 2nd ed. New York: Wiley, 2002.

CHAPTER 17

THE BOUNDARY LAYER

An introduction to the topic of the boundary layer is given in Sec. 16.1. The topic is a very large one. It is an active field of fluid mechanics research, so new results are constantly being published. Here we cannot hope to cover the entire topic; rather, we intend to show by a few examples what types of solution are obtainable and to impart some feeling for the results of the boundary-layer approach. More terminology will be introduced than is necessary for the subject actually treated. This terminology is in common usage in the boundary-layer literature; it is introduced here to show the student how these common terms relate to the other subjects treated in this book.

17.1 PRANDTL'S BOUNDARY-LAYER EQUATIONS

Ludwig Prandtl, the father of the boundary-layer theory, after making the conceptual division of the flow discussed in Sec. 16.1, set out to calculate the flow in the boundary layer. He chose as his starting point the Navier-Stokes equations (Sec. 15.4) and simplified them by dropping the terms he considered unimportant. His simplifications are as follows:

1. The solid surface is taken as the x axis, the boundary layer beginning at the origin; see Fig. 17.1.
2. Gravity is unimportant compared with the other forces acting, so the gravity term can be dropped.

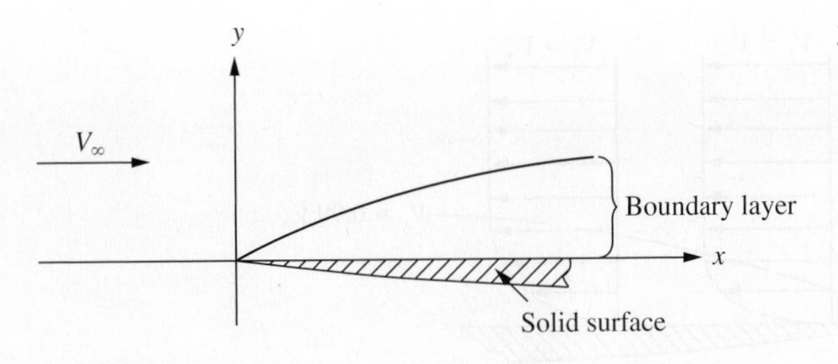

FIGURE 17.1
The boundary layer on a thin, flat plate.

3. The flow is two-dimensional in the x and y directions. This means that V_z (the z component of the velocity) is zero, as are all derivatives with respect to z. These simplifications make the z momentum balance all zeros, so it can be dropped from the list of equations to be solved.
4. Although there is some flow in the y direction within the boundary layer, it is slow enough compared with the flow in the x direction for us not to need to consider the y-directed momentum balance. This does not mean that V_y is zero, but it does mean that $\partial P / \partial y$ is negligible.
5. The $\mu(\partial^2 V_x / \partial x^2)$ term in the momentum balance is small compared with the $\mu(\partial^2 V_x / \partial y^2)$ term and may be dropped.

Making these simplifications in Eq. 15.26, dividing by ρ, and replacing μ / ρ by ν (the kinematic viscosity), Prandtl found

$$\frac{\partial V_x}{\partial t} + V_x \frac{\partial V_x}{\partial x} + V_y \frac{\partial V_x}{\partial y} = -\frac{1}{\rho}\frac{\partial P}{\partial x} + \nu \frac{\partial^2 V_x}{\partial y^2} \tag{17.1}$$

This equation and the two-dimensional, constant-density material balance (see Eq. 15.8)

$$\frac{\partial V_x}{\partial x} + \frac{\partial V_y}{\partial y} = 0 \tag{17.2}$$

are referred to as *boundary-layer equations* or *Prandtl's boundary-layer equations.*

In boundary-layer flows, as in flow in a pipe, the flow can be laminar or turbulent. Prandtl's equations, as shown above, are limited to laminar flows because of the form of the fluid friction term. The analogous form for turbulent flows can be derived [1, Chap. 18] but is of little use.

17.2 THE STEADY-FLOW, LAMINAR BOUNDARY LAYER ON A FLAT PLATE PARALLEL TO THE FLOW

As an example of the use of the boundary-layer equations, we consider the simplest possible boundary-layer problem, the steady flow of a constant-density, Newtonian fluid past a flat plate placed parallel to the flow at velocities low enough for the flow to be laminar everywhere, shown schematically in Fig. 17.2

To find a complete description of this flow, we must find a function, $V_x = V_x(x, y)$, that satisfies Eqs. 17.1 and 17.2 together with the conditions that at $y = 0$ and $x > 0$ the velocity components V_x and V_y are zero (that is, no flow along or across the solid surface) and that, as y becomes large, V_x becomes the same as the perfect-fluid flow (see Chap. 16) outside the boundary layer.

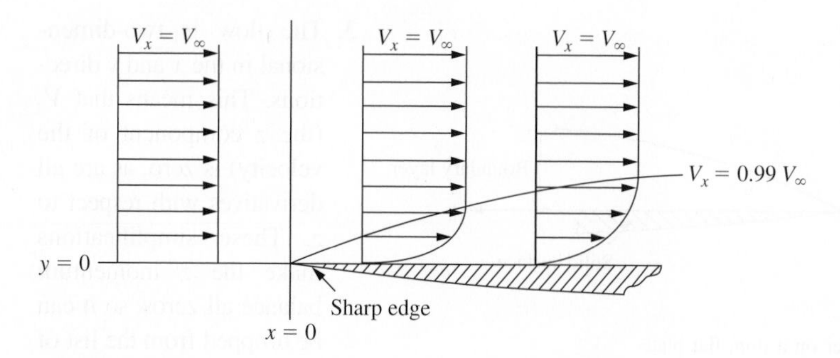

FIGURE 17.2
The velocity distribution in the boundary layer on a thin, flat plate.

For the system shown in Fig. 17.2 the perfect-fluid flow is described by

$$\phi = -V_\infty x \tag{17.3}$$

so that, outside the boundary layer, $V_x = V_\infty$ and $V_y = 0$ for all x and y. Thus, the second condition is that, as y becomes large, V_x must approach V_∞.

From the perfect-fluid solution we find $\partial P / \partial x = 0$. This is true outside the boundary layer; inside the boundary layer it may not be exactly true but, according to Prandtl's third assumption, there is no change in pressure in the y direction inside the boundary layer. Therefore, at any point in this boundary layer the pressure is the same as the pressure at that x in the perfect-fluid flow outside the boundary layer. Boundary-layer experts describe this assumption by saying that the pressure in the perfect-fluid flow outside the boundary layer is *impressed* on the boundary layer. Thus, according to this assumption, the $\partial P / \partial x$ term in Eq. 17.1 can be dropped, giving

$$V_x \frac{\partial V_x}{\partial x} + V_y \frac{\partial V_x}{\partial y} = \nu \frac{\partial^2 V_x}{\partial y^2} \tag{17.4}$$

This problem of determining the flow velocity at every point near the plate is a boundary-value problem, i.e., a set of partial differential equations with specified values on the boundaries. To see how it was solved, we consider first a simpler boundary-value problem. Suppose that an infinite fluid at rest is adjacent to an infinite plane wall at rest. At time zero the wall is suddenly set in motion in the x direction with velocity V_0. If the flow is laminar throughout, then from the momentum balance (Prob. 17.1) for any particle of fluid we find

$$\frac{\partial V_x}{\partial t} = \nu \frac{\partial^2 V_x}{\partial y^2} \tag{17.5}$$

together with a set of boundary conditions. This problem is an approximate description of the flow near the flat side of a ship that starts suddenly.

This boundary-value problem can be completely and rigorously solved by the method of separation of variables or by Laplace transforms. The same equation, with variables renamed, is solved in all heat transfer books and appears in Chaps. 19 and 20. The solution is:

$$V_x = V_0\left(1 - \text{erf}\frac{y}{2(\nu t)^{1/2}}\right) \tag{17.6}$$

Here "erf" is Gauss's error function, defined by

$$\text{erf}\, x = \frac{2}{\pi^{1/2}}\int_0^x e^{-\lambda^2}\, d\lambda \tag{17.A}$$

where λ is a dummy variable. The values of erf are presented in most books of mathematical tables and Fig. 19.5.

The interesting point about Eq. 17.6 is that, although V_x depends on both y and t, the solution indicates that it does not depend on them separately but, rather, depends only on a fixed combination of them, $y/(\nu t)^{0.5}$. If we introduce a new variable, $A = y/(\nu t)^{0.5}$, then the velocity, instead of being a function of y and t, becomes a function of A alone. This is a great simplification. Mathematically, it means that the partial differential equation, Eq. 17.5, can be replaced with an ordinary differential equation. Graphically, it means that instead of representing V_x by a plot of V_x versus y with lines of constant t, we can represent V_x as a single curve on a plot of V_x versus A. This simplification results from the character of the differential equation and its boundary values.

Now we may reconsider the problem of the boundary layer on the flat plate. Blasius [1, p. 156] observed that, although Eqs. 17.4 and 17.5 are not exactly the same, they have a similar form. He also noted the physical similarity between the situations described; if we consider the fluid to be at rest and the boundary layer to be formed by a sharp-edged plate moving through it at constant velocity (that is, Fig. 17.2 as seen by an observer riding with the fluid), then this situation physically has much in common with the flow described by Eq. 17.5. Therefore, Blasius assumed that the solution would be of the same form: that V_x, instead of depending on x and y separately, would depend only on some combination of them. Comparing the physical situation with that in the problem described by Eq. 17.5, he decided that the t in Eq. 17.6 is the time that the fluid has "known" that the plate is moving. For the boundary-layer problem this would be the distance x from the leading edge of the plate divided by the free-stream velocity V_∞. Making this substitution, he defined

$$\eta = y\left(\frac{V_\infty}{\nu x}\right)^{1/2} \tag{17.7}$$

By comparison with Eq. 17.6, he assumed he could write

$$\frac{V_x}{V_\infty} = \text{some function of}\left[y\left(\frac{V_\infty}{x\nu}\right)^{1/2}\right] \tag{17.8}$$

In the case of the other boundary-value problem mentioned, we can demonstrate mathematically that such a substitution is correct; here we cannot make such a demonstration, so the resulting solution rests on this additional assumption. This assumption

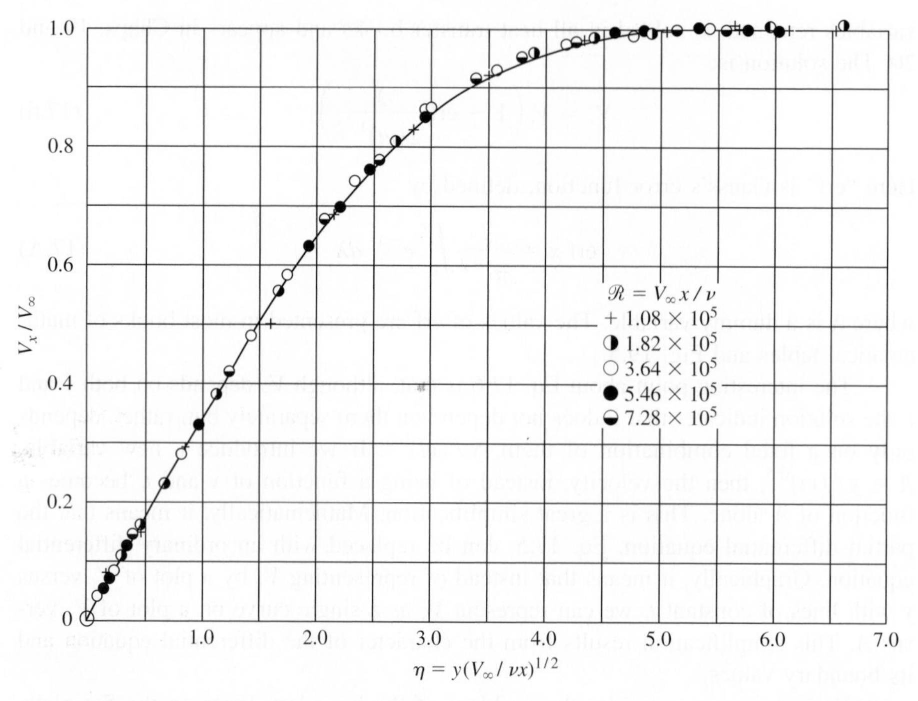

FIGURE 17.3
Blasius's solution for the laminar boundary layer on a flat plate and Nikuradse's experimental tests thereof. (From J. Nikuradse, "Laminar Reibungsschicten an der längsangestroemten Platte"—"Laminar friction layers on plates with parallel flow," *Monograph Zentrale für Wiss. Berichtwesen,* Berlin (1942.))

converts the set of two partial differential equations into a single, ordinary differential equation, which Blasius was able to solve in numerical form. The details of this calculation are shown by Schlichting and Gersten [1, p. 157]; see Prob. 17.4. The result is in the form of a curve of V_x / V_∞ versus η, which is shown in Fig. 17.3.

17.2.1 Boundary Layer Thickness

Blasius's solution for the laminar boundary layer on a flat plate, shown in Fig. 17.3, rests on a considerable string of assumptions and simplifications. However, it has been tested by numerous investigators and found to represent the experimental data very well (Fig. 17.3 shows the comparison between the Blasius's solution and Nikuradse's experimental data). Thus, these assumptions and simplifications seem to be justified.

Blasius's solution to this problem is one of the landmarks of fluid mechanics. From a detailed study of this solution, we can observe most of the principal terms and ideas in boundary-layer theory. Although this solution was found for a flat plate, it provides a guide for the solution for the flow over gently curving surfaces. Thus, although an airplane's wing surface is not flat, most of the calculated values for the laminar boundary layer on a flat plate will be qualitatively true for the laminar flow over an airplane's wing, and much of the quantitative information in Blasius's solution is approximately correct for gently curving surfaces like airplane wings and ship hulls.

From Fig. 17.3 we see that, if the boundary layer is defined as that layer out to $V_x / V_\infty = 0.99$, then the laminar boundary layer extends out to a distance of $\eta \approx 5$. If we let δ be the thickness of the boundary layer, then

$$\delta \approx 5\left(\frac{\nu x}{V_\infty}\right)^{1/2} \tag{17.9}$$

This shows that the laminar boundary layer grows proportionally to the square root of the distance from the front of the plate and that the parabolic shape sketched for it in Fig. 17.2 is correct.

Example 17.1. Calculate the boundary-layer thickness for:

(*a*) A point on an airplane wing 2 ft from the leading edge, when the plane is flying 200 mi / h through air.

(*b*) A point 2 ft from the bow of a ship, when the ship is moving 10 mi / h through water.

For the airplane

$$\delta = 5 \cdot \left(\frac{1.61 \cdot 10^{-4}\ \text{ft}^2/\text{s} \cdot 2\ \text{ft}}{200\ \text{mi}/\text{h} \cdot 5280\ \text{ft}/\text{mi} \cdot \text{h}/3600\ \text{s}}\right)^{1/2}$$
$$= 5.24 \cdot 10^{-3}\ \text{ft} = 0.063\ \text{in} = 1.60\ \text{mm} \tag{17.B}$$

and for the ship

$$\delta = 5 \cdot \left(\frac{1.08 \cdot 10^{-5}\ \text{ft}^2/\text{s} \cdot 2\ \text{ft}}{10\ \text{mi}/\text{h} \cdot 5280\ \text{ft}/\text{mi} \cdot \text{h}/3600\ \text{s}}\right)^{1/2}$$
$$= 6.06 \cdot 10^{-3}\ \text{ft} = 0.073\ \text{in} = 1.8\ \text{mm} \tag{17.C}$$

■

In Sec. 16.1 we discussed Prandtl's assumption that the effect of the transition from the zero velocity at a solid object to the free-stream velocity took place over a very thin layer of fluid. From Example 17.1 we see that this is certainly a very good assumption for laminar flow of air and water and typical ship and airplane velocities.

17.2.2 Boundary Layer Drag

From Blasius's boundary-layer solution we can calculate the drag on any part of a flat plate. Because the solution is based on laminar flow of a Newtonian fluid, we know that the shear stress at any point is given by

$$\tau = \mu \frac{dV_x}{dy} \tag{1.5}$$

Differentiating both sides of Eq. 17.7 with respect to V_x at constant x, we find

$$\left(\frac{d\eta}{dV_x}\right)_x = \left(\frac{V_\infty}{\nu x}\right)^{1/2}\left(\frac{dy}{dV_x}\right)_x \tag{17.10}$$

which can be inverted and rearranged to

$$\left(\frac{dV_x}{dy}\right)_x = V_\infty\left(\frac{V_\infty}{\nu x}\right)^{1/2}\frac{d(V_x/\ V_\infty)_x}{d\eta} \tag{17.11}$$

We want the value of $d(V_x/\ V_\infty)_x/\ d\eta$ at the surface of the plate, that is, at $y = 0$ or, therefore, $\eta = 0$. This we can find from the slope of the curve in Fig. 17.3 to be 0.332. Thus, the shear stress at any point on the surface of the plate is given by

$$\tau_0 = 0.332\mu V_\infty\left(\frac{V_\infty}{\nu x}\right)^{1/2} \tag{17.12}$$

In Sec. 6.13 we showed that the drag force on numerous bodies could be represented in terms of a plot of drag coefficient versus Reynolds number. If we now define a local drag coefficient for some small part of a flat plate as

$$C_f' = \frac{\tau_0}{(1/2)\rho V_\infty^2} \tag{17.13}$$

then from Eq. 17.12 we can calculate the drag coefficient corresponding to Blasius's solution:

$$C_f' = \frac{0.332}{1/2}\cdot\frac{\mu}{\rho}\cdot\frac{V_\infty}{V_\infty^2}\cdot\left(\frac{V_\infty}{\nu x}\right)^{1/2} = 0.664\cdot\left(\frac{\nu}{V_\infty x}\right)^{1/2} \tag{17.14}$$

Here the term in parenthesis on the far right has the same form as $1/\mathcal{R}$. We have seen Reynolds numbers in which the length was a pipe diameter and those in which it was a particle diameter. In boundary-layer theory the natural length to use seems to be the length measured from the leading edge of the solid body. Thus,

$$\mathcal{R}_x = \left(\begin{array}{l}\text{Reynolds number based}\\ \text{on distance from leading edge}\end{array}\right) = \left(\begin{array}{l}\text{boundary layer}\\ \text{Reynolds number}\end{array}\right) = \frac{V_\infty x}{\nu} \tag{17.15}$$

Equation 17.14 says that, according to Blasius's solution, a plot of *local* drag coefficients C_f' versus the Reynolds number should be given by

$$C_f' = \frac{0.664}{(\mathcal{R}_x)^{1/2}} \tag{17.16}$$

Figure 17.4 shows such a plot of experimental local drag coefficients. Those for which the flow is laminar agree very well with Blasius's solution. However, we see that, if the flow is turbulent, the result is quite different. We will discuss turbulent boundary layers in Secs. 17.3 and 17.5. (Elsewhere in fluid mechanics, a drag coefficient is normally C_d. However as shown here, in boundary layers it is normally C_f.)

At the leading edge of the plate ($x = 0$) the Reynolds number is zero; so, according to Eq. 17.16, the drag coefficient should be infinite. This is physically unreal and leads to the conclusion that Blasius's solution is not correct in the

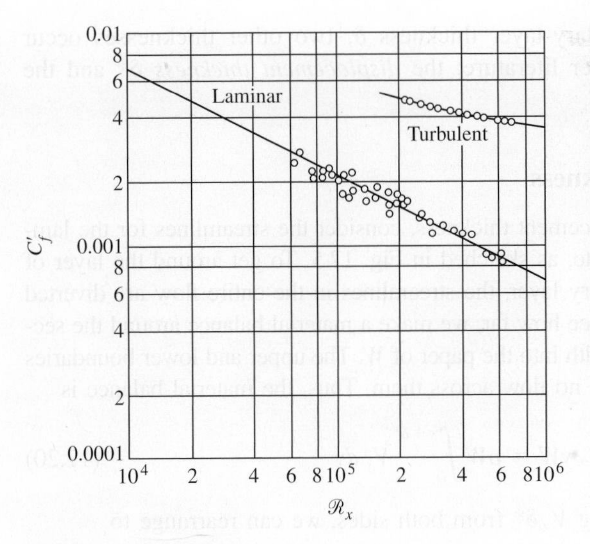

FIGURE 17.4
Local drag coefficient for a flat plate. Experimental data are compared with Blasius's solution (Eq. 17.16) and with Prandtl's equation (Eq. 17.36). (From H. W. Liepmann and S. Dahwan, "Direct measurements of local skin friction in low-speed and high-speed flow," *Proc. First U.S. Nat. Congr. Appl. Mech.*, New York: ASME, (1952) p. 873. Reproduced with the permission of the publisher.)

small region nearest the leading edge [1, p. 162]. This is a minor flaw, which is of little practical concern.

The drag coefficient defined above, C_f', which gives the local drag force, is less convenient for practical calculations than one that gives the drag force on an entire plate. That force for a plate of width W is

$$F = W \int_0^x \tau_0 \, dx \tag{17.17}$$

Now we define a new drag coefficient for the entire surface:

$$C_f = \frac{F}{(1/2)\rho V_\infty^2 A} = \frac{1}{A} \int C_f' \, dA \tag{17.18}$$

For Blasius's laminar boundary-layer solution this becomes

$$C_f = \frac{0.332 \, W \mu V_\infty (V_\infty / \nu)^{1/2} \displaystyle\int_0^x dx / x^{1/2}}{0.5 \, \rho V_\infty^2 \, Wx} = \frac{1.328}{(\mathscr{R}_x)^{1/2}} \tag{17.19}$$

Example 17.2. A square flat plate 1 m^2 is towed behind a ship moving 15 km / h by a long, thin wire that does not disturb the flow. The boundary layers on both sides of the plate are laminar. What is the force required to tow the plate?

The Reynolds number based on the whole length of the plate is

$$\mathscr{R}_x = \frac{15 \cdot 10^3 \text{ m / h} \cdot 1 \text{ m}}{(1.004 \cdot 10^{-6} \text{ m}^2 / \text{s})} \cdot \frac{\text{h}}{3600 \text{ s}} = 4.15 \cdot 10^6 \tag{17.D}$$

so that

$$C_f = \frac{1.328}{(4.15 \cdot 10^6)^{1/2}} = 6.52 \cdot 10^{-4} \tag{17.E}$$

and

$$F = 6.52 \cdot 10^{-4} \cdot \frac{1}{2} \cdot 998.2 \frac{\text{kg}}{\text{m}^3} \cdot \left(\frac{15{,}000}{3600} \cdot \frac{\text{m}}{\text{s}} \right)^2 \cdot 2 \text{ m}^2 \cdot \frac{\text{N} \cdot \text{s}^2}{\text{kg} \cdot \text{m}}$$
$$= 11.3 \text{ N} = 2.54 \text{ lbf} \tag{17.F}$$

■

In addition to the boundary-layer thickness δ, two other thicknesses occur frequently in the boundary-layer literature: the *displacement thickness* δ^* and the *momentum thickness* θ.

17.2.3 Displacement Thickness

To see the meaning of the displacement thickness, consider the streamlines for the laminar boundary layer on a flat plate, as sketched in Fig. 17.5. To get around the layer of slow-moving fluid in the boundary layer, the streamlines in the entire flow are diverted away from the solid surface. To see how far, we make a material balance around the section marked in Fig. 17.5 for a width into the paper of W. The upper and lower boundaries are along streamlines, so there is no flow across them. Thus, the material balance is

$$\rho V_\infty y W = \rho W \int_0^{y+\delta^*} V_x \, dy \tag{17.20}$$

Dividing out ρW and subtracting $V_\infty \delta^*$ from both sides, we can rearrange to

$$-V_\infty \delta^* = -V_\infty (y + \delta^*) + \int_0^{y+\delta^*} V_x \, dy \tag{17.21}$$

which is then rearranged to

$$\delta^* = \frac{\int_0^{y+\delta^*} (V_\infty - V_x) \, dy}{V_\infty} = \int_0^{y+\delta^*} \left(1 - \frac{V_x}{V_\infty}\right) dy \tag{17.22}$$

Here δ^*, the displacement thickness, is the distance the streamlines are moved in the direction perpendicular to the plate. Since $(V_\infty - V_x)$ is 0.01 at the edge of the boundary layer and rapidly goes to zero as y increases, the upper limit of integration in Eq. 17.22 is normally shown as infinity, although any large number will do. Equation 17.22 is correct for any constant-density boundary layer, whether laminar or turbulent.

For Blasius's solution we may perform the integration in Eq. 17.22 graphically on Fig. 17.3 by noting that the region above and to the left of the solid curve is

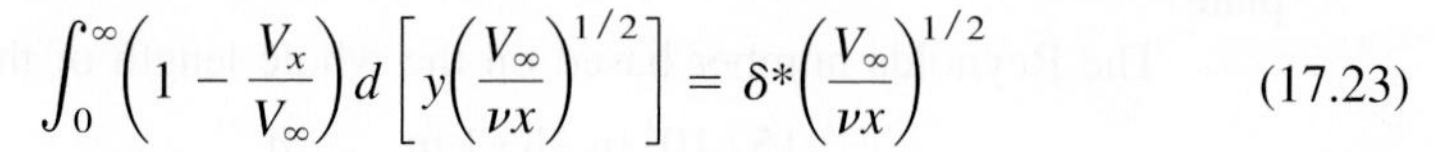

$$\int_0^\infty \left(1 - \frac{V_x}{V_\infty}\right) d\left[y\left(\frac{V_\infty}{\nu x}\right)^{1/2}\right] = \delta^* \left(\frac{V_\infty}{\nu x}\right)^{1/2} \tag{17.23}$$

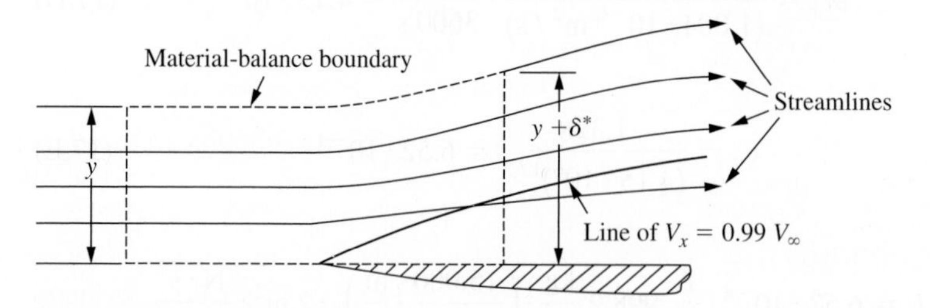

FIGURE 17.5
Displacement thickness.

By graphical integration this region has area 1.72; so that

$$\delta^* = 1.72\left(\frac{\nu x}{V_\infty}\right)^{1/2} \tag{17.24}$$

Comparing Eq. 17.24 with Eq. 17.9, we see that the displacement thickness for a laminar boundary layer is 1.72 / 5, or about one-third of the boundary layer thickness.

17.2.4 Momentum Thickness

The solid body slows down a layer of the fluid without bringing any but the fluid actually touching the wall completely to rest. From all the fluid slowed down it extracts momentum in the form of a drag force on the solid plate. This same amount of momentum and hence the same force could be obtained by stopping completely some layer of the oncoming stream; the thickness of such a layer is called the momentum thickness θ. If such a layer were stopped by a plate W wide, the force required would be $W\theta\rho V_\infty^2$. The force actually exerted by the plate is given by Eq. 17.17; so

$$\theta = \frac{1}{\rho V_\infty^2}\int_0^x \tau_0\, dx \tag{17.25}$$

$$\tau_0 = \rho V_\infty^2 \frac{d\theta}{dx} \tag{17.26}$$

which will be used later in the study of turbulent boundary layers. For Blasius's laminar boundary-layer solution we can substitute for τ_0 in Eq. 17.25, finding

$$\theta = \frac{1}{\rho V_\infty^2}\, 0.322\, \mu V_\infty \left(\frac{V_\infty}{\nu}\right)^{1/2} \int_0^x \frac{dx}{x^{1/2}} = \frac{0.664}{\mathcal{R}_x^{1/2}} \tag{17.27}$$

Comparing Eqs. 17.27 and 17.9, we see that for the laminar boundary layer on a flat plate the momentum thickness is 0.664 / 5, or about one-eighth of the boundary-layer thickness.

We can also show (Prob. 17.9) by a momentum balance around the same region we made the material balance around in Fig. 17.5 that for any boundary layer, laminar or turbulent

$$\theta = \int_0^\infty \frac{V_x}{V_\infty}\left(1 - \frac{V_x}{V_\infty}\right) dy \tag{17.28}$$

This equation also is used in treating turbulent boundary layers.

Blasius's steady-flow, laminar, flat-plate, boundary-layer solution is a numerical solution of his simplification of Prandtl's boundary-layer equations, which are a simplified, one-dimensional momentum balance and a mass balance. This type of solution is known in the boundary-layer literature as an *exact solution.* Exact solutions can be found for only a very limited number of cases. Therefore, approximate methods are available for making reasonable estimates of the behavior of laminar boundary layers (Prob. 17.8).

17.3 TURBULENT BOUNDARY LAYERS

Like the flow in a pipe (Sec. 6.2), the flow in a boundary layer can be laminar or turbulent. In a pipe the transition takes place at a Reynolds number of about 2000, although it may be delayed to higher Reynolds numbers by extreme care to avoid pipe roughness or vibration. In a constant-diameter pipe the flow has the same character over the entire length of the flow, except for a small region near the entrance. The same is not true of a boundary-layer flow. In a boundary-layer flow the characteristic dimension is the distance from the leading edge. As we have seen, the appropriate Reynolds number for boundary-layer calculations is based on this length. As in pipe flow, the Reynolds number furnishes the criterion for transition from laminar to turbulent flow in a boundary layer. For a flat plate, the transition takes place in the Reynolds number region from $3.5 \cdot 10^5$ to $2.8 \cdot 10^6$. This transition is strongly influenced by turbulence in the stream outside the boundary layer and by roughness of the surface. A typical boundary layer on a smooth surface of sufficient length might look like Fig. 17.6. The figure shows not only laminar, transition, and turbulent boundary layers but also a "laminar sublayer" beneath the turbulent boundary layer (to be discussed later).

For laminar boundary layers, as for laminar flow in a pipe, it was possible to calculate the flow behavior from a set of plausible assumptions and then to show experimentally that the flow behaved as calculated. For turbulent boundary layers, as for turbulent flow in pipes, no one is yet able to calculate the flow behavior without starting with experimental data. However, from experimental measurements it has been possible to make some generalizations, which can then be used to extrapolate to other conditions.

In a turbulent flow the velocity at any point fluctuates randomly with time. One may speak of any such velocity as consisting of a time-average component, $V_{\text{time avg}}$, and a fluctuating components, v, so that, at any point, at any instant

$$V = V_{\text{time avg}} + v \tag{17.29}$$

These are defined such that $V_{\text{time avg}}$ is the average reading over some time interval of a velocity meter at the point

$$V_{\text{time avg}} = \frac{1}{t}\int_0^t V\,dt \tag{17.30}$$

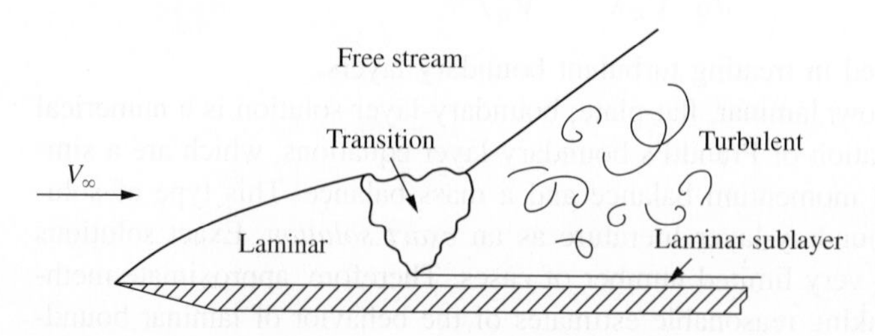

FIGURE 17.6
Laminar-turbulent transition in a boundary layer.

and such that the average value of v over some time interval is zero, because it is positive and negative for equal parts of the total time. In all the following discussions the velocities are the time-average velocities, V_{avg}.

Most of the results available for turbulent boundary layers have been found by measuring time-average velocities at various points in flow in pipes or over flat plates and by attempting to generalize the velocity profiles. For various experimental reasons it is easier to make such measurements in pipes, so most of the results are pipe results. We now consider the turbulent flow in pipes for one section and then return to the turbulent boundary layer.

17.4 TURBULENT FLOW IN PIPES

As discussed in Sec. 6.4, turbulent flow differs from laminar flow in that the principal cause of the shear stresses between adjacent layers of the fluid is the interchange of masses of fluid between adjacent layers of fluid moving at different velocities. This gives rise to additional stresses, called *Reynolds stresses.* The most dramatic effect of these stresses is the large increase of friction heating in turbulent flow over that found in laminar flow. The other dramatic effect is the change in the shape of the velocity profile from laminar to turbulent flow. This is shown in Fig. 17.7, where the experimental turbulent velocity profiles measured by Nikuradse are compared with the profile for laminar flow.

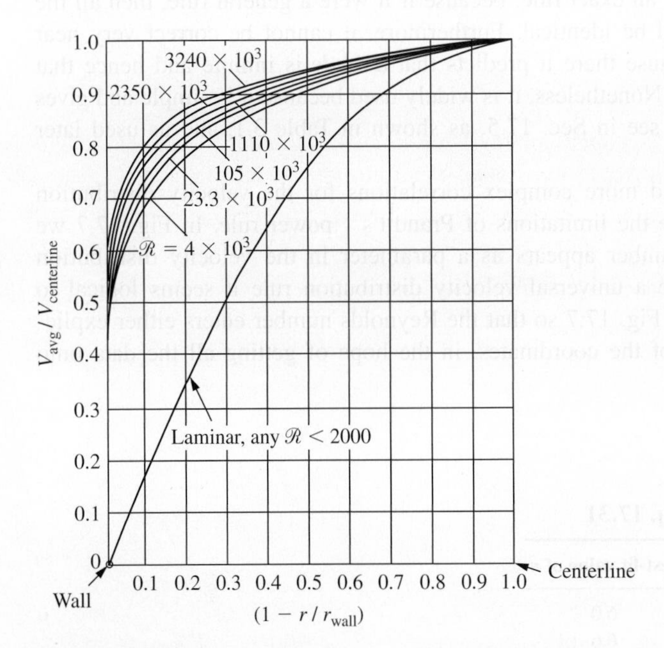

FIGURE 17.7
Velocity distribution in laminar and turbulent flow in smooth circular pipes. (From J. Nikuradse, "Gesetzmaessigkeiten der turbulenten Stroemung in glatten Rohren,"—"Regularities of turbulent flow in smooth tubes"—*Forschungsheft 356* (1932). Reproduced by permission of the publisher.)

As discussed in Sec. 6.3, the velocity profile for laminar flow in a tube is parabolic. For turbulent flow it is much closer to *plug flow,* that is, to a uniform velocity over the entire pipe cross section. Furthermore, as seen from Fig. 17.7, as the Reynolds number is increased, the velocity profile approaches closer and closer to plug flow. At the wall the turbulent eddies disappear; so the shear stress at the wall for both laminar and turbulent flow of Newtonian fluids is given by $\tau_0 = \mu(dV_x/dy)$. Although it is very difficult experimentally to measure velocity gradients very close to the wall, it is clear from Fig. 17.7 that at the wall the velocity gradient is steeper for turbulent flow; hence, the shear stress and friction heating must be larger for turbulent flow than for laminar flow.

Prandtl showed that each of the different turbulent-flow curves in Fig. 17.7 could be represented *fairly well* by an equation of the form

$$\frac{V_x}{V_{x,\text{centerline}}} = \left(1 - \frac{r}{r_{\text{wall}}}\right)^n \tag{17.31}$$

For the curves shown in Fig. 17.7 the value of n that gives the best representation of the experimental curves varies from $\frac{1}{6}$ for the lowest Reynolds number to $\frac{1}{10}$ for the highest Reynolds number. The best-fit values of n in Eq. 17.31 for all the experimental curves are shown in Table 17.1.

Prandtl selected $\frac{1}{7}$ as the best average, deducing "Prandtl's $\frac{1}{7}$ power velocity distribution rule." This is not an exact rule, because if it were a general rule, then all the curves in Fig. 17.7 would be identical. Furthermore, it cannot be correct very near the wall of the tube, because there it predicts that dV/dy is infinite and hence that the shear stress is infinite. Nonetheless, it is widely used because it is simple and gives useful results, as we will see in Sec. 17.5, as shown in Table 3.1, and as used later in Chap. 18.

It is possible to find more complex correlations for the velocity distribution in a pipe that do not have the limitations of Prandtl's $\frac{1}{7}$ power rule. In Fig. 17.7 we see that the Reynolds number appears as a parameter in the velocity distribution plot. In trying to produce a universal velocity distribution rule it seems logical to change the coordinates in Fig. 17.7 so that the Reynolds number enters either explicitly or implicitly in one of the coordinates, in the hope of getting all the data onto one curve.

TABLE 17.1

Best-fit values* of n in Eq. 17.31

Reynolds number	Best-fit value of n
$4.0 \cdot 10^3$	6.0
$2.3 \cdot 10^4$	6.6
$1.1 \cdot 10^5$	7.0
$1.1 \cdot 10^6$	8.8
$2.0 \cdot 10^6$	10
$3.2 \cdot 10^6$	10

*From Schlicting [3, p. 563].

The most successful method of doing this has been to define a new quantity called the *friction velocity*, u^*,

$$\begin{pmatrix}\text{Friction}\\ \text{velocity}\end{pmatrix} = u^* = \left(\frac{\tau_{\text{wall}}}{\rho}\right)^{1/2} = V_{x,\text{avg}}\left(\frac{f}{2}\right)^{1/2} \tag{17.32}$$

where f is the Fanning friction factor used in Chap. 6. The friction velocity is not a physical velocity, which one could measure at some point in the flow, but a combination of terms that has the dimensions of a velocity and hence is called a velocity. Using this "velocity" as a parameter, we can prepare a universal plot of pipe velocities, as shown in Fig. 17.8.

Figure 17.8 shows that in making up the universal velocity distribution it is necessary to introduce two new combinations of variables, which are in common use in the fluid mechanics literature. The ratio of the local velocity to the friction velocity is called u^+ (spoken of as "u plus"). This is also the ratio of the local time-average velocity to the average velocity in the entire flow times $\sqrt{2/f}$. The combination of the distance from the pipe wall and the friction velocity divided by the kinematic viscosity is called y^+. This can be understood as the product of a kind of Reynolds number that is based on distance from the wall rather than on pipe diameter and $\sqrt{f/2}$.

Figure 17.8 shows that the flow can be divided conceptually into three zones: a laminar sublayer nearest the pipe wall, in which the shear stress is principally due to viscous shear; a turbulent core in the middle of the pipe, in which the shear stress

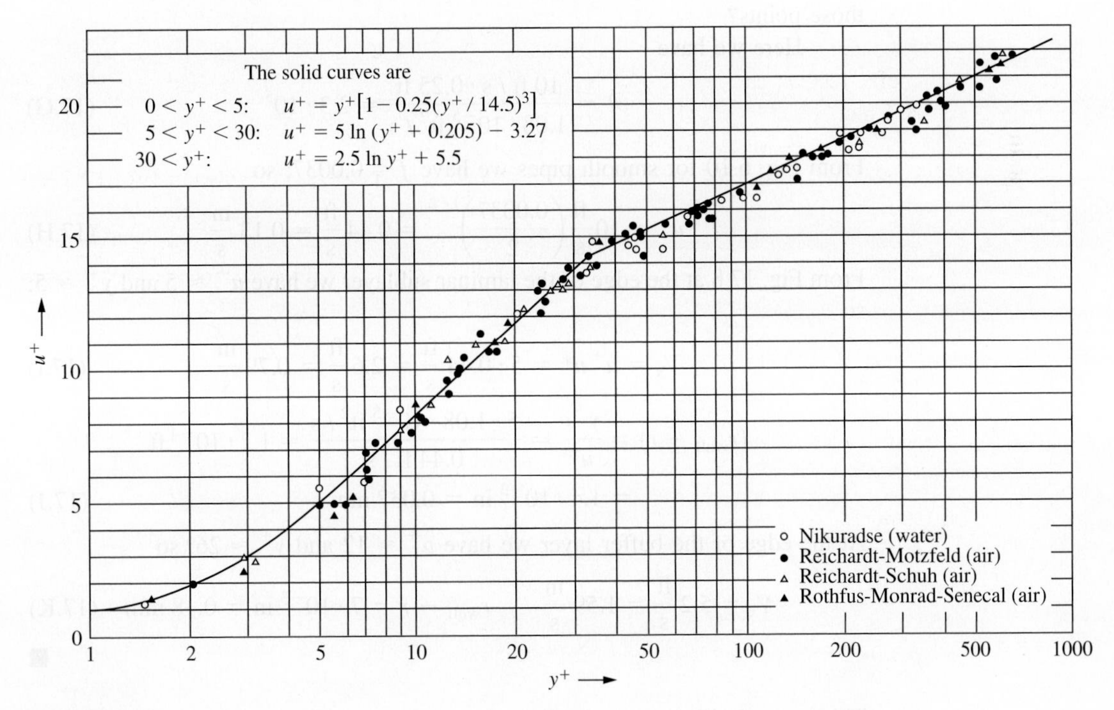

FIGURE 17.8

Universal velocity distribution for turbulent flow in smooth tubes.

is principally due to turbulent Reynolds stresses; and a layer between them, called the buffer layer, in which both viscous and Reynolds stresses are of the same order of magnitude. Good experimental measurements are difficult to make in the laminar sublayer and buffer layer, so there is some controversy over the best location for the boundaries shown in Fig. 17.8. Most investigators place the buffer layer at a y^+ of approximately 5 to 30; some [1, p. 523] use 5 to 70. Furthermore, current work seems to indicate that the location of the edge of these layers is not fixed in place but fluctuates up and down; so these values indicate only the mean locations of these edges [2]. Thus, Fig. 17.8 may be too simple a picture of the actual behavior. Nonetheless, it provides a reasonable conceptual model and is able to correlate most of the available data with reasonable accuracy.

Figure 17.8 is for smooth pipes. As shown in Fig. 6.10, increasing the roughness of the pipe wall in turbulent flow generally leads to an increase in the friction factor. In Fig. 17.8 we see that increasing the friction factor will increase y^+ and decrease u^+; so increasing the roughness while holding everything else constant will move a point on the curve downward and to the right. Schlichting [3, p. 584] presents a plot like Fig. 17.8 with a smooth-pipe line identical to that in Fig. 17.8 and other lines below and to the right of it for various relative roughnesses.

Example 17.3. Water is flowing in a 3 in ID smooth pipe, with an average velocity of 10 ft / s. How far from the wall are the edge of the laminar sublayer and the edge of the buffer layer? What is the time average velocity at each of those points?

Here we have

$$\mathscr{R} = \frac{10 \text{ ft / s} \cdot 0.25 \text{ ft}}{1.08 \cdot 10^{-5} \text{ ft}^2 \text{ / s}} = 2.3 \cdot 10^5 \tag{17.G}$$

From Fig. 6.10 for smooth pipes we have $f = 0.0037$; so

$$u^* = 10 \frac{\text{ft}}{\text{s}} \left(\frac{0.0037}{2} \right)^{1/2} = 0.44 \frac{\text{ft}}{\text{s}} = 0.13 \frac{\text{m}}{\text{s}} \tag{17.H}$$

From Fig. 17.8 at the edge of the laminar sublayer we have $u^+ \approx 5$ and $y^+ \approx 5$; so

$$V_x = u^+ u^* = 5 \cdot 0.44 \frac{\text{ft}}{\text{s}} = 2.6 \frac{\text{ft}}{\text{s}} = 0.79 \frac{\text{m}}{\text{s}} \tag{17.I}$$

$$\begin{aligned}(r_{\text{wall}} - r) &= \frac{y^+ \nu}{u^*} = \frac{5 \cdot 1.08 \cdot 10^{-5} \text{ ft}^2 \text{ / s}}{0.44 \text{ ft / s}} = 1.2 \cdot 10^{-4} \text{ ft} \\ &= 1.4 \cdot 10^{-3} \text{ in} = 0.037 \text{ mm}\end{aligned} \tag{17.J}$$

At the edge of the buffer layer we have $u^+ \approx 12$ and $y^+ \approx 26$; so

$$V_x = 5.2 \frac{\text{ft}}{\text{s}} = 1.59 \frac{\text{m}}{\text{s}} \qquad r_{\text{wall}} - r = 7 \cdot 10^{-3} \text{ in} = 0.18 \text{ mm} \tag{17.K}$$

■

This example illustrates why there are so few experimental data in the laminar sublayer and buffer layer; these layers are extremely thin and have very steep velocity gradients.

17.5 THE STEADY, TURBULENT BOUNDARY LAYER ON A FLAT PLATE

There are no known analytical solutions for turbulent boundary layers that are analogous to Blasius's solution for the laminar boundary layer on a flat plate. Prandtl, to describe the steady, turbulent boundary layer on a flat plate, made the following assumptions:

1. The average velocity in the x direction at any point has the same kind of distribution as that found in a pipe and is represented by Prandtl's $\frac{1}{7}$ power rule (Eq. 17.31) in the form

$$\frac{V_x}{V_\infty} = \left(\frac{y}{\delta}\right)^{1/7} \tag{17.33}$$

This presupposes that the velocity profiles at any x are similar to each other; this is the kind of assumption made by Blasius when he assumed that the velocity was a function of η, not of x and y separately.

2. Over the Reynolds number range of $3 \cdot 10^3$ to 10^5 the friction factor plot for smooth pipes can be approximated (see Prob. 17.13) by

$$f = \frac{0.0791}{\mathcal{R}^{1/2}} \tag{17.34}$$

Blasius has shown that this equation fits the smooth-pipes curve in Fig. 6.10 quite well. Prandtl assumed that it could be taken over directly for determining the shear stress at the surface of the plate, understanding the length term in the Reynolds number given above is *twice* the thickness of the boundary layer.

Combining Eqs. 17.33 and 17.34, as shown in Prob. 17.14, Prandtl found

$$\delta = 0.37x\left(\frac{\nu}{V_\infty x}\right)^{1/5} \tag{17.35}$$

From this equation and several other relations (Prob. 17.15) we can compute the drag coefficient for a turbulent boundary layer, finding

$$C_f' = \frac{0.0576}{\mathcal{R}_x^{1/5}} \tag{17.36}$$

$$C_f = \frac{0.072}{\mathcal{R}_x^{1/5}} \tag{17.37}$$

These two equations are based on some very severe assumptions, and their use can be justified only by experimental verification. Most tests have indicated that they give a very good representation of experimental data. For example, Fig. 17.4 shows a comparison of experimental data on the local drag coefficient with Eq. 17.36; the agreement is excellent. Other experimental data can be adduced to show that these equations are at least satisfactory for engineering purposes. Equation 17.37 assumes that the boundary layer is turbulent from the beginning of the plate ($x = 0$) to the end of the plate. Such a situation can exist if the beginning of the plate is artificially roughened. However, more commonly (Fig. 17.6) the first part of the plate has a laminar boundary layer, which then makes a transition to turbulent flow farther down the plate.

To calculate the drag on such a plate, we would calculate the separate contributions from the laminar and turbulent parts of the boundary layer.

Equation 17.35 also indicates that turbulent boundary layers grow with distance as x to the $\frac{4}{5}$ power, compared with the $\frac{1}{2}$ power for laminar boundary layers. Thus, for the same distance the boundary layer will be larger and growing faster if it is turbulent rather than laminar.

Example 17.4. A speedboat is towing a smooth plate 1 ft wide and 20 ft long through still water at a speed of 50 ft / s. Determine the boundary-layer thickness at the end of the plate and the drag on the plate.

At the end of the plate

$$\mathscr{R}_x = \frac{50 \text{ ft / s} \cdot 20 \text{ ft}}{1.08 \cdot 10^{-5} \text{ ft}^2 / \text{s}} = 0.93 \cdot 10^8 \tag{17.L}$$

and the boundary layer is turbulent, so, from Eq. 17.35

$$\delta = \frac{0.37 \cdot 20 \text{ ft}}{(0.93 \cdot 10^8)^{1/5}} = 0.189 \text{ ft} = 2.3 \text{ in} = 58 \text{ mm} \tag{17.M}$$

As a first approximation, we assume that the entire boundary layer is turbulent so that Eq. 17.37 applies. Then

$$\begin{aligned} F &= C_f \frac{1}{2} \rho V_\infty^2 A \\ &= \frac{0.072}{(0.93 \cdot 10^8)^{1/5}} \cdot \frac{1}{2} \cdot 62.3 \frac{\text{lbm}}{\text{ft}^3} \cdot \left(50 \frac{\text{ft}}{\text{s}}\right)^2 \cdot (2 \cdot 20 \text{ ft} \cdot 1 \text{ ft}) \cdot \frac{\text{lbf} \cdot \text{s}^2}{32.2 \cdot \text{lbm} \cdot \text{ft}} \\ &= 178 \text{ lbf} = 790 \text{ N} \end{aligned} \tag{17.N}$$

Now, to ascertain how large an error we made by assuming that the entire boundary layer was turbulent, we assume that transition from laminar to turbulent flow takes place at an $\mathscr{R}_x$ value of 10^6. From the above, we see that this corresponds to a distance of 1 / 100 of the length of the plate; so the boundary layer over the first 0.2 ft presumably is laminar. For this area the drag due to a laminar boundary layer is given by Eq. 17.19,

$$\begin{aligned} F &= \frac{1.328}{(10^6)^{1/2}} \cdot \frac{1}{2} \cdot 62.3 \frac{\text{lbm}}{\text{ft}^3} \cdot \left(50 \frac{\text{ft}}{\text{s}}\right)^2 \cdot (2 \cdot 0.2 \text{ ft} \cdot 1 \text{ ft}) \cdot \frac{\text{lbf} \cdot \text{s}^2}{32.2 \cdot \text{lbm} \cdot \text{ft}} \\ &= 1.3 \text{ lbf} = 5.8 \text{ N} \end{aligned} \tag{17.O}$$

In the calculation that assumed that the entire boundary layer was turbulent, this leading area contributed a force of

$$\begin{aligned} F &= \frac{0.072}{(10^6)^{1/5}} \cdot \frac{1}{2} \cdot 62.3 \frac{\text{lbm}}{\text{ft}^3} \cdot \left(50 \frac{\text{ft}}{\text{s}}\right)^2 \cdot (2 \cdot 0.2 \text{ ft} \cdot 1 \text{ ft}) \cdot \frac{\text{lbf} \cdot \text{s}^2}{32.2 \cdot \text{lbm} \cdot \text{ft}} \\ &= 4.4 \text{ lbf} = 19.5 \text{ N} \end{aligned} \tag{17.P}$$

So the calculation assuming a completely turbulent boundary layer gives a drag force that is too high by 4.4 − 1.3 = 3.1 lbf. This error is small compared with the uncertainties introduced by the approximate nature of Eq. 17.37. ■

17.6 THE SUCCESSES OF BOUNDARY-LAYER THEORY

The foregoing shows that even in the simplest possible boundary-layer problems the mathematics are formidable. For calculating the boundary layer around an airplane or a ship the mathematics are beyond our current abilities. We must resort to approximations and simplifications, mostly based on the results of the simple cases like those shown. Nonetheless, the success of boundary-layer theory in explaining the behavior of ships, airplanes, projectiles, etc. has been very good.

In Sec. 16.8 we discussed separation. In almost every case, separation results in a great increase in drag, which is normally a very undesirable result. From boundary-layer theory it is possible to make some good estimates of when separation will or will not occur. In particular, it can be shown that turbulent boundary layers are less likely to separate than laminar ones. One may visualize why by considering the flow around a cylinder, as shown in Fig. 16.20. There separation occurs because the fluid flowing near the surface has lost momentum due to drag and cannot penetrate the high-pressure layer behind the cylinder. The effect of turbulence is to transfer momentum from the free stream into the boundary layer by means of turbulent eddies. Thus, the layer near the wall is not slowed as much and does not separate as soon.

The most dramatic example of this occurs in the case of flow around a sphere. A laminar boundary layer separates well ahead of the maximum diameter perpendicular to flow, while a turbulent one clings on much further. This is illustrated in Fig. 17.9, which shows two bowling balls moving at the same velocity in water. One of the balls has a patch of sand glued to its leading surface, making the boundary layer turbulent. This holds the separation back much further. The turbulent boundary layer has a higher

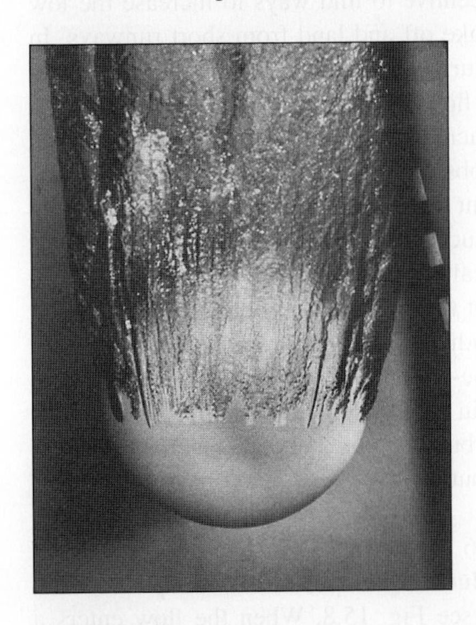

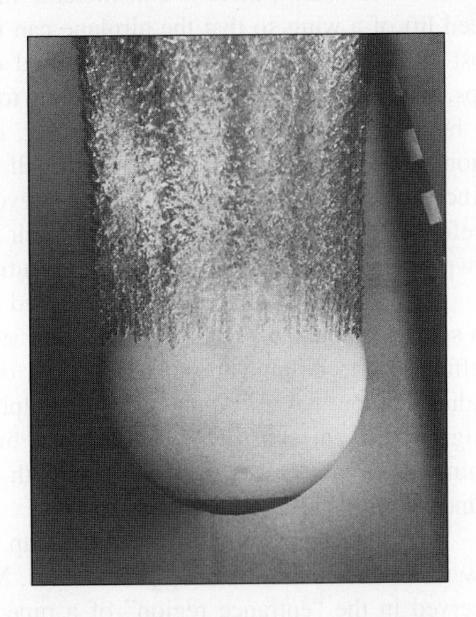

FIGURE 17.9
Two 8.5 in bowling balls entering still water at 25 ft / s. The one on the left is smooth; the one on the right has a patch of sand on its nose. For both photos, $\mathscr{R}_p \approx 1.6 \cdot 10^6$. For the left photo, $C_d \approx 0.4$, for the right photo, $C_d \approx 0.2$. (Official photograph, U.S. Navy).

skin-friction drag than the laminar one but has a much smaller low-pressure wake behind it. Since this low-pressure wake is the principal source of the total drag on the ball, the net drag is greatly reduced by the patch of sand. This principle was discovered long before the idea of the boundary layer by golfers, who found that old, rough balls went farther than new, smooth ones. This discovery led to the invention of the dimpled golf ball, which is rough enough to have a turbulent boundary layer, but still smooth enough to putt well. Its performance was enjoyed by many but understood by no one until the invention of boundary-layer theory [4].

In Fig. 6.24 we showed that the drag coefficient for a sphere moving in a still fluid drops dramatically at a particle Reynolds number of about $3 \cdot 10^5$. Perceptive readers will recognize this as being the lower limit of the transition from laminar to turbulent boundary layers on a flat plate and will conclude that this must be the result of the transition of the boundary layer on the sphere from laminar to turbulent flow. This has been verified experimentally. The function of the dimples on the golf ball is to make this transition occur at a lower Reynolds number.

The lift obtainable from an airplane wing is a strong function of the plane's angle to the horizontal, which is called the *angle of attack*. The higher the angle of attack, the more the lift, up to the point where the flow separates on the top of the wing, causing the lift to decrease dramatically. This phenomenon in airplanes is called *stall*. A stall close to the ground is almost certain to cause a crash. Considerable efforts have been directed at preventing this separation so that airplanes can lift more or have smaller wings.

Stall is normally a problem only on takeoff and landing. One can increase the lift by going faster, but both high takeoff and landing speeds require longer runways than low ones. Thus, there is considerable incentive to find ways to increase the low speed lift of a wing so that the airplane can take off and land from short runways. In most airplanes the wing shape is modified during takeoff and landing by slats and flaps, which allow some high-pressure air to flow from the bottom of the wing into the boundary layer on the top of the wing, increasing its velocity and making separation less likely. Other airplanes use small obstructions on the surface of the wing to increase the turbulence there and thus prevent stall. The function of all these devices is referred to as boundary-layer control; their success indicates that the boundary-layer viewpoint has been very fruitful in aeronautical engineering.

For many years mariners were amazed at the swimming speed of dolphins, who can swim for hours at high speeds. Careful studies showed, for example, that the drag coefficient of a dolphin must be less than one-half that of the best torpedo designs produced by our navies. Apparently, the dolphin can do this because it has a specially-designed resilient skin, which damps out turbulence and keeps his boundary layer laminar in circumstances in which a smooth but rigid surface would have a turbulent boundary layer [5].

In discussing friction in pipes in Chap. 6 we restricted ourselves to the case of flow well downstream of the pipe entrance. Much higher pressure losses per foot are observed in the "entrance region" of a pipe; see Fig. 15.8. When the flow enters a pipe, a boundary layer begins to grow at the wall. From what we have seen here we know that the shear stress is highest at the front of a plate (where the boundary layer is thin). Thus, we would expect the pressure loss per unit length to be greatest at the inlet and to decrease with distance down the pipe. At some distance down the pipe the boundary layers from the opposite walls grow together, filling the pipe. Thereafter the flow is no longer of the boundary-layer type and can be treated as a pipe flow. The distance for the boundary layers to grow together, forming "fully developed" flow, is a function of the pipe diameter and the Reynolds number; see Example 15.7.

Other cases of the success of boundary-layer theory can be cited. The same kind of idea has been taken over into heat-transfer and mass-transfer theory with generally useful results.

Notice also that this entire chapter seems to be a mixture of mathematics and sweeping assumptions. This is typical of boundary-layer studies. The entire problem of describing the flow around some complicated structure like an airplane is far beyond our current abilities, even with supercomputers. However, by using judicious guesswork we can reduce the important parts of the problem to mathematically manageable form. The real genius of Ludwig Prandtl was his ability to guess correctly which terms he could drop out of his equations and still have the calculated result agree with experiments. This inspired guesswork usually can be made only by engineers who have a very good understanding of what every term in an equation means physically. It is strongly recommended that students try to develop this understanding.

17.7 SUMMARY

1. Prandtl started with the Navier-Stokes and material balance equations and discarded enough terms to make his boundary-layer equations, which are the working form of the momentum and continuity equations for boundary-layer problems.
2. By assuming that the velocity was a function of $y(V_\infty / x\nu)^{1/2}$, Blasius was able to solve Prandtl's equations for the steady-flow laminar boundary layer on a flat plate. He found that the laminar boundary-layer thickness is proportional to the square root of the length down the plate.
3. The solution for the turbulent boundary layer is based on results obtained in pipes. Using pipe results plus some strong assumptions, Prandtl calculated that the thickness of a turbulent boundary layer is proportional to the length of the plate to the $\frac{4}{5}$ power.
4. Only a very small number of boundary-layer problems can be solved in any closed or simple mathematical form. However, the boundary-layer viewpoint has been very fruitful in the field of flow around solid bodies and in several other fields.

PROBLEMS

See the Common Units and Values for Problems and Examples, inside the back cover. An asterisk (*) on a problem number indicates that its answer is shown in App. D.

17.1. Derive Eq. 17.5 in each of the following two ways:

(*a*) By writing a momentum balance for a small element of fluid.

(*b*) By canceling the zero terms in the x-directed Navier-Stokes equation, Eq. 15.26. For laminar flow in this geometry it may be assumed that the y and z components of the velocity are everywhere zero.

17.2. List the boundary values for Eq. 17.5.

17.3. Show that the substitution of $A = y/(\nu t)^{0.5}$ converts Eq. 17.5 from a partial differential equation with y and t as variables to an ordinary differential equation with A as a variable.

17.4. Blasius's laminar boundary-layer solution (Fig. 17.3) is an excellent example of the use of the stream function to simplify and solve the partial differential equations of fluid mechanics. To solve Eqs. 17.4 and 17.2 together, he substituted for V_x and V_y from Eqs. 16.24.

(*a*) Show that this leads to

$$\frac{\partial \psi}{\partial y} \cdot \frac{\partial^2 \psi}{\partial y\, \partial x} - \frac{\partial \psi}{\partial x}\frac{\partial^2 \psi}{\partial y^2} = \nu \frac{\partial^3 \psi}{\partial y^3} \tag{17.Q}$$

in which, instead of having two dependent variables, V_x and V_y, we have only one, ψ.

(*b*) We wish ψ to depend on the dimensionless group η, but it cannot depend on η alone, because ψ has the dimensions of (length^2 / time) and η is dimensionless. Therefore, it must depend, not only on η, but also on some combination of the variables x, y, ν, and V_∞. Several such combinations have the required dimensions; Blasius found that the mathematics were simplest if he made the choice

$$\psi = (\nu x V_\infty)^{1/2} f(\eta) \tag{17.R}$$

where $f(\eta)$ is an unknown function, to be determined. Show that on the basis of this choice

$$V_x = -\frac{\partial \psi}{\partial y} = -\frac{\partial \psi}{\partial \eta}\frac{\partial \eta}{\partial y} = V_\infty f'(\eta) \tag{17.S}$$

where f' stands by $df/d\eta$, and that

$$V_y = \frac{\partial \psi}{\partial x} = -0.5\left(\frac{\nu V_\infty}{x}\right)^{1/2}(\eta f' - f)(\eta) \tag{17.T}$$

Hint: For V_y, differentiate directly for $\partial f/\partial x$ and note that $\partial f/\partial x = (df/d\eta)(\partial \eta/\partial x)$.

(*c*) Show that, if we proceed to compute the necessary second and third derivatives the same way as in (*b*) and then substitute them in the first equation in this problem, we find

$$\left(\frac{-V_\infty^2}{2x}\right)\eta f' f'' + \left(\frac{V_\infty^2}{2x}\right)(\eta f' - f) f'' = \nu\left(\frac{V_\infty^2}{2\nu}\right) f''' \tag{17.U}$$

which can be simplified to $ff'' + 2f''' = 0$. This is now an ordinary differential equation, involving only $f(\eta)$ and its derivatives with respect to η. Although it appears simple, it is nonlinear, and no analytical solution of it has been found. However, it has been solved numerically by using an infinite-series method, and tabulated values are available [3, p. 129]. The solid curve in Fig. 17.3 is based on this infinite-series solution.

17.5.* Assuming that the transition from a laminar to a turbulent boundary layer takes place at a Reynolds number of 10^6, what is the maximum thickness for the laminar boundary layer on a flat plate for

(*a*) Air flowing at 10 ft / s?

(*b*) Water flowing at 10 ft / s?

(*c*) Glycerin flowing at 10 ft / s ($\nu_{\text{glycerin}} = 8.07 \cdot 10^{-3}$ ft^2 / s)?

17.6. Calculate the values of the $\mathcal{R}_x$ for the two parts of Example 17.1. Boundary layers on flat plates are laminar only up to $\mathcal{R}_x \approx 3.5 \cdot 10^5$ to $2.8 \cdot 10^6$. Are these values exceeded for the boundary layers in Example 17.1?

17.7. From Blasius's solution (Fig. 17.3) we wish to find V_y (the y component of the velocity) at any point.

(*a*) Starting with the mass-balance equation, Eq. 17.2, show that

$$V_y = -\left[\int_0^y \left(\frac{\partial V_x}{\partial x}\right)_y dy\right]_{\text{all at constant } x} = \frac{V_\infty}{2(V_\infty x/\nu)^{1/2}}\int_0^\eta \frac{d(V_x/V_\infty)}{d\eta}\,\eta\, d\eta \qquad (17.V)$$

Then rearrange to

$$\frac{V_y}{V_\infty}\left(\frac{V_\infty x}{\nu}\right)^{1/2} = \frac{1}{2}\int_0^\eta \frac{d(V_x/V_\infty)}{d\eta}\,\eta\, d\eta \qquad (17.W)$$

and integrate graphically, noting that $d(V_x/V_\infty)/d\eta$ is the slope of Fig. 17.3. Schlichting [3, p. 129] gives a table showing these slopes to five significant figures; he also gives a plot of the result of this graphical integration.

(*b*) The same source shows that at the edge of a laminar boundary layer

$$\left(\frac{V_y}{V_\infty}\right)_{\substack{\text{At the edge of a}\\ \text{laminar boundary layer}}} = \frac{0.8604}{\mathcal{R}_x} \qquad (17.X)$$

Using this equation estimate $(V_y/V_x)_{\substack{\text{At the edge of a}\\ \text{laminar boundary layer}}}$ for $\mathcal{R}_x = 10$, 100, and 1000.

What does this tell you about Prandtl's assumption that we could safely ignore the y component of the momentum balance?

17.8. For boundary layers on curved surfaces, the pressure will change with distance. This greatly complicates the solution of the boundary-layer equations compared with that on a flat plate (in which $\partial P/\partial x = 0$), and as a result very few "exact" solutions are known for such boundary layers. Estimates of the behavior of such boundary layers are given by several methods. To illustrate these methods, we will apply them to the laminar boundary layer on a flat plate, where we can compare the results with Blasius's "exact" solution. These methods begin by assuming a velocity profile of the form $V_x/V_\infty = f(y/\delta)$, where δ is the boundary-layer thickness.

(*a*) Show that any satisfactory assumed function $f(y/\delta)$ must have the following properties: When $(y/\delta) = 0$, then $f = 0$; when $(y/\delta) = 1$, then $f = 1$. In addition, for best results it is desirable that secondary conditions be met: When $(y/\delta) = 0$, then $d^2f/dy^2 = 0$; when $(y/\delta) = 1$, then $df/dy = 0$ and $d^2f/dy^2 = 0$. Show graphically the meaning of these conditions.

(*b*) Several useful choices of $f(y/\delta)$ are $f = (y/\delta)$, $f = \frac{3}{2}(y/\delta) - \frac{1}{2}(y/\delta)^3$, and $f = \sin(\pi/2)(y/\delta)$. Show which of these conditions are satisfied by these functions.

(*c*) Below is the approximate boundary-layer calculation for the assumed function $f = (y/\delta)$. Repeat the calculation for the other assumed functions shown in part (*b*). Substituting $V_x/V_\infty = y/\delta$ in Eq. 17.28 and integrating only from zero to δ, we find

$$\theta = \int_0^\delta \frac{y}{\delta}\left[1 - \frac{y}{\delta}\right]dy = \frac{1}{\delta^2}\left[\frac{y^2\delta}{2} - \frac{y^3}{3}\right]_0^\delta = \frac{\delta}{6} \qquad (17.Y)$$

and, using Newton's law of viscosity in Eq. 17.26, we find

$$\tau_0 = \mu \frac{dV_x}{dy} = \mu \frac{V_\infty}{\delta} = \rho V_\infty^2 \frac{d\theta}{dx} \tag{17.Z}$$

But, as shown above, $\theta = \delta / 6$, and therefore

$$\frac{d\theta}{dx} = \frac{1}{6} \cdot \frac{d\delta}{dx} \qquad \text{or} \qquad \delta \frac{d\delta}{dx} = \frac{6\nu x}{V_\infty} \tag{17.AA}$$

Separating variables and integrating from $x = 0$ and $\delta = 0$ to $x = x$ and $\delta = \delta$, we find

$$\frac{\delta^2}{2} = \frac{6\nu x}{V_\infty} \qquad \text{or} \qquad \delta = \left(\frac{12\nu x}{V_\infty}\right)^{1/2} = 3.46\left(\frac{\nu x}{V_\infty}\right)^{1/2} \tag{17.AB}$$

Comparing this result with Blasius's "exact" solution for the same boundary layer (Eq. 17.9), we find that this approximate solution gives a boundary-layer thickness that is about 3.46 / 5, or approximately 70 percent of the correct one.

Computing the local drag coefficient from Eq. 17.13, we get

$$C_f' = \frac{\mu(V_\infty / \delta)}{\frac{1}{2}\rho V_\infty^2} = \frac{2\nu}{V_\infty}\left(\frac{V_\infty}{12\nu x}\right)^{1/2} = 0.577\left(\frac{\nu}{V_\infty x}\right) \tag{17.AC}$$

Comparing this value with the drag coefficient based on Blasius's solution (Eq. 17.14), we find that this approximate solution gives a drag coefficient of 0.577 / 0.664, or about 87 percent of the correct solution. In the same way, all the other properties of the boundary layer can be computed for this assumed velocity profile or for any other assumed one. The striking thing here is that this very simple assumed profile gives reasonably accurate results with an expenditure of much less effort that does Blasius's solution. For more complicated flows this saving in effort can be vast. More on these approximate methods is given by Schlichting [1, Chap. 8].

17.9. Derive Eq. 17.28 by making an x-momentum balance around the boundaries shown in Fig. 17.5 and solving for the force on the plate. Then eliminate the displacement thickness by means of Eq. 17.22, and equate the force on the plate to $W\theta\rho V_\infty^2$.

17.10. For Prandtl's $\frac{1}{7}$ power velocity rule, calculate the ratio of the maximum velocity at the center of the pipe to the average velocity in the entire pipe. Also calculate the ratio of the kinetic energy of the fluid to the kinetic energy it would have if it were all flowing at the average velocity, and the same relation for the momentum of the flowing fluid. This problem repeats Probs. 3.9 and 3.10. See Table 3.1.

17.11. Any velocity distribution equation that is to represent the flow in the neighborhood of the wall must satisfy the following requirements: $V_x = 0$ at $y = 0$; dV_x / dy is finite and nonzero at $y = 0$. As we saw in the text, Prandtl's $\frac{1}{7}$ power rule satisfies the first condition but not the second. Which of the following kinds of functions satisfy both conditions?

(*a*) $V_x = A + By$.

(*b*) $V_x = By^n$, when n is some power other than 1.

(*c*) $V_x = A \sin y$.

(*d*) $V_x = A \exp y^n$, where n is any power.

17.12. (*a*) Show from the definition of u^+ and y^+ and from the definition of the friction factor that the limiting value of the relation of u^+ to y^+ as y^+ approaches zero must be $u^+ = y^+$, called the *law of the wall.*

(*b*) Figure 17.8 shows that a slightly different function is given as a better fit of the experimental data. It is only intended to be used for $u^+ \leq 5$. At $u^+ = 5$, how much does this function differ from $u^+ = y^+$?

17.13. Show that Eq. 17.34 is approximately correct over the Reynolds number range from $3 \cdot 10^3$ to 10^5 by calculating the friction factor from it and comparing it with the friction factor for smooth pipes in Fig. 6.10 for several $\mathcal{R}$ in this range. How well does this approximation work for $\mathcal{R} = 10^8$?

17.14. Show that, if we make the assumptions shown in Sec. 17.5, we find Prandtl's equation for the drag on a flat plate with a turbulent boundary layer. The procedure is as follows:

(*a*) From the seventh-power distribution rule (Eq. 17.33) deduce the ratio of the momentum thickness to the boundary-layer thickness by using Eq. 17.28. Here the integration is from 0 to δ rather than from 0 to ∞. Answer: $\theta = (\frac{7}{72})\delta$.

(*b*) In Eq. 17.34 the velocity in the Reynolds number and the velocity in the expression for the friction factor are average pipe velocities. From Prandtl's $\frac{1}{7}$ power rule it can be shown that for circular pipes (Prob. 17.10) this average velocity is 0.817 times the maximum velocity. Prandtl rounded this off to 0.8. Making this substitution and recalling from Chap. 6 that

$$f = \tau_0 / \rho \frac{1}{2} V_{x,\text{avg}}^2 \tag{17.AD}$$

convert Eq. 17.4 to

$$\frac{\tau_0}{\rho} \cdot \frac{V_\infty^2}{2} = 0.225 \left(\frac{\nu}{V_\infty \delta} \right)^{1/4} \tag{17.AE}$$

(*c*) Combine this result with Eq. 17.27 and the answer from part 1 of this problem to obtain

$$\frac{7}{72} \cdot \frac{d\delta}{dx} = 0.225 \left(\frac{\nu}{V_\infty \delta} \right)^{1/4} \tag{17.AF}$$

which may be integrated from $\delta = 0$ at $x = 0$ to $\delta = \delta(x)$ at any x, obtaining Eq. 17.35.

17.15. Starting with Eq. 17.35, derive Eqs. 17.36 and 17.37. Use Eq. 17.26 for the shear stress as a function of the momentum thickness and the result obtained in Prob. 17.14 that $\theta = (\frac{7}{72})\delta$. The two different drag coefficients are defined in Eqs. 17.13 and 17.18.

17.16.*In Example 15.1 we considered the laminar flow of a Newtonian fluid between two parallel plates and showed that well downstream from the entrance the velocity distribution was parabolic. At the entrance to such a pair of plates the flow will initially have a uniform velocity, V_0, independent of x and y. Boundary layers will grow from the walls, eventually meeting in the center, as sketched in Fig. 17.10. Show that if we make

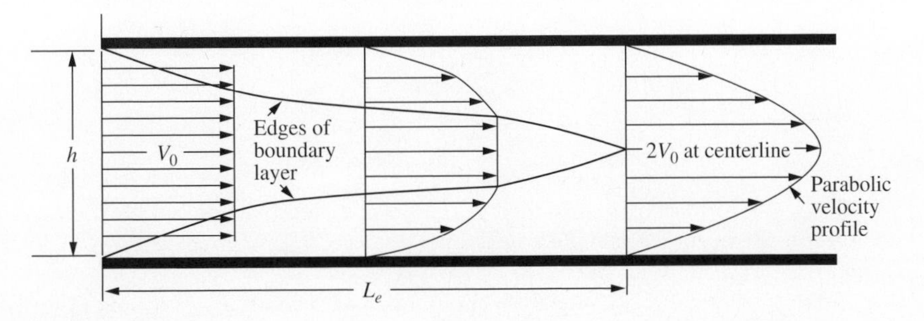

FIGURE 17.10
The entrance length for flow between two parallel plates.

the simplest possible assumption, that the growing boundary layers do not interact with each other and that the fluid between the boundary layers has a constant velocity, then the distance downstream required for the boundary layers to grow together (the "entrance length") would be given by $L_e / h = 0.01\,\mathcal{R}$. These assumptions are gross simplifications; the worst one says that the fluid in the center does not speed up. By material balance we may show that it must reach a velocity of twice the entrance velocity when the layers meet. More complicated analyses that take this into account [3, p. 178] lead to an approximate formula for parallel plates of $L_e / h \approx 0.04\,\mathcal{R}$. To see the magnitude of this entrance length, calculate it for air flowing at 5 ft / s between plates 1.0 in apart. (Here $\mathcal{R}$ is the Reynolds number based on distance between plates, not on distance from the leading edge.)

17.17. The fins on the radiator of your car are about 2 in long. When your car is going 60 mi / h, the free-stream velocity of the air approaching these fins (under the hood of the car and behind the grille) is estimated to be 40 mi / h. Estimate the boundary layer thickness at the downstream end of the fins.

REFERENCES FOR CHAPTER 17

1. Schlichting, H., K. Gersten. *Boundary-Layer Theory.* 8th ed. Berlin: Springer, 2000.
2. Longwell, P. A. *Mechanics of Fluid Flow.* New York: McGraw-Hill, 1966, p.164.
3. Schlichting, H. *Boundary Layer Theory.* 6th ed. transl. J. Kestin. New York: McGraw-Hill, 1968. References are given in the text to the 6th ed. for plots and tables that were not included in later editions.
4. Mehta, R. D. "Aerodynamics of Sports Balls." *Ann. Rev. Fluid Mech. 17,* (1985), pp. 151–189.
5. Kramer, M. O. "Hydrodynamics of the Dolphin." *Adv. Hydrosci. 2,* (1965), pp. 11–130.

CHAPTER 18

TURBULENCE

The detailed behavior of turbulent flows is so complex that in spite of considerable research effort over the past hundred years we do not now possess a comprehensive theory of turbulence or a simple conceptual model of how it works in detail. We can write suitable equations for turbulent flow, but if we attempt to capture the detail in the equations, the resulting computations overwhelm our largest computers. Most of what we know consists of qualitative observations, measurements of various properties of turbulent flows, and some definitions and correlations of these measurements. Turbulence theory helps us understand many of these observations and extend them in some cases. This chapter is largely devoted to descriptions of turbulence, definitions of some of the terms used in describing turbulence, and elementary applications of turbulence models. In the chapter on mixing we will need some ideas from this chapter.

Even though we cannot provide a detailed mathematical description of turbulent flow, we can provide several physical descriptions to help the student form an intuitive picture of turbulent flow.

18.1 NONMATHEMATICAL OBSERVATIONS AND DESCRIPTIONS OF TURBULENCE

Normally we cannot see turbulence in clear fluids like air and water, nor can we see it inside vessels and pipes unless those are transparent. Most of our nonindustrial observations of turbulence are made with clouds, smoke plumes (Fig. 6.3), or mixing of fluids with different colors (e.g., coffee and cream). From those observations, which the reader has certainly made, we can develop some intuitive understanding of turbulence.

18.1.1 Decay of Turbulence

The first observation we can make is that turbulence dies out. If the student vigorously stirs a bowl of soup, the soup will have obvious turbulent eddies in it, but after a few minutes they will be gone and the soup will be motionless. This happens because of viscosity. Large turbulent eddies transfer energy to smaller ones, and they transfer energy to even smaller ones, until the size of the eddy is small enough that viscosity slows and stops it. Thus, the total store of turbulent kinetic energy (ke) that the student put into the soup with a spoon is said to *decay* into internal energy by viscous friction heating. If we want the mass of fluid to have turbulence that does not die out, then we must continue to put in turbulent ke as fast as viscosity is converting that energy to heat.

18.1.2 Production of Turbulence

A shear layer is the most common way to produce turbulence. Stirring the soup most often will be done with a circular motion, inducing a circular flow and/or circular eddies. That is also common in vessels with rotating mixers. But for flows in pipes, ducts, around airplanes or ships, or in the atmosphere there is no such rotating element. Still, we observe that such rotating eddies occur in flows like the flow between two plates, in which no solid surface is rotating. Figure 18.1(*a*) shows one plate sliding relative to another, with fluid in between. If, instead of fluid, the space between the plates had been filled with cylindrical rods, aligned at right angles to the flow, the plates would set the rods in rotation, in the clockwise direction shown. So this *shear flow* sets the fluid between the plates into rotational motion as shown. However, the whole fluid cannot form one rotating cell, so the fluid tends to form parallel threads of rotation, called vortex threads. If the viscosity is high enough or the velocity low enough these rotational threads will be quickly stopped by

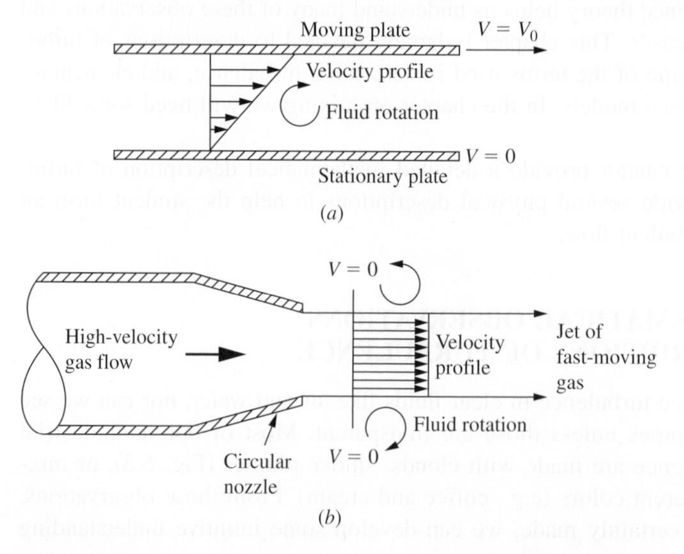

FIGURE 18.1
Fluid rotation produced by (*a*) flow between moving plates and (*b*) a free jet.

viscosity. But for high velocity or low viscosity (actually for high Reynolds number) they will persist.

If the vortex threads were permanently attached to rigid masses of fluid, then they would behave like the solid cylindrical rods described above. But they are not; fluid flows through them, just as fluid flows through the isolated, stationary vortex that forms in a sink drain. The fluid in a vortex thread deforms in response to the forces acting on it. Each vortex thread influences the behavior of its neighbors, because at their boundaries they are generally moving in opposite directions. As a result they twist, kink, stretch, split, and divide. If vortex threads were visible, then a turbulent fluid flow would look like a wriggling mass of interlaced spaghetti, with many different diameters and lengths of spaghetti pieces present, and with large ones constantly being formed, turning into smaller and smaller ones, and the smallest ones eventually disappearing.

The turbulence in pipes, ducts and channels is produced by the shear layers at the walls of the duct. Turbulence is steadily fed into the main flow in the duct and ultimately consumed in viscous heating in the main flow. The turbulence in the wake of ships and airplanes is caused by the shear layer adjacent to the surface of the airplane or ship. It decays with time, due to viscosity, after the airplane or ship passes.

We can also have shear layers not involving solid surfaces as shown in Fig. 18.1(*b*). The jet of hot gas leaving a jet engine is moving very rapidly, relative to the surrounding air; the boundary between the jet of hot gas and the surrounding air forms such a shear layer, which sets the surrounding air into rotation. If the jet is circular, under some conditions it can form circular vortex threads, which form "smoke rings." Similarly, the hot smoke from a cigarette, rising in a still room due to buoyancy (Fig. 6.3) sets up a shear layer between itself and the surrounding air, inducing turbulence.

18.1.3 Free and Confined Turbulent Flows

The turbulent flow in a pipe is quite different in character from the turbulent flow in a wind tunnel or in the lower atmosphere. In the atmosphere or in the central section of a wind tunnel, the nearest wall is so far away that it has little influence on the flow. This kind of flow, substantially uninfluenced by nearby walls, is called *free turbulence*. In a typical long pipe the flow is strongly influenced by the nearby presence of a wall. This kind of flow is often called *shear turbulence* or *wall turbulence*.

In the lower atmosphere or in a wind tunnel the turbulence at any point generally has the same properties in all directions; this is called *isotropic turbulence.* The same term is used with an analogous meaning in the study of matter. Water is isotropic; it has the same properties in all directions. Wood is nonisotropic; it has different properties "with the grain" and "across the grain." (If you do not believe this, try splitting a log across the grain.) Turbulence in pipes is generally not isotropic; it is more intense in the flow direction than at right angles to the flow direction, and most intense in the region about one-tenth of the distance from the pipe wall to the center.

In a pipe, the turbulence character does not change as one moves downstream at a constant radial position. This type of turbulence is called *homogeneous turbulence*. Normally this is not the case in the atmosphere or in a wind tunnel, where the turbulence, although isotropic at a point, tends to become less intense with distance downstream from the source of the disturbance.

18.1.4 Turbulence in the Atmosphere and the Oceans

On a sunny day, the lower atmosphere is normally in turbulent motion, driven by the horizontal shear layer between wind and the ground, and by vertical shear layers caused by rising warm air masses and sinking cold air masses. This turbulence mixes the *troposphere,* up to about 30,000 ft; above that the *stratosphere* has very little turbulence, (see Fig. 2.4). The boundary between the two is often visible to travelers, because modern passenger aircraft fly above that boundary. Below the boundary turbulence slowly mixes fine particles (both natural and man made) up from the ground and makes the air hazy; above the boundary the air is clear.

Students have probably observed, on calm days, that jet airliners leave a condensation trail high in the sky. The trail itself is wrinkled and convoluted, because of the turbulence put into it by the shear layer in the airplane's wake and the turbulence due to the interaction of the rapid jet exhaust and the stationary surrounding air. But after the plane passes, the condensation trail does not grow or disperse; it remains practically constant in size and in place. Eventually molecular diffusion evaporates it. Such trails are formed and visible only in areas of the atmosphere high above the ground, where there is practically no turbulence; if there were turbulence the condensation trails would disappear quickly.

Closer to the ground, turbulence can be fed into the atmosphere by the wind-driven shear layer at the ground or by hot air rising from the solar-heated ground, as in the example of the hot plume from the cigarette in a still room. Students have probably observed that in the lower atmosphere the plumes from smokestacks or from cars on dry dirt roads continually expand as they flow in the downwind direction. This expansion is caused by the turbulence in the lower atmosphere. At night, when the ground is cold, there is very little turbulence in the lower atmosphere and plumes spread very little.

In the oceans turbulence enters mostly through wind-driven waves at the surface and penetrates only a few wavelengths down. There is large-scale buoyant motion in the ocean, caused by solar heating and by temperature and salinity differences. These, and the large-scale ocean currents such as the Gulf Stream, can generate large-scale turbulence at their edges. Away from the surface and from the edges of these large-scale flows, there is very little turbulence in the oceans [1].

In both atmosphere and oceans, density stratification inhibits or destroys turbulence. Atmospheric inversions (cold air under warmer air) damp atmospheric turbulence strongly. When you see a smoke plume that forms a thin, nonspreading streak down the sky, you are almost certainly looking at an atmospheric inversion, [2, Chaps. 5 and 6]. In the oceans density stratification depends on temperature (warm above cold) and salinity (fresh or fresher water over salty water). These stratifications also inhibit turbulence. Such stratifications also occur in flows in pipes and process equipment. If the stratification is unstable (e.g., hot fluid produced at the bottom of a saucepan on a stove) then it will cause vertical flows and turbulence, like water boiling in a saucepan.

18.1.5 Three-Dimensional Turbulence

Turbulence is inherently three-dimensional. If we have a pressure-driven laminar flow between two parallel plates, the velocity will all be in one direction, and the velocity gradients will all be in the direction perpendicular to the parallel plates. There is neither velocity nor a velocity gradient in the direction perpendicular to these two directions. If we increase the Reynolds number enough for the flow to become turbulent, then we will observe fluctuating turbulent velocity components in all three directions.

18.1.6 The Size of Eddies

In liquids or gases at atmospheric pressure, the smallest eddies, which are busily converting turbulent ke into internal energy by viscous heating, are thousands or millions of times as large as the space between molecules, or the mean free path, so that treating turbulence as a phenomenon in a continuous phase and ignoring the existence of molecules and atoms does not introduce any serious error [3, p. 8].

The largest eddies in a confined flow cannot be larger than the dimensions of the confining container. Generally they will be 0.1 to 0.5 times as large as the boundaries of the system or the size of the disturbance causing the turbulence. So in turbulent flow in a pipe or duct, the largest eddies have a length about 25 percent of the pipe diameter. In the wakes of ships and airplanes the largest eddies will be a similar fraction of the size of the ship or airplane. On the scale of major storms in the atmosphere the flow is not turbulent, whereas on the scale of individual parts of the storms it generally is. Thus, the huge eddy of a hurricane or a major storm, seen from an earth satellite and presented by TV weather persons, has little flow and eddying across the main directions of flow, but the cloud edges are ragged, indicating local turbulence.

There is practically no turbulence in small systems. The Reynolds number has a characteristic length (the diameter in pipe flow) and as this becomes small, the Reynolds numbers become small, and the effect of viscosity overwhelms turbulence. There is very little turbulence in the flows in our bodies, and inside most living things.

18.2 WHY STUDY TURBULENCE?

The principal goal of turbulence research is to place turbulent flows on as sound a scientific and computational footing as we now have for laminar flows. For laminar flows we can start with Newton's laws of motion (generally in the form of the Navier-Stokes equations, Sec. 15.4) and from a description of the flow boundaries and the fluid properties deduce a complete mathematical description of the flow. This can be done analytically only for very simple flows, but with computers it is possible to do it numerically for very complex flows by using computational fluid dynamics (CFD). Furthermore, for laminar flows, if we can find the velocity distribution as described above, then we can generally describe the heat transfer and mass transfer in the flow by mathematical analysis without recourse to experimental measurement.

For turbulent flows we are not so fortunate. In general, we cannot calculate such velocity distributions or heat transfer and mass transfer from basic laws but must depend on correlations of experimental measurements. For widely used systems such as the flow inside a straight, cylindrical pipe or the perpendicular flow across the outside of a cylinder, experimental data are available and have been correlated sufficiently well by the methods of dimensional analysis (Chap. 9) to allow us to predict the results of any new experiment with considerable accuracy. However, although this may be satisfactory for the practical engineer, it is unsettling to the theorist. Furthermore, when an extrapolation outside the range of experimental data is required, or application is desired to a shape for which no experimental data are available, the need for a turbulence theory becomes apparent.

Turbulence theories are not so far advanced that they allow us reliably to extrapolate experimental data or calculate flows around new shapes. Rather, they have concentrated on trying to reproduce the existing experimental data from some kind of comprehensive theory. This has not yet been accomplished. On the other hand, the partial results and partial understandings of turbulent flow have been useful in predicting the results of some experiments, e.g., turbulent boundary layers, as discussed in Sec. 17.5.

One of the first historical examples in which turbulence research proved very useful was in the comparison of aircraft tests in wind tunnels with the corresponding results in free flight. The early experimental work in this field indicated that results from one wind tunnel did not necessarily agree with those from another wind tunnel or with the results for free flight. These differences were ultimately explained by the study of the differences in turbulence between various wind tunnels. Some of the first careful measurements of turbulence were made to explain these contradictory experimental results [4].

Heat transfer and mass transfer studies look to fluid mechanics for an understanding of turbulence, because such an understanding of turbulence is very useful in those fields. Even our limited understanding of turbulence already has been of value in those fields. The mixing of fluids plays a significant part in many processes, e.g., the fuel-oxidizer mixing in all combustion processes, the mixing of reactants in most chemical reactors, and the blending of ingredients for foods, plastics, etc. (see Chap. 19). If the fluids being mixed have low viscosities, then turbulence greatly speeds the mixing. Thus, a knowledge of the detailed structure of turbulence is a prerequisite of any scientific understanding of the mixing of low-viscosity fluids. Turbulence plays a major role in low-altitude meteorological processes [5]. Modern theories of astronomy on a galactic scale indicate that turbulence is very important in the evolution of galaxies [6].

Thus, although the study of turbulence is difficult, and the large efforts expended have not resulted in general or comprehensive results, the potential benefits of such a thorough knowledge of turbulence are great enough to justify the effort. For the typical chemical engineer, this chapter is mostly technical-cultural background. However, the terms defined here are in widespread use, and the approximate methods discussed here are used in the two- and three-dimensional CFD programs that typical chemical engineers use.

18.3 TURBULENCE MEASUREMENTS AND DEFINITIONS

As discussed in Sec. 17.3, it is common in discussing turbulence to regard the velocity at any point and any time as consisting of an average component and a fluctuating component:

$$V_x = \overline{V}_x + v_x \tag{18.1}$$

Here V_x is the instantaneous value of the x component of the velocity, $\overline{V}_x$ is the time average of this velocity over some reasonable period, and v_x is the instantaneous fluctuation of this velocity from its time-average value. This is one component of a vector equation

$$\mathbf{V} = \overline{\mathbf{V}} + \mathbf{v} \tag{18.A}$$

which obeys the vector manipulation rules in App. C. Throughout the turbulence literature (and this chapter) a bar over a symbol indicates the average value of that quantity.

Here we must distinguish between time averages and position averages. A time average of some arbitrary function of time, ϕ, is given by

$$\overline{\phi} = \lim_{t \to \infty} \frac{1}{t} \int_0^t \phi(t)\, dt \tag{18.2}$$

which is the average value of the reading of some kind of ϕ-meter located at some fixed point. We can also define position averages as, for example, the average velocity across the cross section of a pipe, as defined in Chap. 3. All average quantities in this chapter are time averages, as defined by Eq. 18.2.

Although the time average given above is shown as the limit as time goes to infinity, in practice it is only necessary to make measurements over a period that is long compared with the frequency of the fluctuations. Thus, for turbulence measurements in pipes the average value found by Eq. 18.2 for t equal to a few seconds is numerically equal to that found for any longer period of time. According to these definitions, the average value of v_x must be zero, because it is negative for as much of the time as it is positive.

The fluctuations in turbulent flow in pipes and channels are mostly so fast that ordinary fluid-flow measuring devices do not detect them at all; those devices record only the values associated with $\overline{V}_x$. For example, the pressures in turbulent pipe flow fluctuate with a very high frequency. However, ordinary pressure gauges do not respond to such high frequencies, so they show a steady average pressure for such a flow. Similarly, the velocities indicated by venturi meters, orifice meters, pitot tubes, etc. for turbulent pipe flows generally do not show the fluctuating component at all; their response is simply too slow. Therefore, a pitot tube in a turbulent pipe flow simply reads $\overline{V}_x$ (subject to slight corrections due to turbulent fluctuations).

To measure v_x we need a flowmeter, which is much more responsive to rapid changes in flow rate. The most successful flowmeter and the one that produced most of the world's measurements of turbulence before about 1980 is the hot-wire anemometer, shown in Fig. 18.2. In such a device a fluid flows over a very thin electrically heated wire whose temperature is much higher than that of the fluid. The wire is generally platinum or tungsten; both metals show a significant increase in electrical resistance with increasing temperature. By suitable electrical circuitry we can measure the fluctuating resistance (and hence the fluctuating temperature) at constant heat input, or the fluctuating heat input required to hold the wire's temperature constant.

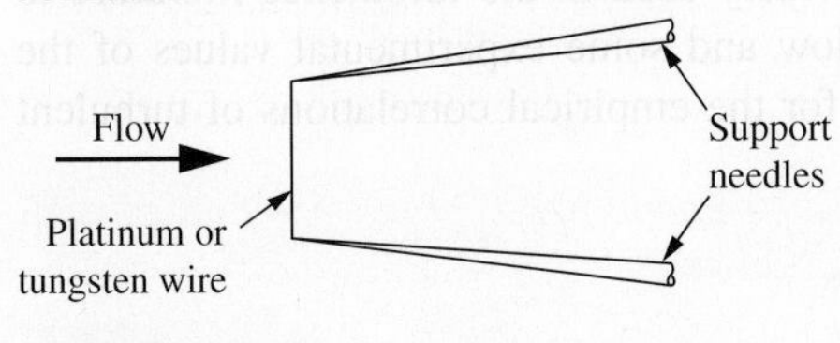

FIGURE 18.2
Hot-wire anemometer. The wire is usually about 0.005 mm in diameter and about 1 mm long.

Over a wide range of flow rates the heat removed from a hot wire by a fluid flowing perpendicular to it is described by

$$\text{Heat-removal rate} \approx A + B \cdot (\text{fluid velocity})^{1/2} \tag{18.3}$$

where A and B are experimental constants. Thus, after suitable calibration we can use a hot-wire anemometer to measure fluid velocities. Furthermore, it can be shown experimentally that, because the wire is very small (the best are about 0.005 mm in diameter, a tenth of the average diameter of a human hair), it can respond very rapidly to changing flow rates. Some hot-wire anemometers can follow velocity fluctuations that are as fast as 50,000 Hz.

The hot-wire anemometer shown in Fig. 18.2 is very sensitive to flow perpendicular to the wire but much less sensitive to flow in the direction parallel to the wire which is much less effective for heat transfer. Thus, by using arrays of hot-wire anemometers placed at different angles to the flow we can determine not only the fluid velocity but also its direction.

Once the basic scheme for hot-wire anemometry was worked out, numerous variants on it were developed [7]. Generally, hot-wire anemometers are delicate, temperamental instruments that require expensive electronic circuitry and considerable care and training to use.

The other procedure for studying turbulence is to inject some tracer into a turbulent flow and measure its concentration distribution at some point downstream in the flow. (This was the method use by Reynolds in his pioneering exploration of turbulence, described in Sec. 6.2). As a tracer we can use a dye, which is injected at a point and then photographed downstream, or an electrically conducting solution (e.g., salt solution in water) whose presence is detected downstream by electric conductivity probes. Both of these methods have the drawback that the "tracer" fluid they introduce may disturb the flow. Another approach is to suspend in the fluid small particles of the same density as that of the fluid and then record their trajectories photographically or by laser-Doppler measurements. Since about 1980 laser-Doppler methods have largely replaced hot-wire methods [8].

18.4 THE EXPERIMENTAL AND MATHEMATICAL DESCRIPTIONS OF TURBULENT FLOWS

The ultimate description we would like to have of turbulent flow would be an explicit expression for $\overline{V}_x$, $\overline{V}_y$, $\overline{V}_z$, and v_x, v_y, and v_z as functions of time and position. Then we could predict the average and the fluctuating velocities at any point and any time. Currently it seems impossible to make such a description; the problem is much too complex. The next best thing is a statistical description of the flow, i.e., what fraction of the time V, v_x, v_y, etc. have certain values. Much of the experimental and theoretical work done on turbulence has been directed at these statistical properties of the flow. Below is a set of definitions that are widely used in the turbulence literature to describe such statistical properties of the flow and some experimental values of the quantities so defined. These form the basis for the empirical correlations of turbulent behavior used in CFD (Chap. 20).

18.4.1 Turbulent Intensity

Turbulent intensity or, simply, *intensity,* is a measure of how strong, violent, or intense the turbulence is. (In the older literature it is often called *level of turbulence* or *degree of turbulence.*) Turbulent intensity is defined by

$$x\text{ – turbulent intensity} = \overline{(v_x^2)}^{1/2} \tag{18.4}$$

Here $\overline{(v_x^2)}^{1/2}$ is the root-mean-square (rms) of the fluctuating x component of the velocity. The relative intensity is defined by

$$x\text{ – relative intensity} = T_x = \frac{\overline{(v_x^2)}^{1/2}}{\overline{V}} \tag{18.5}$$

Here $\overline{V}$ is the average of the absolute magnitude of the vector velocity, equal to $(\overline{V}_x^2 + \overline{V}_y^2 + \overline{V}_z^2)^{1/2}$. We may define the y and z components of the intensity by replacing the subscript x in Eq. 18.4 with a subscript y or a subscript z. If Eq. 18.5 defines the x component of the relative intensity, then the entire relative intensity must be given by

$$T = \left[\frac{1}{3}\left(\frac{\overline{(v_x^2)}}{\overline{V}^2} + \frac{\overline{(v_y^2)}}{\overline{V}^2} + \frac{\overline{(v_z^2)}}{\overline{V}^2}\right)\right]^{1/2} = \left[\frac{1}{3}(T_x^2 + T_y^2 + T_z^2)\right]^{1/2} \tag{18.6}$$

Some writers call T_x, which we call the relative intensity, simply the intensity; others call it the absolute intensity.

As discussed in Sec. 18.2, the average value of v_x is zero, because it is positive as often as it is negative. However, $\overline{(v_x^2)}^{1/2}$, the rms value of v_x, is not zero, because squaring before averaging removes the minus signs. Many authors, to save writing, use the symbols u', v', and w' in place of $\overline{v_x^2}^{1/2}$, $\overline{v_y^2}^{1/2}$, and $\overline{v_z^2}^{1/2}$.

Example 18.1. A turbulent flow at some point is described by the equations

$$V_x = 10\frac{\text{ft}}{\text{s}} + 1\frac{\text{ft}}{\text{s}} \cdot \sin t \qquad V_y = V_z = 0 \tag{18.B}$$

Calculate $\overline{V}_x$, v_x, $\overline{v}_x$, $(\overline{v_x^2}^{1/2})$, T_x, and T.

No real turbulent flow can be described by equations this simple. These equations serve only to show the relations between some defined quantities for turbulent flow. By inspection,

$$\overline{V}_x = 10\frac{\text{ft}}{\text{s}} \qquad v_x = 1\frac{\text{ft}}{\text{s}} \cdot \sin t \tag{18.C}$$

$$\overline{v}_x = \lim_{t\to\infty} \frac{1}{t}\int_0^t 1\frac{\text{ft}}{\text{s}} \cdot \sin t\, dt = 1\frac{\text{ft}}{\text{s}} \lim_{t\to\infty}\left(-\frac{\cos t}{t}\right) = 0 \tag{18.D}$$

$$\overline{(v_x^2)} = \lim_{t\to\infty} \frac{1}{t}\int_0^t \left(1\frac{\text{ft}}{\text{s}}\right)^2 (\sin t)^2\, dt$$

$$= \left(1\frac{\text{ft}}{\text{s}}\right)^2 \lim_{t\to\infty}\frac{1}{t}\left(-\frac{1}{2}\cos t \sin t + \frac{1}{2}t\right) = \frac{1}{2}\frac{\text{ft}^2}{\text{s}^2} \tag{18.E}$$

$$\overline{(v_x^2)}^{1/2} = \left(\frac{1}{2}\frac{\text{ft}^2}{\text{s}^2}\right)^{1/2} = 0.707\frac{\text{ft}}{\text{s}} = 0.22\frac{\text{m}}{\text{s}} \tag{18.F}$$

$$\overline{V} = (\overline{V}_x^2 + 0 + 0)^{1/2} = \overline{V}_x = 10\text{ ft/s} = 3.05\text{ m/s} \tag{18.G}$$

$$T_x = \frac{0.707\text{ ft/s}}{10\text{ ft/s}} = 0.0707 \tag{18.H}$$

and

$$T = \left[\frac{1}{3} \cdot 0.0707^2\right]^{1/2} = 0.0408 \tag{18.I}$$

■

By suitable electronic circuitry a hot-wire anemometer can be made to read directly the rms fluctuating velocity. Thus, with suitable equipment we can directly read the turbulent intensity in the direction perpendicular to the anemometer wire. A typical set of experimental results for turbulent-intensity measurements in a rectangular channel are shown in Fig. 18.3. From this figure we see the following:

1. The intensity varies with location and is generally greatest in a layer near the wall.
2. The intensity falls off to zero right at the wall. This must be, because the solid boundary stops any motion at right angles to the flow direction, and there is zero velocity in the flow direction right at the wall.
3. The maximum observed relative intensity T_x is about 0.19 (measured relative intensities seldom exceed this value).
4. The intensity in the flow direction is greater than the intensity in the direction at right angles to the flow.
5. Near the center of the channel the intensities in the x and y directions are approaching each other (for a very large channel they become practically equal at the center, so that the turbulence there is practically isotropic).

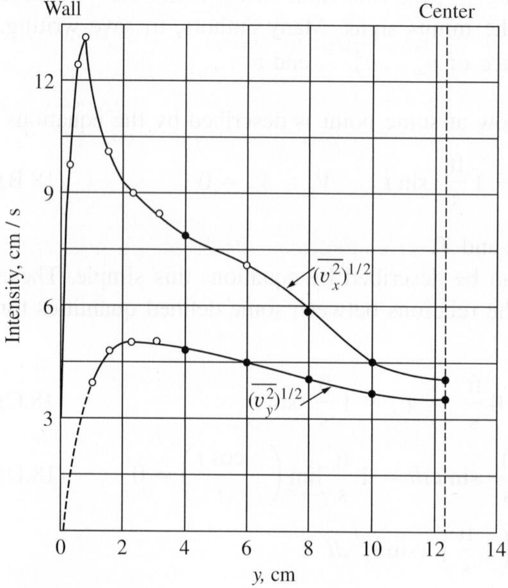

FIGURE 18.3
Turbulent intensity measurements in a rectangular channel 1 m wide and 0.24 m high. The centerline velocity (V_x) is 1 m / s. The value of V_x at the point near the wall where $(\overline{v_x^2})^{1/2}$ is a maximum is about 0.7 m / s, so that the maximum value of T is about 0.19. (From H. Reichardt, "Messungen turbulenter Schwankungen"—"Measurements of turbulent oscillations," *Naturwissenschaften 26*:407 (1938). Reproduced by permission of the publisher.)

The measurements shown in Fig. 18.3 are typical of those made in a channel where no effort has been made to hold the turbulence level low. For wind tunnels we desire as low a turbulent intensity as possible; good ones have $T \approx 0.0005$.

18.4.2 Turbulent ke

A fluid flowing in turbulent flow possesses more ke per unit mass than would the same fluid moving at the same average velocity in nonturbulent flow. Furthermore, the more intense the turbulence, the greater the amount of *turbulent ke.*

The turbulent ke is fed into the flow by some kind of external work, generally a pressure gradient (i.e., injection work) in a pipe or a fan or a blower in a wind tunnel or the sun-driven work of thermal convection currents in the lower atmosphere. This work may first cause an increase in ke in the flow direction or merely a gradient in the velocity that will lead to turbulence. The turbulent ke leaves the flow by viscous conversion to internal energy.

In a wind tunnel the turbulent ke is put in by the blower or an inlet screen, downstream from a blower; no further turbulent ke is put into the flow as it moves through the tunnel (because the tunnel is so short and wide that the walls have little effect in the center, where the tests are made). Thus, the amount of turbulent ke decreases with distance from the blower in a wind tunnel; the turbulent ke decays. Similarly, if we measure the turbulent ke in the wake of a ship or airplane, we find that the turbulent ke is strong close to the ship or airplane but decays farther from the ship or plane, eventually vanishing far enough downstream. In the latter case the turbulent ke is supplied by the engines, which drive the ship or airplane, overcoming drag.

At the other extreme, in "fully developed" turbulent flow well downstream from the entrance of a pipe, turbulent ke is steadily being fed into the flow by the work being expended, overcoming the shear (frictional) resistance at the pipe walls. This ke is steadily being converted to internal energy by viscous friction in the turbulent flow, so that the rate of addition of turbulent ke to the flow exactly balances the rate of destruction of ke by viscous friction, and the amount present stays constant as the fluid moves down the pipe. Thus, the turbulence in a pipe does not decay with distance, as does the turbulence in a wind tunnel or the wake behind a ship or plane.

Example 18.2. Air is flowing in a rectangular channel, 1.0 m wide and 0.24 m high; see Fig 18.3. The centerline velocity is 1.00 m / s. How much nonturbulent ke per unit mass in the flow direction does it have? How rapidly does it degrade to heat?

Based on Table 3.1 we estimate the average velocity as 0.82 m / s. The hydraulic radius is 0.0915 m; the Reynolds number $\approx$ 20,000; and for an assumed relative roughness of $4 \cdot 10^{-5}$ the calculated friction factor is 0.0065, from which the pressure gradient is -0.0286 Pa / m.

The first part is easy:

$$\frac{\begin{pmatrix}\text{Nonturbulent}\\ \text{kinetic energy}\end{pmatrix}}{\text{Mass}} = \frac{V^2}{2} = \frac{(0.82 \text{ m / s})^2}{2} = 0.336\,\frac{\text{m}^2}{\text{s}^2}$$

$$= 0.336\,\frac{\text{J}}{\text{kg}} = 0.000144\,\frac{\text{Btu}}{\text{lbm}} \tag{18.J}$$

If all this ke were converted to internal energy, it would raise the temperature of the air by 0.00060°F = 0.00033°C.

To compute the rate of conversion of some other kind of energy to heat, called the *dissipation rate* (which we will need later in this chapter), we consider a section of pipe with length ΔL and cross-sectional area A:

$$\begin{pmatrix}\text{Dissipation}\\ \text{rate}\end{pmatrix} = \varepsilon = \frac{\begin{pmatrix}\text{power converted}\\ \text{to internal energy}\end{pmatrix}}{\text{mass}} = \frac{-AV\,\Delta P}{A\,\Delta L\rho} = \frac{-\Delta P}{\Delta L}\frac{V}{\rho} \tag{18.7}$$

For this example we calculated that $-\Delta P / \Delta L \approx 0.0286$ Pa / m = 0.00013 psi / 100 ft, so

$$\varepsilon = \frac{-\Delta P}{\Delta L}\frac{V}{\rho} = \frac{0.0286\ \text{Pa}}{\text{m}} \cdot \frac{0.82\ \text{m/s}}{1.20\ \text{kg/m}^3} \cdot \frac{\text{N}}{\text{Pa}\cdot\text{m}^2} \cdot \frac{\text{kg}\cdot\text{m}}{\text{N}\cdot\text{s}^2}$$

$$= 0.0196\,\frac{\text{m}^2}{\text{s}^3} = 0.0196\,\frac{\text{W}}{\text{kg}} = 0.21\,\frac{\text{ft}^2}{\text{s}^3} = 8.4\cdot 10^{-6}\,\frac{\text{Btu}}{\text{lbm}\cdot\text{s}} \tag{18.K}$$

If there were no heat loss, then the external work converted to internal energy would raise the temperature of the air by 0.000035°F / s = 0.0000195°C / s. The values computed here are averages over the whole flow. Later we will see that ε varies from place to place in the flow and that we will need local values. ■

At any point in the flow the x-component of the ke per unit mass is given by

$$\begin{pmatrix}\text{Turbulent ke per unit}\\ \text{mass, } x \text{ component}\end{pmatrix} = \frac{1}{2}\overline{(v_x'^2)} = \frac{1}{2}(x\ \text{intensity})^2 \tag{18.8}$$

and the total ke is the sum of the three component values, so

$$k = \begin{pmatrix}\text{Turbulent ke per unit}\\ \text{mass, three components}\end{pmatrix} = \frac{1}{2}\left(\overline{(v_x'^2)} + \overline{(v_y'^2)} + \overline{(v_z'^2)}\right) = \frac{3}{2}\overline{V}^2 T^2 \tag{18.9}$$

Example 18.3. Estimate the value of k for a point 2 cm from the wall in Fig. 18.3.

Based on chart reading, we have $\overline{(v_x'^2)}^{1/2} \approx 9.5$ cm / s and $\overline{(v_y'^2)}^{1/2} \approx 5.0$ cm / s. No values are shown for $\overline{(v_z'^2)}^{1/2}$; its value will be assumed ≈ 0. Then

$$k = \frac{1}{2}\left[\left(9.5\,\frac{\text{cm}}{\text{s}}\right)^2 + \left(5.0\,\frac{\text{cm}}{\text{s}}\right)^2\right]$$

$$= 57.6\,\frac{\text{cm}^2}{\text{s}^2} = 0.00576\,\frac{\text{m}^2}{\text{s}^2} = 0.00576\,\frac{\text{J}}{\text{kg}} = 2.5\cdot 10^{-6}\,\frac{\text{Btu}}{\text{lbm}} \tag{18.L}$$

■

This is a local value, at one point in the flow.

18.4.3 Scale of Turbulence

The intensity is a measure of "how much" turbulence is in some small mass of matter. We would like to know how big the eddies are as well. Individual eddies are continually changing and do not have fixed or easily defined dimensions. So we refer to the *scale* of an eddy, which represents its characteristic dimension. Scale is like an eddy diameter, but fuzzier. From the discussion above we know that the largest eddies are typically 25 to 40 percent as long as the dimensions of the turbulence-generating system (e.g., pipe diameter or ship hull width). For mixing studies we want to know the size of the smallest eddies. It was shown by Kolmogorov [3] that the scale of the smallest eddies is

$$\begin{pmatrix}\text{Kolmogorov}\\ \text{scale}\end{pmatrix} = \begin{pmatrix}\text{length scale}\\ \text{of smallest eddies}\end{pmatrix} = \left(\frac{\nu^3}{\varepsilon}\right)^{1/4} \tag{18.10}$$

Example 18.4. Estimate the scale of the smallest eddies in Example 18.2. Here we know the average value of ε but not the point values, which vary from place to place in the flow. But, using that average value and the kinematic viscosity of air, we can say that

$$\begin{pmatrix}\text{Kolmogorov}\\ \text{scale}\end{pmatrix} = \left[\frac{(1.613 \cdot 10^{-4}\ \text{ft}^2\,/\,\text{s}\,)^3}{0.21\ \text{ft}^2\,/\,\text{s}^3}\right]^{1/4}$$

$$= 0.0021\ \text{ft} = 0.025\ \text{in} = 0.64\ \text{mm} \tag{18.M}$$

Again, this is based on the average energy dissipation; in some parts of the flow this dissipation will be larger, in others smaller. But because of the $\frac{1}{4}$ power in the definition, the changes in Kolmogorov scale from place to place in any one flow will not be great.

This value is not typical, because it corresponds to a very slow air flow (Fig. 18.3), chosen to make the turbulence measurements there possible. Typical air-conditioning ducts have $V_{\text{average}} \approx 40\ \text{ft}/\text{s} \approx 12\ \text{m}/\text{s}$. If we repeat Example 18.2 and this example for 12 m / s (Prob. 18.3), we find the smallest eddy has a scale of 0.10 mm. For a typical water flow (6 ft / s in a 3 in pipe, Prob. 18.4) the smallest eddy has a scale of 0.033 mm. ■

Observe that the smallest eddies in a typical water flow are about 33 microns large or about 65 percent of the diameter of a typical human hair. There are other defined lengths in the study of turbulence, the *Prandtl mixing length* and the *Taylor scale,* neither of which is widely used by chemical engineers.

18.4.4 Correlation Coefficient

In estimating the effect of eddies, we need to know whether two eddies, or two components of an eddy, work together or are independent of each other. The *correlation coefficient R* (borrowed from statistics, where it is widely used and most often shown as R^2) is a measure of how much of the time two variables coincide with each other. The correlation coefficient of two arbitrary functions of time, $\phi_1(t)$ and $\phi_2(t)$, is defined by

$$\text{Correlation coefficient} = R = \frac{\overline{\phi_1\phi_2}}{\left(\overline{\phi_1^2}\right)^{1/2}\left(\overline{\phi_2^2}\right)^{1/2}} \tag{18.11}$$

This is the average of the product of the instantaneous values of any two functions, divided by the product of their rms averages.

Example 18.5. Determine the correlation coefficients for the following sets of functions: (*a*) $\phi_1(t) = t$ and $\phi_2(t) = t$; and (*b*) $\phi_1(t) = \sin t$ and $\phi_2(t) = \cos t$.

For (*a*) we have

$$R = \frac{\overline{t^2}}{(\overline{t^2})^{1/2}\,(\overline{t^2})^{1/2}} \tag{18.N}$$

Taking the averages as shown by Eq. 18.11, we see that $R = 1$. Thus, for two functions which are the same, $R = 1$.

For (*b*) we have

$$R = \frac{\overline{\sin t \cos t}}{(\overline{\sin^2 t})^{1/2}(\overline{\cos^2 t})^{1/2}} \tag{18.O}$$

Here the numerator is zero (Prob. 18.2), so $R = 0$. Thus, for any two functions that are 90° out of phase with each other, $R = 0$. ■

The correlation coefficient can take values from +1 to −1. Here, it is illustrated for simple analytic functions, where its value is obvious. In the study of turbulence, it is generally applied to randomly fluctuating variables. It can be shown that, if $\phi_1(t)$ and $\phi_2(t)$ are any two randomly fluctuating variables whose average values are zero and whose instantaneous values are not related in any way, then their correlation coefficient is zero. In turbulent flow velocity measurements, suitable electronics let us easily determine the correlation coefficients between simultaneous velocities at two points, or velocities at one point, separated by some small time interval, or two perpendicular components of the velocity at one point.

18.4.5 Spectrum of Turbulence

Another measurable physical property of a turbulent flow is the distribution of frequencies of turbulent oscillations. By suitable electronic filters it is possible to separate the output signal from a hot-wire or laser-Doppler anemometer into various frequency ranges.

If n is the frequency of oscillation in Hz, and we record first the value of $\overline{v_x^2}$ for the entire range of frequencies and then the value of $\Delta\overline{v_x^2}$ for some frequency range Δn, we can form the ratio

$$f(n) = \frac{1}{\overline{v_x^2}} \cdot \frac{\Delta\overline{v_x^2}}{\Delta n} \tag{18.12}$$

In experimental practice we must always use a finite value of Δn, but in principle we can take the limit as Δn approaches zero, finding

$$f(n) = \frac{1}{\overline{v_x^2}} \cdot \frac{d\overline{v_x^2}}{dn} \tag{18.13}$$

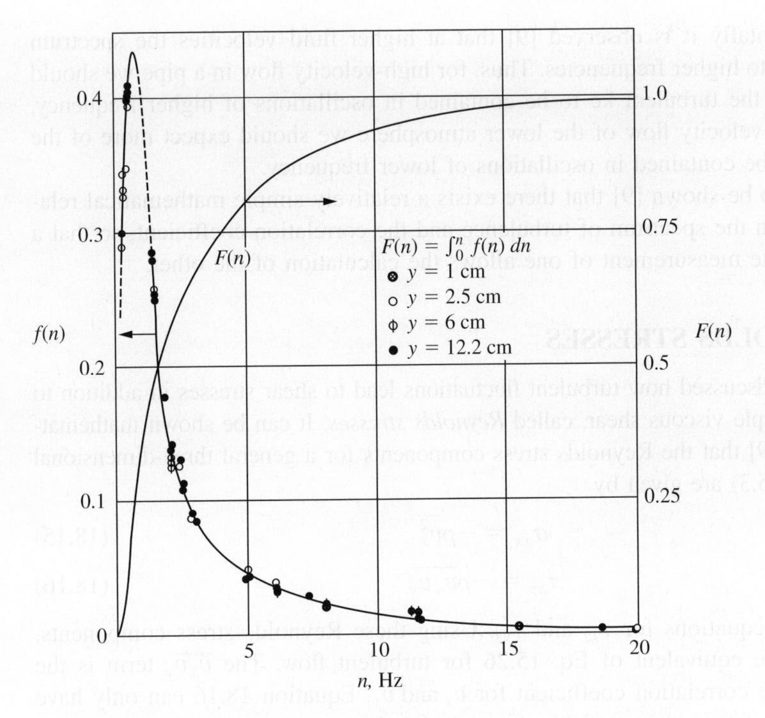

FIGURE 18.4
Spectrum of kinetic energy for the system shown in Fig. 18.3. [From H. Motzfeld, "Frequenzanalyse turbulenter Schwankungen,"—"Frequency analysis of turbulent oscillations," *Z. für Angew. Math. Mech. 18*:362–365 (1938).]

This $f(n)$ is thus the fraction of the total value of $\overline{v_x^2}$ per Hz. Here v_x^2 is twice the x component of the turbulent ke per unit mass, so that this fraction (Eq. 18.13) is really the fraction of the x component of the turbulent ke per Hz. A typical experimental measurement of this fraction (of the total turbulent ke, rather than of the x component) is shown in Fig. 18.4, which also shows the function

$$F(n) = \int_{n=0}^{n=n} f(n)\,dn = \frac{1}{\overline{v_x^2}} \int_{n=0}^{n=n} \frac{d\overline{v_x^2}}{dn}\,dn \tag{18.14}$$

This function, which is simply the area under the $f(n)$ curve up to a given value of n, shows what fraction of the turbulent ke is contained in oscillations of lower frequency. By definition, the value of $F(n)$ must approach 1 as n becomes very large.

A curve like Fig. 18.4 is called a *turbulent kinetic-energy spectrum,* analogous to the energy spectra of light that appear in optics texts. From Fig. 18.4 it is apparent that for this flow half of the turbulent ke is contained in velocity fluctuations that have frequencies between 0 and 2 Hz, and 90 percent of the turbulent ke is contained in fluctuations having frequencies between 0 and about 8 Hz. As a general rule there are more small (high frequency) eddies than large (low frequency) eddies, but most of the turbulent ke is in the large eddies.

Experimentally it is observed [9] that at higher fluid velocities the spectrum curve is shifted to higher frequencies. Thus, for high-velocity flow in a pipe we should expect more of the turbulent ke to be contained in oscillations of higher frequency, whereas in low-velocity flow of the lower atmosphere we should expect more of the turbulent ke to be contained in oscillations of lower frequency.

It can also be shown [9] that there exists a relatively simple mathematical relationship between the spectrum of turbulence and the correlation coefficient, so that a detailed, accurate measurement of one allows the calculation of the other.

18.5 REYNOLDS STRESSES

In Sec. 6.4 we discussed how turbulent fluctuations lead to shear stresses in addition to those due to simple viscous shear, called *Reynolds stresses*. It can be shown mathematically [10, p. 499] that the Reynolds stress components for a general three-dimensional flow (see Sec. 15.3) are given by

$$\sigma_{xx} = -\rho\overline{v_x^2} \tag{18.15}$$

$$\tau_{xy} = -\rho\overline{v_x v_y} \tag{18.16}$$

with analogous equations for τ_{xz} and τ_{yz}. Using these Reynolds stress components, we can form the equivalent of Eq. 15.26 for turbulent flow. The $\overline{v_x v_y}$ term is the numerator of the correlation coefficient for v_x and v_y. Equation 18.16 can only have a nonzero value if these two variables are correlated.

The most interesting of the Reynolds stresses are the shear stresses. From Eq. 18.16 we see that these require that the fluctuations be in two perpendicular directions. If there is no velocity gradient and the turbulence is isotropic, then there will be no such correlation, so that the Reynolds shearing stress will be zero. If there is a velocity gradient (as exists in any flow near a wall or at the boundary of a free jet) then, as shown below, there is always such a correlation, and there will always be a Reynolds stress.

Figure 18.5 represents a flow near a solid wall. The value of $\overline{V}_x$ is shown; $\overline{V}_y$ is 0. Now consider a small mass of fluid that is carried by an eddy upward from A to B. This mass of fluid moves to a larger y over some finite time, so it must have a positive y velocity. However, $\overline{V}_y$ is zero, so this means that for this eddy, v_y is positive.

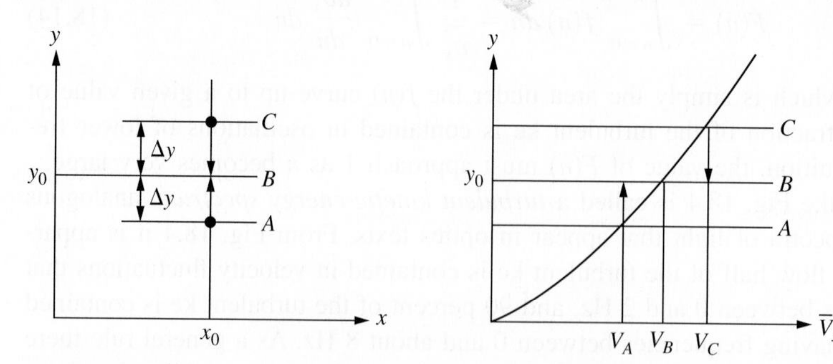

FIGURE 18.5
Diagram showing how Reynolds stresses arise in a turbulent flow with a velocity gradient.

Before this mass of fluid began to move in the y direction, its x velocity was V_A. When it arrives at B, it probably has not had time to change its x velocity, so it still has an x velocity of V_A. This is less than the average x velocity at that point (V_B), so this represents a negative x fluctuation of the velocity at y or a negative v_x.

Now some time later a mass of fluid from C is brought to B by an eddy. By similar arguments, v_y is negative and v_x is positive. Thus, for both kinds of fluctuations $\overline{v_x v_y}$ is negative, and there is, indeed, a correlation between the two velocity fluctuations at right angles to each other and hence a significant Reynolds stress.

18.6 EDDY VISCOSITY

The magnitude and character of these Reynolds stresses can be visualized through the concept of the eddy viscosity, first introduced by Boussinesq [3, p. 23]. He suggested that we retain the form of Newton's law of viscosity,

$$\tau = \mu \frac{dV_x}{dy} \tag{1.4}$$

which only holds for laminar flow, and make it fit the experimental turbulent-flow data by introducing a new quantity called the eddy viscosity ε. The eddy viscosity ε is really an eddy *kinematic* viscosity, but its common name is simply *eddy viscosity.*

$$\tau = (\mu + \rho\varepsilon) \frac{dV_x}{dy} \tag{18.17}$$

To avoid confusion in discussions involving the eddy viscosity, the "ordinary" viscosity is often called the *molecular viscosity*. Observe the notation problem; here we see ε having two meanings in this chapter: the rate of turbulent ke dissipation in Eq. 18.7 and the turbulent eddy kinematic viscosity in Eq. 18.17. These are the standard symbols for these quantities. We will try to make clear which we are using whenever they are used.

This definition of the eddy kinematic viscosity ε has the disadvantage that ε is not a simple property of the fluid, like the molecular viscosity, but is also a function of the flow rate and of position in the flow. It has the advantage that it lets us easily formulate the ratio of the Reynolds stresses to viscous stresses. In addition, in calculations of heat and mass transfer we may introduce a similar eddy thermal conductivity and eddy diffusivity. Under some circumstances these three eddy properties are identical, and under most circumstances they are at least of the same order of magnitude, so this approach helps to apply fluid-flow data to the solution of problems in heat and mass transfer; see Sec. 15.7. This approach to solving problems in turbulent heat and mass transfer through fluid mechanics is called *Reynolds' analogy.*

18.6.1 Finding Eddy Viscosities from Experimental Velocity Distributions

From a measured average velocity profile ($\overline{V}_x$ versus y or r) in some flow and from information on the viscosity and density of the fluid, we can calculate the eddy viscosity for any point in the flow. A typical plot of velocity versus position in turbulent pipe flow is shown in Fig. 17.7. The $\overline{V}_x$ curves there for various $\mathcal{R}$ can each be well represented by an empirical data-fitting equation of the form

$$\frac{\overline{V}_{\text{avg}}}{\overline{V}_{x\text{-centerline}}} = \left(1 - \frac{r}{r_{\text{wall}}}\right)^{1/n} \tag{18.18}$$

Example 18.6. For the flow in Example 18.2, estimate the value of the eddy viscosity at a point half-way between the wall and the center based on Eq. 18.18.

Combining the equation for the shear stress in pipe flow (Eq. 6.4) with Eq. 18.17, we find

$$\tau = \frac{r}{2}\left(\frac{-dP}{dx}\right) = (\mu + \rho\varepsilon)\frac{dV_x}{dy} \tag{18.19}$$

or

$$(\mu + \rho\varepsilon) = \frac{(r/2)(-dP/dx)}{dV_x/dy} = \frac{-(r/2)(-dP/dx)}{dV_x/dr} \tag{18.20}$$

where $y = r_{\text{wall}} - r$ and $dy = -dr$. We know all the terms from Example 18.2 except

$$\frac{dV_x}{dr} = \frac{-V_{\text{centerline}}}{nr_{\text{wall}}}\left(1 - \frac{r}{r_{\text{wall}}}\right)^{(1-n)/n} \tag{18.21}$$

From Example 18.2 we know that $-\Delta P/\Delta L \approx 0.0286$ Pa / m $= 0.00013$ psi / 100 ft and that $V_{\text{centerline}} \approx 1.00$ m / s. From Table 17.1 we estimate that $n \approx 7$. The equivalent of r_{wall} is the distance from the wall to the center of the channel in the short direction, 0.122 m, so that

$$\frac{dV_x}{dr} = \frac{-1.00\text{ m/s}}{7 \cdot 0.122\text{ m}}(1 - 0.5)^{(1-7)/7} = -2.12\frac{1}{\text{s}} \tag{18.P}$$

and

$$(\mu + \rho\varepsilon) = \frac{-(0.122\text{ m}/2)(0.0286\text{ Pa/m})}{-(2.12/\text{s})}$$
$$= 8.24 \cdot 10^{-4}\frac{\text{N}\cdot\text{s}}{\text{m}^2} = 1.72 \cdot 10^{-5}\frac{\text{lbf}\cdot\text{s}}{\text{ft}^2} \tag{18.Q}$$

Most often one sees this as a ratio,

$$\frac{\varepsilon}{\nu} = \frac{\varepsilon}{\mu/\rho} = \frac{\mu + \rho\varepsilon}{\mu} - 1 \approx \frac{\mu + \rho\varepsilon}{\mu}$$
$$= \frac{\nu_{\text{turbulent}}}{\nu_{\text{molecular}}} = \frac{8.24 \cdot 10^{-4}\text{ N}\cdot\text{s/m}^2}{1.21 \cdot 10^{-5}\text{ N}\cdot\text{s/m}^2} = 68 \tag{18.R}$$

which shows that for this approximate velocity distribution equation, at this point in this flow, the Reynolds stresses are ≈ 68 times as large as the viscous stresses. ■

18.6.2 Finding Eddy Viscosities from Semitheoretical Correlations

The previous example shows that if we know the velocity distribution (which we do for simple flows like steady flow in a circular pipe), we can work back from those data to values of the eddy viscosity. But the real problem is the inverse of this one: using the eddy viscosity to estimate the velocity distribution (and other properties like heat transfer and mixing) in complex flows, like that in the industrial furnace in Fig. 1.5.

The CFD programs that make such calculations mostly do so by substituting

$$\begin{pmatrix}\text{Turbulent}\\ \text{viscosity}\end{pmatrix} = \mu_t = \mu + \rho\varepsilon \tag{18.22}$$

in place of μ in the Navier-Stokes equations (Eq. 15.26) and then solving numerically for the resulting velocity distribution. To do so they use semitheoretical correlations for μ_t, of which the most widely used is the k–ε model. The ε in the k–ε model is the turbulent ke dissipation rate (Example 18.2), not the eddy kinematic viscosity. This model calculates the turbulent viscosity by

$$\mu_t = \rho C_\mu \frac{k^2}{\varepsilon} \quad \text{or} \quad \nu_t = \frac{\mu_t}{\rho} = C_\mu \frac{k^2}{\varepsilon} \tag{18.23}$$

where C_μ is a data-fitting constant. The values of k and ε vary from place to place in the flow and obey the general balance equation, rate form (Eq. 3.3), which includes creation and destruction terms.

Example 18.7. Estimate the value of the turbulent kinematic viscosity in a flow of air at a point where $k = 0.00576\ \text{m}^2/\text{s}^2$ (see Example 18.3), $\varepsilon = 0.0196\ \text{m}^2/\text{s}^3$ (see Example 18.2), and $C_\mu = 0.09$ (based on a value taken from a CFD manual).

By direct substitution into Eq. 18.23, we find

$$\nu_t = 0.09 \frac{(0.00576\ \text{m}^2/\text{s}^2)^2}{(0.0196\ \text{m}^2/\text{s}^3)} = 1.53 \cdot 10^{-4}\ \frac{\text{m}^2}{\text{s}} = 1.53 \cdot 10^{-4}\ \frac{\text{N}\cdot\text{s}}{\text{m}^2}$$

$$= 1.64 \cdot 10^{-3}\ \frac{\text{ft}^2}{\text{s}} \tag{18.S}$$

and

$$\frac{\nu_{\text{turbulent}}}{\nu_{\text{molecular}}} = \frac{1.53 \cdot 10^{-4}\ \text{N}\cdot\text{s}/\text{m}^2}{1.21 \cdot 10^{-5}\ \text{N}\cdot\text{s}/\text{m}^2} = 10.3 \tag{18.T}$$

This is approximately $\frac{1}{7}$ of the value in Example 18.6, mostly because we used the available value of k for the whole flow, with the available value of ε at half the distance from the wall to the center. ■

This example only illustrates how a CFD program would compute the turbulent kinematic viscosity at some point, given the input values. The computation of those values is complex and depends on the value of the turbulent kinematic viscosity, so the computation of the three-dimensional velocity at any point must be a circular iterative solution. The k–ε model and related models are rarely if ever used for hand calculations. This example serves only to give some insight into what is going on in CFD models.

18.7 TURBULENCE THEORIES

Most of the best-known theoretical fluid mechanicians of the 20th century attempted to deduce a comprehensive theory of turbulence. The resulting theories each provide some insight into the relations that must exist between various quantities in turbulent flow, but all of them contain undefined constants that must be measured experimentally to make the theory fit the observations. The theories of Prandtl and von Karman are well summarized by Schlichting [11, Chap. 19]. That of G. I. Taylor is summarized by Dryden [12]. The theories of Kolmogorov are discussed by Hinze [3] and Corrsin [6]. The problem of calculating the velocity distribution in turbulent flow in a pipe from the various theories is discussed by Bird, Stewart, and Lightfoot [13, p. 165].

Many of these theories are quite complex mathematically. The more involved mathematics led one fluid mechanician to comment that these theories "confirm one's suspicion that the aim of the statistical theory of turbulence is full employment for mathematicians" [14].

18.8 SUMMARY

1. In analyzing and measuring turbulence it is customary to divide the flow conceptually into steady and fluctuating components.
2. Historically, most turbulence measurements have been made with hot-wire anemometers. In recent years laser-Doppler anemometers have been widely used as well.
3. Free turbulence has a different experimental character from that of turbulence near a solid wall or a free jet.
4. The readily measured experimental properties of a turbulent flow are the time-average velocity and the intensity, scale, and spectrum of the turbulence.
5. Reynolds shear stresses arise out of the correlation of turbulent fluctuations in two perpendicular directions. These are rare in free turbulence but almost always present in turbulent flow near a wall or the edge of a free jet.

PROBLEMS

See the Common Values for Problems and Examples inside the back cover. An asterisk (*) on a problem number indicates that its answer is in App. D.

18.1. Show the relationship between the ke of a stream with velocity $V = \overline{V} + v$ and the ke of a stream with velocity $V = \overline{V}$.

18.2. Repeat Example 18.1 for the same flow, but with $V_y = (0.5 \text{ ft/s})(\sin t)$ instead of $V_y = 0$. Show the values of $\overline{V}_y$, v_y, $\overline{v}_y$, $(\overline{v_y^2})^{1/2}$, T_y, *and* T.

18.3. Repeat Example 18.2 for air flowing 12 m / s in the same duct as Example 18.2.

18.4.*Repeat Example 18.2 for water flowing 6 ft / s in a 3-in schedule 40 pipe.

18.5.*Repeat Example 18.3 for a point 6 cm from the wall in Fig. 18.3.

18.6. In Example 18.4, if we double the energy dissipation rate, how much will that change the Kolmogorov scale?

18.7. Select two functions $\phi_1(t)$ and $\phi_2(t)$ whose correlation coefficient is -1.

18.8. Show that $\overline{\sin t \cdot \cos t} = 0$.

18.9.*If the size of the smallest eddy in a flow is that shown in Example 18.4, the frequency of such an eddy should be about the same as the local stream velocity divided by eddy size. Estimate that frequency for a fluid velocity of 1 m / s and the eddy size from Example 18.4.

18.10. Repeat Example 18.6 for $r / r_{\text{wall}} = 0.9$.

18.11.*Repeat Example 18.6, using the universal velocity distribution plot (Fig. 17.8) instead of Fig. 17.7.

REFERENCES FOR CHAPTER 18

1. Gargett, A. E. "Ocean Turbulence." *Ann. Rev. Fluid Mech. 21* (1989), pp. 419–451.
2. de Nevers, N. *Air Pollution Control Engineering.* 2nd ed. New York: McGraw-Hill, 2000, p. 120.
3. Hinze, J. O. *Turbulence.* 2nd ed. New York: McGraw-Hill, 1975.
4. Dryden, H. L., A. M. Kuethe. "Effect of Turbulence on Wind Tunnel Measurements." *NASA Rept. No. 342* (1929).
5. Priestly, C. B. H. *Turbulent Transfer in the Lower Atmosphere.* Chicago: Chicago University Press, 1959.
6. Corrsin, S. "Turbulent Flow." *American Scientist 49* (1961), pp. 300–325.
7. Perry, A. E. *Hot-Wire Anemometry.* Oxford: Clarendon Press, 1982.
8. Durst, F., A. Melling, J. H. Whitelaw. *Principles and Practice of Laser-Doppler Anemometry.* London: Academic Press, 1981.
9. Taylor, G. I. "The Spectrum of Turbulence." *Proceedings of the Royal Society A 164* (1938), pp. 476–490.
10. Schlichting, H., K. Gersten. *Boundary-Layer Theory.* 8th ed. Berlin: Springer, 2000.
11. Schlichting, H. *Boundary Layer Theory.* 6th ed., Transl. J. Kestin. New York: McGraw-Hill, 1968. References are given to the 6th edition for plots and tables that were not included in later editions.
12. Dryden H. L. "A Review of the Statistical Theory of Turbulence." *Quart. Appl. Math., 1,* 7–42 (1943). This paper and most of G. I. Taylor's basic papers are contained in S. K. Friedlander and L. Topper (eds.), *Turbulence—Classical Papers on Statistical Theory.* New York: Interscience, 1961.
13. Bird, R. B., E. N. Stewart, W. E. Lightfoot. *Transport Phenomena.* 2nd ed. New York: Wiley, 2002.
14. Townsend, A. A. *J. Fluid Mech. 10* (1961), pp. 635–636.

CHAPTER 19

MIXING

19.1 TYPES OF MIXING PROBLEMS

Chemical engineers regularly deal with fluid mixing problems of the following types. Mostly, we mix by generating turbulence in the fluids in which the mixing occurs.

19.1.1 Suspension of Solids

When you put sugar in your coffee, tea, or lemonade, it settles to the bottom. When you stir, the first thing you are doing is *suspending* the solid sugar (SG = 1.58) in the liquid (SG $\approx$ 1.00). In doing so, you increase the rate of dissolution of the sugar, both by increasing the surface area exposed to the liquid and by producing fluid flow over the solid surfaces. This is one of the easiest mixing problems: The requirement is only that most of the solids be carried upward by the fluid so that they do not form a pile on the bottom of the container. Many similar suspension operations occur in chemical engineering, mostly requiring mild stirring. If the local upward velocity is greater than the terminal settling velocity of the particles (Sec. 6.15) then suspension will be practically complete.

19.1.2 Dispersion of Solids

In suspension of solids, the requirement is mostly that the solids not settle to the bottom of the container. In *dispersion* of solids the requirement is normally more difficult: that the solids be uniformly distributed throughout the liquid. The classic example is the dispersion of paint pigment, which consists of opaque, colored particles, normally 0.01 to 5 microns in diameter. To do their job in the paint they must be uniformly dispersed so that, in the paint film placed on the wall, each particle contributes to the paint's ability to hide the underlying surface. Typical pigments have SG $\approx$ 4, so gravity settling in the can during storage was a major problem until 50 years ago. Since then the rheology of the liquid parts of the paint has improved enough that, on a human time scale, pigment settling is negligible.

19.1.3 Blending of Miscible Liquids

When you put cream in your coffee and stir it, the operation you are performing is *blending,* causing two miscible fluids (almost always liquids) to form a single fluid with no perceptible differences in concentration from point to point. In coffee this requirement is not severe; we do not demand uniformity at the scale of a droplet, only at the scale of a mouthful. The corresponding blending of gasoline requires uniformity down to perhaps 1 mm size drops. The various liquids that are blended to make gasoline have different octane numbers; if the blend is not complete at the level of the charge to a single cylinder on a single firing, then some cylinders will receive a higher octane than they need and others will receive less than they need, causing knock. For low-viscosity fluids like coffee and gasoline, blending is done mostly by turbulence, in which the larger eddies repeatedly move parts of one batch of fluid into the midst of the other batch of fluid. For viscous fluids like paints, the flow is mostly laminar, and the blending is accomplished by repeatedly dragging a layer of one fluid through another. The student has perhaps observed in hand-mixing of paints that the first few strokes of the mixer make trails of the pigment into the unpigmented parts of the paint, and over time those trails become thinner and harder to see until they vanish.

19.1.4 Molecular Mixing

For many chemical processes and for combustion, the mixing must be practically complete at the molecular level. This is a more difficult requirement than suspension or blending and generally requires more energetic mixing. As discussed in Chap. 18, turbulent eddies have a minimum size, which is large compared to the dimensions of molecules. Thus, in this kind of mixing we use turbulence to mix down to the size of the smallest eddy, and then rely on molecular diffusion to finish the job.

19.1.5 Blending of Solids

Miscible liquids, once blended, cannot be separated by any ordinary mechanical process. The same is not true of solids. If the individual solid particles have different sizes or different densities, then after mixing they can be separated by screens or gravity devices, or by simple shaking, which will cause the smaller particles to go to the bottom and the larger ones to rise to the top. Tablets of medicines are normally pressed mixtures of a variety of solids; keeping that mixture uniform as it flows from the blender to the tablet presses is not easy. Mixing of solids is a special problem, which will not be discussed further in this book [1].

19.1.6 Mixing in Pipes

Sometimes we wish to blend fluids by flowing them together through a pipe; other times, we want batches of fluid that are following each other down a pipe not to mix.

19.1.7 Emulsification

Many consumer products—mayonnaise, cold cream, salad dressing, homogenized milk—are emulsions, which consist of two immiscible liquids that have been intimately mixed by a high-intensity mixer and whose dispersed droplets do not coalesce, often because the mixtures contains a surface-active emulsifier that prevents coalescence.

19.1.8 Atmospheric or Oceanic Dispersion

Streams of combustion products, often containing pollutants, are regularly emitted from smokestacks to the atmosphere. There they are dispersed by atmospheric mixing, so that their maximum concentration, observed at ground level, is several orders of magnitude less than the concentration in the emitted stream. Similarly treated (or untreated, alas) sewage is dumped into the ocean (or rivers) by a sewage outfall. There it is dispersed to nontoxic concentrations. Before about 1970 the motto of the environmental engineers was "Dilution is the Solution to Pollution." Many cities and states had regulations (e.g., stack height requirements) forcing the dilution of emissions. Current U.S. pollution law is strongly oriented toward treatment and removal of pollutants rather than dilution and disposal in the environment. Nonetheless, such dispersion occurs and is important. Similarly, small amounts of gaseous fuels are released to the atmosphere in many fuel transfers, e.g., filling the gasoline tank of one's car. Atmospheric dispersion reduces the concentration to below the lower flammable limit in a few feet. If it did not, fueling one's car would be very dangerous.

19.2 THE ROLE OF TURBULENCE

We use turbulence for mixing in most mixing applications. The large eddies in turbulence are very effective in coarse mixing, moving the fluid about and folding and turning it. However turbulence is not effective for mixing at the molecular level. We saw in Chap. 18 that the smallest eddies in a turbulent flow will have

$$\begin{pmatrix}\text{Kolmogorov}\\ \text{scale}\end{pmatrix} = \begin{pmatrix}\text{length scale}\\ \text{of smallest eddies}\end{pmatrix} = \left(\frac{\nu^3}{\varepsilon}\right)^{1/4} \tag{18.10}$$

In Example 18.4 we saw that for a typical turbulent flow of water this scale was about $0.03 \text{ mm} = 30$ micron. The intermolecular spacing in liquids is of the order of $10^{-9} \text{ m} = 0.001$ microns. Thus, the smallest turbulent eddies are on the order of 30,000 times the spacing between molecules. If we mix more vigorously, we will increase the value of ε, but it enters Eq. 18.10 to the 0.25 power, so that large changes in ε produce small changes in the size of the smallest eddy. The kinematic viscosity, ν, enters Eq. 18.10 to the 0.75 power, so that as the fluid viscosity increases, the size of the smallest eddy increases significantly. Turbulence mixes well enough that the resultant nonuniformities are sufficiently small that molecular diffusion will finish the mixing job for us if the fluids are not too viscous.

19.3 THE ROLE OF MOLECULAR DIFFUSION

In both the cream- and the sugar-in-coffee examples, once we get to uniformity down to perhaps $10^{-4} \text{ m} = 100$ microns, molecular diffusion finishes the job for us, giving us uniformity down to 10^{-9} m. The role of molecular diffusion in mixing is illustrated in Example 19.1.

Example 19.1. Two layers of fluid, each L thick, are brought into contact. One is pure water, the other is water containing 1 percent acetic acid (HAc). Over time, molecular diffusion causes the concentration in the two layers to become uniform, in this case at 0.5 percent acetic acid throughout. But how long will it take? Figure 19.1 shows the geometry and the change in concentration profile over time, starting as a step and changing over time to a uniform concentration.

This is a classic diffusion problem, which is mathematically identical to a similar problem in heat transfer, whose solution is shown in all heat transfer books (a slab, insulated on all sides but one, with the temperature at that side suddenly changed from 0 to T and maintained at that temperature). The diffusion or heat conduction equation has an infinite-series solution, which is shown in graphical form in all heat-transfer books. Absolute chemical composition uniformity (or temperature uniformity in the heat-transfer problem) requires an infinite time, but 99 percent of all the acetic acid (or heat in the heat transfer problem) that will flow in infinite time flows by the time at which

$$\frac{\text{Fourier number}}{\text{Lewis number}} = \frac{\text{molecular diffusivity} \cdot \text{time}}{(\text{length})^2} = \frac{\mathscr{D} t}{L^2} \approx 2 \tag{19.1}$$

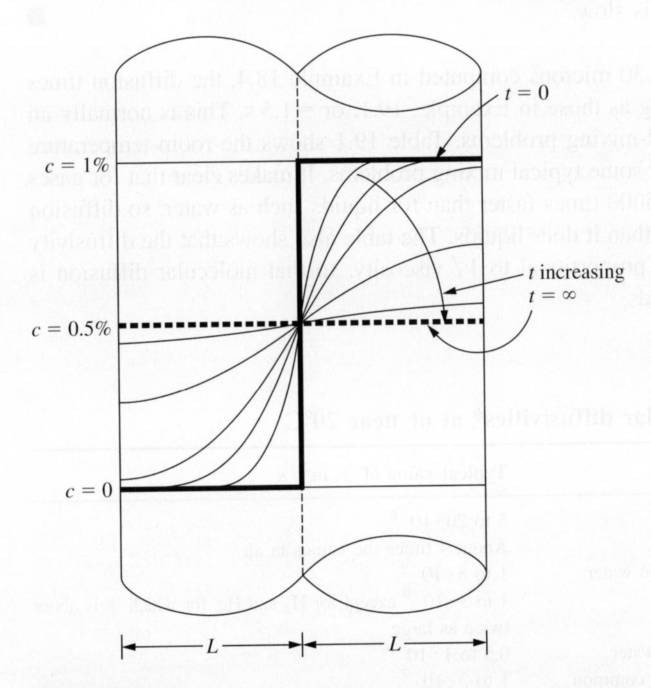

FIGURE 19.1
Schematic for Example 19.1. Two fluid layers, each L thick, are brought together at time zero. The one on the left initially contains no HAc, the one on the right contains 1% HAc. Thus, the initial concentration profile is the step function shown. Over time, diffusion makes the concentration uniform; the profile changes to a horizontal line at 0.5% HAc.

In this problem $\mathscr{D}$ is the molecular diffusivity. In the corresponding heat transfer problem it is the thermal diffusivity, $\alpha = (k / \rho C)$. Both of these quantities (and the kinematic viscosity, which may be thought of as the momentum diffusivity) have the dimension of m^2 / s or equivalent.

For acetic acid in water near room temperature, we have $\mathscr{D} \approx 1.2 \cdot 10^{-9}$ m^2 / s. The time to complete 99 percent of the transfer by molecular diffusion is a very strong function of the thickness L of the layers. If the thickness is 10^{-6} m = 1 micron, then

$$t \approx \frac{2L^2}{\mathscr{D}} = \frac{2 \cdot (10^{-6}\ \text{m})^2}{1.2 \cdot 10^{-9}\ \text{m}^2 / \text{s}} = 1.667 \cdot 10^{-3}\ \text{s} \tag{19.A}$$

One micron is approximately the spacing between individual nerves in our bodies, These nerves pass signals to each other by diffusing acetylcholine across the water-filled spaces between them; this calculation shows that the chemical messages cross the spaces in about 1.7 ms. At this size level, diffusion is fast enough that our brains and nervous systems do not perceive this diffusion delay at all.

However if we repeat the calculation for layers of thickness L = 1 mm or 1 cm, we find that the calculated times for 99 percent completion by diffusion increase by 10^6 and 10^8 to 0.46 h and 46 h. At the size scales of ordinary objects, molecular diffusion is slow. ■

For the eddy size of 30 microns computed in Example 18.4, the diffusion times would be 900 times as long as those in Example. 19.1, or ≈1.5 s. This is normally an acceptable delay for liquid-mixing problems. Table 19.1 shows the room-temperature values of the diffusivity for some typical mixing problems. It makes clear that for gases the diffusivities are about 5000 times faster than for liquids such as water, so diffusion mixes gases much quicker than it does liquids. The table also shows that the diffusivity of liquid is approximately proportional to 1 / viscosity, so that molecular diffusion is slow in highly viscous fluids.

TABLE 19.1

Typical values of molecular diffusivities* at or near 20°C

Substance diffusing	Typical value of $\mathscr{D}$, m^2 / s
Common organic vapors in air	5 to $20 \cdot 10^{-6}$
Same vapors in hydrogen	About 4 times the values in air
Acids and other ionized species in water	1 to $3 \cdot 10^{-9}$
Dissolved gases in water	1 to $3 \cdot 10^{-9}$ except for H_2 and He, for which $\mathscr{D}$ is about twice as large
Sugars and weak electrolytes in water	0.5 to $1 \cdot 10^{-9}$
Common organic liquids in other common organic liquids	1 to $3 \cdot 10^{-9}$
Any of the above in a viscous fluid	Approximately the value in water $\cdot (\mu_{water} / \mu_{fluid})$

*Based on more extensive tabulations in *Perry's Chemical Engineers' Handbook* and the *Handbook of Chemistry and Physics*. Equations for predicting diffusivities and showing the effects of temperature and pressure change on them appear in *The Properties of Liquids and Gases* by Poling, Prausnitz, and O'Connell.

19.4 MIXING IN STIRRED TANKS

Figure 19.2 shows schematics of two versions of a stirred-tank mixer. These are extremely common in chemical engineering. Books on mixing are mostly about this type of mixer and its variants [2–6]. The mixer consists of a vertical, cylindrical vessel,

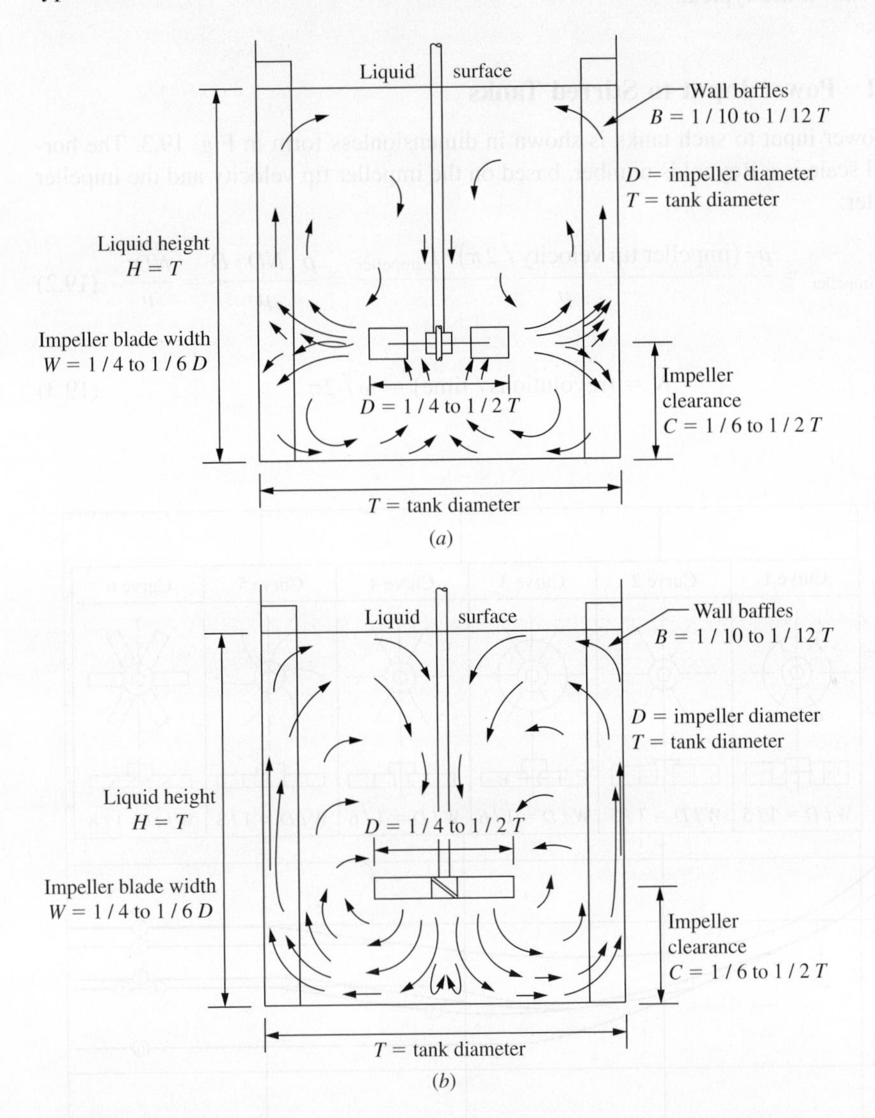

FIGURE 19.2
Schematic for two variations of a cylindrical mixing tank. In both, a shaft, entering from above drives an impeller, which mixes the fluid. Baffles on the wall prevent the whole mass of fluid from rotating. The dimensions shown are typical of these tanks. Part (*a*) shows a radial-flow impeller whose blades are vertical; it produces mostly radial flow, with modest vertical circulation. Part (*b*) shows an axial-flow impeller, whose blades are 45° from the vertical; it produces mostly axial flow, normally downward. [From G. B. Tatterson, *Fluid Mixing and Gas Dispersion in Agitated Tanks,* New York: McGraw-Hill, 1991. Reproduced by permission of the publisher.]

often open at the top. A vertical shaft from the top drives an impeller, which mixes the fluid. Two different impeller designs and their resulting fluid flow patterns are shown. Baffles at the wall (typically four) prevent the whole fluid mass from rotating. These tanks are used in a batch mode or with steady flow in and out. The dimension ratios shown are typical.

19.4.1 Power Input to Stirred Tanks

The power input to such tanks is shown in dimensionless form in Fig. 19.3. The horizontal scale is a Reynolds number, based on the impeller tip velocity and the impeller diameter:

$$\mathscr{R}_{\text{impeller}} = \frac{\rho \cdot (\text{impeller tip velocity} / 2\pi) \cdot D_{\text{impeller}}}{\mu} = \frac{\rho \cdot ND \cdot D}{\mu} = \frac{ND^2}{\nu} \quad (19.2)$$

where

$$N = (\text{revolutions} / \text{time}) = \omega / 2\pi \quad (19.3)$$

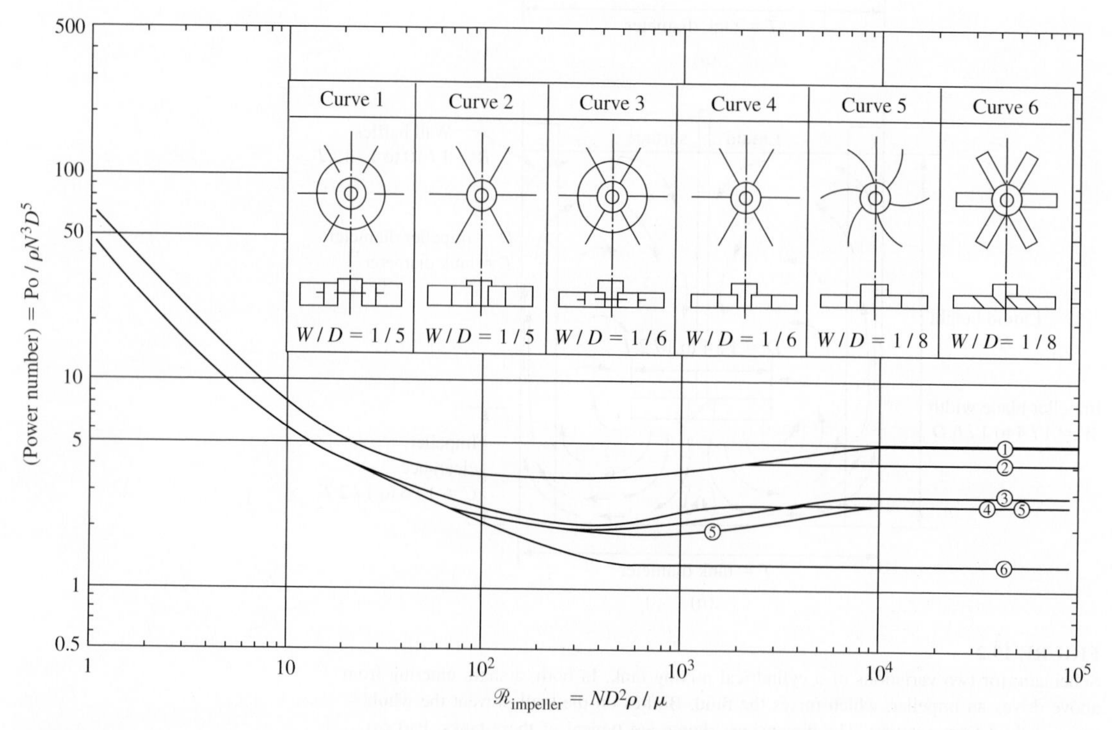

FIGURE 19.3
Power number-Reynolds number correlation for Newtonian fluids using six different impeller designs. [Reprinted with permission from R.L. Bates, P.L. Fondy and R.R. Corpstein, "An Examination of Some Geometric Parameters of Impeller Power", *I & EC Proc. Des. Dev.* 2:310 (1963): Copyright 1963 American Chemical Society.]

and the 2π has been dropped. The vertical scale is the power number,

$$\begin{pmatrix}\text{Power}\\ \text{number}\end{pmatrix} = \frac{\text{Po}}{\rho N^3 D^5} = \frac{(F/D^2)V_{\text{tip}}}{\rho(V_{\text{tip}}/2\pi)^3} = \frac{\Delta P}{\rho(V_{\text{tip}}/2\pi)^2} \tag{19.4}$$

where F is a force, Po is a force times a velocity, and ΔP is a force / area. Except for constants, the rightmost term in Eq. 19.4 is the same as the Fanning friction factor used in Chap. 6. Figure 19.3 is simply Fig. 6.10 for a different geometry.

Example 19.2. A mixing tank with $D_{\text{tank}} = 3$ ft and $D_{\text{impeller}} = 1$ ft has the dimension ratios shown in Fig. 19.2(*a*). The impeller corresponds to Curve 1 in Fig. 19.3 and has $N = 240$ rpm $= 4$ / s. The fluid has the same properties as water. Estimate the power input to the impeller.

$$\mathcal{R}_{\text{impeller}} = \frac{(4\ /\ \text{s}) \cdot (1\ \text{ft})^2}{1.077 \cdot 10^{-5}\ \text{ft}^2\ /\ \text{s}} = 3.7 \cdot 10^5 \tag{19.B}$$

which is well into the turbulent region on Fig. 19.3. From that figure we read that the power number ≈ 5, so that

$$\begin{aligned}\text{Po} = 5\rho N^3 D^5 &= 5 \cdot 62.3\,\frac{\text{lbm}}{\text{ft}^3} \cdot \left(\frac{4}{\text{s}}\right)^3 \cdot (1\ \text{ft})^5 \cdot \frac{\text{lbf} \cdot \text{s}^2}{32.2\ \text{lbm} \cdot \text{ft}} \cdot \frac{\text{hp} \cdot \text{s}}{550\ \text{ft} \cdot \text{lbf}}\\ &= 1.13\ \text{hp} = 0.84\ \text{kW}\end{aligned} \tag{19.C}$$

■

19.4.2 Required Power Input and Mix Time

The preceding calculation is simple. Using Fig. 19.3 or its equivalent for other impellers and tank dimensions, we can easily estimate the power input for any impeller-tank combination. The harder and more interesting question is how much power for how long a time is needed to produce satisfactory mixing. There is no simple answer to that question, because the answer depends on what is "satisfactory," e.g., the level of dispersion needed to suspend sugar? to disperse paint pigments? to make emulsions? for uniformity at the molecular level? The answer also depends strongly on the properties of the fluid(s) and/or solid(s) or gas(es) to be mixed. No simple rule exists. For mixing jobs for which we have no previous data, we normally must test. We can do small-scale tests fairly easily; most engineering laboratories have mixing tanks like that in Example 19.2. But there is no simple set of rules for scaling up from those laboratory tests to the industrial size.

Example 19.3. Tests indicate that the tank-mixer combination in Example 19.2 produces satisfactory mixing for our application in an acceptable time period. If we now build an industrial version that is a scale model with all dimensions increased by 5, using the same N, what should we expect its performance to be?

First, we see that $\mathcal{R}_{\text{impeller}}$ increases by $5^2 = 25$. This keeps us on the flat part of Fig. 19.3, so the power number is unchanged. The required power increases by $5^5 = 3125$, so we would need a 3530 hp drive. That is a huge number, probably unacceptable. A common basis for comparing the intensity of mixing is the (power / tank volume). That ratio is proportional to $(D^5/D^3) = 5^2 = 25$, suggesting that simply keeping the dimension ratios constant and keeping the speed constant leads to much more vigorous mixing as we go up in size.

If, instead of keeping N constant, we decided to keep (power / tank volume) constant, then

$$\frac{\text{Po}}{V} \propto \frac{\rho N^3 D^5}{D^3} \tag{19.5}$$

so for constant (power / tank volume) the N^3D^2 product must be a constant. Thus, in this case,

$$\frac{N_2}{N_1} = \left(\frac{D_1}{D_2}\right)^{2/3} = 5^{2/3} = 2.92 \tag{19.D}$$

and $N_2 = 240 / 2.92 = 82$ rpm. ■

The point of this example is that there is no obvious rule for scaling up from laboratory to industrial mixers of this type. One might choose dimensional similarity and constant $\mathscr{R}_{\text{impeller}}$ or constant (power / tank volume) or constant impeller tip speed, or some combination of these, but one cannot match them all. The most common choice seems to be constant (power / tank volume), but this ratio is only a rough estimate of the best design. In spite of this caution, the suggested values in Table 19.2 are widely published. A summary of rules for estimating how much power and time are needed for mixing is given in McCabe, Smith, and Harriott [7, Chap. 9]. One rule that appears useful is for blending. The impeller that mixes the fluid also pumps fluid through itself, leading to the currents sketched in Fig. 19.2. (Such impellers are quite like the impellers of centrifugal pumps, operating without a pump casing.) There are adequate correlations for the pumping rate of common impeller types. For example, for modest-viscosity miscible liquids, blending will be satisfactory when the total volume that has flowed through the impeller is about five times the volume of the tank. Using this idea and a correlation for the flow, McCabe, Smith, and Harriott [7, p. 260] present the following equation for estimating the required time to blend a tank:

$$Nt_{\text{blend}} = 4.3 \cdot \left(\frac{D_{\text{tank}}}{H_{\text{tank}}}\right) \cdot \left(\frac{D_{\text{tank}}}{D_{\text{impeller}}}\right)^2 \tag{19.6}$$

TABLE 19.2

Commonly reported values of (power / tank volume) and tip speed for mixing in baffled tanks

Operation	Power / tank volume, hp / 1000 gal	Rotor tip speed, ft / s
Blending	0.2–0.5	<7.5
Homogeneous reaction	0.5–1.5	7.5–10
Reaction with heat transfer	1.5–5.0	10–15
Liquid-liquid mixing	5–10	15–20
Liquid-gas mixing	5–10	15–20
Slurrying	10	>20

Example 19.4. For the tank in Example 19.2, how long should it take to blend two miscible, low-viscosity liquids?

$$Nt_{\text{blend}} = 4.3 \cdot (1) \cdot (3)^2 = 38.7 \tag{19.E}$$

$$t_{\text{blend}} = \frac{38.7}{4\,/\,\text{s}} = 9.7\ \text{s} \tag{19.F}$$

■

The same source gives a plot of Nt_{blend} versus $\mathcal{R}_{\text{impeller}}$ for a variety of impeller-tank combinations. The presentation in terms of Nt_{blend} is common because Nt_{blend} is dimensionless, sometimes called the *mixing time factor.*

19.5 MIXING IN PIPE FLOW

Often we put two different fluids into a pipe and use pipe flow for mixing, or put fluids into a pipe one after the other and want them to arrive at the other end of the pipe unmixed. The second case is of great practical significance in petroleum product pipelines.

Consider the situation in which two miscible fluids (e.g., two grades of gasoline) are sent sequentially through the same pipeline. We want to know how much of a mixed zone will exist between them at the end of the line. The situation is sketched in Fig. 19.4. We take the viewpoint of a person riding with the interface between the two fluids. If there were absolutely no mechanical mixing between fluids, then this would be a simple molecular diffusion problem, whose solution is well known. The three-dimensional molecular diffusion equation is

$$\frac{\partial c}{\partial t} = \mathcal{D}_x \frac{\partial^2 c}{\partial x^2} + \mathcal{D}_y \frac{\partial^2 c}{\partial y^2} + \mathcal{D}_z \frac{\partial^2 c}{\partial z^2} \tag{19.7}$$

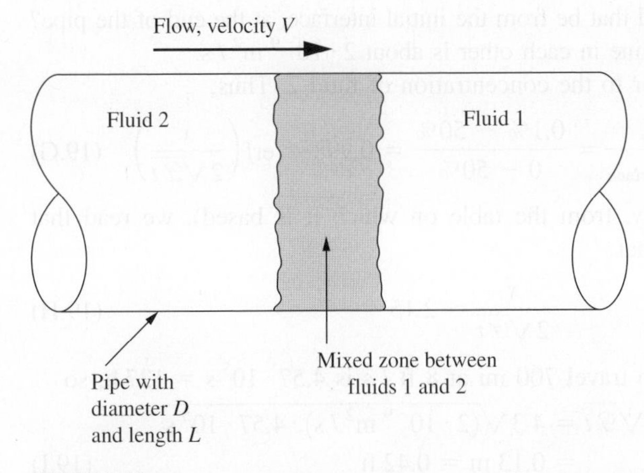

FIGURE 19.4
Schematic of fluid 2 displacing fluid 1 in a long, constant-diameter pipeline. This is seen from the viewpoint of an observer riding with the interface (the Lagrangian viewpoint). The derivation in the text is based on fluid standing still, for a time t equal to the length of the pipeline divided by the fluid velocity $t = L\,/\,V$.

where c is the concentration of the diffusing species and $\mathcal{D}_x$ is the diffusivity in the x direction, etc. For most cases of molecular diffusion, the values of $\mathcal{D}$ do not depend on direction, so we can factor them out on the right side of Eq. 19.7 and find the exact analog of Eq. 16.3 for heat transfer.

The problem sketched in Fig. 19.4 is the exact analog of two semi-infinite slabs initially at different temperatures, brought together at time zero. The plane of contact immediately takes up the average between the two

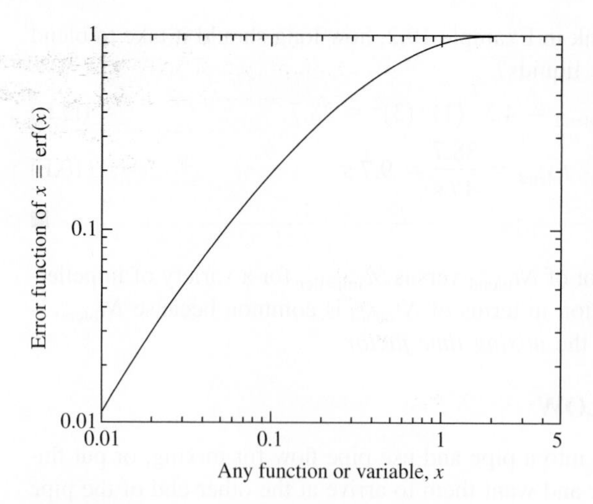

FIGURE 19.5
The error function, erf(x) as a function of x. Excel spreadsheets generate this function by erf(x) = $2 \cdot (\text{NORMSDIST}(x\sqrt{2}) - 0.5)$. In Example 19.5 we use this plot (or this equation) to evaluate $\text{erf}(x / 2\sqrt{\mathscr{D}t})$.

temperatures and remains at that temperature. This differs from the situation in Fig. 19.1, because here the areas into which the diffusion occurs are infinite in depth, not the limited depths in Fig. 19.1. The mathematical solution to this problem (with diffusion only in the x direction) is

$$\frac{c - c_{\text{interface}}}{c_{\text{original}} - c_{\text{interface}}} = \text{erf}\left(\frac{x}{2\sqrt{\mathscr{D}t}}\right) \quad (19.8)$$

where erf is the Gaussian error integral, whose values are given in mathematical tables and shown in Fig 19.5.

Example 19.5. The Chevron Products Pipeline System runs 700 mi from Salt Lake City to Spokane, with intermediate stations at Boise and Pasco. There are two parallel lines, each 8 in in diameter. One line mostly transports different grades of gasoline, the other mostly several grades of diesel fuel and jet fuels. Two different grades of gasoline (fluids 1 and 2) follow each other 700 mi at an average velocity of 8 ft / s. If the only mechanism for mixing were molecular diffusivity (see below), and a concentration of 0.1 percent fluid 2 in fluid 1 is the maximum allowable, how far would that be from the initial interface, at the end of the pipe? $\mathscr{D}$ for two types of gasoline in each other is about $2 \cdot 10^{-9}\ \text{m}^2 / \text{s}$.

Here we let c refer to the concentration of fluid 2. Thus,

$$\frac{c - c_{\text{interface}}}{c_{\text{original}} - c_{\text{interface}}} = \frac{0.1\% - 50\%}{0 - 50\%} = 0.998 = \text{erf}\left(\frac{x}{2\sqrt{\mathscr{D}t}}\right) \quad (19.G)$$

From Fig. 19.5 (actually, from the table on which it is based), we read that 0.998 = erf (2.15), so that

$$\frac{x}{2\sqrt{\mathscr{D}t}} \approx 2.15 \quad (19.H)$$

The time for the fluid to travel 700 mi at 8 ft / s is $4.57 \cdot 10^5$ s = 127 h, so

$$x \approx 2.15 \cdot 2\sqrt{\mathscr{D}t} = 4.3\sqrt{(2 \cdot 10^{-9}\ \text{m}^2/\text{s}) \cdot 4.57 \cdot 10^5\ \text{s}}$$
$$= 0.13\ \text{m} = 0.42\ \text{ft} \quad (19.I)$$

and the mixed zone (from 0.1 to 99.9 percent liquid 2) would have a volume of

$$V_{\text{mixed}} = 2 \cdot 0.42\ \text{ft} \cdot (0.355\ \text{ft}^3\ \text{of liquid / ft of pipe}) = 0.30\ \text{ft}^3 = 2.2\ \text{gal} \quad (19.J)$$

■

This solution is the same as what we would compute if we brought the two fluids into contact and had them stand still for 127 h. If only molecular diffusion were operating, then the mixed zone would be very small. That is not what happens at all. Instead the turbulence in the fluid makes the mixing much more intense. Recall Example 18.6, where we found that the eddy viscosity at one point in a low-speed air flow was 68 times the molecular viscosity. We should expect the same situation here. The experimental data for this type of mixing in turbulent flow are reasonably represented by the semitheoretical relation [8]

$$\frac{\mathscr{D}_{\text{turbulent}}}{VD} = 3.57\sqrt{f} \tag{19.9}$$

where f is the Fanning friction factor for the flow. For turbulent flow, $\mathscr{D}$ is called a *dispersion coefficient* to distinguish it from the molecular diffusivity. In this case, with the dispersion in the flow direction, it is called an *axial dispersion coefficient.*

Example 19.6. Repeat Example 19.5, taking turbulent mixing into account. For this flow the friction factor, determined by the methods of Chap. 6, is $f \approx 0.0039$, so that

$$\mathscr{D}_{\text{turbulent}} = 0.665 \text{ ft} \cdot (8 \text{ ft / s}) \cdot 3.57\sqrt{0.0039} = 1.20 \text{ ft}^2\text{ / s} \tag{19.K}$$

and

$$x \approx 2.15 \cdot 2\sqrt{\mathscr{D}t} = 4.3\sqrt{(1.2 \text{ ft}^2\text{ / s}) \cdot 4.57 \cdot 10^5 \text{ s}} = 3183 \text{ ft} = 970 \text{ m} \tag{19.L}$$

$$\begin{aligned} V_{\text{mixed}} &= 2 \cdot 3183 \text{ ft} \cdot 0.355 \text{ ft}^3\text{ / ft} = 2211 \text{ ft}^3 = 970 \text{ m}^3 \\ &= 16{,}539 \text{ gal} = 394 \text{ bbl} \end{aligned} \tag{19.M}$$

■

Experience with this pipeline [9] indicates that the mixed zone is about $\frac{1}{4}$ of the value calculated above, which reminds us that Eq. 19.9 is only approximate. (But see Probs. 19.9 and 10.) The typical batch size sent through the pipelines is about 10,000 bbl, so from the above calculation we would conclude that approximately 4 percent arrived as the mixed zone, whereas experience suggests about 1 percent. For some products the mixed zone is zero, because, for example, one can put a mixture of regular and premium gasoline in the regular gasoline tank. That mixture will meet regular gasoline performance specifications.

This whole discussion is about axial dispersion in turbulent flow; in laminar flow it is different [10, p. 49] and also interesting.

The previous discussion concerns axial mixing, which we generally would like to avoid. In contrast, we often use flow down a pipe to mix fluids, in which case we want such mixing. Consider Reynold's experiment, in which he introduced a dye streak into a flow (Table 6.1). In turbulent flow, how far down the pipe must we go before we will observe uniform dispersion of the dye? This problem has no well-known solution like the axial dispersion problem discussed above, but we may estimate the distance as follows:

1. First we assume that we can use a turbulent dispersion coefficient and that the semitheoretical value from Eq. 19.9 will serve us. This is a strong assumption,

because axial and radial values of the eddy diffusivity need not be the same; see Fig. 18.3. But it may be the best assumption we can make.

2. We assume that Eq. 19.1 can be used, with the radius or the diameter as the appropriate diffusion length. The choice of the diameter seems conservative.
3. In Eq. 19.1 we can replace t by L/V, where L is the downstream distance. Then

$$\frac{3.57\ VD\sqrt{f}(L/V)}{D^2} = 2 = 3.57\frac{L}{D}\sqrt{f} \tag{19.10}$$

or

$$\frac{L}{D} = \frac{0.56}{\sqrt{f}} \tag{19.11}$$

Example 19.7. In redoing Examples 19.5 and 19.6, we now decide to use the pipe to blend a dye into the fluid by introducing it into the flowing stream. How far downstream does the dye become uniformly distributed throughout the fluid?

From Example 19.6, we know that $f = 0.0039$, so that

$$\frac{L}{D} = \frac{0.56}{\sqrt{0.0039}} \approx 9.0 \tag{19.N}$$

or

$$L = 9.0 \cdot 0.665\ \text{ft} = 6.0\ \text{ft} = 1.8\ \text{m} \tag{19.O}$$

■

This speculative calculation suggests that at $\mathcal{R} \approx 6 \cdot 10^5$ the turbulence is intense enough to mix the dye (or any other soluble substance) into the flowing fluid in a pipe length $\approx$ 9 diameters. Such values are widely reported, but others suggest greater lengths, up to 100 diameters. In laminar flow the eddy diffusivity is zero, so the mixing across the flow is much weaker. When we wish to mix two fluids in laminar flow in a pipe it is common to insert *static mixers* in the pipe, which divide and rotate the flow, repeatedly alternating direction of the rotation. These increase the pressure drop but are effective in laminar flow mixing [3, Chap. 19].

19.6 MIXING IN TURBULENT, CYLINDRICAL FREE JETS

In many cases a jet of fluid passes into a large mass of the same fluid or another fluid of comparable properties, for example, the exhaust from a jet engine or an auto engine, the flow from an air-conditioning duct into a room, or the jet of fluid that is pumped into a tank to stir and mix it. These have been intensely studied for gases, not nearly as much for liquids. The available data are mostly for turbulent flow of jets from a circular orifice, which is the type considered here. Figure 19.6 shows the shape of such a jet for one gas discharging into another, and the nomenclature for describing it.

The figure makes clear that the jet enters from a rounded nozzle from a plenum, in which turbulence should be negligible, so the inlet velocity over the inlet area is practically constant, V_0. As the jet flows into the surrounding gas it generates an annular

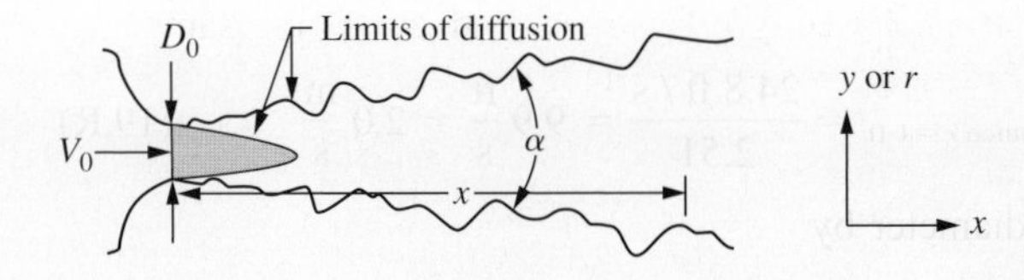

FIGURE 19.6
The behavior of a turbulent free jet, and the notation used to describe it.

turbulent boundary layer that eats into the jet and grows outward from it. The inlet flow extends from the orifice to the point at which the central jet, uninfluenced by the gas around it, is all eaten away. This occurs at $x / D_0 \approx 6.4$ to 7. Up to that distance the velocity along the centerline $= V_0$. Then, from $x / D_0 \approx 7$ to 100 the jet grows steadily in diameter, entrains more and more of the surrounding gas, and slows down. Beyond $x / D_0 \approx 100$ the jet has slowed enough that its velocity is comparable to the random movements in the surroundings, and it ceases to exist as a coherent jet.

Experiments indicate that the momentum flow across any plane perpendicular to the x axis is practically the same; the total mass flowing increases, the velocity falls, and the integral $\int_{\text{all area}} V^2 \rho \, dA$ remains constant. Furthermore, over the range from $x / D_0 \approx 7$ to 100, the velocity profiles are self-similar; they have the same bell-shapes around the axis of the jet. In this region the behavior is reasonably well represented by the following semitheoretical equations [11, pp. 6–21] for air flowing into air:

$$\frac{V_{\text{centerline}}}{V_0} = K \frac{D_0}{x} \tag{19.12}$$

where $K = 5$ for $V_0 = 2.5$ to 5 m / s, and $K = 6.2$ for $V_0 = 10$ to 50 m / s (see Prob. 19.11).

$$\log\left(\frac{V_{\text{centerline}}}{V_{x \text{ at radial distance } r}}\right) = 40\left(\frac{r}{x}\right)^2 \tag{19.13}$$

$$\text{Jet angle} = \alpha = 18° \text{ to } 24° \qquad \text{normally} \approx 20° \tag{19.14}$$

and

$$\begin{pmatrix}\text{Entrainment} \\ \text{ratio}\end{pmatrix} = \frac{(\text{volumetric flow rate of jet})}{(\text{volumetric flow rate of jet})_0} = \frac{Q}{Q_0} = 0.62\sqrt{\frac{x}{D_0}} \tag{19.15}$$

Example 19.8. Air flows from a round, horizontal 1 ft diameter jet into the atmosphere at $V_0 = 40 \text{ ft / s}$ (typical building ventilating system velocity). Estimate the velocity on the centerline at a downstream distance of 10 ft, and the velocity 1 ft away from the centerline at that distance. Estimate the width of the jet and the entrainment ratio at that distance.

From Eq. 19.12, at 10 ft we have

$$\frac{V_{\text{centerline}}}{V_0} = 6.2 \frac{1 \text{ ft}}{10 \text{ ft}} = 0.62 \qquad V_{\text{centerline}} = 24.8 \frac{\text{ft}}{\text{s}} = 7.6 \frac{\text{m}}{\text{s}} \tag{19.P}$$

A plot of $V_{\text{centerline}}$ versus x is a rectangular hyperbola, which falls to 2.5 ft / s at 100 ft. To find the velocity at a downstream distance of 10 ft and a radius of 1 ft, we write

$$\left(\frac{V_{\text{centerline}}}{V_{x \text{ at radial distance } r}}\right) = 10^{40(1/10)^2} = 10^{0.4} = 2.51 \tag{19.Q}$$

and

$$V_{\substack{x \text{ at radial distance } r=1 \text{ ft} \\ \text{and } x=10 \text{ ft}}} = \frac{24.8 \text{ ft / s}}{2.51} = 9.9 \frac{\text{ft}}{\text{s}} = 2.0 \frac{\text{m}}{\text{s}} \tag{19.R}$$

We may calculate the jet diameter by

$$D_{\text{at } x} = D_0\left(1 + \frac{x}{D_0} \sin \alpha\right) = 1 \text{ ft}(1 + \cdot 10 \cdot \sin 20°)$$
$$= 4.42 \text{ ft} = 1.35 \text{ m} \tag{19.S}$$

At 10 ft, the entrainment ratio is

$$\frac{Q}{Q_0} = 0.62\sqrt{\frac{10 \text{ ft}}{1 \text{ ft}}} = 1.96 \tag{19.T}$$

■

This example shows the following:

1. The calculations using these approximate equations are straightforward.
2. The velocities shown are time averages. In any turbulent flow the velocities in simple formulae like these always show time averages.
3. The spreading angle is not clearly defined, and reported values have some variability.
4. The equations are not totally internally consistent. If, for example, we ask for the velocity at the edge of the jet in this example, we find

$$\left(\frac{V_{\text{centerline}}}{V_{x \text{ at radial distance } r}}\right) = 10^{40[(4.42/2)/(10)]^2} = 10^{1.95} = 89.9 \tag{19.U}$$

and

$$V_{\substack{x \text{ at radial distance } r=2.24 \text{ ft} \\ \text{and } x=10 \text{ ft}}} = \frac{24.8 \text{ ft / s}}{89.9} = 0.28 \frac{\text{ft}}{\text{s}} = 0.084 \frac{\text{m}}{\text{s}} \tag{19.V}$$

where perfect consistency of the equations would require a velocity equal to zero.
5. The equations are only reliable to $x / D_0 \approx 100$, at which

$$Q / Q_0 = 0.62\sqrt{100} = 6.2 \tag{19.W}$$

which says that the maximum dilution obtainable in this type of flow is a factor of approximately 6. Beyond the end of the jet, dilution continues by mixing in the surrounding fluid, based on the surrounding fluid turbulence, uninfluenced by the jet.

19.7 MIXING IN ATMOSPHERIC PLUMES

Many chemical engineers are involved with mixing of pollutants into the environment. An important example of this type of calculation is the *gaussian plume* atmospheric dispersion model, sketched in Fig. 19.7. This type of model is widely used in safety and environmental analyses. As shown in that figure, a plume of combustion gases rises from a smokestack and then levels off and flows in the downwind direction.

Such plumes normally rise a considerable distance above the smokestack because they are emitted at temperatures higher than atmospheric and are emitted with

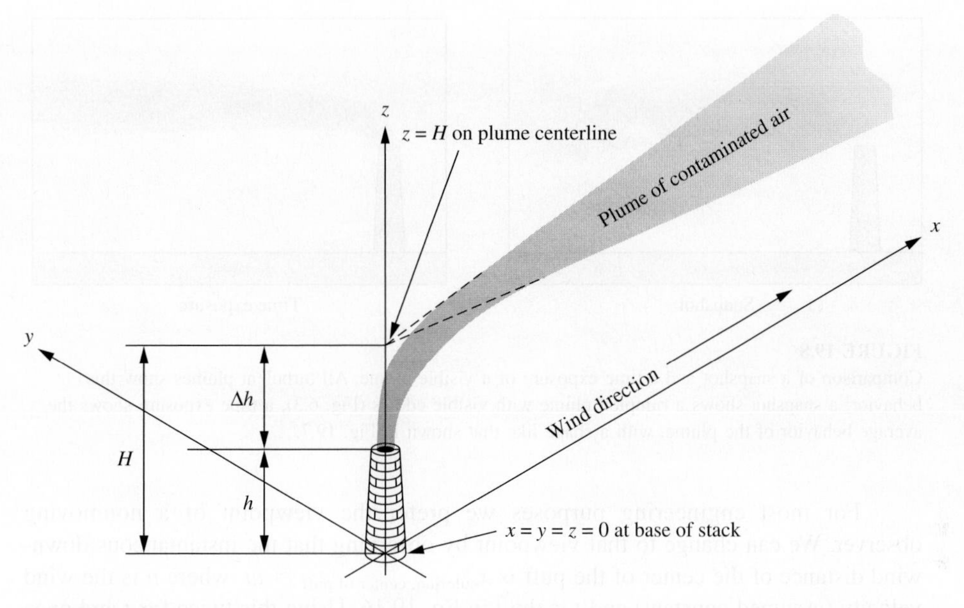

FIGURE 19.7
Coordinate system and nomenclature for the Gaussian plume equation. It replaces the real source with a mathematical point source at 0, 0, *H*, steadily emitting *Q* (g / s) of pollutant with neither buoyancy nor momentum.

a vertical velocity. For a gaussian plume model, the real plume is replaced with a mathematical point source, which has neither upward momentum nor buoyancy, with emission rate Q (normally g / s) emitted from a point with coordinates 0, 0, H, where H is called the *effective stack height,* which is the sum of the *physical stack height* (h in Fig. 19.7) and the *plume rise* (Δh in Fig. 19.7). Physical stack height for any existing plant can be determined with ordinary measuring instruments. Plume rise is discussed in Prob. 19.16. The wind is assumed to blow in the x direction with velocity u, independent of time, location, or elevation. The problem is to compute the concentration due to this source at any point (x, y, z).

Figure 19.8 sketches the instantaneous view of the plume, and a time exposure. Like all turbulent mixing flows, the instantaneous behavior of the plume shows turbulent eddies and snake-like curls (see Fig. 6.3). But a time exposure averages that, and the time-averaged plume is quite smooth and well behaved.

To solve the problem mathematically, we begin with the viewpoint of an observer, riding with the wind (lagrangian viewpoint) passing right over the emission point. We first consider a small "puff"containing X g of pollutant at 0, 0, H, and $t = 0$. Then we apply Eq. 19.7. The solution of this problem is well known [12, p. 225].

$$c = \frac{X}{8(\pi t)^{3/2}(\mathcal{D}_x \mathcal{D}_y \mathcal{D}_z)^{1/2}} \exp - \left\{ \frac{1}{4t} \cdot \left[\frac{x^2}{\mathcal{D}_x} + \frac{y^2}{\mathcal{D}_y} + \frac{(z-H)^2}{\mathcal{D}_z} \right] \right\} \qquad (19.16)$$

Here, t is the time since the release. From the viewpoint of an observer riding with the flow, $x_{\text{lagrangian}}$ represents the downwind or upwind distance from the center of the pollutant cloud, which is assumed to move with local wind velocity.

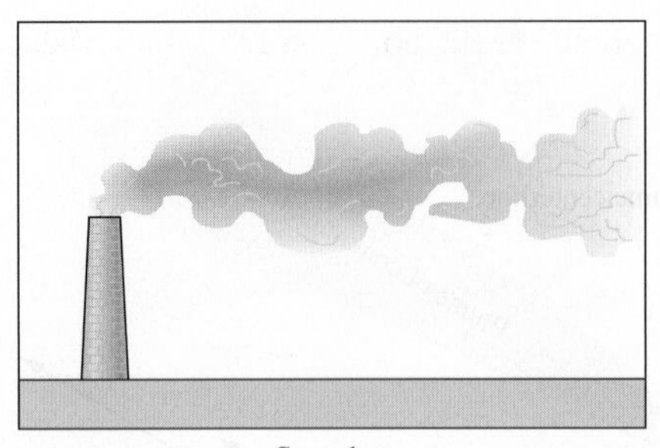

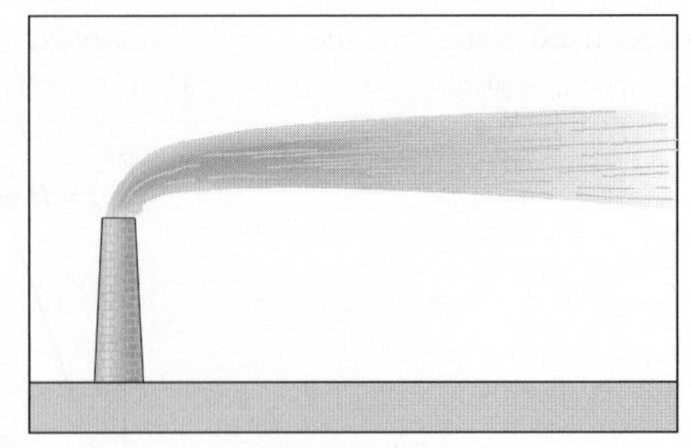

FIGURE 19.8
Comparison of a snapshot and a time exposure of a visible plume. All turbulent plumes show this behavior; a snapshot shows a rumpled plume with visible eddies (Fig. 6.3), a time exposure shows the average behavior of the plume, with a shape like that shown in Fig. 19.7.

For most engineering purposes we prefer the viewpoint of a nonmoving observer. We can change to that viewpoint by observing that the instantaneous downwind distance of the center of the puff is $x_{\text{eulerian, center of puff}} = ut$, where u is the wind velocity (assumed constant) and t is the t in Eq. 19.16. Using this twice for t and once for x, we change Eq. 19.16 to

$$c = \frac{X}{8(\pi x / u)^{3/2} (\mathscr{D}_x \mathscr{D}_y \mathscr{D}_z)^{1/2}} \exp - \left\{ \frac{1}{4x / u} \cdot \left[\frac{(x - ut)^2}{\mathscr{D}_x} + \frac{y^2}{\mathscr{D}_y} + \frac{(z - H)^2}{\mathscr{D}_z} \right] \right\} \tag{19.17}$$

where this x is the distance downwind of the source. As written, this assumes that the $\mathscr{D}$s are molecular diffusivities. But from our previous estimates of turbulent diffusion, we know that in turbulent flow the eddy diffusivities are much larger than the molecular diffusivities. Experimental evidence is strong that while molecular diffusivities are isotropic (which would lead us to simplify Eq. 19.17), eddy diffusivities are not isotropic, so we retain the three different symbols, here and below.

Although Eq. 19.17 would be perfectly satisfactory for our use, for historical reasons the form that appears in the air pollution literature is obtained by making the following three substitutions

$$\mathscr{D}_x = 2\sigma_x^2 \frac{u}{x} \tag{19.18}$$

$$\mathscr{D}_y = 2\sigma_y^2 \frac{u}{x} \tag{19.19}$$

$$\mathscr{D}_z = 2\sigma_z^2 \frac{u}{x} \tag{19.20}$$

where σ_x, σ_y, and σ_z are the three-dimensional *atmospheric dispersion coefficients* or *dispersion factors*. The Greek sigmas are used here to make the formulas look the same as those in the gaussian distribution formulas in statistics. There is little theoretical connection between the two, and some other symbol could just as well have been used here, but the sigmas are used throughout the air pollution literature. Making these substitutions in Eq. 19.17, we find

$$c = \frac{X}{(2\pi)^{3/2}\,\sigma_x\sigma_y\sigma_z} \exp -\left[\frac{(x-ut)^2}{2\sigma_x^2} + \frac{y^2}{2\sigma_y^2} + \frac{(z-H)^2}{2\sigma_z^2}\right] \tag{19.21}$$

the *gaussian puff equation.* It describes the behavior of a "puff"of pollutants. In Eq. 19.21, if we have $x - ut = y = (z - H) = 0$, i.e., the exact center of the moving cloud, then the exp term is $\exp 0 = 1$. This shows that the term before the exp term is the concentration at the center of the cloud, and that the exp term shows how that concentration decreases as we move away from the center of the cloud in three directions. Equation 19.21 (and some variants of it) are used in nuclear and chemical plant safety analysis, where the puff of pollutants is the radioactive or chemical cloud that could be emitted quickly in certain possible types of nuclear or chemical plant accidents. To use it we need the values of the σs, which will be discussed below.

The steady-state equivalent of Eq. 19.21 is used more often than Eq. 19.21. To find it, we assume that the dispersion in the x direction (up and/or downwind) is insignificant compared to that in the two crosswind directions (y and z), so that the basic dispersion equation, Eq. 19.7 is modified by dropping the $\partial^2 c / \partial x^2$ term. In this case the source term X in the dispersion equation becomes Q / u because we are considering spreading by dispersion in the y and z directions, in a slab of air with unit length in the x direction that passes over the source with velocity u. (The reader may verify that Q / u has dimensions of mass / length: i.e., the amount injected per unit length of air passing over the stack.)

Making these substitutions into Eq. 19.21, we find

$$c = \frac{Q}{2\pi u \sigma_y \sigma_z} \exp -\left[\frac{y^2}{2\sigma_y^2} + \frac{(z-H)^2}{2\sigma_z^2}\right] \tag{19.22}$$

where the symbols have the same meaning as before. It is the *basic gaussian plume equation.* It has several widely used variants. It is called the "gaussian" plume equation because the exponential terms have the same form as Gauss's normal distribution function. (If we had chosen our coordinates so that the pollutant source were at some arbitrary point, say (x', y', z') instead of being at $(0, 0, H)$, then the terms in the exponential part of Eqs. 19.21 and 19.22 would be $(x - x')^2, (y - y')^2$, etc. The choice of the origin in Fig. 19.7 simplifies these expressions; one might choose to put the origin of the coordinate system at the top of the plume rise (which would drop the H out of Eq. 19.22), but most of us prefer to have $z = 0$ at ground level.

Example 19.9. A factory emits $Q = 20$ g / s of SO_2. The wind speed is $u = 3$ m / s. At a distance of 1 km downwind, the values of σ_y and σ_z are 30 and 20 m respectively. What are the SO_2 concentrations at the centerline of the plume at $x = 1$ km downwind, and at a point 60 m to the side of and 20 m below the centerline at the same x?

The centerline values are those for which $y = 0$ and $z = H$, so both of the terms in the exp expression are zero. Because $\exp 0 = 1$, the exponential term is unity. Thus, we see that just as in Eq. 19.21, Eq. 19.22 consists of two parts: the first gives the centerline concentration, and the second gives the factor by which the centerline concentration is reduced as we move away from it. At the centerline, 1 km downwind

$$c = \frac{20 \text{ g / s}}{2\pi(3 \text{ m / s})(30 \text{ m})(20 \text{ m})}$$

$$= 0.00177 \frac{\text{g}}{\text{m}^3} = 1770 \frac{\mu\text{g}}{\text{m}^3} \tag{19.X}$$

At the point away from the centerline, we must multiply the above by

$$\exp - \left[\left(\frac{1}{2}\right)\left(\frac{60 \text{ m}}{30 \text{ m}}\right)^2 + \left(\frac{1}{2}\right)\left(\frac{20 \text{ m}}{20 \text{ m}}\right)^2\right] = \exp - \left(2 + \frac{1}{2}\right) = 0.0818 \tag{19.Y}$$

so

$$c = \left(\frac{1770 \ \mu\text{g}}{\text{m}^3}\right) \cdot 0.0818 = \frac{145 \ \mu\text{g}}{\text{m}^3} \tag{19.Z}$$

■

The basic gaussian plume equation predicts a plume that is symmetrical with respect to y and with respect to z. Thus, if we had asked for the concentration 60 m to the other side of and 20 m above the plume centerline, we would have gotten the same answer. The different values of σ_y and σ_z mean that the spreading in the vertical and horizontal directions are not equal. Most often $\sigma_y > \sigma_z$ so that a contour of constant concentration is like an ellipse, with the long axis horizontal. Close to the ground, this symmetry is disturbed, as discussed in Prob. 19.14.

To use Eq. 19.21 or 19.22, one must know the appropriate values of σ_y and σ_z (which were simply guessed in the previous example). From Eqs. 19.18 to 19.20 we would expect them to have the form

$$\sigma_y = \left(\frac{2\,\mathscr{D}_y x}{u}\right)^{1/2} \quad \text{etc.} \tag{19.AA}$$

However, if we reconsider our values of the $\mathscr{D}$s, we see that they are eddy diffusivities, which should depend on wind speed and on the degree of atmospheric turbulence, which is a function of wind speed and degree of solar heating ("insolation"). It is quite plausible to assume that for any given degree of insolation the value of $\mathscr{D}$ will be linearly proportional to the wind speed, i.e., that for any such situation $(\mathscr{D}_y / u)$ and $(\mathscr{D}_z / u)$ are constants. Thus, from Eq. 19.AA we would conclude that for any given meteorological condition, each of the σs should be proportional to the square root of the downwind distance.

Experimental evidence does not agree very well with this prediction. The best available data have been correlated by Turner [13] and others and presented in the form of plots of log σ_y versus log x and log σ_z versus log x. If the above calculation were correct, for each atmospheric condition such plots would be straight lines with slope $\frac{1}{2}$. The best correlations of the experimental results, shown in Figs. 19.9 and 19.10, illustrate that on such plots the horizontal dispersion coefficient, σ_y, does form a family of straight lines (for various atmospheric conditions), but these have a slope

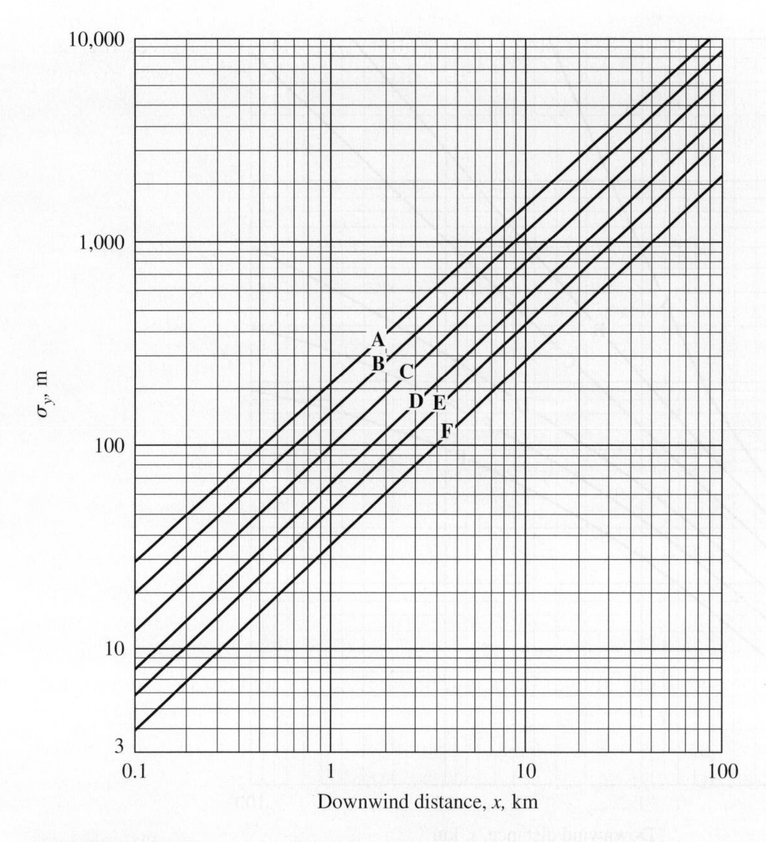

FIGURE 19.9
Horizontal dispersion coefficient, σ_y, as a function of distance downwind from the source, for various stability categories. (From D. B. Turner, "Workbook of Atmospheric Dispersion Estimates," U.S. Environmental Protection Agency Report AP-42, Washington, D.C: U.S. Government Printing Office, (1970).) For puff calculations this plot is often used for σ_x as well. This plot and the next are useful for hand calculations. For computer calculations they are replaced by their *exact equivalents*, $\sigma_y = ax^{0.894}$ and $\sigma_z = cx^d + f$, where x is the downwind distance, expressed in km, the σs are in m, and a, c, d, and f are constants found in the following table.

Stability category		$x \le 1$ km			$x \ge 1$ km		
	a	***c***	***d***	***f***	***c***	***d***	***f***
A	213	440.8	1.941	9.27	459.7	2.094	−9.6
B	156	106.6	1.149	3.3	108.2	1.098	2.0
C	104	61	0.911	0	61	0.911	0
D	68	33.2	0.725	−1.7	44.5	0.516	−13.0
E	50.5	22.8	0.678	−1.3	55.4	0.305	−34.0
F	34	14.35	0.740	−0.35	62.6	0.180	−48.6

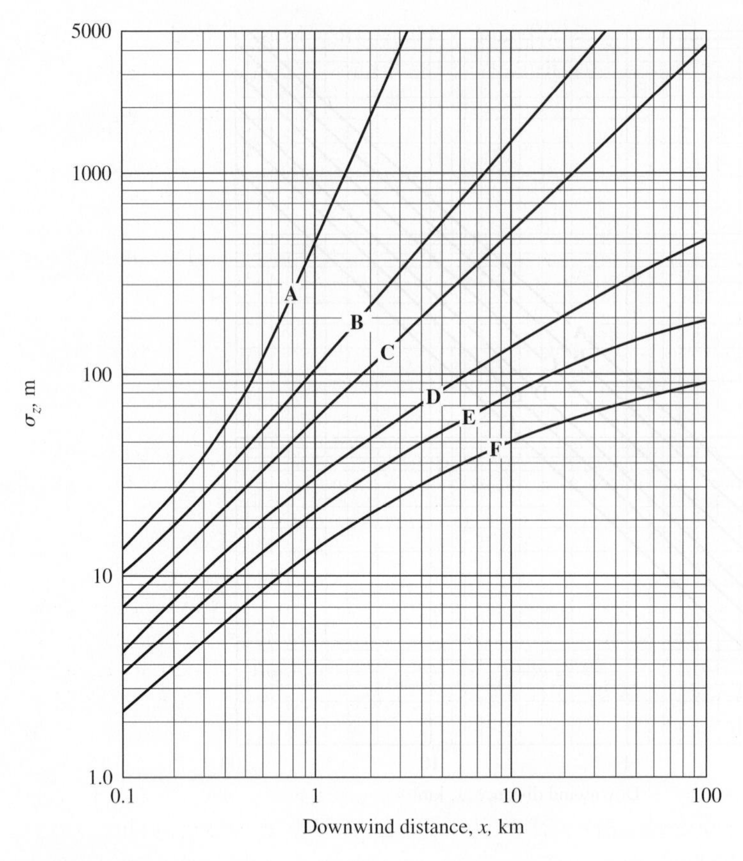

FIGURE 19.10
Vertical dispersion coefficient, σ_z, as a function of distance downwind from the source, for various stability categories, from Turner [13].

of 0.894 instead of the 0.50 that we would expect from the above derivation. The vertical dispersion coefficient, σ_z, forms a fan-shaped pattern for various atmospheric conditions.

Why do the experimental data disagree with our neat theory? They disagree because the assumption that the eddy diffusivity depends only on wind speed and insolation is much too simple to account for all the complicated things that actually go on in the atmosphere, even on days with very simple wind patterns, which are the only ones on which experimental tests of Eq. 19.22 are attempted. Thus, we can say that the preceding derivation shows us a way to obtain a logical material balance for dispersion of a pollutant in the atmosphere, subject to some strong simplifying assumptions, but that we must regard the values of σ_y and σ_z (or the corresponding eddy dispersion values) as experimental quantities that we cannot compute from theory. However, if we accept Figs. 19.9 and 19.10 as adequate representations of the experimental results, we can use them, along with Eq. 19.22, to make predictions of concentrations downwind from point sources. Modern point source atmospheric modeling uses advanced versions of this procedure. The experimental data on which Figs. 19.9 and 19.10 are based are very limited and not necessarily directly applicable to cities. Most of the data were taken for steady flow of winds over grasslands (the Salisbury Plain in England and the grasslands of Nebraska). We use them for cities because we have nothing better.

So far, we have said nothing about the lines labeled A through F on Figs. 19.9 and 19.10. These correspond to different levels of atmospheric stability. On a hot summer afternoon, the sun heats the ground, which in turn heats the air near it causing that air to rise and thus to mix pollutants very well. In this situation the atmosphere is very unstable, and the values of $\mathscr{D}_x$ and $\mathscr{D}_y$ leading to σ_y and σ_z will be large. On a cloudless winter night, the ground cools by radiation to outer space and cools the air near it. The air forms an inversion layer, making the atmosphere very stable and inhibiting the dispersion of pollutants, so the values of σ_y and σ_z will be small.

Figures 19.9 and 19.10 rely on the stability-category classification given by Turner [13], which considers only the incoming solar radiation and the wind speed, reproduced in Table 19.3. There are other systems for estimating σs, but this one is simple and widely used.

Example 19.10. Estimate the values of σ_y and σ_z at a point 0.5 km downwind from a pollution source on a bright summer day with a wind speed greater than 6 m / s. From Table 19.3 we conclude that a bright summer day is one on which the incoming solar radiation is strong, so we would use stability category C. Then, using Figs. 19.9 and 19.10, we would read (for x = 0.5 km) σ_y = 55 m and σ_z = 32 m. ■

This brief section shows only the relation between the atmospheric mixing used in air pollution calculations [14, Chap. 6] and the other fluid mixing problems discussed in this chapter. The analogous problem of the dispersal of wastewater (treated sewage or storm drain water or other wastewater) into other bodies of water (oceans, lakes, rivers) is basically similar, but is made more complex by the significant density differences between the wastewaters and the water into which they are discharged.

TABLE 19.3
Key to stability categories

	Day*			Night*	
	Incoming solar radiation			Thinly overcast or ≥4 / 8 low cloud	≤3 / 8 Cloud
Surface wind speed (at 10 m), m / s	Strong	Moderate	Slight		
0–2	A	A–B	B	—	—
2–3	A–B	B	C	E	F
3–5	B	B–C	C	D	E
5–6	C	C–D	D	D	D
≥6	C	D	D	D	D

*The neutral class, D, should be assumed for overcast conditions, day or night.

19.8 SUMMARY

1. Chemical engineers deal with a great variety of mixing problems.
2. The vast majority of these involve turbulent flow and use turbulence to reduce the scale of the inhomogeneities to fractions of a millimeter.
3. At that scale, molecular diffusion is fairly rapid, so mixing down to the molecular level (as needed for combustion or other chemical reactions) is finished by molecular diffusion.
4. It is common in mixing calculations to use the mathematical forms for molecular diffusion (Fick's law) and replace the molecular diffusivity, which is a function of the fluid but not of the flow, with an eddy diffusivity, which is mostly a function of the flow, and which varies from place to place. The methods for doing this are all approximate; they are widely used.
5. Mixing in laminar flow (high viscosities or small diameters) is different from turbulent mixing; it has an extensive literature [3, 15].

PROBLEMS

See the Common Values for Problems and Examples inside the back cover. An asterisk (*) on a problem number indicates that its answer is in App. D.

19.1.*Gasoline is normally blended from three or four liquid streams. The resulting blend must have approximately the same octane number in every charge sent to the cylinder of an engine. If an auto with a four-cylinder engine is driving 60 mi / h, with an engine speed of 2500 rpm, and fuel use of gal / 25 mi, how large is the charge of fuel going into each cylinder for each combustion? In four-cycle engines fuel is inserted on every second revolution.

19.2. It is a rule of thumb that suspension of solids is easy if their terminal settling velocity is <0.03 ft / s, and difficult if it is >0.15 ft / s. For sugar in water, to what diameters do these settling velocities correspond? Does this rule correspond to your experience stirring sugar into your coffee, tea, or lemonade?

19.3. If paint pigments have SG = 4 and $D = 1\ \mu$, and if the viscosity of the paint is 50 times that of water, how fast should the pigments settle in a stored can of paint? Use Fig. 6.26, and observe that in the Stokes law region the settling velocity is proportional to 1 / viscosity. How far would you expect the particles to settle in a month? This assumes that the liquid part of the paint is a Newtonian fluid. Fifty years ago they all were, and pigment settling was significant. Since then, they have been formulated to be highly non-Newtonian, (thixotropic Bingham plastics), and settling is no longer a problem.

19.4. Repeat Example 19.1 for the mixing of two gases, e.g., CO and O_2.

19.5.*In Example 19.2,

(*a*) What is the impeller tip velocity? How does this compare to the values in Table 19.2?

(*b*) What is the (power / tank volume)? Assume the tank dimension ratios in Fig. 19.2. How does this compare to the values in Table 19.2?

19.6. In Example 19.4, we have now decided to increase all the tank dimensions by a factor of 2. The impeller dimensions and rpm are unchanged. What is the expected time to blend the contents of the tank?

19.7. In Example 19.3, if we want to keep the same $\mathscr{R}_{\text{impeller}}$ in the full-sized tank as in the test tank and also keep geometric similarity, how many revolutions per minute should we use?

19.8. In Example 19.6 we computed an axial dispersion coefficient of 1.2 ft^2 / s. What is the ratio of this value to the value for the molecular diffusion coefficient?

19.9. In Example 19.6 we estimated the size of the mixed zone between two grades of gasoline after a 700-mi pipeline trip. The pipeline operators inform us that this estimate is high by a factor of ≈ 4. Could it be that they use a lower requirement for product purity? Repeat that example, assuming that the requirement is no more than 1 percent fluid 2 in fluid 1, instead of the 0.1 percent in that example. In principle you can read the value you need from Fig. 19.5, but in practice you need the table that Fig 19.5 is based on. From the table we find that $0.98 = \text{erf}(1.65)$.

19.10.*Smith and Schulze [16] present details on laboratory and field tests of product pipelines such as those in Example 19.6. They correlated their results by the totally empirical relation

$$\left(\begin{array}{l}\text{Length of mixed}\\ \text{zone, in ft}\end{array}\right) = \left(\begin{array}{l}\text{pipe length}\\ \text{in ft}\end{array}\right)^{0.62} \cdot (1.075\,\mathscr{R}^{-0.87} + 0.550) \qquad (19.23)$$

where the mixed zone is that from 1 to 99 percent pure product and $\mathscr{R}$ is based on the viscosity of the 50:50 (% / %) mixture of the two fluids. Repeat Example 19.6 using this correlation. Compare the calculated value with that in Prob. 19.9, that in Example 19.6, and that reported by the pipeline operators [9].

19.11. Equation 19.12 shows two values for the constant, corresponding to two different velocity ranges. These values are based on the flow of air jets into still air. Rushton [17] suggests that, for liquid jets flowing into large tanks of liquids, Eq. 19.12 applies, with

$$K = 1.41\,\mathscr{R}^{0.135} \qquad (19.24)$$

How do the two values of K given for air compare with this equation? Assume that the values of K with Eq. 19.12 apply to jets of air at 20°C, 1 ft in diameter.

19.12. Equation 19.W shows that the maximum entrainment ratio from a free jet is about 6. If pure methane ($\approx$natural gas) flows in a free jet into the atmosphere, will the turbulent mixing between the jet and the surrounding air lower the methane content at the end of the jet to below the lower combustible limit, which is about 4 percent?

19.13. Figure 7.13 shows a Bunsen burner, which is a simple example of a confined jet, i.e., a jet inside a tube whose walls are close enough to the jet that the jet is not a free jet. It is different from free jets and from the radial pipe mixing in Sec. 19.5. This type of flow has a considerable literature [18].

(*a*) Typical results indicate that complete mixing of gases occurs in such flows by a downwind distance of $x / D_{\text{pipe or tube}} \approx 5$. What is the value of this ratio for the Bunsen burner in Fig. 7.13?

(*b*) Another recommended approach to estimating the distance for complete mixing is to compute the distance downstream at which the jet diameter, calculated for a free jet of the same dimension as the central jet, becomes equal to the outer tube diameter, and use twice that distance. What would the calculated distance be for that estimate for the jet in Fig. 7.13?

19.14. Equation 19.22 is our best current prediction method for the concentration in steady-state plumes far above the ground. However, we are generally most interested in concentrations at ground level because people and property are exposed at ground level.

The blind application of Eq. 19.22 at or near ground level gives misleadingly low results. The reason is that it indicates that the pollutants continue to disperse at any value of z, even at z less than zero. (Using it alone, we could continue Example 19.9 and compute the concentration underground; the result would bear no relation to what we would observe in nature.) For this reason, it is necessary to account for the effect of the ground.

The ground damps out vertical dispersion. The upward and downward turbulent eddies that spread the plume in the vertical direction cannot penetrate the ground. Thus, the vertical spreading terminates at ground level. The method commonly used to account for this in calculations is to assume that the pollutants that would have carried below $z = 0$ if the ground were not there are "reflected" upward as if the ground were a mirror. Thus, the concentration at any point is that due to the plume itself, plus that reflected upward from the ground. This is equivalent to assuming that there is a mirror-image plume below the ground that transmits as much up through the ground surface as the above-ground plume would transmit down through the ground surface if the ground were not there.

The concentrations due to the "mirror-image" plume are exactly the same as those shown by Eq. 19.22, except that the $(z - H)^2$ term is replaced by $(z + H)^2$. At the ground, $z = 0$, both the main plume and the mirror-image plume have identical values. High in the air, for example, at $z = H$, the main plume has a high concentration $[\exp - (0) = 1]$ while that for the mirror-image plume $[\exp - (2H)^2$ etc.$]$ is a small number. The combined contribution of both plumes is obtained by writing Eq. 19.22 and the analogous equation for the mirror-image plume, adding the values for the two plumes and factoring out the common terms to find

$$c = \frac{Q}{2\pi u \sigma_y \sigma_z} \cdot \left[\exp - 0.5\left(\frac{y}{\sigma_y}\right)^2\right] \cdot \left[\exp - 0.5\left(\frac{z - H}{\sigma_z}\right)^2 + \exp - 0.5\left(\frac{z + H}{\sigma_z}\right)^2\right] \quad (19.25)$$

(*a*) Show the form that this equation takes for a point directly downwind of the source ($y = 0$) and at ground level ($z = 0$). This form is the most widely used simple point-source air-pollution modeling equation.

(*b*) Using that equation, estimate the concentration at ground level, directly under the plume centerline, at $x = 1$ km, for $H = 100$ m, $Q = 10$ g / s, $u = 3$ m / s, and C stability.

19.15. See Prob. 19.14.

(*a*) Show that the resulting equation for ($y = z = 0$) can be written as

$$cu / Q = f(x, H)_{\text{for a given stability category}} \quad (19.26)$$

Figure 19.11 shows this function for C stability.

(*b*) Check the point on this plot for $H = 100$ m, $x = 1$ km, using Figs. 19.9 and 19.10 or their equivalent equations.

(*c*) Repeat part (*b*) of the preceding problem, using Fig 19.11.

19.16. Figure 19.7 shows the plume rising a distance Δh above the top of the stack before leveling out. This distance is called the *plume rise,* described in detail by Briggs [19]. Holland's semi-theoretical equation

$$\Delta h = \frac{V_s D}{u} \cdot \left[1.5 + 2.68 \cdot 10^{-3}\, PD \frac{(T_s - T_a)}{T_s}\right] \quad (19.27)$$

is widely used for estimating plume rise. This is a dimensional equation, in which V_S is the stack exit velocity in m / s, D is the stack diameter in m, u is the wind speed in m / s, Δh is the plume rise in m, P is the pressure in millibars (one standard

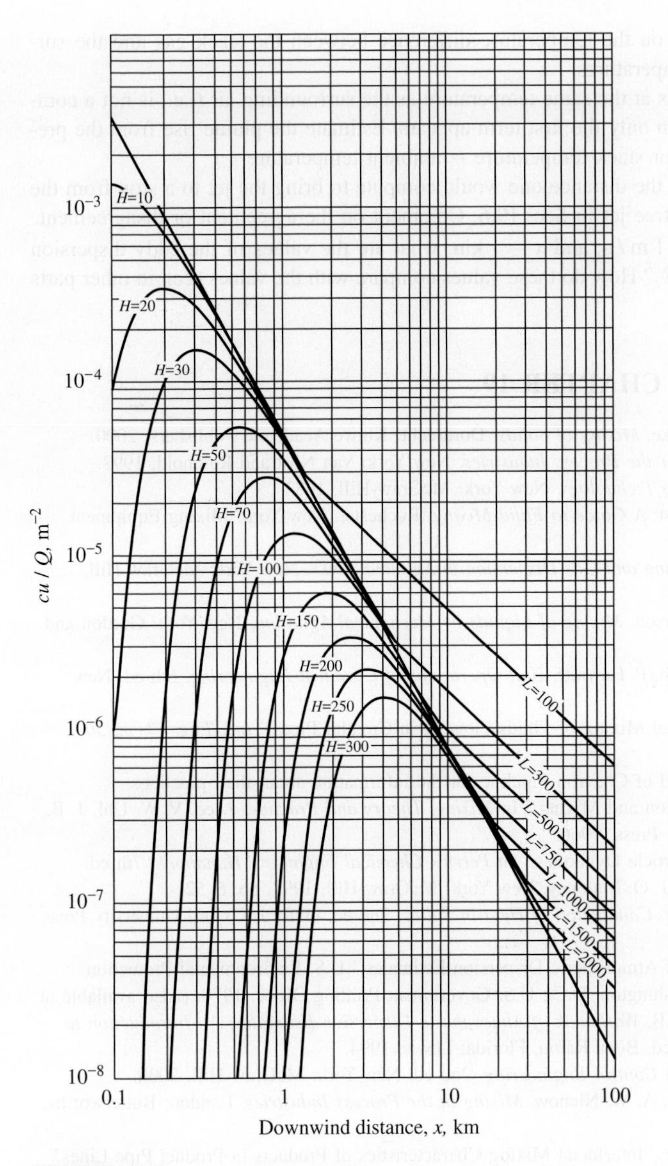

FIGURE 19.11
Ground level cu / Q directly under the plume centerline for C stability only, computed from Eq. 19.25. Similar plots for the other five stability categories are shown by Turner [13]. At the far right the plume has mixed up to the mixing hight (L, m) and can no longer spread upward. The effective stack heights, H, are in m.

atmosphere = 1013 millibars), T_S is the stack gas temperature in K, and T_a is the atmospheric temperature in K.

(*a*) Estimate the plume rise for a 3 m diameter stack with $V_S = 20$ m / s, $u = 2$ m / s, $P = 1$ atm, $T_S = 100°C$, and $T_a = 15°C$ (373 and 288 K).

(*b*) Show that Eq. 19.27 is the sum of two terms, the first of which depends only on V_s, D, and u, and the second of which depends on these and on P and T of the atmosphere and on the temperature difference between the stack gas and the surrounding air temperatures.

(*c*) If the stack gas is at the same temperature as the surrounding air (i.e., is not a combustion gas), then only the first term appears. Estimate the plume rise from the previous example for stack temperature = ambient temperature.

(*d*) Compare this to the distance one would compute to bring the jet to a stop from the equations for a free jet in Sec. 19.6. Comment on the agreement or disagreement.

19.17. For C stability, $u = 1$ m / s, and $x = 1$ km, what are the values of the eddy dispersion coefficients $\mathscr{D}_y$ and $\mathscr{D}_z$? How do these values compare with the values seen in other parts of this chapter?

REFERENCES FOR CHAPTER 19

1. Weinekoetter, R., H. Gericke. *Mixing of Solids.* Dordrecht: Kluwe Academic Publishers, 2000.
2. McDonald, R. J. *Mixing for the Process Industries.* New York: Van Nostrand Reinhold, 1992.
3. Oldshue, J. Y. *Fluid Mixing Technology.* New York: McGraw-Hill, 1983.
4. Oldshue, J. Y., N. R. Herbst. *A Guide to Fluid Mixing.* Rochester, New York: Mixing Equipment Co., 1990.
5. Tatterson, G. B. *Fluid Mixing and Gas Dispersion in Agitated Tanks.* New York: McGraw-Hill, 1991.
6. Ulbrecht, J. J., G. K. Patterson. *Mixing of Liquids by Mechanical Agitation.* New York: Gordon and Breach, 1985.
7. McCabe, W. L., J. C. Smith, P. Harriott. *Unit Operations of Chemical Engineering.* 6th ed. New York: McGraw-Hill, 2001.
8. Levenspiel, O. "Longitudinal Mixing of Fluids Flowing in Circular Pipes." *Ind. Eng. Chem. 50* (1958), pp. 343–346.
9. I thank Mr. Brad Rosewood of Chevron Pipeline for the information about their pipelines.
10. Brodkey, R. S. "Fluid Motion and Mixing." In *Mixing, Theory and Practice, 1.* ed. V. W. Uhl, J. B. Gray. New York: Academic Press, 1966.
11. Tilton, J. N. "Fluid and Particle Dynamics." In *Perry's Chemical Engineers' Handbook.* 7th ed. R. H. Perry, D. W. Green, J. O. Maloney. New York: McGraw-Hill, 1997, pp. 6–52.
12. Carslaw, H. S., J. C. Jaeger. *Conduction of Heat in Solids.* 2nd ed. Oxford: Oxford University Press, 1959.
13. Turner, D.B. "Workbook of Atmospheric Dispersion Estimates." U.S. Environmental Protection Agency Report AP-42. Washington, D.C.: U.S. Government Printing Office, 1970. (Also available at a greater cost as Turner, D.B. *Workbook of Atmospheric Dispersion Estimates: An Introduction to Dispersion Modeling.* 2nd ed. Boca Raton, Florida: Lewis, 1994.
14. de Nevers, N. *Air Pollution Control Engineering.* 2nd ed. New York: McGraw-Hill, 2000.
15. Harnby, N., M. F. Edwards, A. W. Nienow. *Mixing in the Process Industries.* London: Butterworths, 1985.
16. Smith, S. S., R. K. Schulze. "Interfacial Mixing Characteristics of Products in Product Pipe Lines." *Petroleum Engineering 19(3),* 94–103 (Sept. 1948); *20(1),* 330–337 (Oct. 1948).
17. Rushton, J. H. "The Axial Velocity of a Submerged Axially Symmetrical Fluid Jet." *AIChEJ 26* (1980), pp. 1038–1041.
18. Guiraud, P., J. Bertrand, J. Costes. "Laser Measurements of Local Velocity and Concentration in a Turbulent Jet-Stirred Tubular Reactor." *Chem. Eng. Sci. 46* (1991), pp. 1289–1297.
19. Briggs, G. A. *Plume Rise.* Washington D.C.: U.S. Atomic Energy Commission, 1969.

CHAPTER

20

COMPUTATIONAL FLUID DYNAMICS (CFD)

In Chap. 15 we saw that even though we can write the three-dimensional material and momentum balances in a general way, applicable to any fluid flow, we can solve them in closed form (i.e., a set of equations describing the velocities at every point in the flow) only for laminar flows in very simple geometries. With the very restrictive perfect fluid assumptions (Chap. 16) we could solve more complex flows in closed form. But for most flows of practical interest, e.g., the furnace in Fig. 1.15, we do not know how to find such closed-form solutions. Before there were large digital computers, it made no sense to try to find such detailed solutions. With digital computers we can now find solutions to such problems, not in closed algebraic form, but as numerical output in the form of tables or plots. Commercially available computer packages do these calculations, often with very user-friendly input interfaces, so that ordinary chemical engineers can use them without a detailed knowledge of what is going on inside them. This chapter shows what is going on in those computer programs, in simple form, to help such users see how what these packages do relates to the topics covered in this book.

20.1 REPLACING DIFFERENTIAL EQUATIONS WITH DIFFERENCE EQUATIONS

All CFD programs replace the basic differential equations (mass, momentum, sometimes energy, sometimes chemical species balances) with algebraic difference equations, which they then solve numerically. This process (called *discretization*) is illustrated in

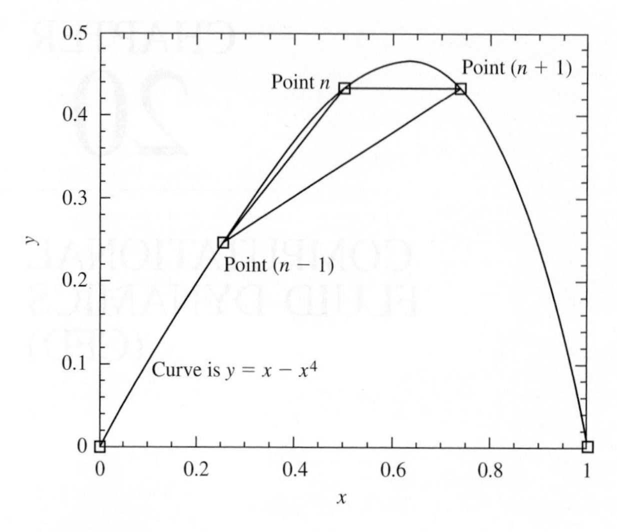

FIGURE 20.1
Geometric representation of forward-backward- and central-difference approximations of (dy/dx) at $x = 0.5$ for simple algebraic expression, $y = x - x^4$. See Example 20.1. The points are spaced at $\Delta x = 0.25$; the example also considers $\Delta x = 0.1$ and 0.01.

Fig. 20.1 where the curve

$$y = x - x^4 \quad \text{(20.A)}$$

(chosen for its simplicity) is drawn, with five equally spaced points that lie on that curve, and three straight lines discussed below. For this equation we know that

$$\frac{dy}{dx} = 1 - 4x^3 \quad \text{(20.B)}$$

and that

$$\frac{d^2y}{dx^2} = -12x^2 \quad \text{(20.C)}$$

If instead of knowing the curve and the equation on which it is based, we knew only the values of the five points, shown as squares on Fig. 20.1, we could estimate the value of dy/dx and d^2y/dx^2 numerically several ways, as shown below. In these estimates, we estimate the values of these two derivatives at point n. The procedure is obviously transferable to point $(n + 1)$ by simply shifting all subscripts upward by 1, etc. At point n we make three estimates of dy/dx, viz.,

$$\left(\frac{dy}{dx}\right)_{\text{at } n} \approx \left(\frac{\Delta y}{\Delta x}\right)_{\substack{\text{forward}\\ \text{difference}}} = \frac{y_{n+1} - y_n}{x_{n+1} - x_n} \quad \text{(20.1)}$$

$$\left(\frac{dy}{dx}\right)_{\text{at } n} \approx \left(\frac{\Delta y}{\Delta x}\right)_{\substack{\text{backward}\\ \text{difference}}} = \frac{y_n - y_{n-1}}{x_n - x_{n-1}} \quad \text{(20.2)}$$

$$\left(\frac{dy}{dx}\right)_{\text{at } n} \approx \left(\frac{\Delta y}{\Delta x}\right)_{\substack{\text{central}\\ \text{difference}}} = \frac{y_{n+1} - y_{n-1}}{x_{n+1} - x_{n-1}} \quad \text{(20.3)}$$

These three estimates are the slopes of the three straight lines drawn on Fig. 20.1. Similarly, we can estimate d^2y/dx^2 by

$$\left(\frac{d^2y}{dx^2}\right)_{\text{at } n} \approx \frac{\left(\frac{\Delta y}{\Delta x}\right)_{\substack{\text{forward}\\ \text{difference}}} - \left(\frac{\Delta y}{\Delta x}\right)_{\substack{\text{backward}\\ \text{difference}}}}{\Delta x} = \frac{y_{n+1} - 2y_n + y_{n-1}}{(x_{n+1} - x_n)^2} \quad \text{(20.4)}$$

Example 20.1. Compute the values of the three estimates of dy/dx and the one estimate of d^2y/dx^2 from the values of y and x at points $(n - 1)$, n, and $(n + 1)$ shown on Fig. 20.1. The x values are 0.25, 0.5, and 0.75, the y values (computed by Eq. 20.A) are 0.2461, 0.4375, and 0.4336, so that

$$\left(\frac{dy}{dx}\right)_{\text{at } n} \approx \left(\frac{\Delta y}{\Delta x}\right)_{\substack{\text{forward}\\ \text{difference}}} = \frac{0.4336 - 0.4375}{0.75 - 0.50} = -0.0156 \tag{20.D}$$

$$\left(\frac{dy}{dx}\right)_{\text{at } n} \approx \left(\frac{\Delta y}{\Delta x}\right)_{\substack{\text{backward}\\ \text{difference}}} = \frac{0.4375 - 0.2461}{0.5 - 0.25} = 0.766 \tag{20.E}$$

$$\left(\frac{dy}{dx}\right)_{\text{at } n} \approx \left(\frac{\Delta y}{\Delta x}\right)_{\substack{\text{central}\\ \text{difference}}} = \frac{\left(\frac{\Delta y}{\Delta x}\right)_{\substack{\text{forward}\\ \text{difference}}} + \left(\frac{\Delta y}{\Delta x}\right)_{\substack{\text{backward}\\ \text{difference}}}}{2}$$

$$= \frac{-0.0156 + 0.766}{2} = 0.375 \tag{20.F}$$

and

$$\left(\frac{d^2y}{dx^2}\right)_{\text{at } n} \approx \frac{\left(\frac{\Delta y}{\Delta x}\right)_{\substack{\text{forward}\\ \text{difference}}} - \left(\frac{\Delta y}{\Delta x}\right)_{\substack{\text{backward}\\ \text{difference}}}}{\Delta x} = \frac{-0.0156 - 0.766}{0.25}$$

$$= -3.125 \tag{20.G}$$

■

None of these are very good estimates of the true values, which we can compute from Eqs. 20.B and 20.C. If we repeat this exercise, spacing the points at $x = 0.1, 0.2, \ldots$, then Fig. 20.1 will be difficult to read, but the numerical results will agree with the true values better, and for $x = 0.01, 0.02, 0.03, \ldots$, they agree very well, as shown in Table 20.1.

TABLE 20.1
Comparison of numerical estimates of first and second derivatives in Example 20.1

	$(dy/dx)_{\text{at } x=0.5}$	$(d^2y/dx^2)_{\text{at } x=0.5}$
True value (Eq. 20.A)	0.50	−3.00
Numerical values for $x = 0, 0.25, 0.50, \ldots$		−3.135
Forward	−0.0156	
Backward	0.765	
Central	0.375	
Numerical values for $x = 0, 0.1, 0.2, 0.3, \ldots$		−3.00
Forward	0.329	
Backward	0.631	
Central	0.480	
Numerical values for $x = 0, 0.01, 0.02, 0.03, \ldots$		−3.00
Forward	0.485	
Backward	0.515	
Central	0.4998	

From this exercise we see the following:

1. The numerical difference formulae are simple, simple to program in computers, and easily visualized as the slopes of lines (and their combinations) on Fig. 20.1.
2. Of the three equations for the first derivative, the central difference gives a better estimate of the correct value at all three values of Δx. The mathematics of this is shown by Anderson [1, p. 132].
3. The accuracy of the approximation increases as Δx decreases. Small intervals mean long computing time. The selection of the proper value of Δx requires a tradeoff between accuracy and computing time.
4. As Δx decreases, the three first-derivative estimates become practically the same. The central difference is always the most accurate, but the other two are widely used in CFD, because they are simpler, and because they work at a solid boundary.
5. This was a simple, one-dimensional example. The examples of technical interest are 2-, 3-, or 4-dimensional (3 space and 1 time).
6. Here we compared our numerical procedure to a known solution. All CFD programs are tested on problems with known analytical solutions. If they can solve those properly we have some confidence they can solve problems for which no analytical solution is known. Similarly most textbook examples, like this one, compare numerical results to known analytical solutions.
7. This replacement of derivatives by divided differences changes a differential equation into a set of algebraic equations. This leads to a lot of bookkeeping, but computers are good at bookkeeping, and we know how to solve large systems of algebraic equations on computers.

20.2 GRIDS

To apply these difference equations to multidimensional problems, we divide the space of interest into many points, located on some kind of a mathematical grid. The simplest possible is the rectangular grid shown in Fig. 20.2 which divides the x-y plane into intervals of Δx and Δy. In two- and three-dimensional fluid mechanics we are always looking for partial derivatives. If we assume that there is a known value of z at every intersection (called a *node*) on Fig 20.2, then we can use the central difference formula to write

$$\left(\frac{\partial z}{\partial x}\right)_{\text{at } n} \approx \left(\frac{\Delta z}{\Delta x}\right)_{\substack{\text{central}\\\text{difference}\\\text{constant } y}} = \frac{z_{(i+1,j)} - z_{(i-1,j)}}{x_{(i+1,j)} - x_{(i-1,j)}} \tag{20.5}$$

and

$$\left(\frac{\partial z}{\partial y}\right)_{\text{at } n} \approx \left(\frac{\Delta z}{\Delta y}\right)_{\substack{\text{central}\\\text{difference}\\\text{constant } x}} = \frac{z_{(i,j+1)} - z_{(i,j-1)}}{y_{(i,j+1)} - y_{(i,j-1)}} \tag{20.6}$$

There is no reason we cannot extend this type of grid to three dimensions (three subscripts instead of two) or three dimensions and time (four subscripts).

The simple rectangular grid and the equations associated with it are the easiest to use; they are used wherever practical. However, we often want the flow to

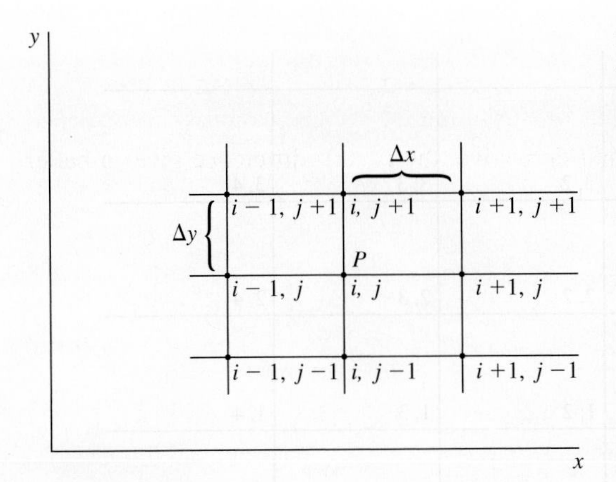

FIGURE 20.2
A simple rectangular grid, centered on point *P*, where the subscripts are *i*, *j*. The *i* subscript increases in the *x* direction, the *j* subscript increases in the *y* direction.

follow the shape of some nonrectangular body, so we must make one of the grid lines in Fig. 20.2 match the shape of that boundary. To capture the detail of the flow in the boundary layer near a solid surface we normally want a much closer grid spacing near such a surface than far away from it. The CFD computer packages normally have a grid-generation routine that will select a suitable grid for your problem; advanced users make up their own grids. Cylindrical or spherical coordinate grids are used for problems with cylindrical or spherical symmetry.

20.3 CFD EQUATIONS

Using the grids and difference equations, we can transform the working differential equations into algebraic form. The example chosen to illustrate this is the unsteady-state one-dimensional laminar flow equation,

$$\frac{\partial V_x}{\partial t} = \nu \frac{\partial^2 V_x}{\partial y^2} \tag{17.5}$$

which we saw in Chap. 17. If we rename the variables, replacing V_x by T and ν (the kinematic viscosity) by α (the thermal diffusivity), then this becomes the one-dimensional unsteady-state heat flow equation, and if we use c (the concentration) and $\mathscr{D}$, the molecular diffusivity or the eddy diffusivity, then it becomes the one-dimensional unsteady-state diffusion equation used in Chap. 19. This is a favorite example equation in CFD, heat-transfer, and mass-transfer texts, because it is simple and useful.

Example 20.2. We wish to represent Eq. 17.5 on *t*-*y* coordinates, as shown in Fig. 20.3. There we see that each node (intersection of two lines) is described by two indices, *i* and *j*, and that the intervals between lines are constant, equal to Δt and Δy. (Other choices are used in CFD, but these are the simplest.) Show the data difference (discretized) algebraic equivalent of Eq. 17.5 for this set coordinate grid.

First we drop the x subscripts because Eq. 17.5 is only about the x component of the velocity. Then we observe that $V_{i,j}$ represents the value of the x component of the velocity at time $= i\,\Delta t$ and distance in the y direction $= j\,\Delta y$. Then we replace the two derivatives in Eq. 17.5 with their approximate values from Eqs. 20.1 and 20.4 to get

$$\frac{V_{i,j} - V_{i-1,j}}{\Delta t} = \nu \frac{V_{i-1,j-1} - 2V_{i-1,j} + V_{i-1,j+1}}{(\Delta y)^2} \tag{20.7}$$

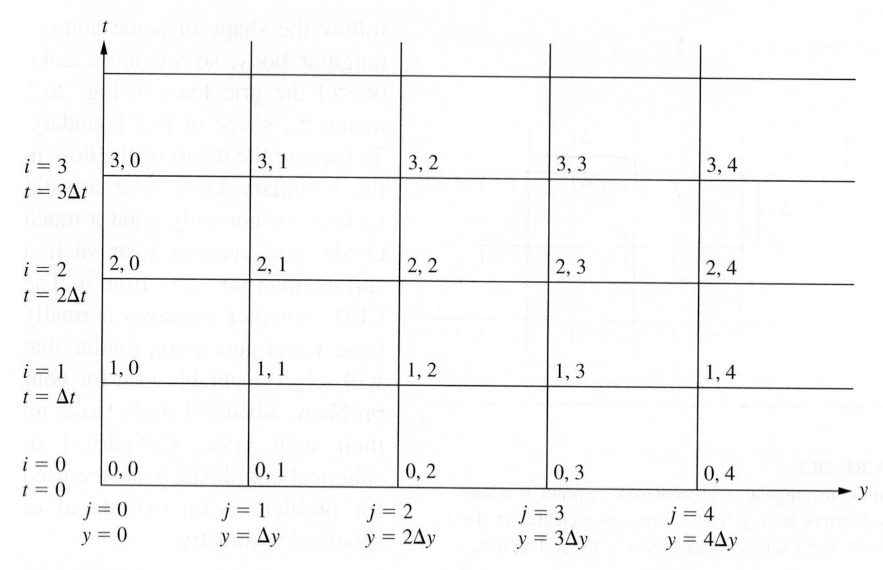

FIGURE 20.3
t-y grid for Example 20.2. The nodes (intersections of lines) are spaced Δt apart in the vertical direction and Δy apart in the horizontal direction. The subscripts increase with distance as shown in Fig. 20.2.

which we can rearrange to

$$V_{i,j} = V_{i-1,j} + \left[\frac{\nu\,\Delta t}{(\Delta y)^2}\right] \cdot (V_{i-1,j-1} - 2V_{i-1,j} + V_{i-1,j+1}) \tag{20.8}$$

There is one such algebraic equation for every node (every combination of i and j). ■

This equation has the property that we can compute the value of $V_{i,j}$ at any point, one at a time, based on the values of V at the $(i-1)$ time and $[\nu\,\Delta t/(\Delta y)^2]$, which is a constant for any given kinematic viscosity and any regular rectangular grid arrangement. For any formulation of this type, called an *explicit formulation,* we can simply march from one time step to the next, solving for all the y points at each time step, based on the values from the end of the preceding time step.

Example 20.3. Solve for V_x as a function of t and y, using Eq. 20.8, for the problem in which, at $t = 0$, the velocity is zero everywhere, and for $t \geq 0$, V_x at $y = 0$ is 5 ft / s, and the fluid is water. This corresponds physically to the boundary layer near the side of a ship that starts suddenly. The heat transfer analog is in all heat transfer books, and the mass transfer analog is Example 19.5.

Here we arbitrarily choose time steps of 2 s, and distance steps as 0.01 ft. The results for the first several time and distance steps are shown in Table 20.2. We see that the column for $y = 0$ ft has value 5 for $t \geq 0$. The row for $t = 0$

TABLE 20.2
Numerical results of Example 20.2*

	y, ft →						
t, s, ↓	**0**	**0.01**	**0.02**	**0.03**	**0.04**	**0.05**	**0.06**
0	5	0	0	0	0	0	0
2	5	1.077	0.000	0.000	0.000	0.000	0.000
4	5	1.690	0.232	0.000	0.000	0.000	0.000
6	5	2.089	0.496	0.050	0.000	0.000	0.000
8	5	2.373	0.743	0.135	0.011	0.000	0.000
10	5	2.588	0.963	0.239	0.035	0.002	0.000
12	5	2.757	1.157	0.351	0.072	0.009	0.000
14	5	2.896	1.328	0.465	0.119	0.021	0.002
16	5	3.011	1.480	0.576	0.172	0.038	0.006
18	5	3.110	1.615	0.684	0.230	0.060	0.011
20	5	3.195	1.737	0.787	0.291	0.086	0.019
Analytical solution		3.150	1.676	0.742	0.270	0.080	0.019

*The first column gives time value(s), the second row the position values (ft). The remainder of the table gives the values of V_x, in ft / s, at that time and location. The last row gives the analytical solution for V_x at $t = 20$ s, corresponding to the numerical solution in the row above.

is all zeros except at $y = 0$. For the chosen set of values,

$$\left[\frac{\nu\,\Delta t}{(\Delta y)^2}\right] = \frac{(1.077 \cdot 10^{-5}\ \mathrm{ft^2/s}) \cdot 2\ \mathrm{s}}{(0.01\ \mathrm{ft})^2} = 0.2154 \tag{20.H}$$

Then, the V value at $i = 1, j = 1$ ($t = 2$ s, $y = 0.01$ ft) is

$$V_{1,1} = 0 + (0.2154) \cdot [(5\ \mathrm{ft/s}) - 2 \cdot 0 + 0] = 1.077\ \mathrm{ft/s} \tag{20.I}$$

This value is shown at 2 s and 0.01 ft in Table 20.2. Repeating Eq. 20.I for the rest of that row shows all zeros. For the next row ($t = 4$ s), the value at 0.01 ft ($i = 2, j = 1$) is

$$V_{2,1} = 1.077 + (0.2154) \cdot [(5\ \mathrm{ft/s}) - 2 \cdot 1.077 + 0] = 1.690\ \mathrm{ft/s} \tag{20.J}$$

and for 0.02 ft

$$V_{2,2} = 0 + (0.2154) \cdot [(1.077\ \mathrm{ft/s}) - 2 \cdot 0 + 0] = 0.232\ \mathrm{ft/s} \tag{20.K}$$

Once this is set up on a spreadsheet, the remaining values in the table are produced in seconds. ■

From this example we see:

1. The procedure is simple. We can start at $t = 0$ and solve for one Δt and then the next, one point at a time (one equation at a time).
2. Table 20.2 shows values rounded to three figures. To minimize round-off error the actual computations carry as many significant figures as the computer allows.

3. This is a favorite problem for CFD examples because its analytical solution is known (if we include the additional assumption that y extends to infinity). That solution (see Eq. 19.8, with variables renamed) is

$$\frac{V_x - V_{x@y=0}}{V_{x@t=0} - V_{x@y=0}} = \text{erf}\left(\frac{x}{2\sqrt{\nu t}}\right) \tag{20.9}$$

or

$$V_x = V_{x,\,y=0} + (V_{x,\,t=0} - V_{x,\,y=0}) \cdot \text{erf}\left(\frac{x}{2\sqrt{\nu t}}\right)$$
$$= 5\frac{\text{ft}}{\text{s}}\left[1 - \text{erf}\left(\frac{x}{2\sqrt{\nu t}}\right)\right] \tag{20.L}$$

The values of this analytical solution for $t = 20$ s are shown as the last row of Table 20.2.

4. Comparison of the last two rows of Table 20.2 reminds us that the numerical solution is an approximation, not an exact solution. In this case the agreement is good; none of the numerical solution values differs from the analytical solution value by more than 0.06 ft / s.

5. This example (and most CFD examples in textbooks) is for one partial differential equation. In real CFD problems there is almost always more than one such equation. Recall Prandtl's boundary layer equations (Chap. 17), which require the simultaneous solution of the mass and momentum equations. In this example we do not bother with the mass equation, Eq. 15.8, because all the terms in it are zero. If all the streamlines are straight and parallel, as they are in this problem, then all the terms in Eq. 15.8 are zero. That is rarely the case in problems of economic interest, so CFD programs are almost always solving more than one partial differential equation simultaneously.

20.4 STABILITY

The previous example also allows us to explore the issue of numerical stability. The process of replacing a derivative by its numerical approximation introduces an error, and round-off in computers introduces an error. If these errors decrease in size from step to step, then the solution is stable. If they increase from step to step, then the solution is unstable. It is proven in CFD texts, e.g., Anderson [1, p. 161], that a solution of this type is stable only if

$$\frac{\nu\,\Delta t}{(\Delta x)^2} \leq \frac{1}{2} \tag{20.10}$$

In Example 20.3 (see Eq. 20.H), this quantity was 0.2154, so stability was not a problem. We may illustrate this difficulty by repeating Example 20.3, taking $\Delta t = 10$ s, which makes this ratio = 1.08. Calculating the first point, we find $V = 5.385$, which is an impossible result (faster than the wall is moving)! The next few entries at $y = 0.01$ ft are -0.83, 12.59 and -17.31 ft / s. The oscillation grows with time, and occurs at all values of y, becoming weaker as y increases. (See Problem 20.7).

If the real problem allows choices of Δt and Δx that satisfy Eq. 20.10, then there is no stability difficulty. For many problems it is a difficulty, and a different approach is needed. In Examples 20.2 and 20.3, we computed the value of $(V_{i,j} - V_{i-1,j}) / \Delta t$ based entirely on the values at $(i - 1)$. That is one of the causes of the instability. If instead we use the values halfway from $(i - 1)$ to i, then Eq. 20.7 becomes

$$\frac{V_{i,j} - V_{i-1,j}}{\Delta t} = \nu \frac{0.5(V_{i-1,j-1} + V_{i,j-1}) - 2[0.5(V_{i-1,j} + V_{i,j})] + 0.5(V_{i-1,j+1} + V_{i,j+1})}{(\Delta y)^2} \tag{20.11}$$

This solves the stability problem, but makes the solution more difficult, because we cannot simply march forward solving one equation at a time as we did in the previous solution. With Eq. 20.11 we must solve the equivalent of Eq. 20.11 for all the points in the problem simultaneously. This is called an *implicit* solution. There will be a set of algebraic equations, one for every point in (i, j) space, to be solved *simultaneously*. These equations are expressed as matrices. They are much too large to solve by hand, or with spreadsheets. CFD programs use matrix-solver routines on them. For some problems this approach is faster and more satisfactory than the explicit approach in Examples 20.2 and 20.3.

20.5 CFD APPLICATIONS

The various ways CFD are used, in order of increasing complexity, are

1. Material balance only. If one starts with the two-dimensional stream function (Chap. 16) in some arbitrary space, one can use CFD to map the flow, and then use that flow map to find other properties [2, p. 541; 3, p. 566]. This approach is used to model the flow around complex structures such as airplanes.
2. Material and momentum balances, laminar flow. Equations 15.8 and 15.27 are solved simultaneously, for arbitrary geometries. These solutions have all the assumptions of the Navier-Stokes equations. Many CFD packages offer this option.
3. Same as item 2, with turbulence. As discussed in Chap. 18, the kinematic viscosity in the Navier-Stokes equations is replaced by an eddy kinematic viscosity, normally based on the k-ε method. This option is offered in most CFD packages.
4. Same as item 3, but the three-dimensional energy balance is solved simultaneously. This addition accounts for changes in physical properties with temperature changes. It is used in aircraft simulations where, as shown in Chap. 8, there are major changes in temperature in adiabatic flows. It is used by chemical engineers in heat transfer problems.
5. Same as item 4, but with chemical reaction, e.g., combustion. This is the version that would be used to model the behavior of the furnace in Fig. 1.15 This type of program is widely used, aiming to improve efficiency of furnaces, product quality, and yield of the materials processed, and to reduce pollutant formation.
6. Even more complex situations, e.g., two-phase flows, flows with drops or particles, flows with freezing or melting, non-Newtonian flows, and porous medium flows [4].

20.6 SUMMARY

1. The basic mathematics of CFD was mostly developed before the advent of fast computers. Using it, one could solve some simple problems by hand, but applications to most problems of engineering interest only became possible when fast computers were developed.
2. CFD solves the basic fluid mechanics equations numerically.
3. This is done by dividing the space (two- or three-dimensional) of the solution into small intervals on some kind of grid and replacing derivatives with divided differences.
4. The result is a set of algebraic equations. In explicit formulations, these can be solved one at a time. In implicit formulations they must all be solved simultaneously.
5. The combination of CFD with heat transfer and chemical reactions, e.g., combustion equations, provides chemical engineers with powerful tools for analyzing problems of great practical interest.
6. This chapter shows only the simplest ideas of CFD. Many more details are presented in Anderson [1].

PROBLEMS

See the Common Values for Problems and Examples inside the back cover. An asterisk (*) on a problem number indicates that its answer is in App. D.

20.1. Show the calculations at all three x intervals ($\Delta x = 0.25, 0.1, 0.01$) for Example 20.1.

20.2. Repeat Example 20.1 for $x = 0.75$.

20.3. Repeat Example 20.1 for $y = \sin x$ at $x = 40°$, using intervals of $\Delta x = 10°$ and $1°$. Compare the numerical derivatives to the analytical derivatives. *Hint:* Spreadsheets normally evaluate sines of angles expressed in radians.

20.4. Show the two second derivatives corresponding to Eqs. 20.5 and 20.6.

20.5. In Example 20.3,

(*a*) Set up the spreadsheet to make the calculations.

(*b*) Rerun Example 20.3 to show that you get the same answers as are presented in Table 20.3.

(*c*) Repeat the calculation out to 20 s, using $\Delta t = 1$ s. Compare the computed values with those in Example 20.3.

20.6. Show the calculation of the analytical values at 20 s in Example 20.3. See Fig. 19.5.

20.7. Repeat Example 20.3 using $\Delta t = 10$ s, out to 100 s, showing the overshoot and oscillation in the values at $y = 0.01$ ft.

REFERENCES FOR CHAPTER 20

1. Anderson, J. D., Jr. *Computational Fluid Dynamics; The Basics with Applications.* New York: McGraw-Hill, 1995.
2. White, F. M. *Fluid Mechanics.* 4th ed. New York: McGraw-Hill, 1999.
3. Wilkes, J. O. *Fluid Mechanics for Chemical Engineers.* Upper Saddle River, NJ: Prentice-Hall, 1999, Sec. 10.4.
4. *Fluent 4.5, Getting Started.* Lebanon, NH: Fluent Inc., 2000.

APPENDIX A TABLES AND CHARTS OF FLUID PROPERTIES, PIPE DIMENSION AND FLOWS, AND HIGH-VELOCITY GAS FLOWS

A.1 VISCOSITIES OF VARIOUS FLUIDS AT 1 ATM PRESSURE

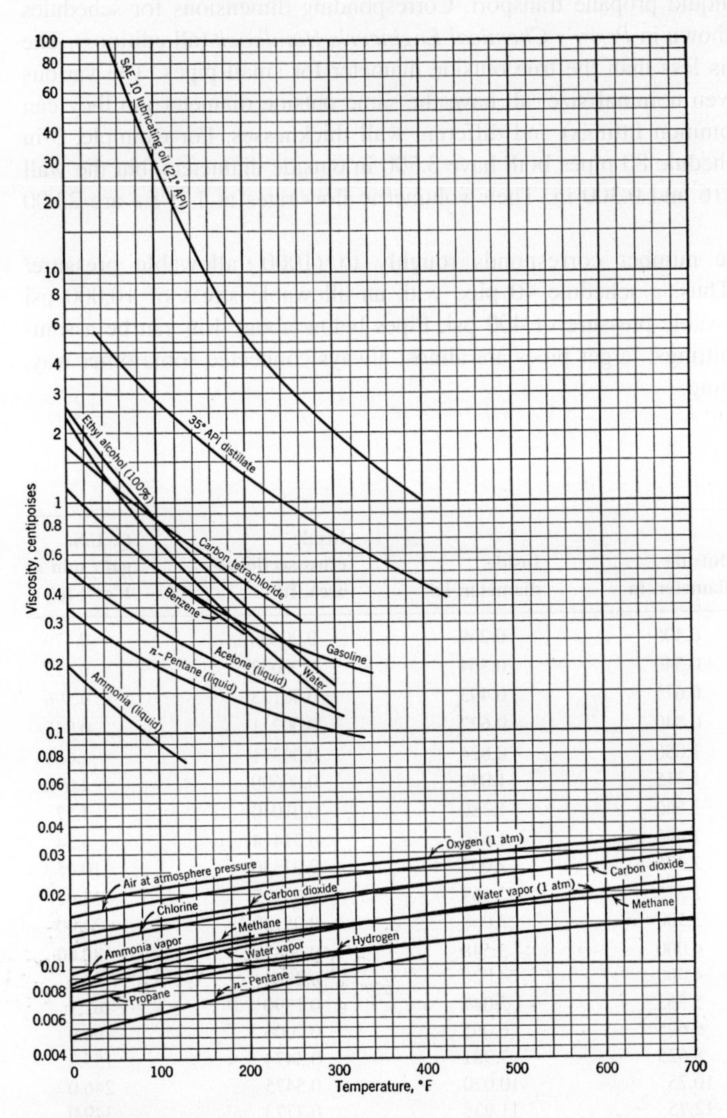

FIGURE A.1
Viscosity as a function of temperature for a variety of gases and liquids. Changes in pressure have little effect on the viscosity, both for gases and liquids, except near the critical point. (From G. G. Brown et al., *Unit Operations,* New York: Wiley (1951), p. 586. Reproduced by permission of the publisher.)

A.2 DIMENSIONS OF U.S. SCHEDULE 40 STEEL PIPE

Schedule 40 is the common pipe in the United States; in your local hardware store the steel pipes for sale are schedule 40. (These are called "galvanized iron" and "black iron" in the hardware store but are actually steel, with or without a rust-preventing zinc coating.) The next thickest wall pipe is Schedule 80, which is required in many applications, e.g., liquid propane transport. Corresponding dimensions for schedules from 5 to 160 are shown in *Perry's Chemical Engineer's Handbook* (all editions). The nominal pipe size is less than the true outside diameter for small pipes. The various schedules, for a given nominal size, all have the same outside diameter (so they can be connected to common fittings) and different wall thicknesses. For example, 3-in schedule 40 and schedule 80 pipes both have 3.500 in outside diameters, but the wall thicknesses are 0.216 and 0.300 in. Their volumetric flow rates at 1 ft / s are 23.00 and 20.55 gpm.

The schedule number corresponds roughly to (1000 · allowable pressure/allowable stress). Thus, a schedule 40 pipe with an allowable stress of 10,000 psi would have an allowable pressure of 400 psi. Pipes below about 4-in can be assembled with screwed fittings; larger pipes are almost always connected some other way, most often by welding.

Nominal pipe size, in	Outside diameter, in	Inside diameter, in	Inside cross-sectional area, ft^2	Q, U.S. gal / min at $V = 1$ ft / s
$\frac{1}{8}$	0.405	0.269	0.00040	0.179
$\frac{1}{4}$	0.540	0.364	0.00072	0.323
$\frac{3}{8}$	0.675	0.493	0.00133	0.596
$\frac{1}{2}$	0.840	0.622	0.00211	0.945
$\frac{3}{4}$	1.050	0.824	0.00371	1.665
1	1.315	1.049	0.00600	2.690
$1\frac{1}{4}$	1.660	1.380	0.01040	4.57
$1\frac{1}{2}$	1.990	1.610	0.01414	6.34
2	2.375	2.067	0.02330	10.45
$2\frac{1}{2}$	2.875	2.469	0.03322	14.92
3	3.500	3.068	0.05130	23.00
$3\frac{1}{2}$	4.000	3.548	0.0687	30.80
4	4.500	4.026	0.0884	39.6
5	5.563	5.047	0.1390	62.3
6	6.625	6.065	0.2006	90.0
8	8.625	7.981	0.3474	155.7
10	10.75	10.020	0.5475	246.0
12	12.75	11.938	0.7773	349.0
14	14.0	13.126	0.9397	422.0
16	16.0	15.00	1.2272	550.0
18	18.0	16.876	1.5533	697.0
20	20.0	18.814	1.9305	866.0

A.3 FLOW OF WATER THROUGH SCHEDULE 40 STEEL PIPE*

Discharge		Pressure drop per 100 ft and velocity in schedule 40 pipe for water at 60°F															
gal / min	ft^3 / s	Veloc-ity, ft / s	Press. drop, lb / in^2	Veloc-ity, ft / s	Press. drop, lb / in^2	Veloc-ity, ft / s	Press. drop, lb / in^2	Veloc-ity, ft / s	Press. drop, lb / in^2	Veloc-ity, ft / s	Press. drop, lb / in^2	Veloc-ity, ft / s	Press. drop, lb / in^2	Veloc-ity, ft / s	Press. drop, lb / in^2	Veloc-ity, ft / s	Press. drop, lb / in^2
		$\frac{1}{8}$ in		$\frac{1}{4}$ in		$\frac{3}{8}$ in		$\frac{1}{2}$ in									
0.2	0.000446	1.13	1.86	0.616	0.359												
0.3	0.000668	1.69	4.22	0.924	0.903	0.504	0.159	0.317	0.061								
0.4	0.000891	2.26	6.98	1.23	1.61	0.672	0.345	0.422	0.086								
										$\frac{3}{4}$ in							
0.5	0.00111	2.82	10.5	1.54	2.39	0.840	0.539	0.528	0.167	0.301	0.033						
0.6	0.00134	3.39	14.7	1.85	3.29	1.01	0.751	0.633	0.240	0.361	0.041						
0.8	0.00178	4.52	25.0	2.46	5.44	1.34	1.25	0.844	0.408	0.481	0.102						
												1 in		$1\frac{1}{4}$ in		$1\frac{1}{2}$ in	
1	0.00223	5.65	37.2	3.08	8.28	1.68	1.85	1.06	0.600	0.602	0.155	0.371	0.048	0.429	0.044		
2	0.00446	11.29	134.4	6.16	30.1	3.36	6.58	2.11	2.10	1.20	0.526	0.743	0.164	0.644	0.090		
3	0.00668			9.25	64.1	5.04	13.9	3.17	4.33	1.81	1.09	1.114	0.336	0.858	0.150	0.473	0.043
		2 in															
4	0.00891			12.33	111.2	6.72	23.9	4.22	7.42	2.41	1.83	1.49	0.565	1.073	0.223	0.630	0.071
5	0.01114					8.40	36.7	5.28	11.2	3.01	2.75	1.86	0.835			0.788	0.104
				$2\frac{1}{2}$ in													
6	0.01337	0.574	0.044			10.08	51.9	6.33	15.8	3.61	3.84	2.23	1.17	1.29	0.309	0.946	0.145
8	0.01782	0.765	0.073			13.44	91.1	8.45	27.7	4.81	6.60	2.97	1.99	1.72	0.518	1.26	0.241
10	0.02228	0.956	0.108	0.670	0.046			10.56	42.4	6.02	9.99	3.71	2.99	2.15	0.774	1.58	0.361
						3 in											
15	0.03342	1.43	0.224	1.01	0.094					9.03	21.6	5.57	6.36	3.22	1.63	2.37	0.755
20	0.04456	1.91	0.375	1.34	0.158	0.868	0.056			12.03	37.8	7.43	10.9	4.29	2.78	3.16	1.28
								$3\frac{1}{2}$ in		4 in							
25	0.05570	2.39	0.561	1.68	0.234	1.09	0.083	0.812	0.041			9.28	16.7	5.37	4.22	3.94	1.93
30	0.06684	2.87	0.786	2.01	0.327	1.30	0.114	0.974	0.056			11.14	23.8	6.44	5.92	4.73	2.72
35	0.07798	3.35	1.05	2.35	0.436	1.52	0.151	1.14	0.704	0.882	0.041	12.99	32.2	7.51	7.90	5.52	3.64
40	0.08912	3.83	1.35	2.68	0.556	1.74	0.192	1.30	0.095	1.01	0.052	14.85	41.5	8.59	10.24	6.30	4.65
45	0.1003	4.30	1.67	3.02	0.668	1.95	0.239	1.46	0.117	1.13	0.064			9.67	12.80	7.09	5.85
												5 in					
50	0.1114	4.78	2.03	3.35	0.839	2.17	0.288	1.62	0.142	1.26	0.076			10.74	15.66	7.88	7.15
60	0.1337	5.74	2.87	4.02	1.18	2.60	0.406	1.95	0.204	1.51	0.107			12.89	22.2	9.47	10.21
70	0.1560	6.70	3.84	4.69	1.59	3.04	0.540	2.27	0.261	1.76	0.143	1.12	0.047			11.05	13.71
80	0.1782	7.65	4.97	5.36	2.03	3.47	0.687	2.60	0.334	2.02	0.180	1.28	0.060			12.62	17.59
90	0.2005	8.60	6.20	6.03	2.53	3.91	0.861	2.92	0.416	2.27	0.224	1.44	0.074			14.20	22.0
														6 in			
100	0.2228	9.56	7.59	6.70	3.09	4.34	1.05	3.25	0.509	2.52	0.272	1.60	0.099	1.11	0.036	15.78	26.9
125	0.2785	11.97	11.76	8.38	4.71	5.43	1.61	4.06	0.769	3.15	0.415	2.01	0.135	1.39	0.055	19.72	41.4
150	0.3342	14.36	16.70	10.05	6.69	6.51	2.24	4.87	1.08	3.78	0.580	2.41	0.190	1.67	0.077		
175	0.3899	16.75	22.3	11.73	8.97	7.60	3.00	5.68	1.44	4.41	0.774	2.81	0.253	1.94	0.102		
200	0.4456	19.14	28.8	13.42	11.68	8.68	3.87	6.49	1.85	5.04	0.985	3.21	0.323	2.22	0.130		
																8 in	
225	0.5013	...	...	15.09	14.63	9.77	4.83	7.30	2.32	5.67	1.23	3.61	0.401	2.50	0.162	1.44	0.043
250	0.557	...	...	...	...	10.85	5.93	8.12	2.84	6.30	1.46	4.01	0.495	2.78	0.195	1.60	0.051
275	0.6127	...	...	...	...	11.94	7.14	8.93	3.40	6.93	1.79	4.41	0.583	3.05	0.234	1.76	0.061
300	0.6684	...	...	...	...	13.00	8.36	9.74	4.02	7.56	2.11	4.81	0.683	3.33	0.275	1.92	0.072
325	0.7241	...	...	...	...	14.12	9.89	10.53	4.09	8.19	2.47	5.21	0.797	3.61	0.320	2.08	0.083

(continued)

A.3 (*continued*)

Discharge		Pressure drop per 100 ft and velocity in schedule 40 pipe for water at 60°F															
gal / min	ft^3 / s	Veloc-ity, ft / s	Press. drop, lb / in^2	Veloc-ity, ft / s	Press. drop, lb / in^2	Veloc-ity, ft / s	Press. drop, lb / in^2	Veloc-ity, ft / s	Press. drop, lb / in^2	Veloc-ity, ft / s	Press. drop, lb / in^2	Veloc-ity, ft / s	Press. drop, lb / in^2	Veloc-ity, ft / s	Press. drop, lb / in^2	Veloc-ity, ft / s	Press. drop, lb / in^2
350	0.7798			...	...	...	...	11.36	5.41	8.82	2.84	5.62	0.919	3.89	0.367	2.24	0.095
375	0.8355			...	...	...	...	12.17	6.18	9.45	3.25	6.02	1.05	4.16	0.416	2.40	0.108
400	0.8912			...	...	...	...	12.98	7.03	10.08	3.68	6.42	1.19	4.44	0.471	2.56	0.121
425	0.9469			...	...	...	...	13.80	7.89	10.71	4.12	6.82	1.33	4.72	0.529	2.73	0.136
450	1.003			...	...	...	...	14.61	8.80	11.34	4.60	7.22	1.48	5.00	0.590	2.89	0.151
		10 in															
475	1.059	1.93	0.054			...	...	...	...	11.97	5.12	7.62	1.64	5.27	0.653	3.04	0.166
500	1.114	2.03	0.059			...	...	...	...	12.60	5.65	8.02	1.81	5.55	0.720	3.21	0.182
550	1.225	2.24	0.071			...	...	...	...	13.85	6.79	8.82	2.17	6.11	0.861	3.53	0.219
600	1.337	2.44	0.083			...	...	...	...	15.12	8.04	9.63	2.55	6.66	1.02	3.85	0.258
650	1.448	2.64	0.097			...	...	...	...	...	...	10.43	2.98	7.22	1.18	4.17	0.301
				12 in													
700	1.560	2.85	0.112	2.01	0.047			...	...	...	...	11.23	3.43	7.78	1.35	4.49	0.343
750	1.671	3.05	0.127	2.15	0.054			...	...	...	...	12.03	3.92	8.33	1.55	4.81	0.392
800	1.782	3.25	0.143	2.29	0.061			...	...	...	...	12.83	4.43	8.88	1.75	5.13	0.443
						14 in											
850	1.894	3.46	0.160	2.44	0.068	2.02	0.042	...	...	...	...	13.64	5.00	9.44	1.96	5.45	0.497
900	2.005	3.66	0.179	2.58	0.075	2.13	0.047	...	...	...	...	14.44	5.58	9.99	2.18	5.77	0.554
950	2.117	3.86	0.198	2.72	0.083	2.25	0.052			...	...	15.24	6.21	10.55	2.42	6.09	0.613
1,000	2.228	4.07	0.218	2.87	0.091	2.37	0.057			...	...	16.04	6.84	11.10	2.68	6.41	0.675
1,100	2.451	4.48	0.260	3.15	0.110	2.61	0.068			...	...	17.65	8.23	12.22	3.22	7.05	0.807
								16 in									
1,200	2.674	4.88	0.306	3.44	0.128	2.85	0.080	2.18	0.042	...	...	...	...	13.33	3.81	7.70	0.948
1,300	2.896	5.29	0.355	3.73	0.150	3.08	0.093	2.36	0.048	...	...	...	...	14.43	4.45	8.33	1.11
1,400	3.119	5.70	0.409	4.01	0.171	3.32	0.107	2.54	0.055					15.55	5.13	8.98	1.28
1,500	3.342	6.10	0.466	4.30	0.195	3.56	0.122	2.72	0.063					16.66	5.85	9.62	1.46
1,600	3.565	6.51	0.527	4.59	0.219	3.79	0.138	2.90	0.071					17.77	6.61	10.26	1.65
										18 in							
1,800	4.010	7.32	0.663	5.16	0.276	4.27	0.172	3.27	0.088	2.58	0.050			19.99	8.37	11.54	2.08
2,000	4.456	8.14	0.808	5.73	0.339	4.74	0.209	3.63	0.107	2.87	0.060			22.21	10.3	12.82	2.55
2,500	5.570	10.17	1.24	7.17	0.515	5.93	0.321	4.54	0.163	3.59	0.091					16.03	3.94
												20 in					
3,000	6.684	12.20	1.76	8.60	0.731	7.11	0.451	5.45	0.232	4.30	0.129	3.46	0.075			19.24	5.59
3,500	7.798	14.24	2.38	10.03	0.982	8.30	0.607	6.35	0.312	5.02	0.173	4.04	0.101			22.44	7.56
														24 in			
4,000	8.912	16.27	3.08	11.47	1.27	9.48	0.787	7.26	0.401	5.74	0.222	4.62	0.129	3.19	0.052	25.65	9.80
4,500	10.03	18.31	3.87	12.90	1.60	10.67	0.990	8.17	0.503	6.46	0.280	5.20	0.162	3.59	0.065	28.87	12.2
5,000	11.14	20.35	4.71	14.33	1.95	11.85	1.21	9.08	0.617	7.17	0.340	5.77	0.199	3.99	0.079	...	...
6,000	13.37	24.41	6.74	17.20	2.77	14.23	1.71	10.89	0.877	8.61	0.483	6.93	0.280	4.79	0.111	...	...
7,000	15.60	28.49	9.11	20.07	3.74	16.60	2.31	12.71	1.18	10.04	0.652	8.08	0.376	5.59	0.150	...	...
8,000	17.82	...	...	22.93	4.84	18.96	2.99	14.52	1.51	11.47	0.839	9.23	0.488	6.38	0.192	...	...
9,000	20.05	...	...	25.79	6.09	21.34	3.76	16.34	1.90	12.91	1.05	10.39	0.608	7.18	0.242	...	...
10,000	22.28	...	...	28.66	7.46	23.71	4.61	18.15	2.34	14.34	1.28	11.54	0.739	7.98	0.294	...	...
12,000	26.74	...	...	34.40	10.7	28.45	6.59	21.79	3.33	17.21	1.83	13.85	1.06	9.58	0.416	...	...
14,000	31.19	...	...	...	...	33.19	8.89	25.42	4.49	20.08	2.45	16.16	1.43	11.17	0.562	...	...
16,000	35.65	...	...	...	...	...	...	29.05	5.83	22.95	3.18	18.47	1.85	12.77	0.723	...	...
18,000	40.10	...	...	...	...	...	...	32.68	7.31	25.82	4.03	20.77	2.32	14.36	0.907	...	...
20,000	44.56	...	...	...	...	...	...	36.31	9.03	28.69	4.93	23.08	2.86	15.96	1.12	...	...

*From Crane Technical Paper No. 410; reproduced by permission of the Crane Company.

A.4 ISENTROPIC COMPRESSIBLE-FLOW TABLES FOR $k = 1.4$

These tables were made up from Eqs. 8.17, 8.20, 8.21, 8.24, and B.6-12. In examples in the text that call for interpolation in this table, the values shown are found from the original equations by using the program that generated this table.

$\mathcal{M}$	T/T_R	P/P_R	ρ/ρ_R	A/A^*	V/c^*
0.00	1	1	1	∞	0.0000
0.05	0.9995	0.9983	0.9988	11.5914	0.0548
0.10	0.9980	0.9930	0.9950	5.8218	0.1094
0.15	0.9955	0.9844	0.9888	3.9103	0.1639
0.20	0.9921	0.9725	0.9803	2.9635	0.2182
0.25	0.9877	0.9575	0.9694	2.4027	0.2722
0.30	0.9823	0.9395	0.9564	2.0351	0.3257
0.35	0.9761	0.9188	0.9413	1.7780	0.3788
0.40	0.9690	0.8956	0.9243	1.5901	0.4313
0.45	0.9611	0.8703	0.9055	1.4487	0.4833
0.50	0.9524	0.8430	0.8852	1.3398	0.5345
0.55	0.9430	0.8142	0.8634	1.2549	0.5851
0.60	0.9328	0.7840	0.8405	1.1882	0.6348
0.65	0.9221	0.7528	0.8164	1.1356	0.6837
0.70	0.9107	0.7209	0.7916	1.0944	0.7318
0.75	0.8989	0.6886	0.7660	1.0624	0.7789
0.80	0.8865	0.6560	0.7400	1.0382	0.8251
0.85	0.8737	0.6235	0.7136	1.0207	0.8704
0.90	0.8606	0.5913	0.6870	1.0089	0.9146
0.95	0.8471	0.5595	0.6604	1.0021	0.9578
1.00	0.8333	0.5283	0.6339	1.0000	1.0000
1.05	0.8193	0.4979	0.6077	1.0020	1.0411
1.10	0.8052	0.4684	0.5817	1.0079	1.0812
1.15	0.7908	0.4398	0.5562	1.0175	1.1203
1.20	0.7764	0.4124	0.5311	1.0304	1.1583
1.25	0.7619	0.3861	0.5067	1.0468	1.1952
1.30	0.7474	0.3609	0.4829	1.0663	1.2311
1.35	0.7329	0.3370	0.4598	1.0890	1.2660
1.40	0.7184	0.3142	0.4374	1.1149	1.2999
1.45	0.7040	0.2927	0.4158	1.1440	1.3327
1.50	0.6897	0.2724	0.3950	1.1762	1.3646
1.55	0.6754	0.2533	0.3750	1.2116	1.3955
1.60	0.6614	0.2353	0.3557	1.2502	1.4254
1.65	0.6475	0.2184	0.3373	1.2922	1.4544
1.70	0.6337	0.2026	0.3197	1.3376	1.4825
1.75	0.6202	0.1878	0.3029	1.3865	1.5097
1.80	0.6068	0.1740	0.2868	1.4390	1.5360
1.85	0.5936	0.1612	0.2715	1.4952	1.5614
1.90	0.5807	0.1492	0.2570	1.5553	1.5861
1.95	0.5680	0.1381	0.2432	1.6193	1.6099
2.00	0.5556	0.1278	0.2300	1.6875	1.6330
2.05	0.5433	0.1182	0.2176	1.7600	1.6553
2.10	0.5313	0.1094	0.2058	1.8369	1.6769
2.15	0.5196	0.1011	0.1946	1.9185	1.6977
2.20	0.5081	0.0935	0.1841	2.0050	1.7179
2.25	0.4969	0.0865	0.1740	2.0964	1.7374
2.30	0.4859	0.0800	0.1646	2.1931	1.7563
2.35	0.4752	0.0740	0.1556	2.2953	1.7745
2.40	0.4647	0.0684	0.1472	2.4031	1.7922
2.45	0.4544	0.0633	0.1392	2.5168	1.8092
2.50	0.4444	0.0585	0.1317	2.6367	1.8257

A.5 NORMAL SHOCK WAVE TABLES FOR $k = 1.4$

These tables were made up from Eqs. 8.37, 8.38, 8.39, B.6-20, and 8.41. In examples in the text that call for interpolation in this table, the values shown are found from the original equations, using the program that generated this table.

$\mathcal{M}_x$	$\mathcal{M}_y$	T_y / T_x	P_y / P_x	ρ_y / ρ_x	P_{R_y} / P_{R_x}
1.00	1	1	1	1	1
1.05	0.9531	1.0328	1.1196	1.0840	0.9999
1.10	0.9118	1.0649	1.2450	1.1691	0.9989
1.15	0.8750	1.0966	1.3763	1.2550	0.9967
1.20	0.8422	1.1280	1.5133	1.3416	0.9928
1.25	0.8126	1.1594	1.6563	1.4286	0.9871
1.30	0.7860	1.1909	1.8050	1.5157	0.9794
1.35	0.7618	1.2226	1.9596	1.6028	0.9697
1.40	0.7397	1.2547	2.1200	1.6897	0.9582
1.45	0.7196	1.2872	2.2863	1.7761	0.9448
1.50	0.7011	1.3202	2.4583	1.8621	0.9298
1.55	0.6841	1.3538	2.6363	1.9473	0.9132
1.60	0.6684	1.3880	2.8200	2.0317	0.8952
1.65	0.6540	1.4228	3.0096	2.1152	0.8760
1.70	0.6405	1.4583	3.2050	2.1977	0.8557
1.75	0.6281	1.4946	3.4063	2.2791	0.8346
1.80	0.6165	1.5316	3.6133	2.3592	0.8127
1.85	0.6057	1.5693	3.8263	2.4381	0.7902
1.90	0.5956	1.6079	4.0450	2.5157	0.7674
1.95	0.5862	1.6473	4.2696	2.5919	0.7442
2.00	0.5774	1.6875	4.5000	2.6667	0.7209

A.6 FLUID DENSITIES

For gases at low pressures the density is given to satisfactory accuracy by the ideal gas law,

$$\rho = MP / RT \tag{A.1}$$

Values of M for common gases are shown in App. A.7. R is a universal constant whose values in various units are shown inside the back cover of this book. For higher pressures, one must account for departures from the ideal gas law. For many gases there are published data on the behavior of that specific gas, e.g., the "steam tables." For quick estimates the best procedure is to use the compressibility factor z, defined so that

$$\rho = MP / zRT \tag{A.2}$$

Clearly, z is identically 1 for ideal gases. For real gases, it can be shown that z for all gases is given *approximately* by

$$z = z(P_R, T_R) \tag{A.3}$$

where $P_R = P / P_{crit}$ and $T_R = T / T_{crit}$. This relation is shown graphically in App. A.8. This simple relationship is accurate to $\pm$ a few percent for values of z near 1, but may have errors of ± 10 percent for values of z less than about 0.6. If greater accuracy is needed, either find specific data for the gas in question, or see Poling, Prausnitz, and O'Connell [1] for more details on estimating the value of z. Values of T_{crit} and P_{crit} for some common gases are also given in App. A.7.

For any fluid (or solid) the density can be written as a Taylor series:

$$\rho = \rho_0 + \frac{\partial \rho}{\partial T}(T - T_0) + \frac{\partial \rho}{\partial P}(P - P_0) + \frac{\partial^2 \rho}{\partial T^2}\left[\frac{(T - T_0)^2}{2}\right] + \frac{\partial^2 \rho}{\partial P^2}\left[\frac{(P - P_0)^2}{2}\right]$$
$$+ \frac{\partial^2 \rho}{\partial P \, \partial T}(P - P_0)(T - T_0) + \cdots \qquad \text{(A.4)}$$

This equation is correct for solids, liquids, and gases under all conditions, if one uses an *infinite* series of terms. For liquids at temperatures well below the critical temperature (say, 200°F below the critical), and also for solids, one may neglect all but the first three terms on the right and write the equation as

$$\rho \approx \rho_0\left[1 + \frac{1}{\rho_0}\left(\frac{\partial \rho}{\partial T}\right)(T - T_0) + \frac{1}{\rho_0}\left(\frac{\partial \rho}{\partial P}\right)(P - P_0)\right]$$
$$= \rho_0[1 - \alpha(T - T_0) + \beta(P - P_0)] \qquad \text{(A.5)}$$

where

$$\alpha = -\frac{1}{\rho_0}\left(\frac{\partial \rho}{\partial T}\right) = \text{coefficient of thermal expansion} \qquad \text{(A.6)}$$

$$\beta = -\frac{1}{\rho_0}\left(\frac{\partial \rho}{\partial P}\right) = \text{isothermal compressibility} = 1 / \text{bulk modulus} \qquad \text{(A.7)}$$

Values of the density, isothermal compressibility, and coefficient of thermal expansion for some common liquids are listed in App. A.9. Notice that for most liquids the effect of a change in temperature is more significant than the effect of a change in pressure. Normally, a temperature decrease of 1°F will have the same effect on the density as a pressure increase of 100 psi.

Generally, the bulk modulus is practically constant over a wide range of pressures, but the coefficient of thermal expansion increases with increasing temperature. Therefore, the single value given in App. A.9 should not be used for temperature changes above 100°F. For a more complete set of values, which includes the effect of increasing temperature on α, see *Perry's Chemical Engineer's Handbook* (all editions). The behavior of liquids near their critical states can be best estimated from App. A.8.

REFERENCE FOR APPENDIX A.6

1. Poling, B. E., J. M. Prausnitz, J. P. O'Connell. *The Properties of Gases and Liquids.* 5th ed. New York; McGraw-Hill, 2001.

A.7 SOME PROPERTIES OF GASES

Much more extensive tables are in *Perry's Chemical Engineer's Handbook* (all editions) and various other reference books.

Gas	*M*, g / mol or lbm / lbmol	T_{crit}, °R	P_{crit}, atm
Hydrogen	2	59.5	12.8
Helium	4	9.47	2.26
Nitrogen	28	227	35.5
Oxygen	32	278	47.9
Air (≈79% N_2, 21% O_2)	29	≈240	≈37
Water	18	1165	218.5
Carbon monoxide	28	239	34.5
Carbon dioxide	44	548	73.0
Freon 12	121	694	40.7
Methane	16	343.4	45.8
Typical gasoline vapor	≈100	≈1000	≈25

A.8 COMPRESSIBILITY FACTOR

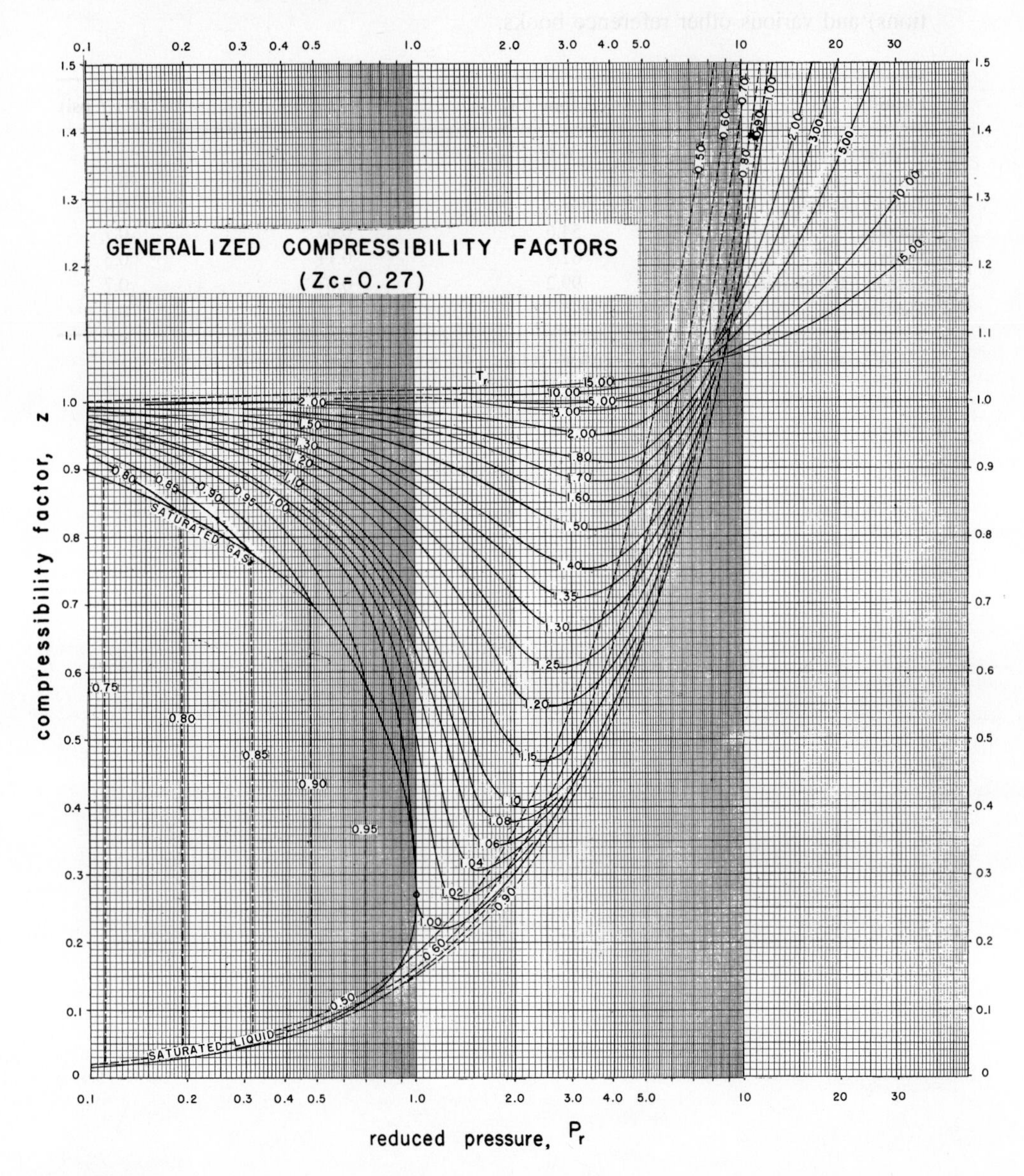

FIGURE A.2
Compressibility factor as a function of reduced temperature and reduced pressure. This plot gives good estimates for fluids with $z_c \approx 0.27$, less accurate estimates for other values of z_c. (From O.A. Hougen et al., *Chemical Process Principles,* 2nd ed., New York: Wiley (1959). Reproduced by permission of the publisher.)

A.9 SOME PROPERTIES OF LIQUIDS

Much more extensive tables are in *Perry's Chemical Engineer's Handbook* (all editions) and various other reference books.

Liquid	Density,* lbm / ft^3	$10^3\ \alpha(1\ /\ °F)$	$10^5\ \beta(1\ /\ psi)$
Hydrogen	4.4	34	11
Helium	9.1	15	48
Typical gasoline	≈44.8	0.7	0.7
Benzene	54.6	0.67	0.7
Water	62.3	0.11	0.3
Carbon tetrachloride	99.2	0.67	0.7
Mercury	845	0.10	0.3

*At 20°C and 1 atm, except H_2 and He, which are at their 1 atm boiling points, 20 and 2.1 K, respectively.

APPENDIX B

DERIVATIONS AND PROOFS

In many places in the text there are proofs and derivations that seem to interrupt the flow of the material. To keep the main text simple and readable, those are placed here, and only referred to in the main text.

B.1 PROOF THAT IN A FLUID AT REST THE PRESSURE IS THE SAME IN ALL DIRECTIONS (SEE SEC. 1.6)

The laws of fluid mechanics work perfectly well in any gravity situation, including the zero gravity of an earth satellite. The following proof is shown for zero gravity, because that makes it simple. The result is then extended for a finite gravity field. The proof rests on the definition of a fluid: "A fluid, when subject to any shear stress, moves."

Consider a prism of fluid, as shown in Fig. B.1. For the fluid to be at rest, there can be no shear forces on any of the surfaces of the prism. Furthermore, because it is at rest, there are no unbalanced forces; i.e., the sum of the forces in any direction is zero. For the sum of the forces in the z direction to be zero, the pressure force on *ABCD* must equal the z component of the pressure force on *BCEF;* or

$$P_{\text{bottom}}\Delta x\,\Delta y = P_{\text{sloping face}}\Delta y(\text{length } BE)\cos\theta \tag{B.1-1}$$

But length *BE* is exactly $\Delta x/(\cos\theta)$, so Eq. B.1-1 reduces to

$$P_{\text{bottom}} = P_{\text{sloping face}} \tag{B.1-2}$$

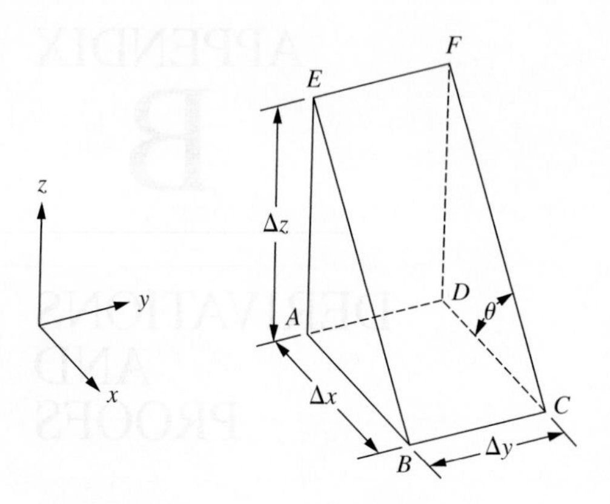

FIGURE B.1
Prism in fluid, for showing force balance.

Equation B.1-2 is true for any angle θ. It shows that the pressure in any direction is the same as the pressure vertically upward and hence that the pressure is the same in all directions.

In the foregoing it is shown that for the zero gravity situation the pressure in an entire body of fluid at rest is the same in all directions. To apply the same reasoning to the prism in the case in which there is significant gravity, one includes in Eq. B.1-1 a term for the force of gravity on the fluid element. One then lets the size of the prism decrease, i.e., $\Delta x \to 0$, and observes that the gravity term is proportional to Δx^3 whereas the pressure terms are proportional to Δx^2; so in the limit (i.e., at a point) the same argument holds as that given above for any value of the acceleration of gravity.

B.2 THE HYDRAULIC JUMP EQUATIONS (SEE SEC. 7.5.3)

We start with the continuity equation,

$$V_1 z_1 = V_2 z_2 \tag{7.AX}$$

and the one-dimensional momentum balance,

$$0 = l\rho z_1 V_1(V_1 - V_2) + \frac{l\rho g}{2}(z_1^2 - z_2^2) \tag{7.BA}$$

We then divide Eq. 7.BA by $l\rho g / 2$ and substitute for V_2 from Eq. 7.AX to find

$$0 = \frac{2V_1^2 z_1}{g}\left(1 - \frac{z_1}{z_2}\right) + (z_1^2 - z_2^2) \tag{B.2-1}$$

We next multiply the far-right term in Eq. B.2-1 by (z_2^2 / z_2^2) and regroup, finding

$$0 = \frac{2V_1^2 z_1}{g}\left(1 - \frac{z_1}{z_2}\right) - z_2^2\left(1 - \frac{z_1^2}{z_2^2}\right) \tag{B.2-2}$$

We factor $(1 - z_1^2 / z_2^2)$ into $(1 + z_1 / z_2)(1 - z_1 / z_2)$ and divide by $(1 - z_1 / z_2)$, finding

$$0 = \frac{2V_1^2 z_1}{g} - z_2^2\left(1 + \frac{z_1}{z_2}\right) \tag{B.2-3}$$

which can be multiplied out to give

$$0 = z_2^2 + z_2 z_1 - \frac{2V_1^2 z_1}{g} \tag{B.2-4}$$

which is a standard quadratic equation with solution

$$z_2 = \frac{-z_1}{2} \pm \sqrt{\left(\frac{z_1}{2}\right)^2 + \frac{2V_1^2 z_1}{g}} \quad (7.52)$$

B.3 THE PROPERTIES OF AN IDEAL GAS

In several derivations in the text, we use the properties of an ideal gas. All of these properties are derived here.

B.3.1 Definitions

An ideal gas is one whose pressure, density, and absolute temperature are related by

$$\frac{P}{\rho} = \frac{RT}{M} \quad (B.3\text{-}1)$$

Here R is the universal gas constant, whose values in various units are shown inside the back cover. The density in this equation is the mass density (mass / volume). If we wish the molar density (moles / volume) we simply drop the M from Eq. B.3-1 and from all the equations in this appendix. There is rarely any question whether the mass density or molar density is the one used in any of the equations in this book.

It is shown in any standard text on thermodynamics that for an ideal gas the enthalpy per unit mass, h, and the internal energy per unit mass, u, are functions of temperature alone; they do not depend on pressure. (The same is *not* true for real gases, or solids or liquids). Thus, for an ideal gas we can define the heat capacity at constant pressure, C_P, and the heat capacity at constant volume, C_V, as follows:

$$C_P = \frac{dh}{dT} \qquad dh = C_P\, dT \quad (B.3\text{-}2)$$

$$C_V = \frac{du}{dT} \qquad du = C_V\, dT \quad (B.3\text{-}3)$$

For any material, the enthalpy per unit mass is defined by $h = u + P/\rho$. Combining these definitions with Eqs. B.3-1, B.3-2, and B.3-3, we find

$$dh = du + d\left(\frac{P}{\rho}\right) \quad (B.3\text{-}4)$$

$$C_P\, dT = C_V\, dT + \frac{R}{M}\, dT \quad (B.3\text{-}5)$$

$$C_P = C_V + \frac{R}{M} \quad (B.3\text{-}6)$$

We now introduce the definition of k, using Eq. B.3-6:

$$k = \frac{C_P}{C_V} = \frac{C_V + R/M}{C_V} = 1 + \frac{R}{MC_V} \quad (B.3\text{-}7)$$

Rearranging produces

$$\frac{R}{MC_V} = k - 1 \tag{B.3-8}$$

or

$$C_V = \frac{R}{M(k-1)} \tag{B.3-9}$$

and

$$C_P = C_V + \frac{R}{M} = \frac{Rk}{M(k-1)} \tag{B.3-10}$$

B.3.2 Isentropic Relations

In any standard text on thermodynamics, one may find the relation among entropy per unit mass, pressure, enthalpy per unit mass, temperature, and density shown below:

$$dh = T\,ds + \frac{1}{\rho}\,dP \tag{B.3-11}$$

(which is true for any pure substance, solid, liquid, or gas; sometimes called *the property equation*). For a constant entropy process. (i.e., an isentropic process), $T\,ds = 0$. We then substitute for dh from Eq. B.3-2 and substitute for C_P from Eq. B.3-10 to find

$$\frac{Rk}{M(k-1)}\,dT_S = \frac{1}{\rho}\,dP_S \tag{B.3-12}$$

The subscript is to remind us that this equation applies only to isentropic processes. Now we replace ρ by its equivalent from Eq. B.3-1:

$$\frac{Rk}{M(k-1)}\,dT_S = \frac{RT}{PM}\,dP_S \tag{B.3-13}$$

Canceling the Rs and Ms and rearranging, we find

$$\frac{k}{k-1}\,\frac{dT_S}{T} = \frac{dP_S}{P} \tag{B.3-14}$$

which is readily integrated to

$$\frac{k}{k-1}\ln T_S\Big]_{T_1}^{T_2} = \ln P_S\Big]_{P_1}^{P_2} \tag{B.3-15}$$

or

$$\left(\frac{T_2}{T_1}\right)_S^{k/(k-1)} = \left(\frac{P_2}{P_1}\right)_S \tag{B.3-16}$$

Substituting $\rho RT/M$ for P_1 and for P_2, and canceling Rs and Ms, we find

$$\left(\frac{T_2}{T_1}\right)_S^{k/(k-1)} = \left(\frac{T_2}{T_1}\right)_S \left(\frac{\rho_2}{\rho_1}\right)_S \tag{B.3-17}$$

or

$$\left(\frac{\rho_2}{\rho_1}\right)_S = \left(\frac{T_2}{T_1}\right)_S^{k/(k-1)} \left(\frac{T_1}{T_2}\right)_S = \left(\frac{T_2}{T_1}\right)_S^{[k/(k-1)]-1} = \left(\frac{T_2}{T_1}\right)_S^{1/(k-1)} \tag{B.3-18}$$

Returning now to Eq. B.3-17, we substitute $PM/R\rho$ for T_1 and for T_2 and cancel, finding

$$\left(\frac{P_2}{P_1}\right)_S = \left(\frac{P_2}{P_1}\cdot\frac{\rho_1}{\rho_2}\right)_S^{k/(k-1)} \tag{B.3-19}$$

Taking both sides to the $(k-1)/k$ power produces

$$\left(\frac{P_2}{P_1}\right)_S^{(k-1)/k} = \left(\frac{P_2}{P_1}\right)_S \left(\frac{\rho_1}{\rho_2}\right)_S \tag{B.3-20}$$

so that

$$\left(\frac{\rho_1}{\rho_2}\right)_S = \left(\frac{P_2}{P_1}\right)_S^{(k-1)/k} \left(\frac{P_1}{P_2}\right)_S = \left(\frac{P_2}{P_1}\right)_S^{-1/k} \tag{B.3-21}$$

This we rearrange to

$$\left(\frac{\rho_1}{\rho_2}\right)_S^{k} = \left(\frac{P_2}{P_1}\right)_S^{-1} = \left(\frac{P_1}{P_2}\right)_S \tag{B.3-22}$$

or

$$\left(\frac{P}{\rho^k}\right)_S = \text{constant} \tag{B.3-23}$$

Now we differentiate this:

$$\frac{dP_S}{\rho^k} + P_S\left(\frac{-k d\rho_S}{\rho^{k+1}}\right) = 0 \tag{B.3-24}$$

or

$$\frac{dP_S}{d\rho_S} = \left(\frac{\partial P}{\partial \rho}\right)_S = \frac{kP_S\rho_S^k}{\rho_S^{k+1}} = \frac{kP_S}{\rho_S} \tag{B.3-25}$$

B.3.3 Entropy Change

Solving Eq. B.3-11 for ds, we find

$$ds = \frac{dh}{T} - \frac{dP}{\rho T} \tag{B.3-26}$$

Substituting from Eqs. B.3-2 and B.3-1 produces

$$ds = C_P \frac{dT}{T} - \frac{R}{M}\left(\frac{dP}{P}\right) \tag{B.3-27}$$

Then, inserting Eq. B.3-10 and factoring, we get

$$ds = \frac{R}{M}\left[\frac{k}{k-1}\left(\frac{dT}{T}\right) - \frac{dP}{P}\right] \tag{B.3-28}$$

which integrates to

$$\frac{M(s_2 - s_1)}{R} = \ln\left(\frac{T_2}{T_1}\right)^{k/(k-1)} - \ln\frac{P_2}{P_1} = \ln\left[\left(\frac{T_2}{T_1}\right)^{k/(k-1)}\left(\frac{P_1}{P_2}\right)\right] \tag{B.3-29}$$

B.4 THE AREA RATIO (EQ. 8.24, SEE SEC. 8.2)

Starting with Eq. 8.23,

$$\frac{A_1}{A^*} = \frac{\rho^* V^*}{\rho_1 V_1} \tag{8.23}$$

we substitute from Eq. B.3-18:

$$\frac{\rho^*}{\rho_1} = \left(\frac{T^*}{T_1}\right)^{1/(k-1)} \tag{B.4-1}$$

and

$$\frac{V^*}{V_1} = \frac{\mathcal{M}^* c^*}{\mathcal{M}_1 c_1} = \frac{1}{\mathcal{M}_1}\left(\frac{c^*}{c_1}\right) \tag{B.4-2}$$

because $\mathcal{M}^* \equiv 1$. But, from Eq. 8.11,

$$\frac{c^*}{c_1} = \left(\frac{T^*}{T_1}\right)^{1/2} \tag{B.4-3}$$

Substituting Eqs. B.4-1, B.4-2, and B.4-3 in Eq. 8.23, we find

$$\frac{A_1}{A^*} = \frac{1}{\mathcal{M}_1}\left(\frac{T^*}{T_1}\right)^{1/2}\left(\frac{T^*}{T_1}\right)^{1/(k-1)} = \frac{1}{\mathcal{M}_1}\left(\frac{T^*}{T_1}\right)^{(k+1)/2(k-1)} \tag{B.4-4}$$

However,

$$\frac{T^*}{T_1} = \frac{T_R/T_1}{T_R/T^*} = \frac{\mathcal{M}_1^2[(k-1)/2] + 1}{1^2[(k-1)/2] + 1} \tag{B.4-5}$$

so that

$$\frac{A}{A^*} = \frac{1}{\mathcal{M}_1}\left\{\left(\mathcal{M}_1^2[(k-1)/2] + 1\right)\Big/\left([(k-1)/2] + 1\right)\right\}^{(k+1)/2(k-1)} \tag{8.24}$$

B.5 HIGH-VELOCITY ADIABATIC FLOW WITH FRICTION

We begin with

$$0 = -\rho V\, dV - dP - \frac{4\tau_{\text{wall}}\, dx}{D} \tag{8.28}$$

and

$$\tau_{\text{wall}} = f\rho \frac{V^2}{2} \tag{8.29}$$

Dividing through by $-AP$ and substituting for τ produces

$$\frac{\rho V\, dV}{P} + \frac{dP}{P} + \frac{\rho V^2}{P} \cdot \frac{2f\, dx}{D} = 0 \tag{B.5-1}$$

The next few steps replace all terms in P, V, and ρ with terms involving the Mach number and constants. First, we rewrite the definition of the Mach number and the speed of sound in an ideal gas as

$$V^2 = \frac{kP}{\rho} \mathcal{M}^2 \tag{B.5-2}$$

or

$$\frac{\rho V^2}{P} = k\mathcal{M}^2 \tag{B.5-3}$$

which we substitute into the third term in Eq. B.5-1. Then we multiply both sides of Eq. B.5-2 by dV / V to find

$$\frac{\rho V}{P} dV = k\mathcal{M}^2 \frac{dV}{V} \tag{B.5-4}$$

We differentiate both sides of Eq. 8.15 and divide left and right sides by

$$V^2 = \frac{kRT}{M} \cdot \mathcal{M}^2 \tag{B.5-5}$$

and simplify, finding

$$\frac{dT}{T} = -\mathcal{M}(k-1) \frac{dV}{V} \tag{B.5-6}$$

Next we take the natural logarithm of both sides of Eq. B.5-5 and differentiate, finding

$$2\frac{dV}{V} = \frac{dT}{T} + 2\frac{d\mathcal{M}}{\mathcal{M}} \tag{B.5-7}$$

Eliminating dT / T between the last two equations and rearranging produces

$$\frac{dV}{V} = \frac{d\mathcal{M} / \mathcal{M}}{\mathcal{M}^2[(k-1)/2] + 1} \tag{B.5-8}$$

Substituting this value of dV/V into Eq. B.5-4 produces

$$\frac{\rho V}{P} dV = \frac{k\mathcal{M}\, d\mathcal{M}}{\mathcal{M}^2[(k-1)/2]+1} \tag{B.5-9}$$

Next we write the mass flow rate per unit area (which must be a constant in this flow) as

$$\frac{\dot{m}}{A} = \rho V = \frac{MP}{RT} V \tag{B.5-10}$$

Taking the natural logarithm of this equation and then differentiating (at constant $\dot{m}/A$) produces

$$\frac{dP}{P} = \frac{dT}{T} - \frac{dV}{V} \tag{B.5-11}$$

Eliminating dT/T with Eq. B.5-6 and dV/V with Eq. B.5-8 produces

$$\frac{dP}{P} = -\frac{(k-1)\mathcal{M}^2+1}{[(k-1)/2]\mathcal{M}^2+1} \cdot \frac{d\mathcal{M}}{\mathcal{M}} \tag{B.5-12}$$

Then returning to Eq. B.5-1, inserting Eqs. B.5-3, B.5-9, and B.5-12 and simplifying, we find

$$\frac{4f\,dx}{D} = \frac{2(1-\mathcal{M}^2)}{[(k-1)/2]\mathcal{M}^2+1} \cdot \frac{d\mathcal{M}}{k\mathcal{M}^3} \tag{B.5-13}$$

which can be expanded by partial fractions to

$$\frac{4f\,dx}{D} = \frac{2d\mathcal{M}}{k\mathcal{M}^3} - \frac{k+1}{k} \cdot \frac{d\mathcal{M}}{\mathcal{M}\{[(k-1)/2]\mathcal{M}^2+1\}} \tag{B.5-14}$$

and integrated from $\mathcal{M} = \mathcal{M}_1$ at $x = 0$ to $\mathcal{M} = \mathcal{M}_2$ at $x = \Delta x$, finding

$$\frac{4f\,\Delta x}{D} - \frac{1}{k}\left(\frac{1}{\mathcal{M}_1^2} - \frac{1}{\mathcal{M}_2^2}\right) + \frac{k+1}{2k} \ln\left\{\frac{\mathcal{M}_2^2}{\mathcal{M}_1^2} \cdot \frac{1+[(k-1)/2]\mathcal{M}_1^2}{1+[(k-1)/2]\mathcal{M}_2^2}\right\} = 0 \tag{8.30}$$

B.6 NORMAL SHOCK WAVES (SEE SEC. 8.5)

In this appendix we derive the equations for a normal shock wave in a ideal gas, starting with the continuity equation,

$$\rho_x V_x = \rho_y V_y \tag{8.36}$$

the momentum balance,

$$V_x - V_y = \frac{P_y}{\rho_y V_y} - \frac{P_x}{\rho_x V_x} \tag{8.37}$$

and the energy balance,

$$\frac{T_R}{T_1} = \mathcal{M}_1^2\left(\frac{k-1}{2}\right) + 1 \tag{8.17}$$

From the ideal gas law and the equation for the speed of sound in a ideal gas,

$$\frac{P}{\rho} = \frac{RT}{M} = \frac{c^2}{k} \tag{B.6-1}$$

Substituting this twice in Eq. 8.37, we find

$$V_x - V_y = \frac{c_y^2}{kV_y} - \frac{c_x^2}{kV_x} \tag{B.6-2}$$

Now we write Eq. 8.17 twice: once for any arbitrary state 1, and once for the critical state (*), at which $\mathcal{M} = 1$. For these states T_R is the same, because the flow is assumed adiabatic, so, dividing one energy balance by the other, we find

$$\frac{T_R / T_1}{T_R / T^*} = \frac{T^*}{T_1} = \frac{\mathcal{M}_1^2[(k-1)/2] + 1}{[(k-1)/2] + 1} \tag{B.6-3}$$

but $T^* / T_1 = c^{*2} / c_1^2$ and $\mathcal{M}_1^2 = V_1^2 / c_1^2$, so

$$\frac{c^{*2}}{c_1^2} = \frac{(V_1^2 / c_1^2)[(k-1)/2] + 1}{[(k-1)/2] + 1} \tag{B.6-4}$$

Multiplying through by c_1^2, simplifying the denominator, and solving for c_1^2 gives

$$c_1^2 = c^{*2}\left(\frac{k+1}{2}\right) - V_1^2\left(\frac{k-1}{2}\right) \tag{B.6-5}$$

Here we have let 1 be an arbitrary state, so we can let it be either x or y. Thus, we will use Eq. B.6-5 twice to eliminate c_x^2 and c_y^2 from Eq. B.6-2:

$$V_x - V_y = \frac{c^{*2}[(k+1)/2] - V_y^2[(k-1)/2]}{kV_y} - \frac{c^{*2}[(k+1)/2] - V_x^2[(k-1)/2]}{kV_x} \tag{B.6-6}$$

Now we multiply both sides by $2kV_x / (k+1)$, finding

$$\frac{2kV_x}{k+1}(V_x - V_y) = \frac{V_x}{V_y}\left[c^{*2} - V_y^2\left(\frac{k-1}{k+1}\right)\right] - c^{*2} + V_x^2\left(\frac{k-1}{k+1}\right) \tag{B.6-7}$$

Regrouping gives us

$$\frac{2kV_x}{k+1}(V_x - V_y) = c^{*2}\frac{V_x - V_y}{V_y} + \frac{k-1}{k+1}V_x(V_x - V_y) \tag{B.6-8}$$

Canceling $(V_x - V_y)$, multiplying by V_y, rearranging, and simplifying, we find

$$V_x V_y = c^{*2} \tag{B.6-9}$$

which is known as the *Prandtl relation* or the *Prandtl-Meyer relation*. Now, returning to Eq. B.6-4, we multiply by $(c_1 / c^*)^2[(k-1)/2+1]$, finding

$$\frac{k-1}{2}+1=\left(\frac{V_1}{c^*}\right)^2\left(\frac{k-1}{2}\right)+\left(\frac{c_1}{c^*}\right)^2 \tag{B.6-10}$$

Substituting for $(c_1 / c^*)^2$ from Eq. B.6-4 and simplifying, we find

$$\frac{k+1}{2}=\left(\frac{V_1}{c^*}\right)^2\left(\frac{k-1}{2}\right)+\frac{(k+1)/2}{\mathcal{M}_1^2(k-1)/2+1} \tag{B.6-11}$$

Multiplying by $(k-1)/2$ and rearranging gives

$$\left(\frac{V_1}{c^*}\right)^2=\frac{k+1}{k-1}\left[1-\frac{1}{\mathcal{M}_1^2(k-1)/2+1}\right]=\frac{\mathcal{M}_1^2(k+1)}{\mathcal{M}_1^2(k-1)+2} \tag{B.6-12}$$

Equation B.6-12 is not specific for shock waves but applies to any steady, adiabatic flow of an ideal gas. The reader may check a few V / c^* values in App. A.4 to see that they have indeed been made up from Eq. B.6-12.

We may now use Eq. B.6-12 twice in Eq. B.6-9, finding

$$\left[\frac{\mathcal{M}_x^2(k+1)}{\mathcal{M}_x^2(k-1)+2}\right]\left[\frac{\mathcal{M}_y^2(k+1)}{\mathcal{M}_y^2(k-1)+2}\right]=1 \tag{B.6-13}$$

from which

$$\mathcal{M}_y^2=\left[\frac{\mathcal{M}_x^2(k-1)+2}{\mathcal{M}_x^2(k+1)^2}\right]\cdot[\mathcal{M}_y^2(k-1)+2] \tag{B.6-14}$$

Collecting $\mathcal{M}_y^2$ terms and using a common denominator gives

$$\mathcal{M}_y^2\left[\frac{\mathcal{M}_x^2(k+1)^2-(k-1)\,[\mathcal{M}_x^2(k-1)+2]}{\mathcal{M}_x^2(k+1)^2}\right]=2\left[\frac{\mathcal{M}_x^2(k-1)+2}{\mathcal{M}_x^2(k+1)^2}\right] \tag{B.6-15}$$

Dividing out the term on the left and simplifying, we find

$$\mathcal{M}_y^2=\frac{1+\mathcal{M}_x^2[(k-1)/2]}{k\mathcal{M}_x^2-[(k-1)/2]} \tag{B.6-16}$$

Now that we have a relationship between the upstream and downstream Mach numbers, we can compute the other relations easily. Writing Eq. 8.17 twice,

$$\frac{T_R/T_x}{T_R/T_y}=\frac{T_y}{T_x}=\frac{\mathcal{M}_x^2[(k-1)/2]+1}{\mathcal{M}_y^2-[(k-1)/2]+1} \tag{B.6-17}$$

Substituting for $\mathcal{M}_y$ from Eq. B.6-16,

$$\frac{T_y}{T_x}=\frac{\mathcal{M}_x^2[(k-1)/2]+1}{\left(\frac{k-1}{2}\right)\left[\frac{1+\mathcal{M}_x^2(k-1)/2}{k\mathcal{M}_x^2-(k-1)/2}\right]+1} \tag{B.6-18}$$

To find the velocity ratio, we write

$$\frac{V_x}{V_y} = \frac{V_x^2}{V_y V_x} = \frac{V_x^2}{c^{*2}} = \frac{\mathcal{M}_x^2(k+1)}{\mathcal{M}_x^2(k-1)+2} \tag{B.6-19}$$

Here we have substituted for $V_x V_y$ from Eq. B.6-9 and then for V_x / c^* from Eq. B.6-12.

From Eq. 8.36 we know that

$$\frac{\rho_y}{\rho_x} = \frac{V_x}{V_y} = \frac{\mathcal{M}_x^2(k+1)}{\mathcal{M}_x^2(k-1)+2} \tag{B.6-20}$$

Finally, we find P_y / P_x from

$$\frac{P_y}{P_x} = \frac{\rho_y}{\rho_x}\frac{T_x}{T_y} = \frac{\left[\dfrac{\mathcal{M}_x^2(k+1)}{\mathcal{M}_x^2(k-1)+2}\right]\left(\dfrac{[(k-1)/2]\{1+\mathcal{M}_x^2[(k-1)/2]\}}{k_x^2-[(k-1)/2]}+1\right)}{\mathcal{M}_x^2[(k-1)/2]+1} \tag{B.6-21}$$

which after some algebra simplifies to

$$\frac{P_y}{P_x} = \frac{2k}{k+1}\mathcal{M}_x^2 - \frac{k-1}{k+1} \tag{B.6-22}$$

Thus, we now have all the pertinent property ratios in terms of $\mathcal{M}_x$, which allows us to tabulate these ratios in App. A.5 and thus to solve normal shock wave problems very conveniently.

B.7 EQUATIONS FOR ADIABATIC, ZERO-CLEARANCE, ISENTROPIC COMPRESSORS PROCESSING IDEAL GASES (SEE SEC. 10.2)

Starting with

$$PV^k = \text{constant for an isentropic process} \tag{B.3-24}$$

we write

$$V = V_1\left(\frac{P_1}{P}\right)^{1/k} = \frac{nRT}{P}\left(\frac{P_1}{P}\right)^{1/k} \tag{B.7-1}$$

Then

$$\int_{P_1}^{P_2} V\,dP = \frac{nRT_1}{P_1}P_1^{1/k}\int_{P_1}^{P_2}\left(\frac{1}{P}\right)^{1/k} dP = \frac{nRT_1}{P^{(k-1)/k}}\int_{P_1}^{P_2}\left(\frac{1}{P}\right)^{1/k} dP \tag{B.7-2}$$

But

$$\int_{P_1}^{P_2}\left(\frac{1}{P}\right)^{1/k} dP = \int_{P_1}^{P_2} P^{-1/k}dP = \frac{1}{1-1/k}P^{(1-1/k)}\Big]_{P_1}^{P_2}$$

$$= \frac{k}{k-1}\left(P_2^{(k-1)/k} - P_1^{(k-1)/k}\right) \tag{B.7-3}$$

so that

$$\int_{P_1}^{P_2} V\,dP = \frac{k}{k-1} nRT_1 \frac{P_2^{(k-1)/k} - P_1^{(k-1)/k}}{P_1^{(k-1)/k}} = \frac{k}{k-1} nRT_1 \left[\left(\frac{P_2}{P_1} \right)^{(k-1)/k} - 1 \right] \tag{9.15}$$

B.8 PROOF THAT THE CURVES OF CONSTANT ϕ AND THE CURVES OF CONSTANT ψ ARE PERPENDICULAR (SEE SEC. 16.5)

Starting with

$$V_x = -\left(\frac{\partial \phi}{\partial x}\right)_y = -\left(\frac{\partial \psi}{\partial y}\right)_x \qquad V_y = -\left(\frac{\partial \phi}{\partial y}\right)_x = \left(\frac{\partial \psi}{\partial x}\right)_y \tag{16.26}$$

we now introduce the identity proven in all calculus books,

$$\left(\frac{\partial A}{\partial B}\right)_C = -\left(\frac{\partial A}{\partial C}\right)_B \left(\frac{\partial C}{\partial B}\right)_A \tag{B.8-1}$$

which is true for any A, B, and C. Using this identity, we find

$$\left(\frac{\partial \phi}{\partial x}\right)_y = -\left(\frac{\partial \phi}{\partial y}\right)_x \left(\frac{\partial y}{\partial x}\right)_\phi \tag{B.8-2}$$

$$\left(\frac{\partial \psi}{\partial y}\right)_x = -\left(\frac{\partial \psi}{\partial x}\right)_y \left(\frac{\partial x}{\partial y}\right)_\psi \tag{B.8-3}$$

Substituting these in Eq. 16.26 produces

$$\left(\frac{\partial \phi}{\partial y}\right)_x \left(\frac{\partial y}{\partial x}\right)_\phi = \left(\frac{\partial \psi}{\partial x}\right)_y \left(\frac{\partial x}{\partial y}\right)_\psi \tag{B.8-4}$$

Substituting for $(\partial \phi / \partial y)_x$ from Eq. 16.26 and canceling,

$$-\left(\frac{\partial y}{\partial x}\right)_\phi = \left(\frac{\partial x}{\partial y}\right)_\psi \tag{B.8-5}$$

or

$$\left(\frac{\partial y}{\partial x}\right)_\phi \left(\frac{\partial y}{\partial x}\right)_\psi = -1 \tag{B.8-6}$$

Equation B.8-6 says that the product of the slopes of a curve of constant ϕ and a curve of constant ψ is -1. This is the condition that the two curves be perpendicular, proven in all calculus books. This was shown for any x and y, not for a specific one, which indicates that at any point x the lines of constant ϕ and constant ψ passing through that point are perpendicular.

APPENDIX C

EQUATIONS FOR TWO- AND THREE-DIMENSIONAL FLUID MECHANICS

C.1 SUMMARY OF VECTOR NOTATION

Please reread Secs. 7.1 and 15.1. This topic is covered in much more detail in mathematics and physics books, and in Bird, Stewart, and Lightfoot [1] and Prandtl and Tietjens [2]. In this book we define the velocity vector as

$$\mathbf{V} = V_x\mathbf{i} + V_y\mathbf{j} + V_z\mathbf{k} \tag{7.A}$$

Here V_x, V_y, and V_z are the scalar components of $\mathbf{V}$, and $\mathbf{i}$, $\mathbf{j}$, and $\mathbf{k}$ are unit vectors in the x, y, and z directions. This can also be written in cylindrical (polar) coordinates as

$$\mathbf{V} = V_r\mathbf{e_r} + V_\theta\mathbf{e_\theta} + V_z\mathbf{k} \tag{C.1}$$

where V_r, V_θ, and V_z are the scalar components of $\mathbf{V}$ and $\mathbf{e_r}$, $\mathbf{e_\theta}$, and $\mathbf{k}$ are the unit vectors in the r, θ, and z directions. For spherical coordinates, the corresponding definition is

$$\mathbf{V} = V_r\mathbf{e_r} + V_\theta\mathbf{e_\theta} + V_\phi\mathbf{e_\phi} \tag{C.2}$$

The unit vectors in the θ and ϕ directions are not nearly as intuitive as the rectangular unit vectors; see any advanced calculus book for their properties.

One of the great merits of the vector notation is that if we can write out the component values in Eq. 7.A, then by straightforward (if tedious) mathematics, we can rewrite the same vector velocity in cylindrical or spherical coordinates. This gives the velocity components in those coordinate systems. This property is also true for vector equations. In the text the equations for conservation of mass and momentum are derived in rectangular coordinates. The following text shows the values of those equations in rectangular, and cylindrical coordinates. In making up or using those equations, we follow the following definitions:

Vector addition or subtraction consists of adding or subtracting the scalar components.

Multiplying a vector by a scalar is equivalent to multiplying all the scalar components of the vector by that scalar.

There are two types of vector multiplication, the dot product and the cross product. The dot product of two vectors, written $\mathbf{V} \cdot \mathbf{r}$, is a scalar whose value is the product of the length of the two vectors, $\mathbf{V}$ and $\mathbf{r}$, times the cosine of the angle between them. The cross product of two vectors, written $\mathbf{V} \times \mathbf{r}$, is a vector whose magnitude is the product of the length of the two vectors, $\mathbf{V}$ and $\mathbf{r}$, times the sine of the angle between them, multiplied by the unit vector normal to the plane that contains the two vectors. The direction of the normal chosen is based on the "right-hand rule."

The dot product of a vector with itself, $\mathbf{V} \cdot \mathbf{V}$, sometimes written $\mathbf{V}^2$, is a scalar whose value is the square of the length of the vector. (The cosine of the angle between a vector and itself is $\cos 0 = 1.0$). The dot product of any unit vector ($\mathbf{i}$, $\mathbf{j}$, or $\mathbf{k}$) with itself is 1.0. The dot product of any unit vector with any of the other unit vectors is zero.

The cross product of a vector with itself is a vector of zero length (the same as a scalar zero). The sine of the angle between a vector and itself is $\sin 0 = 0$. The cross product of any unit vector ($\mathbf{i}$, $\mathbf{j}$, or $\mathbf{k}$) with itself is zero. The cross product of any two unit vectors is $\pm$ the third unit vector.

The derivative of a vector is a vector consisting of the derivatives of the scalar components, multiplied by the corresponding unit vectors, e.g.,

$$\frac{d\mathbf{V}}{dt} = \frac{dV_x}{dt}\mathbf{i} + \frac{dV_y}{dt}\mathbf{j} + \frac{dV_z}{dt}\mathbf{k} \tag{C.3}$$

The *Hamiltonian operator* $\boldsymbol{\nabla}$ called "del" or "nabla" is defined as

$$\boldsymbol{\nabla} = \mathbf{i}\frac{\partial}{\partial x} + \mathbf{j}\frac{\partial}{\partial y} + \mathbf{k}\frac{\partial}{\partial z} \tag{C.4}$$

The dot product of del with a vector is a scalar called the divergence. For example,

$$\begin{aligned} \boldsymbol{\nabla} \cdot \mathbf{V} = \mathbf{div} \cdot \mathbf{V} &= \left(\mathbf{i}\frac{\partial}{\partial x} + \mathbf{j}\frac{\partial}{\partial y} + \mathbf{k}\frac{\partial}{\partial z}\right) \cdot (\mathbf{i}V_x + \mathbf{j}V_y + \mathbf{k}V_z) \\ &= \frac{\partial V_x}{\partial x} + \frac{\partial V_y}{\partial y} + \frac{\partial V_z}{\partial z} \end{aligned} \tag{C.5}$$

(See Fig. 15.2.)

The dot product of del with a scalar is a vector called the gradient. (One often sees $\boldsymbol{\nabla} \cdot$ any scalar written as **grad** $\cdot$ scalar). For example

$$\boldsymbol{\nabla} \cdot P = \mathbf{grad} \cdot P = \left(\mathbf{i}\frac{\partial}{\partial x} + \mathbf{j}\frac{\partial}{\partial y} + \mathbf{k}\frac{\partial}{\partial z}\right) \cdot P = \left(\mathbf{i}\frac{\partial P}{\partial x} + \mathbf{j}\frac{\partial P}{\partial y} + \mathbf{k}\frac{\partial P}{\partial z}\right) \tag{C.6}$$

This vector points "downhill" (i.e., perpendicular to contour lines) and has a magnitude equal to the "steepness" of the slope in the "downhill" direction (Fig. C.1).

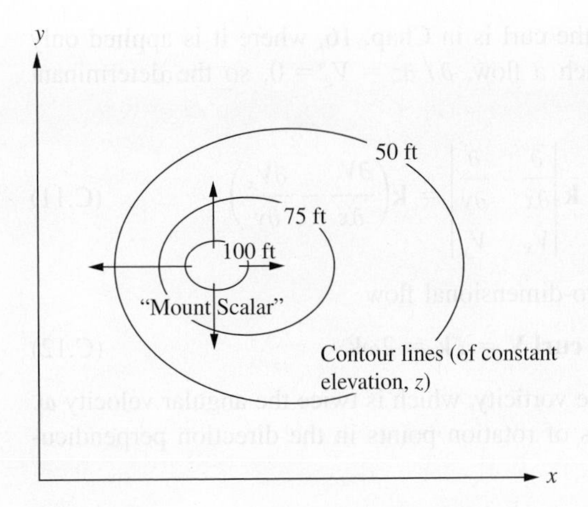

FIGURE C.1
On a two-dimensional contour map, the contours represent lines of equal elevation (which is a scalar); there is a gradient line (a vector) through every point. Its direction is downhill, perpendicular to the contour lines, and its magnitude is proportional to the steepness of the gradient (i.e., how close together the contour lines are). Only four such vectors are shown, as representative examples of the vectors through every point in the scalar field.

Multiplication of a vector by del, $\nabla\mathbf{V}$ (not the same as the dot product $\nabla \cdot \mathbf{V}$), produces a tensor (a nine-member array of derivatives). It is seldom seen by itself, but most often seen in the following dot product:

$$\begin{aligned}\mathbf{V}\cdot\nabla\mathbf{V} &= V_x\frac{\partial \mathbf{V}}{\partial x}+V_y\frac{\partial \mathbf{V}}{\partial y}+V_z\frac{\partial \mathbf{V}}{\partial z}\\ &= \mathbf{i}\left(V_x\frac{\partial V_x}{\partial x}+V_y\frac{\partial V_x}{\partial y}+V_z\frac{\partial V_x}{\partial z}\right)\\ &\quad+\mathbf{j}\left(V_x\frac{\partial V_y}{\partial x}+V_y\frac{\partial V_y}{\partial y}+V_z\frac{\partial V_y}{\partial z}\right)\\ &\quad+\mathbf{k}\left(V_x\frac{\partial V_z}{\partial x}+V_y\frac{\partial V_z}{\partial y}+V_z\frac{\partial V_z}{\partial z}\right)\end{aligned} \tag{C.7}$$

This product appears in the substantive derivative and is widely used in fluid mechanics.

The dot product of del with itself, called the *laplacian operator,* normally written as ∇^2, is

$$\begin{aligned}\nabla^2 = \nabla\cdot\nabla &= \left(\mathbf{i}\frac{\partial}{\partial x}+\mathbf{j}\frac{\partial}{\partial y}+\mathbf{k}\frac{\partial}{\partial z}\right)\cdot\left(\mathbf{i}\frac{\partial}{\partial x}+\mathbf{j}\frac{\partial}{\partial y}+\mathbf{k}\frac{\partial}{\partial z}\right)\\ &= \left(\frac{\partial^2}{\partial x^2}+\frac{\partial^2}{\partial y^2}+\frac{\partial^2}{\partial z^2}\right)\end{aligned} \tag{C.8}$$

Applying it to a scalar produces a three-term scalar. If that scalar is zero, then the three terms of that scalar obey Laplace's equation (Eq. 15.7). Applying ∇^2 to a vector produces a vector of the form

$$\begin{aligned}\nabla^2\mathbf{V} &= \mathbf{i}\left(\frac{\partial^2 V_x}{\partial x^2}+\frac{\partial^2 V_x}{\partial y^2}+\frac{\partial^2 V_x}{\partial z^2}\right)+\mathbf{j}\left(\frac{\partial^2 V_y}{\partial x^2}+\frac{\partial^2 V_y}{\partial y^2}+\frac{\partial^2 V_y}{\partial z^2}\right)\\ &\quad+\mathbf{k}\left(\frac{\partial^2 V_z}{\partial x^2}+\frac{\partial^2 V_z}{\partial y^2}+\frac{\partial^2 V_z}{\partial z^2}\right)\end{aligned} \tag{C.9}$$

The cross product of del with a velocity vector is a vector, called the *curl* (or *rot* in the German literature), that shows the angular velocity of the fluid at any point. The direction of the vector is the axis about which the particle of fluid is rotating. The definition is in terms of a 3-by-3 determinant,

$$\nabla\times\mathbf{V} = \mathbf{curl}\,\mathbf{V} = \begin{vmatrix}\mathbf{i} & \mathbf{j} & \mathbf{k}\\ \dfrac{\partial}{\partial x} & \dfrac{\partial}{\partial y} & \dfrac{\partial}{\partial z}\\ V_x & V_y & V_z\end{vmatrix} \tag{C.10}$$

In this book the only application of the curl is in Chap. 16, where it is applied only in two-dimensional (*x*-*y*) flow. In such a flow, $\partial / \partial z = V_z = 0$, so the determinant simplifies to

$$\nabla \times \mathbf{V} = \mathbf{curl\ V} = \mathbf{k} \begin{vmatrix} \dfrac{\partial}{\partial x} & \dfrac{\partial}{\partial y} \\ V_x & V_y \end{vmatrix} = \mathbf{k}\left(\frac{\partial V_y}{\partial x} - \frac{\partial V_x}{\partial y}\right) \tag{C.11}$$

It is shown in Chap. 16 that for a two-dimensional flow

$$\nabla \times \mathbf{V} = \mathbf{curl\ V} = \zeta \mathbf{k} = 2\omega \mathbf{k} \tag{C.12}$$

where ζ is a defined quantity called the vorticity, which is twice the angular velocity ω. The unit vector $\mathbf{k}$ shows that the axis of rotation points in the direction perpendicular to the *x*-*y* plane.

C.2 THE MASS BALANCE IN THREE COORDINATE SYSTEMS

The three-dimensional mass balance equation (Eq. 15.7) in vector notation is

$$-\frac{\partial \rho}{\partial t} = \nabla \cdot (\rho \mathbf{V}) \tag{15.7}$$

This appears to be a vector equation, but it is not. The only term in it in vector notation is del of a vector, which is a scalar. Equation 15.7 does not have directional components as does the momentum balance. Equation 15.7 is

$$-\frac{\partial \rho}{\partial t} = \frac{\partial(\rho V_x)}{\partial x} + \frac{\partial(\rho V_y)}{\partial y} + \frac{\partial(\rho V_z)}{\partial z} \qquad \text{[rectangular coordinates]} \tag{15.7}$$

$$-\frac{\partial \rho}{\partial t} = \frac{1}{r}\frac{\partial(\rho r V_r)}{\partial r} + \frac{1}{r}\frac{\partial(\rho V_\theta)}{\partial \theta} + \frac{\partial(\rho V_z)}{\partial z} \qquad \text{[cylindrical coordinates]} \tag{C.13}$$

$$-\frac{\partial \rho}{\partial t} = \frac{1}{r^2}\frac{\partial(\rho r^2 V_r)}{\partial r} + \frac{1}{r \sin\theta}\frac{\partial(\rho V_\theta \sin\theta)}{\partial \theta} + \frac{1}{r \sin\theta}\frac{\partial(\rho V_\phi)}{\partial \phi} \qquad \text{[spherical coordinates]} \tag{C.14}$$

C.3 THE MOMENTUM BALANCE FOR LAMINAR, NEWTONIAN, CONSTANT-DENSITY FLOWS (THE NAVIER-STOKES EQUATIONS)

Equations. 15.27,

$$\frac{D\mathbf{V}}{Dt} = \mathbf{g} - \frac{\nabla \cdot P}{\rho} + \frac{\mu}{\rho}\nabla^2\mathbf{V} \quad \text{or} \quad \rho\frac{D\mathbf{V}}{Dt} = \rho\mathbf{g} - \nabla \cdot P + \mu\nabla^2\mathbf{V} \tag{15.27}$$

have the following component equations in rectangular coordinates:

$$\rho\left(\frac{\partial V_x}{\partial t} + V_x\frac{\partial V_x}{\partial x} + V_y\frac{\partial V_x}{\partial y} + V_z\frac{\partial V_x}{\partial z}\right) = \rho g_x - \frac{\partial P}{\partial x} + \mu\left(\frac{\partial^2 V_x}{\partial x^2} + \frac{\partial^2 V_x}{\partial y^2} + \frac{\partial^2 V_x}{\partial z^2}\right) \quad \text{(C.15)}$$

$$\rho\left(\frac{\partial V_y}{\partial t} + V_x\frac{\partial V_y}{\partial x} + V_y\frac{\partial V_y}{\partial y} + V_z\frac{\partial V_y}{\partial z}\right) = \rho g_y - \frac{\partial P}{\partial y} + \mu\left(\frac{\partial^2 V_y}{\partial x^2} + \frac{\partial^2 V_y}{\partial y^2} + \frac{\partial^2 V_y}{\partial z^2}\right) \quad \text{(C.16)}$$

and

$$\rho\left(\frac{\partial V_z}{\partial t} + V_x\frac{\partial V_z}{\partial x} + V_y\frac{\partial V_z}{\partial y} + V_z\frac{\partial V_z}{\partial z}\right) = \rho g_z - \frac{\partial P}{\partial z} + \mu\left(\frac{\partial^2 V_z}{\partial x^2} + \frac{\partial^2 V_z}{\partial y^2} + \frac{\partial^2 V_z}{\partial z^2}\right) \quad \text{(C.17)}$$

Here g_x is the x component of the gravity vector $= g \cos \theta$. If the gravity vector points in the negative z direction, then $g_x = g_y = 0$ and $g_z = -1$.

For cylindrical coordinates, the component equations are

$$\rho\left(\frac{\partial V_r}{\partial t} + V_r\frac{\partial V_r}{\partial r} + \frac{V_\theta}{r}\frac{\partial V_r}{\partial \theta} - \frac{V_\theta^2}{r} + V_z\frac{\partial V_r}{\partial z}\right) = \rho g_r - \frac{\partial P}{\partial r} + \mu\left\{\frac{\partial}{\partial r}\left[\frac{1}{r}\frac{\partial}{\partial r}(rV_r)\right] + \frac{1}{r^2}\frac{\partial^2 V_r}{\partial \theta^2} - \frac{2}{r^2}\frac{\partial V_\theta}{\partial \theta} + \frac{\partial^2 V_r}{\partial z^2}\right\} \quad \text{(C.18)}$$

$$\rho\left(\frac{\partial V_\theta}{\partial t} + V_r\frac{\partial V_\theta}{\partial r} + \frac{V_\theta}{r}\frac{\partial V_\theta}{\partial \theta} + \frac{V_\theta V_r}{r} + V_z\frac{\partial V_\theta}{\partial z}\right) = \rho g_\theta - \frac{1}{r}\frac{\partial P}{\partial \theta} + \mu\left\{\frac{\partial}{\partial r}\left[\frac{1}{r}\frac{\partial}{\partial r}(rV_\theta)\right] + \frac{1}{r^2}\frac{\partial^2 V_\theta}{\partial \theta^2} - \frac{2}{r^2}\frac{\partial V_r}{\partial \theta} + \frac{\partial^2 V_\theta}{\partial z^2}\right\} \quad \text{(C.19)}$$

and

$$\rho\left(\frac{\partial V_z}{\partial t} + V_r\frac{\partial V_z}{\partial r} + \frac{V_\theta}{r}\frac{\partial V_z}{\partial \theta} + V_z\frac{\partial V_z}{\partial z}\right) = \rho g_z - \frac{\partial P}{\partial z} + \mu\left[\frac{1}{r}\frac{\partial}{\partial r}\left(r\frac{\partial V_z}{\partial r}\right) + \frac{1}{r^2}\frac{\partial^2 V_z}{\partial \theta^2} + \frac{\partial^2 V_z}{\partial z^2}\right] \quad \text{(C.20)}$$

For spherical coordinates, the component equations are shown in Bird, Stewart, and Lightfoot [1]. They are used much less often in chemical engineering than the rectangular or cylindrical forms and are not needed for any problem in this book.

REFERENCES FOR APPENDIX C

1. Bird, R. B., E. N. Stewart, W. E. Lightfoot. *Transport Phenomena*. 2nd ed. New York: Wiley, 2002.
2. Prandtl, L., O. G. Tietjens. *Fundamentals of Hydro- and Aeromechanics*. transl. L. Rosenhead. New York: Dover, 1957.

APPENDIX D

ANSWERS TO SELECTED PROBLEMS

1.3. 102 lbm / ft^3
1.5. 1.110 g / cm^3, an error of 0.1%
1.11. Sphere, 6 / D; cube, 6 / E; cylinder, 6 / D
1.15. 1.14 mi^3
1.18. toof = 32.2 ft; dnoces = s / $\sqrt{32.2}$
1.21. 9669 ft / s
1.23. $4.6 \cdot 10^5$
1.25. 0.0145 N = 0.0033 lbf = 1.48 g (force)
1.28. Air pollution models work in g, auto usage data are in mi.

2.1. $5.29 \cdot 10^{-5}$ ft = 0.018 mm
2.3. 998.3 kgf / m^3 = 9.792 kN / m^3
2.8. 16,115 psia = 1096 atm = $1.11 \cdot 10^5$ kPa
16,100 psig = 1095 atm, g = $1.109 \cdot 10^5$ kPa, g
2.12. 0.2236 ft
2.18. 96,783 ft; $P = 0$
2.20. −0.00356 °F / ft, −0.00536 °F / ft
2.22. $1.186 \cdot 10^{19}$ lbm
2.24. 88.92 kPa, g = 12.81 psig
2.27. $F = 2\rho g r^3 / 3$
2.30. 0.60 in = 1.52 cm
2.39. 89.6 vol %

2.40. 62.5 lbf = 278 N
2.43. (*a*) 2573 lbf; (*b*) 2.49991 lbf
2.46. 2000 kg
2.49. 5.2 ft = 1.58 m
2.52. $\Delta h = z \dfrac{\rho_w}{\rho_{\text{Hg}} - \rho_w}$
2.54. (*a*) 2.77 in = 70 mm; (*b*) 10.72 in = 272 mm
2.58. (*a*) 4.808 ft = 1.465 m; (*b*) 0.15%
2.60. 0.017 psi = 0.47 in H_2O = 0.117 kPa
2.62. (*a*) 1234 psia = 1219 psig = 8400 kPa, g
2.65. 127.3 psig
2.68. $14.3 \text{ ft / s}^2 = 4.36 \text{ m / s}^2$, upward
2.70. 1.779° = 0.031 rad
2.72. 0.153 lbf = 0.683 N

3.7. $500 \text{ ft}^3 \text{ / s} = 14.16 \text{ m}^3 \text{ / s}$; 0.467 ft / s = 0.145 m / s
3.13. 4.8 mi / h
3.14. $2.25 \text{ ft}^3 \text{ / s}$; 159 lbm / s; 13 ft / s
3.16. 0.0133 atm
3.19. Rising, 72 mm / h = 2.8 in / h
3.21. 0.0051 lbm / min = 0.0023 kg / min
3.26. About 50,000. Depends on some assumptions.

4.3. 3.75 ft lbf = 5.05 J
4.7. 412 kJ / kg
4.9. −1.602 MW (negative because of sign convention)
4.11. 14.8 Btu = 15.62 kJ
4.15. $-1.44 \cdot 10^{-3}$ lbm = −0.65 g
4.21. −2.24 Btu / lbm = −1.07 kJ / kg

5.1. 11°F, 1.3°F
5.3. (*a*) 31.9 kPa = 4.64 psi, 0
(*b*) 28.67 kPa = 4.28 psi, 3.20 J / kg
(*c*) 0; 32.0 J / kg
5.7. $55.6 \text{ ft}^3 \text{ / s} = 1.57 \text{ m}^3 \text{ / s}$
5.9. $60.6 \text{ ft}^3 \text{ / s} = 1.71 \text{ m}^3 \text{ / s}$
5.13. 127 ft / s = 39 m / s
5.16. 11.3 ft / s = 3.46 m / s
5.21. 5.58 m / s = 18.3 ft / s
5.23. 1341 in / s = 112 ft / s = 34 m / s
5.25. 25.4 ft / s = 17.3 mph = 7.7 m / s
5.27. 0.46 ft / s = 0.14 m / s
5.34. 24.0 ft / s = 7.33 m / s; $4.72 \text{ ft}^3 \text{ / s} = 0.13 \text{ m}^3 \text{ / s}$
5.36. 123 ft / s = 37.5 m / s; $123 \text{ ft}^3 \text{ / s} = 3.48 \text{ m}^3 \text{ / s}$
5.42. 60.3 ft / s = 19.2 m / s; 41.1 ft / s = 12.5 m / s
5.46. 49.1 ft / s = 15.0 m / s; 1.04 rps = 62 rpm

5.49. 36.5 s
5.52. $9.97\ (\mathrm{s}/\sqrt{\mathrm{ft}})\sqrt{h_1}$
5.60. (*a*) $1.0\ \mathrm{lbf/ft^2} = 0.007\ \mathrm{psi} = 0.0479\ \mathrm{kPa}$
(*b*) $29.3\ \mathrm{ft/s} = 8.93\ \mathrm{m/s}$; $Q = 146.5\ \mathrm{ft^3/s} = 8760\ \mathrm{cfm} = 4.15\ \mathrm{m^3/s}$
(*c*) 0.344 hp = 0.264 kW

6.3. 0.0174 psi / ft = 0.92 psi / mi = 0.0039 Pa / m
6.7. 0.068 ft / s = 0.0185 m / s
6.10. 0.114 lbf = 0.507 N
6.14. 20 psi / 1000 ft; ≈40 psi / 1000 ft
6.15. 11.2 psi / 1000 ft = 0.253 Pa / m
6.19. 20.5 psi = 141.6 kPa
6.23. 1.73 psi / 1000 ft = 0.392 Pa / m
6.25. $0.038\ \mathrm{ft^3/s} = 17.2\ \mathrm{gpm} = 0.065\ \mathrm{m^3/s}$
6.31. 1600; 4500
6.42. 0.910 psi; 1.103 psi
6.44. $1032\ \mathrm{gpm} = 0.0623\ \mathrm{m^3/s}$
6.46. 0.0166 = 1.66 ft / 100 ft
6.53. $2.22 \cdot 10^{-6}\ \mathrm{ft^3/s} = 6.3 \cdot 10^{-8}\ \mathrm{m^3/s}$
6.56. 7.31 in = 0.185 m; 22.4 ft / s = 6.83 m / s
6.59. 295.7 ft / s = 92.4 m / s
6.61. 48.1 ft / s = 14.7 m / s
6.63. $0.126\ \mathrm{ft^3/s} = 56.7\ \mathrm{gpm} = 0.0036\ \mathrm{m^3/s}$
6.65. $\mathrm{Max} = 402\ \mathrm{gpm} = 0.025\ \mathrm{m^3/s}$
$\mathrm{Min} = 238\ \mathrm{gpm} = 0.015\ \mathrm{m^3/s}$
6.70. Far end; near end; don't bet much.
6.81. $5.44 \cdot 10^{-7}\ \mathrm{m/s} = 1.79 \cdot 10^{-6}\ \mathrm{ft/s} = 0.15\ \mathrm{ft/day}$
6.83. 1.48 ft / s = 0.45 m / s
6.87. 384 ft / s = 117 m / s; 31.6 s; 8954 ft = 2778 m
107 ft / s = 32.7 m / s; 8.8 s; 696 ft = 212 m

7.1. $-2 \cdot 10^{-24}\ \mathrm{ft/s}$
7.10. $F_x = -586\ \mathrm{N} = -131.7\ \mathrm{lbf}$
$F_y = 1414\ \mathrm{N} = 318\ \mathrm{lbf}$
7.11. (*a*) $V_1 = 19.9\ \mathrm{ft/s}$; $V_2 = 79.7\ \mathrm{ft/s}$; $\dot{m} = 103.4\ \mathrm{lbm/s}$
(*b*) $F = 288\ \mathrm{lbf}$
7.15. 828 lbf = 3.68 kN
7.17. −232.2 lbf = −1.04 kN
7.19. 3.25 kN / (kg / s) = 331.4 lbf / (lbm / s)
7.22. −2480 lbf = −11.0 kN
7.26. (*a*) 48,300 lbf = 215 kN
(*b*) $1.65 \cdot 10^5\ \mathrm{lbf} = 736\ \mathrm{kN}$
7.38. 5947 m / s = 19,500 ft / s
7.45. 816 kPa = 118 psi
7.48. 16.65 ft = 5.07 m; 6.01 ft = 1.83 m
7.50. 25.4 ft / s = 7.73 m / s

7.54. 0.014°F = 0.0077°C
7.56. Cold day; air density greater

8.1. 807 psi / ft = 18.2 MPa / m; 0.97 psi / ft = 22.0 kPa / m
8.4. 10,800 ft / s = 2.0 mi / s = 3.28 km / s
8.6. $3 \cdot 10^5$ psi = $2.07 \cdot 10^4$ bar
8.8. 334 ft / s = 102 m / s
8.10. 226.3°R = −233.7°F = −147.6°C
8.12. 226.3°R = −233.7°F = −147.6°C
3.60 psia = 24.9 kPa
0.0059 lbm / ft^3 = 0.095 kg / m^3
8.14. 500°R = 40°F = 4.6°C
42.5 psia = 311 kPa; 1931 ft / s = 589 m / s
8.16. 7.16 psia = 49.4 kPa
420°R = −40°F = −40°C
2950 ft / s = 899 m / s
0.00636 lbm / ft^3 = 0.102 kg / m^3
8.18. 6.3 psia = 43.6 kPa
440°R = −20°F = −28.7°C; 1.1697
8.20. 662.5°R = 202.5°F = 94.9°C
6.52 psia = 45.0 kPa; 3085 ft / s = 940 m / s
8.22. 0.69 lbm / s = 0.315 kg / s
8.25. 3.69 lbm / s = 1.67 kg / s
8.31. 3.03 psia = 20.9 kPa
0.0265 lbm / ft^3 = 0.42 kg / m^3
8.33. 38,000 lbf = 0.26 MN
8.53. 62,700°R ≈ 62,200°F ≈ 34,500°C
8.54. 167 lbm / s = 75.7 kg / s
8.56. 967.5 ft / s = 295 m / s
631°R = 171°F = 77.3°C
18.05 psia = 124.5 kPa

9.3. $$\pi_2 = \frac{\mu^2}{F\rho} = \frac{2}{(\mathcal{R} \cdot \text{Pressure coefficient})^2}$$
$$\pi_3 = L_1 / L_2$$
9.9. 40.2 mi / h
9.12. $$\pi_4 = \frac{F}{\rho L^4 \omega^2} = \frac{F}{\rho L^2 V^2} \cdot \frac{V^2}{L^2 \omega^2}$$
$$= 0.5 \cdot \begin{pmatrix}\text{pressure} \\ \text{coefficient}\end{pmatrix} \cdot \begin{pmatrix}\text{Strouhal} \\ \text{number}\end{pmatrix}^{-2}$$

10.1. 13 gpm = $8.2 \cdot 10^{-4}$ m^3 / s
10.7. 18.98 hp = 14.16 kW; 31.16 hp = 23.1 kW
24.11 hp = 18.0 kW

10.11. 2860 Btu / lbmol
10.19. 106 hp = 79 kW
10.22. 0.514 ft = 6.2 in = 0.157 m
10.24. 11.4 ft = 3.47 m

11.5. 315 darcies
11.7. $Ca = 4.17 \cdot 10^{-5}$; residual saturation ≈ 0.16
11.9. $\mathscr{R}_{\text{P.M.}} = 1913$; turbulent,
$dP / dx \approx 0.52$ in H_2O / ft
11.13. 0.64 ft / s = 0.195 m / s;
0.0067 ft / s = 0.020 m / s

12.3. $\mathscr{F}_{\text{liquid}} = 0.021\, gh$; $\mathscr{F}_{\text{gas}} = 2.3\, gh$
12.10. $Q_{\text{g-top}} / Q_{\text{g-bottom}} = 1.15$

13.4. $K = 0.44$ lbf / ft; $n \approx 0.31$
13.13. $-dP / dx = 6.22$ psi / 100 ft
13.16. $-dP / dx = 7.81$ psi / 100 ft
13.19. (*a*) 1.583 ft / s; 1.437 ft / s
(*b*) $r_b / r_w = 0.7816$; $V = 1.159$ ft / s

14.2. 1.155 *A*
14.5. 0.084 in; 49 drops / cc
14.7. $\Delta z = 2\sigma A / (g\rho Bx)$; a hyperbola
14.14. 1.6 psi

15.8. 0.6667
15.13. 862 cP
15.20. 1.78 s

16.12. 14.82 psia
16.15. ≈0.50 h
16.16. $\frac{1}{2}$

17.5. 0.081 ft; 0.0054 ft; 4.03 ft
17.16. 8.6 ft

18.3. 22.7 ft^2 / s^3 = 2.108 J / (kg · s)
18.4. 34.6 cm^2 / s^2 = $1.5 \cdot 10^{-6}$ Btu / lbm
18.10. $dV_x / dr = -2.13$ / s

19.1. 0.03 mL per injection
19.5. 12.6 ft / s; 7.1 hp / 1000 gal
19.10. 6492 ft

INDEX

COMMON UNITS AND VALUES FOR PROBLEMS AND EXAMPLES

For all problems and examples in this book, unless it is stated explicitly to the contrary, assume the following:

The acceleration of gravity is $g = 32.17 \text{ ft/s}^2 = 9.81 \text{ m/s}^2$

The surrounding atmospheric pressure is the "standard atmospheric pressure", $P_{\text{surroundings}} = P_{\text{atmospheric}} = 1$ atm $= 14.696 \approx 14.7$ lbf / in^2 $= 33.89$ ft of water $=$ 10.33 m of water $=$ 29.92 in of mercury $=$ 760 mm of mercury $=$ 760 torr $=$ 101.3 kPa $=$ 1.013 bar $=$ 1.033 kgf / cm^2.

If the fluid in the problem or example is water, then it is water at 1 atm pressure and 20°C = 68°F = 293.15 K = 528°R, for which

$$\rho = 62.3 \text{ lbm/ft}^3 = 998.2 \text{ kg/m}^3 = 3.46 \text{ lbmol/ft}^3 = 55.5 \text{ kgmol/m}^3$$

$$= 55.5 \text{ mol/L}$$

$$\mu = 1.002 \text{ cP} = 1.002 \cdot 10^{-3} \text{ Pa} \cdot \text{s} = 6.73 \cdot 10^{-4} \text{ lbm/ft} \cdot \text{s}$$

$$= 2.09 \cdot 10^{-5} \text{ lbf} \cdot \text{s/ft}^2$$

$$\nu = \mu / \rho = 1.004 \cdot 10^{-6} \text{ m}^2/\text{s} = 1.004 \text{ cSt} = 1.077 \cdot 10^{-5} \text{ ft}^2/\text{s}$$

$$M = 18 \text{ g/mol} = 18 \text{ lbm/lbmol}$$

$$\sigma = 0.000415 \text{ lbf/in} = 72.74 \text{ dyne/cm} = 0.07274 \text{ N/m}$$

If the fluid in the problem or example is air, then it is air at 1 atm pressure and 20°C = 68°F = 293.15 K = 528°R for which

$$\rho = 0.075 \text{ lbm/ft}^3 = 1.20 \text{ kg/m}^3 = 2.59 \cdot 10^{-3} \text{ lbmol/ft}^3 = 41.6 \text{ mol/m}^3$$

$$\mu = 0.018 \text{ cP} = 1.8 \cdot 10^{-5} \text{ Pa} \cdot \text{s}$$

$$\nu = \mu / \rho = 1.488 \cdot 10^{-5} \text{ m}^2/\text{s} = 14.88 \text{ cSt} = 1.613 \cdot 10^{-4} \text{ ft}^2/\text{s}$$

$$C_P = 3.5\,R = 6.95 \text{ Btu/lbmol °R} = 6.95 \text{ cal/mol K} = 29.1 \text{ J/mol K}$$

$$M = 29 \text{ g/mol} = 29 \text{ lbm/lbmol}$$

$$k = 1.40$$

Any unspecified gas will be assumed to have the properties of air at 1 atm and 20°C shown above. Standard temperature and pressure (stp) means 1 atm and 20°C = 68°F.

For any ideal gas, the volume per mole is given by $V_{\text{molar}} = 1 / \rho_{\text{molar}} = RT / P$. Wherever R appears in this book it is the universal gas constant, shown on the next page. For real gases at 1 atm pressure the ideal gas equation is an excellent approximation.